Coastal and Estuarine Studies

Coastal and Estuarine Studies

40

David Prandle, Editor

Dynamics and Exchanges in Estuaries and the Coastal Zone

American Geophysical Union
Washington DC

Editor

David Prandle
Proudman Oceanographic Laboratory
Bidston Observatory, Birkenhead
Merseyside L43 7RA, United Kingdom

Library of Congress Cataloging-in-Publication Data

Dynamics and exchanges in estuaries and the coastal zone / David Prandle, editor.
p. cm. — (Coastal and estuarine studies)
Proceedings of a 5th conference, which was held 9–13 July 1990 at the Gregynog study centre.
Includes bibliographical references and index.
ISBN 0-87590-254-5
1. Estuarine oceanography—Congresses. 2. Coasts—Congresses.
I. Prandle, David. II. Series.
GC96.5.D96 1992
551.46'09—dc20 92-6818
CIP

ISBN 0-87590-254-5

Printed in the United States of America.

Contents

Part I. Introduction

Part II. Baroclinic Dynamics

Part III. Circulation

Part IV. Sedimentation

Part V. Modelling (Sediment)

Part VI. Applied Studies

Acknowledgements

The organising committee for the Gregynog Conference was as follows:

R.T. Cheng, U.S.G.S., San Francisco, USA
A.G. Davies, U.C.N.W., Bangor, UK
C.B. Pattiaratchi, C.W.R., Perth, Australia
D. Prandle, P.O.L., Birkenhead, UK
C. van de Kreeke, U. Miami, USA

As organiser of the preceding conference, Ralph Cheng's advice and encouragement was particularly helpful. The success of the conference was, of course, primarily due to the efforts of the participants. In particular I should like to mention Dr. Cath Allen, who, making little of a long-standing illness, both presented and discussed at length her important work on random-walk techniques. Sadly, Cath died in March 1991.

In the preparation of this volume, the painstaking effort of the referees is duly recognised, likewise the help and assistance of the publishers.

Finally, neither the conference nor this volume would have happened without the efforts of Jackie Hardcastle and Mike Blackley of the Proudman Oceanographic Laboratory. I have been warned endlessly by, seemingly, everyone who has ever organised such a conference and volume that it involves a minimum of "one year out". Since my contribution has been minimal, I can only guess at the extent of Mike and Jackie's efforts - sincere thanks.

David Prandle
Proudman Oceanographic Laboratory
Bidston Observatory, Birkenhead
Merseyside L43 7RA, United Kingdom

List of Contributors

D G Aubrey
225 Clark Lab
Woods Hole Oceanographic Institute
Woods Hole
MA 02543
USA

J O Backhaus
Instiut fuer Meereskunde
der Universitaet Hamburg
Troplowitzstr. 7
D-2000 Hamburg 54
Germany

M L Barton
Department of Civil Engineering
University of Birmingham
Birmingham B15 2TT
United Kingdom

V Casulli
Mathematics Department
University of Trento
Trento
Italy

W Chen
Second Institute of Oceanography
PO Box 1207
Hangzhou
Zheijiang 310012
China

R T Cheng
US Geological Survey
Menlo Park
California
USA

P D Craig
CSIRO Division of Oceanography
GPO Box 1538
Hobart Tas. 7001
Australia

G C van Dam
Rijkswaterstaat
Tidal Waters Division
PO Box 20907
2500 EX The Hague
The Netherlands

E J Darbyshire
School of Civil Engineering
University of Birmingham
Birmingham B15 2TT
United Kingdom

J Darbyshire
Unit for Coastal and Estuarine Studies
Marine Science Laboratories
Menai Bridge
Gwynedd LL59 5EY
United Kingdom

A G Davies
School of Ocean Sciences
University College North Wales
Menai Bridge
Gwynedd LL59 5EY
United Kingdom

W P M De Ruijter
University of Utrecht
Institute of Meteorology and Oceanography
Princetonplein 5
3584 CC Utrecht
The Netherlands

S P Dinnel
Center for Marine Science
University of Southern Mississippi
Stennis Space Center
MS 39529
USA

K R Dyer
Institute of Marine Studies
Polytechnic South West
Drake Circus
Plymouth PL4 8AA
United Kingdom

A J Elliott
Unit for Coastal and Estuarine Studies
Marine Science Laboratories
Menai Bridge
Gwynedd LL59 5EY
United Kingdom

C T Friedrichs
MIT-Woods Hole Ocean Institute
Woods Hole Ocean Institute
Woods Hole, MA 02543
USA

A Van der Giessen
National Institute of Public Health
and Environmental Protection
(RIVM), Centre for Mathematical Methods
PO Box 1
3720 BA Bilthoven
The Netherlands

P A Gillibrand
Department of Agriculture, Fisheries
and Food
Marine Laboratories
PO Box 101
Victoria Road
Aberdeen AB9 8TB
United Kingdom

F C Groenendijk
Rijkswaterstaat
Tidal Waters Division
Koningskade 4
2696 AA Den Haag
The Netherlands

P J Hardcastle
Proudman Oceanographic Laboratory
Bidston Observatory
Birkenhead
Merseyside L43 7RA
United Kingdom

P E Holloway
Department of Geography and
Oceanography
Australian Defence Force Academy
Campbell ACT 2600
Australia

F Jiang
Coastal and Oceanographic
Engineering Department
University of Florida
Gainesville
Florida 32611
USA

B King
Australian Institute of Marine Science
Private Mail Bag No 3
Townsville Mail Centre 4810
Australia

R Kirby
Ravensrodd Consultants Ltd
6 Queens Drive
Taunton, Somerset TA1 4XW
United Kingdom

J van de Kreeke
Rosenstiel School of Marine and Atmospheric Science
Division of Applied Marine Physics
4600 Rickenbacker Causeway
Miami, Florida 33149-1098
USA

A Y Kuo
School of Marine Science
Virginia Institute of Marine Science
College of William and Mary
Gloucester Point VA 23062
USA

J L Largier
Center for Coastal Studies, 0209
Scripps Institution of Oceanography
La Jolla, CA 92093
USA

R A Louwersheimer
Rijkswaterstaat
Tidal Waters Division
PO Box 20907
2500 EX The Hague
The Netherlands

D R Lynch
Bedford Institute of Oceanography
PO Box 1006
Dartmouth
Nova Scotia
Canada B2Y 4A2

O Matsuda
Faculty of Applied Biological Science
Hiroshima University
Higashi-Hiroshima 724
Japan

A J Mehta
Coastal and Oceanographic Engineering Department
University of Florida
Gainesville
Florida 32611
USA

J Nicholson
Department of Civil Engineering
University of Liverpool
Brownlow Street
PO Box 147
Liverpool L69 3BX
United Kingdom

B A O'Connor
Department of Civil Engineering
University of Liverpool
Brownlow Street
PO Box 147
Liverpool L69 3BX
United Kingdom

K Park
School of Marine Science
Virginia Institute of Marine Science
College of William and Mary
Gloucester Point VA 23062
USA

T Pohlmann
Instiut fuer Meereskunde
der Universitaet Hamburg
Troplowitzstr. 7
D-2000 Hamburg 54
Germany

D Prandle
Proudman Oceanographic Laboratory
Bidston Observatory
Birkenhead
Merseyside L43 7RA
United Kingdom

Part I

Introduction

1

Introduction

D Prandle

In establishing the peculiar niche of a volume of conference proceedings within the scientific literature, the editor has a responsibility (beyond the culling and processing) to inform readers of the background to the conference, the subject areas and specific foci. More tentatively, underlying directions and common themes should be discerned to perhaps indicate areas of research likely to be fruitful and topical in the next few years. Since many readers will only study selected contributions, the volume is divided into five sections namely:- II. Baroclinic Dynamics; III. Circulation; IV. Sediment Dynamics; V. Sediment Modelling; VI. Applied Studies. This introduction incorporates sub-titles to facilitate the same selectivity.

1.1 The Gregynog Conference

The conference series began in Hamburg in 1978 with the aim of encouraging cooperation between coastal engineers and coastal oceanographers. Proceedings have been published by Springer-Verlag in their section on Coastal and Estuarine Studies. Gregynog (9-13 July 1990) marked the 5th, in what is now a biennial series, to be followed by Perth, Australia in 1992 and the east-coast USA in 1994. Some 70 scientists attended with 40 oral and 20 poster presentations. While participation is primarily by direct invitation, no one is excluded and care is taken to ensure a balance in international representation, between physicists, engineers and mathematicians, and across the spectra from 'grey beards' to research students.

Likewise, residential venues (Greygnog is a remote study centre kindly made available by the University of Wales) have been carefully selected to promote informal contacts across the divides of both age and discipline.

1.2 The Proceedings

All papers have been reviewed by at least two referees including second reviews where major revisions were indicated. These referees were urged both to maintain 'Journal standards' and to ensure coherence within the scope of the volume. As a result of adhering strictly to these requirements, the volume is slimmer but of a uniformly higher standard.

Part II - Baroclinic Dynamics

The order of the five papers in this section conveniently follows a sequence of (first) some large scale observed features of such dynamics thence in the next three papers studies of successively finer scale processes. A final paper then attempts to model the associated turbulent dynamics at a specific location. Craig and Holloway describe how oscillations in the thermal structure of a bay on Australia's east coast can be associated with coastally trapped waves extending to the edge of the adjacent shelf seas. Elliott et al develop a numerical model to simulate the influence of a sill in a Scottish sea loch on the saline stratification - similar dynamics may be found in many fjords. Sturley and Dyer observed the motions of internal waves in a small UK estuary and estimated their contribution to vertical mixing. Darbyshire and West describe direct measurements of turbulent motions and salt fluxes in the same estuary and infer that additional mechanisms must contribute to the observed vertical mixing rate. Finally, Simpson and Sharples present a model study that incorporates a Mellor and Yamada Level 2 turbulence closure scheme to reproduce vertical density differences observed in Liverpool Bay.

Part III - Circulation

This section includes 7 papers concerned with circulation both horizontal and vertical. The scale of circulation varies from internally within a specific estuary, between estuaries (or bays) to circulation between adjacent seas. The first paper by Van de Kreeke determines the flushing rate of a small Florida bay using limited current meter observations in conjunction with data on fresh water inflows and salinity distribution. Largier et al describe the saline exchange in a small South African estuary where gravitational circulation is constrained by a hydraulic control created by the restricted cross sectional area of the mouth. Schroeder et al use extensive observations of salinity distributions to determine the influence of tide, wind and river discharge on the salinity balance between a sub-bay and connecting bay in the Gulf of Mexico. Su and Chen use spectral analysis of sea level oscillations to separate the frequency domains within which local wind forcing versus external sea levels predominate in a large shallow bay in China. Swenson and Saner use sea level data extending as long as 15 years to relate the influence of tides, winds and river flow on flooding (and thereby vegetation) in a Louisiana Delta. Tee developes an x-z numerical model of the St Lawrence estuary to explain 40-50 day axial oscillations observed in both current meter and salinity observations from this long (400 km) and deep (300 m) system). Winkel-Steinberg et al describe numerical model results of exchanges between a deep and shallow sea with an estuarine-like baroclinic circulation.

Part IV - Sediment Dynamics

The extent of the emphasis on sedimentation within this conference reflects the increasing requirement to study estuarine physics for application to water quality problems within which the motion of fine sediments is often crucial. Uncles et al present a comprehensive study of the turbidity maximum in the Tamar; using a numerical model in conjunction with measurements of velocity, salinity and suspended sediment concentration to indicate the controlling mechanisms. Friedrichs et al examine the interaction between dynamics and morphology in short (<20km), strongly tidal embayments, emphasising the predominant role of friction in determining ebb or flood dominance in net sediment movement. Reflecting the character of the conference, these results from a complex analytical study may be contracted with a related study based on analysis of observational data. Kirby examines the profiles of inter-tidal muddy-shores, noting the influence of both tide and waves and extrapolating to consider the impact of any future rise in mean sea level. Thorne et al describe recent developments of an Acoustic Backscatter Probe for measuring sand movement, comparing data from both field and laboratory with theory. Fian and Mehta develop a model to simulate the propagation of surface waves over a layer of soft mud, results explain observations made within the mud layer and emphasise the importance of longer period side-bands in the wave spectra. Darbyshire shows how a triangular array of bottom pressure gauges can be used to estimate wave reflectance on a beach.

Part V - Sediment Modelling

This section includes six papers that address various aspects of modelling dispersion with particular application to sediment movement. Cheng and Casulli examine the accurate definition of parameters in the tidally-averaged dispersion equation, their hypotheses are then rigorously tested using a numerical model embracing the mouth of San Francisco Bay. Smith extends his series of theoretical papers on dispersion in estuaries to consider the effect of bends on particles of varying densities. Davies uses an x-z model to examine the influence of surface waves transported by a (tidal) current indicating increases of up to two orders of magnitude. Wang applies a mathematical model of sediment transport in the Yantze estuary in an attempt to simulate morphological development. Van Dam and Louwersheimer incorporate analytical functions for both vertical and horizontal (sub-grid scale) motions within a vertically averaged coastal zone model to simulate particle dispersion. O'Connor and Nicholson incorporate within a depth-averaged tidal model, the influence of surface waves to estimate sediment motion following dredging in an estuary. This engineering application reminds us that some people must provide answers now.

Part VI - Applied Studies

The six papers in this section emphasise the diversity of regimes, with observations from four continents and from an estuary with (almost) the world's largest tides to one where tidal exchanges are secondary. De Ruijter et al describe the coastal plume from the Rhine which persists for up to 80km alongshore and for 25km offshore; both the localised stratification and the overall geographic configuration are shown to vary strongly with both wind and river discharge. Visser develops a linearised analytical model of the cross-shore circulation within the Rhine Plume, thereby explaining surface observations of a minimum in suspended sediment concentrations along its offshore edge. King and Wolanski use a vertically-averaged tidal model of Australia's Central Great Barrier Reef to study the exchange of coastal waters. Kuo and Park use measurements of currents, salinity and dissolved oxygen to calculate exchange rates between Chesapeake Bay and a sub-estuary concluding that tidal motions are of secondary importance. From observations of water quality parameters close to the edge of a (temperature and salinity) front in Japan's Seto Inland Sea, Yanagi et al illustrate the complexity of the interactions between physical, biological and chemical processes. In the final paper, Shaw describes the range of studies required for the planning of a tidal power scheme in the Bristol Channel.

1.3 Conclusions

Hopefully the above single sentence summaries help in bridging the gap between the titles and abstracts that accompany each paper. The publication deadline forced me to read all 30 papers in one (extended) weekend. While not recommending this, it does yield a useful overview. The niche that commends this volume is, perhaps inadvertently, supplied by the form of presentation assumed by almost all the authors. Specifically, they have written primarily for a broad spectrum of readership reflecting the conference audience. Thus, in contrast to specialised papers for 'narrow' journals, the authors include clearer introductions, more thoughtful justifications of their aims and broader interpretations of their results, while relying on referenced papers for technical details. Consequently the papers are more easily digested and more readily provide both an overview and an insight into the specific discipline.

Lest anyone has missed the discernible trend in estuarine research away from traditional engineering applications towards water quality issues, the question now is to detect any loss of momentum in this veritable band-wagon. Having spent the last five years collaborating with biologists, chemists and sedimentologists in the U.K. North Sea Project, my overwhelming conclusion is for the need to adopt protocols to facilitate such co-operation. The international nature of the commercial market for both instruments and computer hardware and software effectively enforces such protocols in these areas. Likewise the establishment of both national and international data centres avoids much duplication of effort, simplifies modelling simulations and reduces the complexities of inter-comparison exercises. While the conference series deliberately avoids over-representation of narrow numerical modelling studies, approximately 20 discrete modelling modules can be identified within these 30 papers. Almost all of these modules and many more, need to be incorporated in any comprehensive simulation of water quality parameters in estuarine systems. Establishing the protocols for easy inter-connections between such modules (and the related data inputs and outputs) poses a major challenge. It is important that the linking framework adopted is sufficiently flexible to minimise constraints and encourage individuality within co-operative research programmes.

Part II

Baroclinic Dynamics

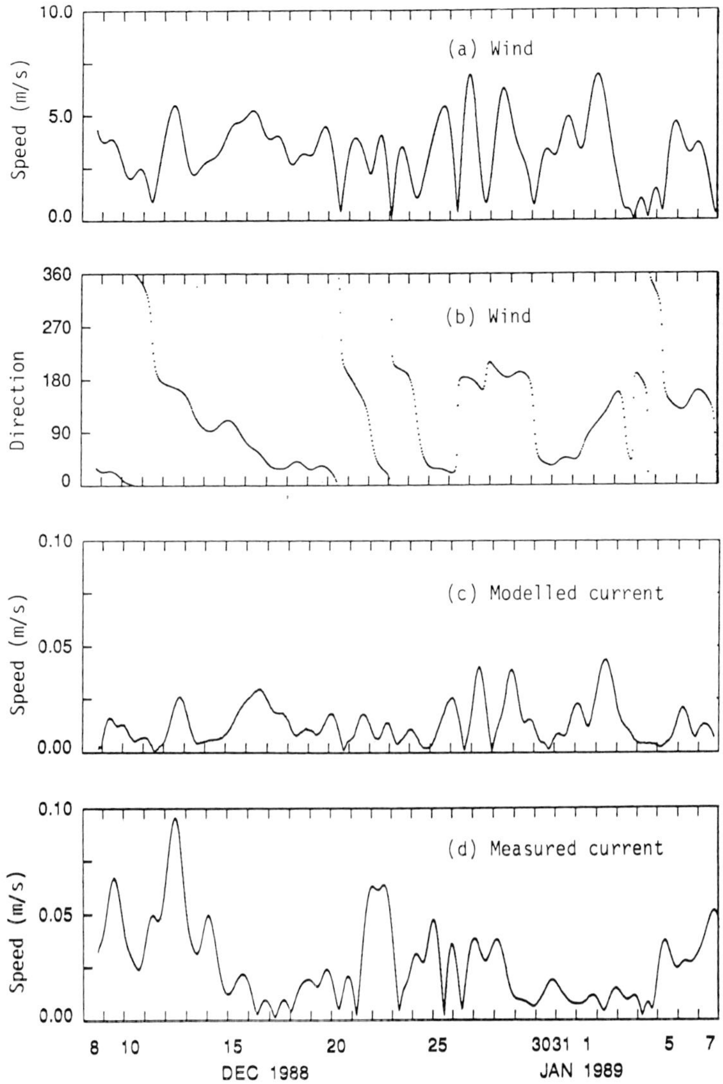

Figure 2: Wind speed (a) and direction (b) for the measurement period, together with simulated (c) and measured (d) current speeds from 6.5 m depth at the Red station (Records are filtered to remove diurnal and higher frequencies).

oscillatory period of 7 to 10 days. Fig. 4 also shows, for comparison, filtered records of the sea-surface elevation from the Black station, and water temperatures from 11 m at Black and 20 m at Yellow. The correspondence between the four curves is obvious. The surface elevation signal was recorded by a submerged pressure sensor, and so does not contain the "inverse barometer" signal. The range of surface movement indicated in the signal is approximately 0.2 m. The Yellow thermistor is 9 m deeper that that at Black, and consistently recording temperatures about 1° cooler. On this basis, the temperature oscillations of around 2° can be converted to a vertical isotherm movement of about 20 m, or approximately 100 times the amplitude of the surface oscillation. The isotherm movement is out-of-phase with the surface: as the surface rises, the isotherms fall, characteristic of baroclinic motion. The atmospheric pressure lags the surface elevation slightly, with a delay of approximately 2 days.

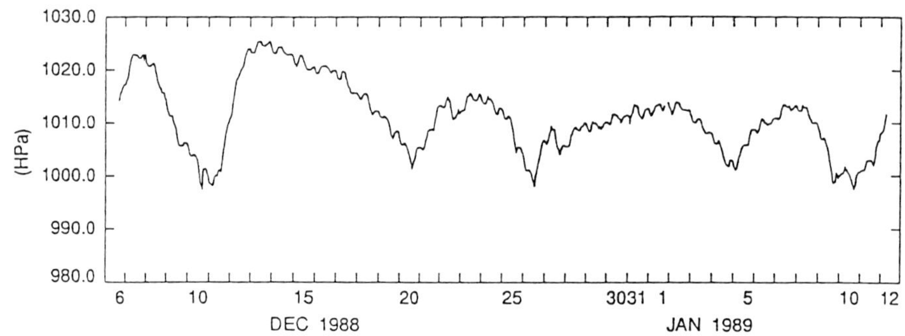

Figure 3: Atmospheric pressure record for 6 December to 12 January

Despite the strong correlation in Figure 4, the atmospheric pressure is unlikely to be directly responsible for the motion indicated by the temperature and surface elevation curves. On the scale of Jervis Bay, the only response to changing atmospheric pressure is likely to be an inverse barometer effect, with small velocities (of order 10^{-3} ms^{-1}) and no baroclinic signal.

The final curve on Figure 4 is the northward wind component, filtered in the same way as the other parameters. At these low frequencies, there is a close correlation between the north-south wind and the atmospheric pressure, with the former leading the latter by about two days. This wind component appears to be the major agent forcing the response seen in the other parameters in Figure 4. There is no obvious reason why Jervis Bay, whose north-south and east-west dimensions are similar, should respond preferentially to winds blowing along the north-south axis. However, the northward wind component is the longshore wind for the open continental shelf outside Jervis Bay. The open shelf waters can be expected to be significantly more responsive to longshore than to cross-shore winds (e.g. Gill and Clarke, 1974).

The most likely explanation for the correlation evident in Figure 4 is that it is indicative of coastally trapped wave activity over the shelf and slope region external to Jervis Bay. During the Australian Coastal Experiment in 1983/84, three mooring lines were deployed across the south-eastern Australian continental shelf and slope to document the propagation of coastally trapped waves (Freeland *et al.*, 1986; Church *et al.*, 1986a, b; Huyer *et al.*, 1988). The first three wave modes were in fact found to account for approximately half the sub-inertial current energy along this part of the coast. The correlation between surface and isotherm displacement, as seen in Figure 4, is an established property of the waves, and the displacement ratio of 1:100 is compatible with theory (e.g. Gill and Clarke, 1974).

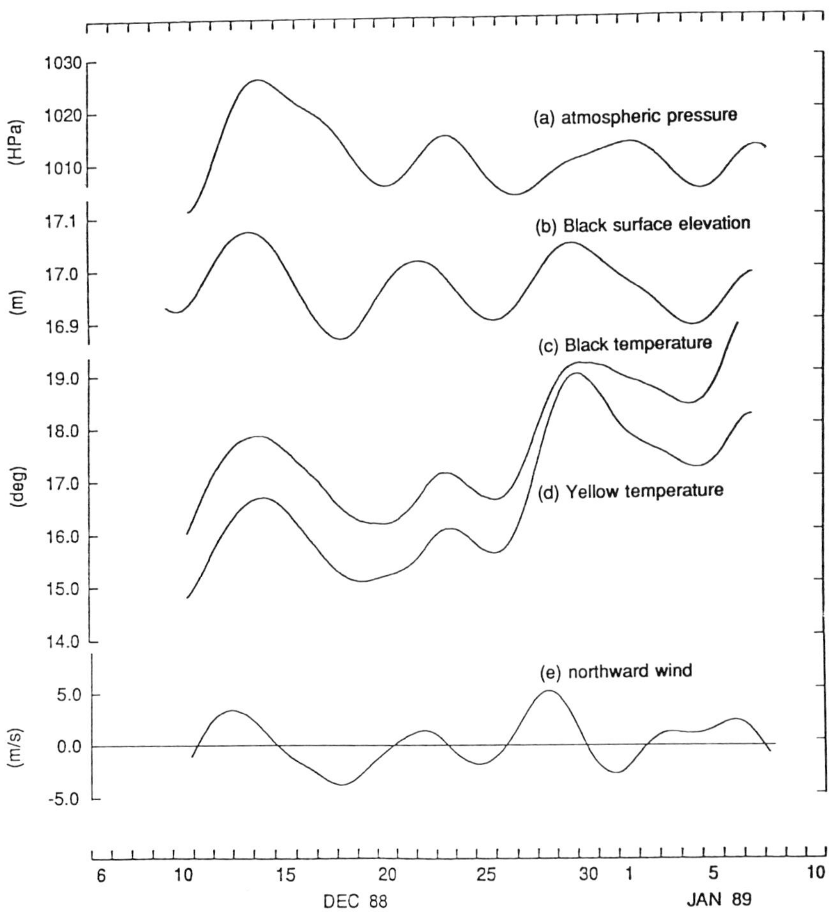

Figure 4: Filtered records of (a) atmospheric pressure, (b) surface elevation from the Black station, (c) water temperature at 11 m depth from Black, (d) water temperature at 20 m from Yellow and (e) northward wind component.

There is ongoing discussion about the origin of the coastally trapped waves in the ACE region, with much of the attention focussed on the role of Bass Strait (e.g. Church and Freeland, 1987; Clarke, 1987; Middleton and Viera, 1991). Local wind will also generate wave activity. The coastally trapped waves are generally described as "free" or "forced" depending upon whether they are generated remotely or locally (e.g. Church *et al.* 1986b). There is no point pursuing this distinction in the present context. However, in Figure 4, it is clear that northward, that is, downwelling favourable, winds are related to rising water temperatures, while southward winds relate to falling temperatures. There is a consistent phase delay of approximately two days from the time of peak wind speed to the corresponding temperature maximum or minimum.

The wavelength of the coastally trapped motion is of the order of a thousand kilometres, much wider than the mouth of Jervis Bay. From the perspective of the bay, then, there is a strongly baroclinic sea-surface and thermal oscillation at the mouth, which forces a response inside the bay. Figure 5 shows the temperature signal recorded by thermistors at approximately 10 m depth on each of the current-meter moorings. At least according to the Green and Blue moorings, the response across the bay is closely correlated to that at the Black station. The response through the water column is demonstrated by the Red mooring (Figure 6), where the deeper thermistors, particularly, show the same temperature signature. In both Figures 5 and 6, the Red 10.5 m record appears to be anomalous, but we have no indication whether the anomaly is real, or due to instrument malfunction.

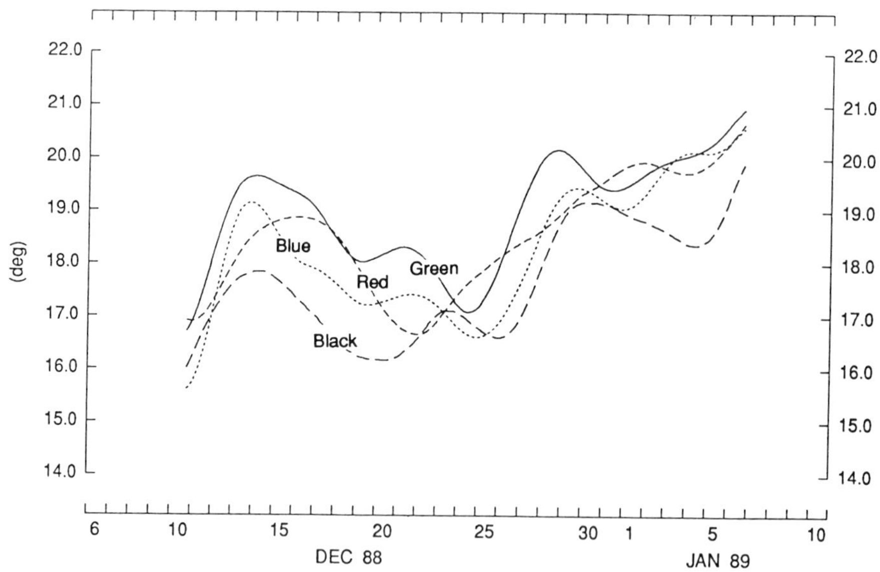

Figure 5: Filtered temperature records from Black at 11 m depth, Blue at 11.5 m, Red at 10.5 m and Green at 9.5 m.

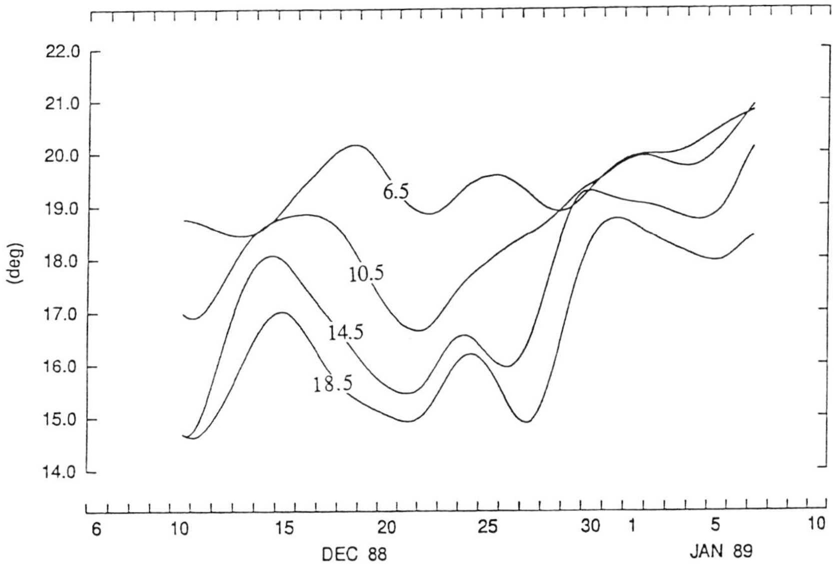

Figure 6: Water temperatures from the Red station at depths of 6.5, 10.5, 14.5 and 18.5 m.

Measured currents have a distinct, though not so clean, response, similar to that of the temperatures. In Figure 7, current components are plotted together with temperatures, for each station in the bay. For the Green and Blue stations, the current components are rotated 45° clockwise, and 30° anticlockwise, respectively, from north, to give a direction approximately parallel to the bathymetry. For the Red station, the eastward component is plotted for meters both at 6.5 m and 18.5 m depth, that is, at both the top and bottom of the water column. As is to be anticipated with baroclinic motion, the two traces are 180° out of phase, that is, almost equal in magnitude but opposite in sign. The Green currents and 6.5 m Red currents, both measured near the surface, are approximately in-phase with the temperatures, while the currents at Blue and at 18.5 m at Red are close to 180° out-of-phase with temperature. For each station in Figure 7, the temperature plot has been moved on the time-axis to maximise, approximately, the visual correlation between the temperatures and the currents. The size of this displacement in time provides an indication of the phase relationship between the signals and is listed in the figure caption.

$$P_z = 0 \qquad at\ z=0. \tag{2.8}$$

This is, of course, the rigid lid condition which is a statement that, on the time-scale of internal wave propagation around the bay, the surface gravity-wave speed is effectively infinite.

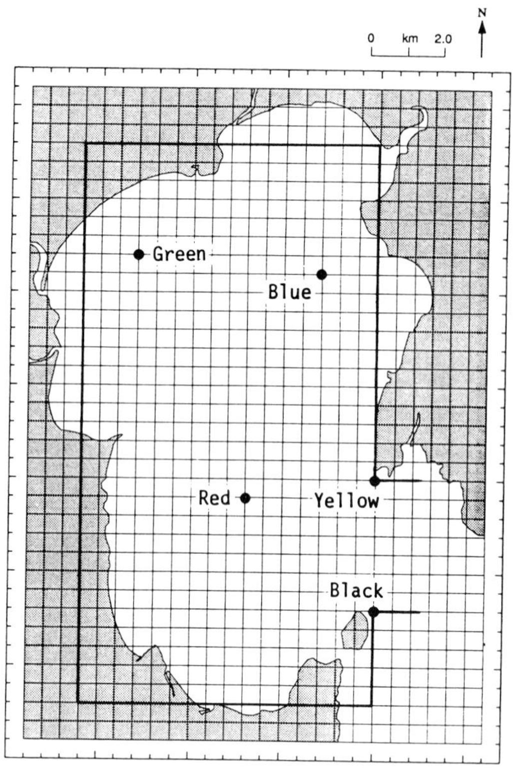

Figure 8: Diagram of the rectangular modelled bay, showing the computational grid and the model station locations.

A final boundary condition is required on the open boundary at the mouth of the bay, where the solution must be stated in a way that represents the (known) movement of the surface and isotherms. The exact specification of this condition will be postponed for the present.

We are considering motion forced at the mouth with periods of the order of five to ten days, For simplicity, we will restrict attention to a single frequency ω, and assume a time-dependence, for the variables in (2.1) to (2.4), of the form $exp(-i\omega t)$. With this time-dependence, all time derivatives in (2.1) to (2.4) can be replaced by the multiplicative factor $-i\omega$. It should be noted that ω will be an order of magnitude smaller than f.

Equations (2.1) to (2.4) can be expressed as a single equation in P,

$$f^2 P_{zz} + N^2 (P_{xx} + P_{yy}) = 0 \; , \tag{2.9}$$

in which ω^2 has been neglected by comparison with f^2. The surface boundary condition on P is (2.8), and the bottom boundary condition, by (2.3), is

$$P_z = 0 \quad on \; z = -H \; . \tag{2.10}$$

On the walls, the zero normal-flow conditions are expressed using the relationships

$$u = \frac{1}{f}\left(\frac{i\omega}{f} P_x - P_y\right) \; . \tag{2.11}$$

and

$$v = \frac{1}{f}\left(\frac{i\omega}{f} P_y + P_x\right) \; . \tag{2.12}$$

By the hydrostatic approximation, implicit in (2.3), the relationship between the pressure P and temperature perturbations T is

$$P_z = g\alpha T \quad , \tag{2.13}$$

where g is the acceleration due to gravity and α is the coefficient of expansion of seawater. The relationship (2.13) assumes that changes in density are due strictly to temperature, a reasonable assumption for Jervis Bay in the summer.

The nominal position of the Black station on the rectangular basin (Figure 8) is $x = X$, $y = y_1$. Using (2.5) and (2.13), we can specify a pressure structure at this location, based on observations of the sea-surface elevation and the water-column temperature structure. The response to a specified sea-level at Black is immediately obvious. If ζ in (2.5) is given at $x = X$, $y = y_1$, then the solution satisfying (2.9) and all the boundary conditions is

$$P = g\zeta , \qquad u = v = w = 0$$

over the whole domain. That is, because of the rigid-lid assumption, the bay responds uniformly and instantaneously to changes in sea-level. We will thus restrict attention, henceforth, only to the dynamics induced by changes in the temperature structure at the mouth.

We can predict the response to temperature changes with some accuracy simply by inspection. In (2.11) and (2.12), the ω/f terms are small. If they are neglected, the pressure behaves as a streamfunction. In particular with this assumption, the pressure at any set depth will be constant all the way around the boundary, taking the same value as that specified at the Black station. The pressure contours, and thus the streamlines, will form closed loops inside the bay, with the maximum numerical value on the wall, decaying away towards the centre of the bay. If the vertical scale is taken as H/π (see (2.14) below)), the horizontal exponential decay scale, by (2.10), is $NH/|f|\pi$. Under this approximation, there will be no flow through the mouth of the bay, and the motion will be in phase across the whole bay.

Neglect of the ω/f terms in (2.11) and (2.12) is similar to the rigid-lid approximation in the sense that it assumes that, on the time-scale of the prescribed oscillations at the mouth, internal waves effectively transmit information instantaneously across the bay. Retention of the ω/f terms will introduce phase delays.

If N is constant, the solution to (2.9) will have the form

$$P = Q(x,y)\cos\frac{n\pi z}{H} , \qquad (2.14)$$

where n is an integer, representing the mode number. Taking $n = 1$, the equation for Q is

$$Q_{xx} + Q_{yy} - \left(\frac{f\pi}{NH}\right)^2 Q = 0 . \qquad (2.15)$$

On an open coastline, the system of equations developed thus far describes internal Kelvin waves of the type treated in detail by Gill and Clarke (1974). On a long straight coastline given by, say, $y = 0$, the free-wave solutions have the form

$$Q = \exp(\pm ikx - \ell y) , \qquad (2.16)$$

where

$$k = \frac{\omega\pi}{NH} , \qquad (2.17)$$

and

$$\ell = \frac{|f|\pi}{NH} . \qquad (2.18)$$

The expressions for k and ℓ are derived using the requirement that v, given by (2.12), is zero on the coast, and again make use of the assumption that $\omega^2 << f^2$. The waves described by (2.16) to (2.18) have velocities strictly parallel to the coast. For typical values, $\omega = 10^{-5}\ s^{-1}$, $|f| = 10^{-4}\ s^{-1}$, $N = 10^{-2}\ s^{-1}$, $H = 20$ m, the offshore length-scale $1/\ell$ is about 1 km, while the longshore wavelength, $2\pi/k$, is about 40 km. According to (2.15), the formal picture of the density-driven dynamics in Jervis Bay is that of a wave propagating into the bay along the south coast of the inlet channel (Figure 8), clockwise around the bay, and then out along the northern coast of the channel. The total solution in the channel is given by

$$Q = A\exp(-ikx - \ell(y-y_1)) + B\exp(ikx - \ell(y_2-y)) \ , \tag{2.19}$$

where A is the (complex) amplitude of the incoming wave, and B that of the outgoing wave. Equation (2.19) which can be applied at $x = X$, for $y_1 \leq y \leq y_2$, in fact becomes the final boundary condition for the problem.

The total problem is specified by the field equation (2.15) subject to boundary conditions

$$-\frac{i\omega}{f}Q_x + Q_y = 0 \tag{2.20}$$

on the walls at $x = 0$ and $x = X$, and

$$\frac{i\omega}{f}Q_y + Q_x = 0 \tag{2.21}$$

on the walls $y = 0$ and $y = Y$. The boundary condition (2.19) is applied as follows. The amplitude of the incoming wave is set so that A exp (ikx) = 1. Equation (2.19) is then differentiated with respect to x, and the unknown coefficient B is eliminated between the differentiated and undifferentiated versions of (2.19) to give

$$Q_x - ikQ = -ik\exp\ell(y_1-y) \tag{2.22}$$

on $x = X$, $y_1 \leq y \leq y_2$. After solving the system for Q, the total solution can be scaled to give the correct amplitude at the Black station.

The solution to (2.15) was obtained by discretising the equation using standard central differences. The numerical grid is shown on Figure 8, with $\Delta x = \Delta y$ = 500m. The boundary conditions were approximated with one-sided, rather than central differences, because central differences in the longshore direction caused a $2\Delta x$-wavelength instability around the boundary. The full discretised problem forms a matrix equation with non-zero right-hand side arising from (2.22). The matrix is amenable to inversion using either direct or iterative techniques.

2.5 Model results

As we have already noted, Jervis Bay was strongly stratified during the measurement program. Figure 9 shows temperature profiles collected before and after the program at the central station shown in Figure 1. We will concentrate on the December 5 profile. The average buoyancy frequency, calculated by fitting a straight line between the surface and bottom

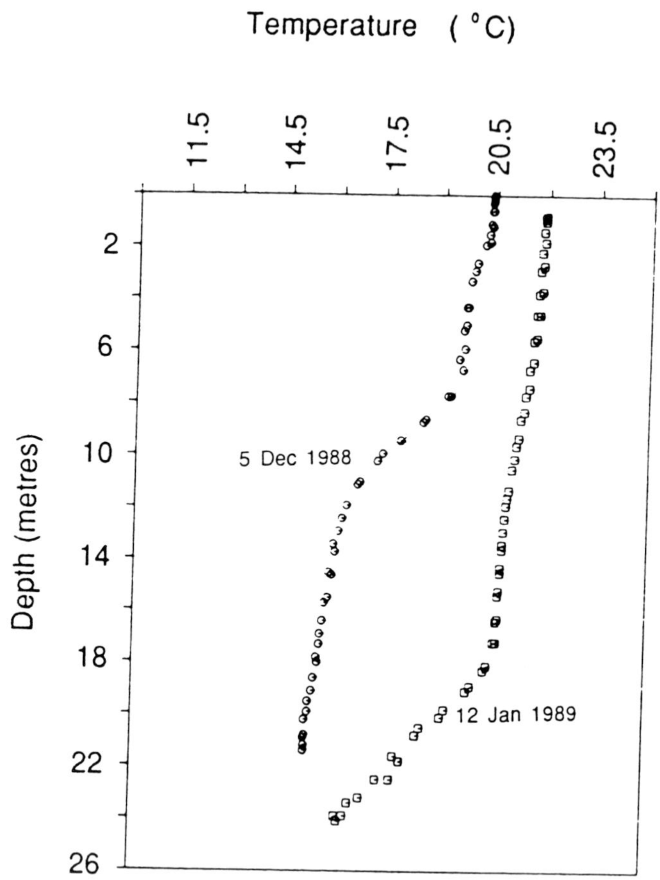

Figure 9: Temperature profiles collected in the central bay on 5 December 1988 and 12 January 1989.

temperatures, is 0.022 s^{-1}. The maximum value, calculated for the temperature gradient in the centre of the water column is close to double this value.

The factor π/NH in (2.15) is an eigenvalue derived, for the case of constant N, from solution of the eigenvalue problem

$$\frac{d}{dz}\left(\frac{1}{N^2(z)}\frac{dZ}{dz}\right)+\kappa^2 Z=0 \ ,$$

with

$$\frac{dZ}{dz}=0 \quad on \quad z=0,-H.$$

With N = 0.022 s^{-1} and H = 20 m, the eigenvalue is

$$\kappa=\frac{\pi}{NH}=7.1 m^{-1}s.$$

For the actual temperature profile (Figure 9), the eigenvalue (calculated numerically) is κ = 4.6 $m^{-1}s$. This is the same eigenvalue as would occur for a constant buoyancy frequency of

$$N=\frac{\pi}{\kappa H}=0.034 s^{-1},$$

although the structure of the eigenvectors for the constant and variable N cases will obviously be different. As a demonstration of the sensitivity to density, model results will be produced for constant N values of both 0.022 and 0.034 s^{-1}.

If P has the vertical structure given by (2.14) then, by (2.13), the temperature at the Black station will be given by

$$T=-T_0 \sin\frac{\pi z}{H},$$

where T_0 is the amplitude, and the minus sign is included because z takes strictly negative values. By (2.13), the pressure at the station is then

$$P=\frac{H}{\pi} g\alpha T_0 \cos\frac{\pi z}{H}.$$

The amplitude T_o and frequency to be used in the models runs were determined as follows. The temperature time-series from the Black station was linearly detrended, then least-squares fitted with a single harmonic of the form

$$T=T_0 \cos(\omega t-\phi), \tag{2.23}$$

where ϕ is the (constant) phase. The correlation between the actual and harmonic time-series was tested over a range of frequencies at intervals of 0.5 x 10^{-6} s^{-1}, and showed a clear best fit at ω = 9 x 10^{-6} s^{-1}, corresponding to an oscillation period of 8 days. The best-fit amplitude was T_0 = 0.56°C. Figure 10 shows a comparison between this harmonic and the detrended temperature signal for the Black station.

A full list of the parameter values used in the model runs is shown in Table 2. The results of the model simulation with $N = 0.034\ s^{-1}$ are shown in Figures 11 and 12. Figure 11 shows the (normalised) pressure field both in-phase and out-of-phase with the temperature forcing; that is, at the time of maximum temperature at Black, and quarter of a period later. For this value of N, the offshore decay scale (baroclinic radius of deformation), $1/\ell$, is 2 km and the longshore wavelength is 136 km, approximately three times the circumference of the basin. Figure 12 shows the surface velocity vectors simultaneous with the pressure fields in Figure 11. By (2.11), (2.12), and (2.14), the velocity varies vertically as $\cos\pi z/H$. Thus bottom velocities, for example, will have the same magnitude but opposite direction to those shown in Figure 12.

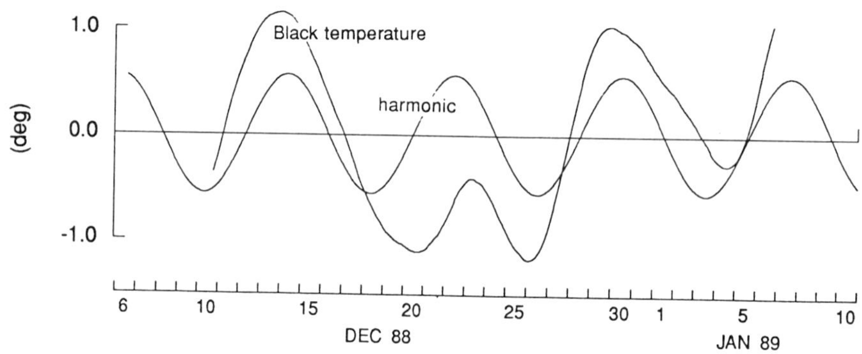

Figure 10: Comparison between the linearly detrended temperature signal from Black, and the best-fit harmonic, having amplitude 0.56°C and frequency $9 \times 10^{-6}\ s^{-1}$

Table 2: Parameters used in model simulations

Name	Symbol	Value
buoyancy frequency	N	$0.034\ s^{-1}$
		$0.022\ s^{-1}$
Coriolis frequency	f	$-10^{-4}\ s^{-1}$
imposed frequency	ω	$9 \times 10^{-6}\ s^{-1}$
accn. due to gravity	g	$9.8\ ms^{-2}$
imposed temp. amplitude	T_o	0.56°C
depth	H	20m
bay width	X	8×10^3 m
bay length	Y	15×10^3 m
grid spacing	Δx	500 m
mouth coordinates	y_1	2.5×10^3 m
	y_2	6.0×10^3 m

The model results all confirm anticipated behaviour in the model basin. The signals for both temperature and velocity decay in magnitude away from the walls. There is a progressive phase delay clockwise around the bay, with peak temperatures, for example, occurring approximately one day later at Yellow, on the north side of the entrance channel, than at Black, where the signal is initiated.

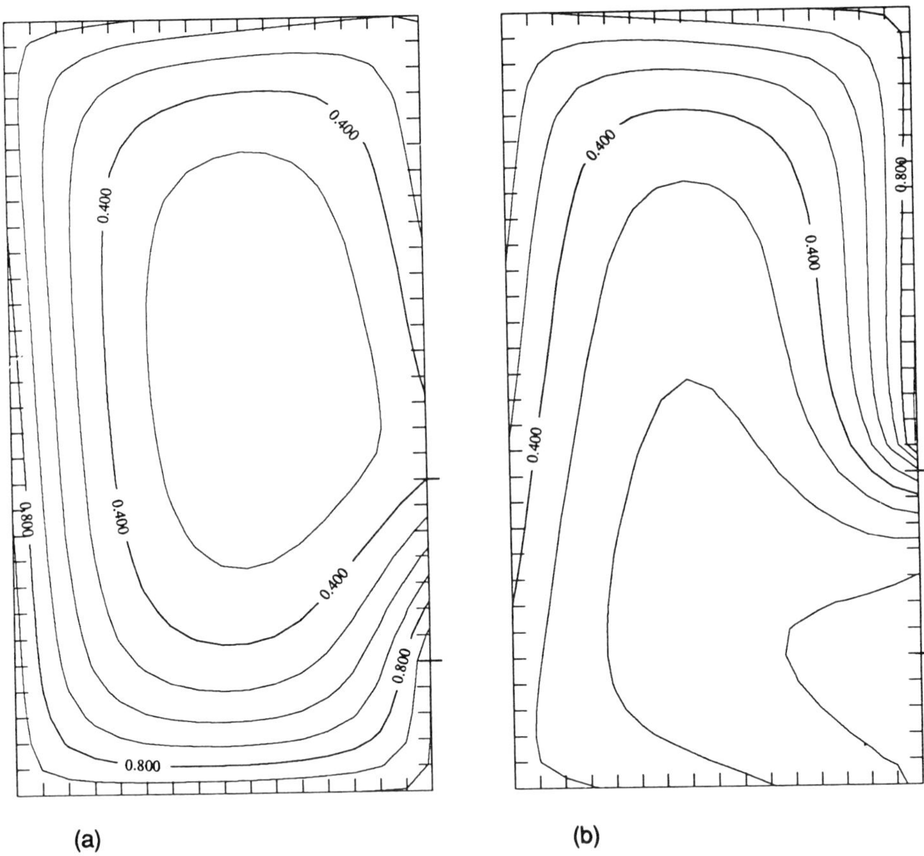

Figure 11: The modelled (normalised) surface pressure field for N = 0.034 s^{-1}, (a) in-phase and (b) out-of-phase with the specified temperature at Black.

For the weaker density structure, N = 0.022 s^{-1}, the solution structure is not dramatically altered (Figure 13). Because the horizontal decay scale is shorter in this case then, for the same amplitude temperature forcing, currents near to the coast are marginally stronger, but the decay is more rapid towards the centre of the bay.

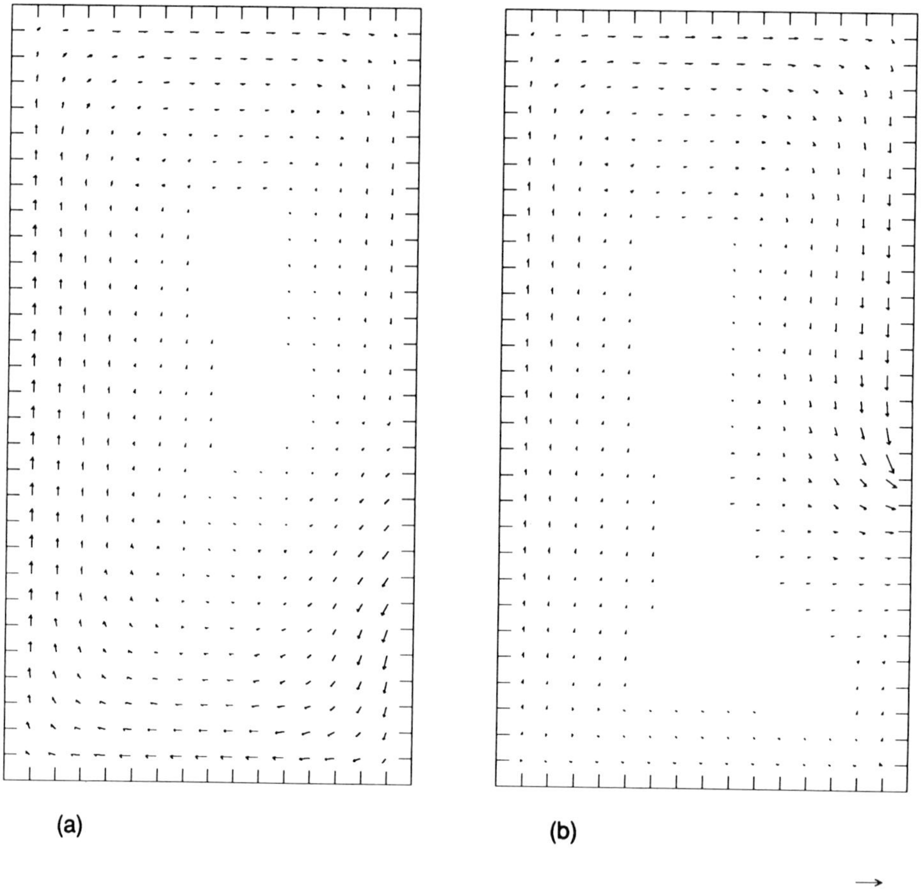

(a) (b)

5 m/s

Figure 12: Modelled surface current fields for N = 0.034 s^{-1}, (a) in-phase and (b) out-of-phase with the specified temperature at Black.

2.6 Discussion

To enable a quantitative comparison with the model results, the measured time-series were each least-squares fitted with a single harmonic having a fixed frequency $\omega = 9 \times 10^{-6}$ s^{-1}, as described for the Black temperature in the previous section. The time-series analysed in this way were those from station depths shown in Figs 5 and 7, except that, for the Red station, the 14.5 m temperature (Fig. 6) and the 6.5 m current were used. The analysis determines an amplitude and phase for each time-series.

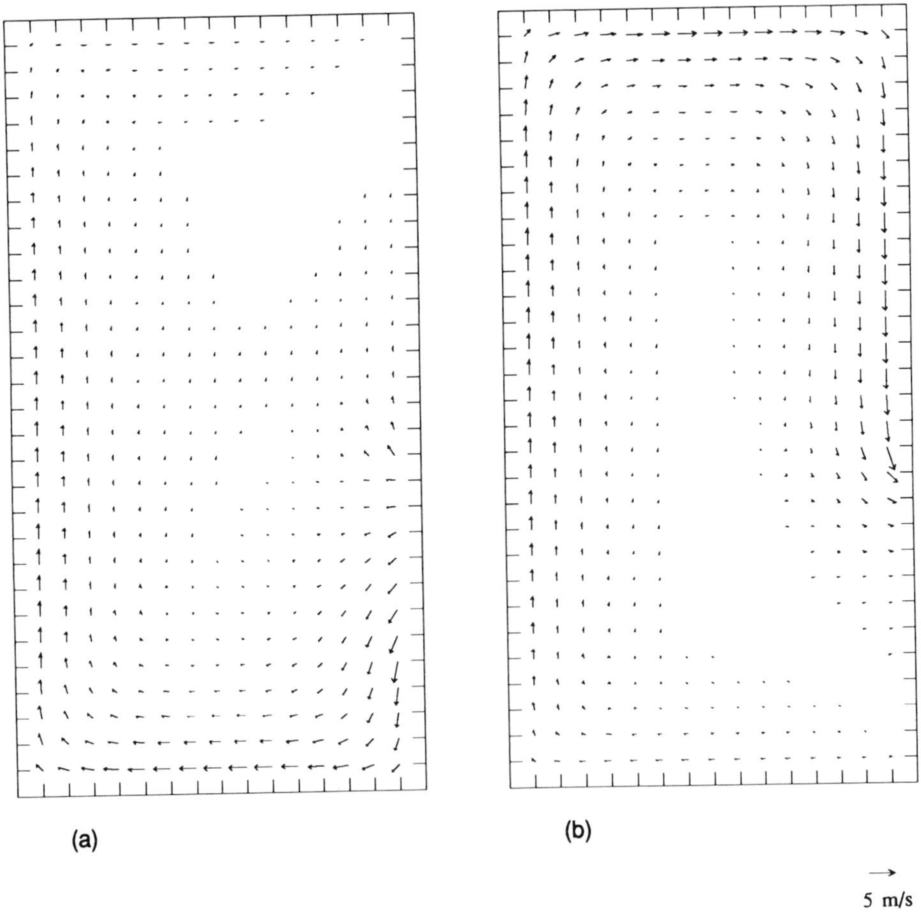

Figure 13: Modelled surface current fields as in Fig. 12, but for N = 0.022 s^{-1}.

Results for both the model and measurements are summarised in Tables 3 and 4. Table 3 lists amplitudes of both temperatures and currents. The current amplitudes quoted are those in the direction of the semi-major axis of the current ellipse. The model amplitudes are maxima over the water column, that is, mid-water values for temperature and surface values for currents. The agreement between the model and the measurements indicated by Table 3 is well within a factor of 2, except at the Red station. The measured velocities and, particularly, temperatures at Red suggest little, if any, decay in signal strength with distance from the coast of the bay.

Table 3: Comparison of modelled and measured temperature and current amplitudes

Station	Temperature (°C)		Current (ms^{-1})	
	Model	Measured	Model	Measured
Black	0.56	0.56		
Red	0.16	0.67	0.003	0.015
Green	0.34	0.44	0.013	0.020
Blue	0.33	0.53	0.014	0.020
Yellow	0.40	0.66		

Table 4 lists phase delays in the temperature signal between adjacent stations, moving clockwise around the bay. There is good agreement in the phase delays between most pairs of stations, with the measurements confirming the model prediction of clockwise signal propagation around the bay. As with the amplitudes, however, there is one fundamental difference. The model is formulated so that the temperature oscillation enters the bay at Black, propagates around, and exits at Yellow. For this reason, the Yellow to Black phase delay in Table 4 is negative rather than positive. The measured signal, however, appears to begin at the Green station and propagate progressively through Blue, Black and Red. In contrast to the model, there is no phase difference between the signal at Yellow and Black, but there instead appears to be some form of signal discontinuity between Red and Green. Interestingly, if the Green time-series had been used to calculate the forcing amplitude T_0 in (2.23), the accuracy of the modelled amplitudes in Table 3 would improve considerably.

Table 4: Comparison of modelled and measured temperature phase delay between adjacent stations

Stations	Phase delay (Days)	
	Modelled	Measured
Black to Red	0.7	0.7
Red to Green	0.2	-2.3
Green to Blue	0.4	0.3
Blue to Yellow	0	0.7
Yellow to Black	-1.2	0

In summary, then, the model appears to predict approximately the correct amplitude and phase relationship in the bay, but the observed lack of decay away from the coast, and the apparent appearance and disappearance of signals in the north-western region are anomalous. Given the simplicity of the model, however, the level of agreement with the data is encouraging.

An obvious shortcoming of the model is the formulation of the forcing at the mouth of the bay. As we found, the prescription of a temperature oscillation at the nominal Black station (Figure 8) leads to an internal Kelvin wave propagating clockwise around the bay. For consistency, to ensure that energy does not accumulate within the bay, it is then necessary to allow the wave to exit, by specification of the radiation condition (2.22) at the mouth. This formulation ignores all complication associated with matching the bay motion to the larger scale motion over the shelf. As just discussed, it also leads to inconsistent phase behaviour across the mouth.

In terms of the behaviour at the mouth, an apparently more consistent model is the even simpler one discussed briefly in section 2.4. This model ignores the ω/f terms in (2.11) and (2.12), so that the bay boundary, including the mouth, becomes a streamline. While the amplitude of the modelled dynamics (Table 3) is not greatly affected by this additional approximation, the model provides no information about phase, since temperature oscillations across the bay are exactly in phase. This is not a bad approximation, since the measured phase delay from Green to Red is only 1.7 days (Table 4), less than a quarter-period.

Retention of the ω/f terms in (2.11) and (2.12), however, allows for internal Kelvin waves, and indicates that some aspects, at least, of the phase structure in the bay are consistent with wave propagation. Since, in reality, the wave does not appear to propagate in and out through the mouth, the model needs, at the next stage, to incorporate dissipation to ensure that the wave energy in the bay does not grow without bound.

Amongst other obvious omissions from the model, realistic bathymetry would modify the structure and propagation rates of the waves inside the bay. Direct surface wind stress is probably capable of generating significant currents at these frequencies (Figure 2). Further, lateral density gradients, probably caused by surface heating, may significantly modify the dynamics. The temperature plot, Figure 5, indicates a temperature rise of up to 2° from the Black to the Green station. Given these omissions, the model has given a useful insight into Jervis Bay's dynamics. Further investigation is likely to require considerable complication of the model.

Acknowledgements

We wish to thank: Graham Symonds and Rick Nunes Vaz for their contributions to this work; Marc Jeffrey and John Mathias for their assistance with the field work; John Andrewartha for preparation of the data; John Church for his comments on the manuscript;and Cathy Flanagan and Nikki Pullen for seeing the manuscript into typescript.

This project was partially funded by the Australian Department of Defence, as a component of a baseline study supervised by the CSIRO Division of Fisheries.

2.7 References

Church, J. A., H. J. Freeland and R. L. Smith, 1986a: Coastal-trapped waves on the east Australian continental shelf. Part I: propagation of modes. *Journal of Physical Oceanography,* 16, 1929-1943.

Church, J. A., N. J. White, A. J. Clarke, H. J. Freeland and R. L. Smith, 1986b: Coastal-trapped waves on the east Australian continental shelf. Part II: model verification. *Journal of Physical Oceanography,* 16, 1945-1957.

Church, J. A. and H. J. Freeland, 1987: The energy source for the coastal-trapped waves in the Australian Coastal Experiment region. *Journal of Physical Oceanography,* 17, 289-300.

Clarke, A. J., 1987: Origin of the coastally trapped waves observed during the Australian Coastal Experiment. *Journal of Physical Oceanography,* 17, 1847-1859.

Craig, P. D., 1989: Water circulation in Jervis Bay. In *Jervis Bay Baseline Studies: Third Progress Report,* December 1989, CSIRO Division of Fisheries, 305-328.

Freeland, H. J., F. M. Boland, J. A. Church, A. J. Clarke, A. M. G. Forbes, A. Huyer, R. L. Smith, R. O. R. Y. Thompson and N. J. White, 1986: The Australian Coastal Experiment: a search for coastal-trapped waves. *Journal of Physical Oceanography,* 16, 1230-1249.

Gill, A. E. and A. J. Clarke, 1974: Wind-induced upwelling, coastal currents and sea-level changes. *Deep-Sea Research,* 21, 325-345.

Hamon, B. V., 1962: The spectrums of mean sea level at Sydney, Coff's Harbour and Lord Howe Island. *J. Geophys. Res.* 67, 5147-5155 (Correction, 1963 *J. Geophys Res.* 68, 4635)

Hamon, B. V., 1966: Continental shelf waves and the effects of atmospheric pressure and wind stress on sea level. *J. Geophys. Res.*, 71, 2883-2893.

Holloway, P. E., G. Symonds, R. Nunes Vaz, M. Jeffrey and J. Mathias, 1989: Oceanographic measurements in Jervis Bay, December 1988 to January 1989. Working paper 1989/2, Department of Geography and Oceanography, Australian Defence Force Academy, 44 pp.

Holloway, P. E., G. Symonds, R. Nunes Vaz, and M.Jeffrey, 1990: Temperature and salinity surveys in Jervis Bay and on the adjacent shelf,. February 1988 to April 1990. Working paper 1990/1, Department of Geography and Oceanography, Australian Defence Force Academy, 59 pp.

Huyer, A., R. L. Smith, P. J. Stabeno, J. A. Church and N. J. White, 1988: Currents off south-eastern Australia: results from the Australian Coastal Experiment. *Australian Journal of Marine and Freshwater Research,* 39, 245-288.

Middleton, J. F. and F. Viera: 1991: The forcing of low frequency motions within Bass Strait. *J. Phys. Oceanogr*, 21, 695-708.

Thompson, R. O. R. Y., 1983: Low-pass filters to suppress inertial and tidal frequencies. *Journal of Physical Oceanography*, 13, 1077-1083.

3

Tidal Mixing near the Sill of a Scottish Sea Loch

A J Elliott, P A Gillibrand and W R Turrell

Abstract

Current and CTD measurements were made during the summer of 1989 near the inner sill of Loch Sunart, a sea loch on the west coast of Scotland. The inner basin, which is around 7km in length and contains water depths of around 90m, is separated from the outer basins by a shallow sill that reaches to within 10m of the surface. Current observations from the inner basin show a striking 180° phase difference between the surface and near-bottom tidal currents. This has been interpreted in terms of the vertical modal structure of the internal tide. Circulation and mixing in the loch is being modelled by the laterally-averaged semi-implicit estuarine model of Wang and Kravitz (1980) which has been modified to allow an irregular vertical grid spacing and to permit the sea surface to move vertically through the grid. Numerical experiments have shown that the baroclinic motion is generated as a result of the sill being located near the depth of the halocline. The model results suggest that an increase in the near-surface stratification following a run-off event may cause the internal tide to strengthen and generate an increased shear across the halocline thus maintaining the vertical mixing between the surface and bottom waters.

3.1 Introduction

Fish farming in Scottish sea lochs has expanded rapidly during the past decade. The annual production of salmon has increased from 500 tonnes in 1979 to 28,500 tonnes in 1989, with a turnover of £100 million. This expansion, however, has raised concern over possible ecological damage caused by the increasing quantities of waste materials produced by the farms. A significant proportion of the food dispensed sinks and settles on the sea bed, and the uneaten food and material excreted by the fish may enhance the levels of nutrients and organic substances in the water column and sediments (Duff, 1988). Secondly, the use of toxic chemicals to treat the caged salmon stock for sea lice has been controversial (Anon, 1989). A third problem relates to bacterial infection of the caged salmon caused by water borne pathogens. These can be carried between adjacent cages by the tidal currents and residual flow and thus spread the disease from an infected cage to its neighbours.

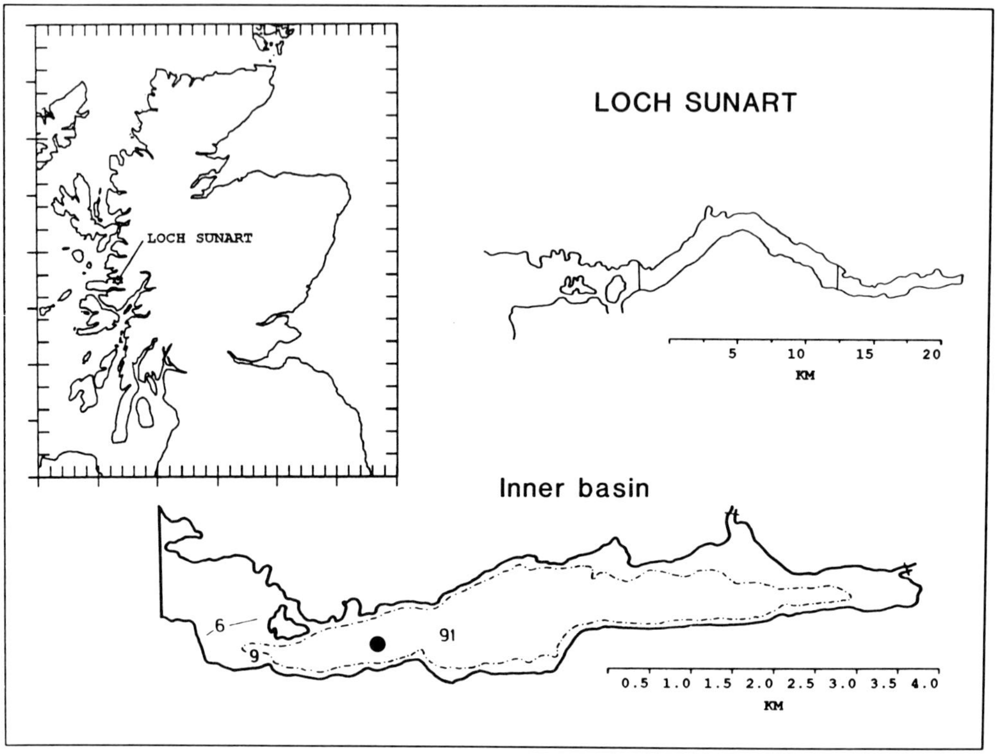

Figure 1: Loch Sunart showing the entire system (top) and inner basin (bottom), the location of the mooring is marked by a circle and bottom depths are shown in metres.

Such problems have prompted studies of the dynamics and flushing characteristics of the fjordic sea lochs on the west coast of Scotland, there being particular interest in the mechanisms that cause the renewal of the deep water of the inner basins and which influence the retention times of pollutants within the system. One particular fjord selected for study is Loch Sunart (Figure 1); a 35km long loch that possesses two major sills, the outer sill being at a depth of 30m while the inner sill is only 10m beneath mean sea level. Maximum water depths are around 120m and 90m in the main and inner basins, respectively. A tidal amplitude at springs of about 2m at the inner sill generates strong tidal currents and turbulence which is evident as a 'boiling' of the surface water above the sill. A surface convergence front, similar to those found in shallower estuaries (Turrell and Simpson, 1988), is observed near the sill during a spring flood tide. Hydrographic surveys during the summer months have shown strong stratification in the inner basin, with low salinity water in a surface layer about 5m thick overlying the deeper more saline water. However, dissolved oxygen values were comparable in the deep water to those found at the surface which suggests that vigorous vertical mixing takes place throughout the summer (DAFS, unpublished data).

In contrast to Sunart, other lochs show an intermittent renewal of the deep water due to the effect of fresh water run-off in controlling the salinity near the sill. For example, in Loch Etive (Edwards and Edelsten, 1977) renewal of the deep water within the basin is prevented at times when the density of the water near the sill, which is controlled by run-off, is less than that of the deep water. For many practical problems, such as those associated with fish farms, it would be useful if a criterion could be developed by which sea lochs could be easily classified into those for which the deep water renewal occurs either continuously throughout the year or else intermittently. Such a classification is likely to require consideration of the work done by the tidal streams near the sill in relation to the potential energy of the salinity induced stratification within the basin (Edwards and Sharples, 1986). Details of the tidal flows and stratification levels are unknown in many lochs and would need to be estimated by modelling. In such cases, therefore, parameterization of the vertical mixing within the loch may be more readily obtained by extending an existing tidal model than by resorting to an argument based on energy considerations. One goal of the Sunart study was to assess the ability of an estuarine model to reproduce observed current and salinity features within the inner basin of a sea loch.

3.2 Observational Results

For 6 weeks from July 8 to August 24 (days 189-236) of 1989 a current meter mooring was placed about 1km to the east of the narrows at the entrance to the inner basin (Fig.1). The sill is located at the western end of the narrows where the depth is less than 10m beneath chart datum. Water depth at the mooring was about 60m, the deepest part of the basin being located about 1km to the east of the mooring. The mooring supported two Aanderaa current meters, one at a depth of 5m and the other at 50m. A pressure gauge to record sea level changes was deployed near the bank of the loch to the south of the mooring. A bottom mounted ADCP instrument was also deployed in the loch; this instrument was placed to the south of the main mooring and collected data during a 10 day period of intensive measurements from August 15-24 (days 227-236). A CTD survey of the loch was made during August 14-15 (days 226-227) and again during August 24-25 (days 236-237).

Figure 2 shows the salinity distributions observed along the centre line of the loch, the inner basin being located between 25-32km from the mouth. (The temperatures were similar during each of the surveys, with typical values of about 13.8°C. The horizontal temperature gradients were weak and vertical contrasts were generally less than 0.5°C in the inner basin, the vertical stratification being dominated by salinity.) During the first survey the surface water within the basin had a value of around 20 psu, while values of around 33 psu were recorded at a depth of about 20m. The stratification was weaker in the outer basins, but still reached values of 4 psu per 20m in the near-surface layers. There was heavy rainfall and strong winds directed into the loch between the two surveys, and Figure 2b shows the resulting marked reduction in the surface salinities near the head of the loch. Surface values of around 5 psu were recorded in the inner basin during the second survey when the strongest halocline was at a depth of about 5m. Temperature was more well-mixed vertically at the time of the second survey, however this may be a result of the reduced data coverage (the CTD failed during the second survey and the results shown are due to bottle data).

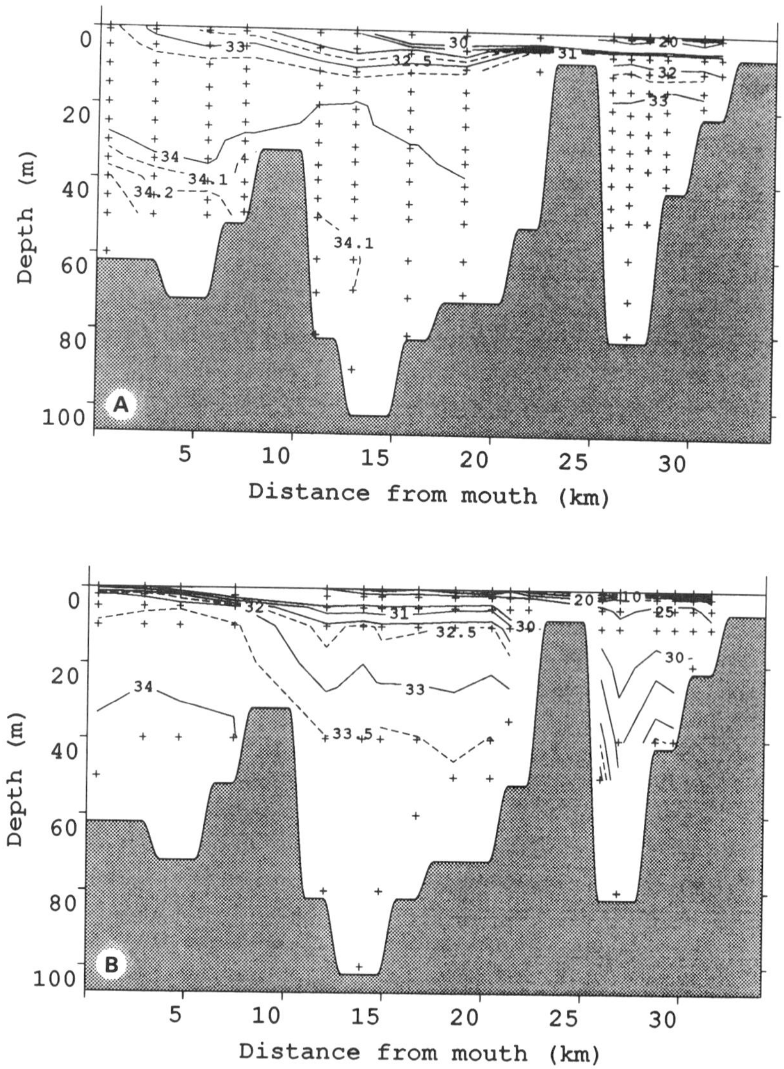

Figure 2: Observed salinity distributions (psu): (a) August 14-15 (days 226-227), (b) August 24-25 (days 236-237).

The mooring data have been analysed to extract information on the propagation of tidal energy into the inner basin. The data series were also low pass filtered to remove the tides and, although the results are not presented here, it is appropriate that some comments should be made about the non-tidal fluctuations. Sea level was recorded by measuring pressure near the sea bed and then converting the pressure signal to water height. Consequently, the elevation signal was automatically corrected for the inverse barometer effect and the non-tidal elevation record showed relatively little energy. The dominant fluctuations were of magnitude 0.2m and occurred over timescales of 2-5 days. Significantly more energy was shown by the low pass current at a depth of 5m. This contained pronounced energy at the 2-day timescale, although the fluctuations had a magnitude of only about 0.05m s^{-1}. The fluctuations at 50m were even weaker, reaching values of only 0.02m s^{-1}, however the 2-day timescale was again apparent. The long-term mean flow was close to zero at both instrument depths. In contrast to shallow estuarine systems which can be dominated by wind forcing, e.g. Chesapeake Bay (Wang, 1979), there was no obvious coherence between the low frequency variability in the sea level, near-surface and near-bottom currents.

Figure 3 shows examples of the mooring data from the period July 28 to August 8 (days 210-220). The non-linear character of the tidal currents is apparent, and results in the near-surface flood currents being stronger and of shorter duration than the ebb. In contrast, the near-bottom current was stronger during the ebb which near the bottom was of shorter duration than the flood.

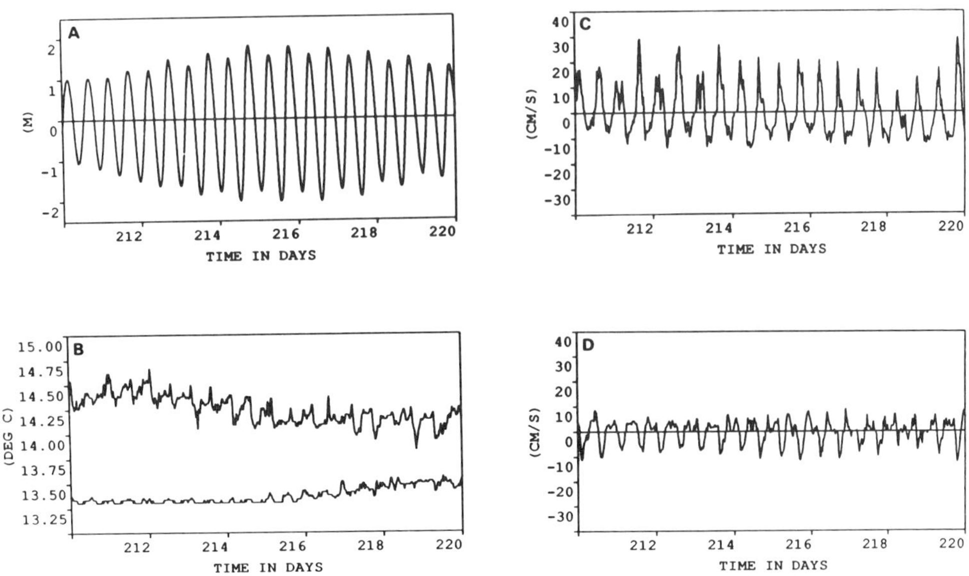

Figure 3: Examples of the data from the period July 29- August 8 (days 210-220): (a) sea level elevation, (b) near-surface (5m) and near-bottom (50m) temperatures, (c) east component of near-surface (5m) current, (d) east component of near-bottom (50m) current.

Table 1 shows the results of a tidal analysis of the elevation and current data. If the tide has the character of a progressive wave as it enters the inner basin we would expect the elevation and currents to be in phase. However, Table 1 shows that the surface current led the elevation by a phase of about 55° at the M_2 frequency while the S_2 current led the elevation by nearly 80°.

Table 1: Results of a tidal analysis of the elevation and current data. Amplitudes are in cm (elevation) and cm s^{-1} (current), phases are with respect to the equilibrium tide at Greenwich, orientation of the ellipse major axis is in degrees anti-clockwise from east.

Elevation								
	H	G						
M_2	136.6	178.0						
S_2	54.8	232.5						
M_4	5.4	201.8						
MS_4	4.7	331.6						
5m								
	East		North		Major	Minor	Orient.	Phase
	H	G	H	G				
M_2	9.1	124.5	1.9	130.0	9.3	0.2	11.8	124.7
S_2	4.5	155.4	1.0	172.9	4.6	0.3	12.3	156.2
M_4	1.3	236.7	0.5	233.5	1.4	-0.0	22.4	-123.8
MS_4	1.4	284.3	0.5	280.0	1.5	-0.0	18.4	-76.2
50m								
	East		North		Major	Minor	Orient.	Phase
	H	G	H	G				
M_2	2.6	317.9	0.5	308.6	2.7	-0.1	11.2	-42.5
S_2	1.5	276.9	0.3	286.0	1.5	0.1	12.4	-82.7
M_4	1.6	63.4	0.4	55.7	1.6	-0.1	13.2	63.0
MS_4	0.9	90.7	0.3	103.5	1.0	0.1	17.9	91.9

This suggests that a significant amount of tidal energy was being reflected from the head of the loch, causing the tidal wave to have some of the character of a standing wave. Both surface and bottom M_2 currents were rectilinear and directed towards 11° (with respect to magnetic east) which implies a true bearing of about 23° anti-clockwise from east. This agrees well with the orientation of the basin axis (Figure 1). A striking feature of the tidal analysis is the nearly 180° phase difference between the M_2 currents near the surface and bottom (Table 1). Figure 4 shows a short portion of the data, taken from a period of spring tides, that illustrates the phase difference in the currents and the non-linear character of the currents and temperature signal.

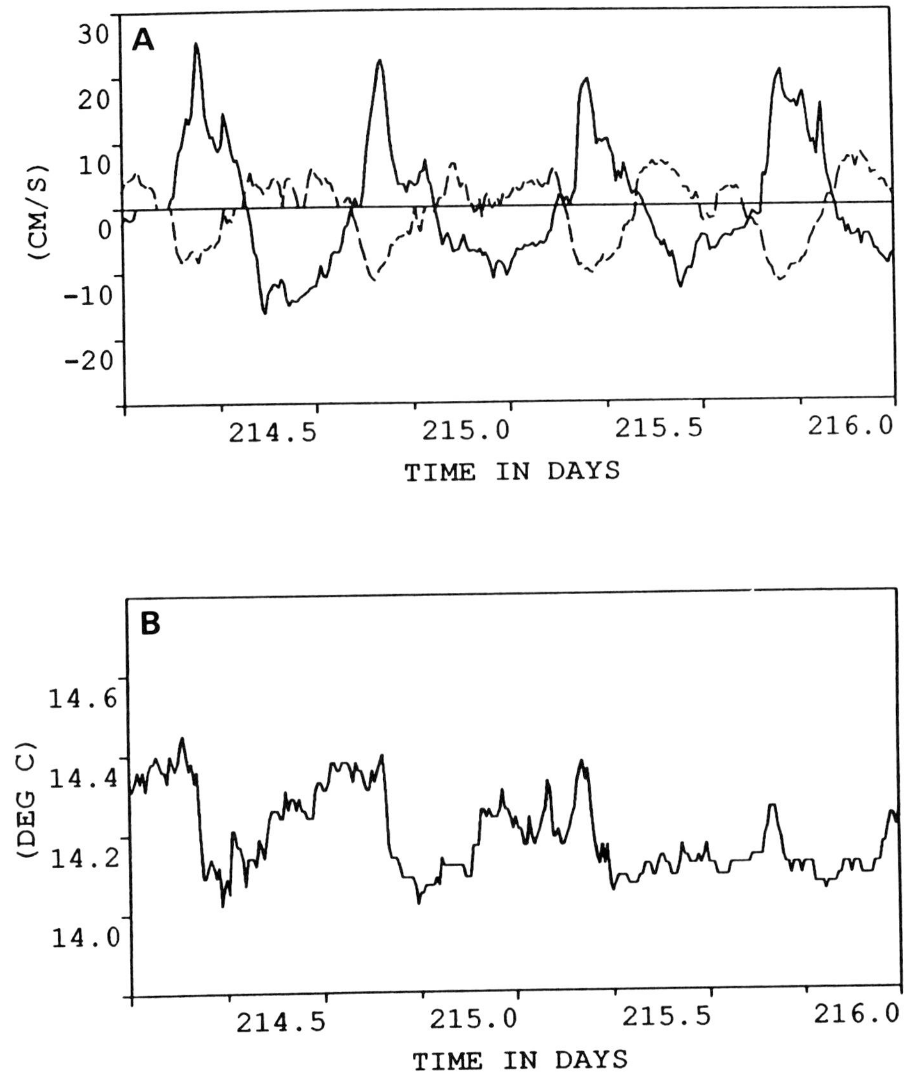

Figure 4: Current and temperature details during a period of spring tides (August 2-3): (a) 5m (solid curve) and 50m (dashed curve) currents, (b) 5m temperature.

Further evidence is provided by Figure 5 which shows the vertical structure of the currents recorded by the ADCP during two tidal cycles; these data suggest that the direction of the tidal current reversed between the depths of 20-30m at the semi-diurnal time scale.

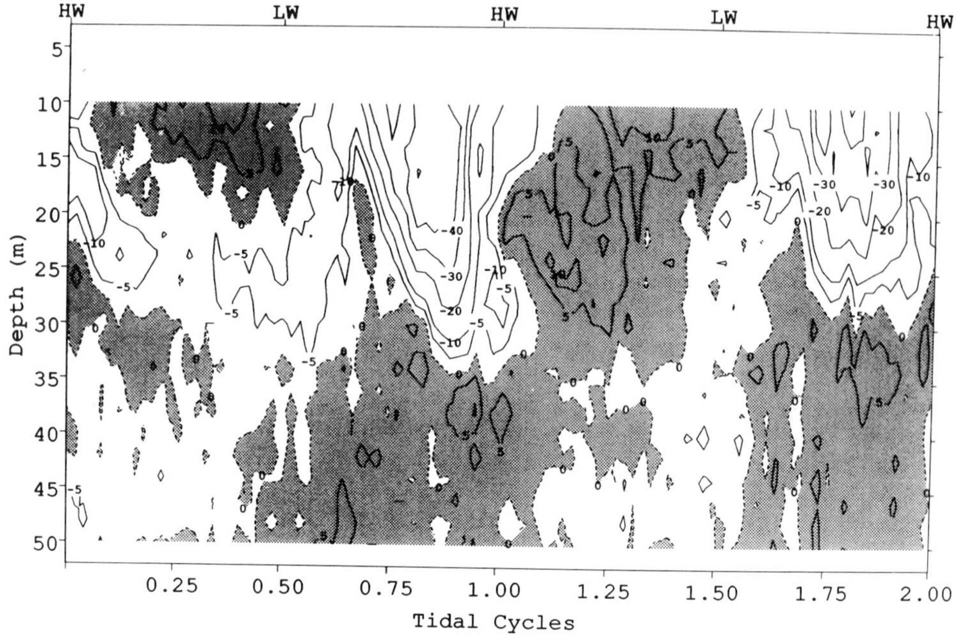

Figure 5: Depth-time contour plot of the ADCP current profile, ebb flow is shown by shading (the data shown cover the period from 1900 on August 18 (day 230) to 2000 on August 19 (day 231)).

Table 1 shows that the shallow water constituents M_4 and MS_4 contributed significantly to the non-linear character of the tide. The quarter-diurnal M_4 was particularly strong in relation to the other constituents at the 50m level where it exceeded the amplitude of S_2. As with M_2, the M_4 constituent showed a 180° phase difference at the two depth levels.

The vertical modal structure can be calculated if it is assumed that the motion is oscillatory in the horizontal with frequency ω and has an amplitude that depends only on depth, i.e. the horizontal and vertical velocities can be written in the form u = U(z).cos(kx-ωt) and w = W(z).sin(kx-ωt) where ω is the known tidal radian frequency and k is an unknown horizontal wavelength (Cushman-Roisin and Svendsen, 1983). Determination of the vertical modes then requires finding the solution of

$$\frac{d^2W}{dz^2} + k^2\left(\frac{N^2}{\omega^2} - 1\right)W = 0 \tag{3.1}$$

where N=N(z) is the Brunt-Väisälä frequency, subject to the boundary conditions W(0)=W(-H)=0 where H is the water depth which is independent of x. The vertical structure of the amplitude of the horizontal velocity, U(z), can then be determined from

$$U = \frac{1}{k} \cdot \frac{dW}{dz} \tag{3.2}$$

If the second derivative in (3.1) is written in finite difference form then a matrix equation for the values of W at each depth can be obtained in which the unknown wavenumber scale k is related to the eigenvalues of the matrix. A more intuitive solution can be found by writing (3.1) in finite difference form, setting W(0)=0 and then marching the solution for W(z) from the surface to the bottom. If W(-H)=0 then a modal solution has been determined, otherwise a new value must be selected for k and the procedure repeated. The method can be made to converge rapidly by careful choice of the parameter k, the number of zero crossings of the solution determining the order of the mode.

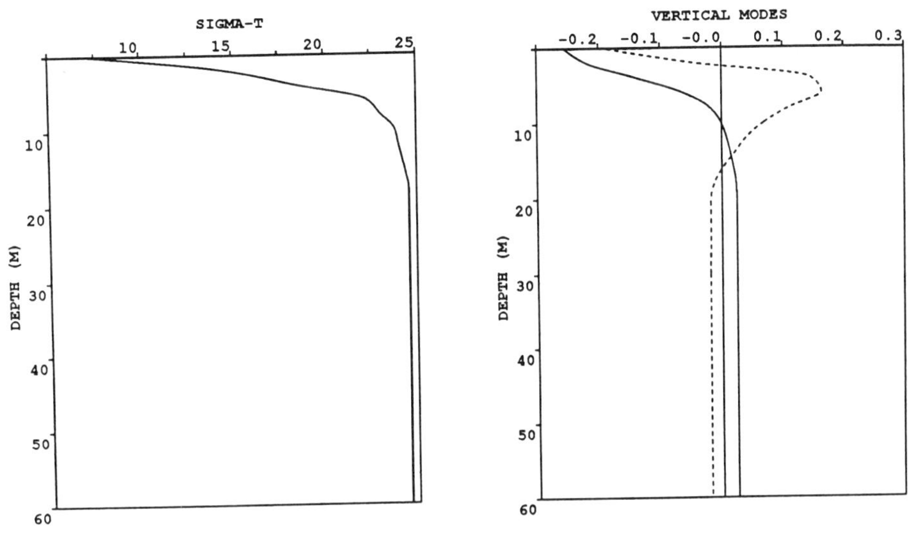

Figure 6: Vertical profile of sigma-t at the mooring position and the computed vertical modes.

Figure 6 shows a representative profile of sigma-t from the mooring position during the survey of August 14-15. This was used to calculate the $N^2(z)$ profile that appears in (3.1) with ω set to the frequency of the M_2 tide. The figure also shows the vertical structure of the amplitude of the horizontal current, the solid curve showing the first mode and the second mode being shown by the dashed curve. The first and second modes had horizontal length scales of 27km and 12km, respectively (Table 2). Both modes show the observed 180° phase difference between the depths of 5m and 50m. It should be noted that the 27km length scale of the first M_2 mode is about four times the length of the inner basin and that a resonant response may therefore exist. The calculations were also made for the frequency of the M_4 tidal constituent and the results are again summarized in Table 2.

Table 2: Results of an analysis of the vertical modal structure. The length scale L is given by L=2π/k where k is the horizontal radian wave number.

Calculated	mode 1		mode 2	
	M_2	M_4	M_2	M_4
L (km)	27.3	13.7	12.2	6.1
$\frac{\lvert U_5 \rvert}{\lvert U_{50} \rvert}$	4.2	4.2	8.1	8.1
Observed	M_2	M_4		
	3.5	0.9		
$\frac{\lvert U_5 \rvert}{\lvert U_{50} \rvert}$				

Although the vertical shapes of the modes were similar at the higher frequency, the length scales were reduced to around 14km and 6km for the first two modes. Although the 6km length scale of the M_4 second mode agrees well with the 7km basin size, the 8:1 ratio of the surface:bottom current speed does not agree with the observed 1:1 ratio. In contrast, the computed first mode ratio of 4:1 for the M_2 frequency agrees reasonably well with the observed ratio of 3.5:1. Both computed modes (Figure 6) predict strong vertical shear in the horizontal flow beneath the halocline in the depth range 5-15m, suggesting that the internal tide can generate an enhanced shear beneath the halocline which may lead to increased vertical mixing rates.

3.3 Modelling Results

The laterally-averaged equations can be written in the form (Wang and Kravitz, 1980):

Equation of continuity

$$\frac{\partial}{\partial x}(uB) + \frac{\partial}{\partial z}(wB) = 0 \qquad (3.3)$$

$$\frac{\partial}{\partial t}(B_0\eta) + \frac{\partial}{\partial x}\int_{-H}^{\eta}(uB)dz = 0 \tag{3.4}$$

Momentum equation

$$\begin{aligned}&\frac{\partial}{\partial t}(uB) + \frac{\partial}{\partial x}(uuB) + \frac{\partial}{\partial z}(uwB)\\ &- \frac{\partial}{\partial x}\left(BN_x\frac{\partial u}{\partial x}\right) - \frac{\partial}{\partial z}\left(BN_z\frac{\partial u}{\partial z}\right) + C_D u|u|\left|\frac{\partial B}{\partial z}\right|\\ &+ gB\frac{\partial \eta}{\partial x} + \frac{gB}{\rho_o}\frac{\partial}{\partial x}\int_z^0 \rho\, dz' = 0\end{aligned} \tag{3.5}$$

Salt conservation

$$\begin{aligned}&\frac{\partial}{\partial t}(sB) + \frac{\partial}{\partial x}(suB) + \frac{\partial}{\partial z}(swB)\\ &- \frac{\partial}{\partial x}\left(BK_x\frac{\partial s}{\partial x}\right) - \frac{\partial}{\partial z}\left(BK_z\frac{\partial s}{\partial z}\right) = 0\end{aligned} \tag{3.6}$$

Equation of state

$$\rho = \rho_0(\alpha + \beta s) \tag{3.7}$$

where x and z are the longitudinal and vertical coordinates; u and w are the corresponding velocity components; B is the width; η is the surface elevation; H is the mean water depth; C_D is the bottom friction coefficient; K_x, N_x are the longitudinal diffusivity and viscosity; and K_z, N_z are the vertical diffusivity and viscosity; ρ_0 is the density of fresh water, and β (7.6 x 10^{-4}) represents the dependence of density on the salinity, s.

The dynamical boundary conditions are given by

$N_z\, \partial u/\partial z$ = wind stress at the surface and $N_z\, \partial u/\partial z = C_d u\,|\,u\,|$ at the bottom. The condition that there should be no vertical flux of salt at the surface and bottom is satisfied providing that $K_z\, \partial s/\partial z = 0$ at $z = \eta$ and $z = -H$.

At the head of the loch the river flow is specified and the salt flux is set to zero, while at the mouth the sea level elevation is specified. When inflow occurs at the mouth the ocean salinity is specified, otherwise an advective balance is assumed and horizontal diffusion is neglected on outflow.

The set of equations (3.3)-(3.7) and the given boundary conditions are standard. One of the first solutions was obtained by Hamilton (1975) who used a two time step integration scheme and allowed the free surface to move vertically through the grid. The model was applied to the Rotterdam Waterway, and then to the Mersey estuary (Bowden and Hamilton, 1975). A similar model was developed by Blumberg (1977) in a study of the circulation of Chesapeake Bay. His model used the leapfrog time stepping scheme and solved equations that had been integrated within each vertical layer. A further model was formulated by Elliott (1976) who did not integrate the equations within each layer but evaluated the system (3.3)-(3.7) at the grid points, and allowed the estuary to branch so that interactions within an estuarine system could be studied. Wang and Kravitz (1980) then developed an implicit solution to the equations (3.4) and (3.5), which allowed them to reduce the computational time by an order of magnitude. An implicit solution of estuarine dynamics had also been derived by Hamilton (1977).

The code of the Wang and Kravitz (1980) model has been applied to Loch Sunart after the following modifications:

(a) The vertical grid spacing can vary with depth. This is an important modification for loch application which requires close grid spacing in the near-surface layers and above sills with less resolution in the deeper water of the basins.

(b) The code has been modified to allow the surface to move through the computational grid (Figure 7). This facility was included in the model of Hamilton (1975) and is necessary when applying models to European waters where tidal ranges are large. Without this modification the near-surface vertical grid spacing cannot be less than the tidal range.

(c) The hybrid advection scheme of James (1986) has been used to treat the horizontal advection of salt. A purely centered scheme generated negative salinities in the regions of strong advection near the head of the loch.

The Wang and Kravitz (1980) model could not be run with a time step very much greater than the one used in a purely explicit model. In initial tests, the explicit code of Elliott (1976) provided realistic tidal behaviour when used with a time step of 15 s while the longest time step that could be used with the implicit code was around 45 s. The rapid depth changes and the strongly non-linear character of the loch circulation are obviously in contrast with those in shallow, dissipative regions such as the Chesapeake Bay in which the non-linear effects are less important.

Three formulations for the vertical eddy coefficients were tested (in cgs units):

(i) Wang (1983) allowed the turbulent fluxes to depend on depth by using a Munk-Anderson parameterization:

$$N_z = N_o + N_1(1+10\ Ri)^{-1/2} \tag{3.8}$$

$$K_z = K_0 + K_1(1+3.33\ Ri)^{-3/2} \tag{3.9}$$

in which Ri is the local Richardson number given by

$$Ri = -\frac{g}{\rho}\frac{\partial \rho}{\partial z} \Big/ \left(\frac{\partial u}{\partial z}\right)^2 \tag{3.10}$$

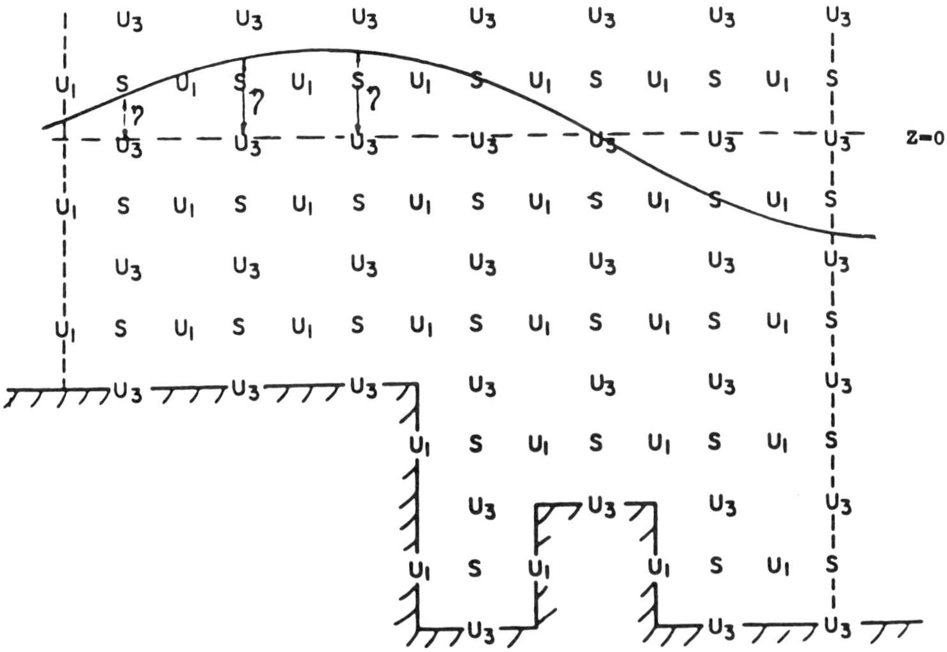

Figure 7: Grid scheme used in the model, the surface is allowed to move vertically through the grid points.

(ii) The model of Bowden and Hamilton (1975) was unstable when the mixing coefficients were depth dependent so they related the eddy coefficients to an overall Richardson number defined in terms of the top to bottom current and density differences. However, with the present integration scheme their formulation remains stable when the Richardson number is made depth dependent, consequently the following expressions were tested in which Ri was given by (3.10):

$$N_z = N_0 + \frac{N_1}{10^2} H|u|.(1+7\ Ri)^{-1/4} \tag{3.11}$$

$$K_z = K_0 + \frac{K_1}{10^2} H|u|.(1+Ri)^{-7/4} \tag{3.12}$$

It should be noted that these expressions are empirical and were derived from data collected in the Mersey Estuary (Bowden and Gilligan, 1971).

(iii) Using the results of Kent and Pritchard (1959), Blumberg (1977) represented the dependence of K_z on the Richardson number as

$$K_z = K_0 + k_1^2\, z^2 \left(1 - \frac{z}{H}\right)^2 \left|\frac{\partial u}{\partial z}\right| \left(1 - \frac{Ri}{Ri_c}\right)^{1/2} \tag{3.13}$$

where Ri_c is a critical value of the Richardson number above which mixing is suppressed. By assuming that the gradient Richardson number (Ri) and the flux Richardson number (Rf) are related by

$$\frac{Rf}{Ri} = \frac{1}{1+Ri}$$

he derived an expression for N_z of the form

$$N_z = K_z(1+Ri) \text{ for } 0<Ri<Ri_c$$
$$\text{and } N_z = K_z = 0 \text{ when } R_i \geq Ri_c \tag{3.14}$$

Figure 8 shows the external forcing that was used when simulating the observed salinity section shown in Figure 2b. The model was spun-up for 3 tidal cycles by specifying and holding constant the salinity observed on August 14-15 (Figure 2a) while forcing with the tidal elevation at the mouth, wind stress and run-off. The integration was then stepped forward for 21 tidal cycles, over the period August 14-25, with the salinity distribution being updated dynamically using (3.6). The final tidal cycle was then averaged to produce the distributions shown in Figure 9. Following the strong run-off event on August 23 the observations showed a very thin (less than 1m thick) surface layer with a salinity of about 5psu. This layer could not be resolved by the 2m grid spacing used in the model; the lowest salinities predicted by the model in the surface layer of the inner basin were of the order of 10psu.

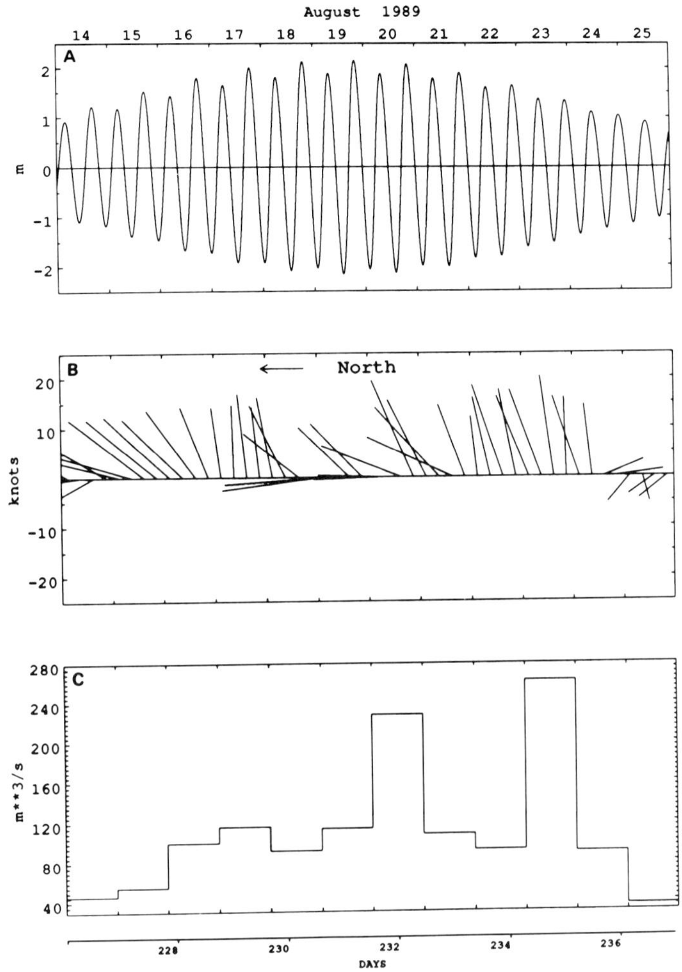

Figure 8: External forcing used to drive the model when simulating the salinity distribution observed on August 24-25.

The Munk-Anderson formulae given by (3.8)-(3.9) produced good agreement with the observations (Figure 9a), both in the near-surface layers of the inner basin and in the deep water of the main basin. However, the low values of the coefficients that were necessary to reproduce the sharp near-surface gradients in the inner basin resulted in reduced mixing at mid-depth in the middle basin. For example, the 33psu contour was reproduced at a depth of about 15m which can be compared to the observed value of about 25m (Figure 2b). For large values of Ri, the Munk-Anderson formulae have the effect of reducing the eddy coefficients to values less than those given by the other methods. The best simulations were obtained using $N_0 = K_0 = 0.05$, $N_1 = 35.0$ and $K_1 = 20.0$.

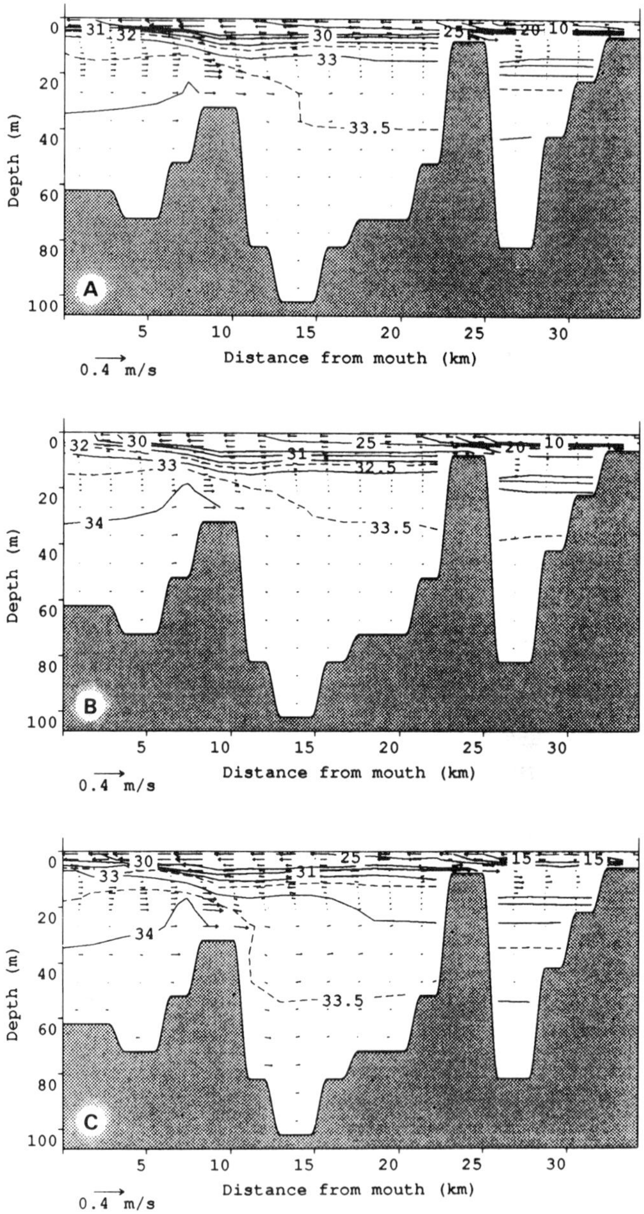

Figure 9: Salinity simulations for different mixing parameterizations: (a) Wang (1983), (b) Bowden and Hamilton (1975), (c) Blumberg (1977).

The formulae (3.11)-(3.12), adapted from Bowden and Hamilton (1975), allow the maximum values of the mixing coefficients to depend on water depth and the instantaneous current speed. The eddy diffusivity, K_z, given by (3.11)-(3.12) for a chosen Ri, is higher than the value obtained using (3.8)-(3.9) due to the strong tidal currents within the loch. In the inner basin this has the effect of reducing the salinity values in the deep water, although the overall agreement with the observed salinities is good (Figure 9b). The best simulations were obtained using $N_0 = K_0 = 0.05$, $N_1 = 0.20$ and $K_1 = 0.10$.

Realistic salinities in the inner basin could not be reproduced when using expressions (3.13)-(3.14) due to Blumberg (1977). Although depth resolving models are usually insensitive to the values chosen for K_x and N_x provided they are less than 10^5 - 10^6 cm^2 s^{-1}, the model would not remain stable unless they were both set to 10 cm^2 s^{-1} when (3.14) was used. The scheme was also unstable unless the critical Richardson number, Ri_c, was set to a value of 5, even though Blumberg (1977) suggests a value of around 10 for the upper limit. With these restrictions, the near-surface salinities in the inner basin reached a minimum value of the order of 20 psu and the observed strong near-surface stratification could not be reproduced (Figure 9c). This suggests that the formulation given by (3.14), which produced realistic salinities in the weekly stratified waters of the Chesapeake Bay, may not be appropriate in highly stratified fjord regions. However, in contrast to the surface waters, reasonable salinity values were computed for the deep water of the outer basins. The best simulations were obtained using $K_0 = 0.05$, $k_1 = 0.035$ and $Ri_c = 2$.

Initial tests aimed at reproducing the vertical phase lag observed in the currents showed that near-surface stratification and the shallow sill at the entrance to the inner basin must both be included in the simulations. Changing the length scale of the inner basin, while holding the sill depth and stratification at chosen levels, did not influence the vertical phase difference. This supports the conclusion made earlier that the internal tide does not result from a resonance of the second mode of the M_4 constituent which has a length scale comparable to the length of the inner basin, and suggests that resonance does not play a part in the baroclinic M_2 response.

Figure 10 shows the results from a series of tests in which the sill at the entrance to the inner basin was set at different depths during high run-off conditions that generate significant near-surface stratification. To make these tests the river flow was set at 65 m^3 s^{-1}, a realistic initial salinity distribution was assigned to the grid points and then the ocean salinity was held fixed while the model was spun-up for 40 tidal cycles without tidal forcing - and without a sill at the entrance to the inner basin. This allowed the internal circulation to establish and for realistic salinity stratification to develop. At the end of the 40 tidal cycles, by which time the solution was approaching steady state, the tidal forcing was applied at the mouth, a sill was inserted at the entrance to the inner basin and the simulation was continued for a further 7 tidal cycles, the last two of which were used to generate the figures. (The results of the 40 tidal cycles spin-up was stored on disk so that the final 7 tidal cycles of simulation could continue without duplicating the initial calculations.) With the sill at 52m (Figure 10a) the simulated tidal currents at the mooring position are fairly barotropic with maximum values at spring tides being reduced to around 10cm s^{-1}. The surface flow ebbs throughout the tidal cycle because the tidal currents during flood are weaker than the river induced surface velocity. Raising the sill to 20m and 14m (Figure 10b,c) creates a jet centered at 10m on the flood tide and introduces the observed vertical phase difference. In Figure 10d are shown results from the simulation with the sill at its actual depth of 8m, and the strong vertical phase signal is in good qualitative agreement with the ADCP velocities shown in Figure 5.

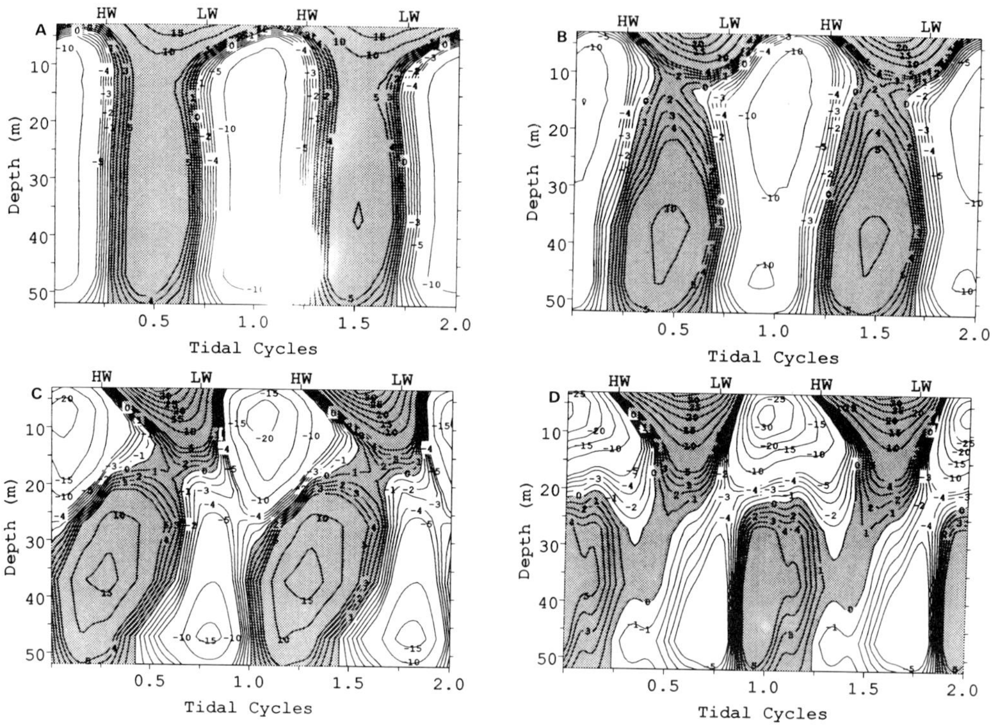

Figure 10: The influence of sill depth on the tidal currents in the inner basin during a period of high run-off: (a) 52m, (b) 20m, (c) 14m, (d) 8m.

A further series of tests was made to investigate the shear induced mixing due to the internal tide. Figure 11 shows the tidal currents at the mooring position simulated during a period of low run-off (10 m^3 s^{-1}) which results in modest near-surface stratification. Although the vertical shear is weak (Figure 11b), well-mixed water in the depth range 10-20m is advected into the basin near the end of the flood tide (Figure 11c) which results in the predicted Richardson number reaching a value of less than 1 (Fig.11d). In contrast, high run-off (65 m^3 s^{-1}) leads to significant shear in the simulated currents (Figure 12a).

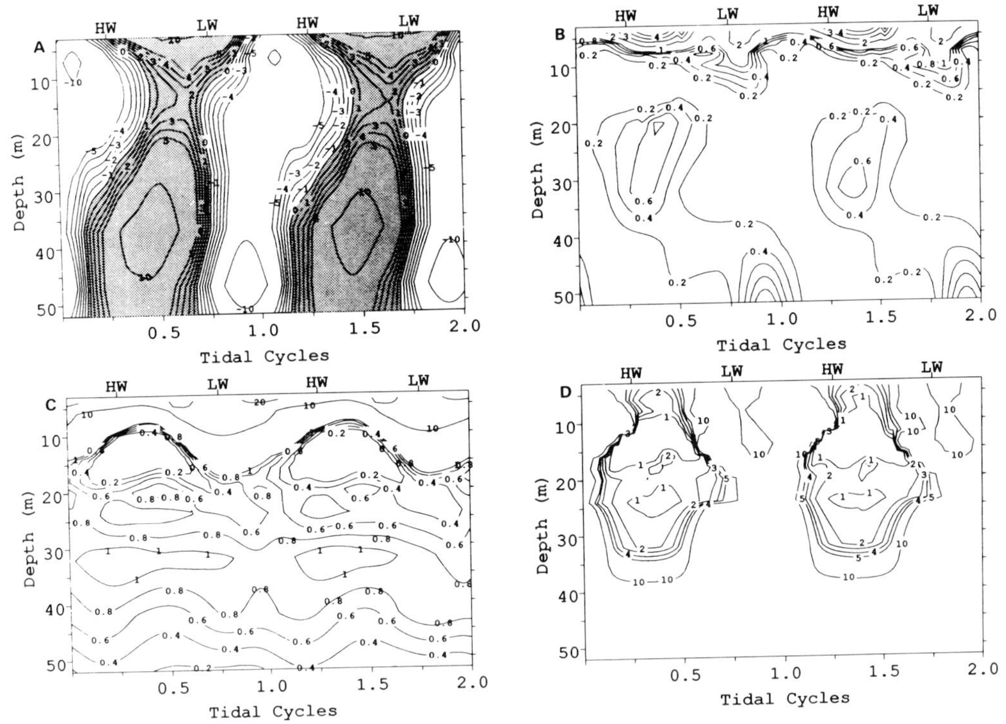

Figure 11: Predicted currents, shear, stratification and Richardson number during a period of low run-off: (a) velocity, (b) $10^4.(\partial u/\partial z)^2$, (c) $-g/\rho(\partial\rho/\partial z).10^4$, (d) Richardson number.

(The currents shown in Figure 12a are slightly different to those shown in Figure 10d because the distributions in Figure 12 were obtained after allowing the model to spin-up for the entire 45 tidal cycles with the sill in position at a depth of 8m.) Figure 12b shows that significant vertical shear is generated in the depth range 10-15m at the middle of the flood. Although the salinity stratification is stronger (Figure 12c) than during low run-off conditions, the increased shear generates Richardson numbers that are as low as 3 in the depth interval between 15-25m. Thus although the vertical stability at 15m is increased by a factor of at least 20 (Figures 11c and 12c), the minimum Richardson number only changes by a factor of about 3 due to the enhanced shear generated by the baroclinic motion.

3.4 Discussion

The observational data collected at the mooring position were characterized by the strongly non-linear nature of the tidal currents and temperature fluctuations. The most striking feature of the semi-diurnal currents was the 180° phase difference between the surface and bottom. An analysis of the vertical modes, based on the observed density profile, showed that both the first and second mode baroclinic motions could produce the observed phase difference. The shallow water M_4 tidal constituent was strong near the bottom, being of comparable magnitude to the M_2 and S_2 constituents, and the length scale of the computed second mode at the M_4 frequency was comparable to the scale of the inner basin. However, it is thought unlikely that a resonance was occurring since the observed ratio of the surface to bottom current speed was closer to that predicted for the semi-diurnal time scale. Inspection of the data series (Figure 4) suggests that the vertical shear was generated by the semi-diurnal tide.

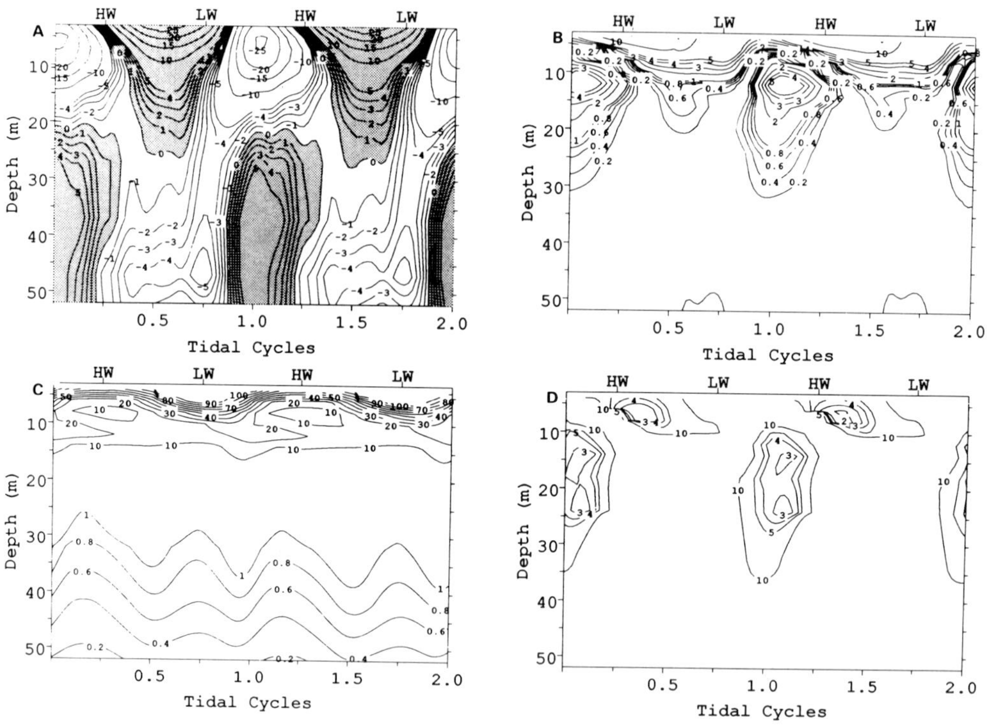

Figure 12: Predicted currents, shear, stratification and Richardson number during a period of high run-off: (a) velocity, (b) $10^4.(\partial u/\partial z)^2$, (c) $-g/\rho(\partial \rho/\partial z).10^4$, (d) Richardson number.

A series of numerical tests have shown that the baroclinic mode is generated by the barotropic tide forcing at the sill of the inner basin, the depth of which is located close to the strong near-surface halocline. The vertical phase difference is only predicted at times of high run-off when the near-surface stratification is well developed. This suggests that the internal mode may act to maintain vertical mixing within the inner basin by increasing the shear beneath the halocline at times of increased vertical stability. Numerical tests made for low and high run-off conditions showed that, although the vertical stability may increase by a factor of 20 at times of high river flow, the Richardson number only increases by a factor of 3 due to the vertical shear generated beneath the halocline by the baroclinic motion.

The highly non-linear character of the loch dynamics limits the size of the time step that can be used with the implicit model. This is in contrast to the shallow, dissipative regions such as the Chesapeake Bay, where the model had been previously applied. The mixing parameterizations of Bowden and Hamilton (1975) and Blumberg (1977), which had been developed for application to the Mersey and Potomac estuaries, respectively, were not successful in simulating the strong near-surface stratification observed in the loch. This suggests that the model could be usefully extended through the use of a closure scheme for the prediction of the turbulent mixing (Smith and Takhar, 1981). A Richardson number probe, consisting of two S4 current meters separated by a CTD was profiled vertically near the sill during the field work and the data are now being analysed to derive estimates of the vertical shear and Richardson number throughout the tidal cycle. This should lead to a better understanding of the shear induced mixing and enable its effective parameterization in future modelling studies.

Acknowledgements

We thank our colleagues in the Hydrography Section of the Marine Laboratory, Aberdeen, in particular R D Adams and R Payne. We also thank the captain and crew of the RV Seol Mara for their help during the field work.

3.5 References

Anon, 1989: Call for pesticide controls. Mar. Poll. Bull., 20, 102.

Blumberg, A.F., 1977: Numerical model of estuarine circulation. J. Hydraul. Div. ASCE, 103, 295-310.

Bowden, K.F. and R.M.Gilligan, 1971: Characteristic features of estuarine circulation as represented in the Mersey estuary. Limn. Oceanogr., 16, 490-502.

Bowden, K.F., and P. Hamilton, 1975: Some experiments with a numerical model of circulation and mixing in a tidal estuary. Estuar. Coast. Mar. Sci., 3, 281-301.

Cushman-Roisin, B., and H. Svendsen, 1983: Internal gravity waves in sill fjords: vertical modes, ray theory and comparison with observations. Coastal Oceanography, H.G. Gade, A. Edwards and H. Svendsen, Eds., Plenum Press, 373-396.

Duff, A., 1988: Tightening controls on fish-farming. Mar. Poll. Bull., 19, 300.

Edwards, A., and D. J. Edelsten, 1977: Deep water renewal of Loch Etive: a three basin Scottish fjord. Estuar. Coast. Mar. Sci.,5, 575-593.

Edwards, A., and F. Sharples, 1986: Scottish sea lochs: a catalogue. Scottish Marine Biological Association/Nature Conservancy Council, 110 pp.

Elliott, A.J., 1976: A numerical model of the internal circulation in a branching tidal estuary. Chesapeake Bay Institute, Spec. Rep.54, The Johns Hopkins University, 85 pp.

Hamilton, P., 1975: A numerical model of the vertical circulation of tidal estuaries and its application to the Rotterdam waterway. Geophys. J. Roy. Astron. Soc., 40, 1-21.

Hamilton, P., 1977: On the numerical formulation of a time-dependent multi-level model of an estuary, with particular reference to boundary conditions. Estuarine Processes II, M. Wiley, Ed., Academic Press, 347-364.

James, I.D., 1986: A front-resolving sigma coordinate sea model with a simple hybrid advection scheme. Applied Math. Modelling, 10, 87-92.

Kent, R.E., and D.W. Pritchard, 1959: A test of mixing length theories in a coastal plane estuary. J. Mar. Res., 8, 67-72.

Smith, T.J., and H.S. Takhar, 1981: A mathematical model for partially mixed estuaries using the turbulence energy equation. Estuar., Coast. and Shelf Sci., 13, 27-45.

Turrell, W.R., and J.H. Simpson, 1988: The measurement and modelling of axial convergence in shallow well-mixed estuaries. Physical Processes in Estuaries, J. Dronkers and W. Van-Leussen, Eds., Springer Verlag, 130-145.

Wang, D.-P., 1979: Wind-driven circulation in the Chesapeake Bay, winter 1975. J. Phys. Oceanogr., 9, 564-572.

Wang, D.-P., 1983: Two-dimensional branching salt intrusion model. J. Waterways, Port, Coastal and Ocean Eng. ASCE, 109, 103-112.

Wang, D.-P., and D.W. Kravitz, 1980: A semi-implicit, two-dimensional model of estuarine circulation. J. Phys. Oceanogr., 10, 441-454.

4

A topographically induced internal wave and mixing in the Tamar Estuary

D R Sturley and K R Dyer

Abstract

The stratified tidal flow over a depression in the Tamar Estuary has been examined on a neap tide. An internal wave formed from this interaction in the manner described by Maxworthy (1979). CTD and velocity measurements from a fixed station, and observations made using a 200 kHz echosounder and the Estuarine Thermistor Spar, have revealed fine details in the wave structure. There is mostly excellent correspondence between regions of acoustic backscatter and temperature gradients, or fine structure. An analysis of the temporal evolution of the wave height shows the formation mechanism to be as Maxworthy (1979) and Lee & Beardsley (1974) proposed, and has highlighted the importance of nonlinear steepening as an evolutionary process.

The onset of vigorous vertical mixing in the seaward end of the wave, after the time of maximum barotropic current, has indicated that this hydraulic phenomenon may be an important contributor to overall estuarine mixing. These observations are the first of their kind for a small partially stratified estuary, and may be evidence of an important transitional phenomenon in estuarine mixing.

4.1 Introduction

Internal waves produced by the interaction of stratified tidal flows with topography, over the continental shelf-break and in fjords, have long been considered important intermittent contributors to vertical mixing and circulation (e.g. Huthnance, 1989). Their importance in estuaries, especially small partially stratified ones, has so far been left relatively unexplored. Intense mixing events, which have been believed to be caused by hydraulic jumps and internal waves, have been reported by Gardner & Smith (1978), Partch & Smith (1978), Partch (1981) and New et al. (1986; 1987). Direct observations of the interaction of stratified tidal flows with topography in estuaries have been rare (e.g. Gardner, 1984; Dyer & New, 1986; New & Dyer, 1987; Pietrzak, 1990). Partch & Smith (1978) have shown that in the Duwamish

Estuary observed intense mixing periods, presumably produced by internal waves, occupied only 20% of the tidal cycle, but accounted for 50% of the vertical turbulent salt flux.

So far accurate estimates in estuaries of the efficiency of conversion of barotropic tidal energy to internal wave energy, and then to increased potential energy through vertical mixing has been lacking. The preliminary results described have shown how the formation of an internal wave occurs on an ebbing tide in the Tamar Estuary, and how subsequent vertical mixing ensues.

New & Dyer (1987) identified a depression in the bed at Cargreen, in the Tamar Estuary, as a promising site for the production of internal waves. Figure 1 shows the bathymetry of the bed at Cargreen. The depression is an elongated elliptical groove, which is central to the main channel. A survey was conducted on 6 November 1989, on the ebb stream of a neap tide (tidal range 2.6m), after a short period of intense rainfall (and riverflow), following a prolonged period of drought. The survey included CTD measurements at a fixed station and observations taken on a longitudinal transect of the estuary using an echosounder and the Estuarine Thermistor Spar, (Sturley & Dyer, 1990). It was thought that these conditions would be most conducive to wave formation.

4.2 Survey

The time of predicted high water at Cargreen, on 6 November 1989, was 10 a.m. A launch was moored at Station "45" (see Figure 1) from an hour and twenty minutes after predicted high water HW+1:20 to HW+5:40, taking vertical profiles of salinity, temperature, current velocity and direction, at twenty minute intervals. These profiles are presented in Figures 2 to 6. The temperature (T) and salinity (S) measurements were made with an MC-5 in-situ T-S bridge (accurate to ± 0.1 salinity units and ± 0.1°C). The current velocity and direction were measured with an NBA (DNC-3) impeller current meter (accurate to ± 2% in velocity, and ± 10° in direction). The depth was calculated from wire length out and angle, with a typical error of 5% to 10%.

From HW+1:20 to HW+6:45 a second launch (RV Catfish) made longitudinal transects over the bed depression with a 200 kHz echosounder and the Estuarine Thermistor Spar deployed. The latter instrument is capable of fine temperature measurements (resolution approx 0.005°C, accuracy ± 0.1°C) at a vertical spacing of 0.25m to a depth of 4m, at a rate of 10Hz (see Sturley & Dyer (1990) for precise details of calibration, design specification and limitations on use). Selected Spar data and echograms are presented in Figures 7 to 12. The Eulerian measurements of temperature and salinity at "45" were used to infer salinity, and hence density, from the Spar data, using the Millero-Poisson equation (Pond & Pickard, 1983).

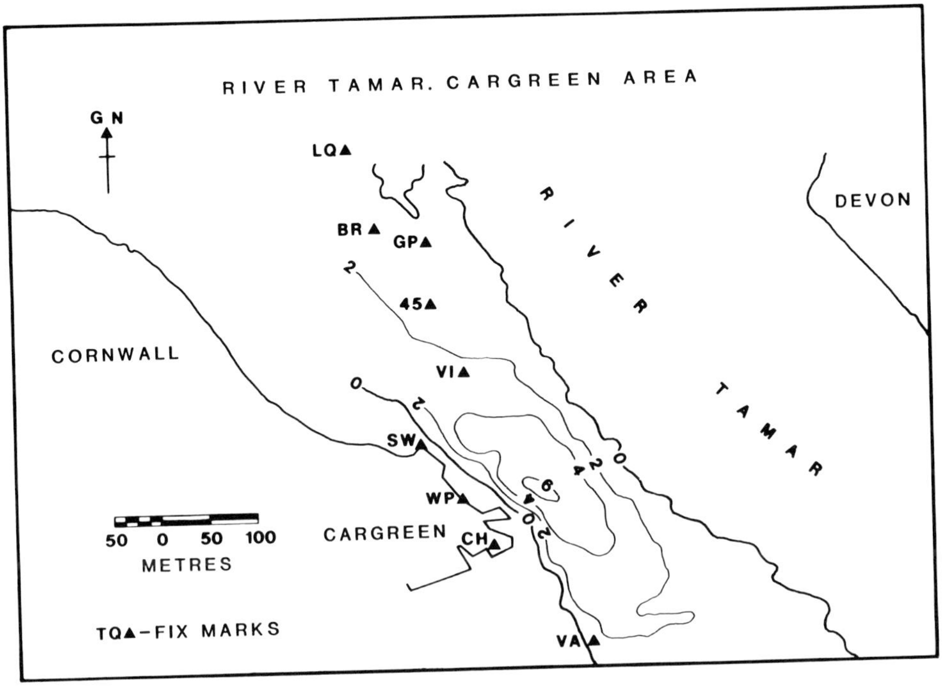

Figure 1: The depression in the Tamar Estuary bed at Cargreen. The Tamar forms a natural boundary between Devon and Cornwall. It is nominally partially mixed and has a mean neap tidal range of 2.3m. Cargreen is 20km from the tidal limit and 10km from the mouth of the estuary. The isobaths are plotted every 1m referenced to chart datum. Station "45" was where the Eulerian measurements were taken.

4.3 Eulerian Measurements

The salinity (Figure 2) and temperature (Figure 3) profiles at "45" show that the estuary was stratified into two layers. They were separated by a region of strong gradient in temperature and salinity. This gradient weakened toward the end of the exercise as the thermocline, and halocline, thickened vertically, with less saline water being incorporated into the thermocline, by entrainment from the upper layer in the seaward end of the wave. However, the difference in density between the two layers remained at about 16 kgm^{-3} throughout the study period, despite each layer becoming marginally less dense, as the ebb tide progressed. T and S were very well correlated in the Eulerian measurements and vertical profiles of T, S and density ρ (not shown) are virtually identical in shape. Figures 2 and 3 show that the vertical migration of the thermocline, which was lowered and then raised, in association with the growth and evolution of the internal wave, greatly influenced the T and S profiles. The position of station "45" was selected to represent the "upstream" conditions of the flow but, as Figures 2 to 5 show, the internal wave distorted the level of the thermocline at this location. The velocity profiles show a similar, but more subtle effect. Figure 6 shows the depth averaged current speed, ū, at "45". It depicts an ebb stream with a double peak in velocity. This behaviour is common in the middle and upper Tamar (George, 1975).

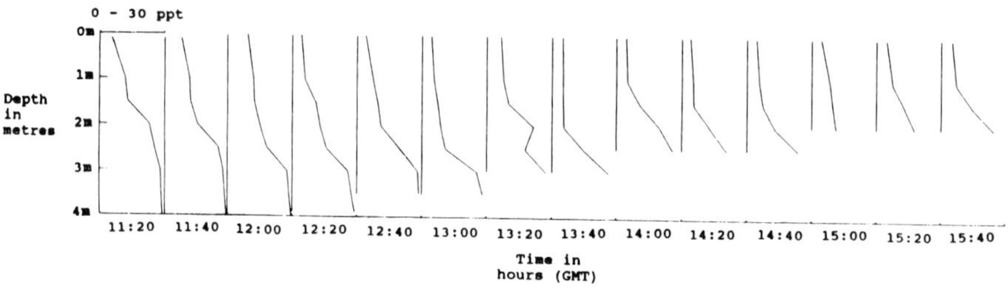

Figure 2: The vertical profiles of salinity at "45" on 6 November 1989, made every 20 minutes, as indicated on the horizontal time axis. The depth is in metres, and the scale for salinity is indicated in salinity units, (ppt on the T-S bridge). The time of predicted high water was 10:00 (GMT).

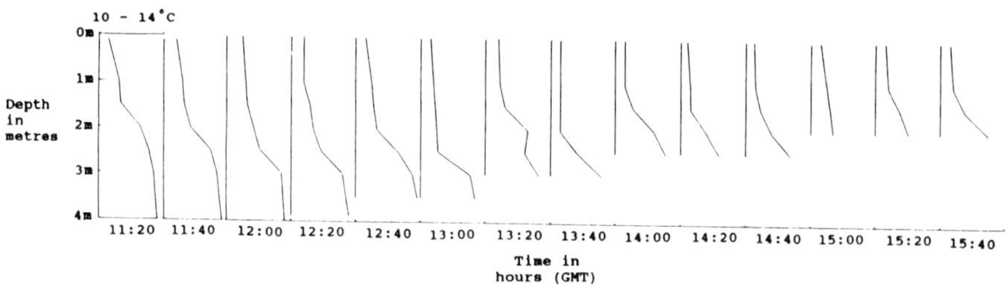

Figure 3. The vertical profiles of temperature at "45" on 6 November 1989, made every 20 minutes, as indicated on the horizontal time axis. The depth is in metres, and the scale for temperature is indicated in the top left hand corner in celsius. The time of predicted high water was 10:00 (GMT).

At the centre of the depression the orientation of the main channel was estimated to be 142° from Chart North. The velocity measured at "45" was divided into vector components, along the orientation of the main channel at "45" (longitudinal velocity-positive velocities are seaward in Figure 4), and in the cross-channel direction (transverse velocity - positive velocities are toward the Cornish bank in Figure 5). Figure 4 indicates maximum seaward flow at mid-depth at, or slightly above, the level of the thermocline. The level of this maximum flow evolved in time in exactly the same way as shown by the T and S profiles. Figure 5 shows a strong transverse component of the flow in the upper layers with negligible flow below the thermocline. The level at which the transverse flow became zero coincided with the bottom of the thermocline, and evolved with it in time. The transverse flow reversed towards the end of the tide. Figures 4 and 5 together show a strongly variable flow associated with the growth and evolution of the lee wave.

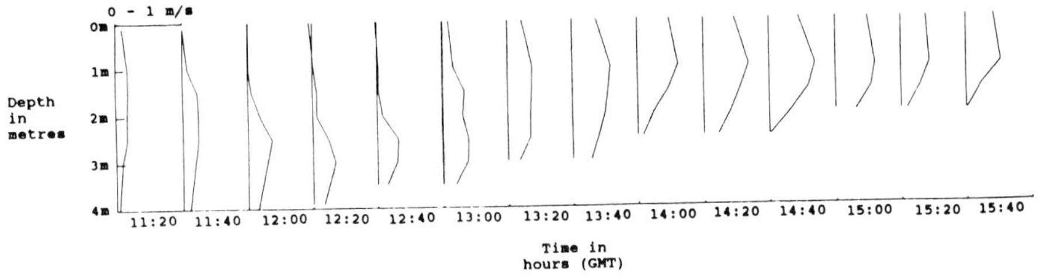

Figure 4: The vertical profiles of longitudinal velocity at "45" on 6 November 1989, made every 20 minutes, as indicated on the horizontal time axis. The depth is in metres, and the scale for velocity is indicated in the top left hand corner, in metres per second. The time of predicted high water was 10:00 (GMT). Positive velocities are seaward.

Calculations of the longitudinal gradient Richardson number, R_i, due to vertical shears in the longitudinal velocity u, where

$$R_i = -\frac{g}{\rho}\left(\frac{\delta\rho}{\delta z}\right)\left(\frac{\delta u}{\delta z}\right)^{-2}$$

show that the thermocline at "45" was stable ($R_i > 0.25$) until 13:20, despite large shears, but R_i is close to 0.25 after this time. (The vertical sampling interval for ρ and u was 0.5m). Using the full velocity vector $\tilde{u} = u + v$ in R_i might make these estimates even lower, showing ample opportunity for shear driven instabilities across the thermocline.

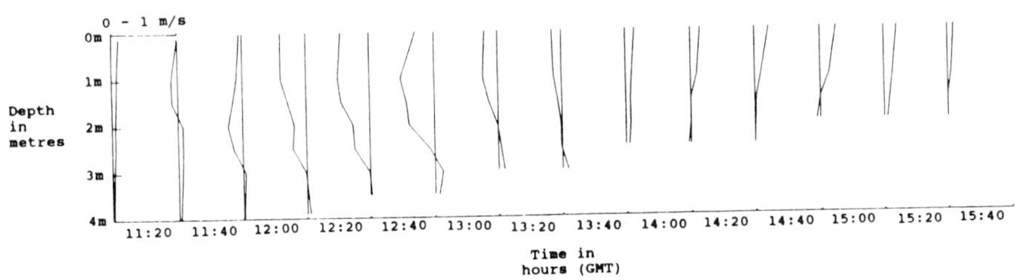

Figure 5: The vertical profiles of transverse velocity at "45" on 6 November 1989, made every 20 minutes, as indicated on the horizontal time axis. The depth is in metres, and the scale for velocity is indicated in the top left hand corner, in meters per second. The time of predicted high water was 10:00 (GMT). Positive velocities are toward the Cornish bank.

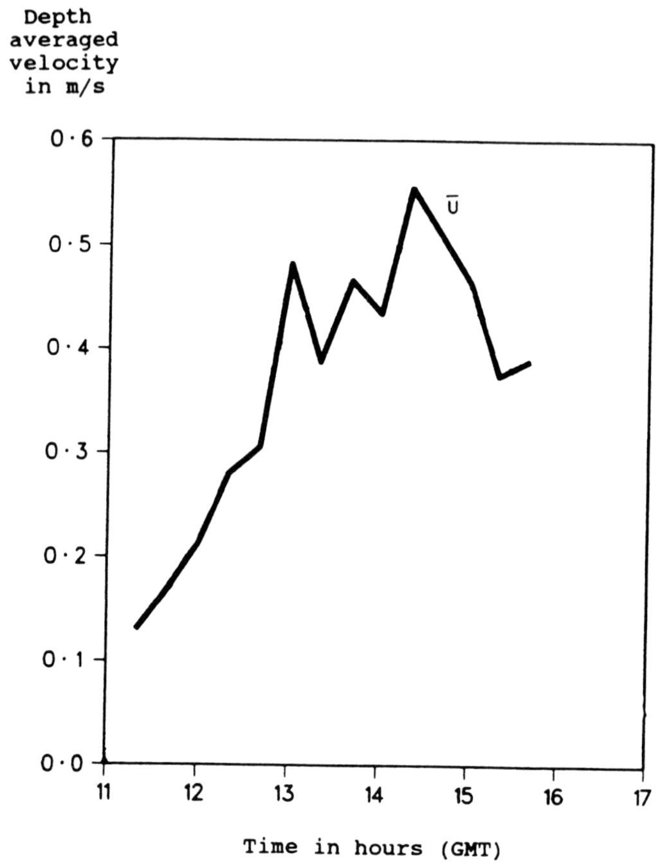

Figure 6: The magnitude of the depth averaged velocity, ū, at "45" on 6 November 1989. The time of predicted high water was 10:00 (GMT).

4.4 Echograms and Spar Data

The Spar data and synchronous echograms presented in Figures 7 to 12 were made on longitudinal transects in the seaward direction. They are in sequence and are annotated with fixed positions (yacht buoys and landmarks). Table 1 shows some of the distances between fixes in metres.

Table 1: Distances between fixed stations shown in Figure 1.

LQ - BR	BR - 45	45 - CH	CH - VA
87m	105m	234m	144m

The Spar data is contoured as isotherms in thermal Maps, representing longitudinal vertical sections of the estuary. The contours are plotted at 0.1°C intervals.

A few general points can be made about the observations:

There is a strong resemblance between the thermal maps and the echograms. The exact cause of acoustic backscatter is uncertain, but density gradients, microstructure and particles, are all good scatterers and are often associated with the thermocline region. Regions where isotherms strongly diverge vertically (e.g. Figure 10 at "VA". from 1 to 3m below the surface), and have associated backscatter in the echograms, are identified as being composed of finestructure, caused by turbulent destratification.

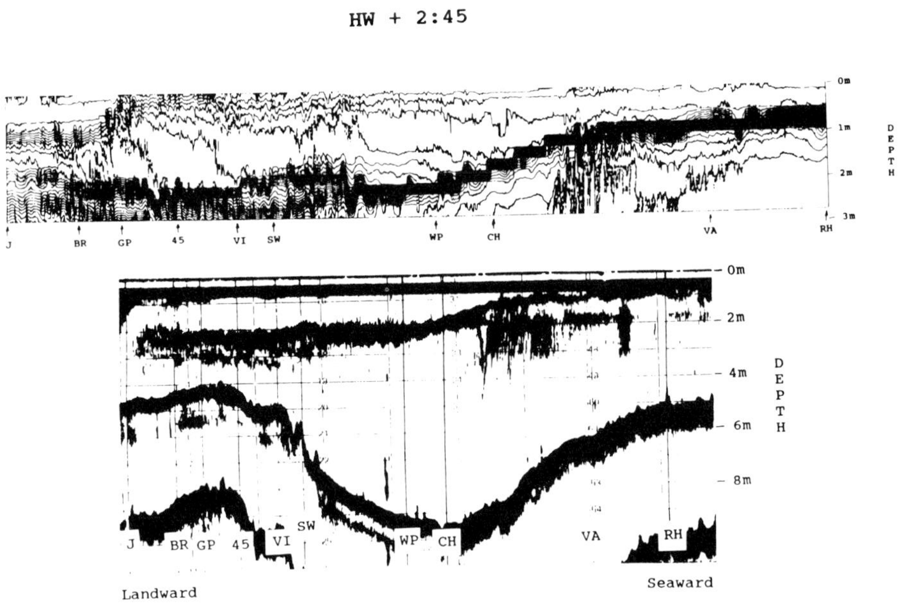

Figure 7: Transect 5 at 12:45 on 6 November 1989.

In Figure 9 (HW+4:45), halfway between "CH" and "VA" the thermocline had an undular structure, which on closer inspection revealed billows.

Propeller mixing in the thermocline is evident (e.g. in Figure 8) seaward of "RH" the isotherms are tangled and there is a patch of associated backscatter. This occurred at a time when RV Catfish was known to have crossed its own wake.

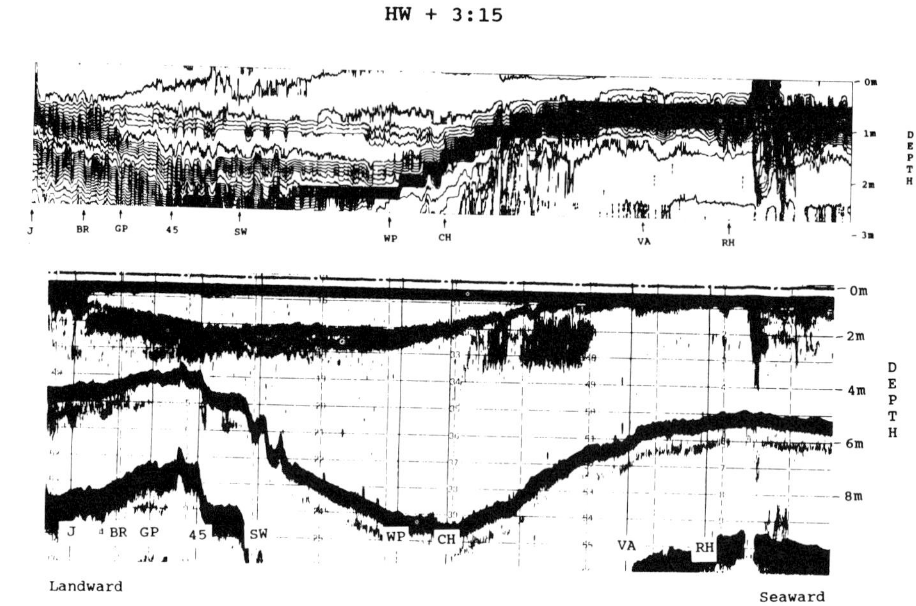

Figure 8: Transect 7 at 13:15 on 6 November 1989.

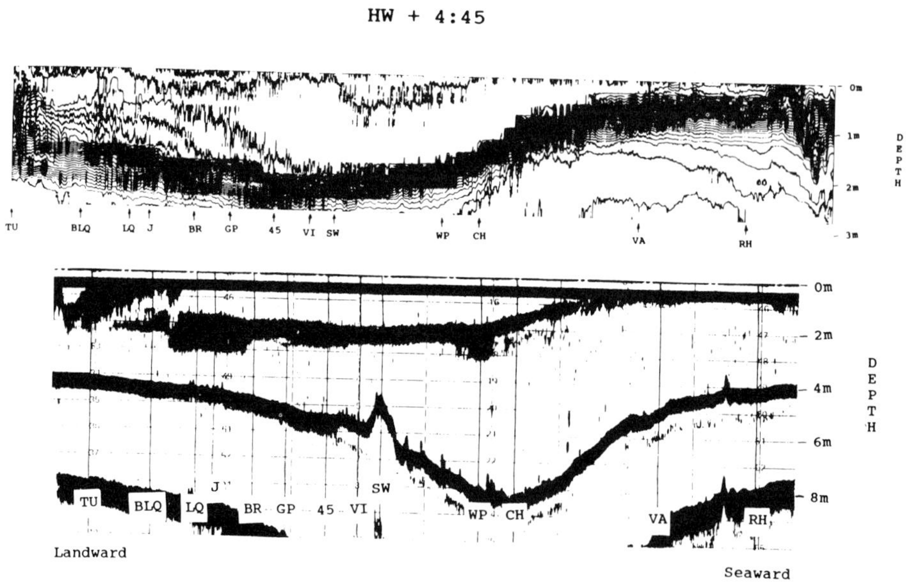

Figure 9: Transect 11 at 14:45 on 6 November 1989.

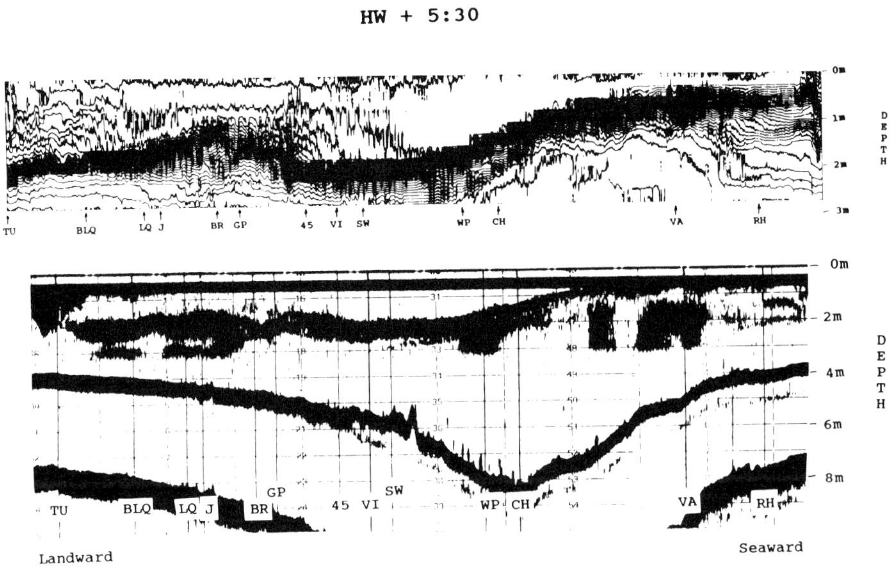

Figure 10: Transect 13 at 15.30 on 6 November 1989.

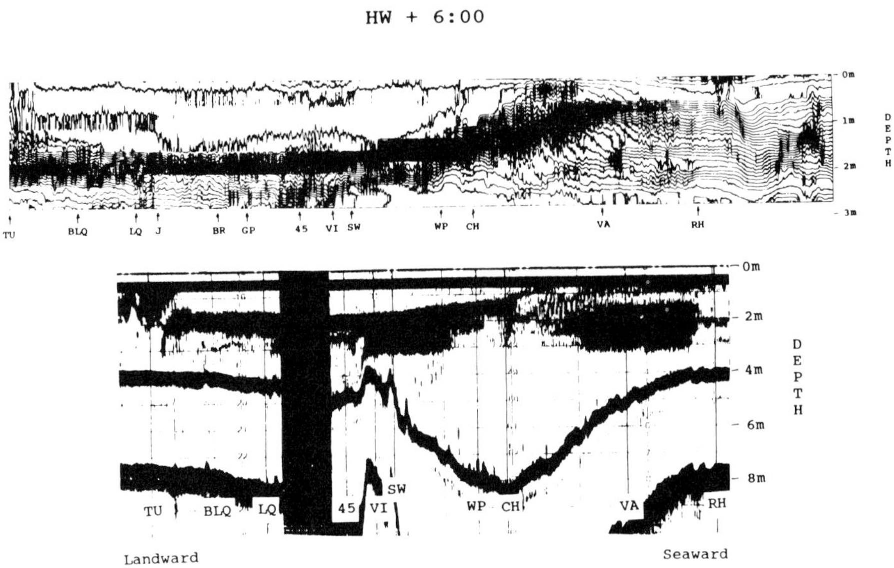

Figure 11: Transect 15 at 16:00 on 6 November 1989.

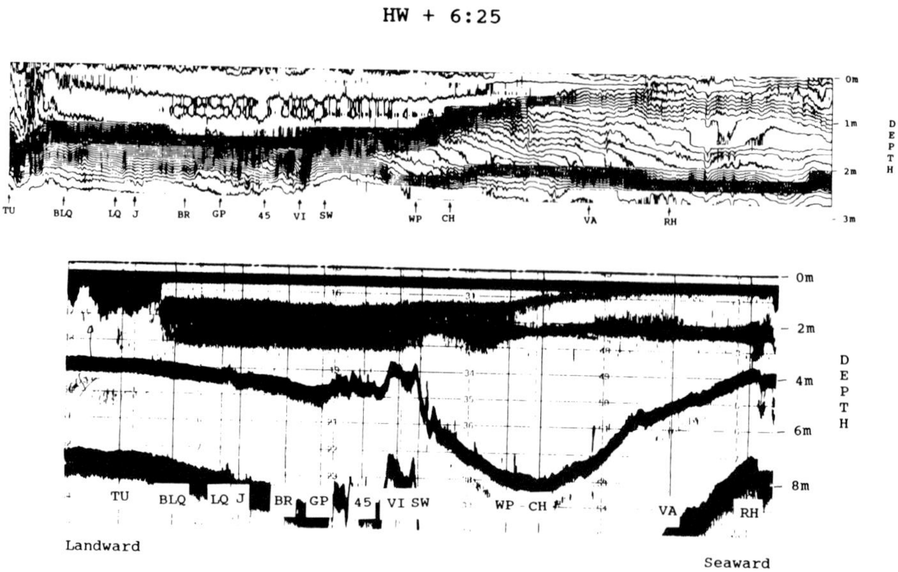

Figure 12: Transect 17 at 16:25 on 6 November 1989.

At about HW+4:45 (Figure 9) the approximate time of maximum stream, the thermocline seaward of the wave thickened vertically reducing the temperature gradient across it. This thickening increased with time, suggesting active mixing in, or across, the thermocline. The region in which this thickening took place spread landward at subsequent times (Figs. 10, 11 and 12), finally reaching the wave trough.

The isotherms often take on a stepped structure, with jumps in the heights of contours corresponding to the spacing between sensors.

4.5 Results

Figures 7 to 12 depict the formation of an internal wave, in the manner described by Maxworthy (1979) and Lee & Beardsley (1974), in the lee of the landward slope of the depression. Figure 13 shows the stages of development of a lee wave, in a conceptual model based on Maxworthy's and Lee & Beardsley's experiments and observations. In this model a lee wave grows as a downward displacement of a density interface (in a two layer fluid) over a bed depression in an accelerating stream, (a), (b) & (c). Just after maximum tidal stream the lee wave, which had developed a large upstream phase velocity in order to remain stationary at maximum stream, now begins to propagate upstream against the slackening tide (d). In doing so it undergoes nonlinear steepening and wave growth due to the nonlinear advective acceleration of the internal tide, by both the barotropic tide and the internal tide itself. There is strong evidence for this steepening mechanism in numerical models of two layer tidal flows over topography (Pingree et al., 1983; Sturley, 1990). Finally, as the tidal stream slackens beyond some critical value, if steepening and dispersive effects become comparable, fission into solitary waves may occur, (e), (f), but this behaviour was not observed in the data presented here.

Figures 7 to 12, themselves, do not allow a comparison with this conceptual model, but an analysis of the wave height shows the behaviour more clearly.

4.5.1 Wave Height

The maximum thermocline displacement, or wave height, has been measured from each transect and these are presented in Figure 14. It was necessary to define the undisturbed thermocline level in each case. Our experience, and numerical modelling (Sturley, 1990), have shown that the downstream (seaward in this case) level is a good indicator of the undisturbed level. The vertical point of maximum gradient, or alternatively, vertical mid-point of the thermocline, was chosen as the boundary between the upper and lower layers.

There is good agreement between the maximum thermocline displacement obtained from the echograms and from the thermal maps. Comparisons between transects suggest that some of the variation is due to the 3-dimensional nature of the wave, and the deviations of the launch from the axis of the estuary, but in general the estimate is reasonable for the whole wave.

Figure 14 shows that the wave was already quite well developed at HW+2:45, in response to the increasing ebb stream, with a displacement of about 1.75m. Between HW+2:45 and the end of the study the depth of the wave trough varies in response to the tidally varying ebb stream. This is shown by the correspondence between the maximum thermocline displacement and the magnitude of the depth averaged velocity, $\bar{u}$, measured at station "45". To remain stationary in the accelerating seaward flow the wave developed a landward phase velocity, continually evolving to match the ebb stream. Between HW+1:20 and HW+3:00, $\bar{u}$ increased and the wave height increased in response. From HW+3:00 to HW+3:20, $\bar{u}$ decreased slightly, but the wave continued to grow through nonlinear steepening, as predicted by the conceptual model. $\bar{u}$ increased again from HW+3:20 to HW+4:20, but the response was a decrease in wave height over this period. This may have been due to increased attenuation at the wave crest, through diffusion and dissipation. Fig. 14 suggests that there may be a "critical" height for this wave, of about 2m. From HW+4:20 to HW+5:20 $\bar{u}$ decreased again. The initial response, from HW+4:45 to HW+5:15, was an increase in height, again as a result of nonlinear steepening. From HW+5:20 to HW+5:40 $\bar{u}$ increased by a small amount, and there is a suggestion that the wave height also increased at this time. The general trend from HW+5:15 to HW+6:15 was a decline in the wave height as it propagated landward losing energy.

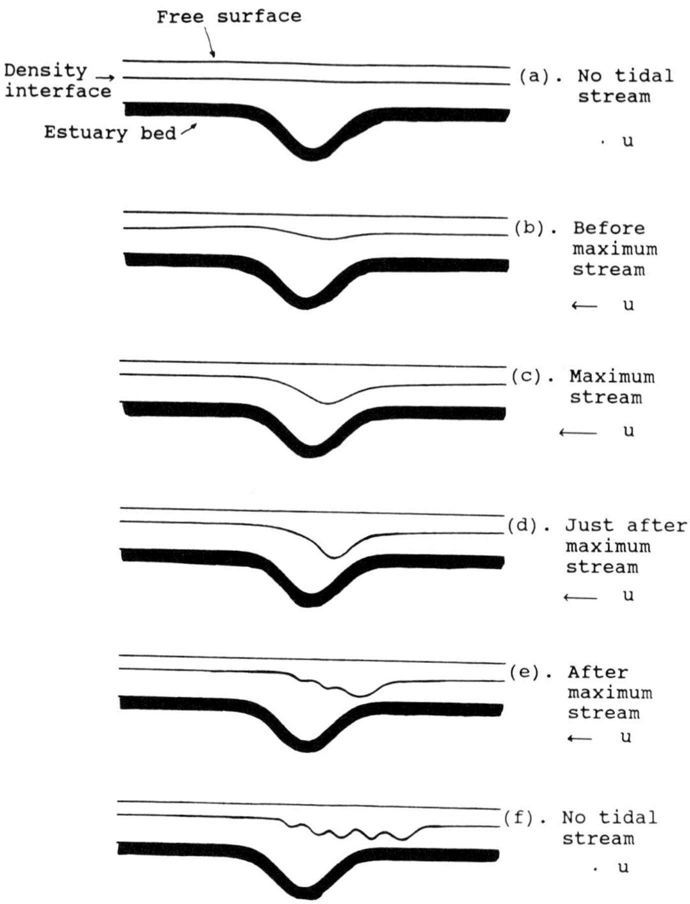

Figure 13: The stages which form the conceptual model of lee wave formation. (a), (b), (c): The wave grows over the landward slope of the bed depression in the accelerating tidal stream. (d): After maximum tide the wave steepens nonlinearly. (d), (e): The wave propagates into the slackening stream. (e), (f): The wave may fission into solitary waves through a balance between dispersion and nonlinear steepening.

At HW+6:45 the thermocline still appeared to be displaced by about 1.3m - 1.4m, but the wave structure and the level of the undisturbed thermocline were difficult to define.

The error incurred in measuring the size of the wave trough is estimated to be of the order $\pm$0.2m. This error is roughly equivalent to a 10% uncertainty in the largest thermocline displacement, and is a result of making essentially subjective judgements about the level of the thermocline. It ignores instrumental errors, which are felt to be small in comparison.

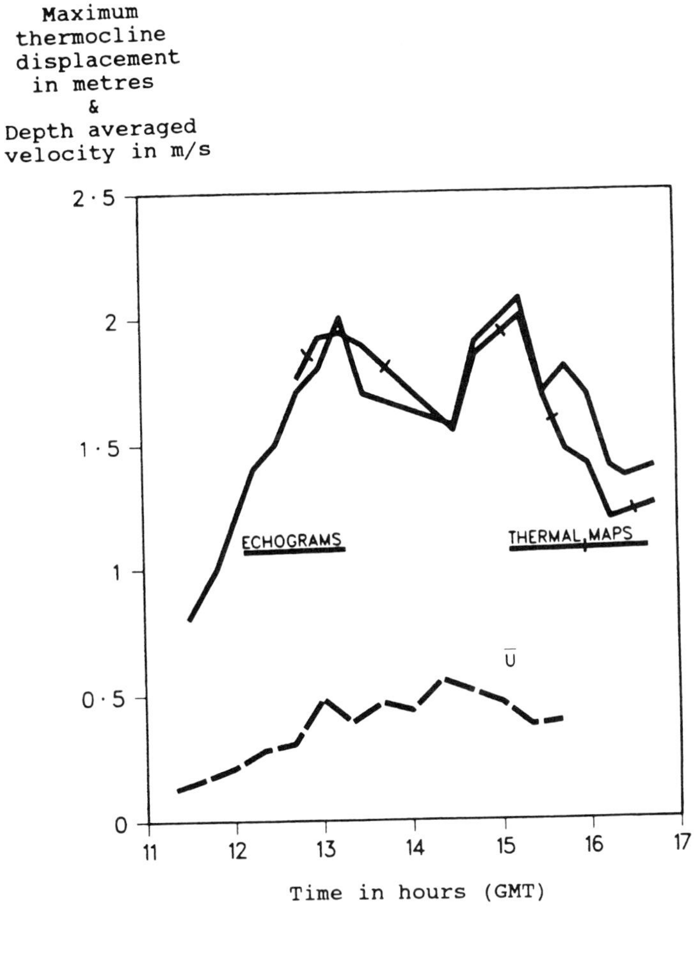

Figure 14: The size of the observed maximum thermocline displacement, in metres, plotted together with the depth averaged speed, ū, at "45" in metres per second. The time of predicted high water was 10:00 (GMT). The observations are derived from both echograms and thermal maps.

4.5.2 Wave Energy

The total wave energy, per unit length in the y (across estuary) direction, E_T, was calculated from (Phillips, 1977):

$$E_T = \int \eta^2(x) \Delta \rho g \delta x$$

over the domain defined by the stations "VA" and "BLQ". This energy represents that of the internal wave only and assumes equipartition of potential and kinetic energy, which is true for linear waves, and good to first order for nonlinear waves. No representation of the barotropic tidal energy is included in this equation, and it assumes the wave is on a thin interface in a two layer fluid, and is 2 dimensional in nature. $\eta(x)$ is the vertical displacement of the density interface, x the longitudinal direction, $\Delta\rho$ the density difference between the layers and $g = 9.8 ms^{-2}$.

In making the calculations it was necessary to define a fixed horizontal domain over which to calculate the energy. Since the surveys did not extend far enough in a landward direction to define completely the upstream thermocline position, "BLQ" was chosen as the landward limit of the waveform. In the cases when the transect did not extend to "BLQ" the thermocline level was extrapolated from the most landward point of the transect. In all cases "VA" was more seaward than the seaward end of the wave. The undisturbed thermocline level was identified in the way already described. The integration was performed in sections, using the fixed station positions, and measuring a mean thermocline displacement between section ends. $\Delta\rho$ was assumed as $16 kgm^{-3}$, which was the mean difference in layer densities over the period HW+2:45 to HW+6:45.

This definition of the boundaries of the wave would, certainly for the latter stages of the study, be restrictive of the landward edge of the wave. However, it is anticipated that the effects of this are not severe, and that the largest uncertainty is in the thermocline displacement, ($\pm$0.2m or $\pm$10%), leading to an estimated error of about 20-25% in the largest energy estimates.

Figure 15 shows the depth averaged speed with the energy measurements derived from the echograms and thermal maps, and these are roughly consistent. The energy follows a similar pattern to the maximum thermocline displacement, but is more exaggerated in its fluctuations, reflecting the dependency on the square of the displacement.

The total wave energy is relatively high, around 160 - $180 kJm^{-1}$, at HW+2.45, before falling to $120\text{-}140 kJm^{-1}$ at HW+3:00 in line with a sharp increase in the ebb stream. The reason for this is unclear. Otherwise, the energy increased with an increasing ebb stream, and continued to increase for a while when the ebb slackened, presumably due to the nonlinear steepening mechanism discussed earlier. The energy diminished when the ebb stream diminished significantly, especially after HW+4:45.

A peak wave energy of about $220 kJm^{-1}$ was measured. From HW+4:45 to HW+6:45 the wave energy dropped by about $85 kJm^{-1}$. Per metre of the length of the wave (about 500m) this amounts to a fall in energy of $170 Jm^{-2}$.

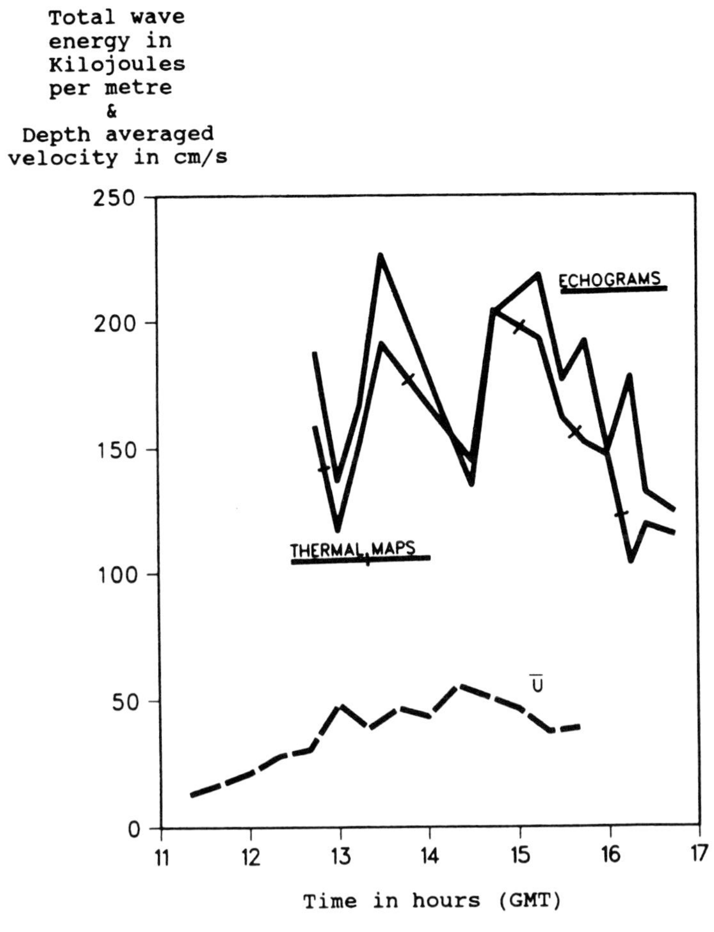

Figure 15: The observed total wave energy in kilo joules per metre, with the depth averaged speed at "45" in cm per second. The time of predicted high water was 10:00 (GMT). The observations are derived from both echograms and thermal maps.

4.5.3 Mixing

Between HW+4:45 and HW+6:45 radical changes in the vertical distribution of density took place at "RH" and "VA". The distortion of the thermocline due to the wave presence enhanced the shear in the seaward end of the wave, near "RH", causing fresher water to be mixed from above the thermocline down into it. Water was then redistributed vertically in the thermocline through mixing. The thermocline thickened, expanding downwards, at the expense of the lower layer. With time this thermocline thickening was manifest along the whole length of the wave. At HW+6:25 (Figure 12) the thermocline from "RH" to "SW" was affected. Mixing in the thermocline had the effect of intensifying the temperature gradient at 2.3m, giving the impression of a "new" thermocline forming in the echograms (Posmentier, 1977). Little mixing appeared to have taken place between the base of the thermocline and the lower layer.

4.6 Discussion

This set of observations has indicated, qualitatively at least, that the formation of an internal wave over topography in a partially mixed estuary is capable of radically altering the stratification locally, through enhanced vertical mixing. This mixing is induced by the increased shear across the thermocline, produced by the distortion of the thermocline by the internal lee wave.

Nonlinear steepening in the slackening tidal stream appeared to play a large role in the wave formation and evolution. There is clear evidence for this role in theoretical studies: Lee & Beardsley (1974), Pingree et al (1983) and Sturley (1990). In particular, non linear advective acceleration of internal waves in a slackening tide can produce steepening and growth. Twice, when the barotropic tidal stream slackened, after a maximum, the wave grew in height as a response. The height of the wave, in relation to the total water depth, itself suggests that nonlinearities might be important in the behaviour of the baroclinic response of the estuary at this location.

A reduction in the wave growth occurred as the wave height approached 2m, and this appeared to be a limiting height. Wave energy was lost through mixing and dissipation and this caused a reduction in wave height. The position at which mixing in, or across, the thermocline was first observed, was in the downstream (seaward) end of the wave. This was also the position at which the shear across the thermocline would have been most enhanced. The region of increased mixing migrated upstream (landward) to meet the trough of the wave, suggesting that it had a larger propagation velocity than the wave itself.

The transfer of energy to mixing was almost certainly the cause of the lee waves decline in height and energy, in the latter stages of the tide (Figures 9 to 12). Future studies aim at improving the quantification of this process.

4.7 References

Dyer, K.R., and A.L. New, 1986: Intermittency in Estuarine mixing. Estuarine variability - proceedings of the Eighth Biennial International Estuarine Research Conference, University of New Hampshire, 1985. Ed. Wolfe, D.A., 321-339, Academic Press, Inc. pp. 509.

Gardner, G.B., 1984: Internal hydraulics and mixing in highly stratified estuaries. PhD Thesis, University of Washington, pp241.

Gardner, G.B., and J.D. Smith, 1978: Turbulent mixing in a salt wedge estuary. Hydraulics of estuaries and fjords, ed. Nihoul, J.C.J., Elsevier Oceanography Series 23, 79-106.

George, K.J., 1975: Tides and tidal streams in the Tamar Estuary. PhD Thesis, University of London.

Huthnance, J.M., 1989: Internal tides and waves near the continental shelf edge. Geophys. & Astrophys. Fluid Dyn., 48, 81-106.

Lee, C., and R.C. Beardsley, 1974: The generation of long nonlinear internal waves in a weakly stratified shear flow. J. Geophys, Res., 79, 3, 453-462.

Maxworthy, T., 1979. A note on the internal solitary waves produced by tidal flow over a 3-D ridge. J. Geophys. Res., 84, C1-6, 338-346.

New, A.L., and K.R. Dyer, 1987: Internal waves and mixing in stratified estuarine flows. Proceedings of International Symposium on Physical Processes in Estuaries, Noordwijkerhout, 1986. Ed. Dronkers, J., Van Leussen, W., 239-254.

New, A.L., K.R. Dyer, and R.E. Lewis, 1986: Predictions of the generation and propagation of internal waves and mixing in a partially stratified estuary. Est. Coast. Shelf Sci., 22, 199-214.

New, A.L., K.R. Dyer, and R.E. Lewis, 1987: Internal waves and intense mixing periods in a partially stratified estuary. Est. Coast. Shelf Sci., 24, 15-33.

Partch, E.N., 1981: Time dependent mixing in a salt wedge estuary, PhD Thesis, University of Washington, 165pp.

Partch, E.N., and J.D. Smith, 1978: Time dependent mixing in a salt wedge estuary. Est. Coast. Shelf Sci., 6, 3-19.

Phillips, O.M., 1977: The Dynamics of the Upper Ocean, 2nd. Ed. Cambridge Univ. Press.

Pietrzak, J.D., C. Kranenburg, and G. Abraham, 1990: Resonant internal waves in fluid flow. Nature, 344, 6269, 844-847.

Pingree, R.D., D.K. Griffiths, and G.T. Mardell, 1983: The structure of the internal tide at the Celtic Sea shelf break. J. Mar. Biol. Assoc. U.K., 64, 99-113.

Pond, S., and G.L. Pickard, 1983: Introductory Dynamical Oceanography. 2nd Ed. Pergamon Press, pp.329.

Posmentier, E.S., 1977: The generation of salinity finestructure by vertical diffusion. J. Phys. Oceanogr., 7, 298-300.

Sturley, D.R.M., 1990: Topographically induced internal waves and enhanced vertical mixing in an estuary. PhD Thesis, Polytechnic South West (CNAA), pp.222.

Sturley, D.R., and K.R. Dyer, 1990: The Estuarine Thermistor Spar: an instrument for making thermal profiles in shallow water. The Hydrographic Journal, 55, Jan. 90, 13-21.

5

Turbulence and shear induced mixing processes in estuaries

E J Darbyshire and J R West

Abstract

During the summer of 1988 data were collected in the upper reaches of the Tamar estuary to investigate vertical turbulent mixing of solute during neap tides. Measurements were made of the vertical variation of turbulent mean longitudinal velocity and salinity and of the turbulent perturbations of the vertical and longitudinal components of velocity and salinity. On the early ebb tide a large vertical salinity gradient developed, the measured turbulent fluxes were found to be small and an extensive internal wave field was observed. Consideration of the temporal evolution of the vertical salinity profiles suggested that the measured fluxes were an underestimate of the required vertical exchange and it was concluded that either undetected small scale transport associated with internal wave breaking or transport due to a transverse secondary circulation was taking place.

5.1 Introduction

The movement of water and solutes in estuaries is governed by the turbulent mean flow structure and turbulent diffusion processes. The mean flow is caused by fluvial, tidal and wind influences and controlled in detail mainly by turbulent diffusion processes. The production of turbulence which causes the diffusion is caused by bed drag characteristics, the mean velocity and density gradients, internal and surface wave characteristics and acceleration. Thus there is strong feedback between the turbulence and mean fields. Given that the bathymetry of estuarine channels is usually complex it may be anticipated that estuarine flows and the associated scalar fields are complex functions of space and time.

In the last two decades extensive measurements have been made in boundary layers in air and water in laboratories, some measurements have been made at sea and a few have been made in estuaries. The paucity of the latter data is surprising considering the anthropogenic pressures on nature leading to numerous mathematical modelling studies of estuarine transport processes. There are, however, considerable physical difficulties in obtaining good field data and it is not easy to develop an understanding of temporally varying three dimensional flow fields from a limited number of point measurements when momentum and solute transport are influencing the flow.

This paper considers some turbulence data collected in the Tamar estuary in July 1988. The data are novel in that the turbulent fluctuations of velocity and salinity have been made over a wide range of relative depths over two ebb tides with ranges close to neap conditions. The resulting turbulence field in a channel whose bathymetry is mainly controlled by the much greater shear stresses induced by spring tide, storm and fluvial flood processes, is relatively weak but the measurements are shown to be adequate for the development of hypotheses concerning transport mechanisms and the partial substantiation of some of these ideas for estuarine channels.

5.2 Previous work

Due to the interaction of vertical velocity shear and longitudinal density gradient, ebb tides in estuaries are frequently associated with significant vertical gradients of salinity and consequently density. Vertical turbulent transport in such environments depends on the maintenance of turbulence which, in stratified flows depends on the relative magnitudes of shear production and vertical buoyancy flux which may be expressed by the local gradient Richardson number,

$$Ri = \frac{-g}{\rho} \frac{\partial \rho / \partial z}{(\partial \bar{U} / \partial z)^2} \tag{5.1}$$

where $\bar{U}$ is the mean longitudinal velocity, ρ is the mean density, g is the acceleration due to gravity and z is positive upwards. Theoretical investigations by Miles (1961) showed that the critical value of Ri in stratified shear flow was 0.25 and that instability was possible below this value.

The vertical turbulent transport of momentum or solute may be related to the gradient of the mean flow via the mixing length given by

$$-\overline{uw} = l_{mz0}^2 \frac{\partial \bar{U}}{\partial z} \left| \frac{\partial \bar{U}}{\partial z} \right| \tag{5.2}$$

for momentum, and by

$$-\overline{sw} = l_{sz0}\, l_{mz0} \frac{\partial \bar{S}}{\partial z} \left| \frac{\partial \bar{U}}{\partial z} \right| \tag{5.3}$$

for a solute. Here $\overline{S}$ is the mean salinity, u, w and s are perturbations from the mean values of longitudinal and vertical velocity and salinity respectively and an overbar indicates a temporal mean value. The subscripts of l, the mixing lengths, are m, momentum, s, salinity, z, vertical transport, and 0, homogeneous conditions. Hereafter vertical transport will be assumed and the subscript z omitted.

Theoretical relationships relating the turbulent mixing lengths in homogeneous and stratified flows to Ri have been proposed by Rossby & Montgomery (1935) and Kent & Pritchard (1959) with

$$l_m = l_{m0}\,(1 + \beta_1 Ri)^{-0.5} \tag{5.4}$$

and

$$l_m = l_{m0}\,(1 + \beta_2 Ri)^{-1.0} \tag{5.5}$$

respectively, where β_1 and β_2 are constants. Values of these constants have been determined by Odd & Roger (1978) by considering a two dimensional mass budget analysis and by West et al. (1985) by direct measurement of the vertical turbulent flux. In both these cases Ri was evaluated from data taken over a period of about 1.0 minute and the interpretation of the results was based on the assumption of turbulence generation in the region of maximum mean velocity shear, i.e. near to the bed. Geyer & Smith (1987) showed that evaluation of Ri over much shorter time scales caused a significant increase in the number of subcritical values of Ri. This reduction was brought about by the additional shear caused by internal wave velocity fluctuations allowing local turbulence production remote from the solid boundary.

The suppression of turbulence by density gradients is also dependant upon the length scales of the turbulence involved. Large length scales may be almost completely suppressed while motions on much smaller scales may be practically unaffected. This can be seen in the steepening of the slope of the salinity variance spectrum in the presence of a density gradient from the value of -5/3 which is observed in homogeneous conditions (Turner (1973)). Various values have been obtained for this steepened slope such as by Dyer (1981) who observed a relationship between the magnitude of the slope and the layer Richardson number.

Internal waves have been cited as the cause of increased turbulent mixing in estuaries by Partch & Smith (1978) and New et al. (1987) although there has been some doubt as to the exact mechanisms of wave generation and breaking. In both these instances the waves have been observed in estuaries where the density structure can be approximated to a two layer situation and the waves propagated along the pycnocline across which the mixing took place.

A further feature of most studies of estuarine mixing has been that they are confined to a two dimensional view. On flood tides in the Conwy estuary Nunes & Simpson (1985) showed that significant transverse currents occur which necessitate the presence of vertical currents within the flow which in this case were downwards towards the centre of the channel and upwards towards the banks. The effect of such currents would be to transport mass and momentum without the need for turbulent exchange. Upward vertical currents were observed during ebb tides on the Great Ouse by Odd & Roger (1978) who commented that the momentum exchange due to them could be as great as that due to turbulence. They did not, however, comment about the mass transfer involved.

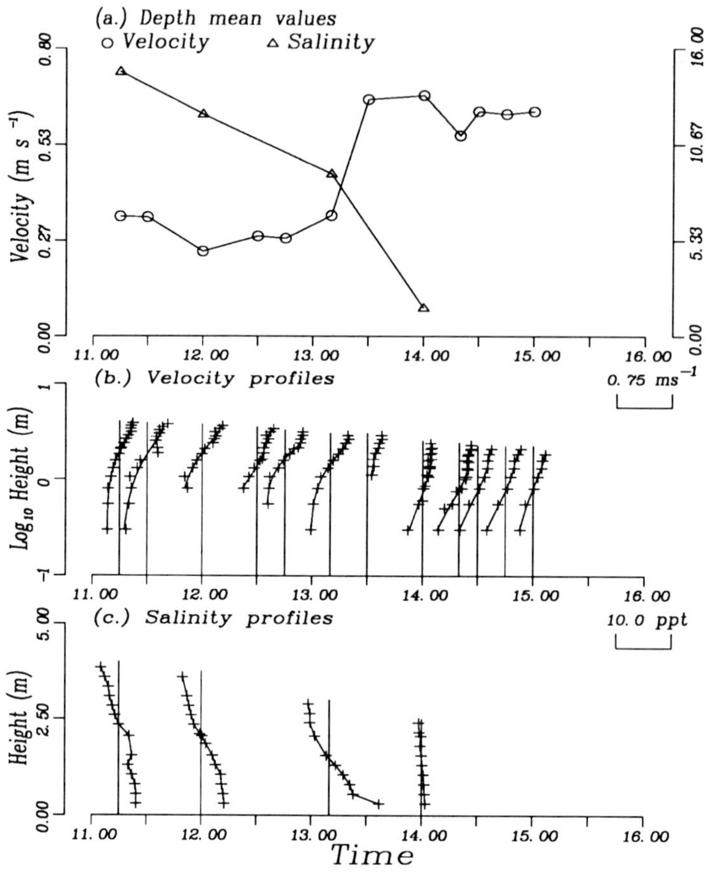

Figure 1: Vertical profiles and depth mean values of velocity and salinity, Cotehele 4/7/88. LSW at 17:45.

5.3 Data collection and analysis

The data were collected near to the centre of the channel of a straight reach of the upper Tamar estuary near to Cotehele Quay. The channel width at this point is about 100m and the depth varies from 1.0m to 5.0m depending on the state of the tide. The measurements were taken on the 4th and 8th of July 1988 when the daily mean fluvial discharges of 6.8 cumecs and 5.6 cumecs respectively were above average for the time of year but low compared with winter values which can exceed 100 cumecs. The 8th was a neap tide and the 4th was an intermediate between neap and spring.

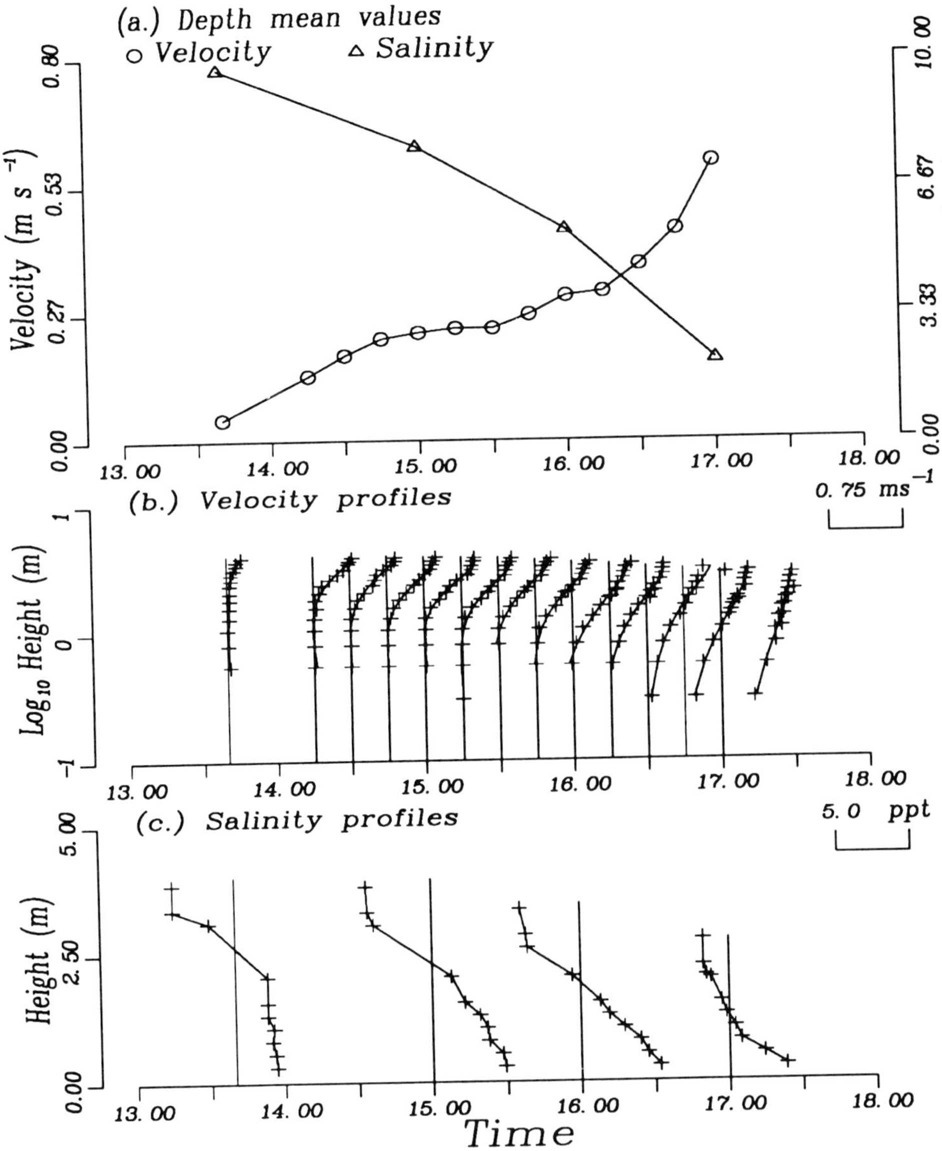

Figure 2: Vertical profiles and depth mean values of velocity and salinity, Cotehele 8/7/88. HSW at 13:15.

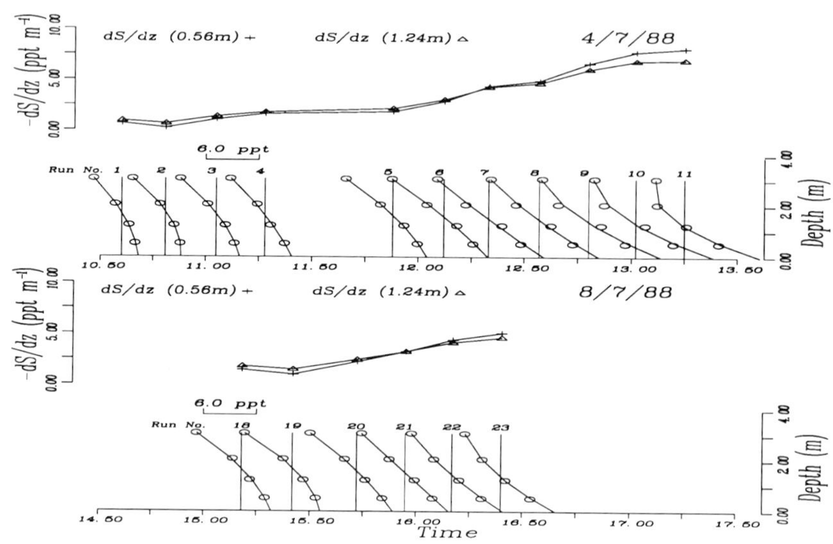

Figure 3: Profiles and vertical gradients of salinity computed from fitted quadratics.

Turbulent mean velocity profiles were measured using eight Braystoke impeller heads mounted on a 2.0m mast at intervals of 0.25m. The mean values were computed over 50s. Turbulent mean salinity profiles were determined from water samples pumped to the surface via pipes whose inlets were attached to the same mast. The salinities were measured with an MC5 salinometer.

Turbulent perturbations of longitudinal and vertical velocity were measured using 5.5×10^{-2}m diameter electromagnetic flowmeters (Colnbrook Instrument Development Ltd.) and those of salinity using an inductive salinometer with four sensor heads. These instruments were mounted on a streamlined mast in such a way that the hydraulic interference between the sensor heads was considered minimal. The flowmeter heads were situated at 0.40m and 1.12m with the salinometer heads at 0.56m, 1.24m, 2.06m and 3.07m above the bed. The separation of the flowmeter and salinometer heads is of relevance to the later discussion. The mast was mounted on a weighted bed frame fitted with orthogonal inclinometers to ensure that it was level on the bed.

Turbulent mean velocities and salinities are shown as vertical profiles in figs. 1 & 2 for the 4th and 8th of July respectively together with the temporal variation of the depth mean values. The velocity profiles are plotted on a logarithmic scale. The points defining each profile have been fitted with a third order polynomial curve in the case of velocities and connected with straight lines in the case of the salinities. Vertical profiles have been plotted as deviation from the depth mean value. The ordinate for each profile is plotted as a vertical line at the time the profile data were collected and with a length equal to the flow depth. The scales for the deviation from the depth mean values are shown as horizontal bars for each parameter.

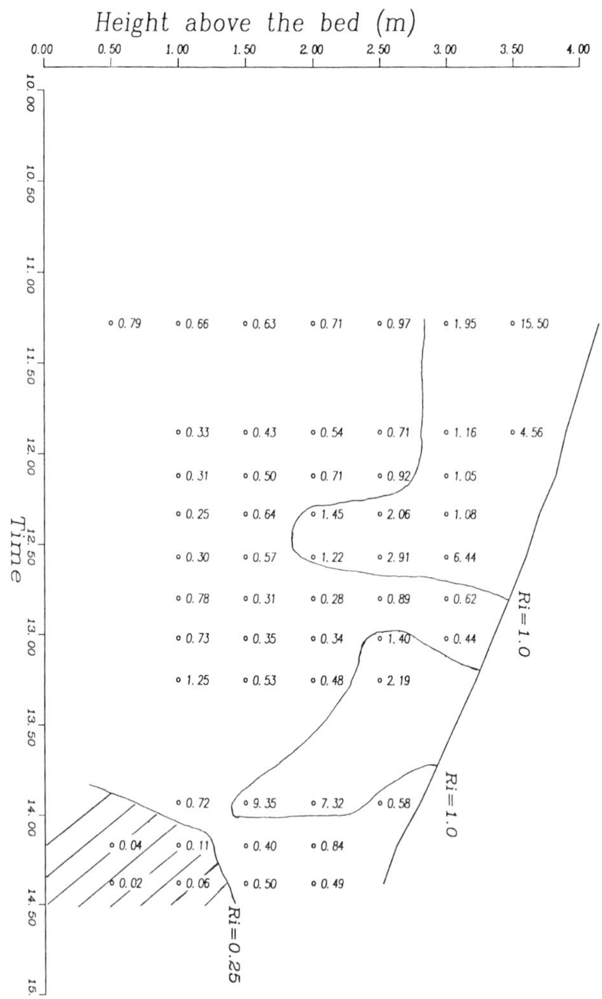

Figure 4: Gradient Richardson Numbers, 4/7/88.

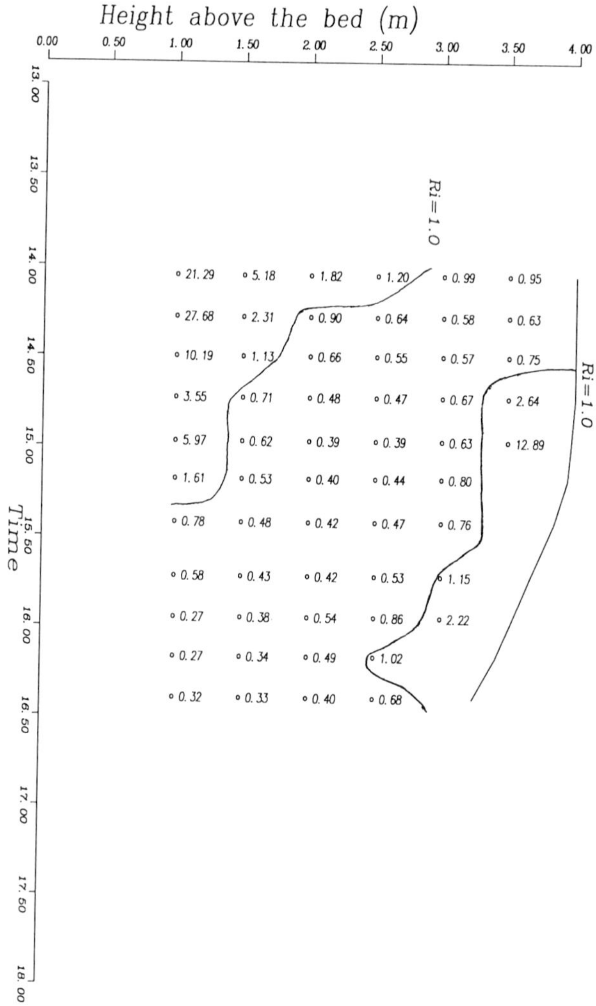

Figure 5: Gradient Richardson Numbers, 8/7/88.

The records of the turbulent perturbations of the velocity and salinity records were divided into runs of 13.6 minutes and digitized at $10s^{-1}$. Large spikes were then removed and a four point smoothing applied to give an effective digitization rate of $2.5s^{-1}$. In order to eliminate longer period non-turbulent contributions to the fluxes the runs were then further divided into sub-sections of 51.2s and the linear trend removed from each record. Each sub-section was then Fourier analysed using a standard FFT routine and the values of the turbulent intensities and turbulent fluxes determined for each one. High and low frequency cut-off errors were computed and applied according to Soulsby (1980). Finally an ensemble mean of the results for each sub-section was evaluated. To assess the lower frequency due to internal waves the spectra of the complete (13.6 minute) salinity records were also computed.

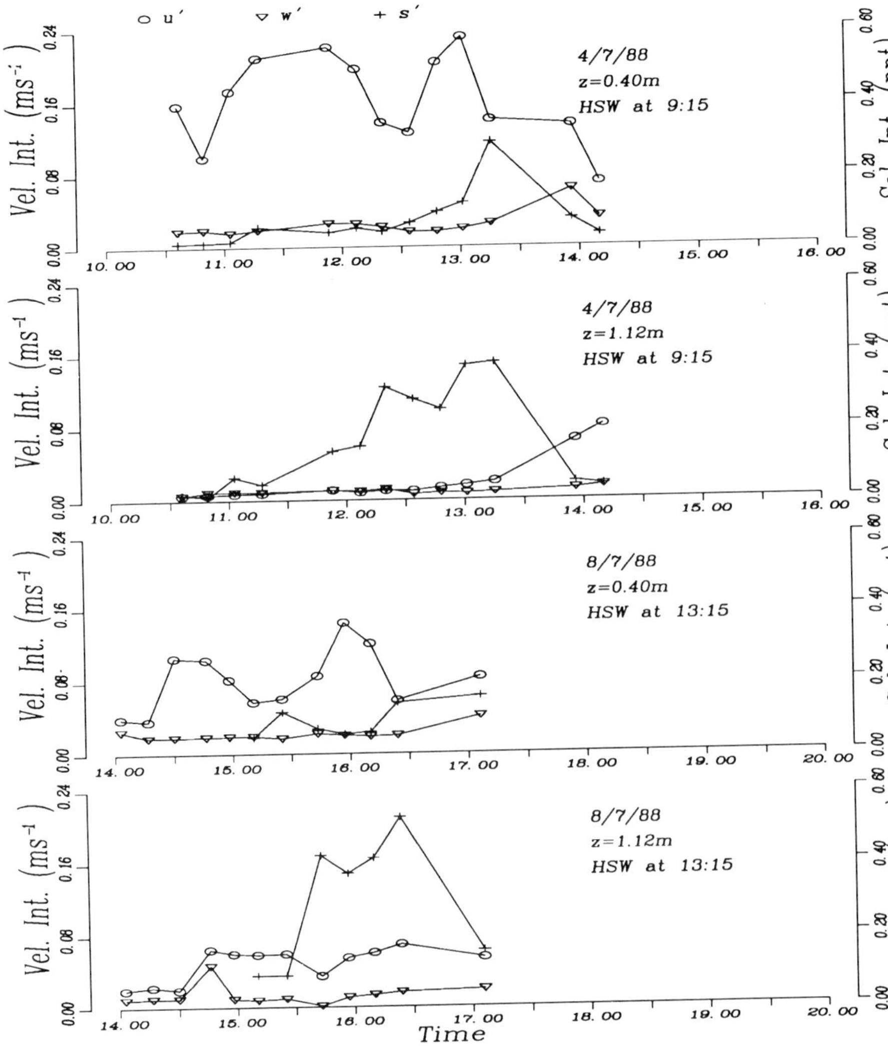

Figure 6: Turbulent intensities of velocity and salinity.

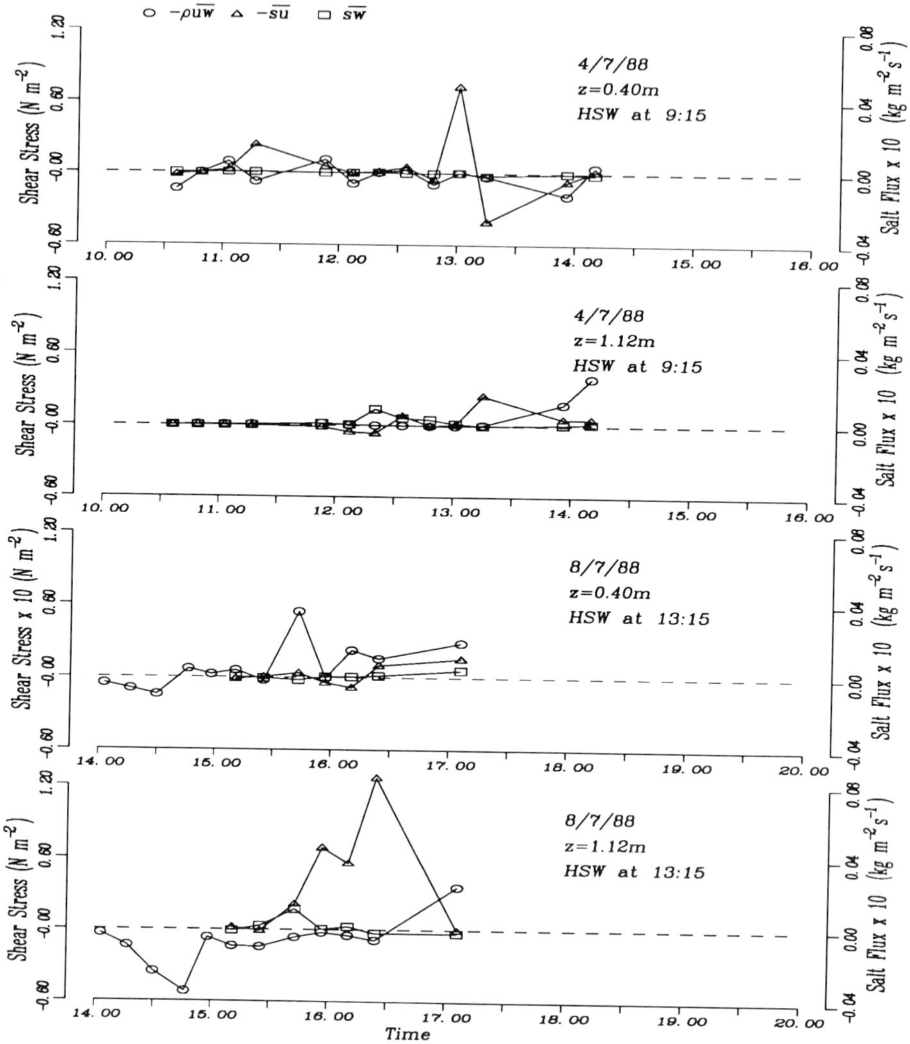

Figure 7: Shear stress and turbulent fluxes of salt.

Computation of the turbulent mean salinity from the inductive salinometer data provided an additional representation of the vertical density structure. Although this view is less detailed in a spatial sense the temporal resolution is much better than for the pumped sample profiles. Profiles computed from these data are shown in fig. 3 with quadratic curves fitted to the four points. Also shown is the temporal variation of the vertical gradient evaluated from the quadratics at z=0.56m and z=1.24m. Each inductive salinometer head had been calibrated against a large MC5 salinometer field data set which included the data which is under consideration here. The correlation coefficients for this linear calibration were 0.9894, 0.9873, 0.9899 & 0.8825 for heads 1, 2, 3 & 4 respectively where head 1 is nearest the bed.

The local gradient Richardson numbers evaluated from the turbulent mean profiles are shown in figs. 4 & 5 for each day. In order to obtain the required vertical gradients vertical interpolations were taken from the profiles to the required heights and then interpolations were made in time assuming linear variation. Values of Ri are not shown where extrapolation of the velocity profiles would have been necessary for their computation which accounts for the scarcity of values at z=0.50m. The linear interpolation is reasonable in the case of the velocity data because the temporal resolution of the profiling is good. In the case of the salinity profiles the resolution is only one per hour. However consideration of the gradients shown in fig. 3 show the assumption of linear variation of the gradients is not unreasonable.

The root mean square values (turbulent intensities) of the velocity (u',w') and of the salinity (s') perturbations about the linear trend in each record are shown in fig. 6 at two depths for the early ebb tide of each day when salinity was present. The associated values of shear stress and vertical and longitudinal turbulent salt fluxes are shown in fig. 7.

5.4 Observations

On both days the ebb tide flow begins at the surface and there is considerable delay before the barotropic forcing due to the surface slope fully overcomes the baroclinic forcing due to the longitudinal density gradient in the lower layers. The velocity profiles remain non-logarithmic for the duration of the saline period. Acceleration of the depth mean velocity to a maximum value of about $0.65ms^{-2}$ is gradual but the process is not uniform throughout the flow with layers at successively lower depths being brought into motion relatively rapidly thus producing local short periods of acceleration.

Significant vertical gradients about $1.0kgm^{-4}$ are present at the commencement of data collection and these are observed to increase, particularly near to the bed, up to about $8.0kgm^{-4}$. Fig. 3 shows the character of the profiles to change from 'concave', where $\partial^2S/dz^2>0$ to 'convex', where $\partial^2S/dz^2<0$, over the period of data collection passing through a period when the profile is near linear.

The gradient Richardson numbers, figs. 4 & 5, are almost always greater than the critical value of 0.25 but also rarely greatly exceed 1.0. The few values which are very large are a result of the use of vertical gradients which are close to the limit of resolution of the instrumentation. The resulting values of Ri are undoubtedly large but their accuracy is debatable.

5.5 Discussion

The turbulent mean profiles of salinity are observed to change over the observation period. This change is a combination of advection of the entire flow structure past the measurement position and the action of the velocity gradient which can have the effect of increasing the vertical gradient and vertical transport of salt. The latter will tend to reduce the vertical solute gradient. The fact that Ri remains above 0.25 for most of the time indicates that turbulent exchange will be severely inhibited. For Ri=0.5 and, using β_1=160 obtained by Odd & Roger (1978) for eqn. 4, the value l_m is reduced to about 10% of its value in homogeneous conditions. However, Ri has been evaluated over timescales of 50s and length scales of 0.25m and conditions for instability may exist on shorter time scales and length scales than these. If this is so then turbulence production must occur at positions remote from the bed. The salinity gradient is large throughout the flow depth and shows no signs of erosion due to vertical mixing, even at the lowest measurement position. This suggests that bed generated turbulence has little, or no, influence on the this stage of the tide.

The observed values of the vertical turbulent flux of salt (fig. 7) are extremely small especially at 0.40m which is consistent with the suggestion of the previous paragraph. In order to compare the effects of velocity shear and vertical turbulent transport the following calculation was performed. Ignoring the effects of advection of the entire flow structure the rate of change of vertical salinity gradient due to the action of vertical velocity shear is given by,

$$\frac{\partial}{\partial t}\left(\frac{\partial \bar{S}}{\partial z}\right) = \frac{\partial \bar{U}}{\partial z}\,\frac{\partial \bar{S}}{\partial x} \tag{5.6}$$

Using the observed values of $\bar{U}$, $\partial\bar{U}/\partial z$ and $\partial\bar{S}/\partial x$ this predicted value was computed for each of the times for which the observed rate of change of $\partial\bar{S}/\partial z$ was available. The value of the longitudinal gradient was obtained by assuming $\partial\bar{S}/\partial x=(1/\bar{U})(\partial\bar{S}/\partial t)$, the temporal gradient obtained from the salinity perturbation records. These values of the predicted and observed rates of change of the vertical gradient are shown in columns 3 & 4 of table 1.

These results show that much of the time the predicted rate of steepening exceeds that which was observed to occur. This suggests that some vertical transport of salt is taking place. If this is entirely due to the vertical turbulent flux then a minimum value of this flux may be calculated. It was assumed that all the salt transported by such a flux into a layer of finite thickness d was accumulated within the layer and that the vertical gradients of salinity and velocity were linear across this layer. This gives a required flux of,

$$\overline{sw} = \frac{d^2}{2}\frac{d}{dt}\left(\frac{\partial \overline{S}}{\partial z}\right) \tag{5.7}$$

The difference between the flux recorded by the sensors and the values of the flux are shown in column 5 of table 1 where a positive value is shown if the predicted flux exceeds that which was observed. Although the number of positive values is not much greater than the number of negative ones their magnitudes are generally greater leading to the conclusion that the vertical turbulent flux is frequently being underestimated. There are three possible reasons for this observation.

Table 1: Predicted and observed rates of change of vertical salinity gradient, vertical turbulent flux deficit and values of $\overline{W}$ needed for mass balance.

Date z	Run	Predicted $\frac{\partial}{\partial t}\left(\frac{\partial S}{\partial z}\right)$ $kgm^{-4}s^{-1}$	Observed $\frac{\Delta(\partial S/\partial z)}{\Delta t}$ $kgm^{-4}s^{-1}$	Flux deficit $-\overline{sw}$ $kgm^{-2}s^{-1}$	$\overline{W}ms^{-1}$
		x 10^{-3}	x 10^{-3}	x 10^{-3}	x 10^{-3}
4/7/88 0.40m	9	0.25	2.00	-0.39	1.50
	10	1.20	1.40	-0.47	1.02
	11	0.25	0.39	0.07	-0.14
4/7/88 1.12m	4	0.02	0.50	-0.16	-1.23
	5	2.00	0.17	0.52	2.54
	6	1.10	1.00	-0.01	-0.74
	7	3.10	1.60	-0.23	5.58
	8	1.50	0.46	0.01	-0.49
	9	1.60	1.60	-0.26	0.82
	10	4.00	1.00	0.90	-1.42
	11	1.50	0.10	0.47	-0.71
8/7/88 0.40m	23	0.64	0.80	-0.04	0.18
8/7/88 1.12m	19	3.50	-0.21	1.23	4.26
	20	4.10	0.91	0.82	6.64
	21	3.10	0.93	-0.18	-5.49
	22	2.10	1.10	0.32	-1.89
	23	0.63	0.57	-0.14	0.49

It is possible that the assumption regarding the effects of advection is incorrect and that they cannot be ignored. Longitudinal sections of estuaries usually show that $\partial\overline{S}/\partial z$ increases in a landward direction towards the limit of salinity intrusion and therefore simple advection of this structure during ebb tides would result in an additional steepening of the profiles which is not observed.

The second possibility is that the vertical transport is occurring but on very small length scales and therefore remains undetected by the turbulence sensors and a third is that transport takes place via some other turbulent mean mechanism.

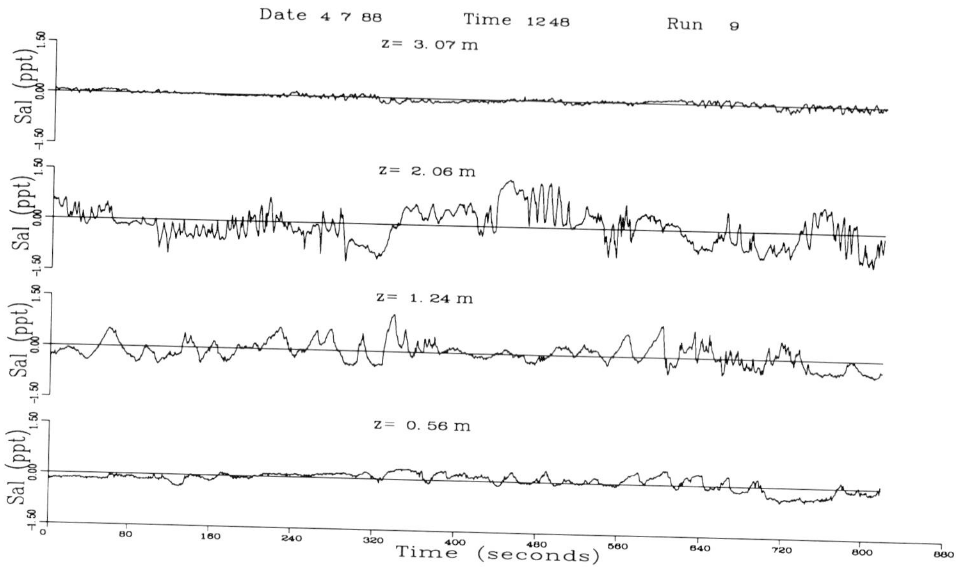

Figure 8: Salinity perturbations at four heights.

Consider first small scale turbulence generation and the possibility that this is associated with internal waves. It is apparent from the salinity time series data shown in fig. 8 that internal waves are present throughout the flow period. It is also apparent that there are many frequencies present and that individual trains of waves do not persist over long periods of time. The frequency spectra of these time series are shown in fig. 9 where inputs of energy due to internal wave activity are apparent particularly for the spectrum computed for z=0.56m.

In order to conduct a simple analytical investigation of the wave field assume an approximate density structure of the flow such that $\partial N/\partial z=0$, where $N^2 = (-g/\rho)(\partial\rho/\partial z)$. Over the depths being considered this approximates to the density profile being a straight line. Fig. 3 shows that this is a realistic first approximation for most of the time. Also ignore the effects of the vertical velocity shear. If the flow is irrotational and the fluid incompressible and is moving past the observer with a velocity $\bar{U}$ then it can be shown, after making further assumptions (appendix 1.), that the first mode internal wave will have the dispersion relation,

$$\bar{U}^2 D^2 k^4 - 2\omega \bar{U} D^2 k^3 + (\omega^2 D^2 + \bar{U}^2 \pi^2 - D^2 N^2) k^2 - 2\omega \bar{U} \pi^2 k + \omega^2 \pi = 0 \tag{5.8}$$

which can be solved numerically for k, the wavenumber where D is the flow depth and ω is the observed wave frequency.

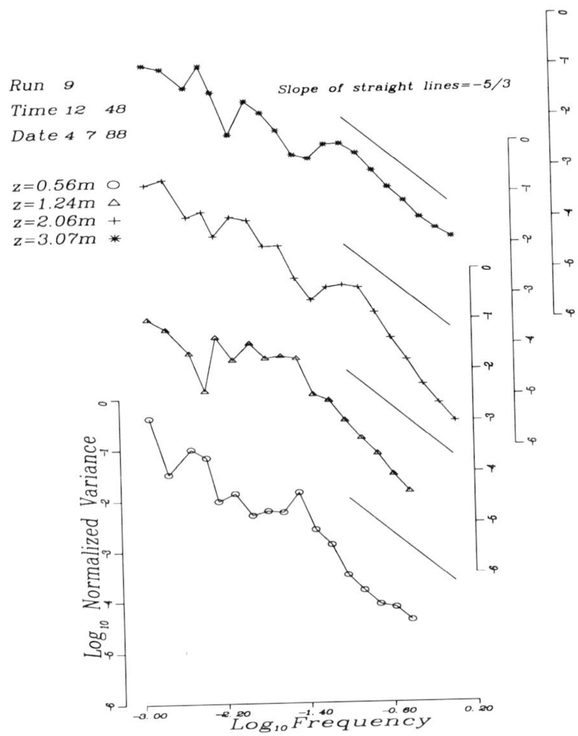

Figure 9: Normalized salinity variance spectra.

On solving for k it is found that two real roots exist, one representing propagation with the flow and one against. The resulting phase velocities are such that the upstream propagating solution cannot propagate against the flow and is therefore not realistic. The waves are therefore assumed to be propagating in the same direction as the mean flow.

The particle velocities associated with the wave propagation are computed by assuming that the observed variance of the salinity signal is due to a single frequency sinusoidally varying wave. The details of assumptions and computation are given in appendix 1 and the resulting maximum longitudinal and vertical particle velocities are given by

$$W_0 = (\partial S/\partial z)^{-1}\, 2\pi\sigma S_0 \tag{5.9}$$

and

$$U_0 = \frac{\delta}{k}\,\frac{W_0}{\tan(\delta z)} \tag{5.10}$$

where $\delta^2 = \omega^2 (N^2 - \omega^2)/k^2$. The results of these computations for one of the sensor heads are shown in table 2 where Z_0, S_0, U_0 and W_0 are the internal wave vertical displacement, salinity, longitudinal and vertical velocity amplitudes. Reference to fig. 8 shows that while the sinusoidal assumption is not very realistic for the z=1.24m time series it also shows that the predicted value of S_0 (table 2, run 9) is not a large overestimate and is sometimes exceeded on timescales of a few wave periods.

Table 2: Salinity, Displacement and Velocity amplitudes, z=1.24m, 4/7/88.

Run	S_0(ppt)	$Z_0 \times 10^2$(m)	$W_0 \times 10^2$(ms^{-1})	$U_0 \times 10^2$(ms^{-1})
1	0.03	1.6	0.6	0.7
2	0.03	1.4	0.3	0.8
3	0.14	6.1	1.7	3.3
4	0.16	6.8	1.9	3.5
5	0.27	8.5	4.7	3.9
6	0.36	9.9	5.3	4.7
7	0.51	13.5	8.0	5.8
8	0.47	12.7	5.2	5.5
9	0.41	10.7	7.1	3.5
10	0.60	16.5	11.8	4.2
11	0.58	20.3	13.3	3.7

These velocities may then be compared with the observed velocity perturbations. In order that the wave effects can be seen more clearly the time series have been smoothed over 6.4s. Simultaneous variations of w,u and s are shown in fig. 10. The first feature which is clearly apparent from the latter part of the run is that positive values of u correlate well in a qualitative sense with positive values of s which is consistent with a wave propagating in the same direction as the mean flow. The effect of advection of the relatively slowly varying wave field past the sensor position is that the observed phase relationship is changed from that which would be observed in still water and positive values of w lag behind positive u and s.

The observed magnitudes of the velocity perturbations are about half those which were predicted by the theory and it remains a matter of speculation whether such waves might break. The presence of vertical velocity shear means that the onset of wave breaking will occur at lower values of U_0 than it would in a non-sheared flow. It is clear however, that should breaking occur, that the resulting turbulence will be on length scales less than the wave displacement amplitude which is generally <0.15m.

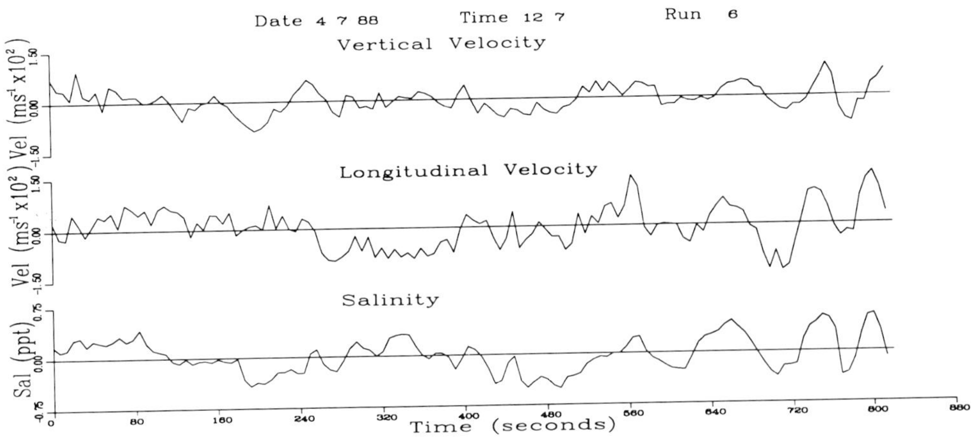

Figure 10: Salinity and velocity perturbations at z=1.12m.

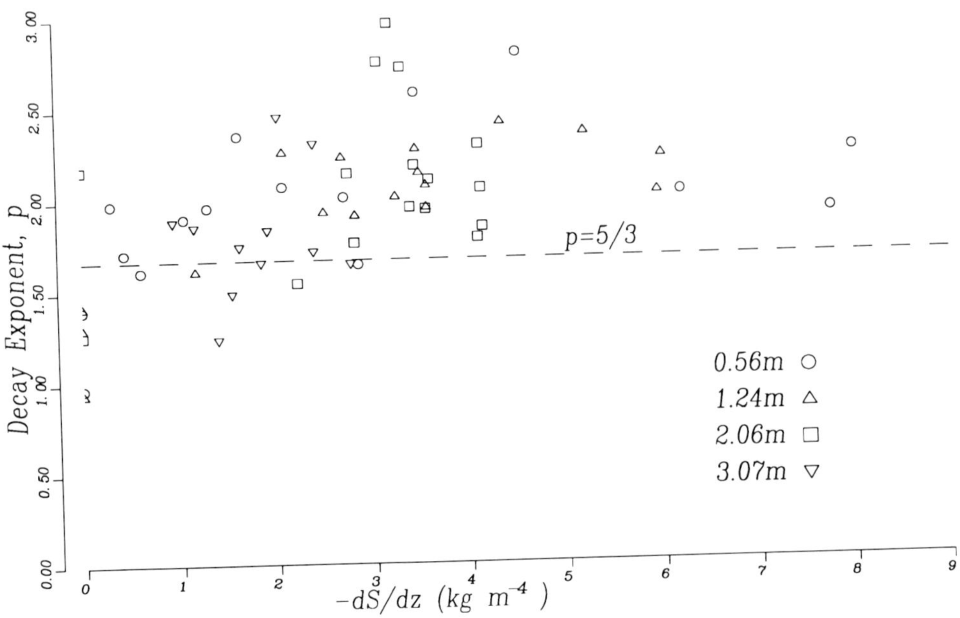

Figure 11: Salinity spectrum decay exponent vs. vertical salinity gradient.

The fact that perturbations of frequencies less than about $1s^{-1}$ are suppressed can be seen from examination of the salinity spectra. A region where the spectrum declines more steeply than -5/3 can be seen extending from frequencies $0.01s^{-1}$ to $1.0s^{-1}$. There is a suggestion that above these frequencies the rate of decrease becomes less steep although there are rather few points upon which to base this speculation. The magnitudes of these slopes have been plotted against the vertical salinity gradient in fig. 11. There is a clear trend to steeper slopes at steeper gradients although there is considerable scatter. The observed values which are <5/3 are generally confined to periods of very low salinity when the presence of random noise may have affected the spectra.

Further evidence of damping as the density gradient increases may be gained from a plot of s' vs. $\partial\bar{S}/\partial z$ (fig. 12). Although not conclusive there appears to be a value of $\partial\bar{S}/\partial z$ above which s' no longer increases. The situation is complicated by the presence of internal wave perturbations which contribute to s' and will not be affected by buoyancy in the same way as turbulent perturbations. Fig. 12 also shows that s' tends to increase with distance from the free surface and the bed which is also due to the fact that the perturbations are largely due to vertical wave perturbations which are smaller near to the flow boundaries.

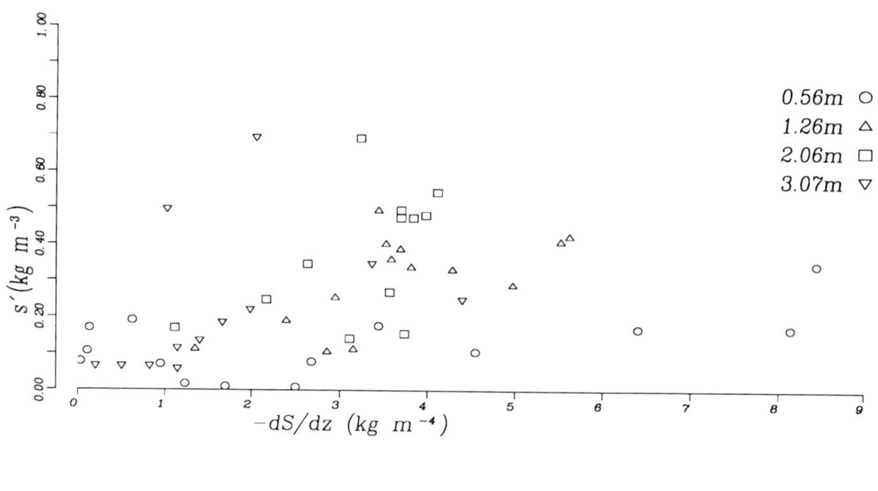

Figure 12: Salinity turbulent intensity vs. vertical salinity gradient.

Finally, consider the possibility of the existence of a flow structure which results in a turbulent mean velocity in the vertical and transverse directions. Although no direct measurements were made it was observed that the axial convergence which was observed to form on flood tides does not survive beyond the early ebb and that the accumulated debris appear to end up on the banks. This is the nature of a transverse flow consistent with upwelling towards the centre of the channel. Such a flow may be anticipated during ebb tides due to the interaction of transverse velocity shear and longitudinal salinity gradient. This should produce, at any given height, lower density water in the main channel and higher density water near to the banks. This transverse density gradient could provide the forcing function for the necessary circulation.

In order to assess whether such a circulation would be realistic the necessary vertical velocity to balance the observed vertical flux deficit was computed in each case. It was assumed that $\partial\bar{W}/\partial z$ was small and that the rate of change of the salinity gradient due to vertical turbulent mean flow is given by $\bar{W}(\partial^2\bar{S}/\partial z^2)$. The computed necessary vertical velocities to balance the flux deficit are shown in table 1. The values of $\partial^2 S/\partial z^2$ were obtained from the quadratic representations of the profiles (fig. 3) which gives a single value for the whole flow depth. This value was observed to change sign during the ebb tide so the necessary value of $\bar{W}$ is not always positive. The values of $\bar{W}$ do not exceed 1.0×10^{-2} ms^{-1} which is small but there is no reason why they should not occur. Given a channel breadth to width ratio of about 20 this could result in transverse velocities of about 10.0×10^{-2} ms^{-1} which may be detectable by surface measurements.

5.6 Conclusions

1. Existing turbulence measuring systems have been shown to have some limitations with respect to the investigation of vertical turbulent transport processes during solute stratified neap ebb tides in a shallow partially mixed estuary. However, preliminary estimates of vertical fluxes were possible and most features of the individual vector and scalar parameters have been determined.

2. The interaction of the longitudinal salinity gradient and velocity shear produce a turbulent mean flow structure which appears to be influenced significantly by local effects. This is true in both a spatial and temporal sense. The main evidence for this local control is that in spite of a tendency for the velocity gradient to increase with time due to the velocity shear the gradient Richardson number remains less than 1.0 over large sections of the flow due to concomitant changes in the salinity gradient. This leads to the conclusion that the velocity and salinity turbulent mean distributions are linked by one or more mechanisms even though steep vertical gradients exist.

3. Measured vertical fluxes of momentum and solutes using instrumentation with sensor separation of about 0.15m gave mixing lengths of 10% of the value found in homogeneous open channel flows. The solute fluxes were generally less than those expected from mass balance calculations based on temporal evolution of the turbulent mean field.

4. Continuous measurements of salinity showed that an incoherent internal wave field was present. The rms. values of the perturbations calculated over 13.6 minutes initially increased with salinity gradient but the trend became less clear at higher gradients. They also showed

a tendency to decrease in the vicinity of upper and lower flow boundaries. An analysis of the wave field for idealized conditions suggested that the waves were propagating with the flow and the predicted particle velocities were not inconsistent with those which were observed. The salinity frequency spectra showed an increasing effect of buoyancy on the higher frequency part of the spectrum as the density gradient increased.

5. Previously other workers have suggested that for turbulent mean gradient Richardson numbers some fine scale turbulence generation may be expected due to internal wave breaking in the regions of maximum shear where, on short time scales, Ri may be reduced below 0.25. This process is probably intermittent and leads to the periodic appearance and decay of the waves.

6. It is conceivable that the secondary circulation caused by a transverse density gradient may contribute to the vertical fluxes. In some cases only a weak circulation is needed for a significant contribution. Such a motion is not physically unrealistic despite the presence of steep density gradients and the absence of surface debris accumulated during the flood tide is circumstantial evidence for its existence.

7. Some progress should be possible in proving some of the above conjecture by using proven techniques. Secondary flow structure may be investigated by drogue tracking using appropriately dimensioned drogues. Continuous point source dye tracing can also provide quantitative estimates of vertical and transverse solute mixing rates and hence diffusion coefficients.

Notation.

d	Layer thickness.
D	Flow depth.
g	Acceleration due to gravity.
k	Wave number.
l	Mixing length.
N	Buoyancy or Brunt-Vaisala frequency.
p	Pressure or Spectrum decay exponent.
Ri	Local gradient Richardson Number.
s	Perturbation of salinity from the turbulent mean value.
s'	Turbulent intensity of salinity.
$\underline{S}$	Salinity.
$\overline{S}$	Turbulent mean salinity.
S_0	Internal wave Salinity amplitude.
t	Time.
u,v,w	Perturbations of longitudinal, transverse and vertical velocities from the turbulent mean values.

u',v',w'	Turbulent intensities of the longitudinal transverse and vertical velocities.
U,V,W	Turbulent mean longitudinal, transverse and vertical velocities.
U_0,W_0	Internal wave longitudinal and vertical velocity amplitudes.
x,y,z	Longitudinal, transverse and vertical coordinate axes.
z	Height above the bed.
Z_0	Internal wave vertical displacement amplitude.
β_1,β_2	Constants.
δ	A parameter.
λ	Wave length.
ρ	Turbulent mean fluid density.
σ	Frequency.
ω	Angular frequency.

Subscripts of l.

m,s	Momentum, salinity.
z	In the vertical direction.
0	Under homogeneous density conditions.

Acknowledgements

The authors gratefully acknowledge the financial support of the Science and Engineering Research Council and the Natural Environment Research Council and the assistance of the members of the Tidal Waters Research Group, School of Civil Engineering, in the collection of data.

5.7 References

Dyer, K.R. 1981: Mixing caused by lateral internal seiching within a partially mixed estuary. Estuarine, Coastal and Shelf Science 15, 443-457.

Geyer, W.G. & Smith, J.D. 1987: Shear instability in a highly stratified estuary. Journal of Physical Oceanography 17, 1668-1679.

Kent, P.E. & Pritchard, D.W. 1959: A test of mixing length theories in a coastal plain estuary. Journal of Marine Research 18, 62-72.

Miles, J.W. 1961: On the stability of heterogeneous shear flows. Journal of Fluid Mechanics 10, 496-508.

New, A.L., Dyer, K.R. & Lewis, R.E. 1987: Internal waves and intense mixing periods in a partially stratified estuary. Estuarine, Coastal and Shelf Science 24, 15-33.

Nunes, R.A. & Simpson, J.H. 1985: Axial convergence in a well mixed estuary. Estuarine, Coastal and Shelf Science 20, 637-649.

Odd, N.V.M. & Rodger, J.G. 1978: Vertical mixing in stratified tidal flows. Proc. ASCE 104(HY-3), 337-351.

Partch, E.N. & Smith, J.D. 1978: Time dependant mixing in a salt wedge estuary. Estuarine and Coastal Marine Science 6, 3-19.

Rossby, C.G. & Montgomery, R.B. 1935: The layer of frictional influence in wind and ocean currents. Pap. Phys. Oceanogr. Meterol. 3(3), 1-101.

Soulsby, R.L. 1980: Selecting record length and digitization rate for near bed turbulence measurements. Journal of Physical Oceanography 10, 208-219.

Turner, J.S. 1973: Buoyancy Effects in Fluids. Cambridge University Press 367pp.

West, J.R, Knight, D.W. & Shiono, K. 1985: A note on the determination of vertical turbulent transport coefficients in a partially mixed estuary. Proceedings of the Institution of Civil Engineers, Part 2 79, 235-246.

Appendix 1 - Internal wave theory

Assuming a flow is irrotational and incompressible, ignoring viscosity and friction and assuming that any waves will be of a period which is short compared with the inertial period, of the order of the period of the earth's rotation, its motion can be described by the four following equations, LeBlond & Mysak (1978).

$$\frac{\partial u}{\partial x} + \frac{\partial w}{\partial z} = 0 \qquad (5.A1)$$

$$\frac{\partial \rho}{\partial t} + w\frac{\partial \rho_0}{\partial z} = 0 \qquad (5.A2)$$

$$\rho_0 \frac{\partial u}{\partial t} = -\frac{\partial p}{\partial x} \tag{5.A3}$$

$$\rho_0 \frac{\partial w}{\partial t} = -\frac{\partial p}{\partial z} - \rho g \tag{5.A4}$$

Here, ρ_0 is the fluid density at the equilibrium position. Eliminating p and ρ from these equations gives the following equation for the vertical velocity, w

$$\frac{\partial}{\partial t^2}\left(\frac{\partial^2 w}{\partial z^2} + \frac{\partial^2 w}{\partial x^2} - \frac{N^2}{g}\frac{\partial w}{\partial z}\right) + N^2 \frac{\partial^2 w}{\partial x^2} = 0 \tag{5.A5}$$

where $N^2 = (-g/\rho_0)(\partial\rho_0/\partial z)$ rads s^{-1}. If it is assumed that the flow can be approximated to a layer of constant buoyancy frequency, i.e. $\partial N^2/\partial z=0$, corresponding to a density profile which is near linear over the water depth, then solutions exist of the form,

$$w = W(z)e^{i(kx-\omega t)} \tag{5.A6}$$

Substituting equation 5.A6 in equation 5.A5 gives

$$\frac{d^2W}{dz^2} + \frac{N^2}{g}\frac{dW}{dz} \mp \lambda^2(\omega^2 - N^2)W = 0 \tag{5.A7}$$

where $\lambda^2 = k^2/\omega^2$, k = the wave number in the horizontal direction and ω = angular frequency in rads s^{-1}.

Solutions to this equation are of the form,

$$W = W_* e^{-N^2 z/2g} \sin(\delta z) \tag{5.A8}$$

where $\delta^2 = \lambda^2 (N^2-\omega^2)-N^4/4g$, and $N^4/4g$ is small compared to the other terms and can be neglected. Imposing the boundary condition of no vertical motion at the free surface, i.e. W=0 at z=D, gives,

$$\delta D = n\pi \tag{5.A9}$$

where n is an integer. Considering the case where n=1 the following dispersion relation is obtained,

$$\omega^2 = \frac{k^2D^2N^2}{\pi^2+k^2D^2} \tag{5.A10}$$

In a moving flow the frequency seen by a fixed observer is, $\omega = \omega_0 + \bar{U}k$. In this equation ω_0 is defined as the frequency relative to the flow which has, until now, been referred to as simply ω. From now on ω will be used to denote the frequency observed by a stationary observer. $\bar{U}$ is the mean longitudinal flow velocity and is positive in the direction of wave propagation. Substituting in equation 5.A10 and rearranging gives,

$$\bar{U}^2D^2k^4 - 2\omega\bar{U}D^2k^3 + (\omega^2D^2 + \bar{U}^2\pi^2 - D^2N^2)k^2 - 2\omega\bar{U}\pi^2k + \omega^2\pi = 0 \tag{5.A11}$$

To calculate particle velocities it is necessary to return to equation 5.A8 and consider the value of W, which from now on will be referred to as W_0, the vertical velocity amplitude. For the purposes of further analysis it is assumed that $e^{-N^2g/2z} = 1$ since both N and z are small. If changes in time of the mean salinity are sufficiently small to be ignored then,

$$\frac{\Delta S}{\Delta t} = \frac{\Delta z}{\Delta t}\frac{\partial s}{\partial z} \tag{5.A12}$$

This allows computation of the wave displacement amplitude in the vertical direction, $Z_0 = S_0 (\partial S/\partial z)^{-1}$. If a single frequency is considered, then equation 5.A12 can be written

$$\frac{dS}{dt} = w\frac{\partial S}{\partial z} \tag{5.A13}$$

To obtain w, sinusoidal variation of the salinity of the form, $S(t) = \bar{S}+S_0 \sin(2\pi\sigma t)$ is assumed where S(t) is the salinity as a function of time, $\bar{S}$ is the mean salinity and S_0 is the amplitude of the salinity variation. This gives

$$w = \left(\frac{\partial S}{\partial z}\right)^{-1} 2\pi\sigma S_0 \cos(2\pi\sigma t) \tag{5.A14}$$

giving,

$$W_0 = \left(\frac{\partial S}{\partial z}\right)^{-1} 2\pi\sigma S_0 \tag{5.A15}$$

thus allowing evaluation of w.

W_0 is a function of z as described by eqn. 5.A8 the simplified form of which is,

$$W_0 = W_* \sin(\delta z) \tag{5.A16}$$

Using eqns. 5.A2, 5.A6 and 4.A16 this leads to

$$w = W_0 e^{i(kx-\omega t)} \tag{5.A17}$$

$$u = -i\, W_0 \frac{\delta}{k} \frac{1}{\tan(\delta z)} e^{i(kx-\omega t)} \tag{5.A18}$$

giving,

$$U_0 = \frac{\delta}{k} \frac{W_0}{\tan(\delta z)} \tag{5.A19}$$

where U_0 is the amplitude of the longitudinal velocity.

6

Dynamically-active models in the prediction of estuarine stratification.

J H Simpson and J Sharples

Abstract

Stratification in estuaries and ROFI's (regions of freshwater influence) exerts a critical environmental control and frequently exhibits large variability. Prediction is inherently difficult because of the localised nature of the buoyancy source and the complex suite of interacting processes that control water column stability.

To advance beyond the first order account of the variability of estuarine stratification given by the prescriptive model (Simpson, Sharples and Rippeth (1991)), it is necessary to have recourse to a dynamically-active model in which the profiles of velocity and density are computed from forcing by surface slopes and density gradients. The model described here employs a Mellor and Yamada level-two closure scheme to represent the interaction between the density-driven and tidal flows via the frictional stresses and their dependence on stratification.

The performance of the model is considered in relation to that of the prescriptive model and compared with recent time series observations from Liverpool Bay. A single component model without rotation will reproduce realistic semidiurnal and semi-monthly cycles, but the latter only when the stirring is artificially increased. As with the prescriptive model, it is found that the inclusion of the effects of rotation in a two component model overcomes this difficulty by reducing the cross-isopycnal flow. Simulations of the springs-neaps cycle with the two-component T-C model using realistic forcing for the tidal flow and the density gradient show encouraging agreement with the observations.

6.1 Introduction

Water column stability is a key factor in determining the environment both in estuaries and in adjacent regions of freshwater influence (ROFI's). Observations of its time-dependence suggest it is frequently highly variable with important cycles on semi-diurnal,semi-monthly and seasonal time scales (e.g. Haas (1977), Simpson et al. (1990)) with more erratic changes associated with episodic wind mixing (Goodrich et al. (1988)).

Stratification inhibits turbulent motions and thus restricts vertical transfer in a way which is known to be effective but is not understood in detail and has proved difficult to parametrise. Episodes of stratification may lead to pronounced hypoxia with important biological and chemical consequences. The level of stratification may also act as a critical control on the supply of nutrients to the upper part of the water column from re-mineralisation sources in the seabed and influence the light climate through its effect on the re-suspension of sediments; both these effects may have a significant bearing on primary production in the areas concerned.

Understanding and, eventually, predicting the development and decay of estuarine stratification must, therefore, be an important goal for physical oceanography requiring complementary programmes of observation and modelling. The difficulty of the problem arises in part from the fact that the primary buoyancy input is localised at one or more sources at the coast rather than being distributed uniformly over the surface as in the heating-stirring competition (which determines seasonal thermal stratification over much of the shelf). In addition the suite of processes involved in the control of stratification interact in a non-linear way generating various feedback effects which may influence the development.

In this paper, we describe new models of the evolution of stratification which allow for some of the important feedback processes that are operating and discuss the results in relation to recent observations of the time history of stratification in a ROFI regime.

6.2 Prescriptive model

In two previous papers (Simpson et al. 1990, Simpson, Sharples and Rippeth (1991)) we have identified the principal processes operating to promote and destroy stratification and attempted to represent these processes in simple models of the time development of stratification. Closely similar approaches have been used by Nunes Vaz et al. (1989) and van Aken (1986),

The stratifying influence of freshwater buoyancy input to an estuary is manifest locally through the shear flow driven by the horizontal density gradient (the estuarine circulation). In addition there is a significant periodic input due to the action of tidal shear on the density gradient (so-called tidal straining) as well as a contribution due to surface heating during spring and summer.

These stratifying influences are opposed by stirring due to the tidal flow and windstress with additional effects from convective cooling during the winter months.

In the prescriptive model (Simpson, Sharples and Rippeth), we specified these processes using analytical solutions for the tidal flow (Prandle (1982)) and density driven flow (Hansen and Rattray (1965) or Heaps (1972)) and used energetics conditions to describe the mixing by tidal and wind-stirring. The model is forced by a depth uniform density gradient and surface slopes which are constrained by a no net flux condition (in the down gradient direction). The justification for the strong assumption of a depth-independent density gradient, which represents an important simplification of the problem, has to be found mainly a posteriori or in reference to observations for the area concerned.

The individual stratifying and mixing inputs are assumed to act independently and so their effects are combined in a simple additive fashion. This linear approach gives a plausible first-order account of observations of the time-dependence of stratification in Liverpool Bay, reproducing both the semi-diurnal and semi-monthly cycles of stability. It, however, explicitly neglects what may be important interactions between the density-driven and tidal flows which occur through the internal frictional stresses and their dependence on the stratification.

When these interactions are introduced, the problem becomes a much more difficult(and interesting) one. For example, instead of being prescribed by the Prandle analytical solution with fixed eddy viscosity, the tidal flow would now be *calculated* for stresses based on an N_z, which varies with stratification, and velocity shear which includes the density driven flow. Likewise the density-driven flow would be modulated with the cycle of tidal stirring as in the laboratory experiments of Linden and Simpson(1988). At the same time the efficiency of mixing, held constant in the presciptive model, would be free to vary with the level of stratification and shear.

6.3 A dynamically active model

We aim for the simplest profile model that can adequately represent these interactions between the density-driven and tidal flows.

Initially we consider flow in the x direction only with no rotation so the dynamics are governed by the equation

$$\frac{\partial u}{\partial t} = -\frac{1}{\rho}\left(\frac{\partial p}{\partial x} + \frac{\partial \tau}{\partial z}\right) \qquad (6.1)$$

where we use conventional notation with z positive upwards and z=0 at the seabed and z=h at the surface.

The pressure p is hydrostatic and represented by

$$p = -g\int_{h+\eta}^{z} \rho\, dz \tag{6.2}$$

where η = surface elevation and $\rho(z)$ = density.

Assuming that $\eta/h \ll 1$, the pressure gradient is approximately

$$\frac{\partial p}{\partial x} = \rho g \frac{\partial \eta}{\partial x} + g(h-z)\frac{\partial \rho}{\partial x} \tag{6.3}$$

where we have further assumed that $\partial\rho/\partial x$ is independent of depth, an assumption which is consistent with the prescriptive model and has the advantage of simplifying the numerical solution.

The shear stress is written in the usual form

$$\tau = -N_z \rho \frac{\partial u}{\partial z} \tag{6.4}$$

where the turbulent eddy viscosity N_z is to be determined from the closure scheme. At the bottom boundary the stress is set equal to

$$\tau_b = -k\rho\, |u_1|\, u_1 \tag{6.5}$$

where u_1 is the near-bottom velocity.

The flow is forced in three ways:

(i) by the imposition of a depth uniform density gradient which may also be (though not necessarily) constant in time.

(ii) by a mean surface gradient of the form:

$$\frac{\partial \bar{\eta}}{\partial x} = \gamma \frac{h}{\rho} \frac{\partial \rho}{\partial x} \tag{6.6}$$

in which the parameter γ may be chosen to make the net transport zero.

(iii) by an oscillating pressure gradient for each tidal constituent giving a sum of the form:

$$g\frac{\partial \eta}{\partial x} = g\frac{\partial \bar{\eta}}{\partial x} + \sum_{1}^{N} A_i \cos(\omega_i t + \lambda_i) \tag{6.7}$$

The density evolution is controlled by horizontal advection and vertical diffusion in the equation:

$$\frac{\partial \rho}{\partial t} = -u\frac{\partial \rho}{\partial x} + \frac{\partial}{\partial z}\left(K_z \frac{\partial \rho}{\partial z}\right) \tag{6.8}$$

where K_z is the eddy diffusivity. Like N_z, the diffusivity is to be determined from closure procedures. To do this we may make use of a level 2 formulation based on Mellor and Yamada(1974) which assumes local equilibrium between the rate of production of turbulent kinetic energy (TKE) and its dissipation in heat and work done against buoyancy forces.

The transfer coefficients are written as:

$$N_z = \phi_m \ell^2 \left|\frac{\partial u}{\partial z}\right| \; ; \; K_z = \phi_h \ell^2 \left|\frac{\partial u}{\partial z}\right| \tag{6.9}$$

where ℓ is a mixing length which has to be specified and the stability functions ϕ_m and ϕ_h are functions of the the flux Richardson number Rf :

$$\phi_m = \left\{\frac{(1-Rf/Rf_c)(1-Rf/Rf_1)}{(1-Rf)^{2/3}(1-Rf/Rf_2)}\right\}^{3/2}$$
$$\phi_h = \frac{1}{Pr}\left\{\frac{(1-Rf/Rf_c)^3(1-Rf/Rf_1)}{(1-Rf)^2(1-Rf/Rf_2)}\right\}^{1/2} \tag{6.10}$$

where Rf_c is a critical value of Rf and Rf_1 and Rf_2 are constants. Pr is the turbulent Prandtl number for homogeneous conditions and is here assumed equal to unity.

The flux Richardson number is found from the local gradient Richardson number Ri according to:

$$Rf = \frac{1}{2}\left\{Rf_1 + Ri\frac{Rf_1}{Rf_2} - \left[\left(Rf_1 + Ri\frac{Rf_1}{Rf_2}\right)^2 - 4Rf_1 Ri\right]^{1/2}\right\} \tag{6.11}$$

We have taken the the required constants (Rf_c,Rf_1,Rf_2)from a recent recalibration of the Mellor-Yamada scheme by J Hamrick(personal communication) which gives the rounded values:

$$Rf_c = \frac{1}{6}\ ;\quad Rf_1 = \frac{1}{4}\ ;\quad Rf_2 = \frac{1}{5} \tag{6.12}$$

The mixing length ℓ must be specified independently and according to Mellor and Yamada is the most arbitrary aspect of the method. We have employed the widely-used form (e.g. Davies (1990)):

$$\ell = Kz(1-z/h)^{1/2} \tag{6.13}$$

which behaves like Kz near the bottom boundary (K = von Karmans k) and falls off rapidly near the surface.

In this level 2 approach, there is no need for a separate differential equation for the TKE which can be obtained explicitly from the equilibrium relation:

$$q = \left\{ N_z B_1 \ell \left(\frac{\partial u}{\partial z}\right)^2 (1-Rf) \right\}^{1/3} \tag{6.14}$$

where q = turbulent intensity and B_1 is a constant.

The model is readily extended to two components of flow so that the effects of rotation may be included where appropriate. The two dynamical equations are now of the form:

$$\begin{aligned} \frac{\partial u}{\partial t} &= fv - g\frac{\partial \eta}{\partial x} - g(h-z)\frac{1}{\rho}\frac{\partial \rho}{\partial x} + \frac{\partial}{\partial z}\left(N_z\frac{\partial u}{\partial z}\right) \\ \frac{\partial v}{\partial t} &= -fu - g\frac{\partial \eta}{\partial y} - g(h-z)\frac{1}{\rho}\frac{\partial \rho}{\partial y} + \frac{\partial}{\partial z}\left(N_z\frac{\partial v}{\partial z}\right) \end{aligned} \tag{6.15}$$

and the forcing includes two components of density gradient and tidal pressure gradient. We retain the condition of no-net-flux normal to the isopycnals and again set the bottom stress in terms of the quadratic drag law now in the form:

$$\begin{aligned} \tau_{bx} &= -k\rho U u_1\ ;\quad U = (u_1^2 + v_1^2)^{1/2} \\ \tau_{by} &= -k\rho U v_1 \end{aligned} \tag{6.16}$$

where u_1 and v_1 are the components of near-bottom velocity.

6.4 Results

6.4.1 Non-rotating model

The unidirectional model was implemented using an explicit finite difference scheme and a time step satisfying the stability condition

$$\Delta t < \frac{1}{2}\frac{(\Delta z)^2}{N_z} \tag{6.17}$$

Two independent versions were written in different languages and tested against each other and against analytical solutions for constant viscosity and diffusivity. (e.g. the Hansen-Rattray estuarine flow solution). The model was then run with driving forces corresponding to the location of the observations in Liverpool Bay (Simpson, Sharples and Rippeth). The result was that the stratification increased continuously and indefinitely. To prevent this run-away stratification, the tidal forcing has been increased somewhat and the density gradient reduced to increase the ratio of stirring to stratifying inputs. (Such an adjustment was found necessary in a previous non-rotating model (Simpson et al. 1990) where it was effected by an increase of N_z).

With these modified inputs, the behaviour of the model in terms of the surface-bottom density difference shows (fig 1b) the same qualitative features as the observations; there is a clear switching between stratified conditions for about 8 days following neaps and almost continuously mixed conditions for the rest of the fortnightly cycle. On top of this pattern is superposed a regular cycle of semi-diurnal fluctuations (clearly apparent in the data) with a maximum amplitude of $\Delta\sigma_t = 0.15$ kg/m ^{3}. The no-net-transport condition was found to be met with a value of γ close to 0.3.

A comparison with the output of the prescriptive model for similar forcing (fig 1b) shows the two models behaving in analagous fashion with similar periods of enduring stratification. There is, however, a significant difference in that the T-C model exhibits a more rapid increase in stratification following neaps and attains a maximum stability more than 4 times that of the prescriptive model, presumably because of the reduced efficiency of mixing in the stratified regime. There are also significant differences between the models in the timing of the onset and breakdown of the enduring stratification.

This single-component model will provide the basis for predicting stratification in the interior of narrow estuaries where the flow is effectively unidirectional. For the presently available data from Liverpool Bay where the flow is certainly influenced by rotation,we shall proceed to a comparison with a two component model in which the Coriolis forces are included.

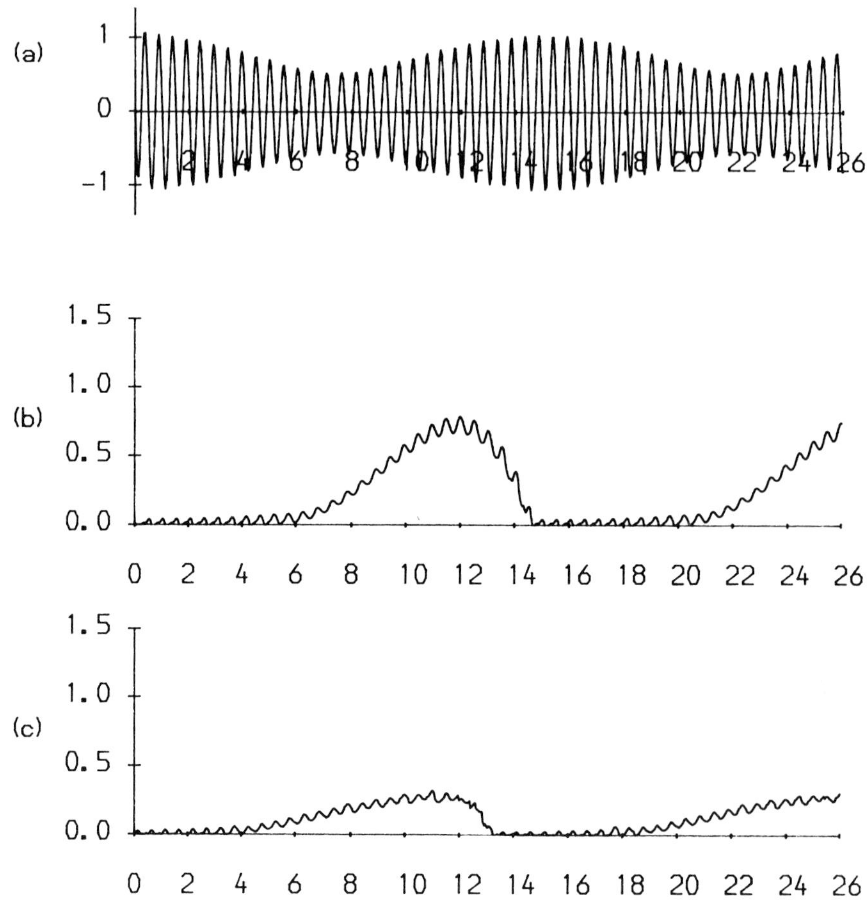

Figure 1: Comparison between turbulence closure and prescriptive models for single component turbulent closure model. (a) Depth mean tidal current (m s^{-1}) versus time (days). (b) Turbulence closure model: bottom-surface density difference (kg m^{-3}) versus time. (c) Prescriptive model: bottom-surface density difference versus time. (N_z=0.08 m^2 s^{-1}). $\frac{1}{\rho}\frac{\partial \rho}{\partial x} = 2.5 \times 10^{-8}$ m^{-1}, depth = 40m, surface wind speed = 5.2 m s^{-1}.

6.4.2 Two component model with rotation

A two component version of the model has also been developed using essentially the same numerical scheme and tested by comparing a constant eddy viscosity form against the results of the analytical calculation of Heaps(1972) for the case of density gradient driving only.

After the decay of inertial oscillations, the model gives a steady state solution very close to the analytical case if the parameter $\gamma = 0.56$. (fig 2).

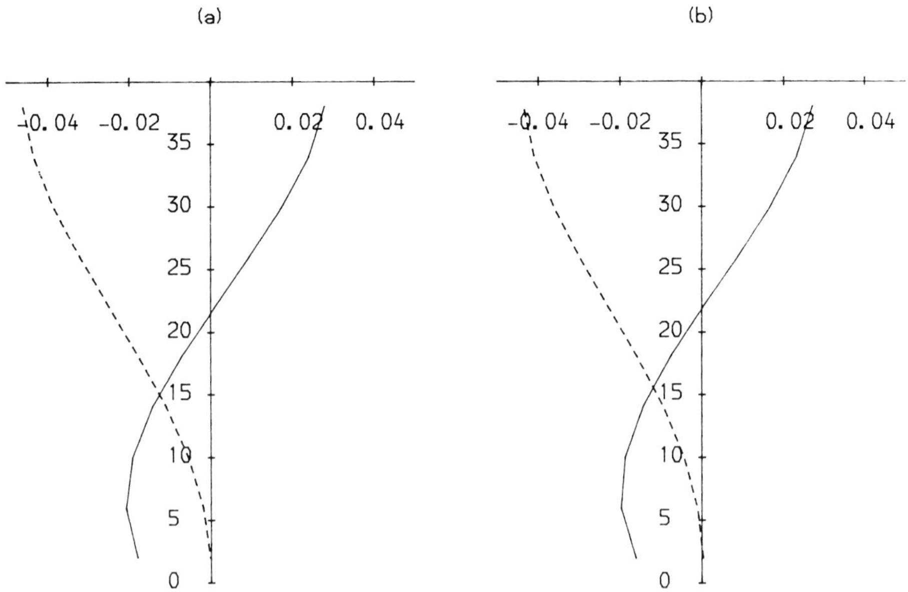

Figure 2: Density current arising from a steady depth-uniform horizontal density gradient in a rotating frame. (a) Heaps (1972) analytical solution. (b) Modelled solution. Horizontal axis is current velocity m s^{-1}), vertical axis is height above seabed (m). Solid line is u component, broken line is v component. $N_z = 0.02$ m^2 s^{-1}, $\frac{1}{\rho}\frac{\partial \rho}{\partial x} = 5.0 \times 10^{-8}$ m^{-1}, and $\gamma = 0.56$.

This two-component model, including the turbulent closure formulation, has been used in an attempt to simulate the springs-neaps cycle in Liverpool Bay for the period March 29 to April 12 1989, for which a time-series of mooring observations is available. Forcing of the model is by two components of surface slope (see table 1 for details of these components and other constants used) and the horizontal,depth-uniform density gradient. The time dependence of the latter for the period concerned (fig 3d) has been determined from observation using a combination of 3 quasi-synoptic CTD sections taken during the deployment period and the mooring time-series from which we have inferred the mean gradient over the tidal excursion.

When run with these realistic inputs, the model reproduces most of the features of the observations. The 8 day period of enduring stratification evident in the model results (fig 3c) closely parallels the observations (fig 3b) with good agreement in the timing and intensity of the maximum stratification. The variation in the magnitude of the semi-diurnal oscillations over the two week period is also reasonably well-represented in the model output.

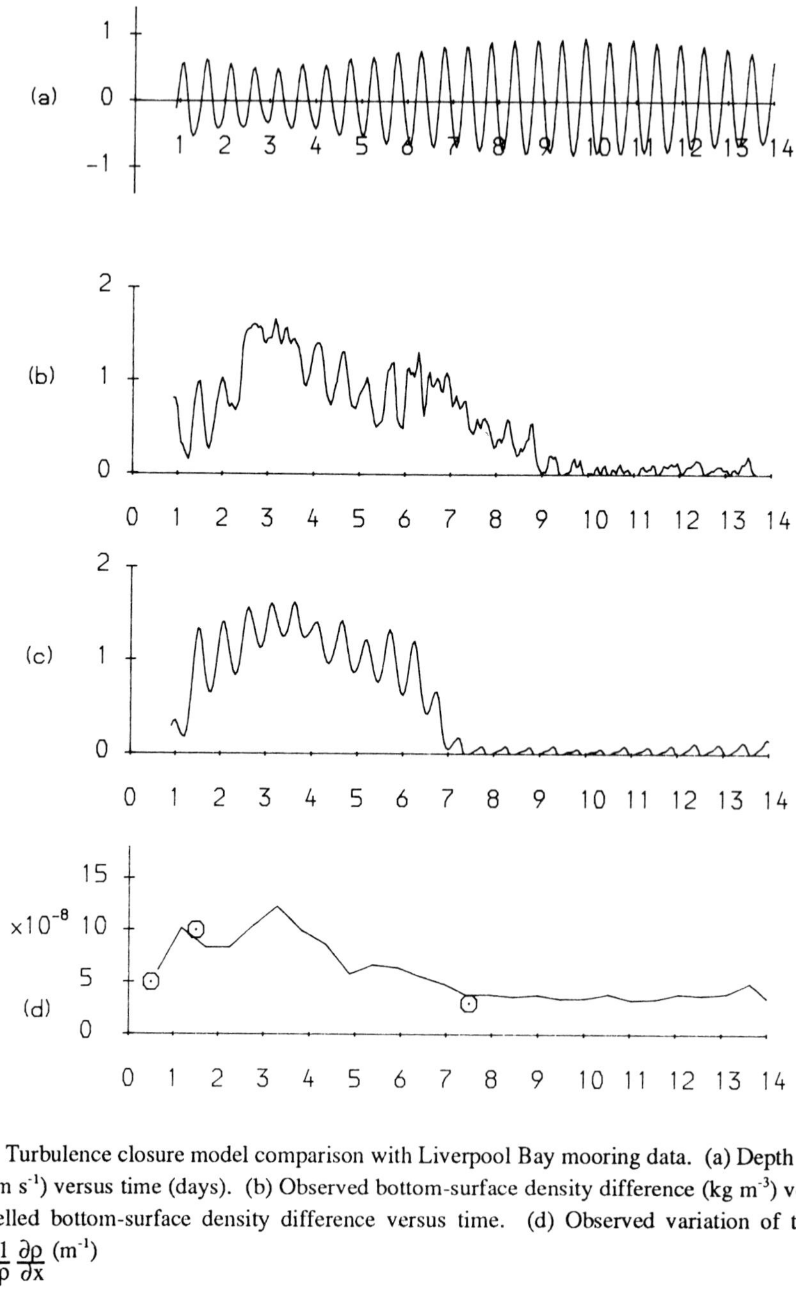

Figure 3: Turbulence closure model comparison with Liverpool Bay mooring data. (a) Depth mean tidal current (m s^{-1}) versus time (days). (b) Observed bottom-surface density difference (kg m^{-3}) versus time. (c) Modelled bottom-surface density difference versus time. (d) Observed variation of the density gradient $\frac{1}{\rho}\frac{\partial \rho}{\partial x}$ (m^{-1})

Table 1: Parameters used by the turbulence closure model in figure 3(c).

Tidal pressure gradient amplitudes m s^{-2}:

Constituent	x-component	phase	y-component	phase
M_2	8.469×10^{-5}	2.415	7.113×10^{-5}	3.749
S_2	3.114×10^{-5}	0.383	2.364×10^{-5}	1.669
N_2	1.497×10^{-5}	0.329	1.211×10^{-5}	1.529

Mean depth = 36.0m.

Depth variation amplitudes:

Constituent	amplitude (m)	phase
M_2	2.52	1.977
S_2	0.83	6.127
N_2	0.45	6.082

Quadratic friction coefficient $k_b = 0.003$
Mean surface slope factor $\gamma = 0.55$
Wind speed from meteorological observations
Number of depth bins= 10
Time step = 0.005 hr.

6.5 Discussion

In this paper, we have proposed single and two component models incorporating a form of the Mellor-Yamada turbulent closure scheme, which, because they incorporate essential feedback mechanisms, offer a more fundamental approach to the problem of predicting estuarine stratification than our previous models. To demonstrate the greater skill of these models than, for example, the prescriptive model, a careful comparison with extensive data sets on estuarine stratification in different situations will be required. A particular priority in this respect is to test the single component model in the narrow estuary situation where rotation can be neglected.

There are, however, encouraging indications from the above results that the T-C model is able to convincingly simulate the time evolution in Liverpool Bay. This is a situation where, as was apparent with the prescriptive model, the effects of rotation must be included in order to limit the cross-isopycnal flow which otherwise induces run-away stratification.

It also appears on the basis of the model trials to date that it is important to allow for variations in the horizontal density gradient which, in previous simplified approaches, we had treated as constant in time as well as depth. Estimates from ship CTD sections and inferences from mooring time series using the tidal excursion indicate changes by a factor of 2-3 over a fortnight. One might expect such changes in gradient to be related to changes in river discharge, but there are indications that the gradients may also be modulated by the springs-neaps cycle in response to the enhanced density current transport at times of low mixing (i.e.neaps).

The parameter γ, which is set to give zero net shorewards transport, accounts in a simple way for the influence of bottom frictional stress on the mean flow. The present condition assumes an instantaneous response of sea level gradient to changes in density gradient and might be replaced,with advantage, by an adjustment with an appropriate time delay. There is also the question of the extent to which γ depends on the forcing parameters; experiments with the current model indicate that its value varies only slowly with changes in the density gradient.

An obvious limitation of the model at present is the lack of provision for surface heat exchange. This omission is not important for the period of the Liverpool Bay observations which is near to the equinox when the net surface exchange is small, but generally surface buoyancy fluxes will need to be included with separate equations for the temperature and salinity replacing the existing density equation.

Similarly, as we move towards models which aim for quantititive predictions of stratification, it will be necessary to utilise a full time history of the wind forcing rather than a single mean vector. With such an input the T-C model should allow a proper description of the way windstress modifies the vertical velocity profile and hence the advection of density.

Acknowledgement

We are grateful to John Hamrick of the Virginia Institute of Marine Science who generously provided guidance and advice in the application of turbulent closure schemes during our stay at VIMS.

6.6 References

van Aken H.M. (1986) The onset of seasonal stratification in shelf seas due to differential advection in the presence of a salinity gradient. Continental Shelf Research, 5 (4), 475-485.

Davies A.G.(1990) A model of the vertical structure of the wave and current bottom boundary layer. Modelling Marine Systems Vol.2, ed. A.M.Davies, CRC Press(in press).

Goodrich D.M., W.C.Boicourt, P.Hamilton, and D.W.Pritchard (1988) Wind induced de-stratification in Chesapeake Bay. Journal of Physical Oceanography, 17 (12), 2232-2240.

Haas L.W. (1977) The effect of the spring-neap tidal cycle on the vertical salinity structure of the James, York, and Rappahannock rivers. Estuarine and Coastal Marine Science, 5, 485-496.

Hansen D.V. and M.Rattray (1965) Gravitational circulation in estuaries. Journal of Marine Research, 23, 104-122.

Heaps N.S. (1972) Estimation of density currents in the Liverpool Bay area of the Irish Sea. Geophysical Journal of the Royal Astronomical Society, 30, 415-432.

Linden P.F. and J.E.Simpson (1988) Modulated mixing and frontogenesis in shallow seas and estuaries. Continental Shelf Research, 8 (10), 1107-1127.

Mellor G.L.and T.Yamada(1974) A hierarchy of turbulence closure models for planetary boundary layers. Journal of Atmospheric Sciences, 31,1791-1806.

Nunes Vaz R.A., G.W.Lennon, and J.R.de Silva Samarasinghe (1989) The negative role of turbulence in estuarine mass transport. Estuarine, Coastal and Shelf Science, 28, 361-377.

Prandle D. (1982) The vertical structure of tidal currents and other oscillatory flows. Continental Shelf Research, 1 (2), 191-207.

Simpson J.H., J.Brown, J.P.Mathews, and G.Allen (1990) Tidal straining, density currents, and stirring in the control of estuarine stratification. Estuaries, 12, 125-132.

Simpson J.H., J.Sharples, and T.P.Rippeth (1991) A prescriptive model of stratification induced by freshwater runoff. Estuarine Coastal and Shelf Science, 33, 23-35.

Part III

Circulation

7

Residual flow in Naples Bay and its effect on constituent concentration, constituent flux and residence time

J van de Kreeke

Abstract

Naples Bay is a shallow, partially mixed estuary in southwest Florida. At its mouth, the estuary branches into two channels, Gordon Pass and the Intra-Coastal Waterway (ICW). Both channels connect to the Gulf of Mexico and therefore allow throughflow or residual flow. The residual rate of flow is determined using recording current meters in a relatively uniform and straight section of the ICW. Measurements extended over a period of one month. The residual rate of flow is defined as the 40 hr low pass filtered instantaneous rate of flow. The monthly mean value is 2.2 m^3s^{-1}. Variations with periods of 5-10 days can be as large as 7 times the mean value. Variations in the residual flow are strongly correlated with wind and atmospheric pressure.

To evaluate the effect of the residual flow on the constituent concentration, constituent fluxes and residence times associated with contaminant sources (marinas) in Naples Bay use is made of a tidally averaged steady state advection-dispersion model. Values of dispersion coefficients are determined from known freshwater inflow and longitudinal salinity distribution and range between 10-100 m^2s^{-1}. Model results show that within the range of observed values, the residual rate of flow has little effect on constituent concentrations and residence times but strongly affects the distribution of the constituent fluxes over Gordon Pass and ICW.

7..1 Introduction

Naples Bay, including Dollar Bay, is a small estuary located on the lower southwest coast of Florida; Fig. 1. It connects to the Gulf of Mexico via Gordon Pass and connects to Rookery Bay via the Intra-Coastal Waterway (ICW). Whereas Naples Bay is an urban estuary surrounded by development and dredged residential canals, Rookery Bay is still in a pristine state and has been designated as an estuarine sanctuary. Naturally, there is concern for waterborne

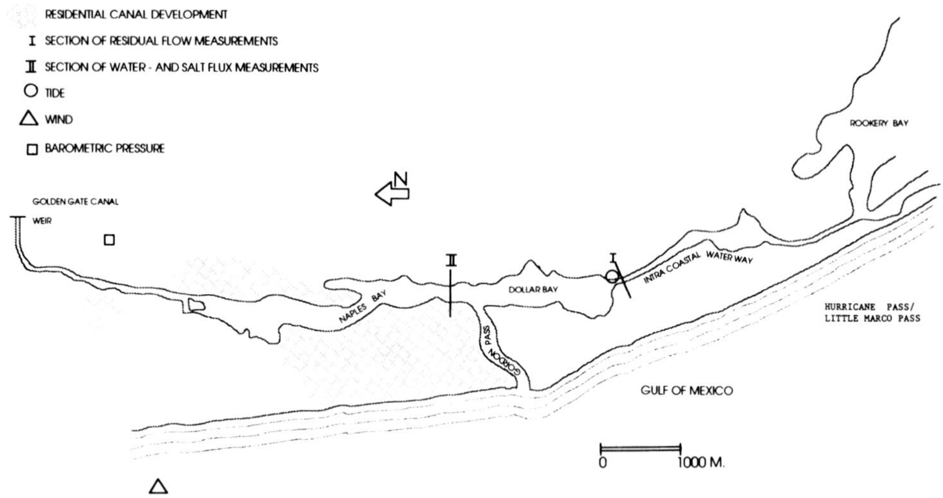

Figure 1: Naples Bay and location of measurement stations

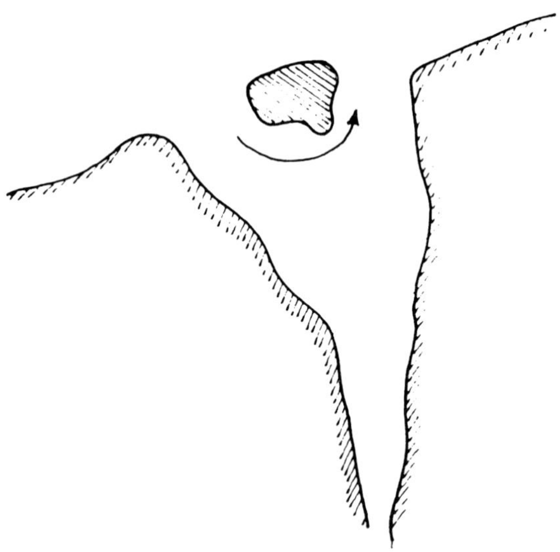

Figure 2: Circulation in Estuary with Multiple Mouths; Plan View

pollutants associated with development to be carried to Rookery Bay via the ICW. In particular, this study is part of the evaluation of a proposed development, including a 600 boat marina at the junction of Naples Bay, Gordon Pass and Dollar Bay. Specific issues to be addressed are the concentrations in Naples Bay and the constituent fluxes towards Rookery Bay resulting from a steady contaminant source at the site of the proposed marina. Also, from a point of view of contaminant concentration, are there better marina locations in the general Naples Bay area?

It seems obvious that potentially the residual flow in the Intra-Coastal Waterway plays an important role in determining the constituent concentrations in Naples Bay and constituent fluxes towards Rookery Bay. Therefore, the focus of this study is on the magnitude and forcing of the residual flow in this channel, its importance as a transport mechanism relative to dispersion and its effect on concentrations, constituent fluxes and residence times. The problem at hand is not unlike that encountered in estuaries that have multiple mouths; see Fig. 2. Examples of the type of estuary are the Hooghly estuary in India and the Ems estuary in The Netherlands. The circulation around the island is a potentially important mechanism for the exchange of water between estuary and ocean.

7.2 Physical characteristics

Waters in the study area are very shallow with typical depths ranging between 2 m and 5 m. As a result of dredging, Naples Bay has some deeper areas up to 8 m.

The primary water movement in the bay is forced by the tide in the Gulf of Mexico. A typical record of water levels observed at a tide station at the southern boundary of Dollar Bay is presented in Fig. 3. For the location of the tide station, see Fig. 1. The tides are of the mixed type. The average tidal range and the average spring tidal range off Gordon Pass respectively are 0.64 m and 0.85 m. (NOAA tide tables). When going towards the head of Naples Bay, tidal ranges increase slightly and reach a maximum in the middle of Naples Bay, and after that slightly decrease (Simpson 1979). This pattern is a result of a sudden change in geometry halfway between Gordon Pass and the head.

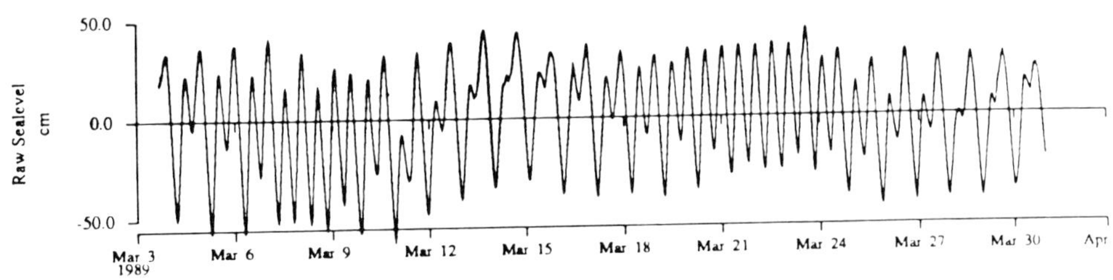

Figure 3: Observed Waterlevels at Tide Station in Dollar Bay

During flood, water enters Gordon Pass and flows into Naples Bay and Dollar Bay. During ebb this pattern reverses. The tidal maximum of the current speed in Gordon Pass varies between 0.6 ms^{-1} and 1 ms^{-1} and in Naples Bay, Dollar Bay, and the Intra-Coastal Waterway varies between 0.2 ms^{-1} and 0.5 ms^{-1}, the exact value depending on the phase of the tide. For average tide conditions, a particle entering Gordon Pass at the beginning of the flood could travel up to the middle of Naples Bay or when entering Dollar Bay could reach the southern boundary of this bay.

The main source of freshwater inflow in Naples Bay is the Golden Gate Canal. Freshwater enters the estuary via a weir. For the period November 1, 1976 - November 1, 1977, the monthly mean values of the freshwater inflow are presented in Fig. 4. During the wet season the rate of flow reaches a maximum of 23 m^3/s. Typically, during the dry season (November-June) the rate of flow is 1 m^3/s. The corresponding steady state longitudinal salinity distribution is presented in Fig. 5. The salinity distribution is based on observations in stations along the axis of the estuary. The salinity distribution shows the typical S-shape. During the dry season the estuary can be classified as mixed to partially mixed; on a tidally averaged basis, differences between bottom and surface salinities do not exceed 2-3 ppt. (Simpson, 1979).

Processes responsible for the salt intrusion and, for that matter, the transport of any dissolved constituent, are dispersion and advection. The dispersion is characterized by a dispersion coefficient. Its value will be estimated using the observed salinity distribution and known freshwater inflow. Advection is associated with the residual flow. For Naples Bay proper the residual flow equals the freshwater discharge. For Gordon Pass/Dollar Bay/ICW, the residual flow is not a priori known. Its value will be estimated from simultaneous current and waterlevel measurements carried out over a period of one month in cross-section I, see Fig. 1.

7.3 Residual flows

7.3.1 Definitions

Assuming a relative uniform distribution over the cross-section

$$Q(t) = A(t)\ V(t) \tag{7.1}$$

in which

Q = rate of flow
A = cross-sectional area (m^2)
V = axial velocity observed by a current meter in the cross-section (ms^{-1})
t = time

Writing

$$A(t) = A_o + W\eta(t) \tag{7.2}$$

in which

A_o = cross-sectional area below mean sea level
W = width of the cross-section
η = waterlevel variation relative to Still Water Level

and substituting in Eq. (7.1), it follows

$$\frac{<Q>}{A_o} = <V> + \frac{<\eta V>}{h_{av}} \tag{7.3}$$

In Eq. (7.3), $h_{av} = A_o/W$ and $< >$ designates low pass filtering with a period of 40 hrs.

In the context of this paper the low pass filtered data is referred to as residuals. The left hand side of Eq. (7.3) represents the transport velocity whereas the first and second term on the righthand side respectively represent the Eulerian mean velocity as observed by the current meter and the vertically averaged Stokes velocity.

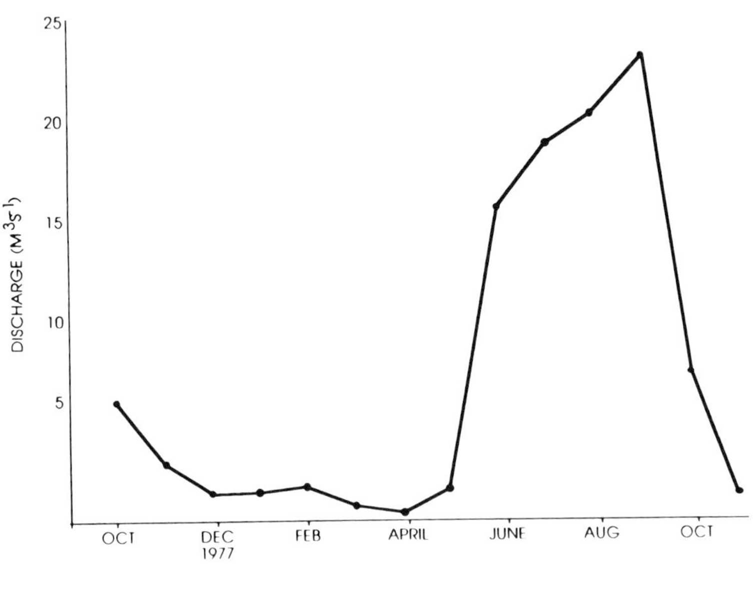

Figure 4: Mean Monthly Discharge of Golden Gate Canal.

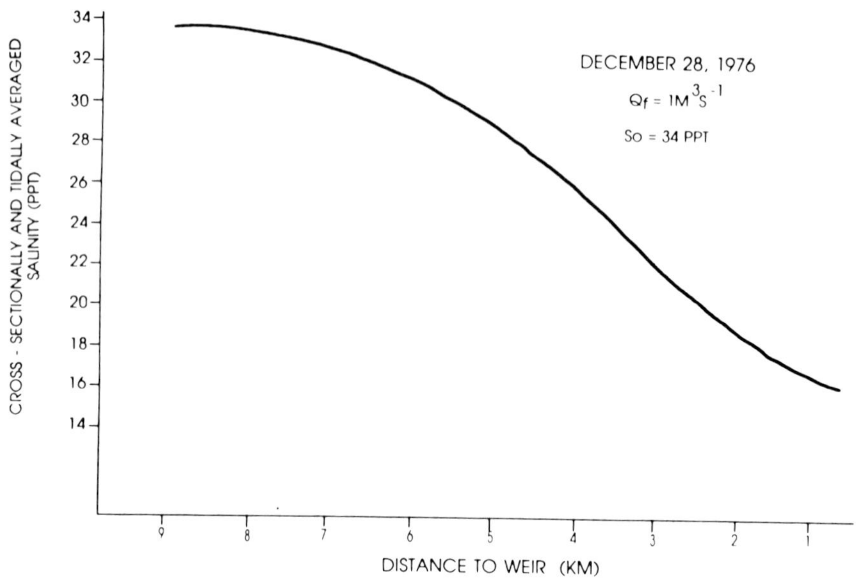

Figure 5: Observed Salinity Distribution in Naples Bay.

7.3.2 Measurements

The determination of the residual flow from a single point measurement is based on the assumption that the observed velocities are representative for the entire cross-section. In practice, this implies that the cross-sectional distribution of the tidal velocities and the Eulerian mean current has to be relatively uniform. To satisfy as much as possible this condition, the cross-section in which the measurements were carried out was selected in a uniform stretch of the Intra-Coastal Waterway; see Fig. 1. A second recording current meter was placed in the same cross-section to gain information on the cross-sectional variability of the current. In addition, measurements with hand held current meters were carried out on March 10, 16, and 24, 1989. Measurements extended over three hours and included ebb- and flood currents. Velocity profiles were measured in four stations in the cross-section located at 20 m, 35 m, 45 m and 60 m; see Fig. 6. Current speed was measured at 30 cm intervals, starting at 30 cm below the surface to 30 cm above the bottom.

To evaluate the Stokes velocity, waterlevels in the cross-section were measured using a pressure sensor. For cross-section and location of the two recording current meters and pressure sensor, see Fig. 6. At the measurement site, the cross-sectional area $A_o = 120$ m^2 the average depth $h_{av} = 1.5$ m and the width W = 80 m. Measurements extended over the period

March 3 - March 31, 1989.

Current meters are NISKIN winged type current meters 6011 MARK II. Meters were operated in the "vector-averaging mode". A burst sample consisting of 16 readings spaced 4 seconds apart was taken every 5 minutes. The 16 readings were vector averaged and recorded as a north and south velocity component. Current meter #348 provided good data for the entire measurement period. Current meter #350 provided good data for the period March 3 - March 7, 1989 and March 16 - March 21, 1989. To determine the axial current speed the NS-EW coordinate system of the #348 current meter was rotated in increments of 1° until the standard deviation of one of the velocity components reached a minimum for a rotation of 336°. The line 336°-156° was taken as the direction of the channel axis. This is close to the 332°-152° direction estimated from the nautical chart.

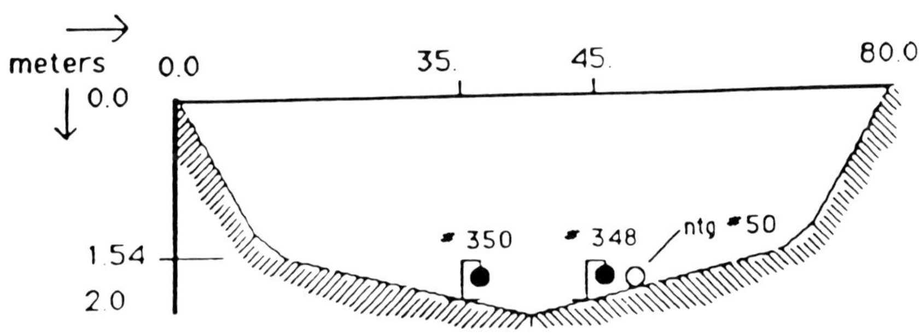

Figure 6: Measurement Section I; Location of Current Meters and Tide Gauge

Waterlevels were derived from the pressure observations of a General Oceanics Model 6041/51 pressure gauge. Pressure was recorded every 15 minutes. In converting to waterlevels, pressures were corrected for atmospheric pressure using hourly observations from the weather station at Naples airport. For location of station, see Fig. 1.

To determine the possible correlation between residual flow and wind, hourly values of wind speed and direction as observed by NOAA, National Weather Service at the Naples Pier were used. Observations refer to an elevation of Mean Sea Level + 10 m. For location of the weather station, see Fig. 1.

7.3.3 Results

Because current meter #348 provided uninterrupted data for the entire measurement period, V(t) in Eqs. (7.1) and (7.3) is taken equal to the axial current speed of this meter. The time

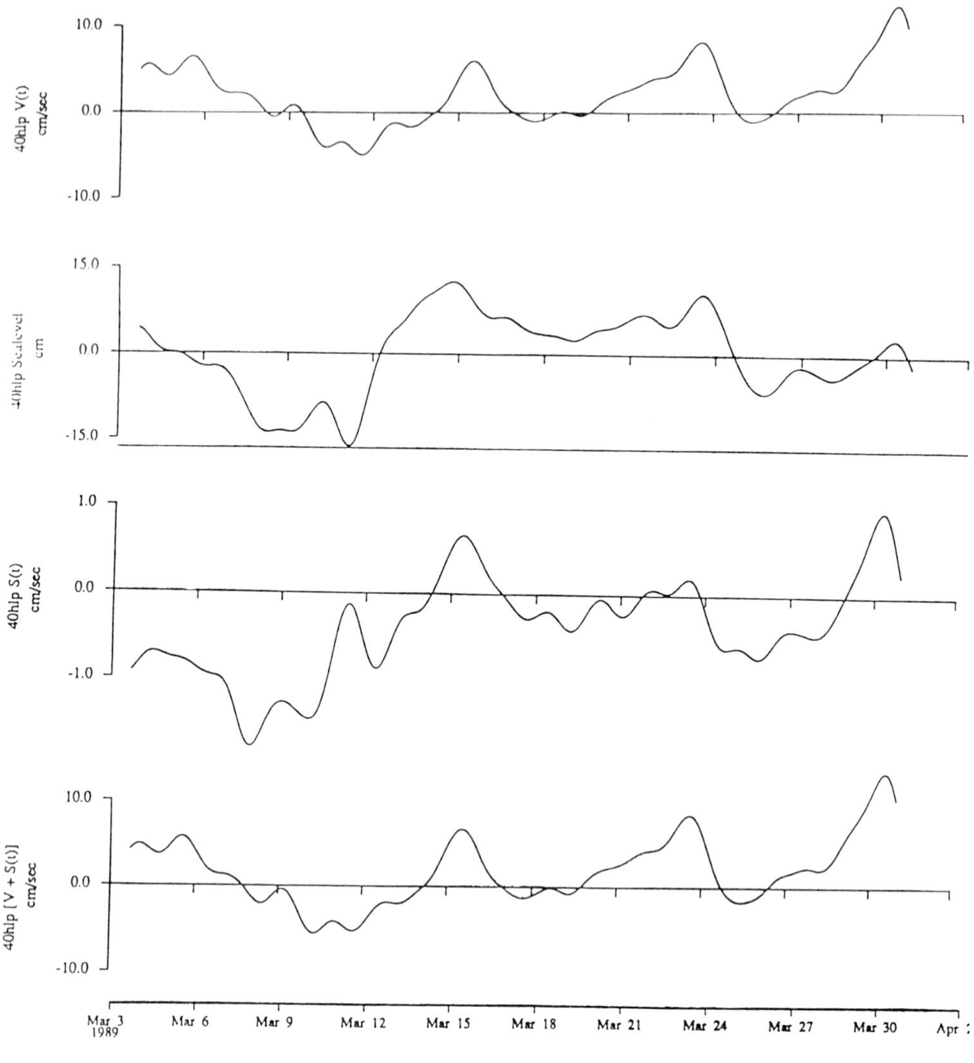

Figure 7. From Top to Bottom, Eulerian Mean Current, Water Level, Stokes Drift and Transport Velocity. Positive direction is towards Naples Bay.

series for the Eulerian mean current obtained by low pass filtering V(t) is presented in Fig. 7. In the same figure, the low pass filtered time series for the waterlevel in the cross-section, the Stokes velocity and the transport velocity are presented. All four time series exhibit variations with a period of 5-10 days. Time series are too short to delineate possible longer period variations in residuals. Because tidal variations in waterlevel and current speed are close to 90° out of phase, the contribution of subtidal variations to the Stokes velocity is substantial. Eulerian current speed is roughly 5 times as large as the Stokes velocity and is the major contributor to the transport velocity. The mean over the time series of the transport velocity is 1.8 cm s^{-1} towards Naples Bay. The maximum value of the transport velocity towards Naples Bay is 13 cm s^{-1} and the maximum value towards Rookery Bay is 5 cm s^{-1}. With a cross-sectional area A_o = 120 m^2, the corresponding values of the residual flow are, 2.2 m^3s^{-1} for the mean, 15.6 m^3 s^{-1} for the maximum flow towards Naples Bay and 6 m^3s^{-1} for the maximum flow towards Rookery Bay.

The 5-10 day variations in the residual flow of the Intra-Coastal Waterway suggest meteorological forcing. Most likely wind and/or atmospheric pressure gradients cause differences in mean sealevel between both end of the waterway. Direct wind forcing is an unlikely contributor as the waterway is protected by mangroves on both sides. The two available meteorological parameters, local wind velocity and local atmospheric pressure show a strong (negative) correlation. Taking the axial wind speed to be representative for meteorological forcing, it follows from Fig. 8 that a strong correlation exists between this parameter and the transport velocity. A linear regression analysis shows a regression coefficient of 0.85 suggesting that 85% of the variance in the transport velocity and therefore in the residence flow can be explained by meteorological forcing.

7.4 Effect of residual flow on residence time, constituent distribution and constituent flux

The time a particle released at a location (x,y) resides in an estuary is referred to as the transit time. The transit time depends on the phase of the tide at which the particle is released. When dealing with a continuous injection of constituent, the average over the tidal period of the transit times of the particles is referred to as the residence time, τ_r. In general, τ_r is a function of x and y. Assuming steady-state conditions, the relation between the residence time τ_r and the tidally and basin averaged concentration $\bar{c}$ is

$$\tau_r = \frac{<V>\bar{c}}{M} \tag{7.4}$$

where <V> is the tidally averaged basin volume and M (kg/sec) is the strength of the source. (van de Kreeke, 1983).

In case of freshwater inflow, it follows from Equation (7.4)

$$\tau_r = \frac{\langle V \rangle}{Q_f T} \frac{s_o - \bar{s}}{s_o} \tag{7.5}$$

in which Q_f is the freshwater discharge, T is the tidal period, s_o is the salt concentration of the receiving waters and <V> $(s_o-\bar{s})/s_o$ represents the volume of freshwater in the estuary. From the longitudinal salinity distribution observed on December 28, 1976, see Fig. 5, and information on the bathymetry presented in Simpson (1979), a value of the tidally and spatially averaged value of $\bar{s}$ = 30.6 ppt was calculated. With <V> = 8.73 10^6 m^3 (= tidally averaged volume of Naples Bay and Gordon Pass), T = 44,712 sec and s_o = 34 ppt it followed that τ_r = 19.3 tidal periods or approximately 10 days.

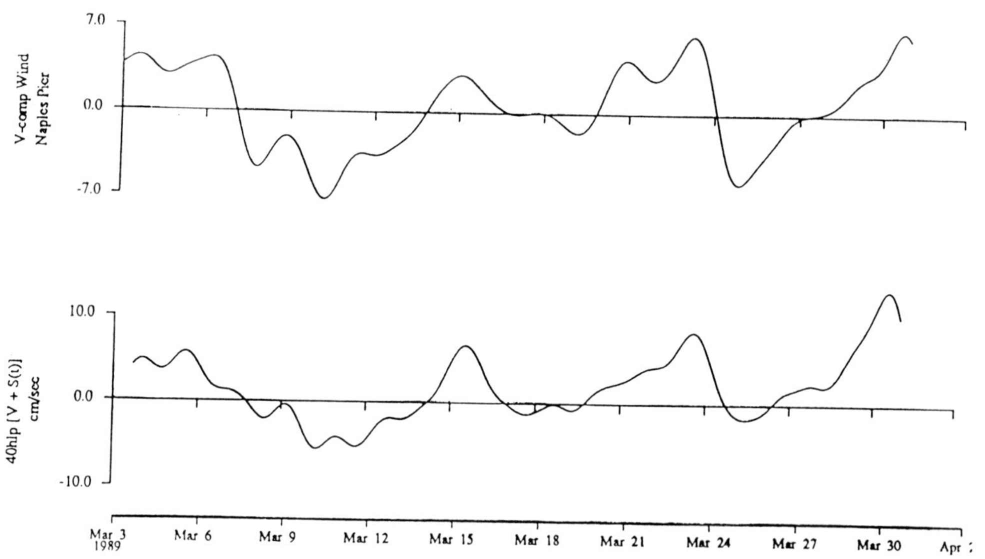

Figure 8: From Top to Bottom. Axial Wind Speed and Transport Velocity. Positive direction is towards Naples Bay.

To evaluate the effect of the residual flow on the residence time, constituent concentration and constituent flux, a link-node model is used; see Fig. 9. For each link the transport equation is

$$\langle Q \rangle \, c - A \, D \, \frac{dc}{dx} = K \tag{7.6}$$

in which

<Q> = residual rate of flow
c = cross-sectionally averaged and low-pass filtered constituent concentration
A = tidal mean cross-sectional area
D = longitudinal dispersion coefficient
K = constituent flux
x = longitudinal coordinate; for positive direction see Fig. 9

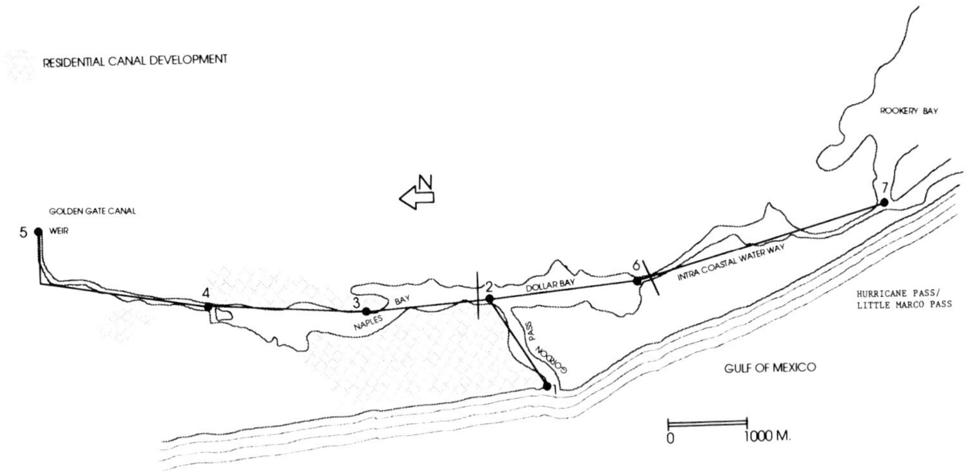

Figure 9: Link-Node Schematization. Arrow indicates positive x-direction.

For each link, <Q>, A, D and K are assumed constant. Solving Eq. 7.6 it follows

$$c_e = c_b \, e^{\frac{\langle Q\rangle L}{AD}} + \frac{K}{\langle Q\rangle}\left(1 - e^{\frac{\langle Q\rangle L}{AD}}\right) \tag{7.7}$$

in which

c_e = constituent concentration at end of link
c = constituent concentration at beginning of link
L = length of link

Making use of the conservation of water and constituent at each node, the values of the concentrations at the nodes follow from Eq. (7.7). Values of concentrations in the Gulf of Mexico, node 1, and in Rookery Bay, node 7, are set equal to zero.

Table 1: Link Characteristics

Link	$A(m^2)$	$L(m)$	$D(m^2s^{-1})$
5-4	150	4000	10
4-3	600	2500	30
3-2	500	2000	100
2-1	450	1500	100
2-6	300	2800	100
6-7	120	4200	50

Table 2: Volumes Assigned to Nodes

Node	$\langle V \rangle$ (m^3)
5	$0.40\ 10^6$
4	$1.30\ 10^6$
3	$2.35\ 10^6$
2	$2.75\ 10^6$
1	$1.40\ 10^6$
6	$0.85\ 10^6$
7	$0.12\ 10^6$

Taking c in Eq. 7.6 to represent the salinity and setting $K = 0$, estimates of the longitudinal dispersion coefficients for Naples Bay were made using the observed steady state salinity distribution of December 28, 1976. Based on these estimates, values of the dispersion coefficients selected for each link are presented in Table 1 together with values of A and L. Values of the dispersion coefficients for Gordon Pass and Dollar Bay were taken equal to the value for lower Naples Bay. As can be inferred from a study on the water and salt fluxes in cross-section II, see Fig. 1, the relatively large dispersion coefficients in the links 3-2, 2-1, and 2-6 can be attributed to trapping of water in the residual canals, phase differences in velocities near the junction and the proximity (less than half the tidal excursion) of the Gulf of Mexico (van de Kreeke, 1982). Similar high values of longitudinal dispersion coefficients were reported by Wolanski and King, (1990) for Hinchinbrook Channel, Australia. This channel has roughly the dimensions of Naples Bay and trapping is associated with storage of water in the mangrove swamp rather than in residential canals.

As a check on the values of the dispersion coefficients the salinity distribution was recomputed using the link-node model. Starting from the computed salinity distribution, the mass of freshwater in Naples Bay was calculated by assigning the volume to each node listed in Table 2. The residence time was then calculated using Eq. 7.4. This resulted in the exact same value as obtained from the observed salinity distribution and known freshwater discharge, i.e., $\tau_r = 10$ days.

The model was then used to calculate the constituent concentrations, constituent fluxes and residence times for continuous releases of dye at various nodes. To test the sensitivity of these parameters to the value of the residual flow, calculations were carried out with $<Q>$ ranging between -2.2 m^3s^{-1} and 6 m^3s^{-1}. $<Q>$ = -2.2 m^3s^{-1} is the mean value over the measurement period and $<Q>$ = 6 m^3s^{-1} is the maximum observed residual flow towards Rookery Bay. The results of the calculations for a continuous dye release at node 5 and for residual flows of -2.2m^3s^{-1} and 6 m^3s^{-1} are presented in Tables 3 and 4. Similarly, for a continuous dye release at node 2 the results are presented in Tables 5 and 6. In all three experiments, the rate of dye release was 1 kg s^{-1} and the discharge of the Golden Gate Canal was 1 m^3s^{-1}.

Concentrating on a constituent discharge at node 5 it follows that a change in residual flow from -2.2 m^3/s^{-1} to 6 m^3/s^{-1} has little effect on concentration distributions and residence times; see Table 3. However, the distribution of the constituent fluxes over the Intra-Coastal Waterway and Gordon Pass differs considerably; see Table 4. For a residual rate of flow of -2.2 m^3s^{-1} the constituent flux in the Intra-Coastal Waterway is 0.02 kg s^{-1} and for a residual rate of flow of 6m^3s^{-1} the constituent flux is 0.18 kg s^{-1}. In both cases the fluxes are towards Rookery Bay. Similar conclusions hold for a constituent discharge at node 2. Concentrations and residence times are not very sensitive to the residual flow in the Intra-Coastal Waterway; see Table 5. However, the constituent flux towards Rookery Bay strongly depends on the residual flow; see Table 6. Comparison of the concentration distributions in Tables 3 and 4 shows that a constituent discharge at node 2 leads to much smaller concentrations and residence times than a constituent discharge at node 5.

Table 3: Concentrations and Residence Times; Source of 1 kg s^{-1} at node 5.

Res. Flow in ICW is -2.2 m^3s^{-1}			Res. Flow in ICW is 6 m^3s^{-1}	
node	c	Residence time	c	Residence time
5	0.944		0.943	
4	0.189		0.188	
3	0.068		0.067	
2	0.031	10.24 days	0.030	10.26 days
1	0.000		0.000	
6	0.024		0.030	
7	0.000		0.000	

Concentrations are in kg m^{-3}

Table 4: Constituent Fluxes; Source of 1 kg s^{-1} at node 5.

Res. Flow in ICW is -2.2 m^3s^{-1}					Res. Flow in ICW is 6 m^3s^{-1}			
Link	<Q>	<Q>c	$-AD\frac{dc}{dx}$	K	<Q>	<Q>c	$-AD\frac{dc}{dx}$	K
5-4	1.0	0.74	0.26	1.00	1	0.79	0.21	1.00
4-3	1.0	0.18	0.82	1.00	1	0.13	0.87	1.00
3-2	1.0	0.05	0.95	1.00	1	0.05	0.95	1.00
2-1	3.2	0.05	0.93	0.98	-5	-0.07	0.89	0.82
2-6	-2.2	-0.06	0.08	0.02	6	0.18	0.00	0.18
6-7	-2.2	-0.01	0.03	0.02	6	0.16	0.02	0.18

Fluxes in kg s^{-1}. Advective and dispersive fluxes pertain to middle of reach

Table 5: Concentrations and Residence Times; Source of 1 kg s^{-1} at node 2.

Res. Flow in ICW is -2.2 m^3s^{-1}			Res. Flow in ICW is 6 m^3s^{-1}	
node	c	Residence time	c	Residence time
5	0.002		0.002	
4	0.025		0.025	
3	0.029		0.029	
2	0.030	2.36 days	0.030	2.40 days
1	0.000		0.000	
6	0.024		0.030	
7	0.000		0.000	

concentrations are in kg m^{-3}

Table 6: Constituent Fluxes; Source of 1 kg s^{-1} at node 2.

Res. Flow in ICW is -2.2 m^3s^{-1}					Res. Flow in ICW is 6 m^3s^{-1}			
Link	<Q>	<Q>c	$-AD\frac{dc}{dx}$	K	<Q>	<Q>c	$-AD\frac{dc}{dx}$	K
5-4	1.0	0.010	-0.01	0.00	1	0.00	0.00	0.00
4-3	1.0	0.025	-0.025	0.00	1	0.03	-0.03	0.00
3-2	1.0	0.029	-0.03	0.00	1	0.03	-0.03	0.00
2-1	3.2	0.050	0.93	0.98	-5	-0.07	0.89	0.82
2-6	-2.2	-0.050	0.07	0.02	6	0.18	0.00	0.18
6-7	-2.2	-0.013	0.03	0.02	6	0.15	0.03	0.18

Fluxes in kg s^{-1}. Advective and dispersive fluxes pertain to middle of reach

7.5 Discussion

The success of measuring residual flow with a single current meter depends on the uniformity of the cross-sectional distributions of the tidal velocities and Eulerian mean current. Experimental data on the distribution of these velocities in a homogeneous tidal channel could not be found. Theoretical studies on the vertical distribution of tidal velocities, Eulerian mean current, Stokes velocity and transport velocity for a sealevel canal are presented by Huang et al. (1986) and Wong (1989). Huang et al. (1986) using a 2DV numerical model calculated the tidal- and mean currents for the Chesapeake and Delaware canal. Wong (1989) presents an analytical solution for the mean currents. The results of both studies suggest that for a typical South Florida coastal channel with a length on the order of 10 km and a depth of 2-3 m, the mean currents do not reverse direction in the vertical.

In the present study, the velocity measurements carried out with the handheld current meters show a practically uniform velocity distribution over the measurement interval, i.e., between the free surface -30 cm and the bottom +30 cm. Velocities of the stations at +35 m and +45m differ less than 5%. Velocities of the stations at +20 m and +60 m are 10-15% lower than those of the two stations in the middle part of the cross-section. Values of the mean of the axial current speed for the period March 3-7, 1989 are 6.5 cm s^{-1} for current meter #350 and 5.8 cm s^{-1} for current meter #348. The means of the axial current speed for the period March 16-21, 1989 are 3.1 cm s^{-1} for current meter #350 and 2.3 cm s^{-1} for current meter #348. Although definitely not proof, the foregoing theoretical and experimental results provide some support for the assumption of a relative uniform velocity distribution and at least suggest that the order of magnitude of the observed residual flow and the period of its fluctuations are correct.

The insensitivity of the constituent concentrations and residence times to the residual flow between Gordon Pass and Rookery Bay in part stems from the fact that in a large portion of the estuary, dispersion is the dominant transport mechanism; see columns 3, 4, 7 and 8 in Table 4 and 6.

The selection of a steady state rather than a time dependent transport equation, Eq. (7.6), is in part based on ease of computation and in part is based on obtaining a worst case with regard to the constituent flux towards Rookery Bay. It is realized that the fluctuations in the residual flow are often too short to reach steady state. Therefore, the calculated constituent flux of 0.18 kg s^{-1} towards Rookery Bay, see Tables 4 and 6 should be considered an upper limit.

7.6 Conclusions

Based on a one month time series of current speed and direction it is estimated that the residual rate of flow in the Intra-Coastal Waterway between Naples Bay and Rookery Bay has a value of 2.2 m^3s^{-1} towards Naples Bay. Maximum values during the measurement period are 15.6m^3s^{-1} towards Naples Bay and 6 m^3s^{-1} towards Rookery Bay. The residual rate of flow exhibits variations with periods between 5 and 10 days. These variations are associated with meteorological forcing. The time series of the current velocity is too short to delineate the forcing of possible variations having periods longer than 10 days.

The contribution of the Eulerian mean current to the residual rate of flow is about five times the contribution of the Stokes velocity.

The influence of the residual rate of flow in the Intra-Coastal Waterway on concentrations and residence times is small. This is a result of the relatively large dispersive transport. The residual flow strongly affects the distribution of the constituent flux between Gordon Pass and Rookery Bay.

Residual rates of flow are derived from one point measurements of current speed and direction. The paper points out the shortcomings in this method and the need for a more sophisticated measurement technique.

From a water quality point of view, i.e., low constituent concentrations, it is better to locate the proposed marina near the junction of Gordon Pass, Dollar Bay, and Naples Bay than further landward in Naples Bay.

7.7 References

Huang, P.S., D.P. Wang, and T.O. Najarian, 1986. Analysis of Residual Currents using a Two-Dimensional Model In: Physics of Shallow Estuaries and Bays, J. van de Kreeke (ed). Lecture notes on Coastal and Estuarine Studies No. 16, 71-80.

Simpson, B. L., Ed. 1979. The Naples Bay Study. Collier County Conservancy. P.O. Box 2136, Naples, Florida.

van de Kreeke, J. 1982. Salt and Water Flux in Naples Bay, Florida. University Miami Rosenstiel School of Marine and Atmospheric Science. Technical Report TR82-2 pp. 28.

van de Kreeke, J. 1983. Residence Time: Application to Small Boat Basins, Journal of Waterway, Port, Coastal and Ocean Engineering. Vol. 109, No. 4, pp 416-428.

Wolanski, E. and B. King, 1990. Trapping in Hinchinbrook Channel. A Giant Mangrove Swamp. To be published in proceedings of 5th International Biennial Conference Physics of Estuaries and Coastal Seas.

Wong, K.C., 1989. Tidally Generated Residual Currents in a Sea Level Canal or Tidal Strait with Constant Breadth and Depth. Journal of Geophysical Research, Vol. 94, No. 6, 8179-8192.

8

The stratified hydrodynamics of the Palmiet - a prototypical bar-built estuary

J L Largier, J H Slinger and S Taljaard

Abstract

Tidal intrusion and mixing in the highly stratified Palmiet Estuary are discussed as an example of a typical small bar-built estuary. The flux of salt through the estuary is dominated by advection owing to the strong gravitational circulation. This vertical circulation pattern is characterized by four processes. The dense saline water enters the estuary as a hydraulically controlled flow through a constricted entrance channel. Beyond an internal hydraulic jump, landward of the flood tidal delta, the saline intrusion propagates as a non-mixing bottom density current and eventually replenishes deep basal reservoirs of saline water in the upper estuary. On the ebb tide this lower saline layer is prevented from flowing seaward by the presence of the sill at the mouth (upstream blocking) - upward entrainment of halocline water occurs via subcritical shear and the sub-halocline water is not removed. The entrained salt is exported to sea by way of the outflowing buoyant surface layer.

This pattern of circulation is expected to be present in countless small bar-built estuaries around the world. While these systems are generally characterized by micro-tidal and high wave-energy coasts, they are observed in a variety of other situations. In the Palmiet Estuary this pattern occurs during both the drier summer and the wetter winter, exhibiting a certain robustness to seasonal changes. These estuaries are generally classified as highly stratified, although the actual stratification varies as a function of season, tide and position in the estuary. Of consequence is the long residence times expected, and observed, for the deep saline water found in the scour holes of the upper estuary. With decreasing river flow and more constricted connections with the sea, tidal mixing decreases and this deep basal water can be resident long enough to become anoxic.

8.1 Introduction

The tidal flow through the mouth of many estuaries is restricted by the presence of a sill or a contraction. Most common is the combined sill and contraction which is provided by a wave-built sand spit (the "bar-built" estuary). The longshore transport of sand seeks to close the mouth and attains a dynamic equilibrium with the tidal scour, which increases as the mouth narrows. A linear relationship has been found between the cross-sectional area of the mouth and the volume of the tidal prism (O'Brien, 1969). The restriction of the tidal flow results in shorter flood tides which are characterized by larger velocities (in the presence of low river flow) and a greater transport of sediment into the estuary. This is enhanced by the suspension of sediments in the surf-zone. There is a net accumulation of sand in the estuary during these periods of low river flow. A flood tidal delta (deposition region) forms landward of the narrow mouth channel and flood and ebb channels develop owing to the strong residual transport of sand. Scouring of the estuary and mouth occur during periods of strong river flow when ebb tidal currents exceed those on the flood tide. This hydrodynamically balanced control of sedimentation is attained in a variety of tidal channels which are subject to scouring and shoaling, both natural and man-made. Given the correct supply of mass and buoyancy, strongly stratified flow and structure occur in the basin landward of this restricted flow through the mouth, as is observed in the Palmiet Estuary.

The Palmiet Estuary is a small, drowned river valley on the south coast of Africa. Its mouth is a narrow channel through a large wave-built sand spit (Figure 1). The height of the sill increases and the width of the contraction decreases through the dry summer season, both being scoured open during the winter rains. The strong river flow during winter storms also scours deeper regions in the bends of the upper estuary (Figure 1). Although the mouth is usually open, on average it closes for about a month at the end of every other summer. As a result of its small size and owing to its undisturbed, natural character, this estuary has been the site of a number of investigations in ecology (Branch and Day, 1984; Taljaard et al., 1986; CSIR, 1987) and hydrodynamics (Largier, 1986; Slinger and Largier, 1989; Largier and Taljaard, 1990). The basin is small (Figure 1) with immediate connections to the sea and river. The river enters the estuary as a fast-flowing mountain stream, low in suspended and dissolved matter.

This paper addresses the nature of the flood-tidal intrusion of seawater and its subsequent propagation up the estuary as a gravity-driven density current. On encountering the scour holes, this dense inflow forms a deep water mass which may exhibit very long residence times. The paper also addresses the upstream blocking and sub-critical entrainment of the lower layer during the ebb tidal outflow. Seasonal examples are discussed and compared - illustrating the general nature of the intrusion and erosion mechanisms in the Palmiet Estuary. It is argued that these dynamics should be common to many of the estuaries on high wave-energy coasts (e.g., southern Africa, southern Australia, western America, etc.). Some of the implications of this stratification and circulation are discussed in the final section.

8.2 The Winter example

Following heavy rain, typically due to the winter passage of mid-latitude frontal storms, the river flows strongly (e.g., 80 m^3s^{-1} on 14 August 1986) and the Palmiet Estuary is completely flushed of any salt. As the river flow decreases and/or the tide increases, new seawater may enter the estuary on the flood tide (Figure 2), as it did on 19 August (river flow 30 m^3s^{-1}, tidal range 1.7 m in the estuary). Approaching the subsequent neap tide (tidal range 0.5 m), the river flow increased again to 40 m^3s^{-1} on 26 August and all the seawater was removed. Of interest here is that period between 19 and 26 August during which the estuary was salt-stratified. We discuss three processes: the tidal intrusion front (plunging flow); the gravity-driven bottom density current (salt wedge), and erosion of the halocline by ebb tidal outflow.

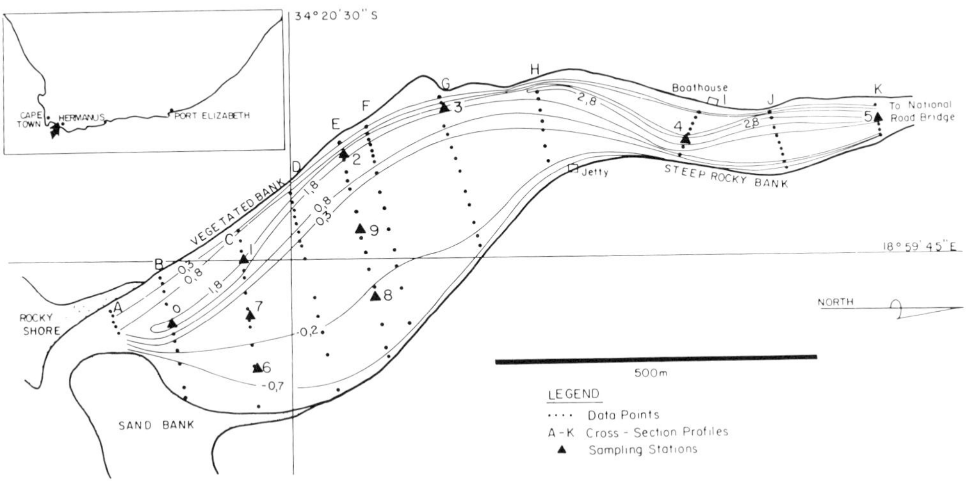

Figure 1: The morphology of the Palmiet Estuary, located on the south coast of Africa. These bathymetric data were collected during winter (August 1986) - isobaths in metres below mean sea level. In summer the shape is essentially the same except for the appearance of a larger flood tidal delta south and west of station 0 and a tidal scour hole around station 0. The presence of fluvial scour holes of 3 to 4-m depth were confirmed by a later survey of the curves in the channel landward of station 3.

The Tidal Intrusion Front: When the net outflow (river outflow minus tidal inflow) drops below a certain value, which is dependent on the density difference of the water types and the size of the constricted mouth (Largier, 1990), the saline water can enter the estuary as a dense underflow. This occurs while the less dense estuarine water is still flowing out. This two-layer exchange flow persists through zero net flow to a net inflow characterized by $Fr_0^2 = 0.3$, where Fr_0^2 is an inflow Froude number

$$Fr_0^2 = \frac{Q_0^2}{g' H_0^3 B_0^2}$$

and Q_0 is the net inflow at the mouth, g' is the reduced gravity, H_0 is the depth and B_0 is the width at the most constricted section (Largier, 1990). The flow is hydraulically controlled by a two-layer composite Froude number $G^2 = F_1^2 + F_2^2$, where F_1^2 and F_2^2 are the densimetric Froude numbers for each layer and $F_i^2 = u_i^2 / g'h_i$, where u is the fluid velocity and h is the layer depth (Armi, 1986). For inflow Froude numbers in excess of 0.3 the ebbing surface layer is blocked ($u_1 = 0$) and a plunge line appears. A surface slick or foam-line marks the surface convergence characteristic of a tidal intrusion front.

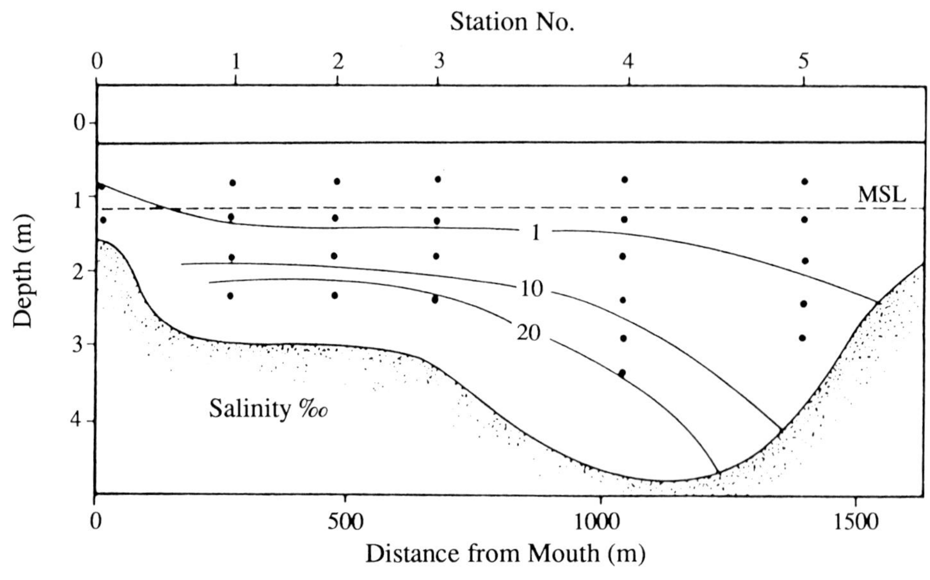

Figure 2: A longitudinal section of the salinity distribution in the water channel during flood tide on 19 August 1986 (see Figure 1 for station locations).

While the first intrusion of seawater was observed on 19 August (Figure 2), we obtained the most detailed measurements on 20 August. In the presence of a 15 m^3s^{-1} river flow, the tide turns at 1235, three hours after low tide in the ocean (0941). Meanwhile, the dense saline water started its subsurface intrusion at about 1200 when the net outflow was 3.4 m^3s^{-1}. Weak surface foam lines (tide lines), which appear to mark the subsurface convergence at the head of the salinity intrusion, were first noticed around station 0 (Figure 1) at 1210. In the narrowest section of the mouth, at about 1230, the upper layer (13.8°C, 1.2°/oo) was flowing out at 0.3 to 0.4 m s^{-1} - varying as a function of surf-zone dynamics (Figure 3). The lower layer (15.2°C, 26°/oo) was flowing in at 0.1 m s^{-1}. The interface occurred at mid-depth (0.5 m), but rose and fell with fluctuations in the net flow as a result of surface gravity and infragravity waves entering the mouth. The two-layer Froude number G^2 is of order unity, indicative of the internal hydraulic control at this section. At station 0 (at 1245) the 0.5 m basal flow is supercritical (F_2^2 = 2.3) with g' = 0.14 m^{2}s^{-1} and u_2 = 0.4 m s^{-1}. Somewhere between stations 0 and 1, where the basal flow is subcritical (Figure 3), the intrusion experiences a hydraulic jump with the attendant mixing. Turbulent mixing is also observed in the mouth and at station 0 as a result of supercritical shear.

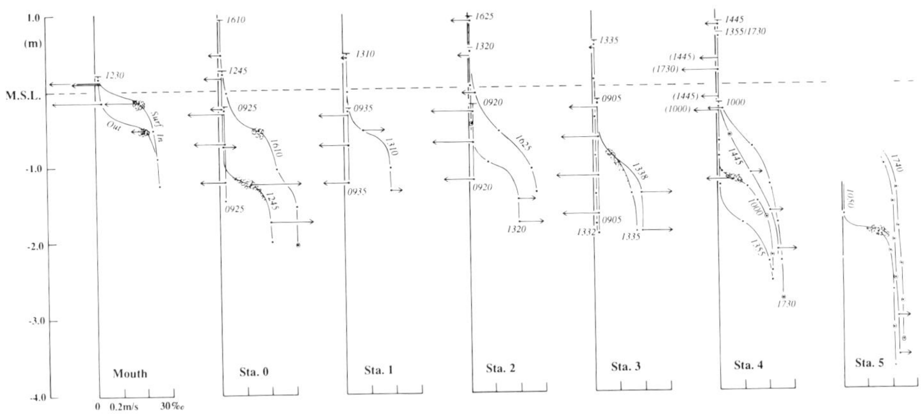

Figure 3: Vertical profiles of salinity and velocity during the flood tide on 20 August 1986. The dots and arrows display the actual measured data. The cloud-like features on the profiles indicate turbulent mixing. Arrows pointing to the left correspond to seaward flow. A circle around a salinity dot represents a near-zero velocity (≤ 0.02 m s^{-1}). Times are noted along the profiles and at the level of the water surface.

By 1300 strong foam lines, characteristic of plunging, are seen at the seaward end of the mouth channel - upstream of the control section. This indicates a net inflow of between 6.7 and 12.2 m^{3}s^{-1}, as observed. However, as Q_0 increases, so does H_0 and *strong* inflow is only attained for several minutes between 1300 and 1330 when Q_0 exceeds 15.7 m^{3}s^{-1} and H_0 = 1.3m. This *strong* inflow is characterized by plunging downstream (landward) of the most constricted section - the plunge line is found at the landward end of the mouth channel (Figure 4). This plunge line, which separates the single layer seawater inflow from the two-layer estuarine flow, is marked not only by the accumulation of foam and flotsam but also by an obvious colour change. These plunging flows have been referred to as tidal intrusion fronts (Simpson and Nunes, 1981) and have been reviewed by Largier (1990). The V-shape seen in Figure 4 is a characteristic shape seen in a number of strong tidal intrusion fronts - the sides of the flow plunge first (near the critical section) whereas the flow in the centre accelerates owing to input from the sides. The centre plunges further downstream (landward). The strong horizontal shear across the front results in shear instabilities which prevent a singularity from developing at the apex of the front.

Figure 4: The V-shaped foam-line associated with the tidal intrusion front during strong inflow on 20 August 1986. The neck of the contraction lies just to the left of the picture. The front is also marked by a colour change between the brown (tannic acid) river water on the right and the foamy green seawater on the left.

The Bottom Density Current: Landward of this tidal intrusion front, beyond the internal hydraulic jump, the seawater continues to propagate as a bottom density current. It arrives at station 3 (Figure 1) at 1334, producing a salinity increase of 18°/∞ in 4 minutes (Figure 3). The most rapid change of 16°/∞ in 100 s indicates a horizontal gradient of 0.8°/∞ m^{-1} through the head, given a speed of advance of 0.2 m s^{-1}. As this basal intrusion moves landward, a weak surface foam line is observed to mark the position of the head. This foam line can be used to estimate the propagation speed of the intrusion. It advanced at about 0.15 m s^{-1} between stations 0 and 3 and accelerated to about 0.20 m s^{-1} on the sloping bed between stations 3 and 4. Similar values of about 0.15±0.05 m s^{-1} are obtained from the arrival times of the salinity front at stations 3 and 4. The empirical result of Simpson and Britter (1979) that $u = 1.15\ c$ where u is the fluid velocity and c is the propagation speed, can be used to obtain smaller values of 0.10 to 0.15 m s^{-1} for the immature density current at stations 0 and 1 at about 1315. Larger values of 0.15 to 0.25 m s^{-1} were obtained at stations 2 and 3.

Theoretical expressions (Benjamin, 1968) and empirical relations (Wilkinson, 1983) have been obtained for the speed of advance of the head of such a gravity current as a function of g', h_2 and the fractional depth h_2/H. The speed obtained for an inviscid intrusion on a flat bed is reduced significantly by viscous drag for small h_2/H (Wilkinson, 1983) and by the presence of a "head wind," an opposing flow in the upper layer (Simpson, 1982). From measurements of g', h_2 and H at stations 1 (1310), 2 (1320) and between stations 3 and 4 (1420), the predicted speeds are 0.20 m s^{-1}, 0.22 m s^{-1} and 0.18 m s^{-1}, respectively. The observations from the water channel seaward of station 3 are less than predicted. It is suggested that this should be expected owing to our strategy of always measuring in the deepest point of the section. Toward the sides h_2 and, particularly, h_2/H are smaller so that c is reduced. The laterally-averaged speed of advance of this basal flow must account for the three-dimensional nature of the spreading of this dense water - c should vary as the width and shape of the channel varies. In contrast, beyond station 3 the theory appears to underpredict, this being due to the downward slope of the bed.

On a sill between stations 4 and 5, this basal intrusion is again subject to internal hydraulic control. The flow accelerates to 0.23 m s^{-1} and $F_2^2 \sim 1.0$ so that the seaward flowing freshwater and the landward flowing saline water are coupled as an exchange flow. An interesting feature is the velocity maximum observed in the halocline, not near the bed as described by Simpson (1987). This velocity maximum, which is also observed at stations 0 and 1 (Figure 3), is due to the absence of turbulence (and momentum diffusion) in the pycnocline as compared with the viscous dynamics of the flow nearer to the bed (Geyer, 1988).

Vertical Exchange: Although the Palmiet is essentially two-layered, this viscous intrusion does exchange water with the overlying layer. During two-layer exchange flow, the turbulent saline flow entrains the less dense ambient estuarine water and is about 20% diluted beyond the hydraulic jump (salinity decreases from 25.5 to 19.5°/∞). The initial dilution from 35.5 to 25.5°/∞ is a result of ebb-plume and surf-zone dynamics. Later, during strong inflow and plunging, water of 31°/∞ is introduced to station 0 which is beyond the plunge line. On flowing from station 0 to 2, this saline layer is diluted a further 13% (to 27°/∞) via the mixing associated with the hydraulic jump. While bulk Richardson numbers of the order of unity are observed at the head of the saline intrusion, the underflow is diluted negligibly (4% to 25.7°/∞) en route to station 4. The products of this turbulence at the head are contained in a third stratified layer between the basal and surface layers. A bulk Richardson number of the order of unity is also observed at plunging.

Even after the tide has turned at the mouth, and after the seawater intrusion is halted, this basal underflow continues to drain into the deepest sections of the estuary. The salt content of the water column at station 4 reaches a maximum between 1600 and 1700, by which time the surface layer is flowing out at 0.3 m s^{-1} (Figure 3) - upward entrainment into the turbulent surface layer begins to remove some of this salt. Richardson numbers were near-unity but supercritical, consistent with the observation that the halocline does not breakdown but is rather eroded slowly from above. The local entrainment velocity at station 4 is 4.2 x 10^{-5} m s^{-1} on the morning ebb tide (between 0930 and 1330), with a maximum of 8.3 x 10^{-5} m s^{-1} observed between 0958 and 1008. In the afternoon (between 1700 and 1900) the entrainment velocity is 3.5 x 10^{-5} m s^{-1}. Following Turner (1973), the observed bulk Richardson numbers of about 3 correspond to entrainment velocities of about 4 x 10^{-5} m s^{-1}. This entrainment in the upper estuary, where the surface flow is contained and strong, is much stronger than in the lower estuary - the overall entrainment velocity is 1.5 x 10^{-5} m s^{-1}. The salinity of the surface water increases from 0 to 1.5°/oo by station 4 but only increases to 2.0°/oo by the mouth.

During most of the year, when river flow in the Palmiet is not strong enough to produce subcritical Richardson numbers, the sills in the estuary are equally important on the ebb tide. The stratified fluid below the sill height and upstream (landward) of the sill is blocked. This water experiences very long residence times consistent with slow vertical entrainment rates. The water below the halocline is unaffected by this entrainment (e.g., below 2 m at station 4, Figure 3). The saline water remaining in the deeper regions (stations 4 and 5) on the following morning (21 August) is evidence that this deep water is resident for longer than a tidal cycle. The average residence time of the saline layer, based on outflow rate versus high-tide volume (Largier and Taljaard, 1990), is 1.6 tidal cycles. The deep water is expected to be resident for substantially longer. The low salinity surface water is effectively non-resident with an average residence time of 0.4 tidal cycles.

8.3 The Summer example

As summer approaches the rain storms are smaller and less frequent, the river flow weakens and the estuary becomes more saline. For much of the summer the Palmiet River flows at about 1 m^3 s^{-1}, with occasional increases after light rains. Tidal flow dominates the estuary with maximum velocities occuring during flood. Stratification remains, however, with the thermal gradients enhancing the salinity gradients in density as the river warms from below 15°C to over 25°C while the ocean generally remains below 15°C. At the same time the reduced river flow allows sand to accumulate in the mouth - the channel becomes narrower and shallower. This serves to protect the estuary from excessive tidal flow and thus, together with the contribution from thermal stratification, the Palmiet continues to be highly stratified most of the time during summer. In February 1985 we obtained some observations of the typical summer version of the stratification structures and processes.

During spring tides which occur in the dry periods (river flow less than 1 m^3 s^{-1}), the shallower channel up to station 3 can become partially mixed following shear-driven ebb tidal mixing. This ebb-tide destratification, due to subcritical Richardson numbers, is well described by comparison with the ebb-tide breakdown of two-layer structure in the Fraser River (Geyer,

1988). Mixing occurred at about 0830 on 6 February, 4 hours before slack water in the mouth. This partially mixed character remains until new saline water of 34.7°/∞ and 17°C intrudes beneath the 32.5°/∞ and 22°C ambient estuarine water at about 1400 (Largier, 1986) - this corresponds to a dilution of seawater (35.5°/∞) of about 25% through a combination of surf-zone and intrusion processes. This 2°/∞ and 5°C stratification remains for the 3 hour flood tide and is eroded again on the ebb tide after 1700 at station 1. The basal intrusion is seen at stations 2, 3 and 4 by about 1430, 1515 and 1630, respectively - corresponding to a realistic speed of propagation of the order of 0.1 m s^{-1} and negligible further dilution. A similar basal density current was observed on the previous day (5 February).

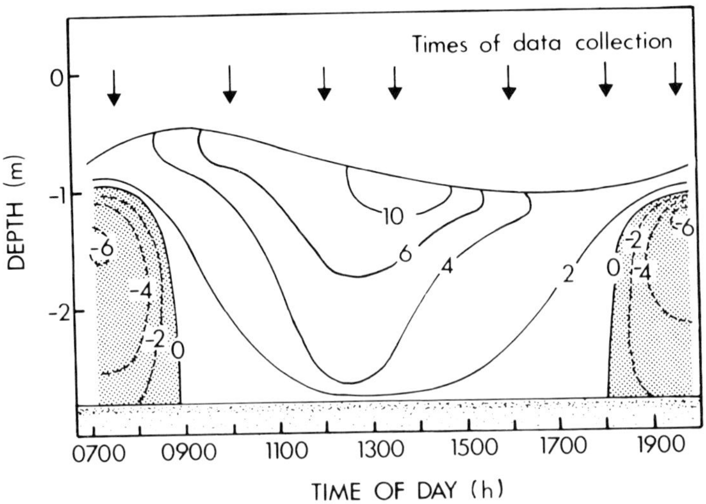

Figure 5: Laterally averaged water velocities at a channel cross-section through station 1 on 12 February 1985. Velocities are in units of cm s^{-1} and negative values correspond to landward flow.

During neap tides, or during slightly enhanced river flow, the two-layer stratification remains through ebb tide. For example, on 12 February (river flow 3 m^3 s^{-1}) bottom salinities of 30°/∞ and surface salinities of 3°/∞ were recorded at station 1 at about 1330, after peak ebb flow (Figure 5) - an interface was found at 0.5 m. Further, during flood tide, curved and V-shaped plunge lines were noted at the landward end of the flood tidal delta. On the earlier flood tide, when inflow was confined to the channel alongside the delta, the plunge line adopted a U-shape which evolved into a V-shape (similar to Figure 4) as the dense inflow strengthened and the centre of the plunge line was pushed further downstream (landward) - these are referred to as type C and D tidal intrusion fronts by Largier (1990). At about 1855, a saline underflow (33.6°/∞ and 0.5 m s^{-1}) was recorded below stationary freshwater at station 0 which was landward of the plunge line (corresponding to a dilution of less than 10% before the hydraulic jump). This supercritical basal flow appears to exceed the critical turbulence velocities required for suspension and to scour the sand bed between the plunge line and the hydraulic jump. A closed area of over 3 m depth is recorded around station 0 in summer. As the water level rises, the nature of the tidal intrusion front changes. The incoming seawater flows over the flood tidal shoals and a new curved plunge line develops along the edge of the shoal as the dense water spills down into the deeper channel landward of the shoal (Figure 6) - this is a type A tidal intrusion front (Largier, 1990).

Figure 6: The curved foam-line and colour change associated with the plunging flow along the edge of the flood tidal delta during late flood on 12 February 1985. Station 0 is to the right of the foam line and the boat is at station 1. The U- and V-shaped plunge lines, which occurred on the earlier flood, were located in the left foreground of the picture. The photographer was standing at the western end of the sand spit, looking north.

Although Figure 5 clearly demonstrates this two-layer velocity structure, it does not display the presence of a transient third layer which is observed as a small basal density current intrusion in summer conditions. It does, however, display a velocity maximum in the lower halocline, as noted for the winter intrusions. The third layer develops through the tidal intrusion process fairly late on the flood tide. Initially the inflowing water plunges below the estuarine water and is substantially diluted by the strong shear, producing a saline layer of about 32.5°/₀₀ which fills most of the estuary. It is suggested that as the flood tide strengthens, so the plunge line is pushed back into deeper water and, eventually, the seawater flowing in along the bottom is sufficiently removed from the interfacial shear that the billows do not dilute it to the same extent as the rest of the inflowing seawater. This appears at station 1 (Figure 7) as a basal intrusion of 33.5°/₀₀ and 13.5°C; it may continue to drain into the deeper channel at station 4. In addition to the temperature-salinity character displayed in Figure 7, this basal intrusion was observed to have enhanced nutrient levels characteristic of the recent coastal upwelling in the ocean.

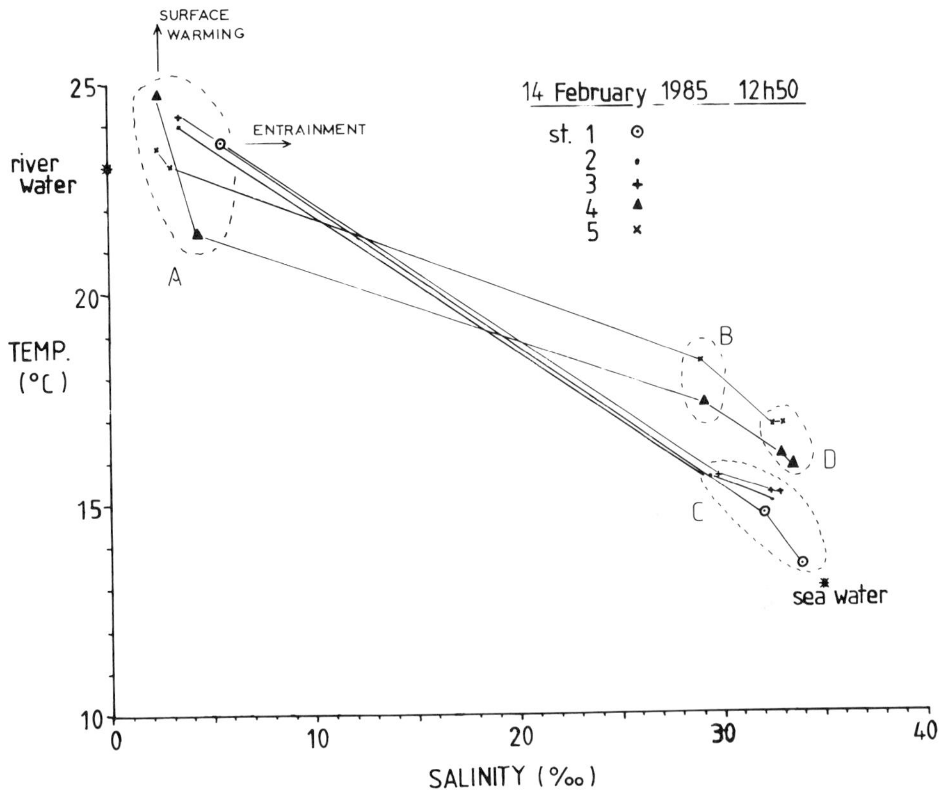

Figure 7: Temperature-salinity data collected at about 1300 (high tide) on 14 February 1985 (data plotted for surface, depth increments of 1 m and bottom at each station). River water was 0°/₀₀ and 23°C whereas seawater was 35°/₀₀ and 13°C. The separation of water into two types is clear. *A* is the surface water, exhibiting surface warming and increasing salinity owing to upward entrainment of halocline water. *B* is the old saline water at middle depth and *D* is the old basal water in the scour holes of the upper estuary. *C* is the newly introduced saline water which is noticeably colder and in the process of intruding under the older water resident at stations 4 and 5. The coldest, most saline point is the bottom density current initiated late on the flood tide.

Through the combined effect of supercritical Richardson numbers and upstream blocking by the sills, the deeper regions continue to play the role of trapping and accumulating the densest water that reaches them. Denser water intrudes beneath the old bottom water and replaces it, similar density water may intrude at some mid-depth and less dense water intrudes along the base of the pycnocline, allowing the old water to remain there for another tidal cycle. It is expected that the residence time of this deep water will be substantially longer than the 1.5 and 5.0 tidal cycles which were calculated for the seawater fraction of the estuary for the dry, spring case and the wet neap case, respectively. In particular, 35‰ water recorded at stations 4 and 5 on 6 February probably remained there until the cold water intrusion on 14 February (note, however, that the deepest points plotted in Figure 7 were not from the deepest spots in the upper estuary, and that some 35‰ water may still have been resident). Average upward entrainment velocities of between 0.5 and 1.0 x 10^{-5} m s^{-1} are calculated during ebb tide, corresponding to a net estuarine circulation of some 0.9 m^3 s^{-1} (Largier, 1986).

In comparison, the freshwater fraction of the estuary has residence times of less than a tidal cycle (as for winter). Typical ebb tide surface velocities average 0.1 m s^{-1} or more for 6 hours, yielding a tidal excursion of 2200 m which is clearly longer than the estuary (Figure 1). There is thus no resident freshwater layer. This outflow of surface water appears to induce a sympathetic seaward flow in the saline lower layer (Figure 5). However, this denser water is blocked by the sill across the mouth and vertical velocities are negligible since bulk Richardson numbers are in excess of 10. It is suggested that the interface is in the process of being set up with an opposing baroclinic pressure gradient growing until it matches the imposed stress of the overflowing layer.

8.4 The general case

The above two observations of stratified flow in the Palmiet Estuary suggest that the summer and winter patterns are variations on a common theme. The significant seasonal changes in river flow and bathymetry serve to counterbalance each other and result in a gravitational circulation pattern which is fairly robust against seasonal effects. The action of wind on the small surface area of the Palmiet appears to be of little consequence in this stratified flow but remote wind effects, through forcing of the coastal ocean, can provide dramatic increases in water level and seawater intrusion - for example, on 7 February 1985 the estuarine water level was 0.39 m higher than for an equal high tide on the previous day.

The vertical gravitational circulation accounts for most of the tidal salt flux in the Palmiet. It is characterized by four processes: hydraulically controlled inflow of dense water through the constricted mouth; basal intrusion of a non-mixing density current; shear-driven upward entrainment into the surface layer and removal by this outflowing layer of buoyant water. This collection of processes appears to be present in different proportions in a variety of tidal inlets - particularly in micro-tidal estuaries, which are characterized by salt-wedge stratification and flood tidal deltas. Tidal plunge lines have been noted in a variety of similar "bar-built" estuaries along the southern African coast (e.g., Sout, Qora, Umgeni), the Australian coast (e.g., Port Hacking - Huzzey, 1982), the California coast (e.g., Mad - Largier, 1991), and the Oregon coast (Pistol - Clifton et al., 1973). Tidal plunge lines have also been noted in Scottish lochs (e.g., Loch Creran - Booth, 1987), in some unconstricted macro-tidal Welsh estuaries (e.g., Seiont - Simpson and Nunes, 1981) and along the coasts of some broader bays (e.g., Humboldt Bay, California, and Chesapeake Bay, Maryland). Field observations of hydraulically controlled exchange flow have also been reported for small boat harbors (e.g., Hilary Harbour in Australia - Schwartz and Imberger, 1988; Hamilton Harbour in Ontario - Hamblin and Lawrence, 1990). The point is that this Palmiet-type system is a common feature, particularly in the countless small "bar-built" estuaries along the high wave-energy coasts of the world.

For the purposes of comparing this Palmiet Estuary with other systems, it is useful to attempt to place it on estuarine classification schemes. We have already noted the geomorphological classifications of the Palmiet as a drowned river valley, a "bar-built" estuary and as a micro-tidal estuary (tidal range less than 2 m) which has a morphology determined primarily by river flow and surf-zone dynamics. In the following hydrodynamic classifications, stratification and flow are compared. Fischer (1972) presents a Estuarine Richardson Number R which expresses the ratio of buoyancy input to tidal mixing. This dimensionless number is intended as an index of stratified versus mixed structure and, thus, of a salt balance dominated by density current advection (e.g., Palmiet) or tidal diffusion. In general, the highly stratified Palmiet exhibits values of the order of 10 (Figure 8) - values in excess of unity taken as representative of stratified estuaries. Smaller values are found towards the mouth (greater tidal velocities) during summer spring tides and low river flow. For example, on 6 February 1985, a small buoyancy input (river flow) and large tidal currents (spring tide) conspire to reduce this Richardson number below unity - consistent with the partially mixed conditions observed on the ebb tide. Similar results are found from an inspection of the Estuarine Stratification Number of Stigebrandt (1988), in which the tidal mixing is scaled by the friction velocity. The numerical values are uncertain but generally much less than unity, indicative of strong stratification. This dimensionless number is location dependent and indicates an increase in stratification towards the river inflow and in the deeper regions as well as a decrease towards the mouth and in the shallower regions. This longitudinal variation in stratification is a common feature of these "bar-built" estuaries. Over the shallows of the flood tidal delta (and over the intertidal sandbanks) the water column is often mixed (single layer saline inflow during flood tide), in the main channel it is stratified, in the deeper regions towards the head the structure is extremely stably stratified (often multi-layered) and beyond the head of the salt wedge the structure is again mixed (fresh water).

The two-criteria classification diagrams of Hansen and Rattray (1966) and Jay and Smith (1988) are more illustrative. Hansen and Rattray (1966) use indices of tidal-average stratification $\Delta S/\bar{S}$ and circulation U_s/U_f (Figure 8), whereas Jay and Smith (1988) use internal (F_B) and barotropic (F_T) Froude numbers. This latter scheme seeks to classify the estuary through evaluation of baroclinic circulation and barotropic tidal forcing. Palmiet values of F_B

are of the order of, but greater than, one - indicating a highly-stratified system and significant baroclinic nonlinearity. In the high-flow winter case (e.g., 25 August 1986), at stations 1, 2 and 3, the saline water is completely removed on the ebb tide and F_B goes to one, corresponding to the most extreme baroclinic nonlinearity. In contrast, at the deeper stations, F_B takes smaller values (going to zero when the saline intrusions are discontinued owing to increased river flow or closing of the mouth) - this indicates increasing stratification strength. The values of F_T, generally greater than 10^{-1}, represent the non-linear nature of the barotropic tide.

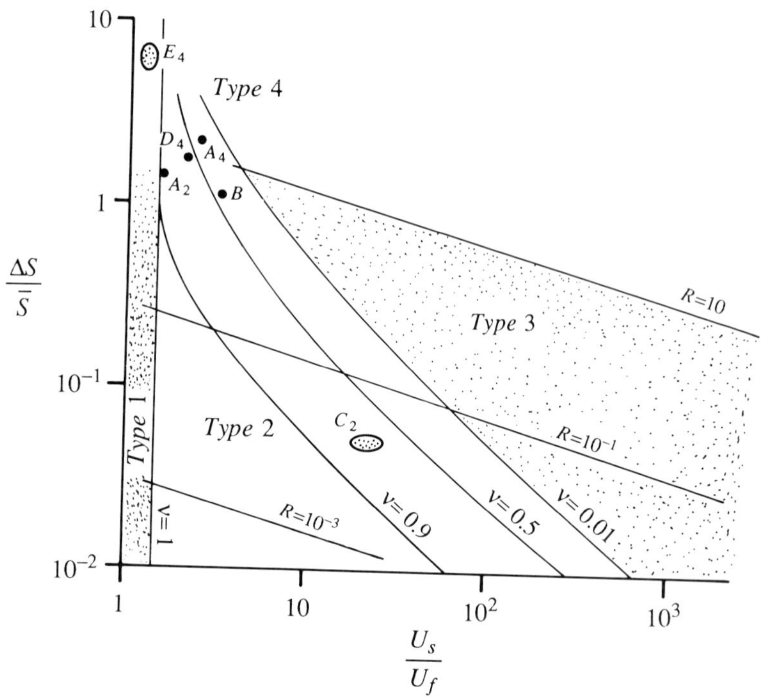

Figure 8: The classification scheme of Hansen and Rattray (1966). The Palmiet is classified as highly-stratified (type 4) and compares with the classification of the Mississippi and Columbia Estuaries. Values for the salt flux diffusion/advection ratio ν and for the Estuarine Richardson Number R (Fischer, 1972) are plotted as functions of the circulation U_s/U_f and stratification $\Delta S/S$ parameters. Points A_2 and A_4 are for stations 2 and 4 during moderate river flow and spring tide on 20 August 1986. Point B is for both stations 1 and 4 during weak flow and neap tide on 12 February 1985. Region C_2 is for station 2 during very weak flow and spring tide on 6 February 1985 - this corresponds to partially mixed conditions. Point D_4 is for station 4 when the mouth was closed and the river flowed very weakly (1 $m^3\ s^{-1}$) in March 1988. Region E_4 is for station 4 when only a thin layer of saline water remained during strong flow (30 $m^3\ s^{-1}$) on 24 August 1986.

In Figure 8, the winter and summer data are plotted on Hansen and Rattray's scheme. While most of the points lie in a small region in the highly-stratified upper left corner, their similarity is a bit deceptive in that this is also a region of large gradients in criteria like the salt flux diffusion/advection ratio ν. It does however provide the essence of the classification: the Palmiet is highly stratified most of the time and the advective salt flux, due to the gravitational circulation, is at least as important as the diffusive fluxes ($\nu \approx 0.5$) - and generally more important in the upper estuary (station 4). Even when the mouth is closed (next section), the data fall into this type 4. The one exception is the data from station 2 on the spring tide during very weak river flow on 6 February 1985. In this case the estuary is partially mixed (type 2) and it is predicted that diffusion dominates the salt flux in the lower estuary.

8.5 A future scenario

Many of the estuaries with constricted mouths also exhibit sills and scour holes further landward. It is this complex bathymetry which is most effective in trapping dense saline water for long periods of time. With increasing depth in the water column, residence times increase (from zero tidal cycles at the surface to several tidal cycles at the base of the deeper regions). With this increase in residence time is a parallel increase in primary productivity and general biomass. However, if the deep water remains resident for too long, the oxygen will be depleted and anoxic conditions may occur - as is reported by CSIR (1982) in the marina canals of the Sandvlei Estuary. Further, decomposition of organic matter also results in high levels of hydrogen sulphide and carbon dioxide (e.g., Swartvlei - CSIR, 1983). This is toxic for animal and plant life and may also be unpleasant for the local society.

Concerned with the potential effects of anthropogenic decreases in river flow (e.g., due to damming of the river) in this sort of system, a field study was conducted in the summer of 1988 when the river flowed so weakly that the mouth closed. We were concerned with how long the water in the deep regions would remain without oxygen ventillation from the surface. As the mouth becomes more constricted, the strength of the tidal mixing in the estuary decreases until it is zero when tidal inflow is prevented. However, the resident buoyancy remains and further buoyancy is input through surface heating and weak river flow - the estuarine classifications indicate strengthening stratification. In the absence of substantial wind forcing or river inflow, would this stratified structure remain, decaying slowly owing to molecular processes?

Following closure of the mouth in January, we visited the Palmiet 8 times from February to April. The temperature-salinity character of the water at station 5 is plotted in Figure 9 for the first 6 visits, evenly spaced over a period of 55 days. The river and surface waters cool by 4°C and 7°C, respectively. This cooling, however, is balanced by a continual input of new fresh water and strong stratification remains. On the first 3 days this new cooler water (mixed with some saline estuary water) is seen to intrude beneath the warm surface water along the top of the halocline. On the fourth day, the affect of a recent freshet is seen in cooler and more saline (owing to increased entrainment of halocline water) surface water. At about this time the estuary begins to overflow the bar across its mouth and larger shear, associated with this throughflow of surface water, accounts for more rapid erosion of the halocline by days 5 and 6. However, the mixing continues to exhibit supercritical Richardson numbers and only the top of the halocline is removed. As the halocline is eroded deeper it retains much of its strength (Figure 9).

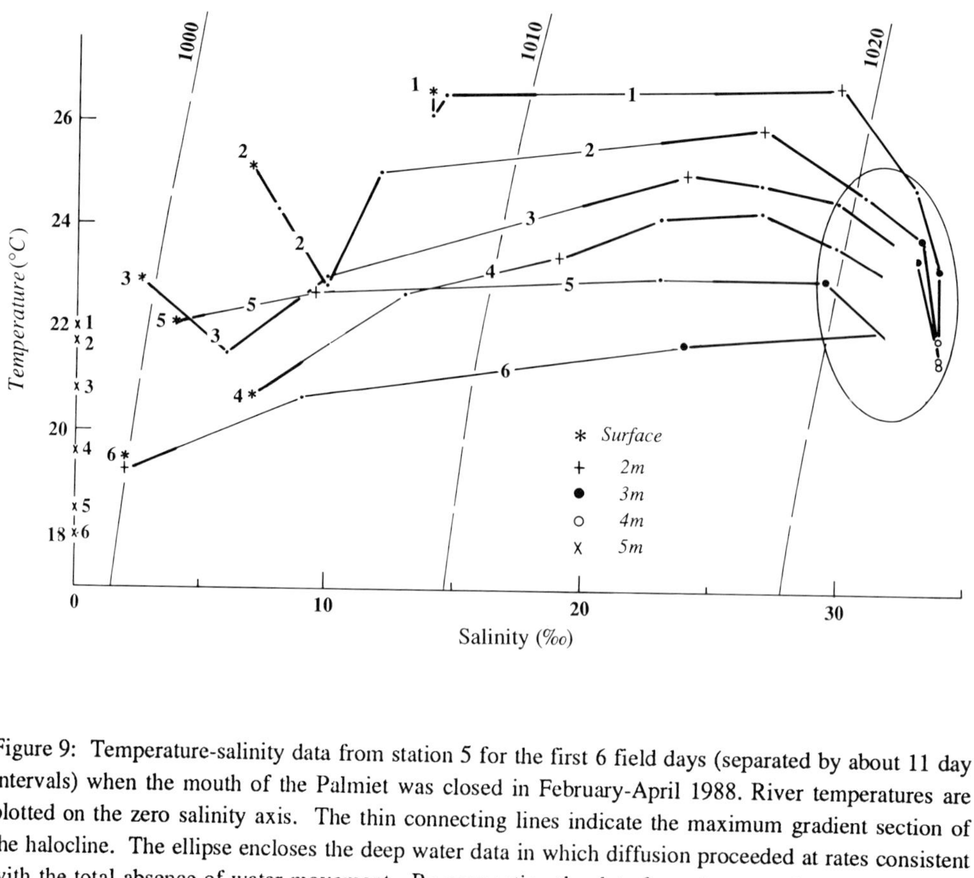

Figure 9: Temperature-salinity data from station 5 for the first 6 field days (separated by about 11 day intervals) when the mouth of the Palmiet was closed in February-April 1988. River temperatures are plotted on the zero salinity axis. The thin connecting lines indicate the maximum gradient section of the halocline. The ellipse encloses the deep water data in which diffusion proceeded at rates consistent with the total absence of water movement. By connecting the data from the same depth and different day, one can obtain an impression of how the temperature-salinity character changed with time.

What is particularly interesting is the apparent absence of mixing below this halocline. Slinger and Largier (1989) calculated diffusion rates based on an assumption of Fickian diffusion. Both salinity and temperature rates indicate that turbulence and advection is absent at depth and that diffusion occurs via molecular action (thermal diffusivity $K_t \sim 10^{-3}$ cm^2 s^{-1}, saline diffusivity $K^s \sim 10^{-5}$ cm^2 s^{-1}). The surface layer turbulence, due to surface wind stress and river inflow (and intrusion), cannot penetrate the strong pycnocline. The deeper observations, which are circled in Figure 9, are representative of this slow molecular diffusion. This sub-pycnocline water is thus resident from when it was introduced in January until the mouth re-opens sufficiently to allow new intrusions of dense saline water (before day 8 in May) - this is a residence time of about 3 months (at only 3-m depth). The accumulation of nutrients and detritus and the depletion of oxygen in this basal water can have severe effects after 3 months. While no such measurements were obtained from the Palmiet, this has been observed in the Sandvlei (CSIR, 1982), Swartvlei (CSIR, 1983), Mtamvuna, Msikaba and Lake Sifungwe (Day, 1981) estuaries along the southern African coast. It can also be expected in many more comparable systems around the world. Of special concern is the potential role of upstream reservoirs and water use in creating these adverse conditions in the Palmiet Estuary and its kin.

Acknowledgements

We gratefully acknowledge the support of the National Research Institute for Oceanology (CSIR) and the South African Committee for Oceanographic Research (FRD, CSIR).

8.6 References

Armi, L., 1986: The hydraulics of two flowing layers with different densities. J. Fluid Mech., 163 , 27-58.

Benjamin, T. B., 1968: Gravity currents and related phenomena. J. Fluid Mech. 31 , 209-248.

Booth, D. A., 1987: Some consequences of a flood tide front in Loch Creran. Estuarine, Coastal and Shelf Science, 24 , 363-375.

Branch, G. M. and J. A. Day, 1984: Ecology of southern African estuaries. XIII. The Palmiet River estuary in the southwestern Cape. S. Afr. J. Zool., 19 (2), 63-77.

Clifton, H. E., R. L. Phillips, and R. E. Hunter, 1973: Depositional structures and processes in the mouths of small coastal streams, southwestern Oregon, p. 115-140. In D. R. Coates (ed.), Coastal Geomorphology , Publications in Geomorphology, New York.

CSIR, 1982: Estuaries of the Cape: Sandvlei CSIR Research Report 413, Stellenbosch, South Africa.

CSIR, 1983: Estuaries of the Cape: Swartvlei CSIR Research Report 421, Stellenbosch, South Africa.

CSIR, 1987: Estuaries of the Cape: Palmiet CSIR Research Report 436, Stellenbosch, South Africa.

Day, J. H. (ed.), 1981: Estuarine Ecology with Particular Reference to Southern Africa , Cape Town, Balkema, 411 pp.

Fischer, H. B., 1972: Mass transport mechanisms in partially stratified estuaries, J. Fluid Mech., 31 (2), 209-248.

Geyer, W. R., 1988: The advance of a salt wedge front: Observations and dynamical model. In Physical Processes in Estuaries (Dronkers/van Leussen, eds.), Springer-Verlag, Heidelberg, pp. 181-195.

Hamblin, P. F. and G. A. Lawrence, 1990: Exchange flows between Hamilton Harbour and Lake Ontario, Proc. 1990 Annual Conf. Canadia Soc. for Civil. Engr.

Hansen, D. V. and M. Rattray, 1966: New dimensions in estuary classification, Limnol. Oceanogr., 11 (3), 319-326.

Huzzey, L. M., 1982: The dynamics of a bathymetrically arrested estuarine front. Estuarine, Coastal and Shelf Science, 15, 537-552.

Jay, D. A. and J. D. Smith, 1988: Residual circulation in and classification of shallow, stratified estuaries. In Physical Processes in Estuaries (Dronkers/van Leussen, eds.), Springer-Verlag, Heidelberg, pp. 21-41.

Largier, J. L., 1986: Structure and mixing in the Palmiet Estuary, South Africa. S. Afr. J. Mar. Sci., 4, 139-152.

Largier, J. L., 1990: Tidal intrusion fronts. Submitted to Estuaries.

Largier, J. L., 1991: Estuarine circulation. Proceedings of the Estuary and Ocean Productivity Workshop - 1990 NE Pacific Chinook and Coho Salmon Symposium (C. Tuss, ed.), Humboldt State University, Sept. 1990, (in press).

Largier, J. L. and S. Taljaard, 1990: Wintertime circulation and stratification in the Palmiet Estuary, South Africa. Submitted to Estuarine, Coastal and Shelf Science.

O'Brien, M. P., 1969: Equilibrium flow areas of inlets on sandy coasts, J. Waterways and Harbors Div., ASCE, No. WWI, Feb. 1969, 43-52.

Schwartz, R. A. and J. Imberger, 1988: Flushing behaviour of a coastal marina, Proc. 21st Inter. Coastal Engr. Conf.

Simpson, J. E., 1982: Gravity currents in the laboratory, atmosphere and ocean, Ann. Rev. Fluid Mech., 14 , 213-234.

Simpson, J. E., 1987: Gravity Currents in the Environment and the Laboratory , Ellis Harwood, Chichester, 244 pp.

Simpson, J. E. and R. E. Britter, 1979: The dynamics of the head of a gravity current advancing over a horizontal surface, J. Fluid Mech., 94 , 477-495.

Simpson, J. H. and R. A. Nunes, 1981: The tidal intrusion front: An estuarine convergence zone. Estuarine, Coastal and Shelf Sci., 13 , 257-266.

Slinger, J., and J. L. Largier, 1989. The evolution of thermohaline structure in a closed estuary. S. Afr. J. Aqua. Sci ., (in press).

Stigebrandt, A., 1988: Dynamic control by topography in estuaries, Hydrodynamics of Estuaries , B. Kjerve (ed.), CRC Press, 17-26.

Taljaard, S., G. A. Eagle and H. F.-K. O. Hennig, 1986: The Palmiet Estuary: A model for water circulation using salinity and temperature measurements over a tidal cycle, Water S. A., 12 (3), 119-126.

Turner, J. S., 1973: Buoyancy Effects in Fluids, Cambridge University Press, Cambridge, 368pp.

Wilkinson, D. L., 1983: Studies into the structure and motion of density currents, Water Research Laboratory Research Report 160, University of New South Wales, Australia, 36 pp.

9

Salinity structure of a shallow, tributary estuary

W W Schroeder, S P Dinnel, and W J Wiseman Jr

Abstract

Analysis of salinity data from 14 field surveys in Weeks Bay (Alabama, USA) indicate that the salinity regime within this shallow, tributary estuary varies significantly both temporally and spatially during periods of minimal tidal range. Horizonal salinity gradients vary from weak to strong and complex, both in the longitudinal and lateral directions. Vertical salinity structure also varies from well mixed to ultra-stratified (up to 14 psu in two meters of water). This variability results principally from subtidal exchanges with the main estuary (Mobile Bay), both in the barotropic and baroclinic modes, and flashy local river runoff. At a secondary level, salinity fields can be influenced by both bathymetry, particularly deeper areas which either channel near-bottom water within the bay or impound bottom waters in scour holes or depression areas, and local winds associated with moderate to high energy events.

Excess fresh water in Weeks Bay, relative to Mobile Bay, can be accounted for by local river inflow. Weeks Bay fill time is one to four days and is dependent on the length of time since the last local freshet. A principal component analysis of stratification resulted in two dominant eigenvectors. The first describes 54% of the total variance and occurs when surface salinities are uniformly fresher than higher salinity bottom waters. The second, limited to the lower bay, describes 17% of the total variance and may indicate gravitational circulation driven by variations in riverine input to the bay.

9.1 Introduction

Results from fourteen months of a two year field study of Weeks Bay, Alabama, provide insight into the structure of the salinity regime in a shallow, microtidal, tributary estuary[1].

---

[1]A tributary estuary is defined herein as an estuary located within a large main estuarine system where the larger estuary serves as the tributary estuary's "coastal ocean" salt source.

Weeks Bay is located on the eastern shore of the Mobile Bay estuary (Fig. 1) in the northern Gulf of Mexico. It is nearly diamond shaped, with a 3.4 km north-south longitudinal axis and a maximum east-west extent of 3.1 km. A small, 5 to 7 m deep scour feature is located in the narrow mouth of the bay and a similar scour feature, 3 to 4 m deep, occurs in the Fish River approximately 200 m upstream of where it empties into the bay. The bay has a mean depth of 1.4 m and a surface area of 6.95 x 10^5 m^2 (Crance, 1971). Tides are principally diurnal with a mean range of approximately 0.4 m. Two rivers discharge into the bay; the Fish River at the northern end, accounting for approximately 73% of the total fresh water input, and the Magnolia River on the eastern shore. Mean combined discharge is estimated at 9 $m^3 s^{-1}$, with freshets up to 4 times larger occurring throughout the year. Short current meter records from the mouth of Weeks Bay are available (Schroeder et al, 1990b). These records are generally from a single meter and, thus, do not resolve the baroclinic component of the flow. A single deployment of two meters does indicate that baroclinic exchanges are occasionally important, but the barotropic component generally is dominant. Tidal currents range in amplitude from 0.15 to 0.3 ms^{-1} and subtidal (40-hour low-passed) currents from 0.1 to 0.3 ms^{-1}.

Previous work in Weeks Bay utilizing water level, wind and runoff data, have shown that subtidal barotropic exchange between Weeks Bay and the Mobile Bay estuary, the adjacent "coastal ocean", is dominated by a co-oscillation of the two bays (Schroeder et al., 1990b). Furthermore, it has been determined that the salinity of Weeks Bay is strongly coupled to the salinity of the Mobile Bay estuary, which is controlled by the extensive seasonal runoff of the Mobile River system, and only secondarily a function of fresh water input from the watershed of Weeks Bay (Schroeder et al., 1989).

If the salinity in Weeks Bay is secondarily a function of the local rivers then the residence time of fresh water in Weeks Bay, relative to that in the coastal ocean, should be rather short. Schroeder et al. (1990b) suggest that normal freshets would refill the bay on the same time scales as wind driven subtidal exchange and normal tidal exchange, and these time scales would be of the order of a few days. In order to estimate the time scale of Weeks Bay response to river flow we use the parameter "fill time". Fill time is defined as the time required for fresh water flow to account for the excess fresh water volume, relative to Mobile Bay salinity, present at the time of each survey. Fill time is a function of time-varying processes, unlike residence time, which is a function of time-averaged processes. The usefulness of the fill time parameter has been documented for fresh water on the Louisiana-Texas continental shelf (Dinnel and Wiseman, 1986).

Schroeder et al. (1990a) have shown that stratification/destratification events can occur in the broad, shallow Mobile Bay estuary at various river stages, and that these events are related to the relative strengths of the winds and river discharge. Very strong stratification variability has also been observed in Weeks Bay: vertical salinity differences from near 0 psu up to 14 psu between surface and 1.5 m depths, in 2.25 m of water. In an attempt to describe the spatial variation in Weeks Bay stratification, a principal component analysis was performed.

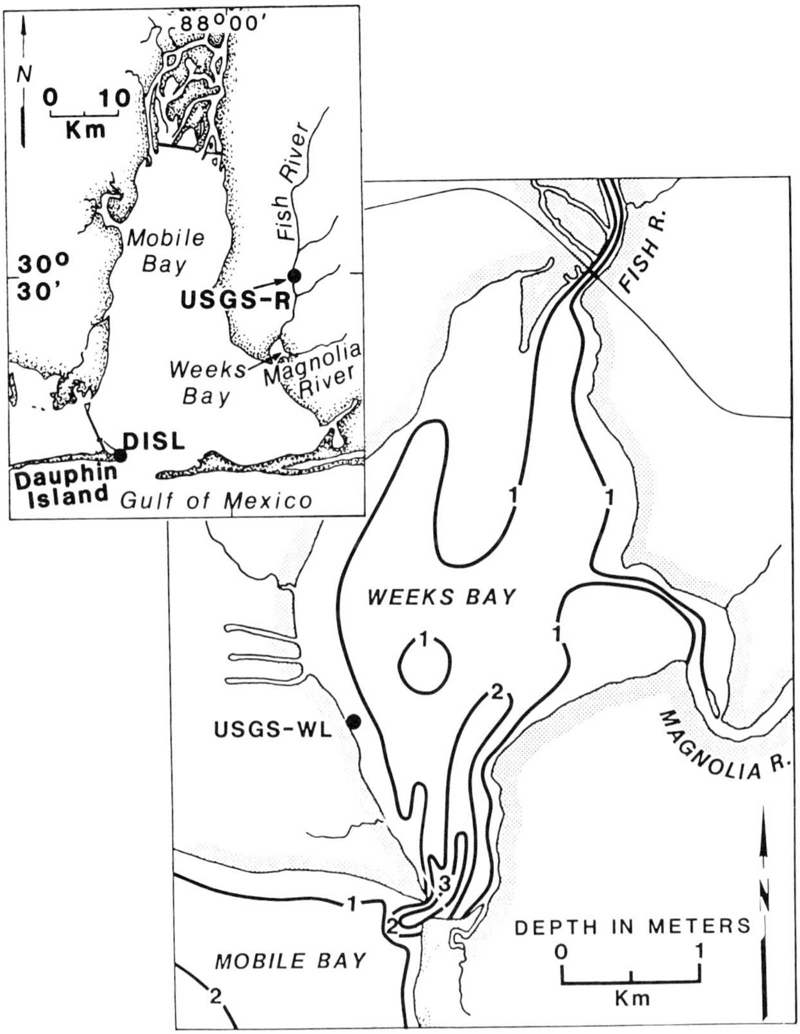

Figure 1: Location map of Weeks Bay, Alabama, showing bathymetry, and location of water level gauge (USGS-WL). Insert shows Mobile Bay, and the locations of the meterological station (DISL) and Fish River discharge gauge (USGS-R).

9.2 Observations and analysis techniques

During the period April 1986 to June 1987 salinity data were collected in Weeks Bay utilizing a Hydrolab Surveyor II. All but one of the surveys were made during periods of low amplitude, equatorial tides. Surveys of the bay were generally completed within six hours or less. Clearly, some aliasing resulted from the non-synoptic nature of the sampling. Since this region of the coast is dominated by diurnal tides, though, we expect that the aliasing will be minimal. Hourly Weeks Bay water levels were obtained from the United States Geological Survey (USGS-WL), and hourly values of wind speed and direction were obtained from the east end of Dauphin Island (DISL) (Fig. 1). DISL winds were converted to north-south and east-west components of wind stress using a drag coefficient that varies with wind speed (Amorocho and DeVries, 1980, 1981). Water levels were demeaned, then both water level and wind stress were low-pass filtered with a 0.6 cycle per day cutoff (Figs. 2b and 2a).

Daily average river discharge measurements for the Fish River were obtained from the USGS (USGS-R) (Fig. 1) starting in November 1986. For the period May to November 1986 Fish River discharge was estimated from local precipitation data. Precipitation was summed over the three reporting stations in the Fish River drainage basin, Bay Minette, Fairhope, and Robertsdale (U.S. Department of Commerce, 1986, 1987, 1988). A linear, multiple regression model, with independent variables of zero-, one-, two-, and three-day lagged precipitation, was fit to 480 days (December 1986 through March 1988) of known gauged Fish River discharge. In effect, we consider the catchment basin to be a linear system with Fish River discharge as the output. The impulse response of the system integrates a weighted version of the runoff over a 4-day period. The impulse response decays rapidly, with time. This simple model, with an R value of 0.815, adequately explained low and moderate discharge values, but tended to underestimate high Fish River discharge. Never-the-less, it was used to estimate the early missing Fish River discharge data. Daily average Fish River discharge is depicted in Figure 2c. To extrapolate from daily averaged Fish River discharge to the total fresh water flow into Weeks Bay, two factors were used. For Fish River discharge, <2.83 m^3 s^{-1}, a factor of 2.76 was used. For higher discharges, >2.83 m^3 s^{-1}, a factor of 3.65 was used (J.L. Pearman, USGS, personal communications).

Maps of the horizontal distribution of salinity were drawn for the sea surface and a depth of 1.5 m below the surface for each survey. When the water depth was less than 1.5 m, the bottom salinity was used at that station. The maps on the deeper level are hereafter referred to as bottom maps. A contouring interval of 2 psu was employed. In addition, north-south longitudinal cross-sections from the Fish River, along the channel on the eastern side of the bay, through the mouth-of-the-bay and out into Mobile Bay were also constructed. An example of such sections is shown in Figure 3. It is clear that most of the variabilty below a depth of 1.5 m is due to ponding of high-salinity water in scour holes. Thus, the maps at 1.5 m depth are representative of the waters below the pycnocline which are not topographically isolated from Mobile Bay.

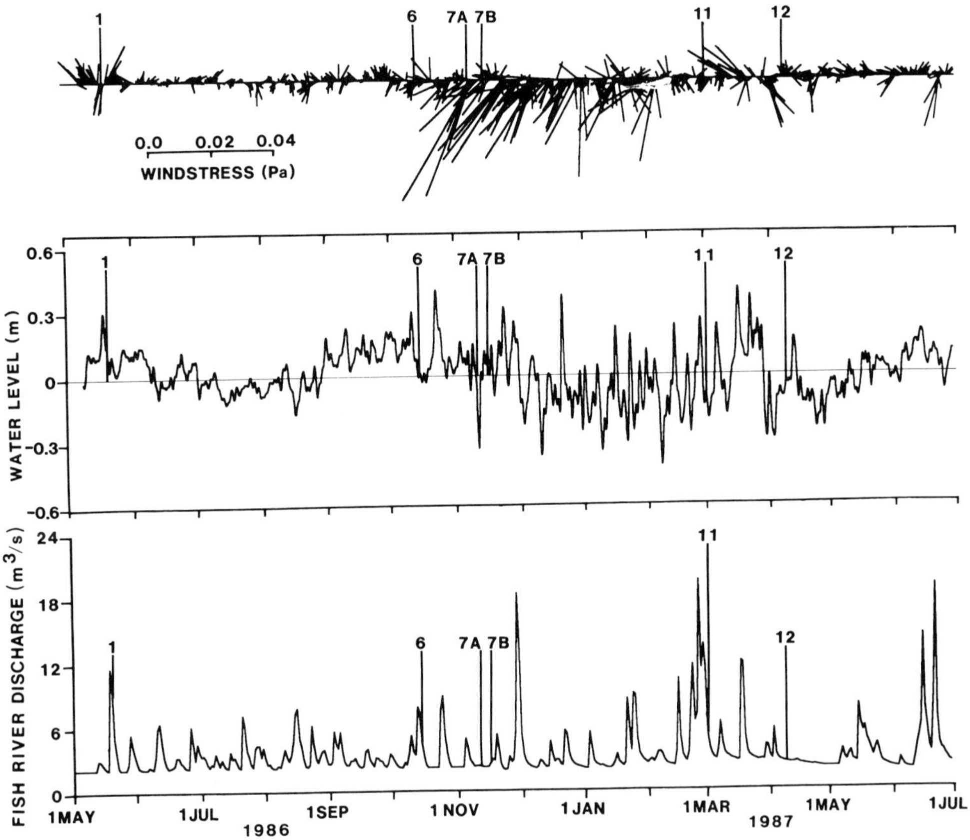

Figure 2: Time series of low-passed wind stress from DISL (a) (top), demeaned, low-passed Weeks Bay water level (b) (center) and daily average Fish River discharge (c) (bottom). Numbers indicate times of individual salinity surveys illustrated in subsequent figures.

For fill time calculations, Weeks Bay was partitioned into 16 segments with one station per segment; segment borders were approximately equidistant between adjacent stations. Two stations in the Fish and Magnolia rivers were excluded. The surface and bottom salinities were averaged and this average salinity was assumed characteristic for each segment. The fresh water fraction, F, for each segment was determined as

$$F = \frac{S_o - S_i}{S_o}$$

Where S_i is the average salinity of the i-th segment and S_o is the salinity of the Mobile Bay station immediately outside the mouth of Weeks Bay. The fresh water fraction multiplied by the station depth and the segment surface area gives the fresh water volume for that segment, and summing over the entire bay produces the excess fresh water volume of Weeks Bay. The total fresh water flow was summed backwards in time, beginning with the day prior to the survey, until the excess fresh water volume in the bay present at the time of the survey, was accounted for. This length of time is the fill time.

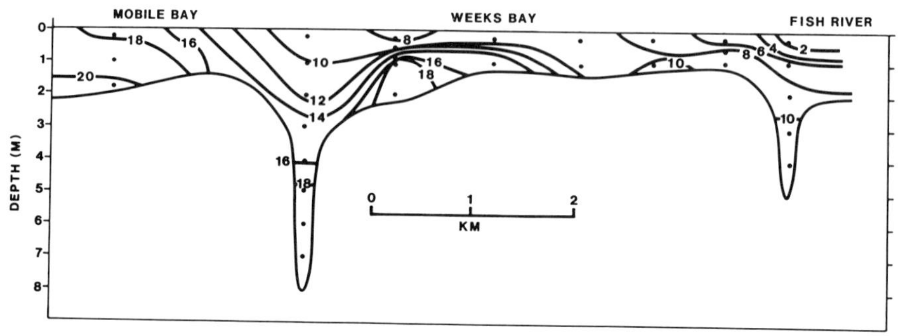

Figure 3: Longitudinal salinity (psu) cross-section from Mobile Bay station, along the channel on the eastern side of Weeks Bay to the Fish River station, survey 1, on 20 May 1986.

A measure of the stratification at each station in Weeks Bay was estimated as the bottom salinity minus the surface salinity divided by the station depth. High values indicate large, vertical salinity gradients, low values indicate small vertical gradients, and a zero value indicates a vertically homogeneous water column. This parameter is also indicative of density stratification as Weeks Bay is nearly isothermal throughout the year. A principal component analysis was performed using this stratification parameter for 16 stations on the 14 surveys. Only those stations within the bay proper were included, two stations, one in both the Fish and Magnolia River mouths and the station in Mobile Bay were excluded. The principal component analysis was performed on the covariance matrix using PRINCOMP (SAS, 1985).

9.3 Horizontal and vertical salinity patterns

In the following section we will hypothesize that specific horizontal salinity patterns and vertical salinity structures, during low amplitude tidal periods, result principally from (1) subtidal exchanges with Mobile Bay both in the barotropic and baroclinical modes and (2) flashy local runoff from the Fish River. At a secondary level, salinity fields can be influenced by (1) bathymetry, particularly deeper areas which either channel near-bottom water within the bay or impound bottom waters in scour holes or depression areas, (2) local winds associated with moderate to high energy events, and (3) high discharge from the Magnolia River freshening the eastern side of the bay. The scenarios which we present are self-consistent and consistent with prior measurements within Weeks Bay (Schroeder et al, 1990b). Some caution is necessary. We are presenting scenarios of time-dependent events inferred from single snapshots of the salinity field and time series of the forcing. When the data might be interpreted in different fashions, we present that which appears most reasonable to us.

9.3.1 -Subtidal Exchange-

The salinity distributions from surveys 1 (Figs. 3 and 4) and 11 (Fig. 5) illustrate typical patterns associated with subtidal exchange events. Survey 1 was undertaken during a falling water level period following a major intrusion event (Fig. 2b) which moved high salinity water from southeastern Mobile Bay into Weeks Bay. The salinity pattern in the longitudinal section (Fig. 3) suggests that at the peak of the intrusion bottom waters with salinities greater than 18 psu were forced up into the middle of the bay as evidenced by the 18 psu water found in the deep scour hole at the mouth of the bay (Fig. 3) and the elongated dome of 18 psu water located in the deep channel on the east side of the lower bay (Figs. 3 and 4b). In addition, 10 psu water would have extended up into the lower reaches of the Fish River as evidenced by the 10 psu water in the scour hole near the mouth of the river (Fig. 3). For many days prior to the survey, winds were weak, but persistent from the southeast. This would tend to drive water into the bay because of Ekman set-up at the coast. Water levels within Weeks Bay were high. Immediately prior to the survey, winds shifted around to blow from the north, causing water levels to drop. As fresh surface water was driven seaward, a deep return flow may have been maintained.

As a result of falling water levels the high salinity bottom waters mixed, spread laterally to the west over the deeper portions of the lower bay and moved seaward. At the same time lower salinity water flowed seaward in the surface layer and covered the shallow bottom areas adjacent to the mouths of both rivers (Figs. 4a and 4b). During survey 1, winds were weak from the north (Fig. 2a) and river discharge was just coming under the influence of recent rains with a moderate to high increase in flow (Fig. 2c) seen as very low surface salinities at the head of the bay (Figs. 3 and 4a).

Survey 11 also occurred during a falling water level period following a major intrusion event (Fig. 2b). Winds were very weak (Fig. 2a), but unlike survey 1, survey 11 followed a 10-day period of moderate to high river discharge (Fig. 2c). High salinity bottom waters extended from Mobile Bay into Weeks Bay (Fig. 5b). However, although water levels are falling (Fig. 2b), the tightly packed isohalines in the bottom salinity field show no sign of mixing or spreading. This appears to be the result of the strong fresh water flow from the

Fish River that dominates the upper and western bay (Figs. 5a and 5b). During both surveys salinity distribution patterns are consistent with a gravitational circulation. Because the salinity surveys were completed within 4 to 6 hours, during low-amplitude equatorial tides, we expect that aliasing of the salinity field was kept to a minimum.

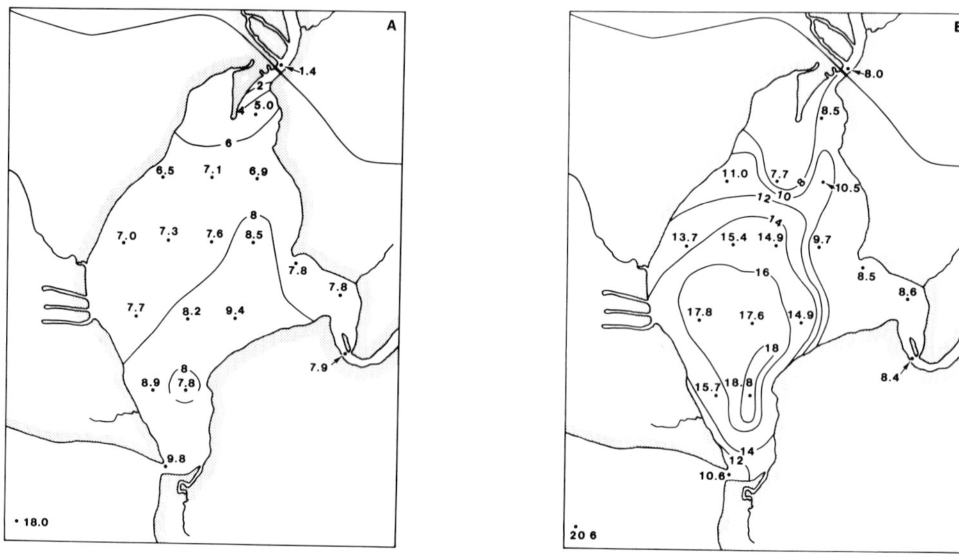

Figure 4: Surface (a) and bottom (b) salinity (psu) for Week Bay for survey 1, on 20 May 1986.

9.3.2 -Fish River Discharge-

The influence of local runoff from the Fish River can be seen in the salinity distributions from survey 11 (Fig. 5) undertaken towards the end of a multiple day high river discharge period (Fig. 2c) and from surveys 1 (Figs. 3 and 4) and 6 (Fig. 6) made at the time of short, moderate river discharge events (Fig. 2c). During survey 11, high salinity waters were confined to the deepest portion of the channel in the lower bay (Fig. 5b), while surface waters, except in the lower bay, and bottom waters in the upper and western bay were nearly fresh (Figs. 5a and 5b). As a result of the large pulse of fresh water (Fig. 2c) the intrusion of high salinity Mobile Bay waters was limited to the deep channel in the lower bay where a very strong halocline formed with a maximum vertical gradient of 14.2 psu between the surface and 1.5 m depths, in 2.25 m of water (Figs. 5a and 5b).

Also of interest is the obvious lack of influence Fish River discharge has on the surface salinity at the Mobile Bay station (Fig. 5a). This is noted by the 4.1 psu surface salinity at the mouth of the bay compared with the 15.1 psu surface salinity at the Mobile Bay station (Fig. 5a). However, low salinity surface waters from Weeks Bay could move into Mobile Bay as a narrow plume, flowing along the shoreline. This condition has been observed in a

Landsat satellite image (not shown) which reveals a ribbon-like plume of turbid water exiting Weeks Bay and oriented to the west and northwest along the shoreline of Mobile Bay.

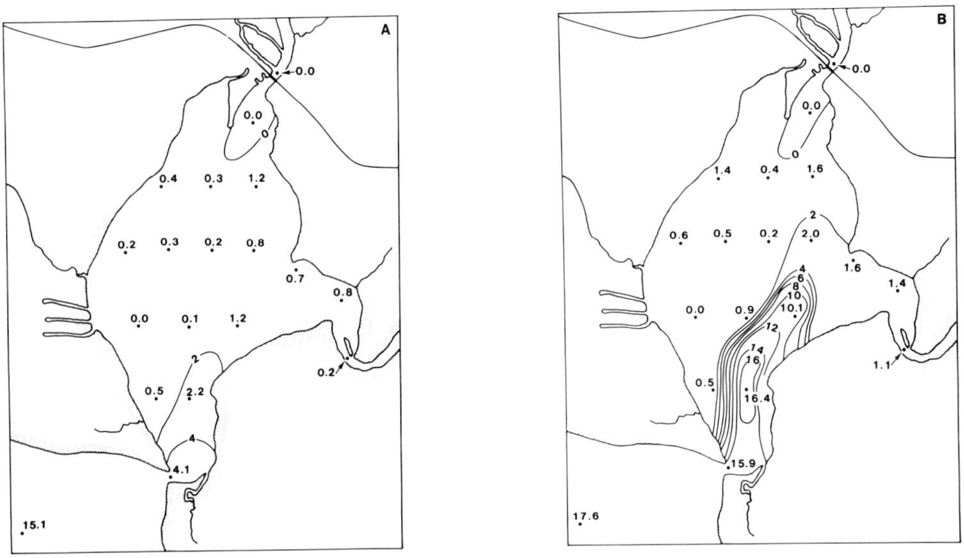

Figure 5: Surface (a) and bottom (b) salinity (psu) for Weeks Bay for survey 11, on 2 March 1987.

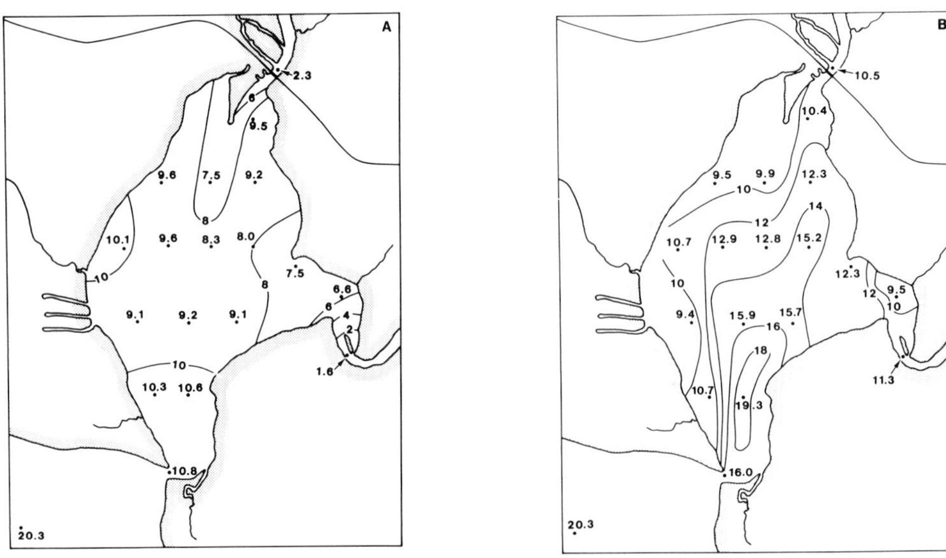

Figure 6: Surface (a) and bottom (b) salinity (psu) for Weeks Bay for survey 6, on 15 October 1986.

During surveys 1 and 6, high salinity bottom waters were observed spread over the deeper, lower and middle bay regions (Figs. 4b and 6b), while lower salinity surface waters occurred throughout the bay. At the mouth of the Fish River during survey 6 (Fig. 6a), very low salinity waters were measured. At the times of both of these surveys the salinity fields exhibited well developed haloclines over most of the bay with the strongest stratification again confined to the region of the deep channel in the lower bay (Figs. 4 and 6). Fish River discharge was insufficient to completely flush salt from surface waters in the lower reaches of the river (Figs. 3, 4a and 6a). It is possible, of course, that maximum river discharges impacted the bay after surveys 1 and 6 were completed.

9.3.3 -Bathymetry-

Bathymetry controls on the salinity distribution might be expected to be slight in waters as shallow as Weeks Bay (Fig. 1). This is not consistently the case. As noted above, the deep channel in lower Weeks Bay directs much of the subtidal exchange of bottom waters with Mobile Bay and permits the development of very strong haloclines (see Fig. 5). In addition, it is clearly seen in Figure 3 that the deep scour features located in the mouth of the bay and near the mouth of the Fish River can readily impound high salinity bottom waters and, at least temporarily, isolate these waters from the remaining bay-river system. At times, much smaller bathymetric variations also influence the structure of bottom salinity fields. For example, tongues of relatively high salinity bottom waters have been observed in the deeper, eastern portions of the middle (Fig. 6b) and middle-upper (Fig. 8b) bay. Also, during survey 7A an elongated pool of 16.0 psu bottom water (Fig. 7b) was found within the finger-shaped depression which lies between the northwestern shore of Weeks Bay and a submerged shoal that extends into the bay from the north (Fig.1). This pool of bottom water apparently remained intact, maintaining a small surface to bottom salinity difference of 2 psu, while the remainder of the bay was nearly vertically homogenized by moderately strong north winds associated with the beginning of a winter cold front passage (Fig. 2a).

9.3.4 -Local Winds-

Subtidal exchange processes are often driven by Ekman convergences or divergences forced by non-local winds (Schroeder et al, 1990b). At certain times, though, local wind forcing is clearly important. Survey 7A was made just as coastal Alabama was coming under the influence of a winter cold front, while survey 7B occurred five days later, 48 hours after the passage of the front. The winds over this period (Fig. 2a) were (1) moderately strong (5 to 10 m s^{-1}) from the north fourteen hours prior to and during survey 7A, (2) strong (8 to 16 m s^{-1}) from the north-northeast to northeast for 44 hours after survey 7A followed by 38 hours of approximately 5 to 8 m s^{-1} winds from the northeast to east, and (3) <5 m s^{-1} winds from the southwest to west 24 hours prior to and during survey 7B. Comparison of Figures 7 and 8 from the two surveys indicates a number of significant changes which were likely due to the winds.

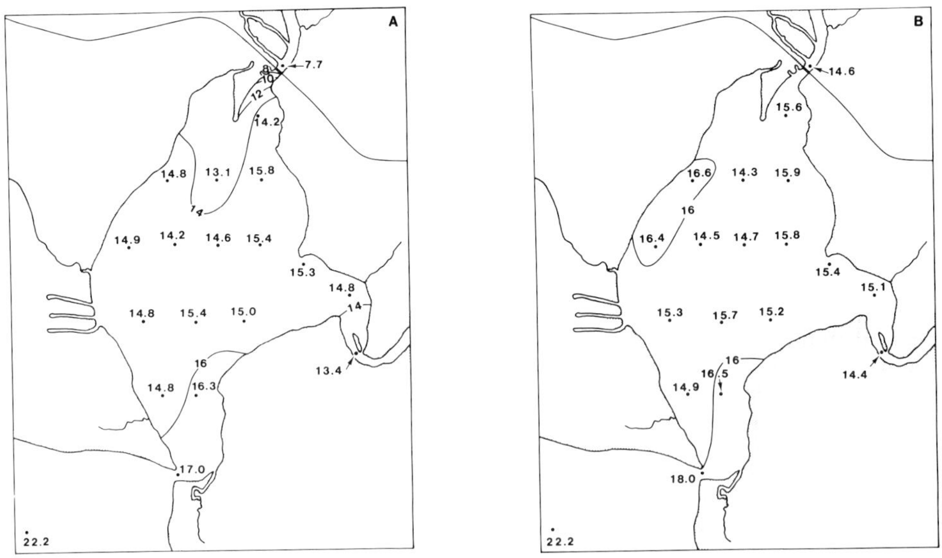

Figure 7: Surface (a) and bottom (b) salinity psu for Weeks Bay for survey 7A, on 12 November 1986.

Weeks Bay waters were first nearly vertically homogenized and also well mixed laterally and longitudinally (Figs. 7a and 7b), presumably due to wind mixing. Strong vertical and horizontal gradients occurred only in the upper bay at the mouth of the Fish River (Figs. 7a and 7b). After survey 7A, at the time of the strongest northerly winds, bay waters were rapidly set-down (Fig. 2b) and waters within Weeks Bay were likely completely mixed. There would have been very strong mixing in Mobile Bay, also. When the winds relaxed (Fig. 2a), bay water level rebounded (Fig. 2b) and Weeks Bay refilled with high salinity Mobile Bay water. These high salinity waters moved up into the deeper areas of Weeks Bay, appear to have reached a maximum intrusion and then subsided some just before the survey 7B. This can be seen in the distribution of bottom salinities (Fig. 8b) where a lens of 20 psu water in the northeast corner of the bay and 21.7 psu water at the mouth of the Magnolia River have both been pinched off from the large pool of >20 psu water in the lower bay (Fig.8b). The pool of 20 psu in the lower bay has likewise been detached from the higher salinity "coastal ocean" in Mobile Bay (Fig. 8b).

High salinity waters were also pooled on the western side of Weeks Bay. This is seen in both the surface and bottom salinity fields and probably resulted from upwelling along the upwind shore of this shallow bay. The effects of the southwest and west winds prior to and during survey 7B can also be seen in the position of the Fish River plume which was oriented along the eastern, downwind side of the bay (Fig. 8a).

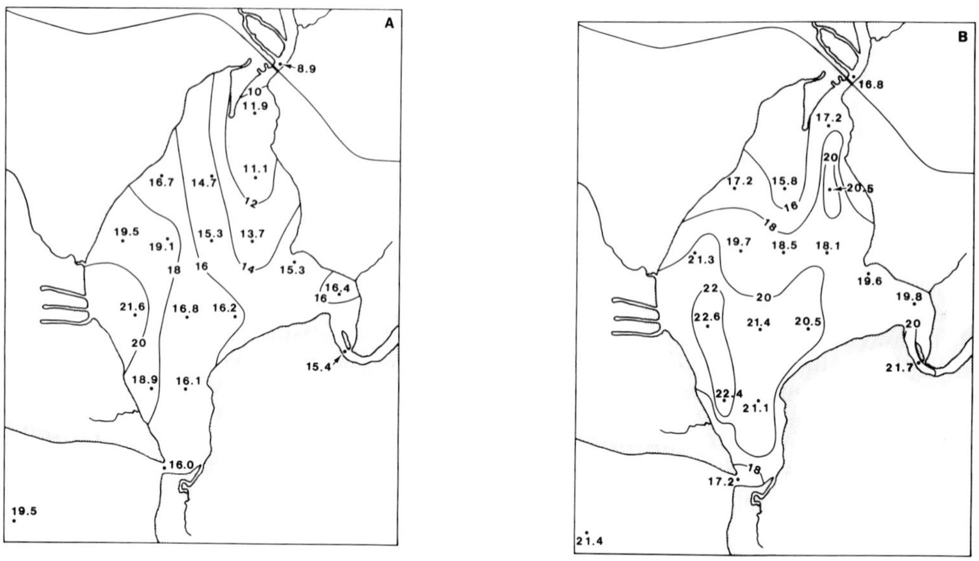

Figure 8: Surface (a) and bottom (b) salinity psu for Weeks Bay for survey 7B, on 17 November 1986.

9.3.5 -Mobile Bay/Mobile River System-

In previous work (Schroeder et al., 1990b), we have indicated that the flushing time of Weeks Bay is very short. Therefore, variations in the salinity of Mobile Bay should be readily transmitted to Weeks Bay. The salinity variations in Mobile Bay are normally of long time scale (weeks to months). Occasionally, when the Mobile River system abruptly floods, shorter time scale changes (one to two weeks) in salinity may occur. Such an event resulted in the low salinities observed within Weeks Bay during survey 12 (Fig. 9). Although the discharge from the Fish River was very low (Fig. 2c), the waters of Weeks Bay are seen to be nearly fresh, as are those of Mobile Bay. Therefore, we assume the bulk of the low salinity water within Weeks Bay originated in Mobile Bay as a result of Mobile River System discharge, not from Fish River discharge.

9.3.6 -Magnolia River Discharge-

The Magnolia River discharge accounts for less than 25% of the total fresh water flow to Weeks Bay. Its plume is, consequently, proportionally smaller than that of the Fish River. Furthermore, since the drainage basins of both rivers are local, the same weather systems will tend to force both river discharges. An example of the Magnolia River freshening the eastern side of the bay can be seen during survey 6 (Fig. 6a).

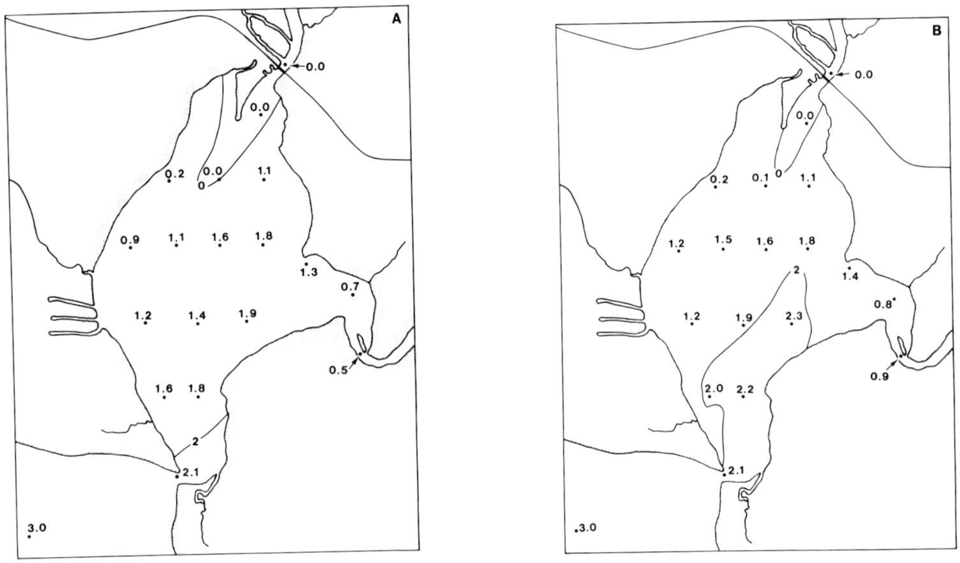

Figure 9: Surface (a) and bottom (b) salinity psu for Weeks Bay for survey 12, on 10 April 1987.

9.4 Fill time

Fish River discharge is flashy (Fig. 2c). Discharges respond rapidly to local precipitation then return to baseline levels in just a few days. Excess fresh water flow into Weeks Bay, relative to Mobile Bay salinity, is assumed to be accounted for by the total fresh water flow into Weeks Bay. Most of the 14 surveys followed a noticeable freshet and fill times were generally between one and four days. The fill time was zero during the time of survey 10 (not shown) when the entire system, Weeks Bay and Mobile Bay, had very low salinities, i.e. 1.5 - 2.5 psu, and there was no excess fresh water. The longest fill time, six days, was associated with survey 11. A multiple-day, high discharge period had occurred a week before and Weeks Bay appeared to have received an influx of higher salinity bottom water from Mobile Bay a few days before survey 11. A brief, but sharp peak in Weeks Bay water level suggests Mobile Bay water was drawn in along the bottom (Fig 2b). Winds at this time were extremely weak and we are probably seeing the results of a significant gravitational circulation. A rapid drop in Weeks Bay water level then trapped the higher salinity water in the deeper, lower bay, and high Fish River discharge created a fresh surface layer. Fill time for the excess water in Weeks Bay is short and is dependent upon the length of time since a locally-driven freshet. The pulses of fresh water could possibly remain in Weeks Bay for more than one week, but no fill time needed more than one preceding freshet to account for the excess fresh water.

9.5 Salinity stratification

Weeks Bay average salinities range from 0 to 20 psu. Strong stratification, as well as strong horizontal gradients, can occur within the bay when salinities are between 5 and 15 psu. Stratification in Mobile Bay shows no significant correlation with the stratification in Weeks Bay.

The principal component analysis of the stratification empirically partitions the variance of the stratification parameter into orthogonal patterns (or eigenvectors) known as principal components. Each eigenvector has an associated eigenvalue that defines the percentage of the total variance explained by that eigenvector. The most important eigenvector describes over half the total variance, 54%, of the stratification parameter. This eigenvector has high amplitudes of similar sign over much of the bay indicating that most of the salinity stations respond coherently to external forcing (Fig. 10a). The salinity station in the bay mouth and those stations nearer the river entrances have low amplitudes and do not contribute much individual variance to the first eigenvector (Fig. 10b). The second most important eigenvector describes 17% of the total variance in the stratification parameter. The highest amplitudes of the second eigenvector describes stratification that is localized to the lower region of the bay. There is little contribution to the second eigenvector from the individual stations farther away from the deeper, lower bay (Fig. 10d). Although the remaining eigenvalues combined describe 39% of the total variance, no additional eigenvalue is statistically different from random noise at the 95% level using the criteria of Overland and Preisendorfer (1982).

The first principal component correlates well with the arithmetic average of the stratification parameter over each survey, an R value of 0.985. This implies that the dominant stratification pattern is consistent bay-wide. The first eigenvector is very similar to the pattern of strong stratification which occurs when the surface salinities are uniformly fresher than the higher salinity bottom water (for example see Fig. 4). The second principal component correlates with one- and two-day old Fish River discharge, with R values of 0.66 and 0.65 respectively. Higher river discharge is associated with stronger stratification in the lower bay. This suggests that local discharge does have some influence on the stratification in Weeks Bay. This principal component is strongest for surveys with high salinity bottom water near the bay mouth and low salinity over the rest of the bay (see Fig. 5b). It may indicate gravitational circulation driven by variations in riverine input to the bay.

The timing of our surveys was biased away from periods of strong tropic tides. The mechanical mixing of the surface and bottom salinities in Weeks Bay, because of weak tidal currents, may be much lower than needed to break down the observed stratification. In Mobile Bay proper, stratification appears to break down when strong tidal currents occur in conjunction with significant wave activity (Schroeder et al., 1990a). It is not yet clear whether a similar relationship holds in Weeks Bay.

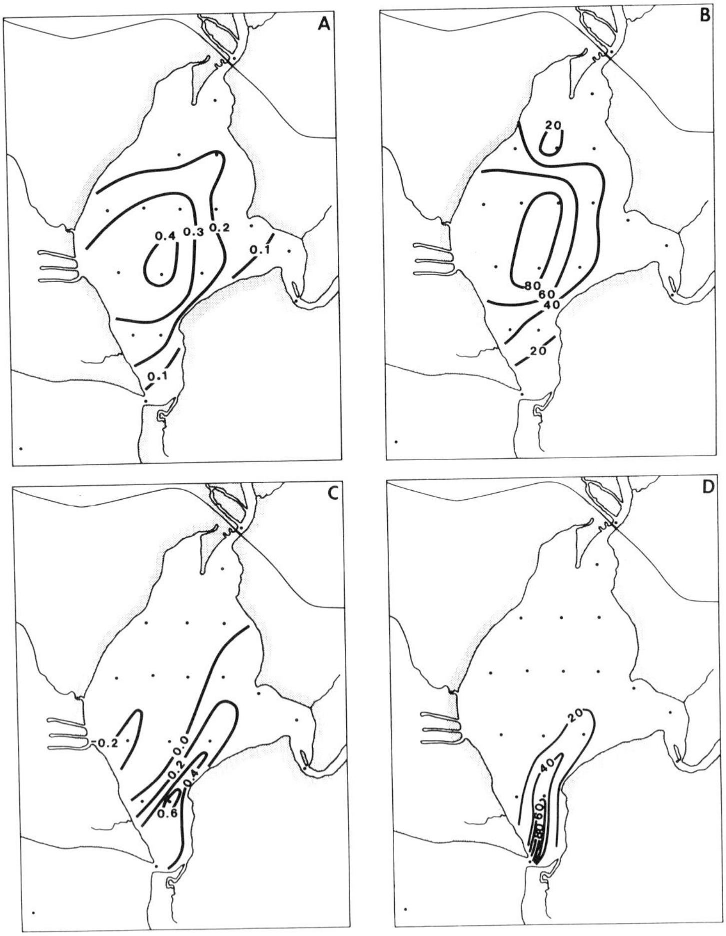

Figure 10: First and second eigenvectors of stratification parameter. Amplitudes for first eigenvector (a) which has an eigenvalue that describes 54% of the total variance; percentage of individual station variance contributed to first eigenvector is shown in panel (b). Amplitudes for second eigenvector (c) which has an eigenvalue that describes 17% of the total variance; percentage of individual station variance contributed to second eigenvector is shown in panel (d).

Acknowledgements

This work is the result of research sponsored by the U.S. Department of Commerce, NOAA, NOS, Office of Ocean and Coastal Resource Management, Marine and Estuarine Management

Division under Grant No. NA86AA-0-CZ017, The University of Alabama Marine Science Program, the Marine Environmental Sciences Consortium of Alabama and the Mississippi-Alabama Sea Grant Program (Grant No. NA85AA-D-SG005; Projects R/ER-16 and E/O-16). Partial support was also received from the U.S. Geological Survey Wetlands Loss Project, Contract No. 14-08-0001-23411. T.R. Matthews served as the Graduate Research Assistant for the project. J.L. Pearman and H.H. Jeffcoat coordinated cooperative monitoring with the USGS.

Contribution No. 162 of the Aquatic Biology Program, The University of Alabama; Contribution No. 193 of the Marine Environmental Sciences Consortium, Dauphin Island, Alabama; and Contribution No. 0117 of the Center for Marine Science, University of Southern Mississippi.

9.6 References

Amorocho, J. and J.J. DeVries, 1980: A new evaluation of the wind stress coefficient over water surfaces. *J. Geophys. Res.*, 85:433-442.

Amorocho, J. and J.J. DeVries, 1981: Reply. *J. Geophys. Res.*, 86:4308.

Crance, J.H., 1971: Description of Alabama estuarine areas - cooperative Gulf of Mexico estuarine inventory. *Alabama Mar. Res.* Bull., 6, 85 pp.

Dinnel, S.P. and Wm. J. Wiseman, Jr., 1986: Fresh water on the Louisiana and Texas shelf. *Cont. Shelf Res.*, 6, 765-784

Overland, J.E. and R. W. Preisendorfer, 1982: A significance test for principal components applied to a cyclone climatology. *Mon. Weather Rev.*, 110, 1-4.

SAS Institute, Inc., 1985: *SAS User's Guide: Statistics, Version 5 edition.* Cary, NC. 956 pp.

Schroeder, W.W., S.P. Dinnel and Wm.J. Wiseman, Jr., 1989: Stratification variability in a tributary estuary. Abstracts, Tenth Biennial International Estuarine Research Federation Conference, Baltimore, Maryland, Oct. 1989. p. 105.

Schroeder, W.W., S.P. Dinnel and Wm.J. Wiseman, Jr., 1990a: Salinity stratification in a river-dominated estuary. *Estuaries*, 13(2):145-154.

Schroeder, W.W., Wm.J. Wiseman, Jr. and S.P. Dinnel, 1990b: Wind and river induced fluctuations in a small, shallow, tributary estuary. *Residual Currents and Longterm Transport*, R.T. Cheng, Ed., Springer-Verlag, pp. 481-493.

U.S. Department of Commerce, 1986: *Climatological Data, Alabama*, Volume 92. National Climatic Data Center, Asheville, NC. 264 pp.

U.S. Department of Commerce, 1987: *Climatological Data, Alabama*, Volume 93. National Climatic Data Center, Asheville, NC. 276 pp.

U.S. Department of Commerce, 1988: *Climatological Data, Alabama*, Volume 94. National Climatic Data Center, Asheville, NC. 276 pp.

10

On meteorologically induced subtidal motions in Hangzhou Bay

J L Su and W Chen

Abstract

Multiple input spectral analysis is applied to subtidal sea level oscillations inside Hangzhou Bay which is large and shallow. Among the three forcing factors for the sea level oscillations: The coastal sea level is the dominant one; for periods less than 10 days the local along-bay wind stress component is also important; the local cross-bay wind stress component has only minor importance. One dimensional linear inviscid model is shown to be useful to describe the subtidal bay oscillations. The analysis shows that, when the inputs are highly correlated multiple input spectral analysis can encounter difficulty.

10.1 Introduction

Hangzhou Bay is a large funnel-shaped bay along the East China Sea coast of China (Fig.1), whose width is about 100 km at the mouth and narrows down to around 20 km some 100 km to the west. Several large cities are located near the bay and a nuclear power plant is being built on the north shore of the bay. Consequently, there have been interests in the understanding of the environmental quality of the bay water.

The bay is shallow with an average depth of about 10 m and the average tidal range at the baymouth is around 2.5 m. Thus tidal flushing is probably the dominant mechanism for the bay-shelf water exchange. Nevertheless, it is also of interest to understand how meteorologically induced oscillations contribute to this water exchange. In this study, meteorologically induced subtidal motions in Hangzhou Bay are investigated, using mainly the tidal and meteorological data.

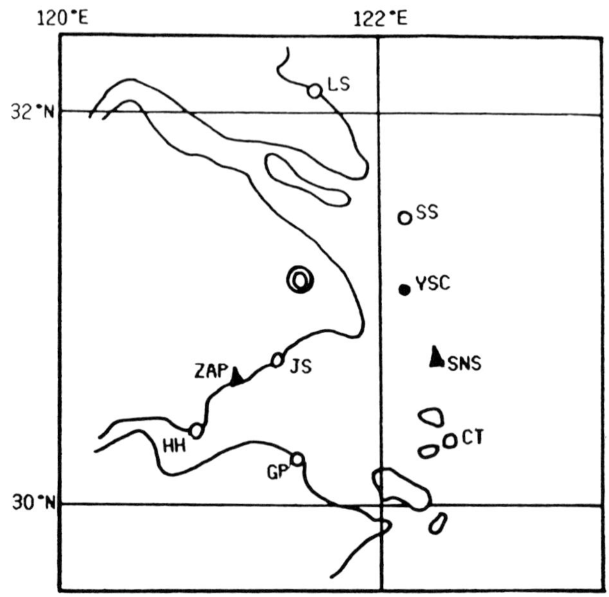

Figure 1: Locations of data stations: tide (circle); meteorology (solid circle);wind (solid triangle). Double concentric circles denote the Shanghai city.

10.2 The Data

The data stations are shown in Fig. 1. The two wind stations, YSC and SNS, provide the remote (over the shelf) wind data and the wind station ZAP gives the local (over the bay) wind data. Throughout this study the term "wind stress" refers to the quadratic wind velocity part of the usual wind stress formula. For the wind stress components the east (cross-shelf) is taken as the positive direction of the x-axis and the north (alongshelf) the y-axis. The sea level station CT provides the data of the coastal sea level which also forces oscillations in the bay. In the following we will use the terms cross-shelf and alongshelf winds when discussing the coastal sea level oscillations. Methods and formulas used in data analysis are the standard ones (e.g., Moores and Smith, 1968; Jenkins and Watts, 1968).

The data period used spans from 12.1973 to 12.1974. All the time series are error detected, low-passed and then decimated with a 24-hour interval before use. A Lanczos-cosine filter (half-power point of 40 hr) is used to low-pass the hourly sea level records and a weighted symmetric filter (half-power point of 54 hr) to low-pass the 6-hourly wind stress and atmospheric pressure records. All sea level records are also barometrically adjusted. Two subsets of the data period, 12.1973 to 3.1974 and 6.1974 to 9.1974, are chosen to represent the winter and the summer characteristics, respectively.

Tukey window is used in the spectral computations. For the complete record the frequency resolution is 0.011 cpd and the equivalent degrees of freedom is 8.2, while for the seasonal records they are 0.045 cpd and 10.7, respectively. Larger degree-of-freedom Tukey windows have also been employed with no effects on the main conclusions.

10.3 Multiple input system

It is known that the effect of meteorological forcing on the sea level oscillations of a narrow estuary can be resolved into the local and remote components (Wang and Elliott, 1978). The former is forced by local winds along the thalweg, setting up a sea level difference along the estuary. The latter is induced by the coastal sea level oscillations at the mouth of the estuary, which propagates with little attenuation into the estuary. Both components are important for oscillations with periods of a few days. For longer period oscillations, however, the remote effect is expected to dominate (Garvine, 1985).

Hangzhou Bay is rather wide and has a gentle bottom topography. Cross-bay winds may also contribute significantly to the sea level oscillations in the bay. Thus for oscillations in Hangzhou Bay with short periods, say, less than 10 days, we have a three-input system and the inputs are the coastal (CT) sea level and the two components of the local wind stress. For long period oscillations in Hangzhou Bay, however, only the coastal sea level is needed as the input. As will be demonstrated later, coastal sea level oscillations with periods of a few days are caused mainly by the alongshelf winds, whereas longer period oscillations are caused by both alongshelf and cross-shelf winds. Therefore, for the long period case, we may also treat the bay system as having two inputs, namely, the two components of the remote wind stress. Because of the large extent of a weather system the local winds and the shelf winds are statistically correlated. The two wind stress components at the same station also correlate with each other. Therefore, in both frequency range the Hangzhou Bay sea level oscillation is the output of a multiple input system whose inputs are correlated.

For subsequent discussions we list a few formulae of multiple input spectral analysis. For simplicity, consider a linear system with one output (u) and two inputs (x_1 and x_2). Assuming that x_1 and x_2 are correlated it can be shown (Chen and Su, 1991) that the output spectrum of this system can be written as

$$S_{uu} = E_{1/2} + E_{12} + E_{2/1} + E_{21} + S_R$$

where $E_{1/2} = H_{u1}\, S_{u1/2} = R^2_{u1/2}\, S_{uu/2}$,

$$E_{12} = H_{u1} (S_{u1} - S_{u1/2}),$$

and likewise for $E_{2/1}$ and E_{21}. "S" denotes either the autospectrum or the cross-spectrum, "H" the frequency response function and "R" the coherence squared. The slash symbol in the subscript of either a spectrum or a coherence squared indicates a "partial" quantity. $E_{1/2}$ is the contribution of the uncorrelated part of x_1 to S_{uu} and E_{12} is the contribution to S_{uu} due to the part of x_1 that is correlated with x_2. We take the modulus of $(E_{1/2} + E_{12})$ as a measure of the contribution of x_1 to S_{uu}. The meanings of $E_{2/1}$ and E_{21} can also be explained accordingly. We note that the sum $(E_{12} + E_{21})$ is a real number.

Similar formulae can be derived for linear systems with three or more correlated inputs (Chen and Su, 1991).

When a system is studied with a multiple input spectral analysis two problems will arise if the inputs are highly correlated. Firstly, unless there are causal relations among the highly correlated inputs and unless these relations are also studied, interpretation of partial cross spectrum analysis results may be difficult. Secondly, a frequency response function depends on the ratio of a partial cross spectrum over a partial autospectrum (Bendat and Piersol, 1980). When two inputs are highly correlated with each other and one of these inputs is highly correlated with the output, the variance of the estimation of the corresponding frequency response function is likely to be large.

10.4 Results and Discussions

10.4.1 Sea level oscillations

Fig. 2 gives the spectra of the sea level oscillations both inside the bay (GP) and at the coast (CT). The general forms of the two spectra are quite similar, especially at the low frequency end. Inside the bay the spectrum has a prominent peak near the 15 day period which is not evident at the coast (Fig. 2). This is due to aliasing of the diurnal and semi-diurnal tides which have not been entirely removed by the filter used (Fang and Yu,1985). Amplification of the semi-diurnal tides by the funnel-shaped bay contributed to this high aliased peak. The spectrum of the along-bay sea level difference between GP and CT is also plotted in Fig. 2. For periods longer than 10 days but away from a narrow band near the 15-day period this spectrum is significantly lower than the sea level spectrum of GP, confirming Garvine's suggestion (1985) that the remote effect dominates the sea level oscillations in estuaries. At all frequencies the spectrum of the transverse-bay sea level difference between JS and HH was found to be lower than the JS sea-level spectrum by a factor around 10, suggesting that the effect of the north-south component of the local wind-stress is probably of minor importance.

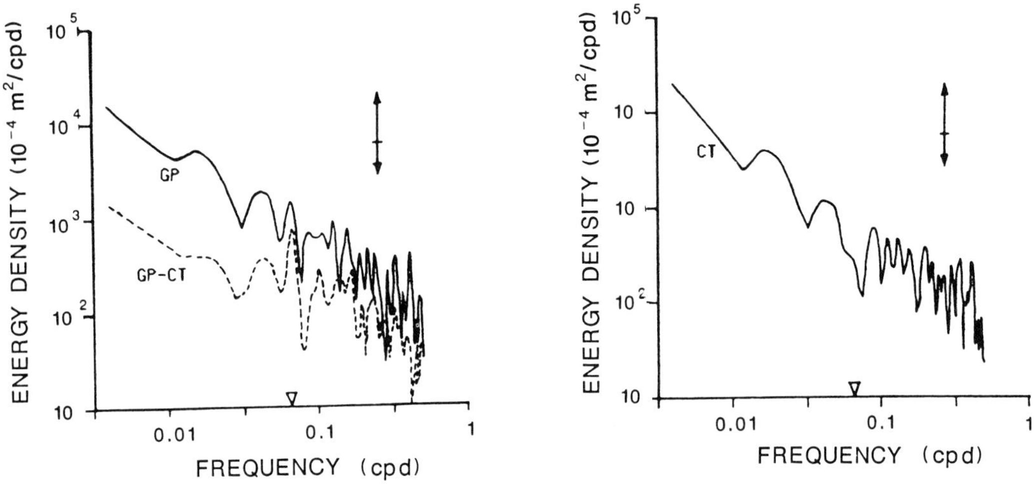

Figure 2: Sea level spectra inside the bay (GP) and at the coast (CT) (solid lines), and sea level difference (GP-CT) spectrum (dashed line). Inverted triangle on the abscissa indicates the 15 day period. (The 95% confidence interval is marked)

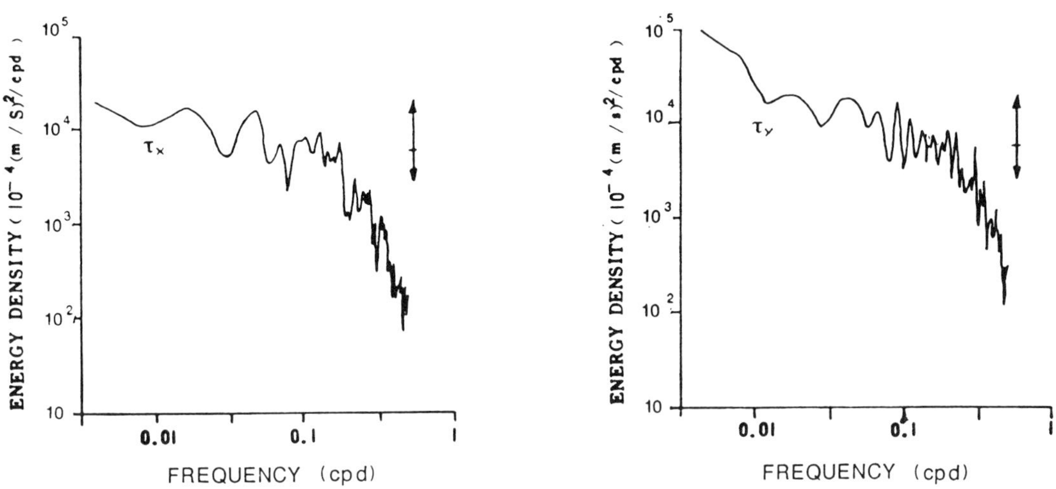

Figure 3: Spectra of the wind-stress components over the shelf (YSC). (The 95% confidence interval is marked)

The wind-stress spectra of YSC are given in Fig.3. The north-south (alongshelf) wind-stress spectrum is in general slightly greater than the east-west (cross-shelf) wind-stress spectrum. Spectral peaks at frequencies corresponding to the peaks of the sea level spectra are also found in one or both of the wind-stress spectra, including the fortnight frequency (Fig. 2). At the high frequency end the wind-stress spectra drops off more sharply than the sea level spectra. This is because, at the high frequency end, the sea level inside the bay is also excited by continental shelf waves (Chen and Su, 1991).

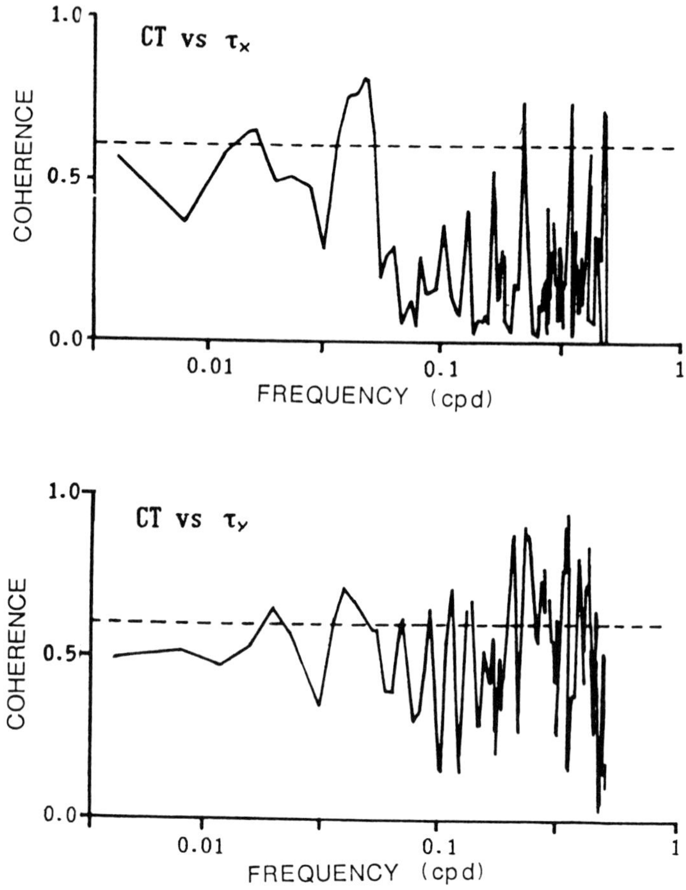

Figure 4: Squared coherences between the coastal sea level at CT and the shelf (YSC) wind-stress components. (The 95% significance level is at the horizontal dashed line)

The squared coherences between the shelf (YSC) wind stress components and sea levels at the coast (CT) and inside the bay (GP) are shown in Figs. 4 and 5. The figures suggest that subtidal motions inside and outside the bay with periods up to 70 days are driven, directly or indirectly, by meteorological forces, except for oscillations near the fortnight period. In the following we will ignore the sea level oscillations near the fortnight period unless otherwise stated. At the coast (Fig. 4) the alongshelf wind-stress component seems to be responsible for most of the sea level oscillations at periods shorter than 10 days, whereas for longer periods both alongshelf and cross-shelf winds are important. Inside the bay (Fig. 5) both wind stress components are influential for the sea level oscillations. For periods longer than 10 days influence of the east-west winds on the bay oscillations is mostly indirect through the coastal sea level oscillations (Figs.2 and 5). For shorter periods this influence also comes via the local winds directly, resulting in sea level differences along the bay (Figs. 2, 4 and 5). As mentioned above direct effect of the north-south components of the local winds is probably of minor importance.

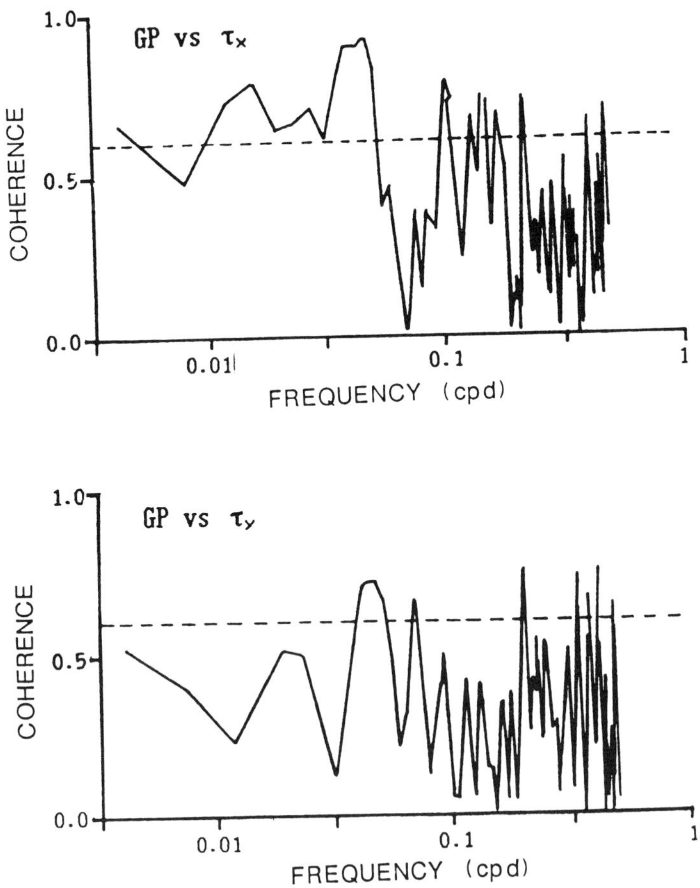

Figure 5: Squared coherences between the bay sea level at GP and the shelf (YSC) wind-stress components. (The 95% significance level is at the horizontal dashed line)

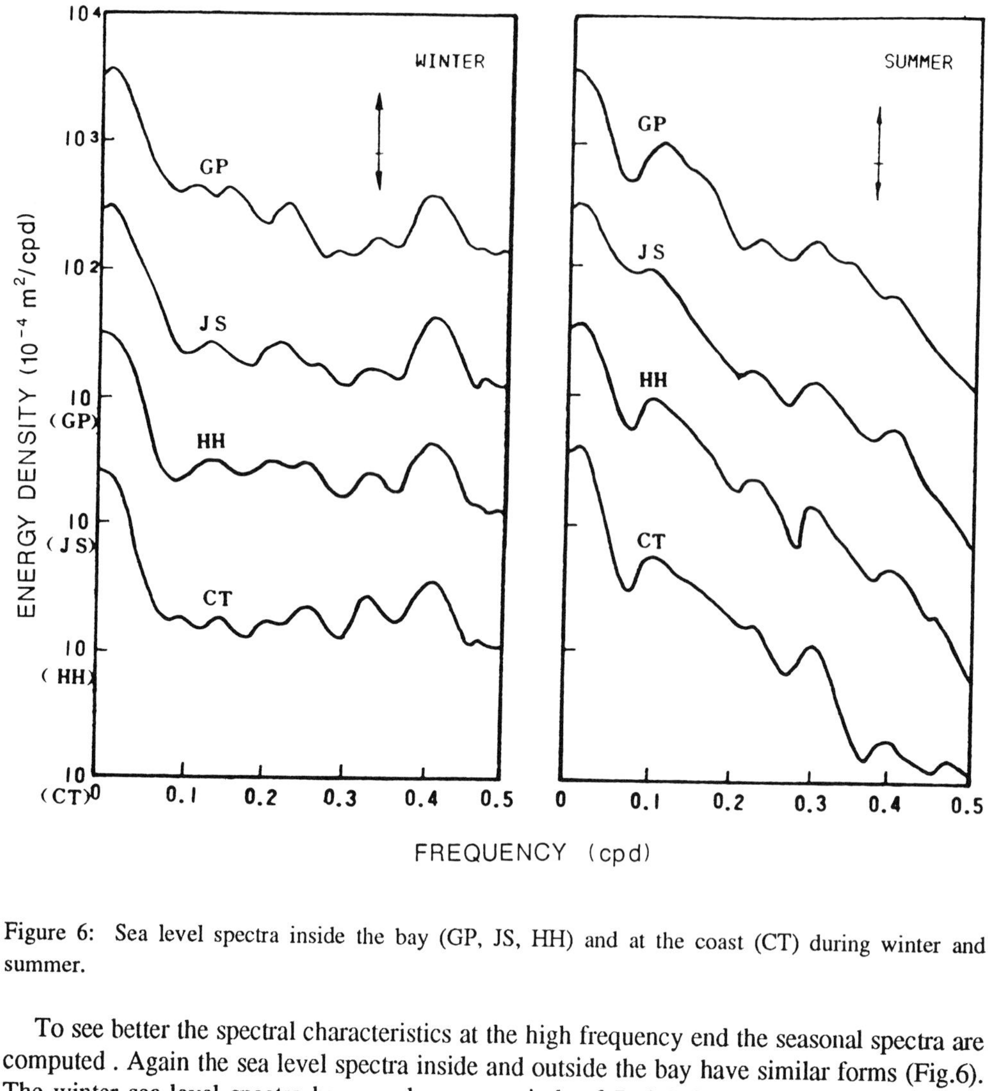

Figure 6: Sea level spectra inside the bay (GP, JS, HH) and at the coast (CT) during winter and summer.

To see better the spectral characteristics at the high frequency end the seasonal spectra are computed . Again the sea level spectra inside and outside the bay have similar forms (Fig.6). The winter sea-level spectra have peaks near periods of 7, 4.5, 3 and 2.5 days, whereas in summer the peaks are close to periods of 10, 4.5 and 3.3 days. Compared with the summer case the winter sea- level spectra are noticeably higher at periods shorter than 5 days. This is probably due to the effect of the shelf waves (Chen and Su, 1991) which are more energetic in winter (Chen and Su,1987). Peaks at periods close to positions of sea level spectra peaks are also found in the respective seasonal spectra of one or more of the four wind stress components (Fig. 7). The local winds are more attenuated than the shelf winds. However, at the coast the east-west winds are more energetic, whereas inside the bay the north-south winds are stronger.

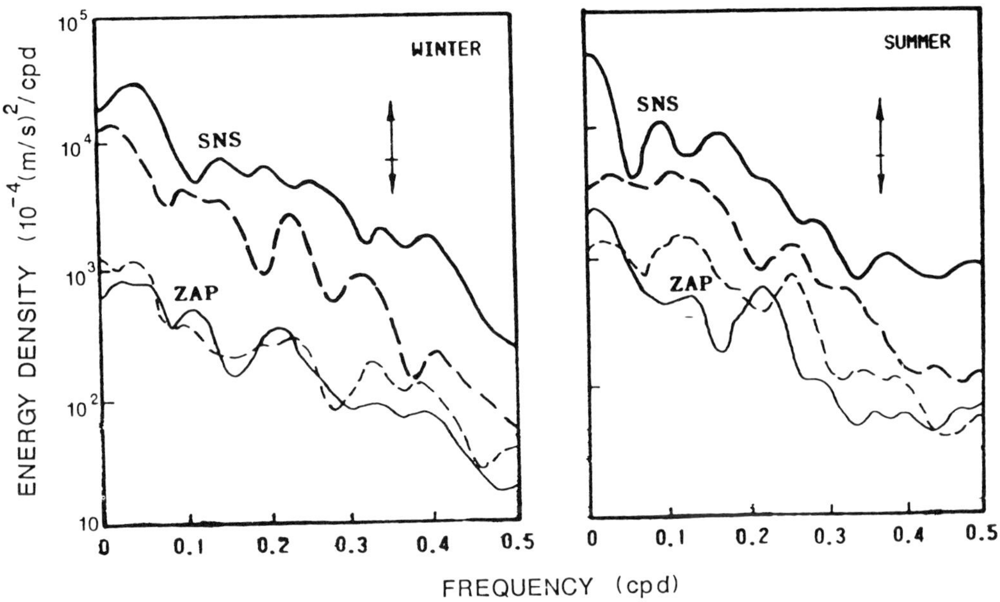

Figure 7: Wind-stress spectra over the shelf (SNS) and inside the bay (ZAP) during winter and summer. (τ_x - solid line; τ_y - dashed line)

Based on the above results we will apply the cross-spectral analysis separately on two bands of the subtidal motions in Hangzhou Bay, one shorter than 10 days and the other longer.

10.4.2 One-dimensional model

Although Hangzhou Bay is rather wide, its axis is aligned more or less in the east-west direction. A one-dimensional vertically integrated model was constructed to study the subtidal sea-level responses of the bay both to the east-west winds over the bay and to the coastal sea-level oscillations (Chen and Su, 1991). Scale analysis showed that the advection, time-dependent and bottom-friction terms can all be neglected from the momentum equation. The solution shows that the variable baywidth affects only the transport.

When only the sea level at the bay mouth is allowed to oscillate with certain frequency the sea level inside the bay oscillates in-phase with the same amplitude. When only a uniform oscillating east-west wind is applied over the bay the sea level inside the bay is found to oscillate at 180 degrees out-of- phase with the wind. In this case the amplitude of the oscillation inside the bay varies as $C\rho_a\tau|x|/\rho gH$. τ is the east-west wind stress (dimension $(m/s)^2$), C the wind stress coefficent, ρ_a the air density, ρ the water density, g the gravitational acceleration, H the average water depth, and x the distance measured from the bay mouth. Therefore, as far as the effect of the local east-west winds is concerned, the model predicts that the amplitude of the sea level response at GP is about twice as large as that at JS.

Equivalently, the sea level inside the bay has a frequency response function of unity with respect to the coastal sea level and a frequency response function of $C\rho_a|x|/\rho gH$ with respect to the east-west wind stress.

10.4.3 Multiple input spectral analysis: periods shorter than 10 days

Analysis given above of the short period oscillations suggests that the effect of the local cross-bay, i.e., north-south, winds is of minor importance. To quantify this statement we first consider the bay oscillations as a system with three inputs, namely, the coastal sea level (η_{CT}) and the two components (τ_x, τ_y) of the local winds. Applying the three-input spectral analysis to the winter data the percentage contributions of η_{CT}, τ_x, τ_y and the residual to the sea level spectrum (0.1-0.5 cpd) at each bay station have been computed (Table 1). We recall that the contributions of η_{CT}, τ_x and τ_y are measured by the moduli of complex numbers. Hence the sum of the four percentage contributions at each bay station can exceed 100 for either season.

Table 1: Percentage contributions to the winter bay sea-level spectra (0.1-0.5 cpd): three-input model.

Station	GP	JS	HH
Local winds			
τ_x	33	14	9
τ_y	14	12	10
Coastal sea level, η_{CT}	78	93	96
Residual	11	3	2

Table 1 shows that at each station the contribution from η_{CT} is overwhelming. It also shows that contributions from τ_x and τ_y are comparable, contrary to the earlier suggestion. However, the three inputs are highly correlated with one another (Fig.8), especially between τ_y and η_{CT}, and with the bay oscillations (e.g., Fig. 9). The composition of the contributions also reflect the effects of these high correlations. At GP, for example, the ratio of the correlated part to the uncorrelated part of each contribution is around 5.4 for τ_y, 1.8 for τ_x, and 1.9 for η_{CT}. As pointed out earlier, for a highly inter-correlated multiple input system large variances may be associated with the estimation of the frequency response functions, which in turn will affect the estimates of the contributions of the related inputs. The three winter gain functions at GP are given in Fig.10. The gain function with respect to η_{CT} has large variations from the one-dimensional result of unity. Similar results are also found at the other two stations. The coastal sea level gradient due to meteorological forcing is usually of the order of 10^{-7} in the alongshelf direction so that the one-dimensional model should be a reasonably good approximation. One possible reason for these large deviations is due to the high correlation among the inputs.

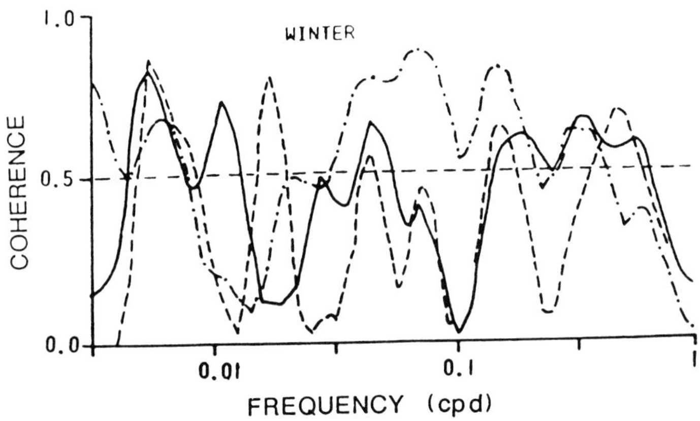

Figure 8: The winter coherence squared between η_{CT} and τ_y (dash-dotted line), η_{CT} and τ_x (dashed line) and τ_x and τ_y (solid line). (The 95% significance level is at the horizontal dashed line)

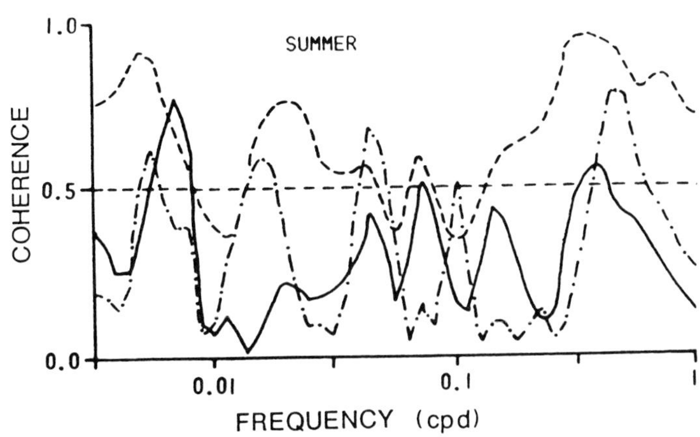

Figure 9: The winter coherence squared between the sea level at GP and (a) η_{CT} (dashed line), (b) τ_x (dash-dotted line) and (c) τ_y (solid line). (The 95% significance level is at the horizontal dashed line)

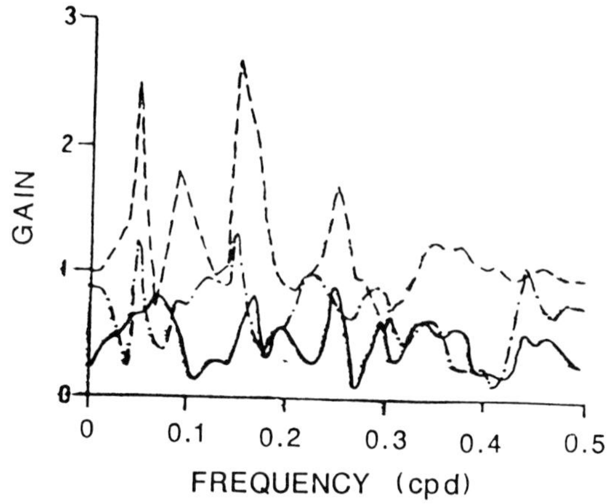

Figure 10: The winter gain function of the sea level at GP with respect to η_{CT} (dashed line), τ_x (dash-dotted line) and τ_y (solid line). (Three-input model results)

Table 2: Percentage contributions to the seasonal bay sea-level (0.1-0.5 cpd): modified three-input system.

	winter			summer		
Station	GP	JS	HH	GP	JS	HH
Local winds						
τ_x	30	17	12	36	31	25
τ_y	10	9	14	6	4	11
Coastal sea level, η_{CT}	64	85	79	65	75	72
Residual	18	7	7	15	11	8

Based on the above arguments we construct a modified three-input system where the frequency response function with respect to η_{CT} is assigned the value of unity a priori. Using slightly modified formulae from those used in computing the values in Table 1 we find, for the modified system, the percentage contributions of η_{CT}, τ_x, τ_y and the residual to the bay sea-level spectra as listed in Table 2. Contribution from η_{CT} remains dominant for all stations. Except during winter at HH, contribution from τ_y is in general less than half of that from τ_x. Station HH is located in shallow water south of an extensive tidal flat. Strong northerly winter winds are likely the reason for, at HH, the contribution from τ_y to be comparable to that from τ_x. At all stations the contribution from the residual is larger than the corresponding value computed from the original system (Table 1). This is expected since the frequency response function related to η_{CT} is predetermined and not by formulae equivalent to a least square regression method.

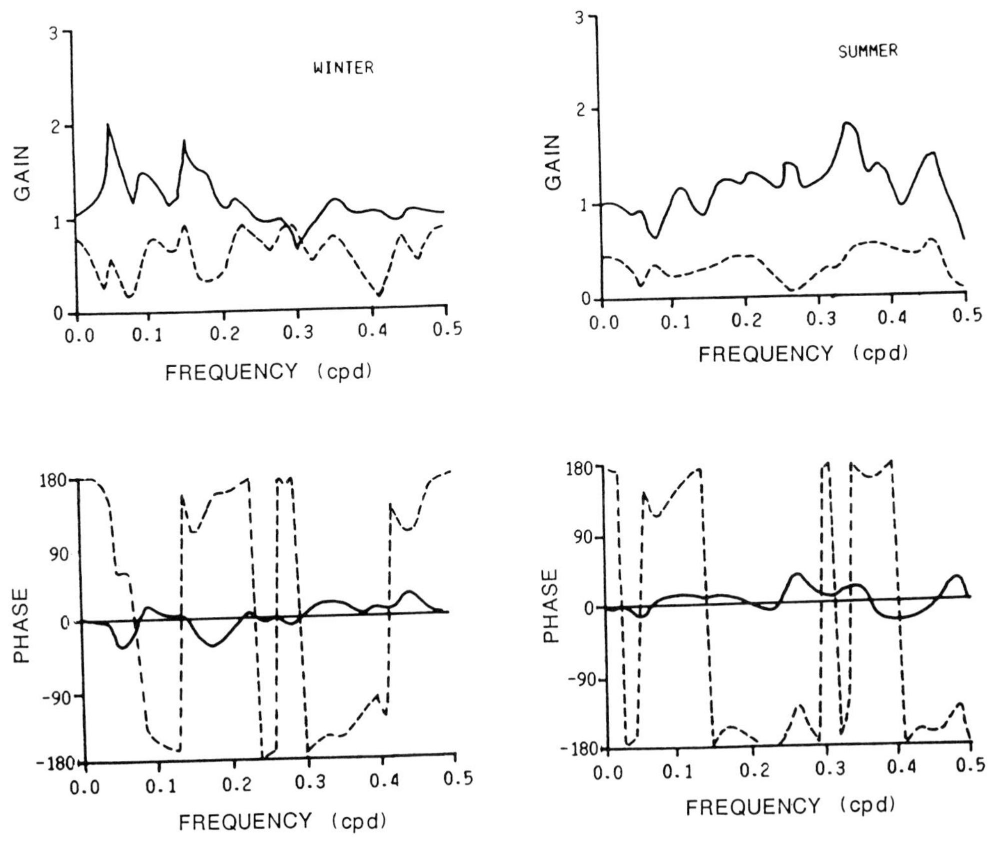

Figure 11: The gain and phase functions of the sea level at GP with respect to η_{CT} (solid line) and τ_x (dashed line). (Two-input model results)

From the above discussions we therefore ignore the effect of the cross-bay wind stress and treat the bay oscillation problem as a two-input systems. The two inputs, η_{CT} and τ_x, are correlated but not too highly. The computed frequency response functions at GP and JS are given in Figs. 11 and 12. The gain functions at GP show reduced oscillations than the previous results obtained from the three-input system (Fig.10). The mean values of the gains over 0.1 to 0.5 cpd are listed in Table 3. The one-dimensional model discussed before predicts a constant unity value for the gain with respect to η_{CT} and $C\rho_a|x|/PgH$ for the gain with respect to τ_x. The computed gain with respect to η_{CT} has average values ranging from 1.03 to 1.24. If we take $C=1.5\times10^{-3}$, $\rho_a/\rho = 1.25\times10^{-3}$, H=10 m and $|x|$ =100 km, the distance of GP from the bay mouth, the predicted value for the gain with respect to τ_x at GP is 1.9×10^{-3} m/(m/s)^2, which is one-third to one-half the values given in Table 3. The predicted gain with respect to τ_x at either JS or HH should be one half that at GP. The computed ratio ranges between 0.35 and 0.83. As to the phase functions the model predicts a constant zero phase with respect to η_{CT} and a constant 180 degree phase with respect to τ_x. The computed results (Figs. 11 and 12) agree well with these predictions. Therefore, overall speaking, the one-dimensional model can be used to describe the response of Hangzhou Bay to the forcing of η_{CT} and τ_x for periods less than 10 days.

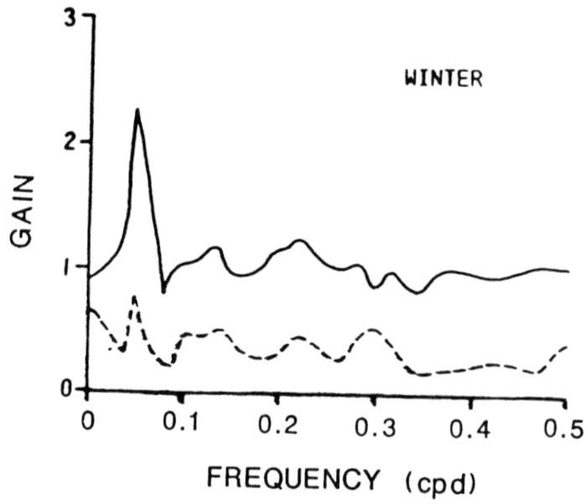

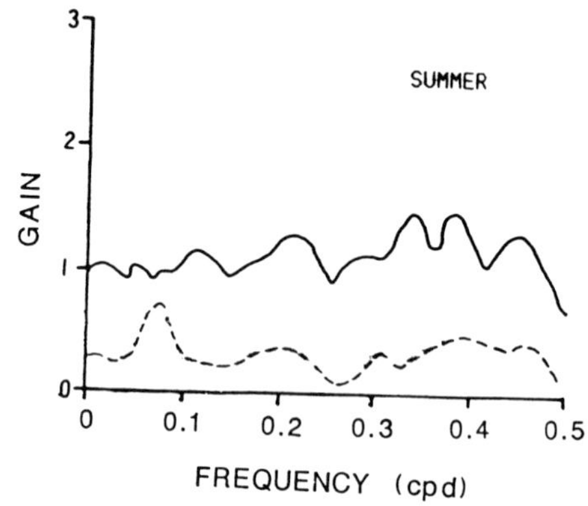

Figure 12: The gain function of the sea level at JS with respect to η_{CT} (solid line) and τ_x (dashed line). (Two-input model results)

Table 3: Mean values of the gains of the bay oscillations over 0.1 to 0.5 cpd.

	Winter			Summer		
Station	GP	JS	HH	GP	JS	HH
Forcing coastal sea level, η_{CT} east-west wind stress, τ_x (10^{-3}m/(m/s)^2)	1.08 6.2	1.03 3.4	1.24 2.2	1.22 3.5	1.15 2.9	1.21 2.7

If we use a larger degree-of-freedom for the spectral computations the oscillations in the cross-spectral results will become smaller. The conclusions remain, however, unchanged.

10.4.4 Multiple input spectral analysis: periods longer than 10 days

The coastal sea level is driven by both the shelf winds and shelf waves propagating down from the north. The latter usually have periods less than 10 days. As discussed earlier, for short period coastal sea level oscillations the alongshelf wind stress component dominates, whereas for long period oscillations the cross-shelf wind stress component becomes equally important. The two-input system used above does not work well here because the two inputs, the coastal sea level and east-west local wind stress component, are now highly correlated. However, as discussed before it is sufficient to use the coastal sea level as the only input to study the long period oscillations inside the bay. Alternatively, we can also use the two components of the shelf wind stress as the inputs, also denoted respectively as τ_x and τ_y in the following discussions. In this way we can also gain some information on the effect of east-west winds over the bay.

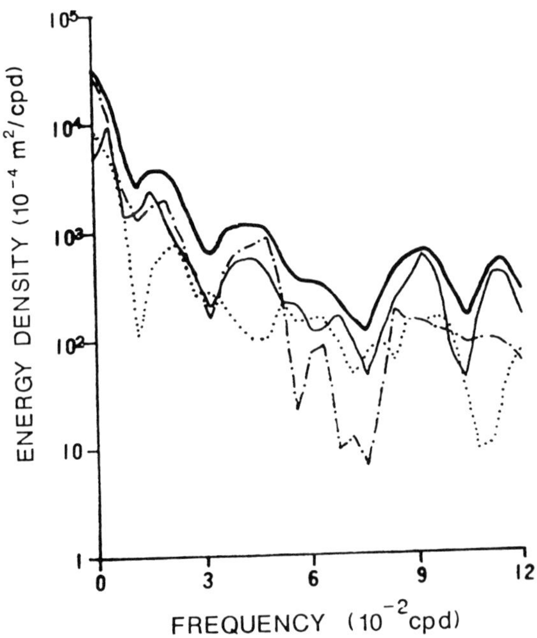

Figure 13: Contributions to the coastal (CT) sea level spectrum (heavy solid line) from τ_x dash-dotted line), τ_y (light solid line) and the residual (dotted line).

It was found from the coherence squared between τ_x and τ_y that the two components are correlated but not too highly. We first apply the two-input model to the coastal sea level oscillations. Contributions from τ_x, τ_y and the residual to the coastal (CT) sea level spectrum are shown in Fig. 13. Evidently the contribution from τ_y dominates over those from both τ_x and the residual at periods shorter than 12 days, while at periods longer than 20 days both wind stress components are important, confirming the arguments given above. Next we apply the two-input model to the bay oscillations. In Fig. 14 we show the results at GP. At periods greater than about 18 days contribution from τ_x to the sea level oscillations at GP is actually more important than that from τ_y. This suggests that, like the short period case studied above, the contribution from the north-south component of the local wind stress is significantly lower than that from the east-west component. We note the dominant contribution from τ_y around the 15-day period. This is an artifact since, as we commented on earlier, the peak of the GP sea-level spectrum at the 15-day period is due mainly to aliasing from the incompletely removed tides. In Table 4 the percentage contributions from τ_x, τ_y and the residual to the sea level spectrum (0.01 to 0.06 cpd) at GP, JS, HH and CT are listed.

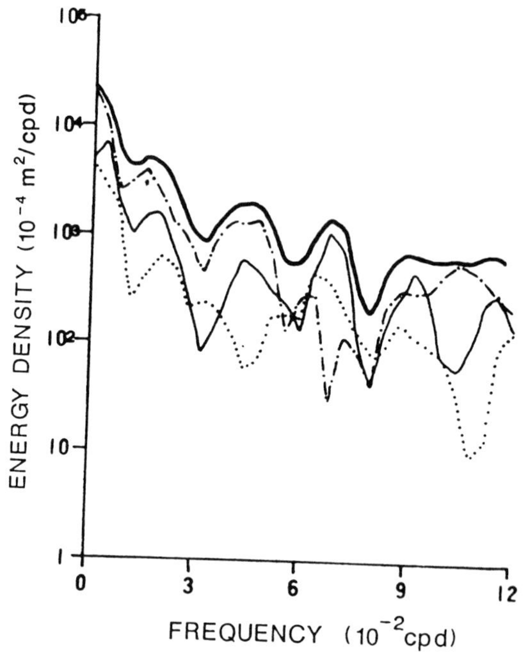

Figure 14: Contributions to the bay (GP) sea level spectrum (heavy solid line) from τ_x (dash-dotted line), τ_y (light solid line) and the residual (dotted line).

Table 4. Percentage contributions to the sea level spectra (0.01 to 0.06 cpd): two-input model.

Station	GP	JS	HH	CT
τ_x	69	69	64	49
τ_y	28	25	33	45
Residual	12	15	13	16

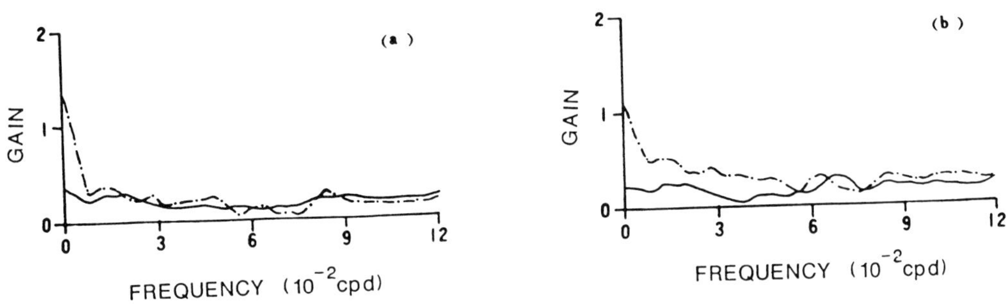

Figure 15. The gain functions of the sea level at (a) CT and (b) GP with respect to τ_x (dash-dotted line) and τ_y (solid line).

The gain functions at CT and GP are shown in Fig. 15. The two gains with respect to τ_y are quite similar, indicating that the contribution from the cross-bay winds to the bay oscillations is of minor importance. The gain with respect to τ_x at GP is higher than that at CT by about 10^{-3} m/(m/s)^2. This difference is about half of the value predicted by the one-dimensional model for the gain at GP due to east-west wind stress over the bay. The phase functions at CT and GP were found to vary close to 180 or -180 degrees. This is expected since both northerly and easterly shelf winds will lower the coastal sea level and the easterly winds over the bay will lower the sea level at GP.

10.5 Summary

Applying multiple input spectral analysis to sea level data of Hangzhou Bay it is found that

a. Coastal sea level has the dominant influence on the subtidal sea level oscillations inside the bay. The local wind stress component along the axis of the bay is also important for the bay oscillations at periods shorter than 10 days. The local cross-bay wind stress component is of minor importance for these oscillations.

b. Linear inviscid one-dimensional model can be used to describe the sea level response of the bay to forcing by both the coastal sea level and the along bay winds.

This study demonstrated that, when the inputs are highly correlated multiple input spectral analysis can encounter difficulty.

10.6 References

Bendat, J.S., and A.G. Piersol, 1980: Engineering Applications of Correlation and Spectral Analysis, John Wiley & Sons.

Chen, D.K., and J. L. Su, 1987: Continental shelf waves along the coast of China, Acta Oceanologica Sinica, 6, 317-334.

Chen, W., and J. L. Su, 1991: Low-frequency sea level fluctuations in the Hangzhou Bay, Acta Oceanologica Sinica, 10, (to appear).

Fang, G. H., and Z. W. Yu, 1985: On calculation and use of daily mean sea level, II. On use of daily mean sea level data, J. Oceanogr. Huanghai & Bohai Seas, 3(2), 1-9 (in Chinese).

Garvine, R., 1985: A simple model of estuarine subtidal fluctuations forced by local and remote wind stress, J. Geophys. Res. 90 (C6), 11945-11948.

Jenkins, G.M., and D.G. Watts, 1968: Spectral Analysis and its Applications, Holden-Day, 525 pp.

Moores, C. N. K., and R. L. Smith, 1968: Continental shelf waves off Oregon, J. Geophys. Res., 73, 549-557.

Wang, D.P., and A. Elliott, 1978: Non-tidal variability in Chesapeake Bay and Potomac River, evidence for non-local forcing, J. Phys. Oceanogr., 8, 225-232.

Walters, R.A., and C. Heston, 1982: Removing tidal-period variations from time-series data using low-pass digital filters, J. Phys. Oceanogr., 12, 112-115.

11

Water level fluctuations in the Atchafalaya Delta, Louisiana: tidal forcing versus river forcing

E M Swenson and C E Sasser

Abstract

The Louisiana coastal marshes were built over the last 8,000 years by the successive formation, abandonment and subsequent subsidence and erosion of delta lobes from the Mississippi River. The Atchafalaya River Delta is one current site of deposition and land building in this process. Sediments of the Atchafalaya River have been filling Atchafalaya Bay since 1950, and new islands first emerged in 1973. This new land is being formed in a coastal region which is currently experiencing land loss at a rate of about 155 km^2 per year. The surrounding marshes are basically fresh due to the presence of the river but also have a water level signal that is strongly influenced by astronomical tides, atmospheric forcing and the Atchafalaya River itself. The resulting water level regime (inundation and frequency) is a key element controlling the marsh vegetation dynamics within this deltaic system.

Data for the analysis came from four recording water level stations located within the system. The general water level pattern was similar for all stations, showing a strong diurnal tidal signal superimposed upon other larger scale events. These larger scale events included both frontal passage events as well as the flooding events of the Atchafalaya River. The river flow explains about 86% of the variation in water levels on an annual time scale and about 45% of the variation on monthly time scales. On time scales of several days the atmospheric forcing becomes important, with the north-south wind component explaining about 62% of the variation in water levels.

The resulting marsh inundation regime shows a logistic shaped curve with low elevations (< -25 cm NGVD) being flooded 100% of the time and high elevations (>50 cm NGVD) only being flooded a few hours a year. The total hours the marsh is flooded, at the mid-elevations (~25 cm NGVD), shows substantial variation among years, with the difference between large flood years and small flood years being on the order of 3,000 hours. In addition, the frequency of flooding shows changes on the order of 250 events per year. Thus, vegetation growing within this system experiences wide ranges in both flooding frequency and duration.

These year to year differences are determined by the flow of the Atchafalaya River.

The hydrologic regime is the major abiotic factor influencing the vegetation distribution within this system. An understanding of the relationship of this regime to the vegetation succession will provide key data on which to base management decisions regarding the implementation of fresh water and sediment diversion projects. Presently, such large scale diversion projects are being proposed for the restoration of the deteriorating Louisiana Coastal Wetlands.

11.1 Introduction

The Louisiana coastal marshes were formed, over the last 8,000 years by the successive formation, abandonment, and subsequent subsidence and erosion of delta lobes from the Mississippi River. The modern delta, which has occupied its present position for about the last 700 years, consists of two depositional areas, (1) the main delta of the Mississippi river and (2) the Atchafalaya river delta. The Atchafalaya delta is the current site of deposition and land building in this process, since the sediments in the main delta are being discharged into the deeper waters of the continental shelf Kolb and Van Lopik (1966); Van Heerden (1983); Baumann et al. (1984). Since 1963, the flow of the Atchafalaya River has been stabilized at 30% of the combined flow of the Mississippi and Red rivers by control structures located in Simmesport, Louisiana. Since 1983, sediments of the Atchafalaya River have been filling Atchafalaya Bay Van Heerden and Roberts (1980). New islands first emerged in 1973, and emergent plants began to colonize as soon as the sediments had accreted to intertidal elevations. These marshes are basically fresh, due to the presence of the Atchafalaya River but also have a water level signal that is strongly influenced by the astronomical tides as well as atmospheric forcing Adams and Baumann (1980). The Atchafalaya Delta represents an area where land building is occurring in a coastal zone which is otherwise experiencing land loss at a rate of about 155 km^2 per year Turner (1990).

One of the key elements in the vegetation dynamics of coastal marshes is the role played by the water level regime. Of particular concern is the duration and frequency of inundation. It has been shown, that the periodic flooding of a marsh system may serve to enhance the productivity of the emergent vascular plants Teal (1962), Hopkinson and Hoffman (1984), to enhance the flux of nutrients between the marsh proper and adjacent waters, and increase the available feeding habitat for small estuarine-dependent fish species Boesch and Turner (1984), Zimmerman and Minello (1984), Minello and Zimmerman (1985). These studies have been largely confined to salt and brackish marsh systems. Field studies investigating the hydrologic regime as it relates to the vegetation, indicate that the degree of inundation may also be an important parameter affecting fresh marsh vegetation. Sasser (1977), Mendelssohn and McKee (1987). Recent work on vegetation within the Atchafalaya islands by Johnson et al. (1985) indicated that the environmental factors which vary most consistently with the vegetation are those which depend directly on hydrology. Their work however, did not include a detailed analysis of the hydrologic characteristics of the islands. In a modeling study of vegetation dynamics within the Atchafalaya Delta Rejmánek et al. (1987) the predicted changes in vegetation distribution were assumed to be strongly related to the flooding regime within the islands. An understanding of the hydrologic regime of this area and its relationship to the vegetation community can provide important management information for use with river diversion projects being designed for the large-scale creation and restoration of Louisiana coastal wetlands.

In this paper, we describe the inundation pattern of the Islands in the Atchafalaya Delta, with emphasis on the frequency and duration of flooding. The relationship between the observed patterns and the major hydrologic forcing functions is also described.

11.2 Study area

The Atchafalaya Bay system (Figure 1) is a shallow estuary fringed by salt and brackish marsh. The delta proper is characterized by a series of lobate islands, which first emerged following the record flood of 1973. The islands became vegetated within two years and have since become important wetland habitat for wildlife and waterfowl. In general, the islands are characterized by an elevation gradient from a high upstream tip, usually vegetated by willow (*Salix nigra*) to a low-lying, arrowhead (*Sagittaria spp.*) marsh. This low-lying marsh is characterized by extensive mud-flats and is vegetated on a seasonal basis. The intermediate elevations are colonized by a mixture of vegetation which is also seasonal in nature. This elevation gradient serves to define a flooding gradient with less flooding at the upstream tip of the island and greater flooding on the mud flats.

11.3 Data and methods

In connection with its mission of Flood Control and Channel Maintenance within the Atchafalaya Delta Complex, the U.S. Army Corps of Engineers (COE) routinely monitors water levels at four locations within this system (Figure 1). The stations of interest included Deer Island (DI), in the Lower Atchafalaya River, the Amerada Hess Platform (AH) in the delta islands proper, and Eugene Island (EI) an offshore station. The fourth station, Belle Island (BI), is located in an oil field on a salt dome, and was not used in this study.

The records of water stage at these stations were obtained from the COE and were machine digitized at one hour intervals. They covered the time period from January 1979 through December 1986. The data were digitized using a "Numonics 2400" digitizing tablet interfaced with an "IBM AT" computer. During digitizing, notes were taken regarding any gaps, anomalous data or time base drift on the records. These notes were used later for data editing.

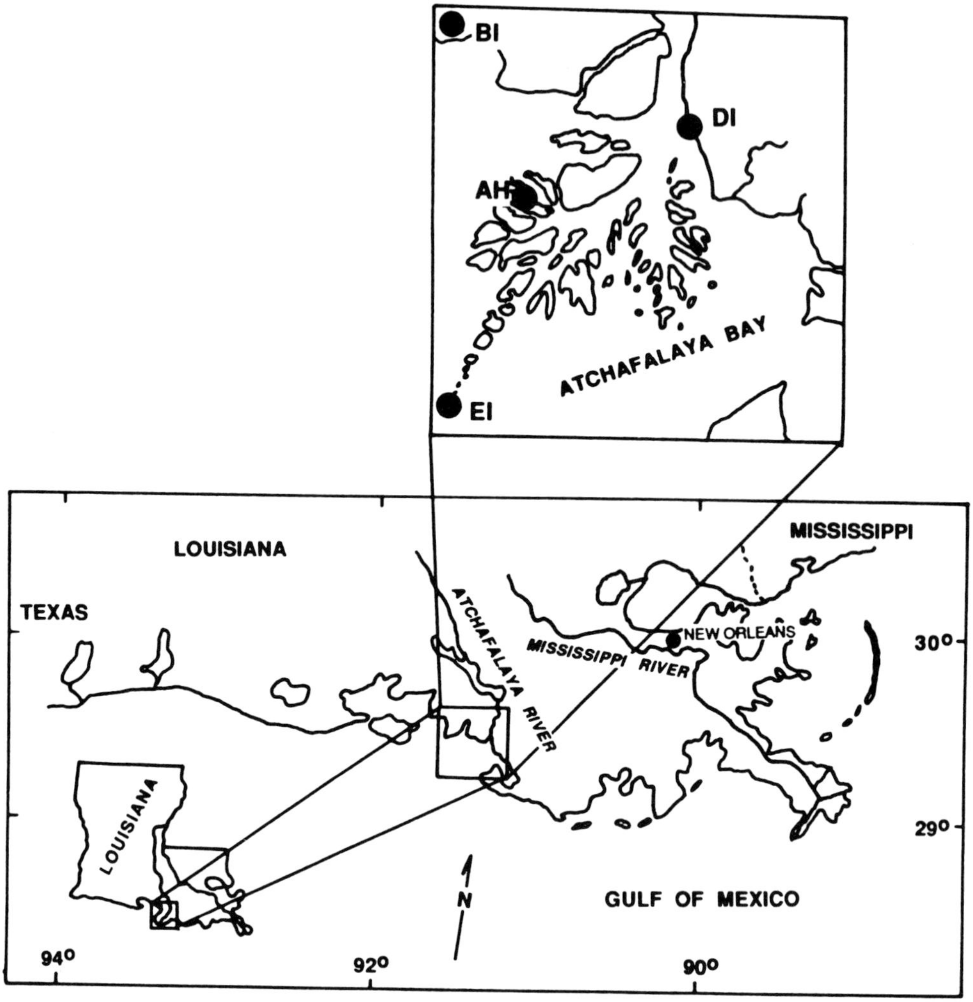

Figure 1. Base map of the Atchafalaya Delta Area showing the locations of the Corps of Engineers tide gages used in the analysis. DI = Deer Island, AH = Amerada Hess Platform and EI = Eugene Island, BI = Bell Isle.

Data analyses were conducted using the LSU Mainframe with "Statistical Analysis System" SAS Institute Inc. (1985 a,b,c). After all of the digitized data had been transferred, a preliminary analysis was performed. During this analysis, the data from all months for any given station were concatenated; any corrections, based upon the digitizing notes, were made; and a date and time variable was created for the data sets. The raw data were then plotted, and any obvious spikes were edited out. In general the data were quite good with few data points edited out, although there were numerous gaps in the records.

In order to average out the slight errors in digitizing, three-hour means were computed from the original one-hour data set. The three-hour means were used in all final analyses.

The final analysis of the data focused on the determination of flooding statistics for the delta islands. Of particular concern were frequency and duration of flooding as a function of elevation, a major controlling factor for the vegetation. The frequency of flooding was calculated by counting the number of individual flooding events within a given elevation class. The duration was calculated by measuring the length of each flooding event. These parameters were computed using 10 cm elevation class intervals. The 5 cm flooding data was then interpolated from the 10 cm flooding data. The flooding statistics were determined on a monthly basis, then summarized as the annual values presented here. Figure 2 presents a schematic representation of the definition of flooding frequency and duration as used in this paper. In this example, there are three flooding events (1, 2 and 3). These flooding events are then subdivided by 10 cm elevation classes. This subdivision yields the following breakdown; three events greater then 0.0, three events greater than 10.0, one event greater then 20.0, and one event greater then 30.0. The duration of each event at a given elevation (in hours) is indicated by the letters (A1, A2, A3, B1, etc).

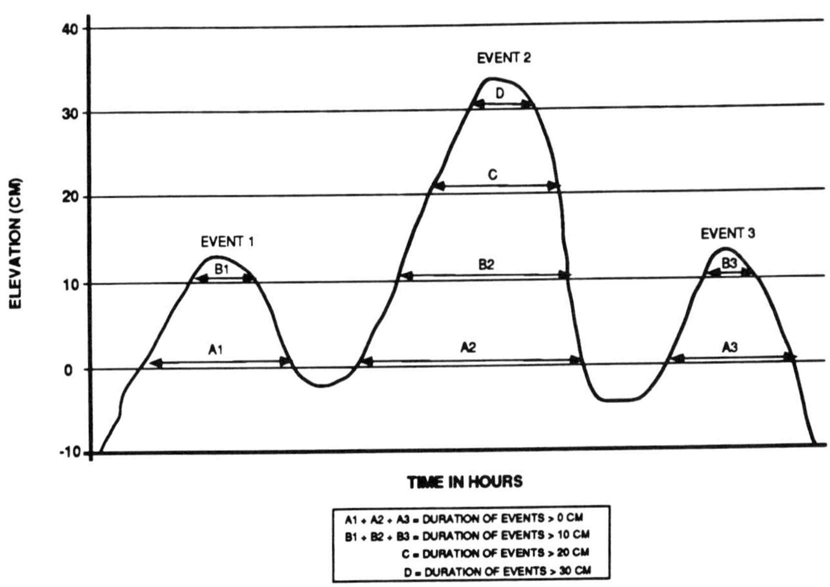

Figure 2: Schematic of a typical water level curve illustrating the measurements made. Each of the flooding events (1, 2, 3) is subdivided into 10 cm elevation classes. The flooding duration for a particular elevation class is indicated by the letters (A1, A2, A3, B1, etc.).

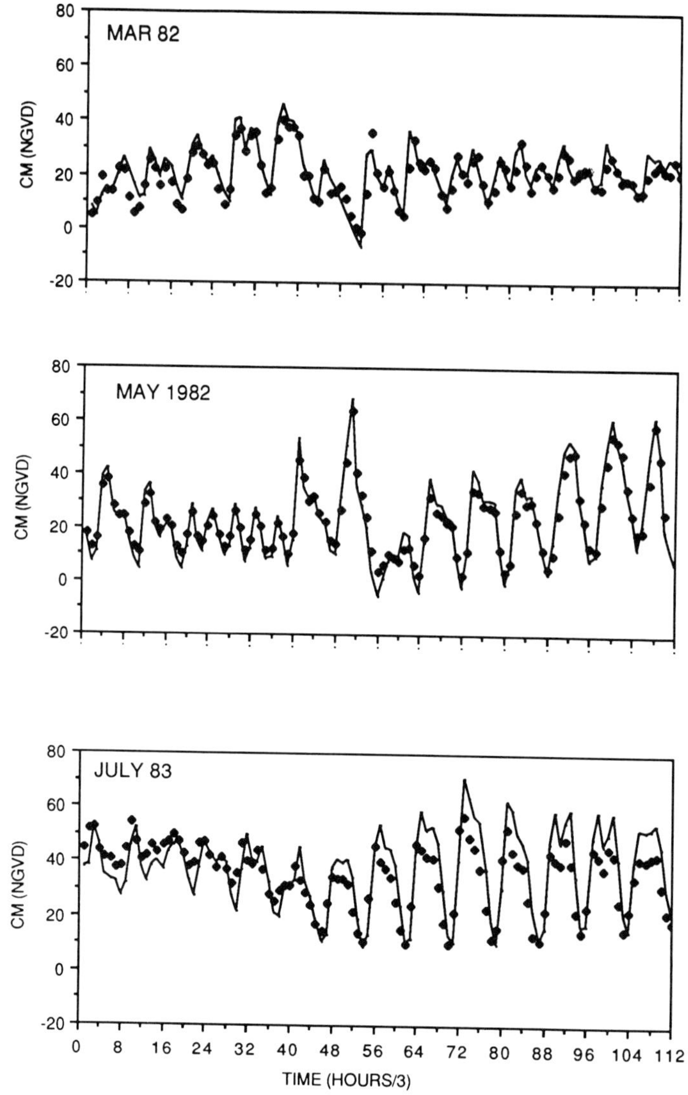

Figure 3: Plot of measured water levels (solid line) from Amerada Hess against water levels predicted from Deer Island data (diamonds), for (top to bottom) March 1982, May 1982 and July 1983. The horizontal axis is time (hours/3) and represents the first two weeks of data from the month. The vertical axis is in centimeters relative to National Geodetic Vertical Datum (NGVD).

In conducting the analysis, it was necessary to have the water level data from the delta islands proper (Amerada Hess) as opposed to offshore (Eugene Island) or from Atchafalaya

Bay (Deer Island). However, there were quite a few gaps in the record from Amerada Hess. In order to fill in these gaps, data from the Deer Island gage was used to predict the missing Amerada Hess data based upon a relationship obtained when data was available from both gages. A total of eight months of data were used in this analysis. A regression analysis between Amerada Hess and Deer Island was conducted using data from five of the eight months to develop a predictive equation. This predictive equation was then tested by predicting the water level curve for the remaining three months data. The resulting predictions were subtracted from the actual measured data to obtain the error (residual) of the prediction. Figure 3 presents examples of the predicted and measured water levels for comparison. The results, indicated that the DI three-hourly values can successfully predict the AH three-hourly values with mean residual errors of less then 10%. However, for the final analysis we were primarily concerned with the calculation of flooding frequency and duration as opposed to the actual point measurements of water levels. In order to further check on the accuracy of the prediction, the flooding frequency and duration were calculated using both the measured data and the predicted data. The results indicated excellent agreement between flooding frequencies and durations calculated from the measured data and the predicted data.

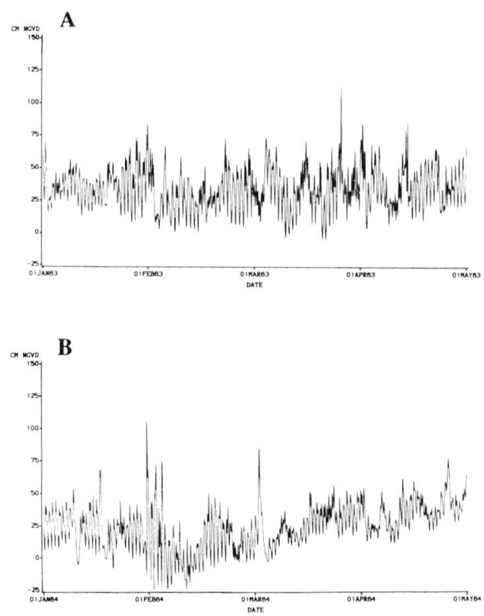

Figure 4: Time series (A) of three-hour mean water levels at the Amerada Hess.gage in centimeters relative to National Geodetic Vertical Datum (NGVD) for January thru April 1983, and (B) time series of three-hour mean water levels at the Amerada Hess.gage in centimeters relative to National Geodetic Vertical Datum (NGVD) for January thru April 1984

11.4 Results and Discussion

Figure 4 presents three hour mean water level data from Amerada Hess for January thru April for 1983 and 1984. Figures 5 and 6 present examples of the daily mean water levels at Amerada Hess and Atchafalaya River flow at Simmesport for 1983 and 1984. The Amerada Hess water level data shown in Figure 3 is typical of the data from all years, which show a strong diurnal tidal signal superimposed upon other larger scale events. These larger scale events include both frontal passages that influence sea level and flooding events of the Atchafalaya River. This type of signal is quite common to water levels measured in coastal Louisiana Byrne et al. (1976), Chuang and Swenson (1981), Swenson and Turner (1987). The river flow data shows a typical pattern with a spring flood, followed by a fairly rapid drop in flow during the summer.

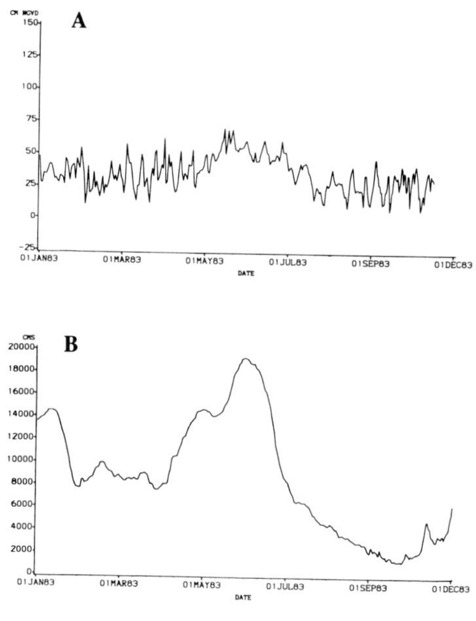

Figure 5: Time series (A) of daily mean water levels at the Amerada Hess.gage in centimeters relative to National Geodetic Vertical Datum (NGVD), and mean daily river flow (B) of the Atchafalaya river at Simmsport in cubic meters per second. Data is from 1983.

Figure 7 presents a plot of mean monthly water levels at AH and mean monthly flow of the Atchafalaya River. It can be seen that the water levels mimic the river flow. Linear regression between the two variables indicate that the river flow accounts for about 35% of the variation in water levels, at this time scale. Figure 8, presents the yearly mean river flow and water level data. There appears to be a close relationship between the water level and

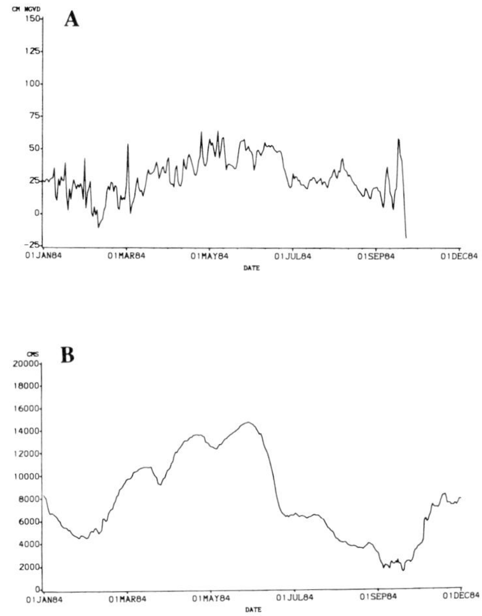

Figure 6: Time series (A) of daily mean water levels at the Amerada Hess.gage in centimeters relative to National Geodetic Vertical Datum (NGVD), and mean daily river flow (B) of the Atchafalaya river at Simmsport in cubic meters per second. Data is from 1984.

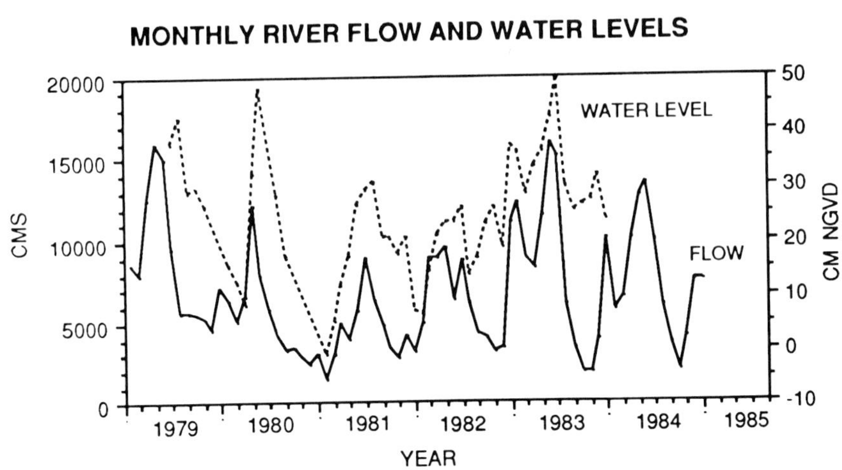

Figure 7: Time series of mean monthly Atchafalaya River flow (solid line) and mean monthly water level (dashed line) at Amerada Hess.

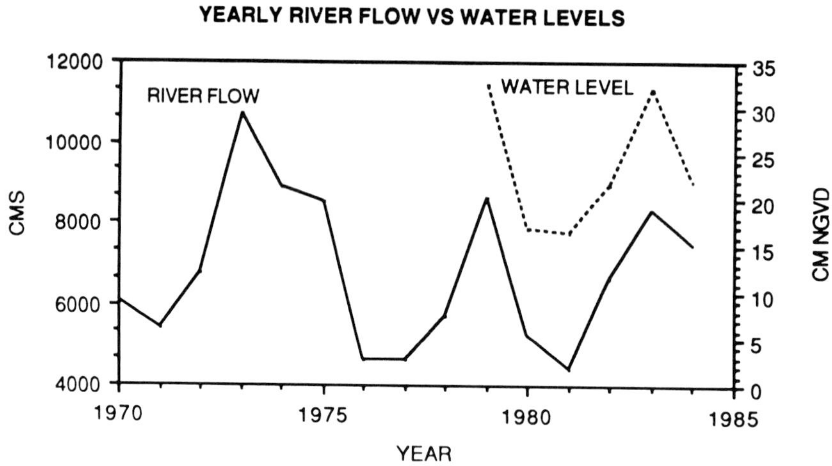

Figure 8: Time series of yearly mean Atchafalaya River flow (solid line) and yearly mean water level (dashed line) at Amerada Hess.

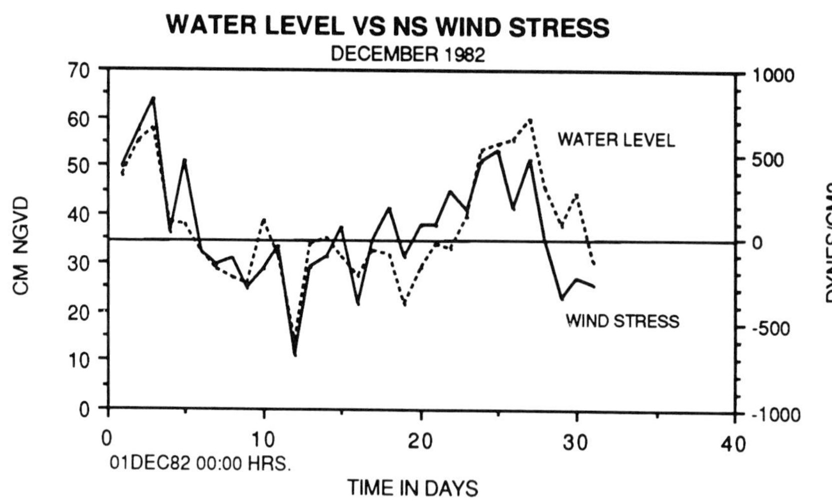

Figure 9: Time series of daily mean north-south wind component (solid line) and daily mean water level (dashed line) at Amerada Hess for December 1982.

the river flow. Although there are only 6 data points (1979 - 1984), regression analysis between the two indicated that river flow may explain about 86% of the water level variation at time scales of a year. At shorter time scales (daily) the river flow explains only 20% of the variation. At these shorter time scales, both the diurnal tide and the effect of frontal passages become important. Figure 9 presents an example of water level and north-south wind component for December, 1982. Linear regression indicates that the wind events explain about 62% of the water level variation, in this particular instance.

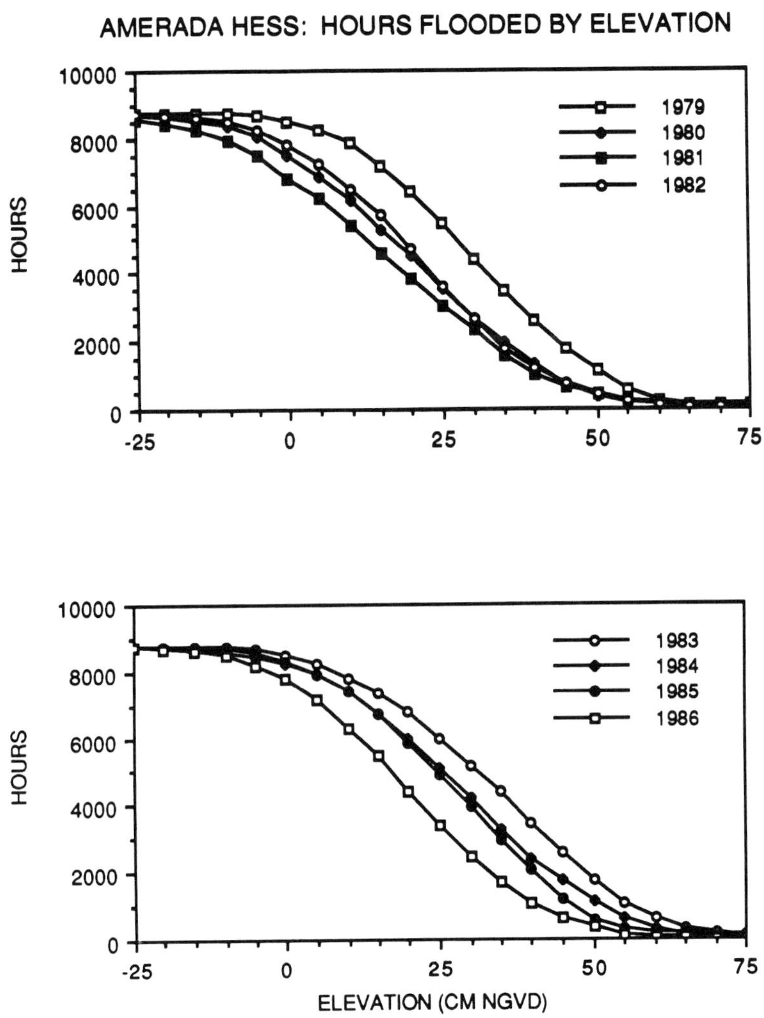

Figure 10. Yearly flooding duration curves for Amerada Hess for 1979 through 1986. Indicated are the number of hours a given elevation was flooded during the year.

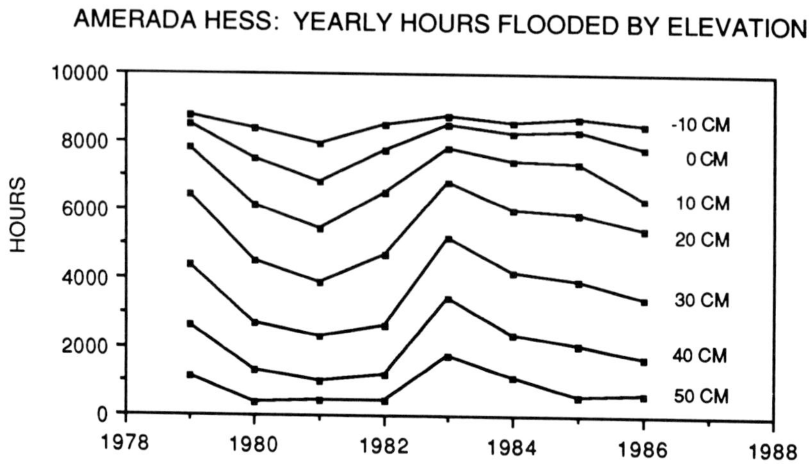

Figure 11: Time series of flooding duration for various elevations at Amerada Hess. Indicated are the number of hours a given elevation was flooded for the year.

Table 1: Yearly flooding duration from the Atchafalaya Delta at Amerada Hess. Indicated are the number of hours flooded for a given elevation (in cm relative to NGVD).

Elevation (cm)	1979	1980	1981	1982	1983	1984	1985	1986
-25	8760	8703	8568	8722	8760	8724	8760	8725
-20	8760	8637	8454	8690	8760	8692	8745	8705
-15	8760	8495	8253	8601	8750	8653	8716	8630
-10	8742	8345	7941	8481	8730	8578	8659	8471
-5	8676	8032	7467	8216	8656	8436	8536	8196
0	8499	7477	6813	7776	8493	8227	8323	7798
5	8238	6878	6198	7204	8227	7908	7934	7165
10	7835	6146	5424	6470	7818	7423	7396	6303
15	7174	5304	4614	5702	7369	6721	6737	5461
20	6409	4520	3867	4709	6787	5992	5867	4419
25	5453	3537	3042	3612	5998	5097	4934	3426
30	4375	2669	2298	2647	5167	4199	3949	2469
35	3457	1940	1602	1740	4371	3268	2950	1682
40	2605	1298	1029	1212	3446	2393	2056	1044
45	1788	717	618	732	2608	1743	1219	656
50	1103	382	411	441	1773	1128	535	357
55	585	193	246	248	1081	659	311	148
60	252	131	192	124	605	296	214	83
65	116	96	144	68	326	147	167	38
70	88	60	132	42	166	95	120	24
75	73	47	67	29	66	44	95	0
80	47	25	0	17	27	13	69	0
85	45	12	0	7	7	13	62	0
90	15	8	0	4	7	13	55	0
95	15	3	0	4	7	13	51	0
100	12	2	0	4	7	9	51	0
125	5	0	0	0	0	0	26	0
150+	0	0	0	0	0	0	3	0

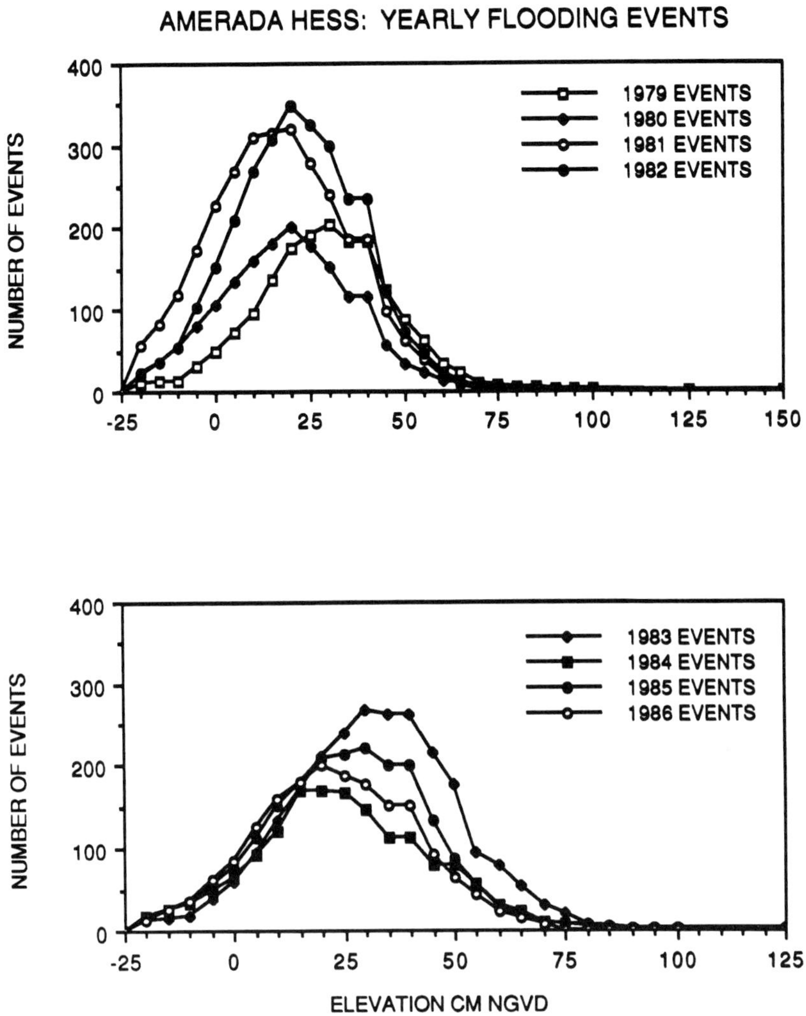

Figure 12: Yearly flooding frequency plots for Amerada Hess for 1979 through 1986. Indicated are the number of times a given elevation was flooded during the year.

The annual flooding duration curves for the Delta Islands (AH) from 1979 through 1986 are presented in Figure 10. These plots present the cumulative distribution of hours flooded as a function of elevation. The elevations are in centimeters relative to the National Geodetic Vertical Datum (NGVD). In general, it can be seen that elevations less than about -25 cm NGVD are flooded 100% of the time (there are 8,760 hours in a common year), the curve then exhibits a logistic shape with very few hours of flooding above elevations of 50 cm NGVD. The variation among years is large, with the differences at mid elevation classes (around 25 cm NGVD) between large flood years (1979, 1983) and smaller flood years (1981) being on the order of 3,000 hours. Thus vegetation growing at these levels would experience a wide range in flooding from year to year. This variation can be seen more clearly in Figure 11. and Table 1 which present flooding duration as a function of elevation.

Figure 12 and Table 2 present the annual flooding frequency data for the Delta Islands (AH) from 1979 through 1986. These plots present the distribution of flooding events as a function of elevation. The curves are bell shaped with the maximum number of events occurring at values of 25 to 50 cm NGVD. As with the flooding duration, the variation among years is large, with the differences at mid elevation classes (around 25 cm NGVD) being on the order of 250 events per year. Thus vegetation growing at these levels would experience a very wide range in both flooding frequency and duration.

Table 2: Yearly flooding frequency from the Atchafalaya Delta at Amerada Hess. Indicated are the number of flooding events for a given elevation (in cm relative to NGVD).

Elevation (cm)	1979	1980	1981	1982	1983	1984	1985	1986
-25	1	1	1	1	1	1	1	1
-20	11	21	56	22	12	17	17	14
-15	13	37	82	36	15	26	26	26
-10	14	54	118	53	18	34	34	37
-5	31	80	172	103	39	49	56	61
0	49	107	226	153	60	65	77	85
5	72	134	268	210	96	93	114	126
10	96	161	309	268	133	121	152	161
15	136	181	315	308	172	171	180	180
20	176	201	321	349	212	170	208	200
25	190	177	280	324	241	169	214	189
30	204	153	240	299	269	147	221	178
35	182	117	187	236	262	113	201	151
40	182	117	187	236	262	113	201	151
45	124	58	97	122	216	80	134	94
50	88	34	61	72	178	80	88	65
55	61	23	39	47	96	56	58	44
60	34	13	18	22	79	31	29	24
65	22	10	9	13	55	22	20	16
70	10	7	2	5	32	11	11	7
75	8	5	1	4	20	8	9	0
80	6	3	0	3	9	4	7	0
85	4	2	0	2	5	3	5	0
90	2	1	0	1	1	1	2	0
95	2	1	0	1	1	1	2	0
100	2	1	0	1	1	1	2	0
125	1	0	0	0	0	0	2	0
150	0	0	0	0	0	0	1	0

11.5 Conclusions

Due to the role of the various forcing functions within the delta system, vegetation growing within the islands must cope with a wide range of flooding frequency and duration, depending upon the elevation at which it is growing. In addition, the major hydrologic forcing changes throughout the year. This results in a pattern of emergence, growth and die-back of the vegetation which follows the yearly water level pattern.

The emergence of the vegetation in the spring is controlled by the duration of the flood peak of the Atchafalaya River. Once the River stage has decreased to a level low enough to allow the mud flats to have some sub-aerial exposure, the vegetation emerges. During the summer months when the vegetation is actively growing, the water level patterns are controlled by the astronomical tides. Recent research Sasser and Fuller (1988) indicated grazing stress within the islands to be an important biotic control of the vegetation. At elevations where the inundation is high, the resulting flooding frequency and duration becomes a major abiotic factor influencing the vegetation growth and distribution. In these instances, the grazing stress, coupled with the physiological (flooding) stress can lead to a reduction in the vegetation cover, particularly for the *Sagittaria* species.

Table 3: Listing of the major hydrologic forces controlling the water levels within the Atchafalaya Delta Islands. Indicated are the major forcing functions, by season, and the hypothesized vegetation response to this forcing function.

SEASON	MAJOR HYDROLOGIC FORCE	VEGETATION RESPONSE
WINTER	FRONTAL PASSAGES	NON-VEGETATED (DORMANT)
SPRING	RIVER INCREASE THEN DECREASE	DROP OF RIVER DETERMINES EMERGENCE OF VEGETATION
SUMMER	ASTRONOMICAL TIDES	VEGETATION DISTRIBUTION INFLUENCED BY FREQUENCY AND LENGTH OF INUNDATION
FALL	ASTRONOMICAL TIDES FRONTAL PASSAGES	ONSET OF STRONG COLD FRONTS DETERMINES DIE BACK OF VEGETATION DUE TO SALINITY INCREASE WHEN SOUTH WINDS MOVE GULF WATER INLAND

In the fall, southerly winds associated with a major front can move more saline Gulf of Mexico water into the delta, flooding the islands. Thus, the vegetation must also cope with periodic salt stress. We have hypothesized that this salinity stress may be responsible for the large-scale die back of vegetation observed during the Fall and early Winter. This die back occurs primarily in the *Sagittaria* and seasonal vegetational communities. The major hydrologic forces and their hypothesized influence on the vegetation are listed in Table 3. In addition, tropical storms and hurricanes, although rare, can serve as important land building forces within the delta Rejmánek, M. et al. (1988).

This research represents the first step in describing the hydrologic regime of the Atchafalaya Delta islands. We have thus far limited our analysis to a statistical description of the flooding characteristics of the islands, the major abiotic factor influencing the vegetation distribution in an area of Coastal Louisiana with a large fresh water and sediment input. A more complete understanding of the relationship between the hydrologic regime and vegetation succession with particular emphasis on the hypothesized salinity stress from fall frontal activity is the topic of proposed future studies. In addition, further detailed analysis of the existing water level data using spectral techniques should be conducted. The long-term time series water level records discussed in this paper when coupled with the existing meteorological data should yield a more complete understanding of the forcing mechanisms within the delta environment. Possibly some sort of predictive model could be developed. Such information will provide a data base on which to base management decision regarding fresh water and sediment diversion projects currently being considered as a means of restoring marshes in Coastal Louisiana.

Acknowledgements

The authors thank the U. S. Army Corps of Engineers, particularly Mr. Robert Cunningham, for supplying us with copies of the raw data records from the gages used in this study. We also acknowledge the assistance of Ms Jessica Kastler who spent countless hours digitizing the raw records while keeping careful notes on the process. Her diligence made transferring and editing the data a much simpler task. We also acknowledge comments made by two anonymous reviewers.

This study was supported through the National Marine Fisheries Service, Habitat Conservation Division, U. S. Department of Commerce Additional support came from the Louisiana Board of Regents Research and Development Program and the Louisiana Sea Grant College Program, a part of the National Sea Grant College program maintained by the National Oceanic and Atmospheric Administration, U.S. Department of Commerce.

11.6 References

Adams, R. D. and R. H. Baumann. 1980. Land Building in Coastal Louisiana: Emergence of the Atchafalaya Bay Delta. Center for Wetland Resources, Louisiana State University, Baton Rouge, LA

70803. Sea Grant Publication LSU-SG-80-02.

Baumann, R. H., J. W. Day and C. A. Miller. 1984. Mississippi Deltaic wetland survival: Sedimentation versus coastal submergence. Science 224: 1093-1095.

Boesch, D. F. and R. E. Turner. 1984. Dependence of fisheries species on salt marshes: the role of food and refuge. Estuaries 7(4A):460-468.

Byrne, P., M. Borengasser, G. Drew, R. Muller, B. L. Smith, Jr., and C. Wax. 1976. Barataria Basin: Hydrologic and Climatologic Processes. Center for Wetland Resources, Louisiana State University, Baton Rouge, Louisiana 70803. Sea Grant Publication LSU-T-76-012. 175 pp.

Chuang, W. S., E. M. Swenson. 1981. Subtidal Water Level Variations in Lake Pontchartrain, Louisiana. Journal Geophysical Research 86(C5):4198-4204.

Hopkinson, C. S. , Jr. and F. A. Hoffman. 1984. The estuary extended-a-recipient-system study of estuarine outwelling in Georgia. Pages 313-330 in V. S. Kennedy ed. The Estuary as a Filter. Academic Press, Orlando, Florida.

Johnson, W. B., C. E. Sasser and J. G. Gosselink. 1985. Succession of Vegetation in an Evolving river delta, Atchafalaya Bay, Louisiana. Journal of Ecology 73:973-986.

Kolb, C. R. and J. R. Van Lopik. 1966. Depositional environments of the Mississippi River deltaic plain - southeastern Louisiana. pp. 17-61 in Shirley, M. L. ed., Deltas, Houston Geologic Society, Houston, Texas

Mendelssohn, I. A. and K. L. McKee. 1987. Experimental field and greenhouse verification of the influence of saltwater intrusion and submergence on marsh deterioration: mechanisms of action. ch. 8 in Turner, R. E. and D. R. Cahoon, editors. Causes of Wetland Loss in the Coastal Central Gulf of Mexico. Volume II: Technical Narrative. Final report submitted to Minerals Management Service, New Orleans, LA. Contract No. 14-12-0001-30252. OCS Study/MMS 87-0120. 400 pp.

Minello, T. J. and R. J. Zimmerman. 1985. Differential selection for vegetative structure between juvenile brown shrimp (*Penaeus aztecus*) and white shrimp (*P. setiferus*), and implications in predator-prey relationships. Estuarine, Coastal and Shelf Science. 20: 707-716.

Rejmánek, M., Sasser, C. E. and Gosselink, J. G. 1987. Modeling Vegetation Dynamics in the Mississippi River deltaic plain. Vegetatio 69, 133-140.

Rejmánek, M. Sasser, C. E. and Peterson, G. W. 1988. Hurricane-Induced Sediment deposition in a Gulf Coast marsh. Estuarine, Coastal and Shelf Science. 27:217-222.

SAS Institute, Inc. 1985a. User's Guide: Basics, Version 5 Edition. Cary, NC: Sas Institute Inc., 1290 pp.

SAS Institute, Inc. 1985b. User's Guide: Statistics, Version 5 Edition. Cary, NC: Sas Institute Inc., 956 pp.

SAS Institute, Inc. 1985c. SAS/Graph User's Guide, Version 5 Edition. Cary, NC: Sas Institute Inc., 596 pp.

Sasser, C. E. 1977..Distribution of vegetation in Louisiana coastal marshes as response to tidal flooding. M.S. Thesis. Louisiana State University, Baton Rouge, Louisiana, 70803. 40 pp.

Sasser, C. E. and D. A. Fuller, eds. 1988. Vegetation and Waterfowl Use of Islands in Atchafalaya Bay. Final report submitted to Louisiana Board of Regents, 150 Riverside Mall, Suite 129, Baton Rouge, LA. Contract No. 86-LBR/018-B04.

Swenson, E. M. and R. E. Turner. 1987. Spoil Banks: Effects on a Coastal Marsh Water Level Regime. Estuarine, Coastal and Shelf Science. 24:599-609.

Teal, J. M. 1962. Energy flow in the Salt Marsh Ecosystem of Georgia. Ecology 43: 614-624.

Turner, R. E. 1990. Landscape Development and Coastal Wetland Loss in the Northern Central Gulf of Mexico. Amer. Zool., 30:89-105.

Van Heerden, I. L. 1983. Deltaic Sedimentation in eastern Atchafalaya Bay, Louisiana. Center for Wetland Resources, Louisiana State University, Baton Rouge.

Van Heerden, I. L. and H. H. Roberts. 1980. The Atchafalaya delta: rapid prodegradation along a traditionally retreating coast (southcentral Louisiana). Zeitschrift fur Geomorphologie. 38: 188-201.

Zimmerman, R. J. and T. J. Minello. 1984. Densities of Penaeus aztecus, Penaeus setiferus and other natant macrofauna in a Texas saltmarsh. Estuaries 7(4A):421-433.

12

Modelling of low-frequency salinity variations in the St. Lawrence Estuary

K T Tee

Abstract

Low-frequency salinity oscillations with the period of 40-50 days in the St. Lawrence Estuary were examined using current meter measurements in 1982, and a two-dimensional (cross-sectional averaged) numerical estuarine circulation model. Forcing of the model includes low-frequency variations of wind, ice and freshwater runoff from January to September, 1982. Results of the model simulation and field observation consistently show that the salinity oscillation (1) is forced mainly by the along-channel wind stress with maximum salinity corresponding to maximum wind forcing in downstream direction, (2) occurs earlier in the middle estuary, and (3) occurs earlier at deeper locations in the middle and lower estuaries. The wind induced upwelling and downwelling are the major factor producing the salinity variations.

12.1 Introduction

The St. Lawrence Estuary is a long and deep estuary (Fig. 1). Its length from Quebec City, where salt intrusion ends, to Pointe-des-Monts, where the coastline suddenly diverges, is about 400 km. An abrupt rise in the Laurentian Channel near Tadoussac in the middle estuary separates the upper estuary, with typical depths of 20-50 m, and the lower estuary, with typical depths of 200-300 m. The bottom topography in the upper estuary is fairly complex with a deep channel in the northern section and a shallow channel in the southern section.

Studies of subtidal variations in other coastal estuaries were carried out mainly in shallow water (generally less than 15 m), and for periods between 2 and 30 days (Weisberg, 1976; Weisberg and Sturges, 1976; Elliott, 1978; Elliott et al., 1978; Wang and Elliott, 1978; Wang, 1979a,b; Wong and Garvine, 1984; Vieira, 1986; Smith, 1977; Schroeder and Wiseman, 1985; Walters, 1982; Goodrich et al. 1987). Seasonal variations of sea level data were discussed by Wang (1979a) and Walters (1982). The response of the St. Lawrence Estuary to atmospheric and buoyancy forcing is very different from those of the shallow estuaries. This is because the St. Lawrence Estuary includes the deep Laurentian Channel where frictional effects are small and stratification is strong. The response is expected to have a strong baroclinic component associated with the vertical density gradient.

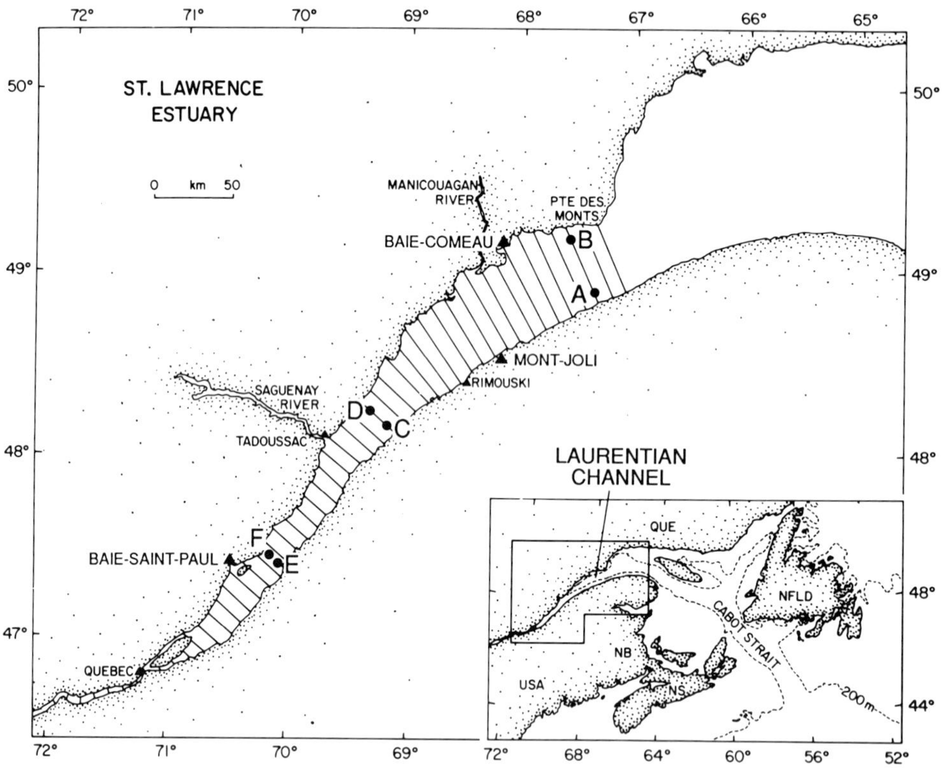

Figure 1: The location and the numerical grid system of the St. Lawrence Estuary. The current meter moorings at stations A, B., C., D, E and F are indicated by solid dots.

For the St. Lawrence Estuary, there exist two sets of long-term (>100 days) current meter data, one near the mouth of the estuary carried out in 1979 (El-Sabh et al., 1982; Mertz et al. 1988, 1989); and another, carried out in 1982, involves simultaneous observations in the upper, middle and lower estuaries (Tee, 1989). Both observations show subtidal oscillations with the periods of 10-15 days and 80 days (Tee, 1990). The 10-15 days oscillation is induced by meteorological forcing. The forcing of the 80 days oscillation is unclear.

In addition to these oscillations, the 1982 observations also show an oscillation with the period of 40-50 days. A simple analysis indicated that the oscillation in the middle estuary occurred earlier than that in the upper and lower estuaries. Because of short current meter records, an accurate description of the oscillation using spectral analysis cannot be carried out.

One of the objectives of this study is to confirm this initial occurrence of the oscillation in the middle estuary using a two-dimensional (cross-sectionally averaged) estuarine circulation model. The observed vertical variation of the oscillation will also be examined and compared with the model result.

Section 12.2 describes the observed low-frequency salinity variations in the St. Lawrence Estuary. The formulation and forcing of the numerical model are described in section 12.3. The comparison between observed and computed low-frequency salinity variations is examined in section 12.4.

12.2 The observation of low-frequency salinity variations

In 1982, fourteen current meters were moored from April to September at six stations along the St. Lawrence Estuary: stations A and B near Pointe-des-Monts in the lower estuary, stations C and D near Tadoussac in the middle estuary, and stations E and F near Baie-Saint-Paul in the upper estuary (Fig.1). There were two Aanderaa RCM-5 current meters at each station except station C, where there were four Aanderaa RCM-5 current meters. At the stations in the middle and lower estuaries (A-D) the meters were moored at 20 and 30 meters from the surface. Another two meters at station C were moored at 40 and 50 m. In the upper estuary, the meters were moored at 14 and 19 m at station E and at 16 and 21 m at station F. All the current meters functioned properly except those at station F, where the lower current meter was lost and the upper current meter contained a short record (62 days) which is not used in this study.

The data were collected at 30 minute intervals. The currents were resolved into along-channel and cross-channel components. The along-channel axis is directed from the head to the mouth of the estuary, and the cross-channel axis from northwest to southeast shores. To study the low-frequency oscillations, the time series of salinity, temperature and current components were first low-passed with a Cartwright filter (half power at 35 h) and subsampled at 6 h intervals. The low-passed data were then fitted (least squares method) to 10th order polynomials. The low-frequency variation of the temperature is closely correlated with the salinity variation, with high (low) temperature associated with low (high) salinity. Because the low-frequency velocity variation has strong signals of spring-neap variation generated by tidal rectification, and/or seasonal variation generated probably by freshwater runoff (Tee, 1989), the 40-50 days signal in the velocity variation cannot be identified clearly. In the following, only the low-frequency oscillation in salinity is discussed and compared with the model prediction.

Figure 2 shows the low-frequency components of the along-channel wind velocity and the atmospheric pressure at Mont Joli (Fig.1), and the measured salinity at current meter stations. An example of the 40-50 day oscillation is the salinity maxima around days 115, 150, 200 and 250 and minima around days 130,180 and 220 at station C. We can see from the figure that

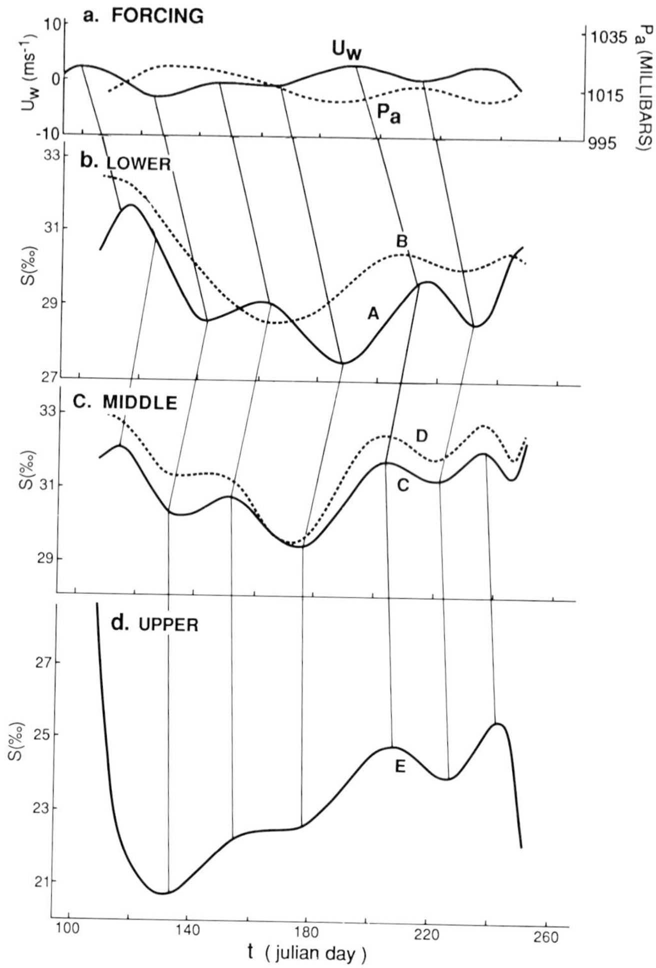

Figure 2: The low-frequency oscillations of the atmospheric forcing and the salinity for the upper current meters in 1982. (a) The along-channel wind velocity u_w, (solid) and the atmospheric pressure p_a (dashed). (b) The salinity at stations A (solid) and B (dashed) in the lower estuary. (c) The salinity at stations C (solid) and D (dashed) in the middle estuary. (d) The salinity at station E in the upper estuary. Positive U_w indicates the wind blowing toward the mouth of the estuary. The forcing and propagation of the 40- to 50-day oscillation is shown. The wind velocity U_w in (a) has a maximum at days 100 which corrects the correspond day of 110 shown previously (Fig.11 of Tee, 1989). This correction is due to the use of longer wind record in the present study (see Fig.5).

there are some apparent correspondences between the salinity variation and the atmospheric forcing. The correlation between the salinity and wind velocity is positive: maximum salinity corresponds to maximum wind velocity. The salinity maxima or minima at stations C and D in the middle estuary occur earlier than those at stations A and B in the lower estuary and at station E in the upper estuary.

The vertical variation of low-frequency salinity data in the St. Lawrence Estuary was not examined in the previous studies (Tee, 1989). Here, it was found that the oscillations at stations A, B, C, and D occurred earlier at deeper current meters, and those at station E occurred slightly later at deeper meter. The magnitudes of the variation decrease with depths. An example of the upward propagation of the maxima or minima, and the decreases of the low-frequency signals at station C is shown in Figure 3.

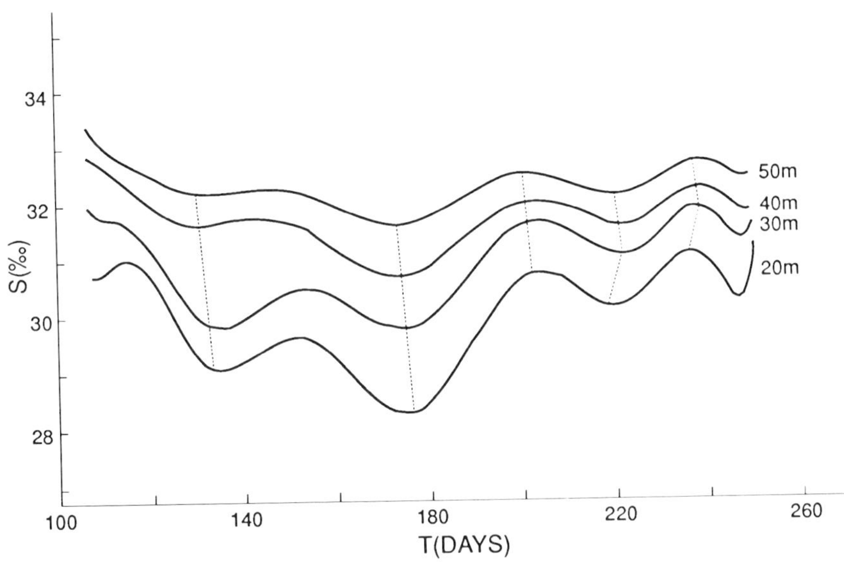

Figure 3: The vertical structure (20, 30, 40, 50 m) of the observed low-frequency salinity variation at station C in the middle estuary. The upward propagation of the salinity variation is shown.

12.3 A numerical model and forcing functions

A two-dimensional (cross-sectional averaged) numerical model was used by Tee and Lim (1987) to study the freshwater pulse in the St. Lawrence Estuary. The model is modified here to include wind forcing and ice formation and melting. For detailed descriptions of the numerical model, the reader is referred to Tee and Lim (1987). Only those aspects that are related directly to this study are outlined here.

12.3.1 Governing equations

In the Cartesian co-ordinates, the governing equations of the model, averaged over the width of the channel, are:

$$\frac{\partial u}{\partial t} + \frac{1}{B}\frac{\partial (Bu^2)}{\partial x} + \frac{\partial}{\partial z}(uw) = -\frac{1}{\rho_0}\frac{\partial p}{\partial x} + \frac{\partial}{\partial z}\left(N\frac{\partial u}{\partial z}\right) \tag{12.1}$$

$$\frac{1}{B}\frac{\partial (Bu)}{\partial x} + \frac{\partial w}{\partial z} = 0 \tag{12.2}$$

$$\frac{\partial s}{\partial t} + \frac{1}{B}\frac{\partial (Bsu)}{\partial x} + \frac{\partial}{\partial z}(sw) = \frac{\partial}{\partial z}\left(K\frac{\partial s}{\partial z}\right) \tag{12.3}$$

where x is the horizontal co-ordinate measured seaward from the head of the estuary; z is the vertical co-ordinate measured upward from the mean sea level; u and w are the velocity components in the x and z directions, N and K are the coefficients of vertical eddy viscosity and diffusivity, B is the width of the estuary, t is time, s is salinity, ρ_0 is the density of fresh water, and p is pressure.

The density (ρ) is related to the salinity by:

$$\rho = \rho_0 (1 + 7.57\times 10^{-4} s) \tag{12.4}$$

where s is expressed in terms of parts per thousand (ppt). The pressure is derived from the hydrostatic approximation,

$$p(z) = p_s + \int_z^0 g\rho dz \tag{12.5}$$

where p_s is the pressure at surface, and g is gravity.

The St. Lawrence Estuary is subdivided into 49 sections with constant horizontal grid spacing (Δx) of 9.5 km. There are twelve levels with variable spacings in the vertical. Beginning from the surface, the vertical grid spacing (Δz) are 10 m each for the first six spacings, 20 m each for the next two spacings, 30 m for the following spacing and 40 m each for the last three spacings.

The vertical eddy coefficients, N and K, are taken to be a function of the Richardson number:

$$N = a + b(1 + 7R_i)^{-1/4} \tag{12.6}$$

and

$$K = (1/2)a + b(1 + R_i)^{-7/4} \tag{12.7}$$

where

$$R_i = - g(\partial\rho/\partial z)/(\partial u/dz)^2 \tag{12.8}$$

is the Richardson number, and a and b are constants. These forms of N and K were chosen by Bowden and Hamilton (1975) for their studies of the estuarine circulation. The constants a and b are chosen to be $3\text{x}10^{-4}$ m^2s^{-1}, the values found to give the best fit between computed and observed salinity in February and May of 1963 in the St. Lawrence Estuary (Tee and Lim, 1987).

The bottom boundary conditions are:

$u_n = 0$ (zero normal velocity) (12.9)
$N\partial u/\partial z = \lambda u_b$ (linear bottom friction) (12.10)
$\partial s/\partial z = 0$ (no salt flux) (12.11)

where u_n is the normal velocity at the bottom, u_b is the horizontal velocity near the bottom taken to be the value at the deepest level, and λ is the linear friction coefficient taken to be 0.001 ms^{-1}. At the surface (z=0), the boundary conditions are:

$w = 0$ (rigid-lid condition) (12.12)
$\rho N\partial u/\partial z = \tau_x$ (wind stress) (12.13)
$\rho K\partial s/\partial z = -[\rho_i(s_i - s)\partial h/\partial t]$ (salt flux) (12.14)

where τ_x is the along-channel component of the wind stress, h is the ice's thickness, ρ_i = 0.0009 kg m^{-3} is the ice density, and s_i is the ice salinity taken simply to be zero. The salt flux is negative if the ice is melting and positive when it is forming. The simple boundary conditions given in (12.13) and (12.14) do not take into account complex interaction between ice and water, which probably affects significantly the circulation in the estuary during part of January and February when a large area of the estuary was covered with ice. However the inclusion of salt flux (12.14) in the model probably produces the first order effect of ice on the salinity variation in the estuary.

Although the comparison of observed and model salinity is carried out in the ice-free season, the ice formation and melting are included in the model because the observation in

April can be affected by ice melting in March. Note that the response of the estuary, estimated by using the ratio of length scale (500 km) to velocity scale (0.1 ms^{-1}), is about two months.

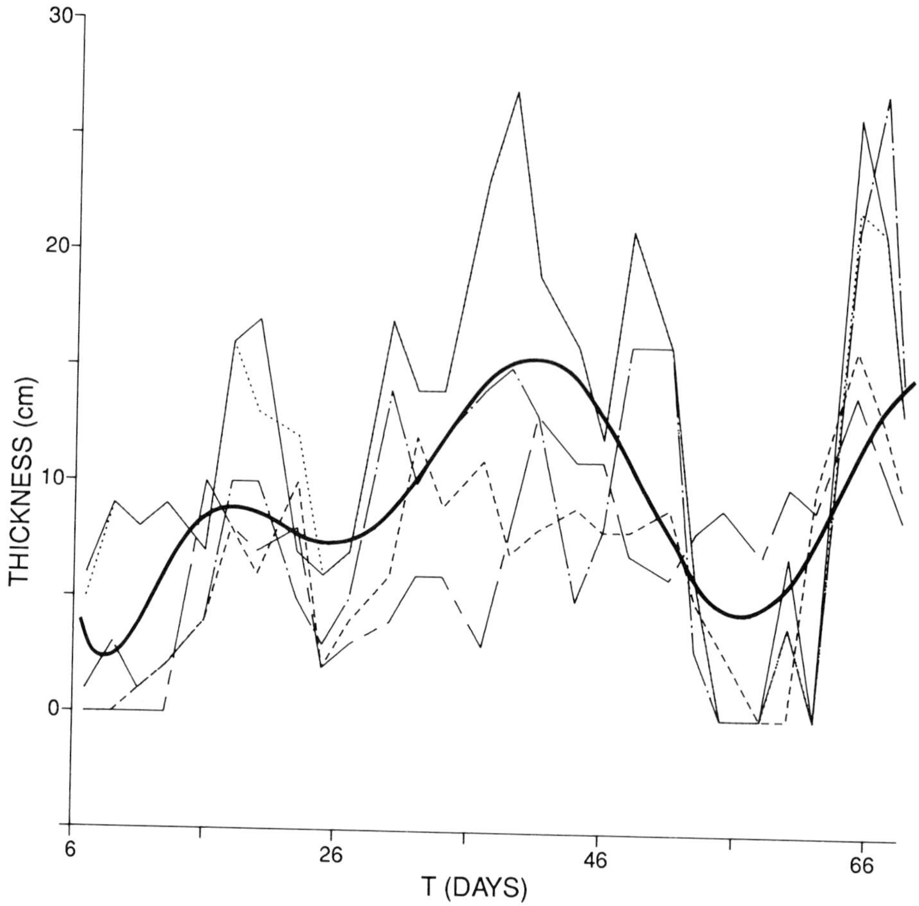

Figure 4: The ice thickness in the St. Lawrence Estuary for the year 1982 estimated from ice charts produced by Ice Forecasting Central, Ottawa, Canada. The estuary is subdivided into five sections of equal horizontal spacing. The section number increases from one near the head to five near the mouth of the estuary. Solid: section 1; dotted: section 2; dash-dotted: section 3; dashed: section 4; long and short dashed: section 5. The low-frequency variation of the ice thickness used in the model is obtained by averaging the thickness over all the sections, and then fitting (least squares method) the averaged data into an 11th order polynomial. The result is plotted as a thick solid curve.

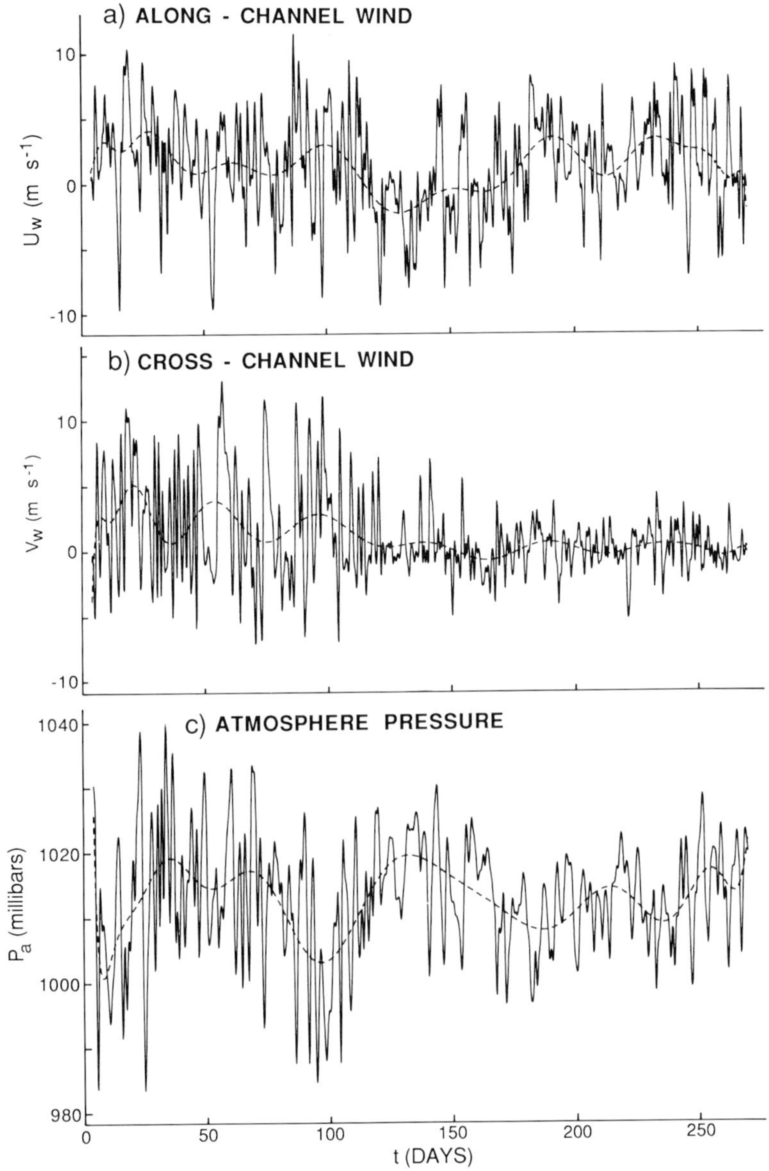

Figure 5: The (a) along-channel (u_w), (b) cross-channel (v_w) component of the wind velocity, and (c) the atmospheric pressure (p_a) at Mont-Joli. Solid: low-passed data; Dashed: low-frequency data obtained by fitting (least squares method) the low-passed data to a 20th order polynomial.

12.3.2 Forcing functions

The ice thickness and coverage for the year 1982 were estimated from ice charts provided by Ice Forecasting Central, Ottawa, Canada. The chart was produced every two to three days. Most areas of the estuary except those near the head were ice free after March 12. The modelling area is subdivided into five sections of equal horizontal spacing. The average thickness is computed for each section and shown in Fig.4. The section number increases from one near the head to five near the mouth of the estuary. The thickness at the two sections near the head of the estuary (solid- and dotted-curves) are basically the same except for the period between days 18 and 25 and near day 65 when the two thicknesses are slightly different. As expected, the ice near the head of the estuary where the salinity is low is generally thicker than downstream. The exception are those between days 47 and 58 when the ice near the mouth of the estuary is significantly thicker than that inside the estuary. The low-frequency variation of the ice thickness (thick solid curve) used for computing salt flux on the surface (Eq. 12.14) is obtained by averaging the thickness over all the sections, and then fitting (least squares method) the averaged data into an 11th order polynomial.

Hourly wind and atmospheric pressure (P_a) data in 1982 at Mont-Joli and Baie-Comeau in the lower estuary and Quebec City in the upper estuary were obtained from the Environment Canada Weather Office. It was found that the oscillations in the atmospheric data at the three stations were similar and occurred more or less simultaneously. This is because the weather system in the Gulf of St.Lawrence area generally has a large horizontal scale (order of thousand kilometers) and propagates rapidly (order of thousand kilometer per day). In the following, only the data at Mont-Joli are applied in the model and analysis.

The hourly wind data were resolved into along-channel (u_w) and cross-channel (v_w) components. The resolved data were low-passed with a Cartwright filter (half power at 31 hours) and subsampled at 6 hour intervals. Figure 5 shows the low-passed (solid-curves) and low-frequency (dashed curves) wind velocity and atmospheric pressure. The low frequency data are obtained by fitting (least squares method) the low-passed data to a 20th order polynomial. As indicated by the figure, the low-frequency data of u_w and v_w are inversely correlated with that of P_a; maxima in u_w and v_w correspond to minima in P_a, or vice versa. Within the period of observation (between days 105 and 250), the along-channel component of the wind velocity was much larger than the cross-channel component. In the 2-D modelling carried out in this study, only the former component is needed.

The freshwater runoff into the St. Lawrence Estuary is contributed mainly from the St. Lawrence River (78-80%, Tee and Lim, 1987; El-Sabh, 1988). The contributions from Saguenay River in the middle estuary and the Manicouagan River System in the lower estuary are respectively about 10 to 13%, and 9 to 10%. The runoff from St. Lawrence River is characterized by seasonal peaks in spring and fall, while that from Saguenay has a series of short pulses (10-20 days) between May and July (Tee, 1989). The runoff from Manicouagan River system, because it is highly regulated for hydroelectric power, is more or less uniform throughout the year. Figure 6 shows the 1982 freshwater runoff through Montreal (solid curve), 240 km upstream of Quebec City. The dashed-curve in the figure is the low-frequency curve obtained by fitting (least square method) the runoff data to a 20th order polynomial. The runoff through Quebec City near the upstream end of the model (dash dotted, Fig.6) is estimated from the mean ratio (1960-1972) of monthly discharges through Quebec City to that through Montreal. A phase lag of 7 days is applied for the runoff to arrive at Quebec City from Montreal (F. Jordan, Bedford Institute of Oceanography, Canada, Private Communication). In the present form of two-dimensional model, only the runoff through the head of the estuary (Quebec City) can be incorporated.

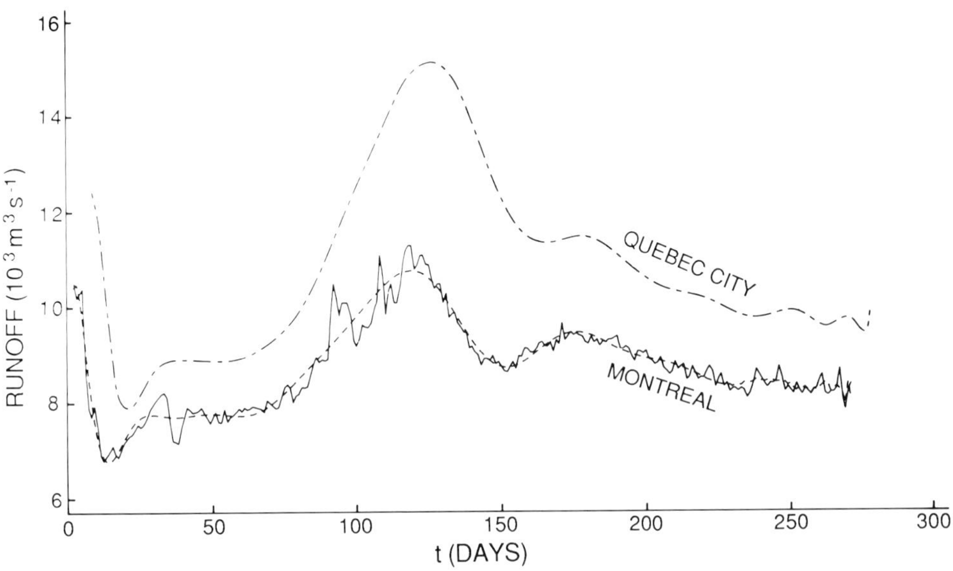

Figure 6: Freshwater runoff through Montreal (about 240 km upstream of Quebec City) and Quebec City. Solid: daily runoff through Montreal; Dashed: low-frequency runoff through Montreal obtained by fitting (least squares method) the daily data to a 20th order polynomial; dash-dotted: low-frequency runoff through Quebec City.

12.4 Comparison between observed and computed salinity variation

In the numerical model, a steady state estuarine circulation was first obtained by using the observed wind, ice and runoff on January 7, 1982. The computation was then continued for nine months using the long period forcing described in the previous section.

Figure 7 shows the comparison between computed (dash-dotted curves) and observed (solid and dashed curves) salinity variation for the upper current meter at stations A, B, C, D and

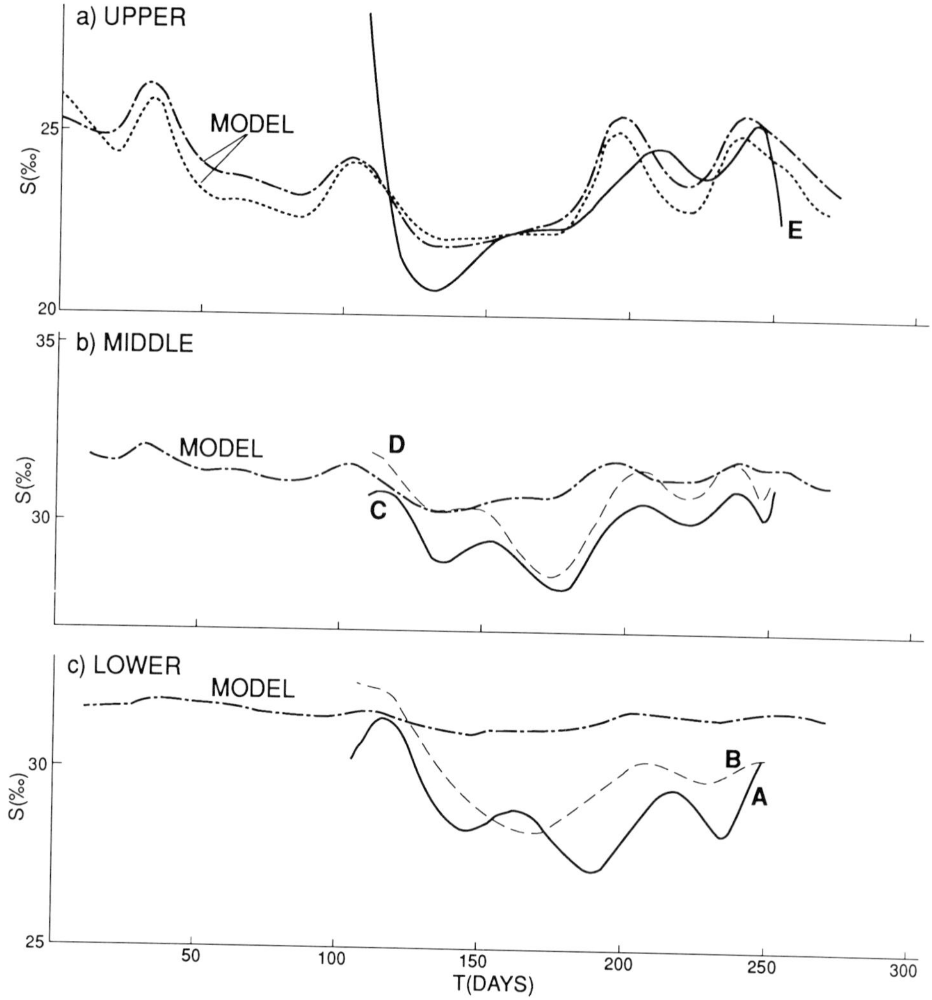

Figure 7: The comparison between computed and observed low-frequency salinity variation at the upper current meters at (a) station E in the upper estuary, (b) stations C and D in the middle estuary, and (c) stations A and B in the lower estuary. Solid: observed salinity at the southern stations (A, C, E); dashed: observed salinity at the northern stations (B, D); dash-dotted: computed salinity with all the forcings; dotted: computed salinity with the wind forcing only.

E. At station E in the upper estuary, we can see from Fig.7a that the observed long-period salinity variation can be reproduced approximately by the numerical model. Note that because of the simplicities of the model (e.g. cross-sectional averaged) and the contamination of the observed 40-50 days salinity variation by spring-neap signals generated by strong tidal rectification process in the upper estuary (Tee, 1989), the observed low-frequency variation of salinity is expected to be reproduced only qualitatively. Figure 7a indicates that the observed magnitudes of the variation, and occurrence of two major maxima around days 200 and 250, and two major minima around days 130 and 220 are simulated reasonably well by the model.

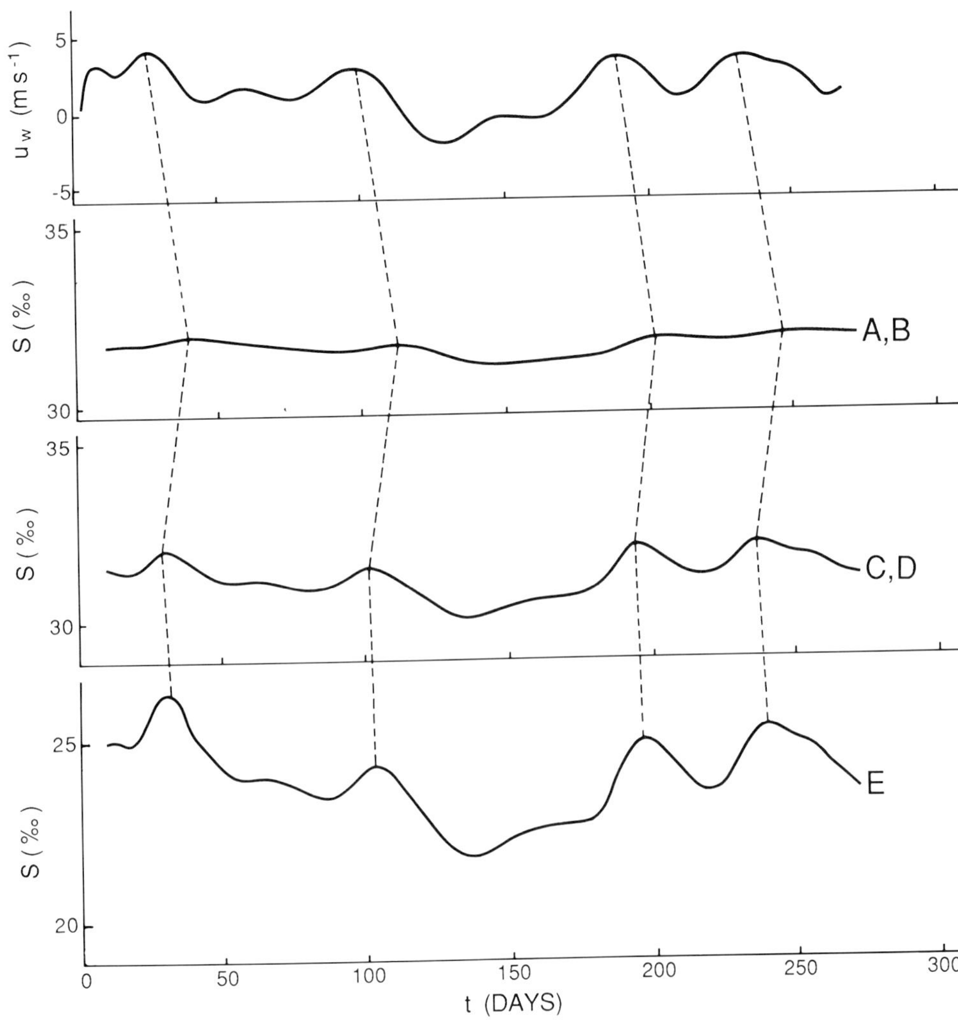

Figure 8: The wind forcing and the model results of 40- to 50-day salinity oscillation at 20 m in the St. Lawrence Estuary. The along-channel wind velocity u_w is shown in the upper curve. The names of the stations (A, B, C, D and E) are indicated on the right-hand side of the salinity curves. The forcing and propagation of the 40- to 50-day oscillation is shown.

At stations C and D in the middle estuary, the observed and simulated salinity variations during the later period between days 190 and 250 have comparable magnitudes [~ 0.8 ppt for the observed data and ~ 0.6 ppt for the model; Fig.7b]. However, the observed salinity variations with maximum at days 150, and minima at days 130 and 180 during the early period are underestimated significantly by the model. This may be because the runoff by Saguenay River near the stations (Fig.1), which has significant outflow during this early period (maximum value of 2.8×10^3 m^3s^{-1} at days 140; see Fig.14 of Tee, 1989), is not included in the model. The absence of this low salinity water in the model during spring results in a small vertical salinity gradient near the surface in the middle estuary, which then produces a small salinity variation from wind-induced vertical displacement of halocline.

At stations A and B in the lower estuary, the simulated low-frequency salinity variation was found to be significantly smaller than those detected in the observed data (Fig.7c). This disagreement may be because the freshwater runoff from the Manicouagan River system near these stations (Fig.1) is not included in the numerical model. The runoff from this river system has been found to reduce significantly the surface salinity in the lower estuary during spring and summer (Therriault and Levassiur, 1985).

Although some of the amplitude of the salinity variation in the middle and lower estuaries cannot be simulated quantitatively, the time lags of the salinity maxima or minima between the stations shown in Fig.2 are reproduced very well. Using the model salinity at 20 m at stations A, B, C, D and E, the evidence of time lags for the salinity oscillation is shown in Fig.8. The upper curve is the low-frequency variation of the along-channel wind. The two middle curves are the salinity variation for stations A and B in the lower estuary, and C and D in the middle estuary. The salinity variation for station E in the upper estuary is shown in the lower curve. By comparing Figs.2 and 8, we can see that the occurrences of the observed salinity maxima and minima which are earlier at stations C and D, and later at stations A, B and E are simulated well by the numerical model.

Figure 9a shows the quantitative comparison of relative time lags at 20m between the stations. The observed time lag of 11.2 days between stations in the lower (A, B) and middle (C, D) estuaries compares well with the computed time lag of 9.2 days from the numerical model. Similarly, the observed time lag of 5.2 days between the stations in the upper (E) and middle (C, D) estuaries agrees with the computed time lag of 3.2 days from the model.

The computed vertical variation of salinity oscillation at stations C and D in the middle estuary is shown in Fig.10. The number on the right-hand side of the curve indicates the depths from the surface. We can see from the figure that the magnitudes of oscillation decrease with depth and the oscillation occurs earlier at deeper locations. The same results were observed in the water column of the upper and lower estuaries. By comparing Figs.3 and 10, we can see that the numerical model also reproduces the decrease of magnitude toward deeper water and the upward propagation of salinity oscillation observed at station C. Figure 9b shows the quantitative comparison of relative time lags at various depths at this station. Between the salinity variation at 20 and 50 m, the observed time lag of 2.2 days compares well with the model time lag of 2.7 days. The same result can be found for the salinity variation at stations A and B in the lower estuary (Fig.9c), where the observed time lag of 3.5 days between the salinity at 20 and 30 m compares well with the computed time lag of 5.2 days. At station E in the upper estuary, the salinity oscillation at the lower current meter (at 19 m) lags those at the upper current meter (at 14 m) by 0.8 days. This small time lag disagrees with the model's results which shows that the salinity variation in the lower current meter leads that of the upper meter by 1.8 days (Fig.9d). The cause of the disagreement is unclear. It may be because the observed time lags (0.8 days) is too small to be estimated accurately.

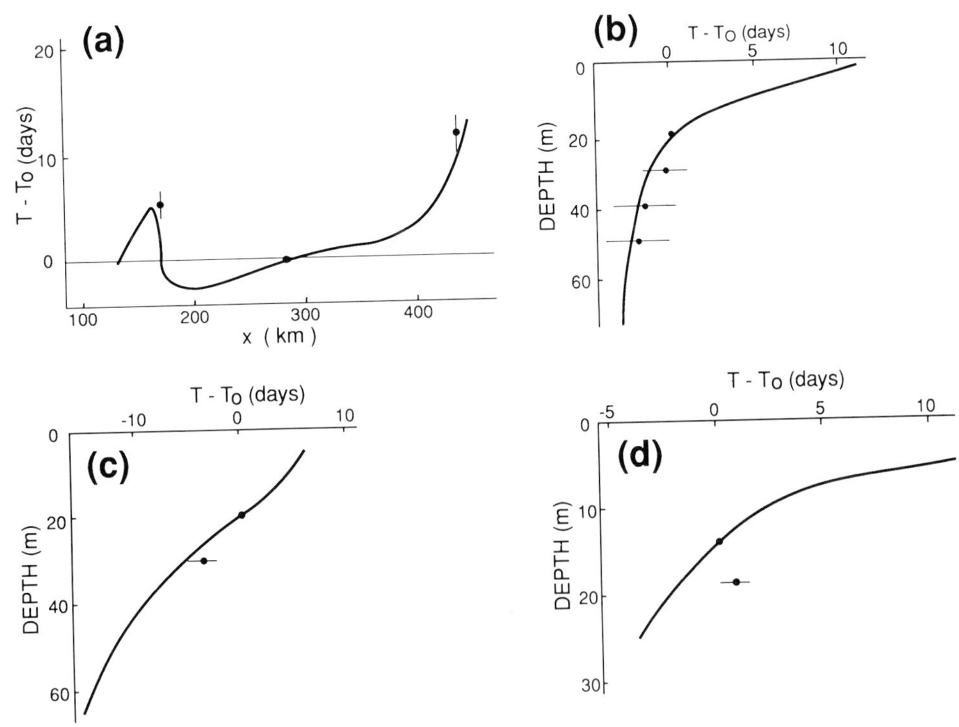

Figure 9: The comparison of relative time lags ($T - T_0$) between observed (solid dots) and model (solid curves) results of the salinity variation in the St.Lawrence Estuary. (a) The horizontal variation of ($T - T_0$) at 20 m, where T_0 is the time lags at stations C and D. (b) The vertical variation of ($T - T_0$) at stations C and D in the middle estuary, where T_0 is the time lags at 20m. (c) The vertical variation of ($T - T_0$) at stations A and B in the lower estuary, where T_0 is the time lags at 20 m. (b) The vertical variation of ($T - T_0$) at stations E in the upper estuary, where T_0 is the time lags at 14 m.

The model results indicate that the contribution of ice on the salinity variation is quite small, less than 5% of the total salinity variation in most of the area, except near the head of the estuary (near Quebec City) where it contributes to a maximum of about 18%. The comparison between observed and simulated low-frequency salinity variation is found to be affected insignificantly by the ice forcing.

The observed 40-50 days salinity variations are induced by wind forcing. This is shown by the close correspondence between wind forcing and salinity variations (Figs. 2 and 8). To examine more accurately the importance of this forcing on the low-frequency salinity variations, a computation was carried out by varying the wind-forcing, but keeping the ice and runoff at their January 7 values. An example of the computation at 14 m at station E is shown in Fig.7a by a dotted curve. By comparing dash dotted (all forcings) and dotted (wind forcing only) curves, we can see that the observed long-period salinity variation is induced mainly by wind forcing. Similar results were observed for the other stations in the middle and lower estuaries.

As the wind blows downstream (positive U_w), water particles in the upper portion of the water column are advected out to the Gulf of St. Lawrence. Because the salinity increases toward the mouth of the estuary, the advection of low salinity from upstream reduces the local salinity in the surface layer. This horizontal advection of salinity cannot explain the positive correlation between wind forcing and salinity in the surface layer throughout the estuary (Figs.5 and 11). Horizontal advection of salinity below surface layer in the deeper portion of the estuary (including stations A, B, C and D) also cannot explain the salinity oscillation because the salinity below 20 m in this region is more or less uniform along the estuary (see Fig.10 of Tee and Lim, 1987).

The observed and simulated positive correlation between wind forcing and salinity is induced mainly by vertical movement of halocline. As the total transport in the water column for the two-dimensional model is given by the freshwater runoff through the head of the estuary, the wind induced downstream flow in the upper layer generates the upstream flow in the lower layer of the water column (Fig.12). This two-layer estuarine circulation produces upwelling in the estuary which advects halocline toward the surface. The salinity is thus expected to correspond with the downstream wind velocity (Figs.2 and 8). The upward movement of halocline is particularly strong near stations C and D in the middle estuary because the stations are near the head of the Laurentian Channel where the steep slope of the bottom topography diverts the upstream flow from deep to shallow waters resulting in strong upwelling. This strong respond of salinity to wind forcing indicates the early formation of salinity maximum near these stations. More detailed explanation of the salinity variations, including the horizontal and vertical propagation of the salinity maxima and minima, requires an extensive analysis of the complex and nonlinear interactions between velocity and salinity. This analysis is the subject of a future study.

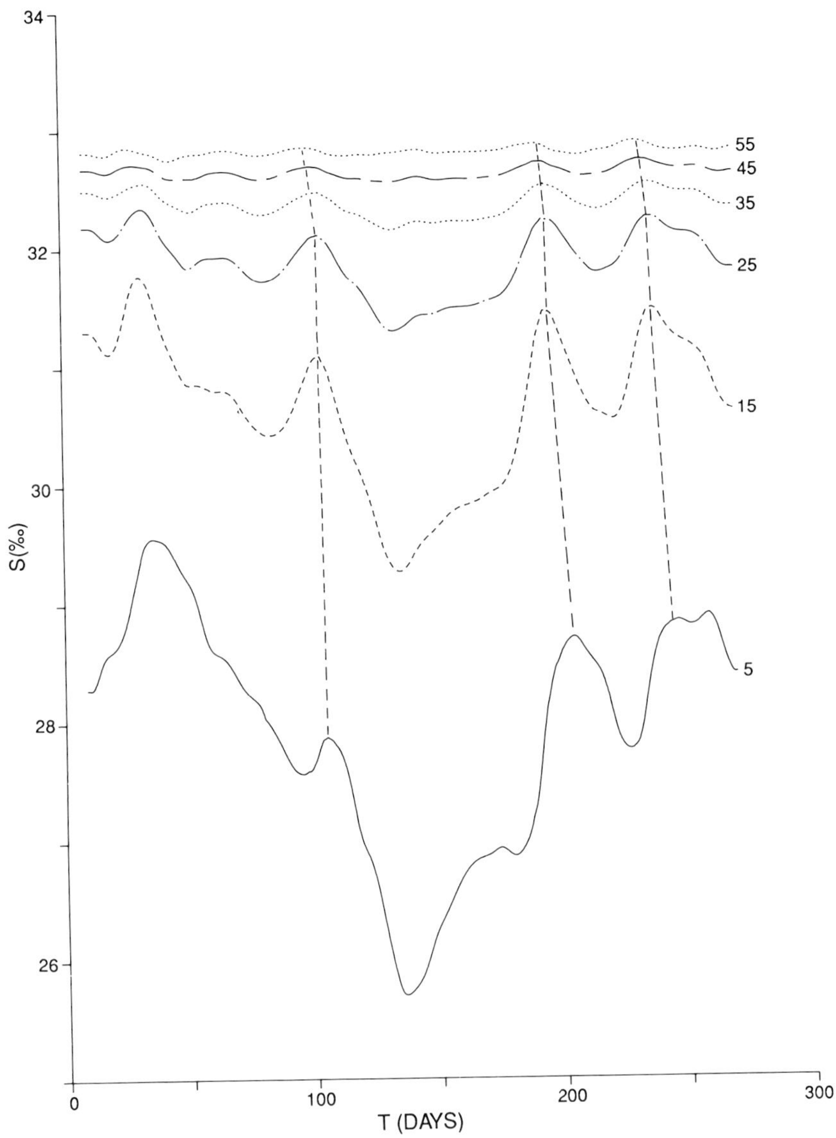

Figure 10: The model results of low-frequency salinity variation at station C and D in the middle estuary. The decrease of magnitudes with depth and the upward propagation of the salinity maxima and minima is shown in the figure.

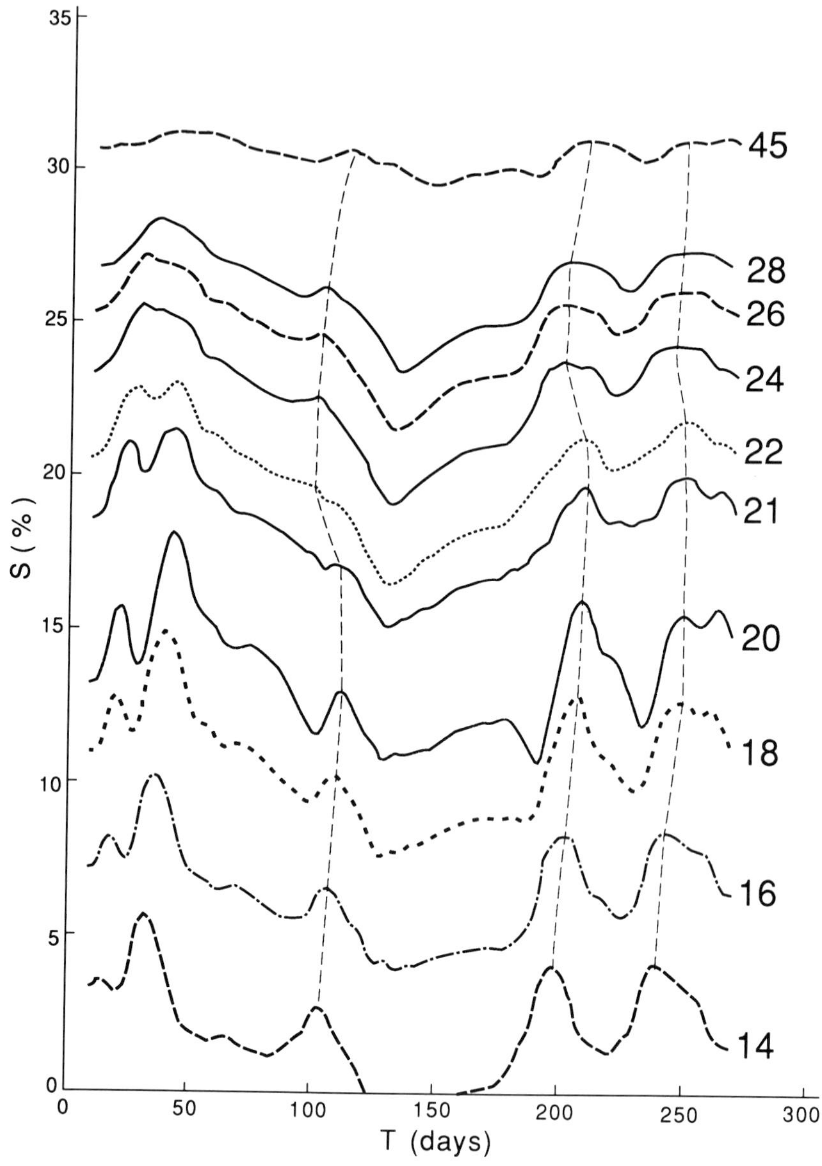

Figure 11: The model results of salinity variations at 5 m on various sections in the St.Lawrence Estuary. The section number measured from the head of the estuary (Fig.1) is shown on the right-hand side of each curve.

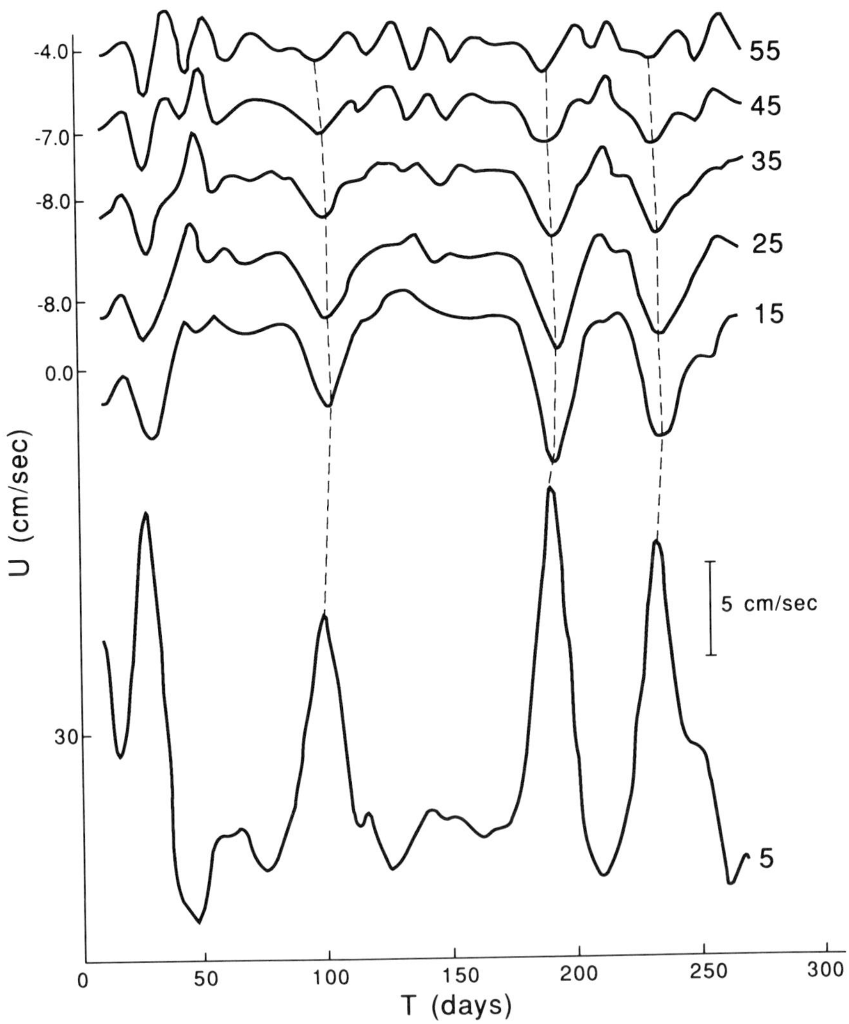

Figure 12: The model results of velocity variation at stations C and D. The depth (m) from the surface is shown on the right-hand side of each curve.

12.5 Conclusion

Between April and September 1982, fourteen current meters were moored at six stations along the St. Lawrence Estuary: two stations each in the lower, middle and upper estuaries (Fig.1).

The experimental data reveal the low-frequency oscillation with the period of 40-50 days. A simple analysis indicated that the oscillation in the middle estuary occurred earlier than that in the upper and lower estuaries. In the vertical water column, the oscillation occurred earlier at deeper locations in the middle and lower estuaries, and slightly earlier at shallower locations in the upper estuary.

Because of the short current meter records (about 150 days), the time lags between the stations cannot be estimated objectively using a spectral method. To confirm the existence and propagation of this oscillation in the St. Lawrence Estuary, a two-dimensional (cross-sectional averaged) estuarine circulation model with the observed low-frequency forcing of wind, ice and freshwater runoff was developed. The observed salinity oscillation in the upper estuary is simulated well but that in the lower estuary is underestimated significantly by the model. In the middle estuary, the summer-time (days 190 to 250) salinity oscillation is simulated well, but the spring-time (days 100 to 190) oscillation is underestimated by the model. This underestimation in the middle and lower estuaries is probably because the runoff from Saguenay and Manicouagan Rivers (Fig.1) are not included in the model.

Although the model underestimates some observed magnitudes of the salinity oscillation, it reproduces many important features of the oscillation including the decrease of magnitude from surface to deeper water, and the earlier occurrence of the oscillation in the middle estuary and at deeper locations. The model also confirms that the most important forcing of the oscillation is the along-channel wind stress. The wind-induced upwelling and downwelling are the major factor producing the salinity variations. Results of this study indicates the important effects of wind forcing on the formation, distribution and propagation of salinity oscillations in coastal estuaries.

12.6 References

Elliott, A. J., 1978: Observation of the meteorologically induced circulation in the Potomac Estuary, Estua. Coastal Mar. Sci., 6, 285-299.

Elliott, D. P. Wang, and D. W. Pritchard, 1978: The circulation near the head of Chesapeake Bay, J. Mar. Res., 36, 643-655.

El-Sabh, M. I., H. J. Lie and V. G. Koutitonsky, 1982: Variability of the near surface residual current in the lower St. Lawrence Estuary, J. Geophys. Res., 87, 9589-9600.

Goodrich, D. M., W. C. Boicourt, P. Hamilton and D. P. Pritchard, 1987: Wind-induced destratification in Chesapeake Bay, J. Phys. Oceanogr., 17, 2232-2240.

Mertz, G., M. I. El-Sabh and V. G. Koutitonsky, 1988: Wind-driven motions at the mouth of the lower St. Lawrence Estuary, Atmosphere-Ocean, 26, 509-523.

Mertz, G., M. I. El-Sabh and V. G. Koutitonsky, 1989: Low frequency variability in the lower St. Lawrence Estuary, J. Mar. Res., 47, 285-302.

Schroeder, W. W. and W. J. Wiseman, Jr., 1985: Low-frequency shelf-estuary exchange processes in Mobile Bay and other estuarine systems on the northern Gulf of Mexico, in Proceedings of the Eight Biennial International Estuarine Research Conference, edited by D. A. Wolfe, 355-367, Acadexic, San Diego, California.

Smith, N. P., 1977: Meteorological and tidal exchanges between Corpus Christi Bay, Texas, and the Northwestern Gulf of Mexico, Estuar. Coastal Mar. Sci., 5, 511-520.

Tee, K. T., 1989: Subtidal salinity and velocity variations in the St. Lawrence Estuary, J. Geophys. Res.,94, 8075-8090.

Tee, K.T., 1990: Meteorologically and buoyancy induced subtidal salinity and velocity variations in the St.Lawrence Estuary, in "Oceanography of a large-scale estuarine system - the St.Lawrence" edited by M. I. El-Sabh and N. Silverberg, Coastal and estuarine Studies, 39, 51-70.

Tee, K.T. and T. H. Lim, 1987: The freshwater pulse - a numerical model with application to the St. Lawrence Estuary, J. Mar. Res., 45, 871-909.

Therriault, J. C. and M. Levasseur, 1985: Control of phytoplankton production in the lower St. Lawrence Estuary : light and freshwater runoff, Naturaliste Can., 112, 77-96.

Vieira, M. E. C., 1986: The meteorologically driven circulation in mid-Chesapeake Bay, J. Mar. Res., 44, 473-493.

Walters, R. A., 1982: Low-frequency variation in sea level and currents in South San Francisco Bay, J. Phys. Oceanogr., 12, 658-668.

Wang, D. P., 1979a: Subtidal sea level variations in the Chesapeake Bay and relations to atmospheric forcing, J. Phys. Oceanogr., 9, 413-421.

Wang, D. P., 1979b: Wind-driven circulation in the Chesapeake Bay, Winter 1975, J. Phys. Oceanogr., 9, 564-572.

Wang, D.P. and A. J. Elliott, 1978: Nontidal variability in the Chesapeake Bay and Potomac River: Evidence for non-local forcing, J. Phys. Oceanogr., 8, 225-232.

Weisberg, R. H., 1976: The nontidal flow in the Providence River of Narragansett Bay: A Stochastic approach to estuarine circulation, J. Phys., Oceanogr., 6, 721-734.

Weisberg, R. H. and W. Sturges, 1976: Velocity observation in the West Passage of Narragansett Bay: A partially mixed estuary, J. Phys. Oceanogr., 6, 345-354.

Wong, K. C. and R. W. Garvine, 1984: Observations of wind-induced subtidal variability in the Delaware Estuary, J. Geophys. Res., 89, 589-597.

13

On the estuarine circulation within the Kattegat

N Winkel-Steinberg, J O Backhaus and T Pohlmann

Abstract

A three-dimensional, prognostic, baroclinic circulation model, based upon the primitive equations is applied to the Kattegat and adjacent sea regions in order to provide an estimate of the estuarine circulation under forcing conditions typical for the region. The model domain includes parts of the Skagerrak and hence it allows to study the communication between a rather deep and a very shallow sea region in the presence of stratification. The circulation is predominantly governed by the influence of winds and stratification. Tidal dynamics play a rather marginal role and were therefore not considered in the model. The discussion and evaluation of the model results is carried out with the intention to explain the influence of the circulation and the entrainment of nutrient-rich bottom water on the distribution of nutrients. These physical processes are considered to be important mechanisms for the preconditioning of the frequently occuring oxygen depletion events in the Kattegat. The model results compare rather well with the known hydrographical features of the Kattegat. Entrainment rates derived from the model results are compared with estimates obtained from an integrated box-model for the Kattegat, developed by Stigebrandt (1983).

13.1 Introduction

The North Sea and the Baltic Sea communicate via a transition-region with complex topography (fig. 1). It consists of the Skagerrak, the Kattegat, the Belt Sea and the Sound and the Western Baltic (Kiel Bight). In the Kattegat, the circulation and transformation of the water masses is crucially determined by both, local processes (winds, stratification) and by the influence of the far-field North Sea and Baltic, respectively. There are weather conditions, which, due to the combined effect of winds and air pressure and due to the shallowness of the sea, can cause strong in- and outflow events. (More extensive descriptions of the hydrography and the dynamics of the transition region are given by Aure and Saetre (1981), Svansson (1984), Larsson and Rodhe (1979), Thorstensson (1978), Knudsen (1899), Dietrich and Schott (1974).) These events caused by the meteorological and hydrodynamical far-field have predominantly synoptic time-scales of 1-5 days and they contribute significantly to the extreme variability, which is typical for the region. As a result the currents may vary between zero and (locally) up to five knots and the salinity can vary within a range of fifteen psu (practical salinity units). In the transition region an increasing abundance of oxygen depletions due to anthropogenic eutrophication is observed for the past, say ten years. These events have caused severe harm to the local fisheries and to the marine environment. They are caused by local nutrient inputs via the atmosphere and via run-off from the mainland and from numerous islands in the domain and, on the other hand, from remote sources located in the southern North Sea and the eastern Baltic, respectively. For the contribution from the west the circulation and advection in the North Sea, in particular along the Danish west coast (the Jutland current), plays an important role for the nutrient balance in the Kattegat. The nutrients carried with the Jutland current enter the Kattegat at depth, via a deep channel along the Swedish west coast, due to the strong density contrast to the much less saline Baltic run-off. If not disturbed by the atmosphere this run-off is a geostrophic buoyant plume which follows the Swedish west coast towards north. The extension of this coastal plume within and beyond the Skagerrak is called the Norwegian coastal current. The salinity and also the nutrient balance within the Kattegat is hence strongly determined by advection (North Sea inflow and Baltic run-off) and by entrainment between the in- and outflowing water masses, respectively. The latter is dominated by local atmospheric disturbances.

In early summer 1988 a very excessive toxic and harmful algae-bloom, which appeared in the press as the "Killer-Algae" episode, occured. Besides political reactions it initiated a number of investigations Backhaus et. al. (1988), Aksnes et. al. (1989). Further, a report of an international group of experts, called upon by the International Council for the Exploration of the Sea (ICES), was edited by Skjoldal and Dundas (1989). It comprises physical and bio-geochemical reasons and interpretations for the dramatic event. It likewise documents the still considerable lack of knowledge about anthropogenic disturbances of the marine environment. Amongst other questions these investigations addressed the relative importance of local versus remote nutrient sources in the context of environmental accidents and of an increased trend in the eutrophication of the north-west European waters.

Only a few modelling studies have as yet been carried out, for the transition region Kattegat in particular, which would help to answer some of the above questions. There is a canal model approach by Svansson (1972), then the well known and widely accepted integrated box-model by Stigebrandt (1983) and a three-dimensional (grid-box) circulation model by Mueller-Navarra (1983). Only Stigebrandt's model so far took into account the two-layer density structure in conjunction with salt entrainment. His model is carefully calibrated and in good agreement with observations. However, due to the laterally integrated boxes, it cannot provide insight into the spatial variability of the circulation and of the advection of substances and of salt and nutrient entrainment. The model of Mueller-Navarra is a diagnostic one and, hence it does not allow the simulation of the space-time evolution of the baroclinic fields. Further it does not contain a term that explicitly describes the vertical mixing of salt and temperature.

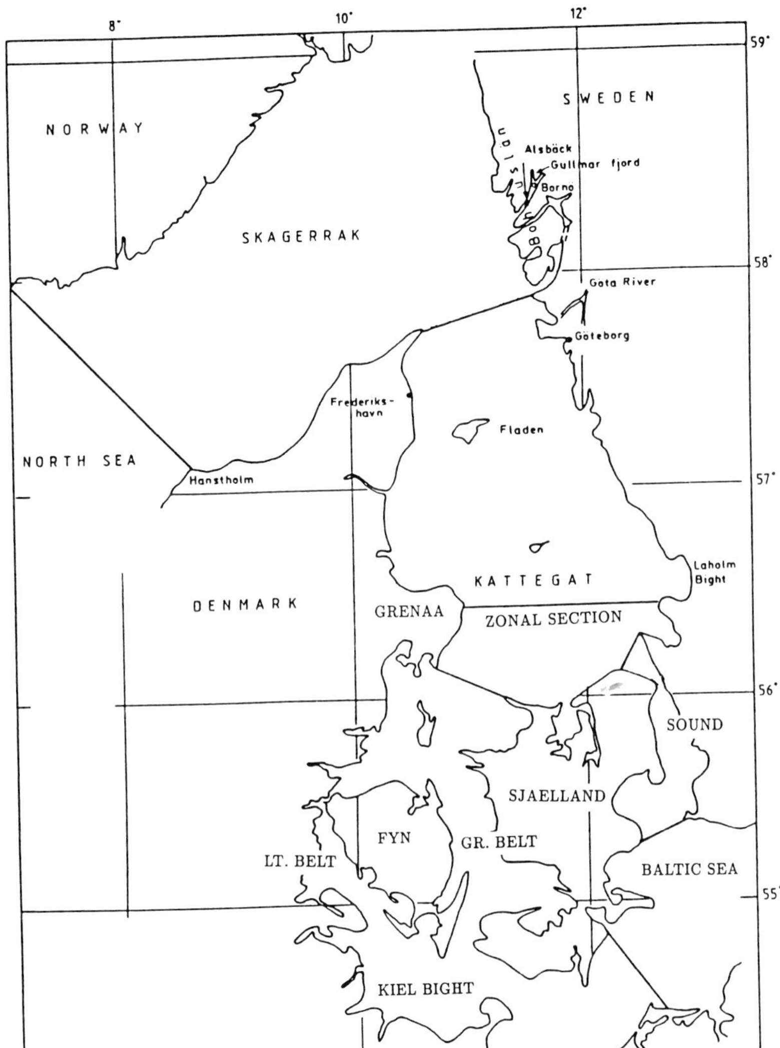

Figure 1: Map of the study area

This contribution is based upon a prognostic baroclinic circulation model, where vertical mixing rates are determined by a space-time dependent algorithm. It is therefore anticipated that the results provide insight into the spatial structure of the baroclinic fields due to the effects of both, advection and entrainment. As a whole this investigation aims at a better understanding of the role of the circulation in regard to the dispersion of nutrients and, in view of the oxygen depletions, in particular to the entrainment of nutrient-rich bottom waters into the euphotic near-surface zone within the Kattegat. This paper is a short and still rather superficial description of the results of a preliminary investigation phase, which in the future, in collaboration with the Swedish Meteorological and Hydrological Institute, will be continued and deepened with the availability of recent field data from the SKAGEX '90 experiment.

13.2 A brief summary of the numerical model scheme

The three-dimensional numerical model is based upon the scheme described by Backhaus (1985) and by Backhaus and Hainbucher (1987). Those terms in the nonlinear primitive equations, which in their discrete numerical approximation may cause numerical instability, are introduced by implicit algorithms, which effectively avoid instability. The scheme as such is semi-implicit, because the remaining terms are treated explicitly. The density is derived from the standard nonlinear oceanographic equation of state. Vertical diffusion (mixing) of momentum, heat and salt is approximated by a space-time dependent approach proposed by Kochergin (1987), who derived the mixing principle from the equation for the turbulent energy. The principle considers shear production and buoyancy; the equation is diagnostic due to the assumption of a local stationarity of the turbulent kinetic energy. For further details see Kochergin (citation above) or Winkel (1989). The vertical momentum exchange coefficient, derived from the Kochergin-relation was set to vary within the range of 0.013×10^{-4} to $500 \times 10^{-4}\ m^2s^{-1}$.

The model is defined on the Arakawa-C grid and it uses a two time-level scheme. For the approximation of the advection of momentum, heat and salt a Lagrangian, "selective vector-upstream scheme" was used. This will be described here in some more detail, because it is missing in the above cited papers. The scheme is, due to the upwind algorithm, a priori diffusive. However, compared to a straight forward, classical upwind scheme it has the advantage of a small cross-flow diffusivity, which is worse whenever the flow direction is aligned with or close to the diagonal of the coordinate system. The improvement was obtained by a Lagrangian extension of the straight forward upwind operator. The idea of the selective vector-upstream scheme is readily explained in a graphical way by means of the below sketch (fig. 2). As an example, and also to compare the different upwind algorithms, we consider a flow from the south-west quadrant, which, in the Lagrangian sense, advects the property P to the central grid-point C within one time-step (for the sake of clarity the other, irrelevant quadrants around C have not been drawn). The position of P is uniquely defined by the advection velocity, i.e. the Lagrangian path during one time-step, which ends at C.

We consider the three cases:

a) component-upwind: it involves the grid-points: W, C, S
b) vector-upwind: it involves the grid-points: SW, W, C, S
c) selective vector-upwind: it involves the grid-points: SW, C, S

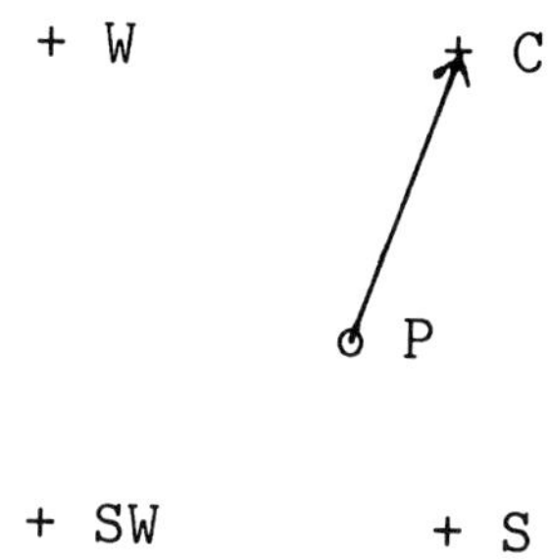

Figure 2: Location of grid-points
(denotation according to compass-rose indexing; i.e. SW means south-west from the central point C).

All three cases become equivalent, whenever the flow is aligned with a coordinate axis. It should, however, be mentioned, that neither case b) nor c) can be backtraced to the analytical form of the advection term, because of the Lagrangian form of the discrete advection approximation. This interpretes and considers that the property P (salt or momentum etc.) is determined by a spatial interpolation (which is true in each of the cases a) to c)). At the end of the time-step the property P arrives at the location C and causes a local advective change at this point. The intention of the versions b) (see Haltiner 1977) and c) is to improve the quality of the spatial interpolation of the property P. This is rather bad for case a) when the flow is diagonal. It is easily verified that the most selective case c) must have the smallest cross-flow diffusivity. In the above example the interpolation for case c) within the triangle (SW, C, S) will give the most precise spatial estimate of P, because it uses the smallest distances between P and the fixed Eulerian grid-points within the south-west quadrant (related to C). Experiments have shown that version c), which was used in the following simulations, gives the most realistic results of advection processes. This is mainly due to the fact, that only in case c) the artificial diffusion behaves isotropic. Using a suitable fine grid, formulation c) helps to effectively avoid the application of complicated and computer-time consuming flux-corrected schemes. For the vertical advection a component-upwind approach is adopted in the model. As we have applied an explicit scheme for all advective terms the time-step is restricted by the CFL-stability criterion.

13.3 Adaption of the model to the transition region

The model domain as a discrete approximation of the topography is depicted in figure 3. The horizontal increments of the north-oriented spherical grid are 5 and 3 minutes in meridional and zonal direction, respectively. This is approximately 5000 m. The dimension of the horizontal matrix is 115 times 61 grid-points. In the vertical, in order to obtain a decent resolution of the vertical stratification in the Kattegat (maximum depth about 90 m), there are eight equidistant layers with 5 m thickness each. Towards increasing depth the vertical increment is gradually increased, leading to a total number of 18 model layers for the model domain. Within the deep portions (maximum depth 700 m) of the Skagerrak the vertical resolution is of the order of 100 to 250 m. The internal Rossby-Radius of deformation for the Kattegat is of the order of five kilometers. Hence the model is not an eddy-resolving one. The time-step of the semi-implicit model was set to 600 s. For an explicit scheme the included deep region of the Skagerrak would have required a time-step of about 50 s. The initial baroclinic fields (T, S) for an early summer situation were interpolated from a coarser North Sea - Baltic model, developed by Boehlich and Backhaus (1987). The spatial resolution of that model is 12 nautical miles and hence the initial fields are only a very crude approximation of the typical density structure of the domain of interest.

Similarly sea levels at the open boundaries were interpolated from the coarser, encompassing model. Due to the estuarine character of the transition region a considerable sea level difference of about 0.40 m exists between the open Baltic boundary and the western boundary within the Skagerrak (fig. 4). The (mean) atmospheric forcing for the early summer situation was obtained from a meteorological atlas (Backhaus et. al. 1985), which was compiled from a data base for the years 1955 - 1982. The 6-hourly, gridded data (150×10^3 m) was kindly provided by the Norwegian Meteorological Institute. The mean wind field applied was almost homogeneous; the direction nearly everywhere is from the west and the rather moderate wind speed varies within the range of 3.5 to 4.5 ms^{-1} (fig. 5).

13.4 The mean Summer circulation

The "Killer-Algae" event occured during the months May to June 1988. The initial model simulation was prepared for a climatological mean of an early summer situation, in order to provide an estimate of a mean summerly estuarine circulation prior to future simulations of hydrographic and meteorological anomalies, which were quite pronounced for the previous winter and early summer 1988 (Backhaus et. al. 1988). The model was started from the above described initial and forcing fields and due to the prognostic equations for temperature and salinity the initially very crude baroclinic fields are eventually adjusted to the topography and to the circulation by geostrophic adjustment and self-advection and mixing. After about 40 days significant changes in the flow fields as well as in the baroclinic fields vanished and it was then assumed that the model had settled down to an equilibrium state under the applied forcing which was held constant. A second experiment was conducted in which the wind forcing was omitted, in order to quantify the solely buoyancy driven circulation. The model was run for the same period. In the following some few results of these two experiments are presented with the intention to provide insight into the mean circulation and T, S-structure, obtained with the model. At the end of both simulations the baroclinic fields are notably more structured than the crude initial fields and in very good agreement to the mean hydrography known from observations (citations in introduction).

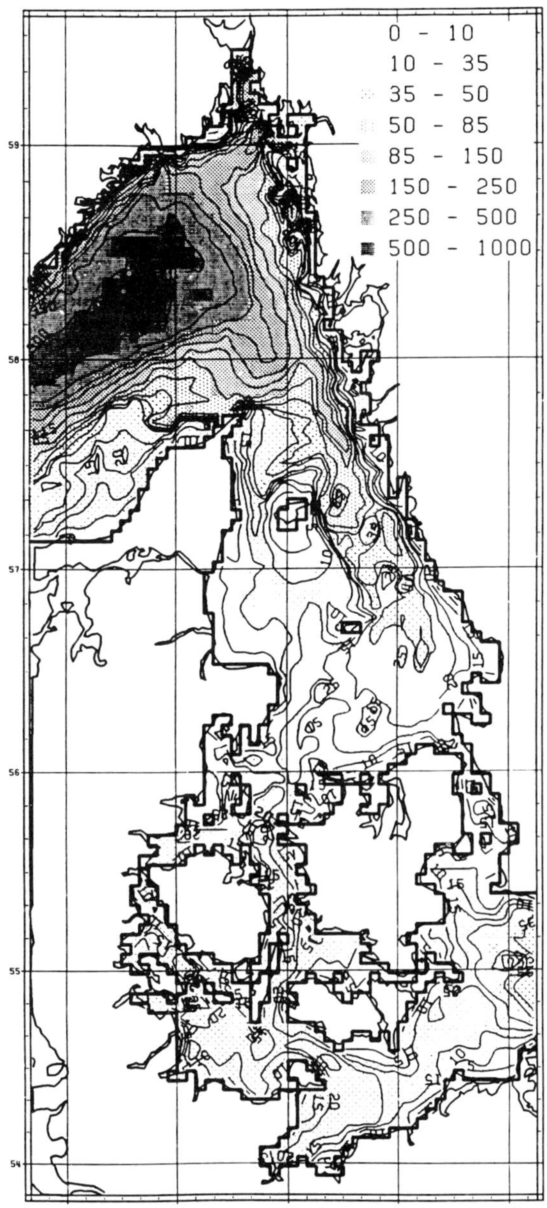

Figure 3: Model domain (discretized topography, depths in m)

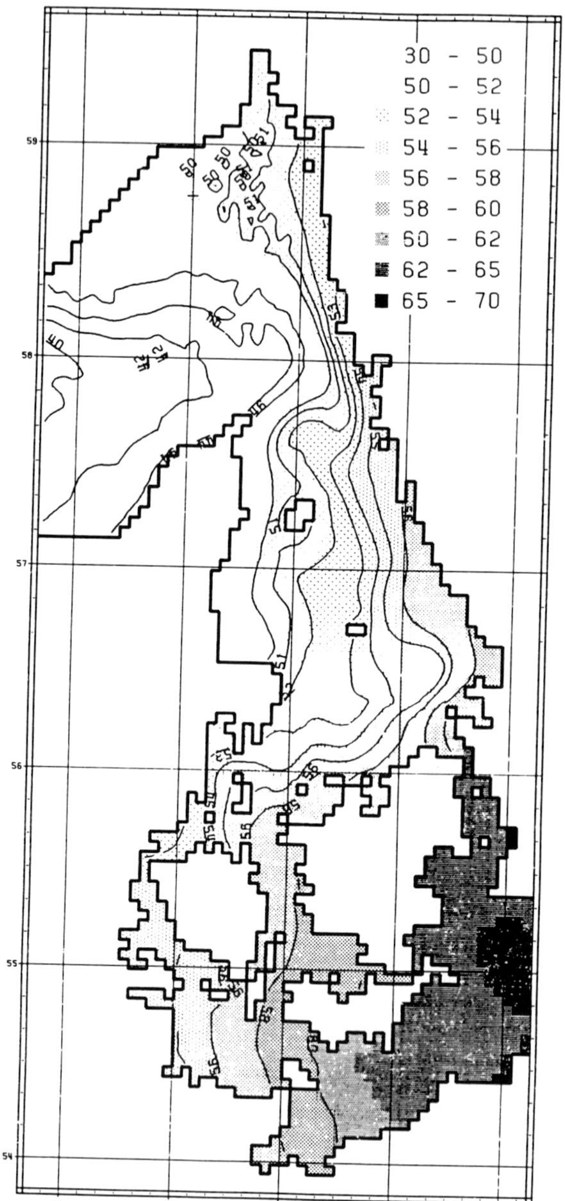

Figure 4: Sea surface topography (m x 10) for mean summer situation

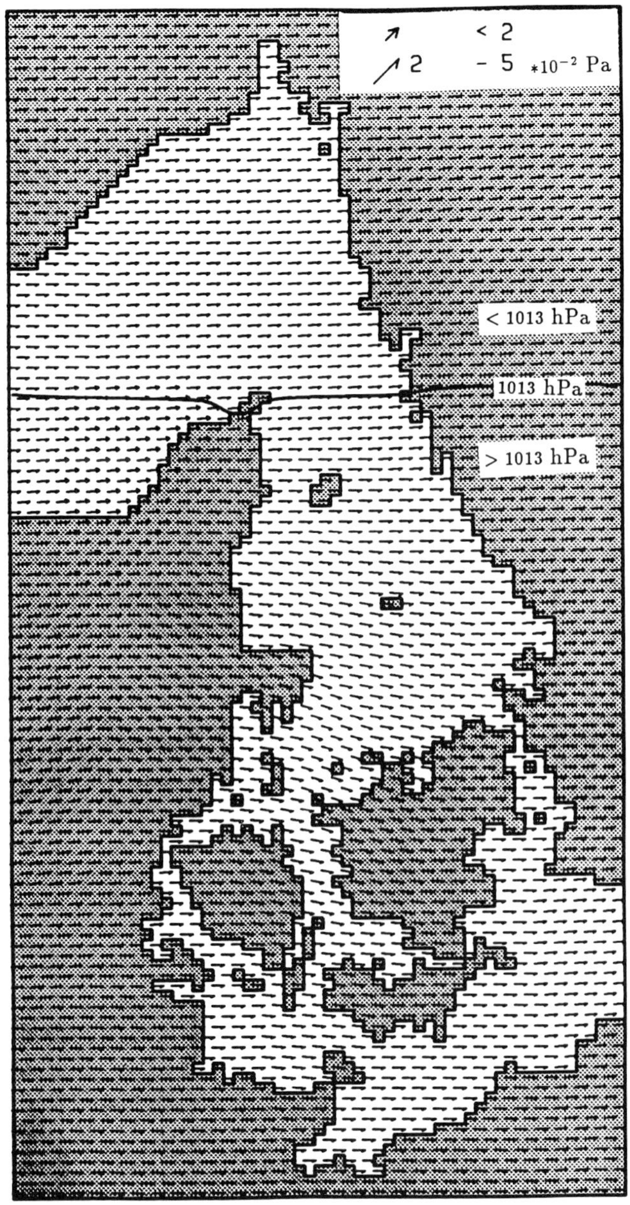

Figure 5: Atmospheric forcing for mean summer situation
windstress (Pa x 10^{-2})
air pressure (Pa x 10^{2})

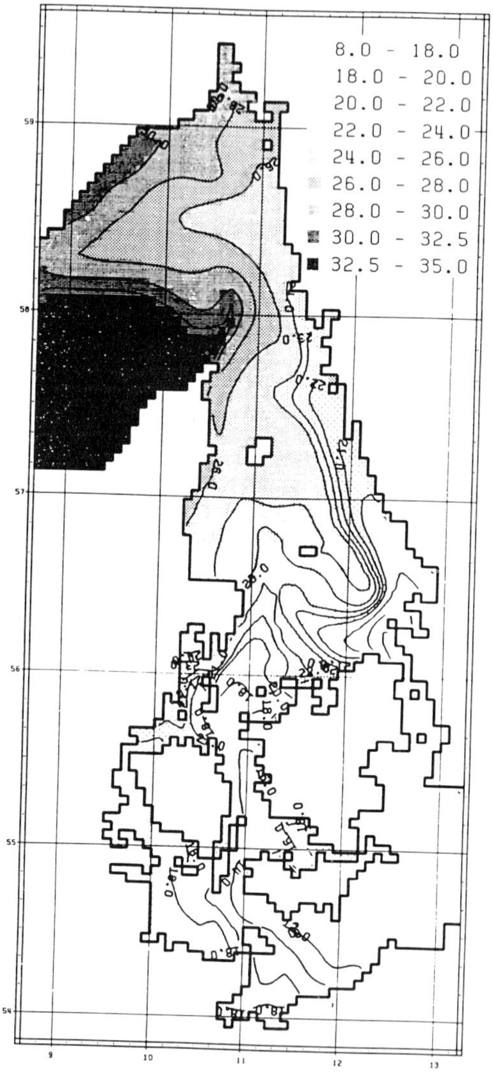

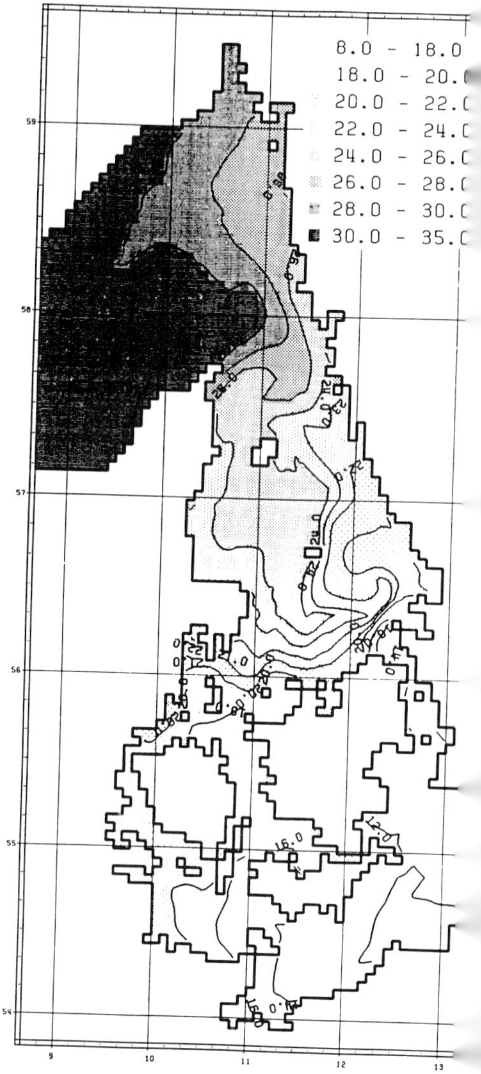

Figure 6: Surface salinity (psu) for mean summer situation
a) without wind forcing
b) forced by mean summer wind

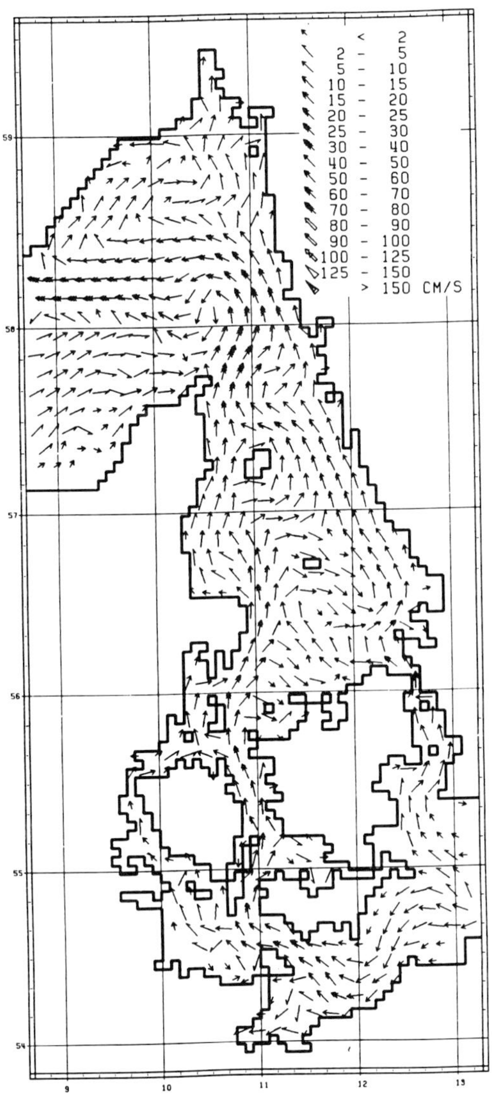

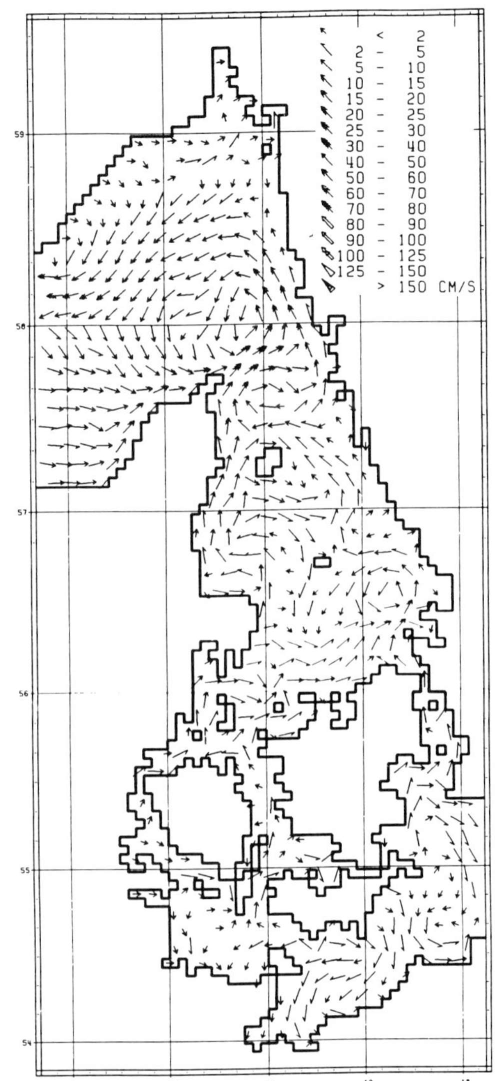

Figure 7: Circulation pattern at different depth

a) near sea surface (0 - 5 m); without wind forcing

b) near sea surface (0 - 5 m); forced by mean summer wind

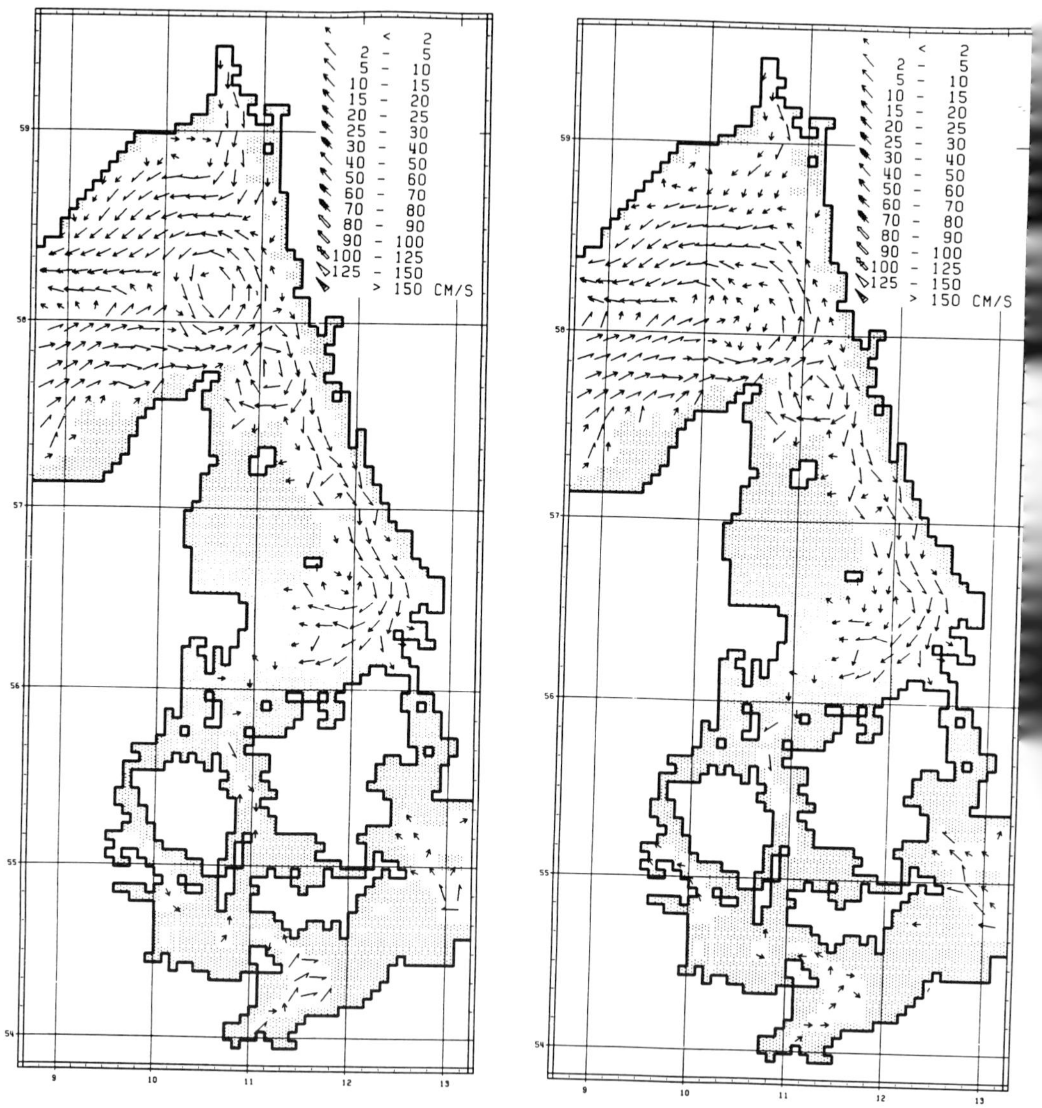

Figure 7: Circulation pattern at different depth
c) at 20 - 25 m depth; without wind forcing
d) at 20 - 25 m depth; forced by mean summer wind

An inspection and comparison of the computed surface salinities from both experiments (fig. 6 a, b) reveals that the buoyant plume that emerges into the Kattegat and flows along the Swedish coast line is largely retarded and almost blocked by the effect of the applied weak wind forcing (cf. fig. 5). This also holds for the Skagerrak, which is in agreement with observations of Aure and Saetre (1981), who report about a blocking of the coastal current for westerly winds. The rather drastic changes in the model results that are induced by the applied weak wind forcing support the observed high sensitivity of the region to wind events. Along the Swedish west coast the salinity increases by about ten psu on its way from the northern entrance of the Sound to the northern boundary of the Kattegat (latitude of Skagen), which is a result of the entrainment of haline bottom water in the deep channel along the Swedish coast.

The sea surface elevation for the mean summer situation, depicted in figure 4 shows the strong sea level gradient between North Sea and Baltic, which is maintained by the estuarine stratification. The near surface circulation for both experiments (fig. 7 a, b) shows that also the flow is sensitive to the applied changes. In particular two counter-clock-wise rotating stationary eddies in the transition Skagerrak/Kattegat, which occur for the wind induced blocking situation, support the assumption that the communication between the two topographically very different regions is very dependent on the wind forcing. The northward (out-) flow through the Great Belt and the Sound is likewise blocked by the wind forcing. The flow regime at greater depth (20 - 25 m) is less sensitive to the wind disturbance and shows the expected inflow of haline water within the channel along the Swedish coast (fig. 7 c, d; note that only half the flow information was drawn for the sake of clarity). In the northern entrance of the Kattegat the stationary eddy reaches down to the bottom.

13.5 Model experiments with different boundary conditions (forcing)

Apart from the mean estuarine circulation during early summer described above, our model configuration was used to estimate the influence of changes in the forcing (winds and in/outflow) on the stratification within the Kattegat. These experiments were conducted by J. Svensson and his colleagues of the SMHI at Gothenburg. Each of their experiments was initialised using the mean summer situation. For a zonal section within the Kattegat, just north of the northern entrance of the Sound (cf. fig. 1), the resulting salinity contours (J. Svensson, pers. communication) are shown for the following cases (fig. 8 a-d):

a) Mean summer situation including mean wind (fig. 8 a) This section agrees well with long term means of the salinity, compiled by Thorstensson (1980) and Szaron (1979). It complements figure 6 a (surface salinity) and serves as a reference mean value for the following cases.

b) Short term outflow situation with strong north-easterly wind (fig. 8 b) After three days with a fairly strong north-easterly wind of 10 ms^{-1} and the sea level at the Baltic boundary risen by 0.50 m only the local effect of the wind tends to broaden the coastal outflow plume. As a result the salinity increases at the Swedish coast in conjunction with a local upward motion (cf. the 28 psu contour in figure 8 a).

c) Long term outflow situation with moderate north-easterly wind (fig. 8 c) In contrast to the case b) the model was run for 26 days with a moderate north-easterly wind of 5 ms^{-1} and the water level in the Baltic risen by 0.50 m likewise. This situation allows that the advection of low salinity water masses from the Baltic cause a significant effect within the Kattegat. As a result rather low salinity values appear at the Swedish coast and the outflowing plume is not aligned with the coast due to the offshore wind forcing. Due to a re-circulation within the Kattegat also low salinity values appear at the Danish side of the section.

d) Inflow situation for south-westerly winds (fig. 8 d) The wind forcing was set to a south-westerly wind of 10 ms^{-1} and the model was run for a period of three days. At the entrance towards the North Sea the Jutland current was increased by a respective rise in the sea level gradient normal to the coast. Although the simulation is rather short an intrusion of haline bottom water appears at the western flank of the section in the southern Kattegat. The coastal plume at the Swedish side has narrowed in comparision to the mean summer reference situation (fig. 8 a).

These experiments demonstrate that the model reacts in a rather sensitive way to changes in the forcing conditions. They also help to demonstrate why the Kattegat is characterized by a very high variability of the circulation and hydrography. The model results are in good qualitative agreement with features observed during similar meteorological conditions.

13.6 Salt entrainment

Nutrients are rather non-conservative tracer due to biological activity. Therefore, in the discussion of the fate of nutrients often the salinity is taken as a tracer. Since nutrients enter the sea via the freshwater run-off they are in a good approximation inversely correlated to the salinity. The entrainment of salt in the Kattegat is hence also a measure for the entrainment of nutrients, which are advected into the region by the Jutland current. The vertical diffusive salt-flux (fig. 9) derived from the model output is a good measure of entrainment. It shows that indeed, as supported by a water mass analysis of observations, entrainment occurs along the entire channel at the Swedish west coast. A new feature that emerges from the result is the strong entrainment at the northern coast of the island Sjaelland, which is the southern coastal boundary of the Kattegat. The salt entrained there is then advected northward with the outflowing water. In contrast to the box model of Stigebrandt (citation in introduction) this model provides insight into the spatial distribution of the entrainment. Due to the complex topography this is far from being uniform. An inspection of the entrainment rates for the entire

Kattegat yields a five-times lower value than that published by Stigebrandt (1983). The main reason for this discrepancy is that he forced his model with transient winds, whereas our estimates are derived from the mean (and rather calm) summer situation.

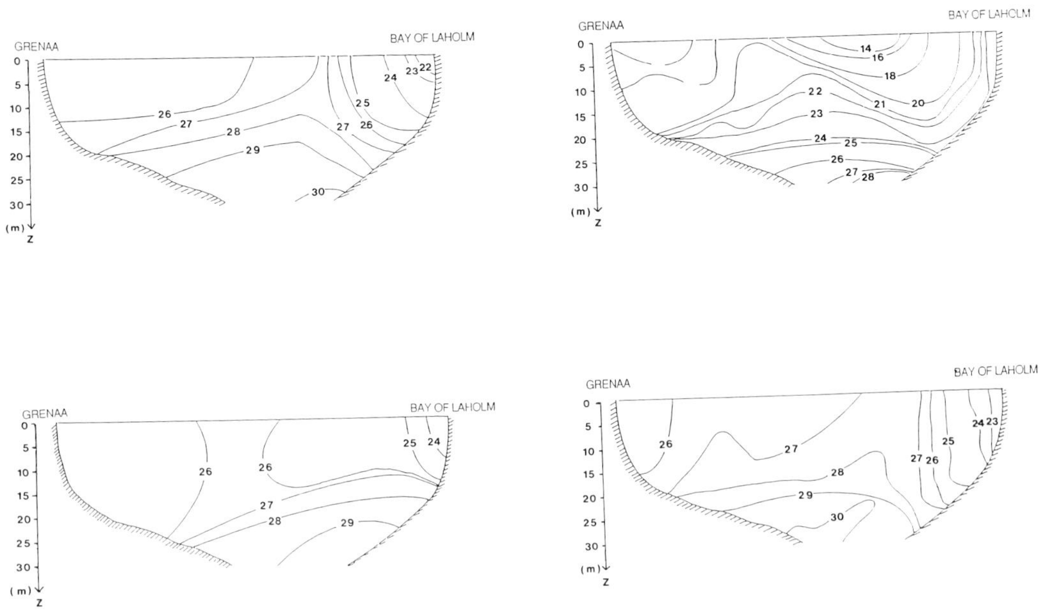

Figure 8: Salinity (psu units) on a zonal section through southern Kattegat (for location cf. fig. 1)

a) reference case (mean summer situation)

b) short-term outflow (3 days), north-easterly wind 10 ms^{-1}

c) long-term outflow (26 days), north-easterly wind 5 ms^{-1}

d) short-term inflow (3 days), south-westerly wind 10 ms^{-1}

Future model experiments, conducted with transient forcing, should allow for a more consistent and realistic comparison. It is then expected that the mixing rates produced by the model are much higher. It is interesting to note that especially the region with strong entrainment at the southern coast of the Kattegat is known for its high abundance of oxygen depletions. There secondary algae-blooms are favoured by the entrainment of nutrient-rich bottom waters into the euphotic surface zone, which, in early summer, often has a low nutrient content due to the regular spring plankton bloom.

13.7 A tracer experiment

The import of water masses from the North Sea into the Kattegat is demonstrated by a (passive and conservative) tracer experiment, where at the western boundary of the model domain a continuous release of arbitrary matter (100 %) was prescribed. The dispersion of this tracer was calculated with an Eulerian transport model which has the same grid-resolution as the circulation model. In this model the horizontal and the vertical advection as well as the vertical diffusion was determined according to the scheme outlined in section 13.2. The input data for this simulation was provided by the circulation model described in section 13.3.

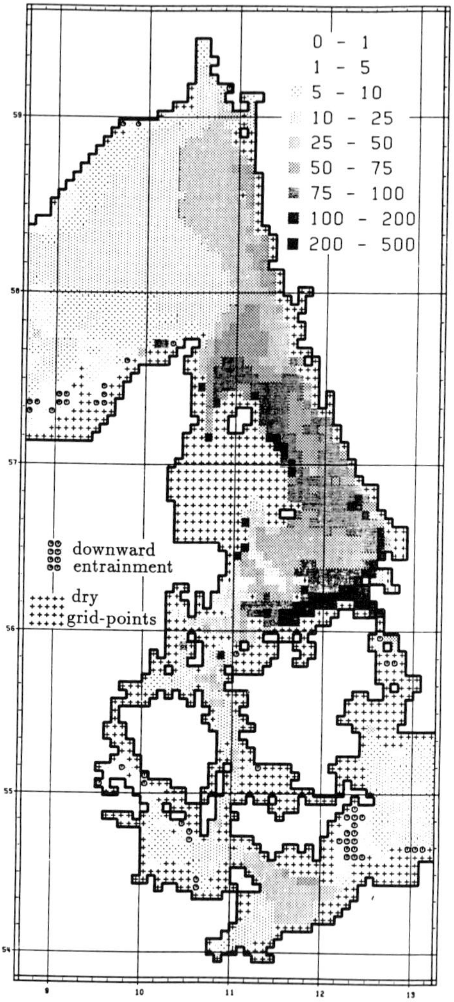

Figure 9: Salt entrainment rates at 15 m depth (ms^{-1} x psu)

As known from water mass characteristics the waters of the Jutland current enter the domain via the deep channel along the Swedish west coast. They change the signature of the buoyant surface plume by entrainment. These features are shown in the figures 10 a-c, which depict tracer concentrations (dissolved arbitrary matter) at three different depth ranges. Clearly the tracer at depth proceeds farthest south and the concentrations closer to the surface are caused by (vertical) entrainment. This dispersion experiment is in perfect agreement with the interpretation of the Kattegat dynamics, as derived from observations.

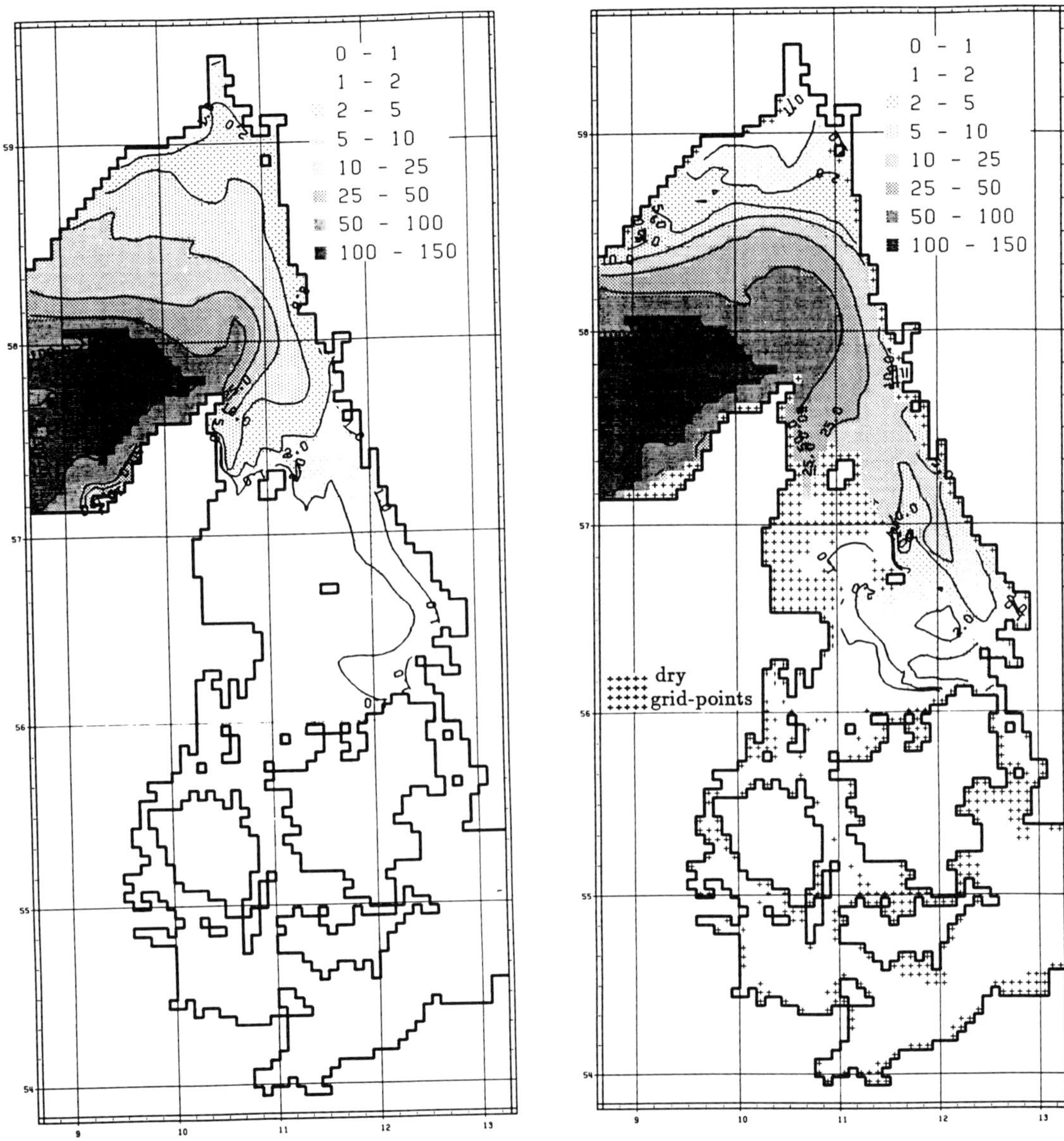

Figure 10: Concentrations of passive and conservative tracers in percent

a) surface distribution after 106 days

b) distribution in depth range: 10-15 m after 106 days

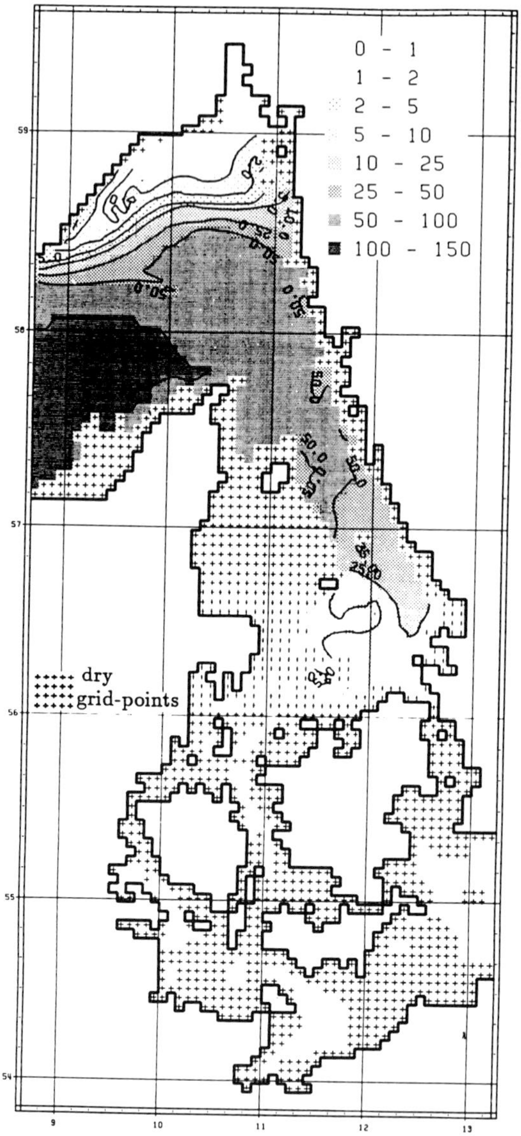

Figure 10: Concentrations of passive and conservative tracers in percent
c) distribution in depth range: 20-25 m after 106 days

13.8 Conclusions and Outlook

Some preliminary results on the adaptation and application of a baroclinic circulation model on the stratified transition and communication region between the North Sea and the Baltic have been presented. In collaboration with Scandinavian colleagues this work shall be continued with the intention to support the investigation of the very complex and considerable range of questions and problems in conjunction with anthropogenic eutrophication of coastal waters in northern Europe. Presently the performance of the model is only confirmed in qualitative way. However, only the long term application of the model with a realistic, transient forcing and in conjunction with suitably sampled field data for its validation, will help to gain confidence in the model and to support its reliability in regard to its application in environmental problems.

Acknowledgement

The authors wish to thank J. Svensson of the SMHI at Gothenburg for conducting all model experiments described in section V.

13.9 References

Aure, J., R. Saetre, 1981: Wind effects on the Skagerrak outflow, in: The Norwegian Coastal Current, Vol. I, ed.: R. Saetre, M. Mork.

Aksnes, D. L., J. Aure, G. K. Furnes, H. R. Skoldal, R. Saetre, 1989: Analysis of the Chrysochromulina Polylepis bloom in the Skagerrak in May 1988, environmental conditions and possible causes, Bergen Scientific Centre, Rep. BSC-89/1.

Backhaus, J. O., 1985: A three-dimensional model for the simulation of shelf sea dynamics. Dt. hydrogr. Z., 38 H. 4, 165-187.

Backhaus, J. O., D. Hainbucher, 1987: A finite difference general circulation model for shelf seas and its application to low frequency variability on the North-European shelf. In: three-dimensional models of marine and estaurine dynamics, 221-244, ed.: J.C.J.Nihoul and B.M. Jamart, Elsevier Science publishers B.V., Amsterdam, NL.

Backhaus, J. O., J. Bartsch, D. Quadfasel, J. Guddal, 1985: Atlas of monthly surface fields of air pressure, wind stress and wind stress curl over the North Eastern Atlantic Ocean: 1955-1982, Techn. Rep. 3-1985, Universitaet Hamburg, Inst. f. Meereskunde.

Backhaus, J. O., J. Bartsch, P. Damm, D. Hainbucher, T. Pohlmann, D. Quadfasel, G. Wegner, 1988: Circulation and hydrographic conditions in the North Sea during winter and spring 1987/88: a physical oceanographic study of the extreme algae bloom of spring 1988, Techn. Rep. 3-1988 (English version), Universitaet Hamburg, Inst. f. Meereskunde.

Boehlich, M. J., J. O. Backhaus, 1987: Simulation windangeregter Stroemungen im System Nordsee/Ostsee zur Interpretation eutrophierungsrelevanter biologisch-chemischer Prozesse in deutschen Kuestengewaessern, Forschungsbericht 102 04 215/18 Umweltbundesamt Berlin.

Dietrich, G., F. Schott, 1974: Wasserhaushalt und Stroemungen, in: Meereskunde der Ostsee, ed.: L.Magaard, G. Reinheimer, Springer Verlag, Berlin, New York, Heidelberg, 31-41.
Haltiner, G. J., 1977: Numerical weather prediction, John Wiley and Sons, Inc.

ICES, 1989: Report on: The Chrysochromulina Polylepis bloom in the Skagerrak and the Kattegat in May-June 1988: environmental conditions, possible causes, and effects, ICES, C.M.1989/L:18, ed.: H.R. Skjoldal, I. Dundas.

Kochergin, V. P., 1987: Three-dimensional prognostic models, in:Three-dimensional coastal ocean models, ed.: N. S. Heaps, Coastal and estuarine sciences 4, American Geophysical Union, Washington, D.C., 201-208.

Knudsen, M., 1899: De hydrografiske forhold i de danske farvande indenfor Skagen i 1849-88, Komm. for Vidensk. Unders. i den Danske farvande, 2:2.

Larsson, A. M., J. Rodhe, 1979: Hydrographical and chemical observations in the Skagerrak 1975-77, Report no. 29, Oceanografiska Intstitutionen, Goeteborgs Universitet, Goeteborg.

Mueller-Navarra, S., 1983: Modellergebnisse zur baroklinen Zirkulation im Kattegat, im Sund und in der Beltsee, Dt. hydrogr. Z., 36 H. 6.

Stigebrandt, A., 1983: A model for the exchange of water and salt between the Baltic and the Skagerrak, J. Phys. Oceanogr., 13, 411-427.

Svansson, A., 1972: Canal models of the sea and salinity variations in the Baltic and adjacent waters. Fisheries board of Sweden, Series Hydrography rep. no. 26.

Svansson, A., 1984: Hydrographic features of the Kattegat, National Board of Fisheries, Inst. of Hydrographic Research, Gothenburg, Sweden.

Szaron, J., 1979: Preliminary transport computation of water, salt and nutrients through the Goeteborg-Fredrikhavn section (GF) in the northern Kattegat, based upon measurements, 1975-77, Medd. Havsfiskelab., Goeteborg, Sweden.

Thorstensson, B., 1978: Decade mean values of hydrographical parameters from Marstrand to Skagen (the P-section), Medd. Havsfiskelab., Goeteborg, Sweden.

Winkel, N., 1989: Salz-Entrainment im Kattegat unter sommerlichen Bedingungen - eine Modellstudie, Diploma Thesis, Universitaet Hamburg, Inst. f. Meereskunde.

Part IV

Sedimentation

14

Observations of fine-sediment concentrations and transport in the turbidity maximum region of an estuary

R J Uncles, J A Stephens and M L Barton

Abstract

It is shown that a mobile stock of fine sediment and a suspended particulate matter (SPM) turbidity maximum can be maintained in the upper reaches of the Tamar Estuary by tidal pumping mechanisms. Tidal pumping is associated with asymmetry in both the flood and ebb, barotropic tidal currents and in the durations of the high and low-water slack periods. The asymmetry between ebb and flood tidal currents leads to an asymmetry in the rates of erosion of bed sediments, owing to the dependence of erosion rates on bed shear stress. The difference in durations of the high and low-water slack periods leads to a difference in the amount of suspended fine sediment which is deposited during slack currents. These barotropic, tidal processes are modified by the existence of buoyancy effects. Data from three types of experiments in the turbidity maximum area of the Tamar are used to illustrate tidal processes and their modifications by buoyancy.

Vertical profiling data show pronounced stratification during the early ebb following high water, with subsequent reduction of vertical mixing. It is not until the stratification has been eroded significantly in the faster currents and with the approach of the freshwater-saltwater interface that SPM can be mixed into the water column. Vertical mixing is, as an average over the water column, more intense during the flood. The mixing maximizes as the interface passes through the site on the flood and SPM is mixed into the near-surface waters where considerable up-estuary transport can occur. These data are complemented by a longitudinal transect experiment. In the stratified region down-estuary of the freshwater-saltwater interface, immediately following high water, a considerable period of time is available for settling of SPM. As the ebb currents increase, the stratification erodes and vertical mixing increases, leading to high SPM loads as the interface passes down-estuary. Much of the SPM is advected down-estuary, where it has little time to settle and deposit at low water, and subsequently advects back up-estuary in close association with the interface. These features are also apparent in data obtained from two longitudinally separated instrument packages deployed at 0.25m above the bed in the turbidity maximum region. SPM concentrations, longitudinal velocity and salinity data recorded at 5 minute intervals over a 10 day period demonstrate the growth of the turbidity maximum over the neap to spring-tide period and its dependence on both local resuspension, and therefore the location of the stock of bed-source sediment, and the presence of the freshwater-saltwater interface.

14.1 Introduction

A suspended particulate matter (SPM, or turbidity) maximum frequently occurs in the low salinity reaches of partly-mixed and well-mixed estuaries. It has been known for many years that this phenomenon can be caused by local resuspension of sediments in the upper estuary (Allen et al., 1980) or by gravitational circulation accumulating riverine and coastal SPM (Festa and Hansen, 1978), or by a combination of these two factors (Officer and Nichols, 1980; Officer, 1981). Numerical solutions have been derived for turbidity maximum formation due to both density-driven circulation (Festa and Hansen, 1978), and to tidally-induced resuspension (Uncles and Stephens, 1989).

Recently, it has been shown quantitatively that intratidal variations in water-column salinity stratification, and therefore stability and vertical mixing, may also play an important role in turbidity maximum formation (Hamblin, 1989). It has also been shown that stratification due to resuspended sediment can damp turbulence and vertical mixing near the bed, which may, in turn, lead to the formation of a lutocline (Sheng and Villaret, 1989; Wolanski et al., 1988; Kirby and Parker, 1977) and affect the behaviour of the turbidity maximum. The supply of fine sediment is also of crucial importance; Grabemann and Krause (1989) concluded from a study of the turbidity maximum in the Weser Estuary that tidally-induced resuspension from limited sources, and deposition at slack water, are the dominant intratidal processes, while the density-driven circulation provided a long-term supply of fine sediment to the turbidity maximum region.

This paper has three main objectives. First, we demonstrate that tidal processes are able to maintain a turbidity maximum and a long-term supply of fine sediment to the upper reaches of the Tamar, a typical macrotidal, partly-mixed estuary (Figure 1) and that the position and magnitude of the bed sediment supply and turbidity maximum depend on seasonal influences. Second, we present vertical-profile and longitudinal-transect data which demonstrate the intratidal behaviour of the turbidity maximum and the effects of water-column stability on its advection and growth. Third, we present data on near-bed currents, salinity and SPM concentrations at 5 minute intervals over a spring-neap cycle for two longitudinally-separated locations in the turbidity maximum region of the Tamar and demonstrate the importance of both advection and the modifying influence of the freshwater-saltwater interface on the resuspension processes and transport of cohesive, fine sediment.

The possible importance of the resuspension of bottom sediments by tidal currents to the formation of the turbidity maximum in mesotidal and macrotidal estuaries has been recognised for some time (Officer 1981 review paper). The following tidal processes can be important (Allen et al., 1980; Uncles et al, 1985 a,b,c; Dronkers, 1986): (a) asymmetry in the tidal currents in which flood currents exceed ebb currents and high-water slack periods exceed low-water slack periods, (b) suspension of eroded bottom sediments, and (c) the existence of an up-estuary maximum in the tidal currents and thus in the erosion of sediments. In this paper we provide some evidence for the importance of tidal processes and demonstrate the modification of these processes by intratidal variations in water-column stability and the presence of the freshwater-saltwater interface.

14.2 Tamar sediments

14.2.1 Bed sediments

The Tamar is a muddy estuary; much of the intertidal deposited and suspended sediment consists of mud (that is, primarily cohesive, silt and clay admixtures, Delo(1988)). The terms, fine sediment or mud, are used in this paper to describe these admixtures. Large areas of intertidal mudflats exist throughout much of the Tamar's length. Fine sediment does not occur on the bed within 3 km of the tidal limit at Weir Head (Figure 1) under winter conditions due to scour by the strong fresh-water flows (Bale, 1987). Mobile bed sediment varies from 50% silt (<63μm grain diameter) at 3km from the weir to 90-95% at 13km during winter. A sandy region then occurs between 13-16km with 30-50% silt. The silt content increases to about 80% near the mouth of the Tavy (Figure 1) and then falls to about 60% at the mouth of the Tamar. Silty, mobile bed sediment occurs within 1km of the weir at Weir Head during summer and the silt content increases from 60-70% at the mouth to 90-95% at the head (Bale, 1987). The water content of bed sediment is about 70% by weight in the areas of high silt content. Typical particle sizes in the silty bed sediment are 20-30μm and the organic content is typically about 10%.

14.2.2 Suspended sediments

A turbidity maximum exists in the upper reaches of the estuary and comprises fine sediment of around 20-30μm median diameter, as determined by in-situ laser diffraction (West et al., 1990). These maximum SPM concentrations occur around maximum flood and ebb current speeds. Minimum concentrations occur around high water when a wide range of SPM particle sizes co-exist but with large, loosely-bound flocs dominating (median diameters 200-300μm).

It is shown in Uncles and Stephens (1989) that purely tidal processes can generate a turbidity maximum in the Tamar. Tidal currents and elevations are computed using numerical solutions of the standard, one-dimensional hydrodynamic equations. Realistic topography is incorporated. Boundary conditions are taken to be the predicted tidal oscillations in water level at the mouth of the Tamar and the observed fresh-water inputs at its head. The space-step in the model is 1km and the variable time-step is of the order of several minutes.

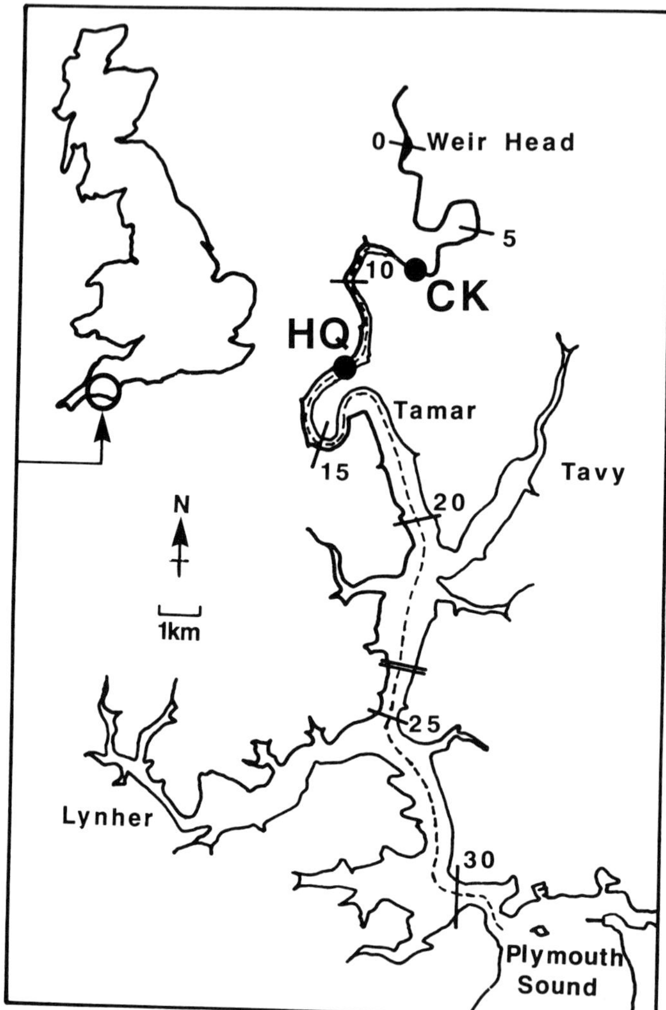

Figure 1: The Tamar Estuary, showing its sub-division into 5km segments and the locations of Stations CK and HQ in the turbidity maximum region which have been used as deployment sites for the long-term instrument packages.

The one-dimensional equation of sediment conservation is solved numerically on the same grid and with the same time-step as the hydrodynamical model. All parameter values are given in Uncles and Stephens (1989). Computed currents are used to define advection of sediment and local resuspension and deposition. Local terms are based on formulations given by Odd and Owen (1972). Odd and Owen (1972) assumed that the rate of erosion is proportional to the excess shear stress (τ-τ_e), where τ_e is the critical erosion stress and τ the bed shear stress. Our model uses a quadratic relationship between bed shear stress and cross-sectionally averaged velocity. Based on limited profiling data and measurements of the shear strength of intertidal mudflats, which indicate increasing strength from head to mouth, we take the critical stress for erosion to increase from 0.14 N m^{-2} near the head to 1.6 N m^{-2} near the mouth. An erosion rate "constant" of $3x10^{-5}$ kg m^{-2} s^{-1} is used.

For deposition, Odd and Owen (1972), following Krone (1962), assumed that the rate of deposition is proportional to the deficit shear stress. On the basis of flume data Odd and Owen, following Krone (1962), took the depositional stress to be 0.06 N m^{-2}. According to Delo (1988), the present state of knowledge of the critical bed shear stress for deposition indicates values between 0.06 - 0.10 N m^{-2}. We use a value of 0.06 N m^{-2}. The rate of deposition is proportional to the aggregate fall velocity, which in turn is taken to be proportional to the SPM concentration with a constant of proportionality of $2x10^{-6}$ m s^{-1} $(ppm)^{-1}$.

The fresh-water input SPM concentration, P_f, is related to the fresh-water flow by a power law deduced from regressional analysis of measured data (Harris et al., 1984). SPM concentrations in coastal waters need not be specified on the ebb. It is assumed that coastal waters lying further offshore than a mean spring-tide tidal excursion have concentrations of 5ppm (consistent with measured values) and that a linear gradient in SPM concentration exists in the offshore direction.

Figure 2(a) shows computed (dashed lines) and measured longitudinal distributions of depth-averaged, high-water SPM at spring tides during 1985. The date on which measurements were made, followed by the fresh-water run-off during that day (m^3 s^{-1}) and the tidal range (m) are given in the top right of each diagram. With reference to intratidal data presented later, the SPM concentrations tend to be lowest at high-water slack.

The model simulates a high-water turbidity maximum in the upper reaches which is roughly of the correct magnitude and in the correct position. A comparison of data on 11/03 and 09/04 clearly shows the effect of high run-off for a similar tidal range (5.0 and 4.9m); the maximum is displaced down-estuary at high run-off (60 as opposed to 9 m^3 s^{-1}) and has a much smaller magnitude. This behaviour is reproduced by the model. A comparison of data on 11/03 and 27/06 clearly shows the effect of tidal range on the magnitude of the turbidity maximum for a similar fresh-water run-off (9 and 8 m^3 s^{-1}); the maximum is located in roughly the same place but is much reduced for the smaller tidal range (3.2 as opposed to 5.0 m). This behaviour is also reproduced by the model.

A feature of the high-water data which is not reproduced by the model is the narrowness of the turbidity maximum peak, the tidal model smears the SPM maximum over a tidal excursion. It is shown later that this narrowness may be associated with the freshwater-saltwater interface and the effect of water-column stability down-estuary of the interface.

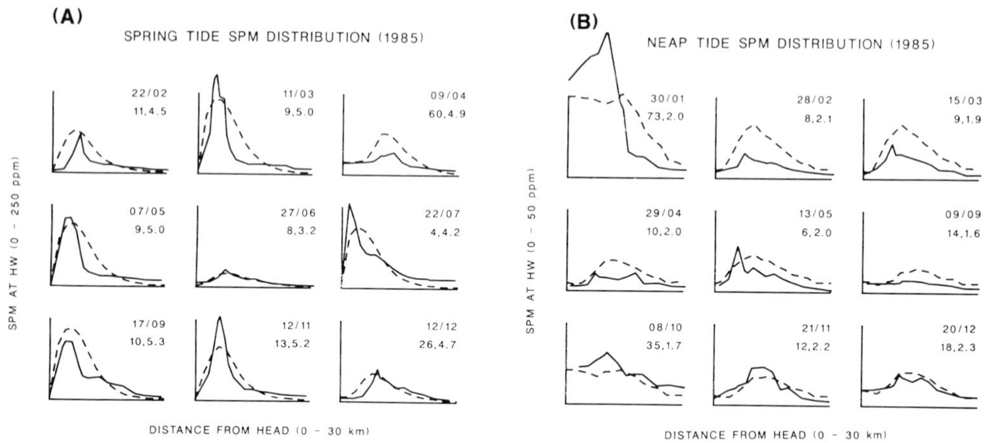

Figure 2: Measured and computed (dashed lines) distributions of SPM at high water of (a): spring tides and (b): neap tides along the estuary during 1985. Each vertical axis represents 0-250ppm in (a) and 0-50ppm in (b). Each horizontal axis represents 0-30km. The date, run-off ($m^3\ s^{-1}$) and tidal range (m) for each period of observation are shown in the top right of each diagram. Reproduced from Uncles and Stephens(1989), with permission of the American Geophysical Union.

Figure 2(b) shows computed (dashed lines) and measured longitudinal distributions of SPM at neap tides during 1985. Concentrations are much smaller than those at spring tides. A comparison of measured data for 30/01 and 29/04, which have the same tidal ranges (2.0m) but very different values for the run-off (73 as opposed to 10 $m^3\ s^{-1}$), demonstrates the much higher fresh-water SPM concentrations at high run-off. The pronounced turbidity maximum on 30/01 was probably due to resuspension in the strong ebb currents; it is reproduced by the model, but is much less pronounced.

14.2.3 Fine-sediment transport

The tidal model and observations establish that the turbidity maximum within the Tamar migrates longitudinally in response to changing river flows. The same occurs in the Gironde Estuary (Allen et al., 1977, 1980); the maximum migrates up-estuary during low fresh-water discharge, and is located down-estuary during high discharge. There is an accompanying shift in the sediment shoaling pattern, which consists of sediment accumulation in the upper estuary in summer and autumn, and in the lower estuary in winter and spring. A similar seasonal pattern of sediment movement has been observed in the Tamar Estuary (Bale et al., 1985) and has been modelled using purely tidal processes (Uncles et al., 1988).

The sediments model was run for 12 years of simulated time. During the simulation, environmental forcing functions of tide and fresh-water run-off repeated themselves three times with a four-year periodicity. This was done to ensure that quasi-periodic solutions for the sediment-mass distributions could be achieved. Observed Tamar forcing functions for the period 1982-1985 were used. The distribution of sediment mass in the upper 20 km of estuary was essentially the same for the second and third four-year periods.

The estimated flux of sediment across the head of the estuary for the period 1982-1985 is shown in Figure 3(a). These data have been subjected to a running two-week average to facilitate presentation. The estimated sediment inputs to the estuary correlate with and amplify the seasonal variations in fresh-water run-off. Inputs are very small during low run-off summer periods and large during high run-off winter periods. This is emphasized in Figure 3(b), which shows the cumulative fluvial input of sediment to the estuary during 1982-1985. The major changes in sediment load occur during 2-4 month winter periods.

Figure 3(c) shows the calculated mass of fine sediment deposited on the bed of the estuary (kg per metre of estuarine length) during 1982-1985 at 2 km and 8 km from the head. The datum of zero sediment mass at a location defines the situation in which all of the sediment which is able to be eroded and transported, under the conditions experienced during 1982-1985, has been moved from the location. Figure 3(d) provides the same information for 2 km and 18 km from the head. Data are running-averages over a spring-neap (14.7 day) period to facilitate presentation. A pronounced seasonal cycle in sediment mass occurs within the upper 20 km of our model estuary, as observed in the Tamar by Bale et al. (1985).

Fine sediment accumulates on the bed of the estuary at 2 km from the head during low run-off summer months. Sediment is transported down-estuary during high run-off winter months, so that this region is subsequently denuded of sediment. The sediment accumulation close to the head is very sensitive to fresh-water flows. Increasing summer accumulation from 1982 to 1984 (Figure 3(c)) is a consequence of longer summer periods of low fresh-water run-off. The small accumulation during the summer of 1985 is a consequence of higher than usual run-off during a summer period.

The accumulation of sediment at 2 km from the head occurs during times when the fluvial sediment input is small (Figures 3(a,b)). The calculated mobile sediment accumulated at 2 km from the head during the low run-off summer of 1984 is 2.2×10^3 kg m^{-1}. If the fluvial input of sediment during this period were deposited in the upper 2 km of estuary, it would amount to 10^2 kg m^{-1}. Therefore, in agreement with the observations of Bale et al. (1985), the accumulated sediment mass must be derived from up-estuary transport of material and not from deposition of fluvial-derived sediment. The removal of sediment during high run-off winter months must be due to erosion in the fast ebb currents. When run-off is low, the tidal flood currents exceed the ebb currents, and sediment is eroded from down-estuary and subsequently transported into the upper reaches. Direct observations of these processes are given in Uncles et al. (1985 a, b, c).

The pattern of sediment accumulation shown in Figure 3(c) at 2 km from the head is repeated down to about 6 km. The summer accumulation of sediment maximizes at 4 km. This pattern becomes less sensitive to run-off further from the head. Accumulation of sediment begins earlier in the summer, and removal is delayed until later in the autumn and winter. Daily fluctuations in run-off have little influence on the accumulated mass of sediment.

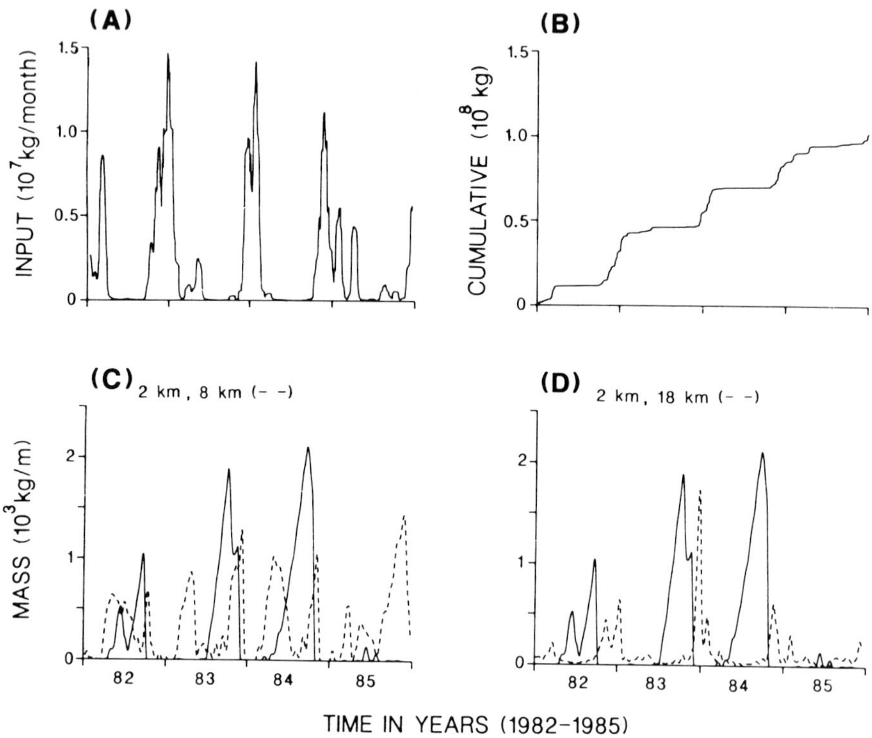

Figure 3: Estimated fluvial sediment inputs and simulated mass accumulation on the bed of the estuary as functions of time during 1982-1985. (a): Fluvial sediment inputs (10^7 kg month^{-1}); (b): Cumulative fluvial sediment inputs (10^8 kg); (c): Mass of mobile bed sediment per unit length of estuary (10^3 kg m^{-1}) at 2 km (full line) and 8 km (dashed line - plotted at twice actual magnitude for clarity) from head; (d): Mass of mobile bed sediment per unit length of estuary (10^3 kg m^{-1}) at 2 km (full line) and 18 km (dashed line - plotted at twice actual magnitude) from head. Reproduced from Uncles et al.(1988), with permission of the Estuarine Research Federation.

The seasonal pattern of sediment accumulation changes further down-estuary. At 8 km from the head, sediment is accumulated and removed twice a year (Figure 3(c)). Sediment is transported down-estuary during highest run-off until the upper region is depleted. Sediment is tidally-pumped up-estuary early in the year, when the run-off is sufficiently low that tidal pumping can dominate. When the run-off is very low during summer, sediment is tidally pumped up-estuary faster than it is being supplied from down-estuary, and sediment is depleted. During the onset of high run-off, sediment is transported down-estuary from the upper reaches. Some of this sediment accumulates at 8 km until the run-off achieves very high values and the sediment is then transported further down-estuary.

The seasonal pattern of sediment accumulation is again different at 18 km from the head (Figure 3(d)). Sediment accumulation is strongly and positively correlated with run-off and fluvial sediment influx (Figure 3(a)). The calculated mobile sediment mass accumulated at 18 km during the winter of 1983/84 is 10^3 kg m^{-1} (Figure 3(d)). If the fluvial sediment influx during this period (Figure 3(b)) were deposited over the upper 20 km of estuary, it would supply 10^3 kg m^{-1}. Therefore, the accumulated sediment could be derived from the down-estuary transport and deposition of fluvial inputs, any additional mass being due to transport and accumulation of mobile estuarine sediment. This estuarine sediment is rapidly accumulated from up-estuary during the onset of high run-off, and is rapidly removed by erosion and up-estuary transport when the rate of supply of mobile estuarine sediment decreases. Very little up-estuary transport occurs below 20 km from the head (most transport is down-estuary), so that only a single period of sediment accumulation occurs during the year.

14.3 Intratidal behaviour of turbidity maximum

14.3.1 Vertical profiling data

The intratidal behaviour of salinity, velocity and SPM transport has been observed on a number of occasions in the upper reaches of the Tamar using direct-reading profiling equipment from anchored vessels. We present one set of data from a station close to CK in Figure 1. The measurements were made during a spring tidal cycle on 7 July 1982 with a tidal range of 3.9m (small springs) and a low fresh-water run-off of 3.5 m^3 s^{-1}.

Figure 4(a) shows contour plots of salinity on a depth-time diagram. The vertical scale is height above bed and a line depicting the water surface is drawn to represent total depth of water. The freshwater-saltwater interface (somewhat arbitrary between 0 and 1 ppt but taken in this diagram as 0.1 ppt) passes through the station on the ebb at roughly 11.5 hours and on the flood at roughly 16.5 hours. The salinity stratification on the ebb between high water slack and peak ebb currents at mid-water is much higher than during the flood and implies inhibited vertical mixing of salt and SPM during the ebb.

Figure 4(c) shows contour plots of SPM concentration on the depth-time diagram. Concentrations are low at high water and, when compared with the salinity contours, show an increase as the ebb progresses and salinity stratification erodes. A turbidity maximum is reached in the fresh water roughly 0.5-1 hour after the passage of the freshwater-saltwater interface. Concentrations drop to 100-150ppm around low water and then increase as the freshwater-saltwater interface approaches on the flood. The turbidity maximum is reached as the freshwater-saltwater interface passes the station. The flood turbidity maximum is mixed more effectively through the water column than the ebb maximum.

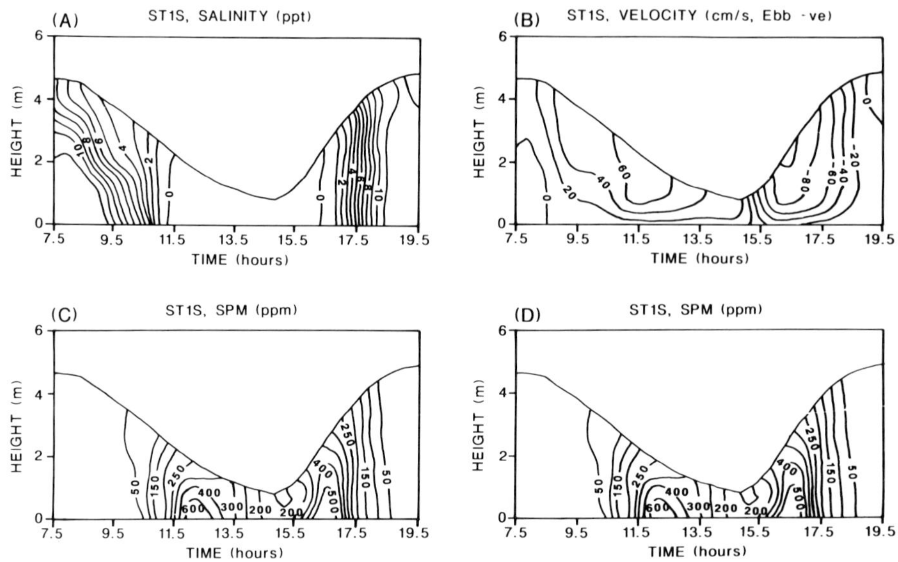

Figure 4: Contour plots of salinity, velocity (ebb positive here) and SPM concentrations on a height-time diagram over a spring tidal cycle at Station ST1S near Station CK (located in Figure 1) on 7 July 1982. (a): Salinity contours (0.1, 1, 2, ..., 12 ppt); (b): Velocity contours (0, - 20, ..., -100 and 0, 20,.., 60 cm s^{-1}); (c): SPM concentrations (50, 100, 150, ..., 600 ppm); (d): Same as (c).

The velocity contours are shown in Figure 4(b) and the SPM concentration contours are reproduced again in Figure 4(d). Ebb currents have maximum surface speeds of about 0.6 m s^{-1} and near-bed shear maximizes at approximately 11.5 hours. The SPM concentrations rapidly increase at this time and maximize roughly 0.5-1 hour later when the near-bed shear is still high but the depth-averaged currents have decreased somewhat (Figure 4(b)). Flood currents have maximum speeds of about 1 m s^{-1}. Maximum near-bed shear occurs at approximately the same time as the maximum surface and depth-averaged currents (16.5 hours) which also corresponds to the passage of the turbidity maximum (Figure 4(d)). The flood turbidity maximum is mixed into the near-surface waters where the SPM is carried up-estuary by the strong currents.

The difference in the behaviour of the turbidity maximum between ebb and flood portions of the tide is clearer in Figure 5(a) where depth-averaged current speed, salinity and SPM concentrations are plotted against time. SPM concentrations lag depth-mean currents on the ebb and coincide on the flood. The estimated vertical eddy diffusivity, K_z, at 0.25m above the bed and the depth-averaged SPM concentrations are plotted against time in Figure 5(b). The estimation of K_z at 0.25m assumes the validity of the mixing-length approximation with Munk-Anderson damping (Munk and Anderson, 1948) and that an approximately linear increase in K_z occurs with height above, and close to, the bed; vertical density gradients due to SPM are ignored. Maximum SPM loads appear to occur after maximum bed-shear stresses are reached and correspond to maximum vertical mixing near the bed when near-bed suspensions are mixed into the flow.

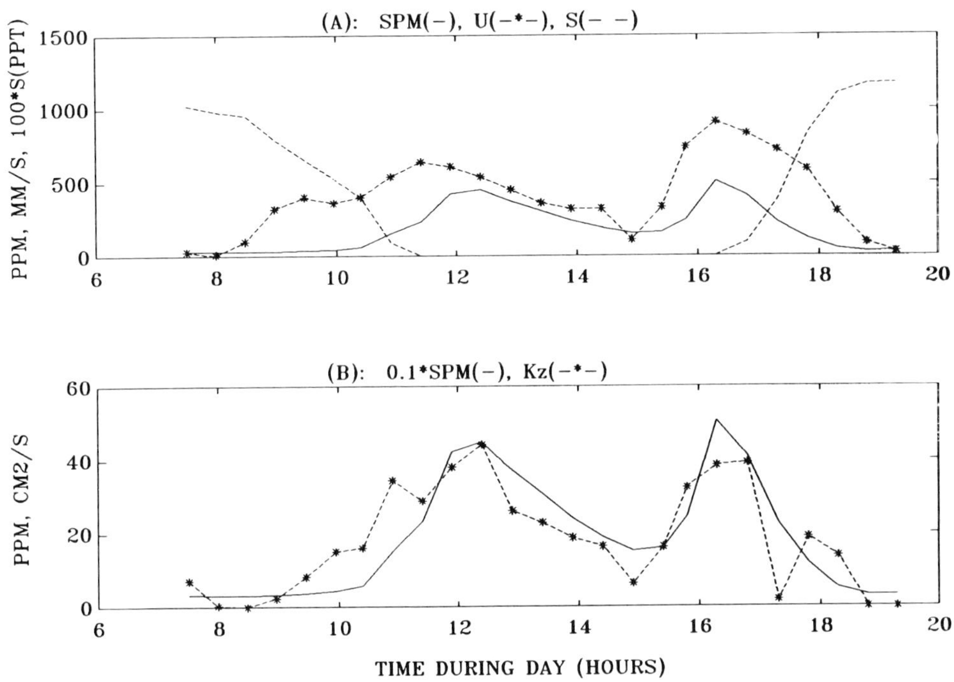

Figure 5: (a): Depth-averaged SPM (-), speed (U, -*-) and salinity (S, - -) and (b): Depth-averaged SPM (-) and the estimated vertical diffusivity (K_z, -*-) at 0.25m above the bed as functions of time for the contoured data shown in Figure 4.

14.3.2 Longitudinal transect data

Figures 6(a,b) show longitudinal transect data of vertical profiles for SPM concentrations and salinity during a spring tide on 8 July 1988 in the turbidity maximum region of the Tamar.

The salinity is depicted by dashed contours. The times of beginning and end for each transect are shown together with the state of the tide relative to high or low water and the longitudinal current at 0.25m above the bed near the down-estuary boundary of the transect.

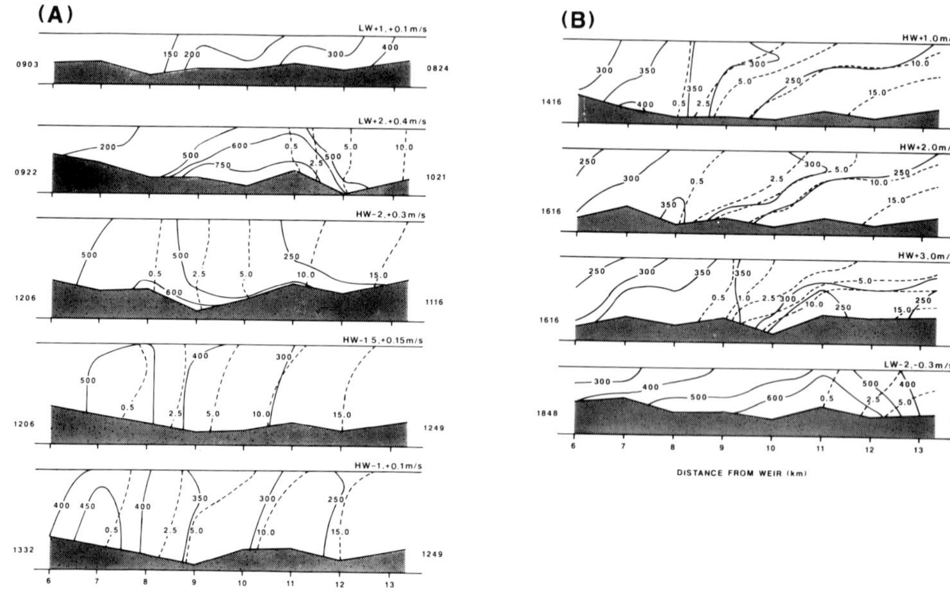

Figure 6: Longitudinal transect data between Stations CK and HQ located in Figure 1 showing SPM (-) and salinity (- -) contours during a spring tide on 8 July 1988. The longitudinal velocity at 0.25m above the bed at 12.5km is shown on the top right of each plot and the start and end times for each transect on the left and right of each plot. (a): Data for low water (LW) plus 1 hour until high water (HW) minus 1 hour; (b): Data for high water (HW) plus 1 hour until low water (LW) minus 2 hours.

The salinity data show the freshwater-saltwater interface advecting into the region on the flood with strong vertical mixing. At high water the up-estuary fresh water is ebbing, as is the surface water in the down-estuary, stratified region. The near-bed water in the stratified region 'stalls', partly because of the opposing longitudinal density gradient, and the stratification becomes intense as fresher water ebbs in the surface layers. The near-bed water in the stratified region begins to ebb roughly two hours before low water and the freshwater-saltwater interface moves down the estuary.

At the start of the flood the SPM concentration increases down estuary, so that the turbidity maximum is, by implication, associated with the freshwater-saltwater interface region. As the interface advects into the region on the flood there is considerable resuspension of near-bed mud at the interface in the strong currents. Some of this SPM moves with the interface but much of it settles and deposits as high-water slack approaches. Some SPM remains and reaches the limit of intrusion of the interface, where it forms a slowly eroding turbidity maximum as particles settle and deposit. At the beginning of the ebb in the fresh water the lower SPM concentrations advect down-estuary. In the stratified region the SPM ebbs in the surface layers. Deposition occurs in the stagnant, near-bed water. As the near-bed currents increase on the ebb and the interface moves down-estuary, strong vertical mixing and resuspension occur in the unstratified water behind the interface and the associated turbidity maximum advects down-estuary.

14.4 Long-term monitoring of SPM concentrations

14.4.1 Instrumentation

A near-bed, instrument package has been developed to monitor long-term behaviour of the SPM concentrations. Measurements of SPM concentration are made using Partech Electronics (UK) light-transmittance turbidity meters (0-10^2 and 0-10^4ppm). These instruments are calibrated using gravimetric determinations of suspended sediment concentrations in the measurement area. Flume-calibrated, Colnbrook Instrument (UK) E/M current meters are used to measure horizontal velocity. Valeport Marine (UK) conductivity/temperature sensors are used to determine temperature and salinity. These measurements are made at 0.25m above the bed and the pressure sensor is located in the top of the frame at 0.65m above the bed. Batteries, electronics boards and data logger are housed at the top of the frame and horizontal stabilising legs are fitted to its base. Data are recorded every 5 minutes. Details of the instruments are given in Uncles et al. (1989), although the frames have since been redesigned to accommodate the shallow waters of the upper estuary.

14.4.2 Deployment

Two of these instrument packages have been deployed in the turbidity maximum region of the Tamar Estuary at Stations CK and HQ in Figure 1 for 10 days at a time for several occasions during 1988 and 1989. Stations CK and HQ are 7 and 12.5km from the head, respectively. The data discussed here were obtained during November 1988 at a time of low fresh-water flows but following a period of spates during October. We therefore expect from Figure 3(c) that much of the mobile mud has moved from the upper reaches, within a few kilometres of the head, and has been deposited between HQ and CK and for some short distance up-estuary of CK.

14.4.3 Spring-neap behaviour of SPM concentrations

SPM concentrations at 0.25m above the bed for Station CK are shown in Figure 7(a), during a 10-day period in November 1988. The data at HQ are very similar, although concentrations are somewhat lower, and are therefore not presented. The near-bed SPM levels fall to low values during the low neap tide on day 4 and subsequently increase on the rising spring tides. Each tidal cycle comprises two main peaks for both stations. The SPM data exhibit striking similarity for all tidal cycles. The increasing SPM concentrations with increasing tidal range demonstrate that resuspension in the tidal currents provides the source of water-column SPM for the turbidity maximum. This SPM is, in turn, derived from the fine sediment deposits on the bed of the estuary.

The fluxes of SPM (F, ppm m s^{-1}) at 0.25m above the bed for Station CK are shown in Figure 7(b). The corresponding data at HQ are very similar and are therefore not presented. The tidally-averaged (25 hour running mean) fluxes, <F>, are drawn as the bold line and demonstrate that the residual fluxes are much smaller than the instantaneous values, so that one would expect the net amount of mud transported during a single tide to be very small. However, the magnitude and direction of transport, based on just one point in each cross-section, has little significance to the estuarine mud budget.

Numerical solutions have been derived for turbidity maximum formation due to density-driven circulation (Festa and Hansen, 1978), and this circulation pattern may be expected to contribute to SPM transport in the presence of tidally-induced resuspension. Figures 7(c,d) show the residual, longitudinal current at 0.25m above the bed for Stations CK and HQ, respectively. An up-estuary flow occurs at HQ during the neap tides and smaller spring tides but reverses during the larger spring tides (Figure 7(d)) when SPM concentrations are largest (Figure 7(a)). At CK the near-bed residual flow is directed down-estuary (Figure 7(c)), so that density-driven circulations alone cannot be invoked as a causative mechanism for turbidity maximum formation in the upper estuary.

14.4.4 Intratidal behaviour of water depths and velocities

We will focus on the single tidal cycle during 9.0-9.5 days at the end of the record, when SPM levels and fluxes are highest (Figures 7(a,b)). The behaviour of the turbidity maximum is similar during the preceding spring tides. Flood and ebb SPM peaks also occur at neap tides, but these have much lower SPM concentrations due to the greatly reduced influence of resuspension processes.

Although the 25 hour averaging has the effect of shortening the time series, as shown in Figures 7(c,d), a 12.5 hour average of the velocity data for the last tidal cycle confirms the trend in Figures 7(c,d) which shows that near-bed residual currents are directed down-estuary when SPM levels are largest.

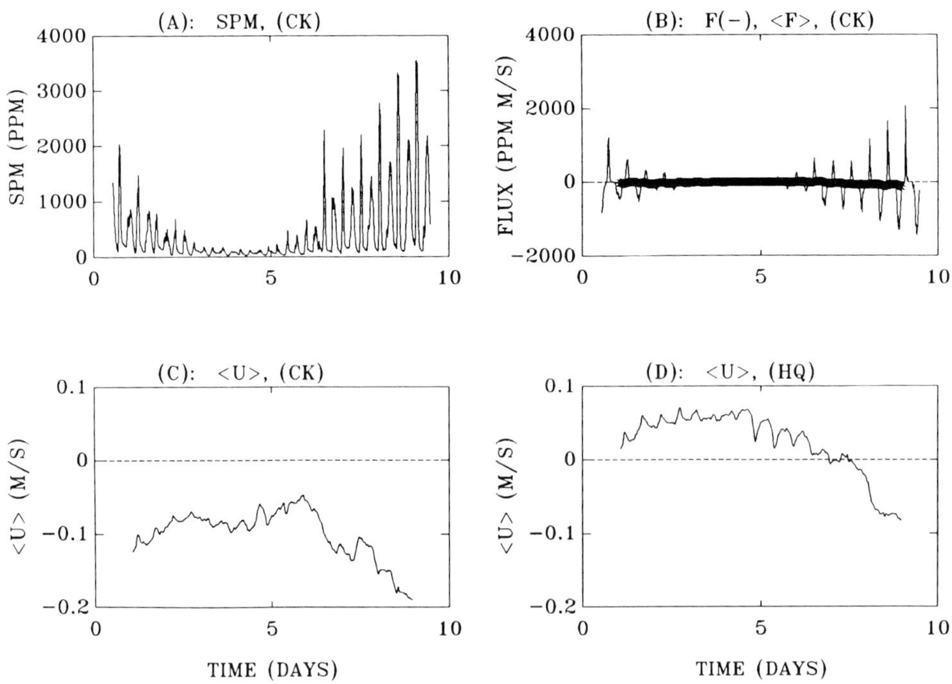

Figure 7: Time series data at 0.25m above the bed for Stations CK and HQ located in Figure 1; times 0.0 to 1.0 days represent November 14, 1988. (a): SPM concentrations at Station CK; (b): SPM fluxes at Station CK (F, flood positive) with a 25 hour running mean, <F>, drawn as the bold line; (c,d): 25-hour averaged longitudinal currents (<U>, flood positive) at Station CK (c) and HQ (d).

Background physical data are shown in Figure 8. Total water depths at both sites are plotted in Figure 8(a) and vary between about 1m and 6m over the tidal cycle. The CK site is roughly 1m deeper than the HQ site, despite being further up-estuary. Low water is approximately 0.5 hour later at Station CK and high water occurs at approximately the same time. Salinity data are shown in Figure 8(b) and demonstrate that the freshwater-saltwater interface intrudes just a short distance up-estuary of CK.

Longitudinal velocities at 0.25m above the bed are shown in Figure 8(c) for CK (dashed line) and HQ. Rapid accelerations occur between maximum ebb and maximum flood currents with only very brief periods of slack currents at low water. Low water slack occurs roughly 0.5 hour later at CK. Following maximum flood currents at HQ there is a slow deceleration until high-water slack at roughly 221 hours. This is followed by a long period of slow currents which are the result of shallow water overtides and density-driven currents opposing the near-bed, ebb-directed tidal flow immediately following high water. The rapid deceleration following maximum flood currents at CK and the long period of near-zero currents on the mid-flood appears to be triggered by resuspension of high SPM concentrations. This behaviour is currently not understood, although it is consistently observed on several earlier tidal cycles and is possibly related to the damping of turbulence and reduction of bed shear stress in the presence of high SPM loads, as modelled by Sheng and Villaret (1989).

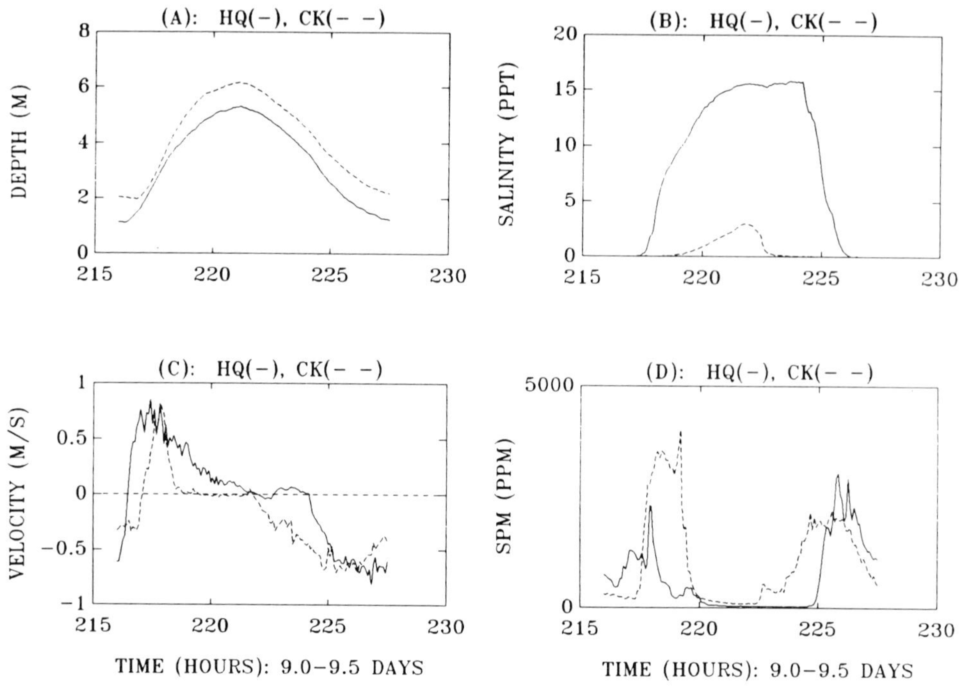

Figure 8: Time series data at 0.25m above the bed for Stations CK (- -) and HQ (-) located in Figure 1; times 216 to 228 hours correspond to times 9.0-9.5 days in Figure 7. (a): Total depths of water; (b): Salinities; (c): Longitudinal currents (flood positive); (d): SPM concentrations.

14.4.5 Intratidal behaviour of SPM concentrations

Figure 8(d) shows the SPM concentrations at HQ (-) and CK (- -) as functions of time for the final tidal cycle covering 9.0-9.5 days of the record. Flood SPM levels are smaller than ebb levels at HQ, the reverse being true at CK, despite similar near-bed, peak current velocities at both sites (Figure 8(c)). This implies a scarcity of source mud on the bed, down-estuary of HQ. Each main flood and ebb SPM peak comprises several sub-peaks, some of which may be the result of patchiness in the bed-source mud, which in turn may lead to patchiness in the resuspended SPM.

Both sites exhibit low SPM levels around low-water slack (roughly 217 hours) and extended periods of very low SPM levels around high-water slack (roughly 221 hours). The main flood SPM peak at HQ leads that at CK and the main ebb peak at CK leads that at HQ. However, this is not unequivocal evidence that advection of SPM dominates local processes because Figure 8(c) shows that flood currents (and possible local resuspension) at HQ lead flood currents at CK, whereas ebb currents at CK (and possible local resuspension) lead those at HQ.

14.4.6 Intratidal behaviour at CK

SPM concentration (-), longitudinal velocity (- -) and salinity (-.-) at 0.25m above the bed for CK during the tidal cycle at 9.0-9.5 days are shown in Figure 9(a). The salinity is zero for most of the tide and reaches only 3 ppt at high-water slack. Considering the ebb portion of the tide after roughly 222 hours, the sudden increase in SPM concentrations corresponds to the increase in vertical diffusivity following the passage of the freshwater-saltwater interface once the critical erosion shear stress has been exceeded. The salinity is zero for most of the ebb.

A marked correlation between ebb velocity and SPM concentrations at 0.25m above the bed is exhibited in Figure 9(b). This scatter plot implies a critical velocity for erosion of 0.2 m s^{-1}, or a critical bed shear stress of about 0.15 N m^{-2} (assuming this to be $3.7u^2$, where u is measured at 0.25m above the bed, West et al., 1990). This is similar to critical shear stresses for erosion of some estuarine and San Francisco Bay muds (Villaret and Paulic, 1986; Mehta, 1988). The strong correlation implies that resuspension and advection of mud from immediately up-estuary of CK are the dominant processes governing SPM loads during the ebb at CK.

The flood SPM levels (times greater than 217 hours in Figure 9(a)) show a dramatic increase as the critical shear stress is exceeded. Salinity is zero at this time and the enhanced mixing on the flood tide associated with stronger current speeds (implied from profiling measurements but only marginally reflected in the near-bed data presented) lead to elevated SPM levels at 0.25m above the bed. The near-bed SPM-induced density gradients appear to 'stall' the near-bed current and leads to a decline in SPM levels as SPM either settles or is replaced by particles mixed down from flood-directed currents higher in the water column where SPM levels will be smaller. The increasing salinity implies that a similar vertical mixing is occurring for salt. The passage of the freshwater-saltwater interface leads to a subsequent peak in SPM levels as SPM associated with the interface advects through CK.

14.4.7 Intratidal behaviour at HQ

SPM concentration (-), longitudinal velocity (- -) and salinity (-.-) at 0.25m above the bed for HQ during the tidal cycle at 9.0-9.5 days are shown in Figure 9(c). The salinity is zero during the peak ebb to peak flood and reaches 15 ppt at high-water slack. Considering the ebb portion of the tide (after roughly 224 hours for currents at 0.25m), the sudden increase in SPM concentrations appears to correspond to the resuspension and down-estuary advection of mud deposited over high water between CK and HQ, some of which is carried in the surface layers of the stable water column. An increase in vertical diffusivity follows the erosion of stratification and mixes near-bed SPM higher into the water column. The passage of the freshwater-saltwater interface results in a maximum in the vertical mixing and advection through HQ of SPM associated with the interface because of resuspension further up-estuary. SPM levels fall following the passage of the interface despite strong near-bed currents, which implies an absence of local source mud and the presence of sufficient vertical mixing through the water column to result in 'dilution' of resuspended SPM advected from much further up-estuary.

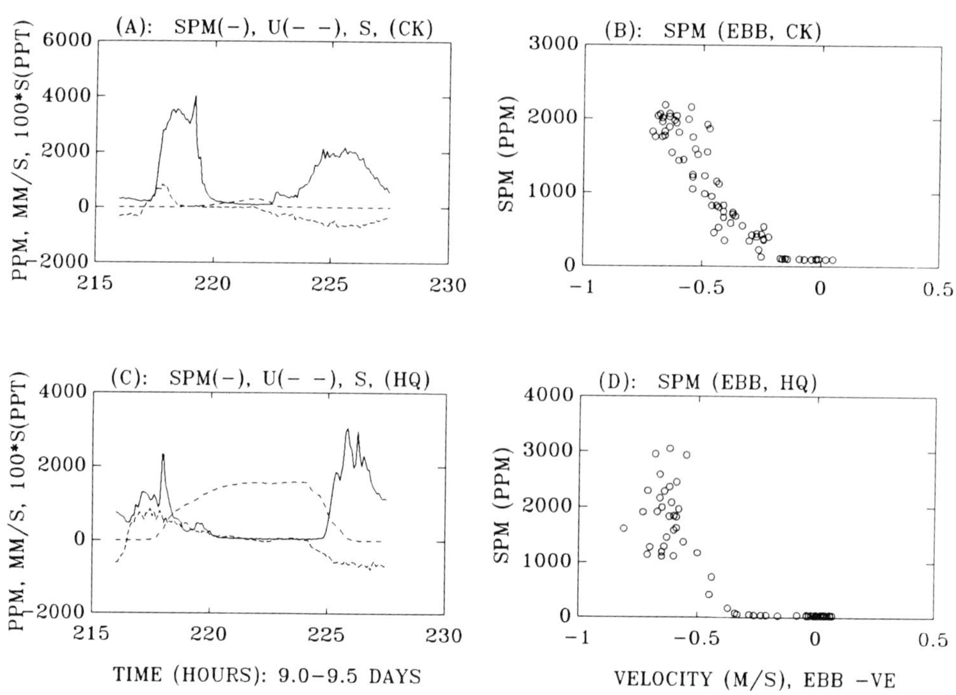

Figure 9: Time series data at 0.25m above the bed for Stations CK and HQ located in Figure 1; times 216 to 228 hours correspond to times 9.0-9.5 days in Figure 7. (a): SPM concentration (-), longitudinal velocity (U, - -) and salinity (S, -.-) at CK; (b): Scatter plot of SPM concentration against longitudinal velocity during the ebb (-ve) flow at CK; (c): SPM concentration (-), longitudinal velocity (U, - -) and salinity (S, -.-) at HQ; (d): Scatter plot of SPM concentration against longitudinal velocity during the ebb (-ve) flow at HQ.

A correlation between ebb velocity and SPM concentrations at 0.25m above the bed is exhibited in Figure 9(d). This scatter plot implies a critical velocity for erosion of 0.35 m s^{-1}, or a critical bed shear stress of about 0.45 N m^{-2} (assuming this to be $3.7u^2$, where u is measured at 0.25m above the bed, West et al., 1990). However, the correlation is clearly poor for current speeds in excess of 0.5 m s^{-1}, which implies that advection of SPM resuspended from some considerable distance up-estuary of HQ is the dominant process governing SPM loads during the ebb at HQ.

The flood SPM levels (times greater than 216.5 hours in Figure 9(c)) show an increase as soon as the near-bed current turns to flood. Therefore, SPM which has not had time to settle during the brief slack-water period returns through HQ. The SPM peak associated with the freshwater-saltwater interface passes through HQ when the near-bed salinity is 3ppt, which is somewhat higher than during the ebb and implies a lag between the movement of high concentration SPM near the bed and currents higher in the water column following low-water slack. SPM levels fall after the passage of the interface, indicating that the bed-source mud is exhausted.

14.5 Conclusions

Numerical modelling shows that a mobile stock of fine sediment can be maintained in the upper reaches of the Tamar Estuary by tidal pumping mechanisms associated with asymmetry in both the flood and ebb tidal, barotropic currents (and rates of erosion) and in the durations of the high and low-water slack periods. The time available for deposition of SPM at high water greatly exceeds that at low water. A turbidity maximum in the SPM can also be maintained by purely tidal mechanisms in the upper reaches.

These modelled tidal processes are modified by buoyancy effects. Data from three types of experiments in the turbidity maximum area of the Tamar have been used to illustrate tidal processes and their modifications by buoyancy. Vertical profiling data show pronounced stratification during the early ebb following high water, with subsequent reduction of vertical mixing. Therefore, although bed shear stresses may be high, it is not until the stratification has been eroded significantly with the approach of the freshwater-saltwater interface that large amounts of SPM can be mixed into the water column. Vertical mixing is more intense, as an average over the water column, during the flood. The mixing maximizes as the interface passes through the site and SPM is mixed into the near-surface waters where considerable up-estuary transport can occur.

These data are complemented by a longitudinal transect experiment, which demonstrates the importance of resuspension. In the stratified region down-estuary of the freshwater-saltwater interface, immediately following high water, a considerable period of time is available for settling of SPM. As the ebb currents increase, the stratification erodes and vertical mixing increases, leading to high SPM loads as the interface passes down-estuary.

Much of the SPM is advected down-estuary, where it has little time to settle and deposit at low water, and subsequently advects back up-estuary in close association with the interface.

These features are also apparent in data obtained from two longitudinally separated instrument packages deployed at 0.25m above the bed in the turbidity maximum area. SPM concentrations, longitudinal velocity and salinity data recorded at 5 minute intervals over a 10 day period demonstrate the growth of the turbidity maximum over the neap to spring-tide period and its dependence on both local resuspension, and therefore the location of the stock of bed-source sediment, and the presence of the freshwater-saltwater interface.

14.6 References

Allen, G.P., J.C. Salomon, P. Bassoulet, Y. Du Penhoat and C. De Grandpre, 1980: Effects of tides on mixing and suspended sediment transport in macrotidal estuaries, Sedimentary Geology, 26, 69-90.

Allen, G.P., G.Sauzay, P. Castaing and J.M. Jouanneau, 1977: Transport and deposition of suspended sediment in the Gironde Estuary, France, In: Estuarine processes, (ed. M Wiley), Academic Press, N.Y., 63-81.

Bale, A. J., 1987: The characteristics, behaviour and heterogeneous chemical reactivity of estuarine suspended particles. Ph D Thesis, Plymouth, UK, 216pp.

Bale, A.J., A.W. Morris and R.J.M. Howland, 1985: Seasonal sediment movement in the Tamar Estuary, Oceanologica Acta, 8, 1-16.

Delo, E.A., 1988: Estuarine muds manual. Report SR 164. Hydraulics Research, Wallingford, U.K.

Dronkers, J., 1986: Tide-induced residual transport of fine sediment, In: Physics of shallow estuaries and bays, (ed. J. van de Kreeke), Lecture notes on coastal and estuarine studies, 16, Springer-Verlag, 228-244.

Festa, J.F. and D.V. Hansen, 1978: Turbidity maxima in partially mixed estuaries: A two-dimensional numerical model, Estuarine and Coastal Marine Science, 7, 347-359.

Grabemann, I. and G. Krause, 1989: Transport processes of suspended matter derived from time-series in a tidal estuary, J. Geophysical Research, 94, No. C10, 14373-14379.

Hamblin, P.F., 1989: Observations and model of sediment transport near the turbidity maximum of the Upper Saint Lawrence Estuary, J. Geophysical Research, 94, No. C10, 14419-14428.

Harris, J.R.W., A.J. Bale, B.L. Bayne, R.F.C. Mantoura, A.W. Morris, L.A. Nelson, P.J. Radford, R.J. Uncles, S.A. Weston and J. Widdows, 1984: A preliminary model of the dispersal and biological effect of toxins in the Tamar Estuary, England, Ecological Modelling, 22, 253-284.

Kirby, R.A. and W.R. Parker, 1977: The physical characteristics and environmental significance of fine-sediment suspensions in estuaries, in: Estuaries, Geophysics and the environment, (ed. C.B. Officer), 110-120, National Academy Press, Washington, D.C.

Krone, R. B., 1962: Flume studies of the transport of sediment in estuarial shoaling processes, Univ. California, Hydraulic Engineering Laboratory.

Mehta, A. J., 1988: Laboratory studies on cohesive sediment deposition and erosion. In: Physical processes in estuaries (ed. J. Dronkers and W van Leussen), Springer-Verlag, 427-445.

Munk, W.H. and E.J. Anderson, 1948: Note on the theory of the thermocline, J. Marine Research, 7, 276-295.

Odd, N.V.M. and M.W. Owen, 1972: A two-layer model of mud transport in the Thames Estuary, Proceedings of the Institution of Civil Engineers, Paper 7517S, 175-205, 1972.

Officer, C.B., 1981: Physical dynamics of estuarine suspended sediments, Marine Geology, 40, 1-14.

Officer, C.B. and M.N. Nichols, 1980: Box model application to a study of suspended sediment distributions and fluxes in partially mixed estuaries. In: Estuarine Perspectives, Academic Press, London, 329-340.

Sheng, Y.P. and C. Villaret, 1989: Modeling the effect of suspended sediment stratification on bottom exchange processes, J. Geophysical Research, 94, No. C10, 14429-14444.

Uncles, R.J., R.C.A. Elliott and S.A. Weston, 1985a: Observed fluxes of water, salt and suspended sediment in a partly mixed estuary, Estuarine and Coastal Shelf Science, 20, 147-167.

Uncles, R.J., R.C.A. Elliott and S.A. Weston, 1985b: Dispersion of salt and suspended sediment in a partly mixed estuary, Estuaries, 8, 256-269.

Uncles, R.J., R.C.A. Elliott and S.A. Weston, 1985c: Lateral distributions of water, salt and sediment transport in a partly mixed estuary, In: Proc. of the 19th Int. Conf. on Coastal Engineering, ASCE, Chapter 204, 3067-3077.

Uncles, R.J., J.A. Stephens and T.Y. Woodrow, 1988: Seasonal cycling of estuarine sediment and contaminant transport, Estuaries, 11, 108-116.

Uncles, R.J. and J.A. Stephens, 1989: Distributions of suspended sediment at high water in a macrotidal estuary, J. Geophysical Research, 94, No. C10, 14395-14405.

Uncles, R.J., J.A. Stephens, R.H. Bruce, C.D. Barrett and N.J. Bloomer, 1989: Observations of temporal variability in a partly mixed estuary using a new instrument package, In: Developments in estuarine and coastal study techniques (eds. J. McManus and M. Elliott), Olsen and Olsen - International Symposium Series, Fredensborg, Denmark, 151-158.

Villaret, C. and M. Paulic, 1986: Experiments on the erosion of deposited and placed cohesive sediments in an annular flume and a rocking flume. Rep. UFL/COEL-86/007, Coastal and Oceanographic Eng. Dept., Univ. Florida, Gainesville.

West, J.R., K.O.K. Oduyemi, A.J. Bale and A.W. Morris, 1990: The field measurement of sediment transport parameters in estuaries, Estuarine, Coastal and Shelf Science, 30, 167-183.

Wolanski, E., J. Chappell, P. Ridd and R. Vertessy, 1988: Fluidization of mud in estuaries, J. Geophysical Research, 93, No. C3, 2351-2361.

15

Velocity asymmetries in frictionally-dominated tidal embayments: longitudinal and lateral variability

C T Friedrichs, D R Lynch and D G Aubrey

Abstract

Asymmetric tides play a central role in net near-bed transport and, ultimately, in the morphologic evolution of small (length <~20 km), well-mixed (tidal amplitude/depth >~0.1) tidal embayments. In this study we examine the physics which cause different peak tidal velocities during the ebb and the flood, and relate these patterns of velocity dominance to geomorphologies common to shallow tidal embayments. Under conditions appropriate to many shallow embayments, friction is one-to-two orders of magnitude larger than acceleration in the cross-sectionally integrated momentum equation. Tidal flow in these systems therefore may be described by a zero-inertia diffusive model which captures the most important non-linearities, while greatly simplifying conceptual interpretation. Linearized solutions are governed by the frictional length scale L_f, which determines both the propagation speed and the rate of amplitude decay. Non-linear behavior is represented by variation in the magnitude of L_f during the tidal cycle which, in turn, reflects variation in total system width and mean channel depth. Non-linear numerical solutions to the zero-inertia equation recreate patterns of velocity distortion observed in the field and are consistent with conceptual extensions of the linearized solutions. Shallower areas tend to be flood-dominant and enhance landward near-bed transport; deeper areas tend to be ebb-dominant and enhance seaward near-bed transport. Changes in morphology are associated with reversals in velocity-dominance. The feedback between hydrodynamics and embayment morphology is strongly positive. We hypothesize that antecedent geometry may dictate the initial overall sense of velocity distortion for an embayment, and that hydrodynamics slowly modulate that morphology in a generally positive fashion.

15.1 Introduction

The study of tidal non-linearities is important for geologic understanding of coastal evolution, especially in shallow embayments which are ephemeral geologically. Asymmetric tidal velocities contribute strongly to the morphodynamics of small (length <~20 km), shallow (tidal amplitude/mean depth >~0.1) tidal embayments. Net bedload sediment transport often occurs in tidal systems having asymmetric tidal currents, even under conditions of zero residual water volume transport (Aubrey, 1986). If bedload transport is assumed to be a geometric function of velocity (e.g., Meyer-Peter and Muller, 1948), then the direction of net coarse sediment transport will tend to be in the same direction as the maximum current velocity. Residual transport of suspended sediment has also been related to asymmetries in tidal velocity (Postma, 1961; Aubrey, 1986; Dronkers, 1986a). If a longer period of slack water occurs near high water than near low water, more sediment may settle out of the water column after the flood than after the ebb. Furthermore, asymmetries in maximum velocity directly contribute to net suspended sediment transport under conditions of significant scour lag; that is, when a larger velocity is required to erode cohesive sediment than that at which it settles back to the bed (Postma, 1967). Thus asymmetric velocity patterns in tidal embayments provide insight into potential patterns of sediment transport and, ultimately, into the evolution of their morphology.

Peak bottom stress is a function of maximum depth-averaged velocity for well-mixed conditions. In this study we examine the physics which cause different peak velocities during the ebb and the flood, and therefore contribute to net near-bed sediment transport. We then relate these patterns of velocity dominance to geomorphologies common to shallow tidal embayments. Here we define a point within an embayment as being flood- or ebb-dominant if the maximum current speed during a tidal cycle at that location occurs during the flood or ebb, respectively. Asymmetries in flood and ebb duration can be determined from observations of tidal elevation alone, suggesting the probable velocity distortion type. However, to determine unambiguously local ebb- or flood-dominance, velocities must be analyzed. In this study we investigate five specific aspects of velocity dominance observed in shallow tidal embayments. The first three concern cross-sectionally averaged flow (i.e., one-dimensional), whereas the last two consider depth-averaged flow (i.e., two-dimensional): 1. Embayments characterized by large changes in mean channel depth during the tidal cycle have flood-dominant main channels. 2. Embayments characterized by large changes in total width during the tidal cycle have ebb-dominant main channels. 3. Systems with flood-dominant main channels tend to have ebb-dominant tidal inlets. 4. Systems with ebb-dominant main channels tend to have flood-dominant shallow margins and intertidal areas. 5. Tidal inlet cross-sections are often composed of separate deep or ebb-dominant and shallow or flood-dominant sections.

To investigate spatial variations in velocity asymmetry in a variety of shallow embayments, a zero-inertia diffusive model of tidal dynamics is developed for well-mixed tidal flow. We show that under conditions appropriate to many small, well-mixed tidal embayments, friction is one-to-two orders of magnitude larger than the acceleration term in the cross-sectionally integrated momentum equation. Linear analytic solutions to the zero-inertia equation are used to identify non-dimensional parameters which govern non-linear behavior. Non-linear numerical solutions to the zero-inertia equation are used to recreate the patterns of velocity distortion observed in the field. A physically-based conceptual model for velocity distortion and near-bed sediment transport results.

We conclude that patterns of velocity asymmetry in these systems reinforce morphology and likely play a central feedback role in its creation and maintenance. Shallower areas tend to be flood-dominant and enhance landward near-bed transport; deeper areas tend to be ebb-dominant and enhance seaward near-bed transport. Changes in morphology are associated with reversals in velocity-dominance. Also, because of the dependence of channel geometry on tidal prism, we hypothesize that antecedent geometry may dictate the overall sense of velocity distortion for an embayment, and, therefore, predetermine evolutionary trends.

15.2 Frictional Dominance of the Momentum Equation

15.2.1 Previous work

Through scaling arguments, field measurement and/or numerical modeling of the individual terms of the 1-d momentum equation, many authors have demonstrated the dominance of friction over the inertial terms in well-mixed, shallow tidal embayments and estuaries. A survey of the literature (Table 1) indicates that in systems of interest (well-mixed, tidal amplitude/mean depth >~0.1, tidal velocities ~0.5 ms^{-1}), the friction term is typically one-to-two orders of magnitude larger than either the local or the advective acceleration term. Furthermore, the local and advective acceleration terms are typically of opposite sign and partially cancel. The relative magnitude of friction depends directly on the choice of bottom drag coefficient, which is 0.009, on average, among the first ten studies listed in Table 1. Other authors who have emphasized observed non-linearities, but who did not measure the momentum terms directly, may have underestimated the effect of friction by assuming a "classical" shelf value of c_d = ~0.002 to be appropriate for systems having mean depths of only a few meters (e.g., Robinson et al., 1983; Uncles and Kjerfve, 1986). The last two entries in Table 1 demonstrate the limits of the frictional-dominance assumption. In the Lower Columbia River Estuary where salinity intrusion is present, the shear stress at the bed is reduced by stratification in the water column, partially decoupling the overlying flow from the bottom (Giese and Jay, 1989). Upriver beyond the intrusion of salinity, the tidal pressure gradient is primarily balanced by friction (Giese and Jay, 1989). In this paper, the frictional-dominance approximation is applied only to tidal embayments having well-mixed watercolumns. In the final example (Pingree and Maddock, 1978), the English Channel is simply too deep and bottom stress too small for friction to dominate the momentum equation. We apply frictional-dominance only to systems having large ratios of tidal amplitude to mean water depth, $\zeta_o / \bar{h} \geq$ ~0.1.

Table 1: Magnitudes of local (*LA*) and advective acceleration (*AA*) relative to the friction term (*F*), along with relevant characteristic scales.

Location	LA/F	AA/F	$\bar{h}$ (m)	ζ_o (m)	u_o (ms^{-1})	c_d	Source
1. 1-d trapezoidal channel	0.04	0.01	2	1	0.5	0.02	Speer (1984)
2. 1-d model of Conwy Estuary, Wales	0.04	0.04	3	2.4	0.5	0.01	Münchow and Garvine (1991)
3. Conwy Estuary, Wales	0.05	0.02	3	2.4	0.5	0.006	Wallis and Knight (1984)
4. Fraser Estuary, BC Canada	0.05	0.05	9	4.5	1	0.005	LeBlond (1978)
5. 1-d model, Stony Brk. Harbor, NY, USA	0.05	0.1	2	0.9	1	0.01	Park (1985)
6. Macquarie Harbor, Tasmania	0.1	0.007	6	0.5	1	0.004	van de Kreeke (1967)
7. Great Bay, NH USA	0.1	0.02	7	1.3	1	0.03	Swift and Brown (1983)
8. St. Lawrence Estuary, Canada	0.1	0.1	7	3.5	1	0.001	LeBlond (1978)
9. Ingram Thorofare, NJ, USA	0.2	0.07	3	0.5	0.7	0.002	Weisman et al. (1990)
10. Delaware Estuary, USA	0.4	0.03	6	0.7	1	0.0025	Parker (1984)
11. *Columbia R. Estuary, WA, USA	1	1	10	1	1	0.0008	Giese and Jay (1989)
12. *English Channel	5	2.5	40	2	0.8	0.0025	Pingree and Maddock (1978)

*These two examples are included to illustrate limits of the frictional dominance assumption. See text for discussion.

It was only recently that the dominant role of friction in the momentum balance of shallow tidally-dominated estuaries was demonstrated to the estuarine community by LeBlond (1978). Despite the work of LeBlond, the more appropriate frictional interpretation of highly non-linear tidal flows in shallow embayments has received relatively little attention. During the last decade only a few papers have used field observations (Aubrey and Speer, 1985; Friedrichs and Aubrey, 1988), numerical modeling (Speer and Aubrey, 1985; Westerink et al., 1989), or scaling arguments (Parker, 1984; Munchow and Garvine, 1991) specifically to address the role of non-linear frictional distortion in shallow tidal embayments. The neglect of non-linear frictional effects may result in part from long-held classical views of depth-averaged estuarine non-linear distortion. A century-and-a-half ago, Airy (1845) noted that in a frictionless estuary where $\zeta_0 / \bar{h}$ is large, the tide propagates as a shallow water wave with depth-dependent phase velocity. Lamb (1932) combined the frictionless non-linear advection and continuity effects into one non-linear expression for frictionless phase velocity:

$$c = \sqrt{g\bar{h}}\left(1 + \frac{3}{2}\frac{\zeta(t)}{\bar{h}} + O\left(\frac{\zeta_0}{\bar{h}}\right)^2\right), \tag{15.1}$$

where $\bar{h}$ is undisturbed mean depth, and $\zeta(t)$ is time-dependent tidal elevation. Because total water depth is significantly greater at the tide crest than at the trough, the crest propagates more quickly than the trough, resulting in a shorter flood, a longer ebb, and occurrence of highest velocity currents during the flood. Since the time of Airy, "shallow water" non-linearity has been synonymous with the non-linear effects of frictionless acceleration and continuity. Discussion of Airy-type shallow water non-linearity has largely overshadowed the enhancement of frictional non-linearities as $\zeta_0 / \bar{h}$ increases. In the following section, a scalar analysis appropriate to shallow tidal embayments of interest is used to demonstrate the dominance of friction over inertia in the 1-d momentum equation.

15.2.2 Scalar analysis

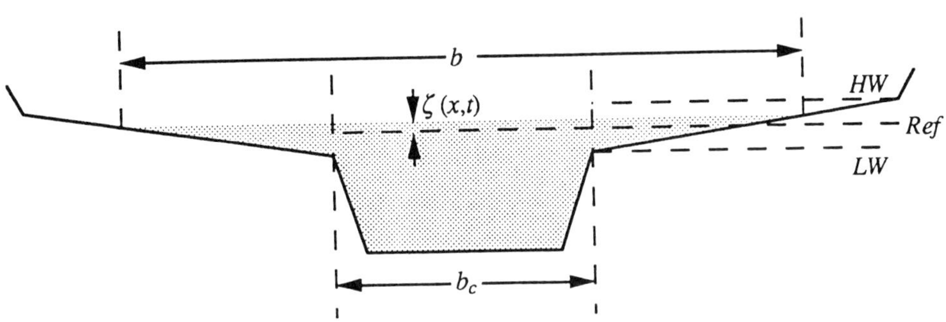

Figure 1: Schematic diagram of a tidal embayment cross-section: $b(x,t)$ is the total width of the embayment cross-section, including tidal flats which act in a storage capacity only; b_c is the width of the momentum-transporting tidal channel; ζ = surface elevation relative to a vertical datum; *HW* = high water; *LW* = low water; *Ref* = datum level.

The cross-sectionally integrated, one-dimensional equations of motion for well-behaved, channelized flow in a tidal embayment with intertidal flats (Fig. 1) may be expressed as:

$$\frac{\partial Q}{\partial x} + \frac{\partial A}{\partial t} = 0, \qquad \textbf{continuity} \tag{15.2}$$

$$\frac{\partial Q}{\partial t} + \frac{\partial}{\partial x}\frac{Q^2}{A_c} + gA_c\frac{\partial \zeta}{\partial x} + c_d\left|\frac{Q}{A_c}\right|\frac{Q}{A_c}P_c = 0, \qquad \text{momentum} \tag{15.3}$$

where Q is along-channel flux (confined to the sub-tidal channel), x is horizontal distance, A is total cross-sectional area (including tidal flats), A_c and P_c are the cross-sectional area and perimeter of the momentum transporting, sub-tidal channel, and other symbols are as previously defined. Restated in terms of characteristic scales, (15.2) and (15.3) become

$$O\left(\frac{\delta Q}{\delta x}\right) + O\left(\frac{\delta A}{\delta t}\right) = 0, \tag{15.4}$$

$$O\left(\frac{\delta Q}{\delta t}\right) + O\left(\frac{Q}{A_c}\frac{\delta Q}{\delta x}\right) + O\left(\left(\frac{Q}{A_c}\right)^2\frac{\delta A_c}{\delta x}\right) + O\left(gA_c\frac{\delta \zeta}{\delta x}\right) + O\left(c_d\left(\frac{Q}{A_c}\right)^2 P_c\right) = 0. \tag{15.5}$$

We assume $O(\delta Q) \approx O(Q)$ and solve for δx in (14.4). Then substituting δx into (15.5) gives:

$$O\left(\frac{\delta Q}{\delta t}\right) + O\left(\frac{\delta A}{A_c}\frac{\delta Q}{\delta t}\right) + O\left(\frac{\delta A}{A_c}\frac{\delta A_c}{A_c}\frac{\delta Q}{\delta t}\right) + O\left(gA_c\frac{\delta\zeta\,\delta A}{\delta t\,\delta Q}\right) + O\left(c_d\left(\frac{Q}{A_c}\right)^2 P_c\right) = 0. \tag{15.6}$$

If we assume $O(\delta A/A_c) \leq O(1)$ (which is the case for all observed and/or modeled embayments considered in this paper), then the sum of the three inertia terms is $O(\delta Q/\delta t)$. Combining the first three terms of (15.6) and dividing (15.6) by the scale of the surface gradient gives:

$$O\left(\frac{Q^2}{gA_c\delta A\,\delta\zeta}\right) + O(1) + O\left(\frac{c_d\,\delta t\,Q^3 P_c}{gA_c^3\,\delta A\,\delta\zeta}\right) = 0. \tag{15.7}$$

The ratio of the friction scale to the inertial scale is then

$$\frac{F}{I} = \frac{\delta t\,u_0 c_d}{R_c}, \tag{15.8}$$

where u_o is the characteristic scale of velocity, averaged over the momentum-transporting portion of the cross-section, and $R_c = A_c/P_c$ is the hydraulic radius of the momentum-transporting channel. In shallow tidal embayments of interest, u_o is of the order 0.5 ms^{-1}, $c_d \approx 10^{-2}$, the semi-diurnal period $\delta t = 4.5 \times 10^4$ sec, and 1 m $\leq R_c \leq$ 10 m. Therefore F will typically be one-to-two orders of magnitude larger than I in these tidally-dominated embayments. Since F/I is frequency-dependent, however, one should use the period of the overtide of interest when considering strongly non-linear flow. This is not a serious limitation: F will still dominate I by one-to-two orders of magnitude, even if one scales (15.8) with the quarter-diurnal tidal period.

Assuming $F/I >> 1$, then from (15.7) for a single embayment:

$$c_d \sim O\left(\frac{gA^3\delta A\,\delta\zeta}{\delta t Q^3 P}\right)\left(\frac{A_c}{A}\right)^3\left(\frac{P}{P_c}\right) \propto \left(\frac{A_c}{A}\right)^3\left(\frac{P}{P_c}\right). \tag{15.9}$$

Thus c_d is sensitive to the fraction of the cross-section that is defined as momentum-transporting. If one compares drag coefficients from several tidal embayments, a consistent definition should be used. In Table 1 (entries 1-10), the momentum-transporting cross-sections used in the balance calculations are all clearly defined, sub-tidal channels with negligible change in width during the tidal cycle. In cases where extensive intertidal flats are immediately adjacent to subtidal channels (Speer, 1984; Park, 1985; Munchow and Garvine, 1991), the authors treated intertidal flats in a storage capacity only, excluding extensive flats from the momentum balance.

15.2.3 Numerical comparison

We also solved (15.2) and (15.3) numerically, with and without the inertial terms, to demonstrate further the validity of the zero-inertia approximation in shallow regimes. The model embayment consists of a seven kilometer long, wide rectangular channel (i.e., $R_c \approx$ channel depth, $P_c \approx b_c$ = constant), with a linearly sloping bottom. We let $c_d = (0.013\ \text{m}^{1/3})\, R_c^{-1/3}$, using the Manning formulation for bottom friction. For $R_c = 2$ m, $c_d \approx 0.01$. The channel bottom rises from two meters below mean sea level at the ocean end to one meter above sea level inland, and is forced by a one meter amplitude ocean tide. Eqs. (15.2) and (15.3) with the inertial terms were solved explicitly with a 1-d finite difference model (Δt = 15 s, Δx = 250 m, from Speer and Aubrey, 1985). Omitting the inertial terms, (15.2) and (15.3) were solved implicitly using Crank-Nicolson in time (Δt = 450 s) and 1-d finite elements in space (Δx = 250 m). Without the inertial terms, the solution is free of advective time-stepping constraints, allowing stable solution at much larger Δt. This may be an important consideration in the solution of highly complex geometries in two dimensions.

The zero-inertia and fully dynamic solutions are essentially identical (Fig. 2). Except at slack water, the pressure gradient and friction terms in the fully dynamic solution dominate the inertial terms by one-to-two orders of magnitude. Furthermore, the local and advective acceleration terms are typically of opposite sign and partially cancel. Clearly the zero-inertia assumption is inappropriate at slack water when $u = 0$ and in the immediate vicinity of sudden channel expansions or tight channel meanders. However, during most of the tidal cycle and for the majority of the tidal embayment, the zero-inertia assumption is well-justified.

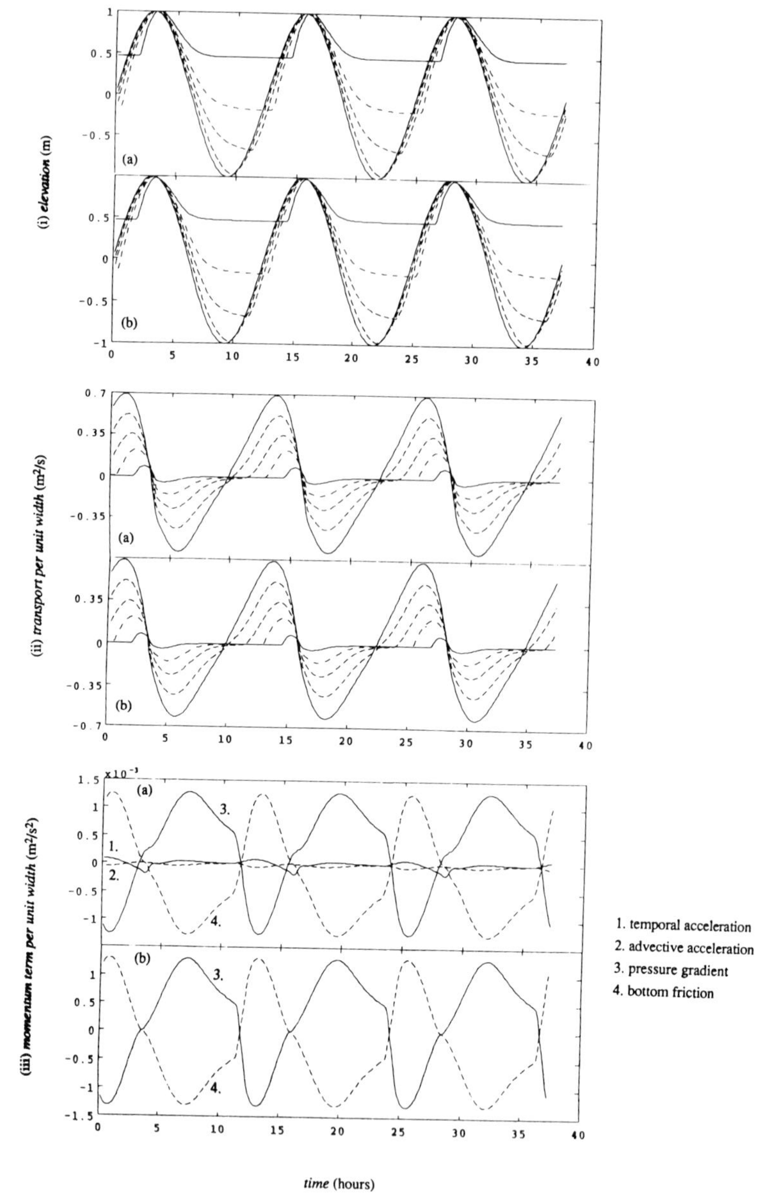

Figure 2: Comparison of numerical solutions to the one-dimensional equations of motion with and without the inertial terms. (i) Surface elevation and (ii) transport per unit width at five tidal stations, and (iii) momentum balance at middle station: (a) with the inertial terms; (b) without the inertial terms.

15.3 Analytical treatment of the linearized zero-inertia equation

15.3.1 Formulation

The zero-inertia momentum equation for well-behaved, cross-sectionally averaged flow may be expressed as (e.g., Henderson, 1966):

$$\frac{\partial \zeta}{\partial x} + n^2 R_c^{-4/3} \, |u| u = 0 \quad \text{or} \quad u = -\frac{1}{n} R_c^{2/3} \left| \frac{\partial \zeta}{\partial x} \right|^{-1/2} \frac{\partial \zeta}{\partial x}, \tag{15.10a,b}$$

where u is velocity averaged over the momentum-transporting portion of the cross-section (Fig. 1), $n = R_c^{1/6}(c_d/g)^{1/2}$ is the Manning friction coefficient, and the other symbols are as previously defined. Cross-sectionally averaged continuity (15.2) may be re-expressed, with momentum-transport confined to the channel, as

$$b \frac{\partial \zeta}{\partial t} + \frac{\partial}{\partial x} b_c h u = 0, \tag{15.11}$$

where b and b_c are as defined in Fig. 1, and h is the cross-sectionally averaged depth of the momentum-transporting channel. Substituting (15.10b) into (15.11) gives

$$\frac{\partial \zeta}{\partial t} - \frac{1}{b} \frac{\partial}{\partial x} \left(D \frac{\partial \zeta}{\partial x} \right) = 0, \tag{15.12}$$

where D is a non-linear diffusion coefficient given by

$$D = \frac{h}{n} b_c R_c^{2/3} \left| \frac{\partial \zeta}{\partial x} \right|^{-1/2} \tag{15.13}$$

Eq. (15.12) may be solved to first order analytically if we assume that D is finite and slowly varying in time and space during most of the tidal cycle. This analysis is inappropriate for tidal flow at slack water because D is undefined for zero sea-surface gradient. However this linearized analysis still provides valuable insight into the behavior of frictionally-dominated tidal propagation during the majority of the tidal cycle. Under these assumptions, (15.12) may be restated as

$$\frac{\partial \zeta}{\partial t} = \kappa \frac{\partial^2 \zeta}{\partial x^2}, \tag{15.14}$$

with a "constant" diffusion coefficient, κ, defined (with $R_c \approx h$) as

$$\kappa = \frac{b_c}{b}\frac{h^{5/3}}{n}\left|\frac{\partial\zeta}{\partial x}\right|^{-1/2} \tag{15.15}$$

The boundary conditions for the problem are no-flow at $x = L$, where L is the length of the channel, and sinusoidal forcing of amplitude ζ_o at $x = 0$. The analytic solution for $\zeta(x,t)$ is

$$\zeta_o \frac{\cos(x/L_f - \omega t - \psi)e^{-x/L_f} + \cos(x/L_f + \omega t)e^{x/L_f - \psi} + \cos(x/L_f - \omega t)e^{-x/L_f + \psi} + \cos(x/L_f + \omega t - \psi)e^{x/L_f}}{2(\cos\psi + \cosh\psi)}, \tag{15.16}$$

where ω is angular tidal frequency, $L_f = (2\kappa/\omega)^{1/2}$ and $\psi = 2L/L_f$.

15.3.2. Length scales and zeroth-order results

Eq. (15.16) describes an exponentially decaying, partially progressive, partially standing wave governed by the frictional length scale L_f. L_f determines both the propagation speed ($c = L_f\omega$) and the rate of amplitude decay (e-folding length = L_f) of each of the four terms in (15.16). If $L/L_f << 1$, then (15.16) reduces to an x-independent standing wave. If $L/L_f >> 1$, then (15.16) reduces to an equation for a diffusive propagating wave:

$$\zeta(x,t) = \zeta_o \cos(x/L_f - \omega t)e^{-x/L_f}. \tag{15.17}$$

Eq. (15.17) is analogous to the infinite tidal channel solution considered by LeBlond (1978). The infinite channel solution of LeBlond, however, does not capture the full range of zeroth-order (i.e., amplitude decay and phase lag) behavior found in the small ($L < \sim 20$ km) systems of interest to the present study. Significant variation is seen in (15.16) over a small range of L/L_f (Fig. 3).

We estimate L_f as a function of known parameters by first evaluating $|\partial\zeta/\partial x|$ at $x = 0$. From (15.16), it may be shown that

$$\frac{\partial\zeta}{\partial x}(x=0) = \frac{2^{1/2}\zeta_o}{L_f(\cos\psi + \cosh\psi)}\left(\sin\psi\cos(\omega t - \pi/4) + \sinh\psi\sin(\omega t - \pi/4)\right). \tag{15.18}$$

For $L/L_f \geq 1.25$, $\sinh\psi$ is at least an order of magnitude larger than $\sin\psi$, and $\sinh\psi \approx \cosh\psi$. Thus for L and L_f of the same order, we may determine $|\partial\zeta/\partial x|$ using these simplifying approximations. An estimate for the time-averaged absolute value of (15.18) is then given by

$$\left|\frac{\partial \zeta}{\partial x}\right|(x=0) \sim \frac{2^{1/2}\zeta_o}{L_f}\ \overline{\sin(\pi/4-\omega t)} = \frac{2^{3/2}\zeta_o}{\pi L_f} \tag{15.19}$$

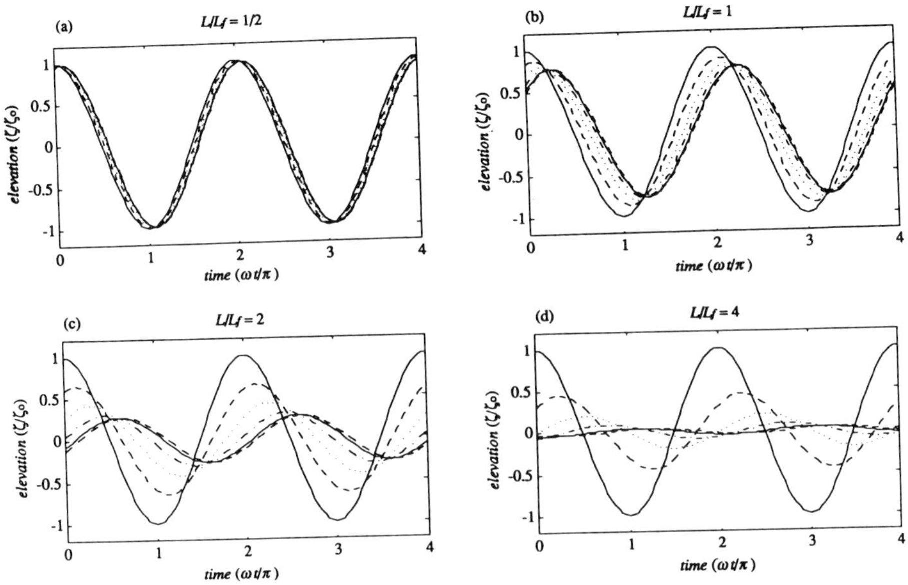

Figure 3: Time series of Eq. (15.16) during two tidal cycles recorded at x/L = 0, 0.2, 0.4, 0.6, 0.8 and 1.0: (a) L/L_f = 1/2; (b) L/L_f = 1; (c) L/L_f = 2; (d) L/L_f = 4.

We solve directly for L_f by substituting this relation into (15.15) along with the definition $L_f = (2\kappa/\omega)^{1/2}$:

$$L_f \sim 2^{1/6}\pi^{1/3}\left(\frac{b_c}{b}\right)^{2/3} h^{10/9}\zeta_o^{-1/3}(n\,\omega)^{-2/3}. \tag{15.20}$$

According to (15.20), propagation speed decreases and amplitude decay rate increases as depth is reduced, friction is increased, or total embayment width is increased relative to the width of the channel. Eq. (15.20) is also consistent with observations that indicate shallow tidal embayments act as low-pass filters (Aubrey and Speer, 1985).

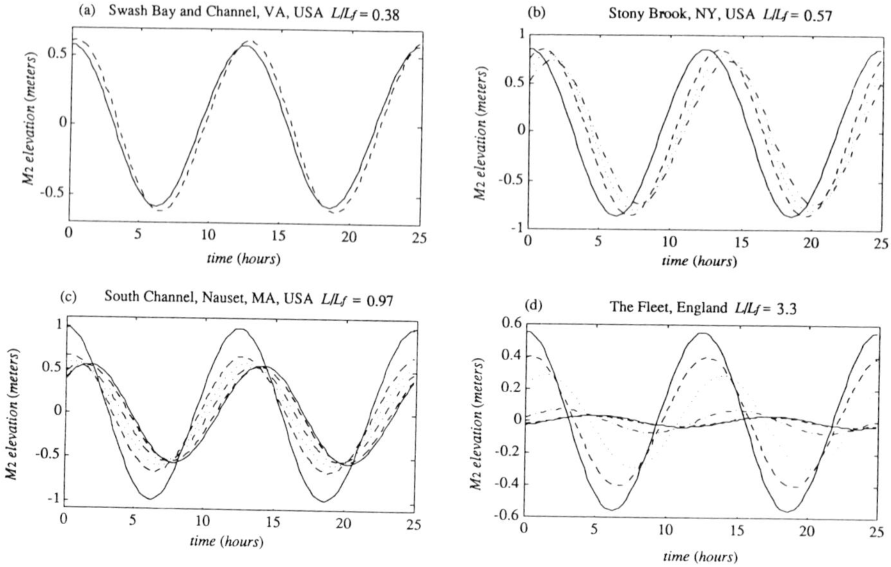

Figure 4: Time series of mean observed M_2 surface elevation during two tidal cycles: (a) Swash Bay and Channel (x/L = 0, 1); (b) Stony Brook Harbor (x/L = 0, 0.2, 0.4, 0.8); (c) South Channel, Nauset (x/L = 0, 0.2, 0.3, 0.5, 0.7, 0.9); (d) The Fleet (x/L = 0, 0.2, 0.4, 0.6, 0.8, 1).

The zeroth-order decay and delay of the M_2 tide in four tidal embayments (Fig. 4) are consistent with behavior predicted by simple, linear solutions to the zero-inertia approximation (Fig. 3). Table 2 summarizes the parameters used in calculating L_f for four real systems. We choose b_c/b equal to embayment area at mean low water divided by embayment area at mean tide level (except for The Fleet, for which Robinson et al., 1983, provide b_c and b). The values given for Manning's n (Table 2) are based on the n or c_d values used by the authors in 1-d numerical models of the embayments. Robinson et al. (1983) assumed a typical shelf value for the drag coefficient (c_d = 0.002), possibly underestimating the frictional effects in the tidal channel. They then matched observations of tidal elevation by tuning the values of b_c/b in their numerical model, a model which limits momentum transport to the central width, b_c. Eq. (15.20) demonstrates that decreasing b_c/b is analogous to increasing n in a frictionally-dominated system. Since Robinson et al. tuned b_c/b, the L/L_f value calculated for The Fleet (Table 2) should still be accurate, even if Robinson et al. underestimated n. Robinson et al. noted that low values of b_c/b strongly increase amplitude decay and phase lag, but they mistakenly assumed that the tide in The Fleet propagates as an Airy-type frictionless long-wave. With c_d, b_c, Q and h as given by Robinson et al., $F/I \approx 40$ for a semi-diurnal tide in The Fleet. By assuming the dominance of acceleration over friction, they derived a phase speed of $(ghb_c/b)^{1/2}$, which becomes smaller at low b_c/b. Robinson et al. also neglected friction in their discussion of tidal non-linearity and attributed observed flood-dominance only to

Airy-type behavior (see Eq. 15.1). The following discussion will demonstrate, however, that frictional-dominance provides a more appropriate, physically-based model for the non-linear behavior of shallow tidal embayments of interest here.

Table 2: Geometric and frictional length scales (and the parameters used for calculation) for four shallow tidal embayments.

Location	L (km)	$\overline{h}$ (m)	$M2_o$ (m)	$\overline{b_c/b}$	Manning's n ($m^{-1/3}s$)	L/L_f	Sources
1. Swash Bay and Channel, Wachapreague, VA	6.4	2.5	0.58	0.61	0.025	0.38	Boon and Byrne (1983);
2. Stony Brook Harbor, NY, USA	5.2	1.7	0.86	0.77	0.035	0.57	Park (1985); Friedrichs and Aubrey (1988)
3. South Channel, Nauset, MA, USA	8.2	2.1	0.98	0.76	0.051	0.97	Aubrey and Speer (1984, 1985); Friedrichs and Aubrey (1988)
4. The Fleet, England	12.5	0.7	0.57	0.32	0.015	3.3	Robinson et al. (1983)

15.3.3. Non-linearities predicted by the linearized equation

Simple analytic solutions such as those displayed in Fig. 3 assume L/L_f to be constant throughout the tidal cycle. However, non-dimensional parameters important to non-linear behavior can still be identified by considering variations in the magnitude of L_f during the tidal cycle. Since the tidal wave propagation speed is proportional to the frictional length scale ($c = L_f\,\omega$), the tidal wave will propagate most quickly during the portion of the tidal cycle where L_f is greatest. If L_f is significantly greater around high water than low water ($L_{fhigh} >> L_{flow}$), then the propagation speed will be greater near high water, and the crest of the tide will propagate more quickly than the trough. The result will be a shorter flood, a longer ebb, and occurrence of highest velocity currents during the flood. Analogously, if $L_{fhigh} << L_{flow}$, the trough will propagate more quickly, and ebb-dominant currents will result. From (15.20) the ratio of high- to low-water frictional length scales is dependent on two non-dimensional parameters, b_{high}/b_{low} and h_{high}/h_{low}:

$$\frac{L_{f\,high}}{L_{f\,low}} \sim \left(\frac{b_{high}}{b_{low}}\right)^{-2/3} \left(\frac{h_{high}}{h_{low}}\right)^{10/9} \qquad (15.21)$$

This conceptual extension of linear theory may be tested by measuring tidal lags relative to boundary forcing in real systems. Tidal lag, Θ, is inversely proportional to the tidal propagation speed and, therefore, inversely proportional to L_f. Thus the ratio of tidal lags at high and low water is given by

$$\frac{\Theta_{high}}{\Theta_{low}} \approx \left(\frac{L_{f\,high}}{L_{f\,low}}\right)^{-1} . \tag{15.22}$$

The ratio of Θ_{high} to Θ_{low} provides a measure of the degree of tidal asymmetry and, by our conceptual extension of linear theory, a rough measure of the ratio of frictional length scales near high and low water. Table 3 compares ratios of Θ_{high} to Θ_{low} observed in four shallow embayments with values predicted by (15.22). In all four cases the general trend (>1 or <1) predicted by (15.22) is in agreement with observations.

Table 3: Predicted ratio of tidal lags (and the parameters used for calculation) along with measured tidal lag ratios at innermost gauge, based on M_2 and M_4 only. ΔD = rise duration - fall duration.

Location	predicted $\overline{h}_{high}/\overline{h}_{low}$	predicted $\overline{b}_{high}/\overline{b}_{low}$	predicted $\Theta_{high}/\Theta_{low}$	observed $\overline{\Theta}$ (hours)	observed ΔD (hours)	observed $\Theta_{high}/\Theta_{low}$	Sources
1. Swash Bay and Channel, Wachapreague, VA	1.6	4.0	1.49	0.65	0.28	1.55	Byrne et al. (1975); Boon and Byrne (1981)
2. Stony Brook Harbor, NY, USA	2.8	1.6	0.44	1.7	−1.1	0.53	Park (1985); Friedrichs and Aubrey (1988)
3. South Channel, Nauset, MA, USA	1.9	1.6	0.67	1.6	−1.3	0.43	Aubrey and Speer (1985); Friedrichs and Aubrey (1988)
4. The Fleet, England	2.1	1.0	0.44	4.9	−1.6	0.72	Robinson et al. (1983)

15.4 Numerical treatment of the non-linear zero-inertia equation

Linear solutions to the zero-inertia equation identify parameters (L_{fhigh}/L_{flow}, b_{high}/b_{low}, h_{high}/h_{low}) which likely govern non-linear behavior. We now utilize fully non-linear solutions to confirm the importance of these non-dimensional parameters and to investigate five specific patterns of velocity asymmetry observed in the field: (a) Embayments characterized by large changes in mean channel depth during the tidal cycle have flood-dominant main channels. (b)

Embayments characterized by large changes in total width during the tidal cycle have ebb-dominant main channels. (c) Systems with flood-dominant main channels tend to have ebb-dominant tidal inlets. (d) Systems with ebb-dominant main channels tend to have flood-dominant shallow margins and intertidal areas. (e) Tidal inlet cross-sections are often composed of separate deep or ebb-dominant and shallow or flood-dominant sections. The following sub-section addresses the first two observed patterns of velocity distortion and discusses empirical relationships between tidal prism and channel geometry. Sub-sections 2 - 4 individually discuss patterns (c) - (e).

Table 4: Results of least-squares linear regressions for channel area (A_c), width (b_c) and mean depth ($\bar{h}$) as exponential functions of spring tidal prism (Ω), of the form $\chi = \alpha\Omega^{\beta}$, where χ is the dependent variable and α and β are constants. The linear regressions are performed on the natural logs of the variables, therefore the linear equation is of the form $\ln \chi = \alpha + \beta \ln \Omega$. () = standard error.

	χ	$\ln \alpha$	β
1. Along channel (data from Chantler, 1974)	A_c	- 6.8 (0.4)	0.87 (0.02)
	b_c	-3.2 (0.4)	0.56 (0.02)
	$\bar{h}$	-3.8 (0.3)	0.32 (0.02)
2. At tidal inlet (data from Jarrett, 1976)	A_c	-11 (0.9)	1.1 (0.05)
	b_c	-8.3 (1.4)	0.86 (0.08)
	$\bar{h}$	-2.8 (1.2)	0.25 (0.07)

15.4.1 Velocity-dominance of the main tidal channel

Linear solutions to the zero-inertia equation suggest that either flood- or ebb-dominant velocities will be produced in the main tidal channel by non-linear solutions to (15.12) depending on the values of b_{high}/b_{low} and h_{high}/h_{low}. Therefore we wished to include a range of these non-dimensional geometric parameters. It was practical to investigate only a finite number of fully non-linear solutions, and we wished to limit investigations to realistic morphologies. Thus it was useful to follow empirical relations between tidal discharge and channel geometry. Two of the most systematic compilations of empirical data are those by Jarrett (1976), who related channel cross-sectional parameters at tidal inlets to spring tidal prism, Ω, and by Chantler (1974), who related cross-sectional parameters along the length of tidal channels to maximum spring discharge, Q_m. Here we utilize both sets of data by approximating maximum spring prism as $\Omega = 2Q_m/\omega$, and fitting exponential relationships of the form $\alpha\Omega^\beta$ (α and β constant) to channel cross-sectional area, width and mean depth at mean tide level (Fig. 5; Table 4). Following these empirical relations, channel width and depth become unique functions of tidal prism, and reasonably realistic model geometries can be produced easily by first specifying the tidal range and area of intertidal storage.

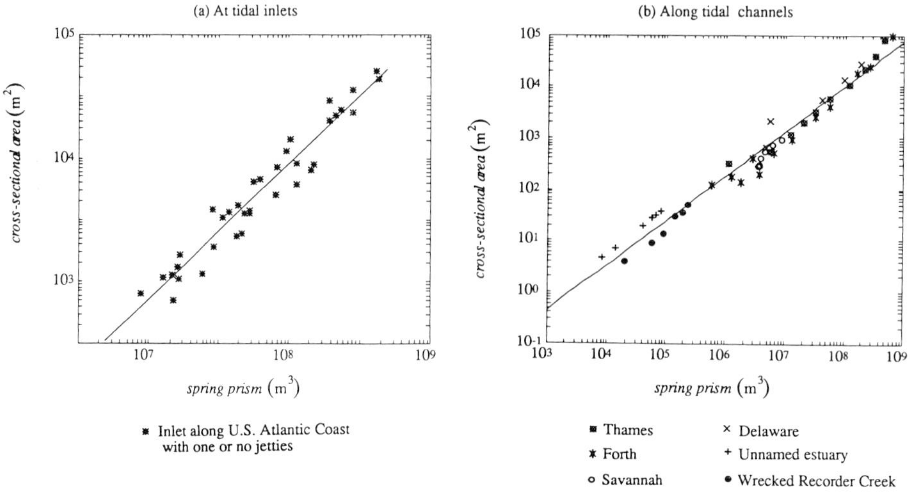

Figure 5: Cross-sectional area as a function of spring tidal prism: (a) Measured at 37 tidal inlets along the Atlantic Coast of the U.S. with one or no jetties (data from Jarrett, 1976); (b) Measured at stations along the length of six tidal channels (data from Chantler, 1974).

In solving (15.12) we chose mean tide conditions of $\zeta_o = 1$ m, $L = 7000$ m and $n = 0.036$. Manning's $n = R_c^{1/6}(c_d/g)^{1/2} = 0.036$ corresponds to $c_d = 0.01$ for $R_c = 2$ m or $c_d = 0.006$ for $R_c = 10$ m, a rough approximation of the range seen in Table 1. The distribution of intertidal storage for the model is shown in Fig. 6. We assumed maximum spring discharge to be one-and-a-half times that of a mean tide and then determined channel geometries on the basis of the empirical relations summarized in Table 4. Jarrett's data were used to determine channel cross-sections at the inlet, whereas Chantler's data were used to determine cross-sections within the inner portion of the basin. A linear combination of the two relations was used in-between. Basin widths ranging from 100 m to 10 km resulted in 10^5 m^3 $< \Omega <$ 10^8 m^3 and the distribution of embayment cross-sections shown schematically in Fig. 7. Small tidal prisms are associated with shallow channels and narrow intertidal areas, i.e., $h_{high}/h_{low} >> 1$. Large tidal prisms are associated with deep channels and wide intertidal areas, i.e., $b_{high}/b_{low} >> 1$. For the range of tidal prisms examined, Fig. 8 displays Lf_{high}/Lf_{low} (averaged along-channel) as predicted by linear theory. This extension of linear theory to highly non-linear solutions is accurate to lowest order only. Nonetheless, results are generally consistent with the fully non-linear solutions to be discussed. As linear theory suggests, small prisms with $L_{fhigh} > L_{flow}$ are associated with flood-dominant main channels, and large prisms with $L_{fhigh} < L_{flow}$ are associated with ebb-dominant main channels.

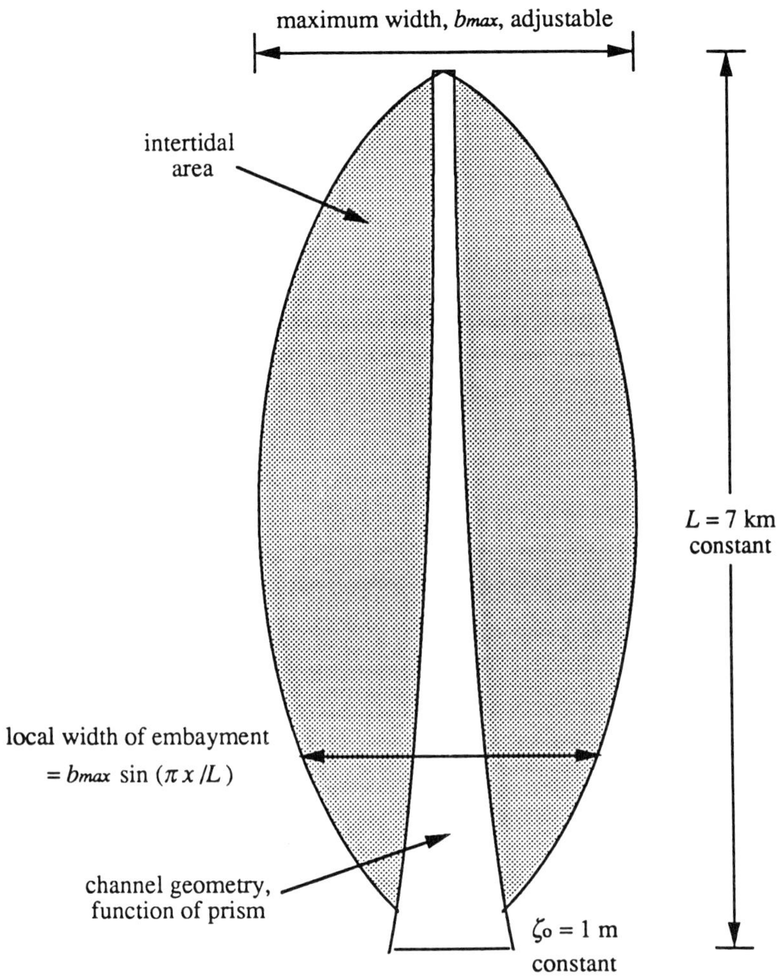

Figure 6: Schematic diagram of model tidal embayment. Tidal basin area is specified by a sine function. With the forcing tidal amplitude and embayment length fixed, the choice of maximum embayment width determines the spring tidal prism, which, in turn, empirically determines the channel geometry.

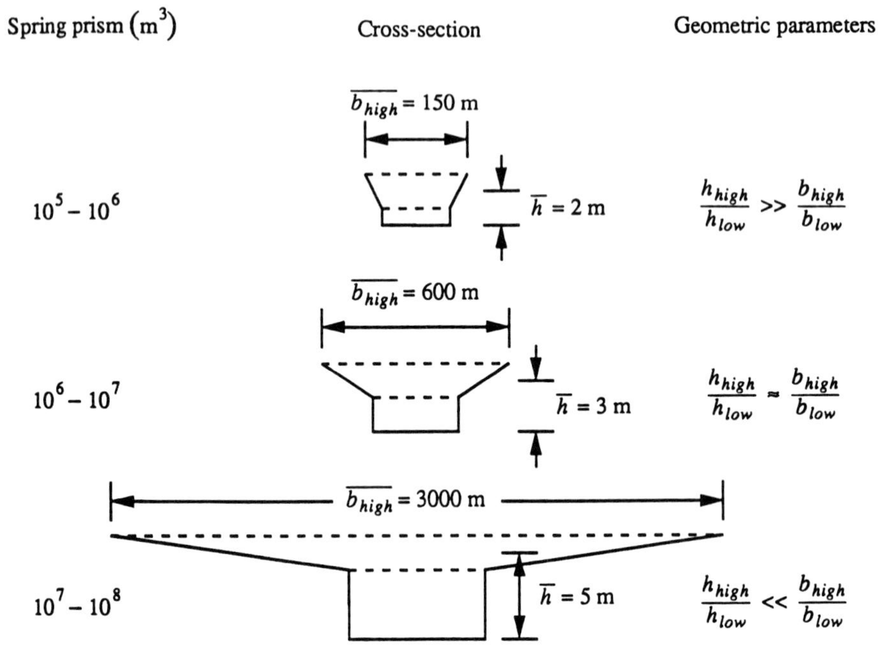

Figure 7: Schematic diagram of the relationships (with ζ_o and L fixed) between spring tidal prism, embayment cross-sectional geometry and the non-dimensional geometric parameters which govern tidal asymmetry.

Several observational studies of tidal embayments indicate deep main channels and large intertidal flats are associated with ebb-dominance, whereas shallow channels and small intertidal storage are associated with flood-dominance (e.g., Boon and Byrne, 1981; Aubrey and Speer, 1985; Friedrichs and Aubrey, 1988). Friedrichs and Aubrey examined tidal elevations from 54 gauges in 26 embayments. In systems characterized by large changes in total width during the tidal cycle ($b_{high} >> b_{low}$, $h_{high} \approx h_{low}$), they observed ebb duration to decrease as V_s/V_c (volume stored in intertidal areas/volume in channels) increased. In systems with large changes in mean depth during the tidal cycle ($b_{high} \approx b_{low}$, $h_{high} >> h_{low}$), they observed flood duration to decrease as ζ_o/h increased. Solutions to the non-linear zero-inertia equation are in agreement with these field observations (Fig. 9). Velocity asymmetry is parameterized by the ratio of maximum flood speed to maximum ebb speed, calculated along the axis of the embayment. Under conditions of well-mixed tidal flow, this ratio captures the differences in peak bottom stresses which are important to net near-bed sediment transport. Non-linear solutions to (15.12) indicate that low water propagates fastest ($Lf_{high} < Lf_{low}$) in ebb-dominant channels (Fig. 10a), whereas high water propagates fastest

($Lf_{high} > Lf_{low}$) in flood-dominant channels (Fig. 10b).

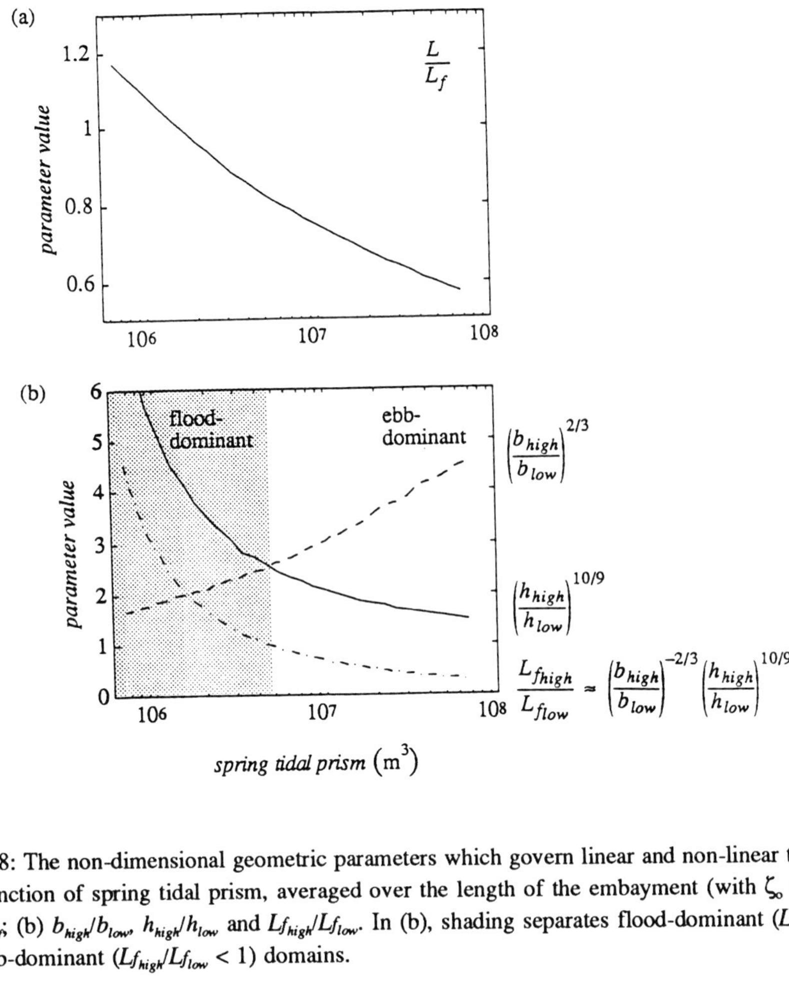

Figure 8: The non-dimensional geometric parameters which govern linear and non-linear tidal behavior as a function of spring tidal prism, averaged over the length of the embayment (with ζ_o and L fixed): (a) L/L_f; (b) b_{high}/b_{low}, h_{high}/h_{low} and Lf_{high}/Lf_{low}. In (b), shading separates flood-dominant ($Lf_{high}/Lf_{low} > 1$) and ebb-dominant ($Lf_{high}/Lf_{low} < 1$) domains.

Eq. (15.12) was solved implicitly using Crank-Nicolson in time ($\Delta t = 450$ s) and 1-d finite elements in space ($\Delta x = 250$ m). Solutions for (15.12) were compared to solutions for (15.2) and (15.3), which include the acceleration terms in the momentum equation (Fig. 10c,d). Eqs. (15.2) and (15.3) were solved explicitly with a 1-d finite difference model ($\Delta t = 15$ s, $\Delta x = 250$ m, see Fig. 2). The zero-inertia solution is virtually identical to the solution which includes the acceleration terms, except immediately around slack water when the surface gradient is zero.

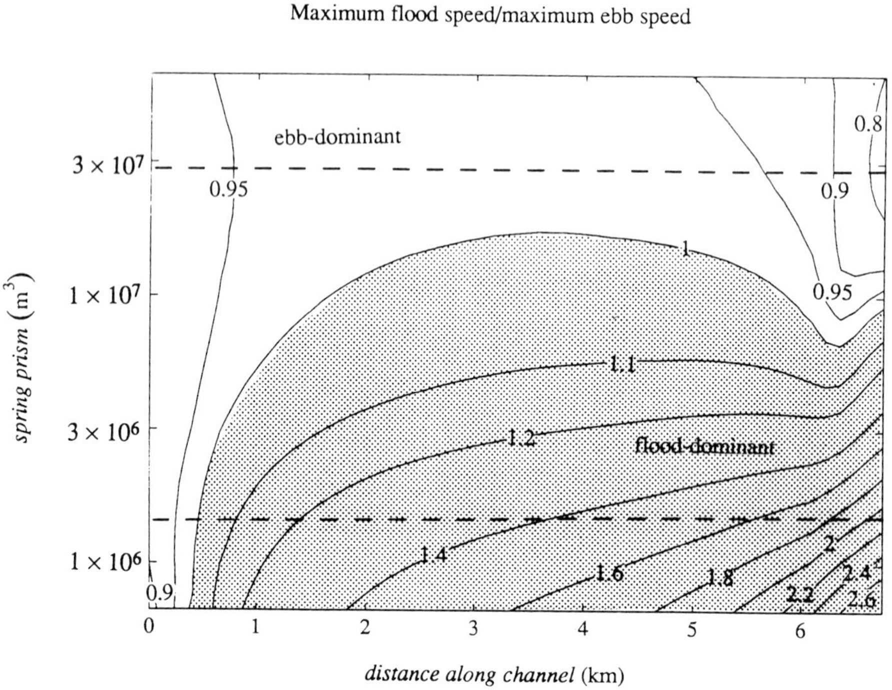

Figure 9: Results of 30 numerical solutions to (15.11). The ratio of cross-sectionally averaged maximum flood to maximum ebb speed during a mean tidal cycle is contoured as a function of local position along the channel and total embayment spring prism. If maximum flood speed/maximum ebb speed is greater than one, the channel is flood-dominant; if the ratio is less than one, the channel is ebb-dominant. The dashed lines indicate two models examined in more detail, results from which were used to force other models (see Figs. 10, 12, 13).

15.4.2 Ebb-dominant inlets to flood-dominant main channels

Ebb-dominant velocities are often observed in the immediate vicinity of inlets to otherwise flood-dominant embayments (e.g., Murrells Inlet, SC, Perry et al., 1978; New Chatham Inlet, MA, Giese et al., 1989). Solutions to the non-linear zero-inertia equation are consistent with these observations (Fig. 9), indicating systems with flood-dominant main channels contain ebb-dominant cross-sectionally averaged velocities in the immediate vicinity of the inlet. In embayments with $h_{high}/h_{low} >> 1$, Boon and Byrne (1981) attributed local ebb-dominant velocities at inlets to continuity effects. In frictionally-dominated systems, changes in elevation within the inner embayment may lag sea surface forcing by up to several hours. According to Boon and Byrne, if inlet velocity is in phase with discharge from the inner embayment, while inlet cross-sectional area is in phase with the sea tide, average inlet area will be smaller during ebb than during flood. Continuity then forces stronger ebb than flood velocities to occur at the inlet.

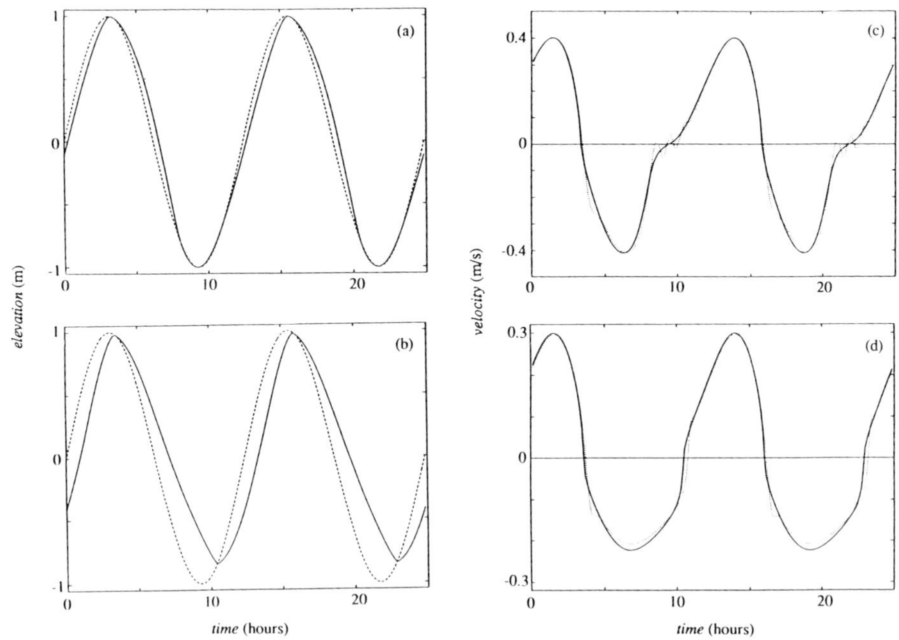

Figure 10: Time series of tidal elevations at the forcing end and 5500 m up the tidal channel: (a) ebb-dominant main channel; (b) flood-dominant main channel (see Fig. 9). Time series of tidal velocities produced by the zero-inertia equation and a geometrically identical model which includes the acceleration terms, 5500 m up the tidal channel: (c) ebb-dominant main channel; (d) flood-dominant main channel.

It may be inappropriate, however, to consider velocities at the inlet to be forced by the phase of velocities within the embayment. Advective effects are negligible in these systems, so velocities at the inlet cannot be viewed as forced by velocities elsewhere. An explanation more appropriate to frictionally-dominated embayments results from consideration of set-up. In systems with $L_{fhigh} > L_{flow}$, low tide decays more quickly than high tide, i.e., differences between ocean and embayment elevations are significantly larger near low water than near high water (Fig.10a). Near the inlet a much steeper surface gradient occurs near low water than near high water. By solving (15.10) for velocity,

$$|u| = \frac{R_c^{2/3}}{n} \left|\frac{\partial \zeta}{\partial x}\right|^{1/2}, \tag{15.23}$$

we see that a much steeper surface gradient near low water leads to highest velocities at the inlet during the ebb (assuming the effect of the increase in surface gradient is larger than the effect of the accompanying decrease in hydraulic radius).

15.4.3 Flood-dominant intertidal areas adjacent to ebb-dominant main channels

Observations of intertidal embayments indicate that tidal channels which drain completely at low tide are flood-dominant (Lincoln and FitzGerald, 1988; Speer et al., 1991). Intertidal areas bordering relatively deep ebb-dominant tidal channels have also been observed to be flood dominant. Swift and Brown (1983) presented harmonic analyses of the Great Bay, NH, that indicate a landward transition from ebb-dominance in the main channel to flood-dominance in the marginal shallows and intertidal areas (Table 5). The Great Bay Estuary has ζ_o = 1.3 m, mean channel depths on the order of 7 m, and extensive areas of intertidal flats (Fig. 11). A conceptual explanation for the landward transition from ebb- to flood-dominance observed at Great Bay follows from a consideration of spatial variations in the frictional length scale, L_f, in two dimensions. In 2-d, (15.12), (15.13) and (15.20) become

$$\vec{\nabla}.(D\vec{\nabla}\zeta) - \frac{\partial \zeta}{\partial t} = 0, \tag{15.24}$$

$$D = \frac{H^{5/3}}{n} |\vec{\nabla}\zeta|^{-1/2}, \tag{15.25}$$

$$L_f \propto H^{10/9} \zeta_o^{-1/3} (n\omega)^{-2/3}, \tag{15.26}$$

where $H(x,y)$ is total local depth. Over all but the innermost portion of the Great Bay system, deep channels are found adjacent to large intertidal areas. Thus average depths over most of the system are shallowest around high water, and high tide propagates more slowly than low tide ($Lf_{high} < Lf_{low}$). Over the innermost marginal shallows, however, average depths are much shallower around low water ($Lf_{high} > Lf_{low}$). Increased frictional losses around low water slow the propagation of low tide, and the tide becomes locally flood-dominant.

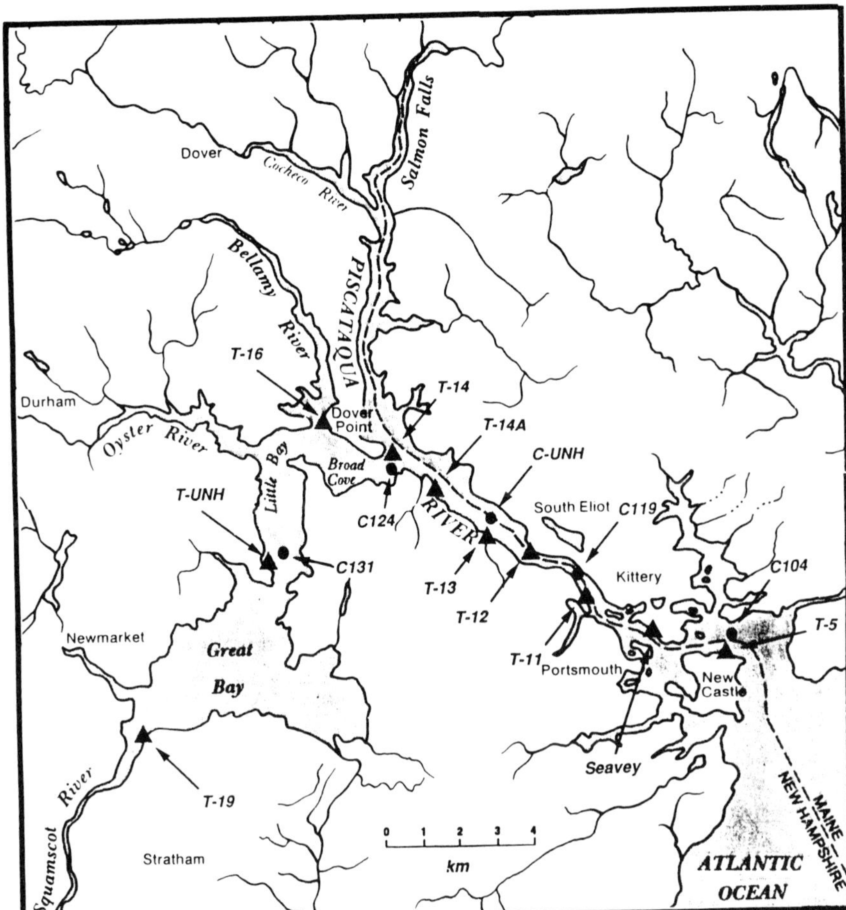

Figure 11: Location map of the Great Bay Estuary, New Hampshire, indicating locations of tidal elevation stations (▲) and current moorings (•). (After Swift and Brown, 1983).

To investigate the physics which result in flood-dominance over marginal shallows, we solved (15.24) for a linearly sloping intertidal flat, rising from one meter below mean tide level at the forced boundary to one meter above mean tide level at the landward end, over a total horizontal distance of 2500 m. We performed two experiments, each using $\Delta t = 90$ s and $\Delta x = 62.5$ m, and each forced with different time-series taken from previous 1-d solutions (see Fig. 10). The first forcing time-series had a shorter duration flood and the second had a shorter duration ebb. The first experiment resulted in flood-dominant currents across the entire width of the intertidal flat. In the second case currents adjacent to the distorted forcing were ebb-dominant, but half way across the flat the velocity-dominance reversed and the currents became flood-dominant (Fig. 12). Across the intertidal flat average depths are much shallower around low water ($Lf_{high} > Lf_{low}$), and high water propagates more quickly than low water. The faster propagation of high water further enhances flood-dominance when the tide at the edge of the flat is forced by a shorter duration flood (Fig. 13a,c). If the boundary is forced by a shorter duration ebb, however, the faster propagation of high water first must overcome the pre-existing shorter ebb before eventually producing a shorter flood and flood-dominance farther out across the flat (Fig. 13b,d). The sharp peak in flood velocity which develops with distance over the intertidal flat is consistent with observations of sediment resuspension at the Great Bay Estuary, NH. Anderson (1973) indicated that flood tide sediment resuspension is at at least twice that of the ebb over flats at the Great Bay.

Table 5: Characteristics of the M_2 and M_4 tide observed at tide gauges and current meters throughout the Great Bay, NH tidal embayment (from Swift and Brown, 1983). Velocity data are in bold italics. A shorter duration ebb, suggesting probable ebb-dominance, is described by a sea-surface relative phase ranging from 180°-360°. Neglecting residual currents, ebb-dominance is indicated more directly by a velocity relative phase of 90°-270°. The remaining ranges of relative phase characterize flood-dominance (Friedrichs and Aubrey, 1988).

Gauge or meter	distance (km)	M_2 (m or ms^{-1})	M_2 phase (deg)	M_4/M_2	relative phase (deg)	dominance
1. T-5	4.2	1.29	106	0.016	238	ebb
2. C 104	***4.2***	***0.47***	***248***	***0.085***	***107***	***ebb***
3. Seavey	5.5	1.20	114	0.017	207	ebb
4. T-11	7.3	1.12	117	0.026	243	ebb
5. C 119	***7.8***	***0.70***	***248***	***0.029***	***129***	***ebb***
6. T-12	9.3	1.00	128	0.030	231	ebb
7. T-13	10.7	0.95	133	0.042	235	ebb
8. T-14A	12.5	0.93	140	0.043	258	ebb
9. T-14	13.3	0.94	145	0.032	213	ebb
10. C 124	***13.3***	***1.48***	***262***	***0.027***	***222***	***ebb***
11. T-16	15.0	0.83	166	0.012	156	flood
12. C 131	***19.2***	***0.31***	***261***	***0.097***	***333***	***flood***
13. T-UNH	20.0	0.87	171	0.034	260	ebb
14. T-19	26.0	0.92	176	0.033	103	flood

15.4.4 Inlets with distinct flood- and ebb-dominant sections

In a series of articles, Kjerfve and others (Kjerfve, 1978; Kjerfve and Proehl, 1979; Kjerfve et al., 1981) discussed variability of velocity-dominance across cross-sections of tidal channels at North Inlet, SC. North Inlet is a well-mixed tidal embayment with $\zeta_o = 0.7$ m, $\bar{h} = 2$-3 m and little fresh water inflow (Nummedal and Humphries, 1978; Friedrichs and Aubrey, 1988). Kjerfve and others found velocity-dominance to be related to a characteristic bimodal distribution of cross-sectional bathymetry near the tidal inlet. High velocity, strongly ebb-dominant flow was typically observed in the deeper of two sub-parallel channels, whereas the shallower channel typically contained lower velocity, weakly flood-dominant flow. These observations are consistent with qualitative descriptions of tidal inlet flood- and ebb-channels by Hayes (1975; 1980). Kjerfve and Proehl speculated about the observed velocity distribution, including the influence of channel curvature, the shape of the flood-tide delta, and the nearby junction of two tidal creeks. From the consideration of the zero-inertia equation, however, it appears that the velocity pattern observed at North Inlet may be a direct result of the distribution of cross-sectional bathymetry.

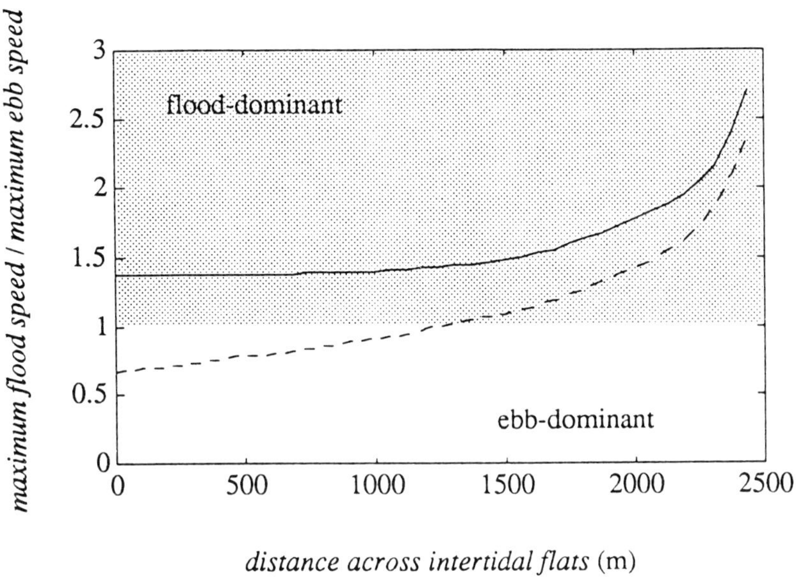

Figure 12: The ratio of depth-averaged maximum flood to maximum ebb speed during a mean tidal cycle, plotted as a function of distance across an intertidal flat. The forcing elevations were produced by the two models highlighted in Fig. 9, recorded 6500 m into the tidal channels. The solid (dashed) line is forced by a time series with a shorter duration flood (ebb).

If flow through a channel cross-section is approximately straight and parallel, it is reasonable to assume sea-surface elevation and gradient are relatively constant across that given channel cross-section at any one time (e.g., Henderson, 1966). Therefore, by the two-dimensional equivalent to (15.23)

$$|u| = \frac{H^{2/3}}{n}|\vec{\nabla}\zeta|^{1/2}, \tag{15.27}$$

depth-averaged velocities will consistently be smaller over the shallower portions of the channel cross-section. Furthermore, the difference in velocity between the deep and shallow channels will be enhanced at lower water levels. Since elevation changes within the embayment at North Inlet lag behind ocean forcing (Nummedal and Humphries, 1978), depths near the inlet are less, on average, during the ebb than during the flood. Because $h_{ebb}/h_{flood} <$ 1, frictional drag is significantly greater in the shallow section during the flood than during the ebb. Therefore ebb-currents in the shallow section are much weaker than ebb-currents in the deep section, whereas flood-currents in the shallow section are only slightly weaker than flood-currents in the deep section. It follows that the deep and shallow portions of the cross-section should be ebb- and flood-dominant, respectively, if the total cross-sectionally integrated flow at North Inlet were neither ebb- nor flood-dominant. Considered as a whole, however, the embayment at North Inlet is a typical ebb-dominant system with moderately deep channels (relative to tidal amplitude) and large areas of intertidal storage (Nummedal and Humphries, 1978; Friedrichs and Aubrey, 1988). Thus the deeper channel is strongly ebb-dominant, whereas the shallow channel is only weakly flood-dominant.

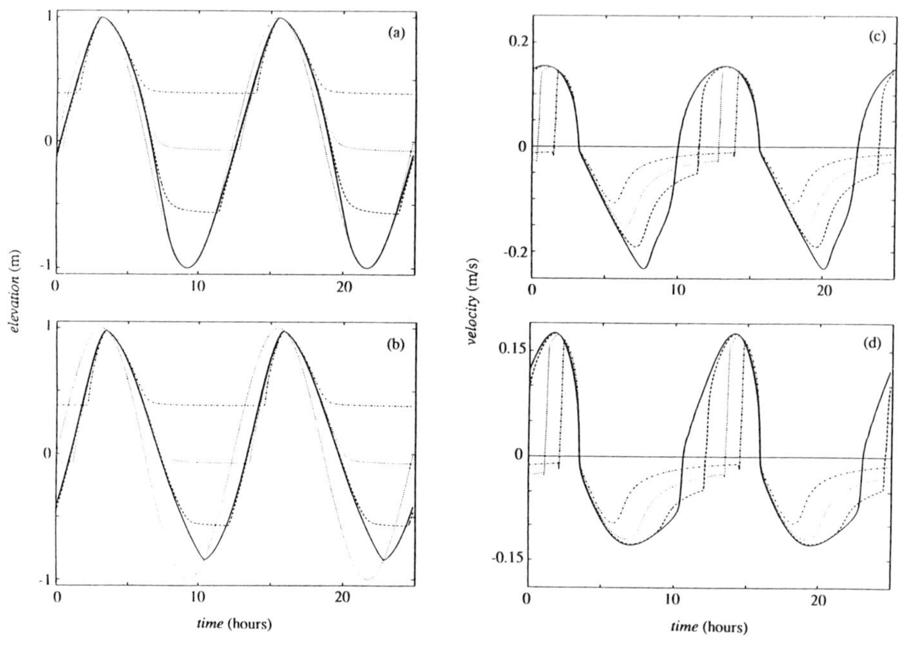

Figure 13: Time series of an undistorted sine curve, tidal elevations used for boundary forcing, and tidal elevations recorded at 500 m intervals across the intertidal flat: (a) forced by a time series with a shorter duration ebb; (b) forced by a time series with a shorter duration flood (see Fig. 9). Time series of tidal velocities recorded at 500 m intervals across the intertidal flat: (c) forced by a time series with a shorter duration ebb; (d) forced by a time series with a shorter duration flood.

Uncles and Kjerfve (1986) attempted to reproduce the residual velocity distribution at North Inlet with a 1-d numerical model that included the cross-sectional distribution of bathymetry. The model, which discretized the cross-section into twelve elements, neglected advection but included the effects of local acceleration. The model was forced by three periodic time series of axial surface gradient and elevation, each based on the M_4 and M_2 components of a single tidal cycle observed by Kjerfve and Proehl. Uncles and Kjerfve observed that depth-averaged velocity from the model was closely related to topography and was larger in the deep water where the effect of friction was smaller. Their model partially reproduced the depth-dependent distribution of residual velocity observed in the field. However the field observations were generally more depth dependent than the model results. Uncles and Kjerfve speculated that their error was due to advection effects neglected by their numerical formulation. However, we have argued that advection is negligible in shallow, relatively straight channels within well-mixed embayments such as North Inlet. Their error instead may be due to an underestimate of the drag coefficient appropriate to North Inlet. Uncles and Kjerfve applied an open-shelf value of c_d = 0.0025, even though Finley (1976) had already published observations from North Inlet that indicated $c_d \approx 0.01$.

To investigate the physics which result in a flood-dominant shallow section and an ebb-dominant deep section within a single inlet, we solved (15.24) for a one-kilometer length of channel composed of distinct shallow and deep sections. We performed two experiments, each using Δt = 450 s and a two-dimensional finite element grid composed of approximately 1000 elements (Fig. 14). In each case the seaward end of the channel was forced by a sine curve with ζ_o = 1 m, and the landward end was forced by time-series recorded 1 km into the previously solved 1-d channels. The mean depth and width of the 2-d channels were also identical to the geometry of the first kilometer of the 1-d channels which produced the forcing time-series. The channel depths in the two cases differ only by a constant scaling factor. The first model is narrower and shallower and was forced at the landward end by a time-series with a shorter duration flood. The second model has a greater width and mean depth and was forced at the landward end by a time-series with a shorter duration ebb. The first experiment resulted in strongly ebb-dominant currents in the deep section of the channel and strongly flood-dominant currents in the shallow section. In the second case the deeper section was strongly ebb-dominant, and the shallower section was weakly ebb-dominant (Fig. 14). Consistent with the conceptual discussion, the division of a single inlet into ebb- and flood-dominant sections is enhanced as h_{high}/h_{low} is increased. One can also expect enhancement of this effect as L/L_f is increased because h_{ebb}/h_{flood} will decrease as the tidal wave becomes more progressive.

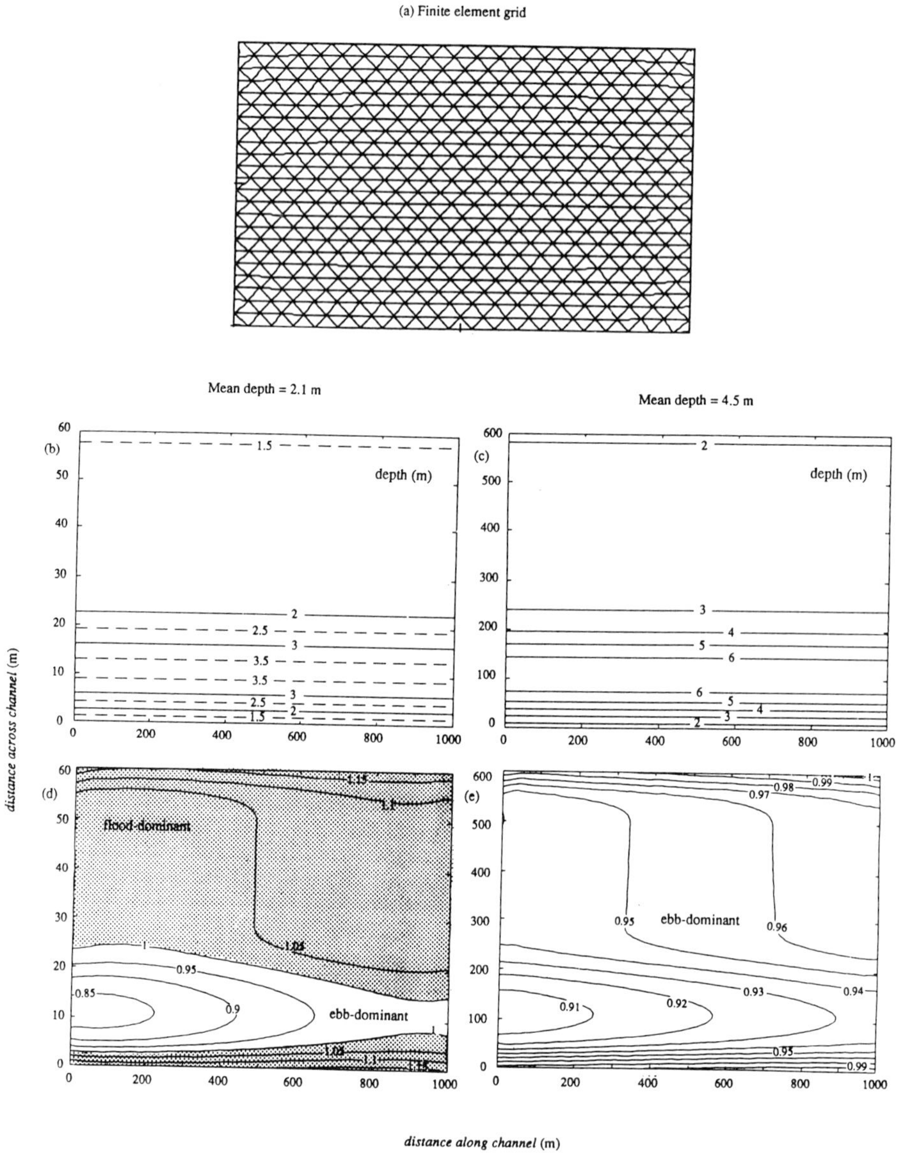

Figure 14: Two-dimensional finite element models of tidal channels with distinct deep and shallow sections. The models are forced by a sine curve at one end and elevations recorded from the two models highlighted in Fig. 9 at the other end, a distance of one kilometer upstream: (a) The finite element grid used for both models; (b), (c) the bathymetry of the two channels; (d), (e) the ratio of depth-averaged maximum flood to maximum ebb speed in the two channels.

15.5 Implications for sediment transport and morphologic evolution

From the preceding examination of velocity asymmetries produced by the zero-inertia equation, systematic patterns are identified which reinforce the underlying embayment morphology: Shallow and deep areas enhance flood- and ebb-dominance, respectively. Considering the embayment as a whole, shallow systems ($h_{high} >> h_{low}$, $b_{high} \approx b_{low}$) are largely flood-dominant, except in their deepest portions near the tidal inlet, which are primarily ebb-dominant. Deep systems ($h_{high} \approx h_{low}$, $b_{high} >> b_{low}$) are largely ebb-dominant, except in their shallowest portions over intertidal flats, which are primarily flood-dominant. Also, sharp changes in morphology are associated with reversals in velocity dominance. Across the width of tidal inlets, a local change from ebb- to flood-dominance is typically associated with a marked decrease in channel depth. Similarly, the transition from ebb- to flood-dominance observed in the inner embayment is associated with a morphologic break from deep tidal channel to shallow intertidal flat. The velocity patterns reinforce the morphologies: flood-dominance enhances landward near-bed transport, channel shoaling, and further reductions in depth; ebb-dominance enhances seaward near-bed transport, channel erosion, and further increases in depth.

Several authors have separated tidal embayment morphologies into two categories. Dronkers (1986b) discussed two fundamental types of basins: Type 1 basins have deep channels (ζ_o / h << 1) and high tidal flat areas, on the average above mean sea level. Type 2 basins have shallow channels ($\bar{h}$ at most a few times larger than ζ_o) and low tidal flat areas, which at high water become part of the stream cross section. Dronkers' Type 1 and 2 systems are equivalent to typical ebb- and flood-dominant geometries discussed by others (e.g., Boon and Byrne, 1981; Speer and Aubrey, 1985; Friedrichs and Aubrey, 1988). Boon and Byrne, (1981) suggested that in a system with flood-dominant main channels and lacking extensive tidal flats, high velocity floods could fill an initially deep basin to a level just flooded at high water, increase the area of flats, and eventually produce an ebb-dominant main channel. But results from Speer and Aubrey (1985) and Friedrichs and Aubrey (1988) indicated flood-dominance of the main channels does not lead to eventual ebb-dominance. Net bed-load transport more likely will cause shallow, flood-dominant channels to shoal, whereas under ebb-dominant conditions, deeper channels may be maintained.

It can be misleading, however, to label entire embayments as ebb-dominant or flood-dominant. As this paper shows, different portions of a single embayment typically contain distinct patterns of velocity distortion. For example, local flood-dominance over peripheral shallows and intertidal areas may occur at the edges of otherwise ebb-dominant systems. During the evolution of these systems, a morphologic transition may be enhanced between the ebb-dominant central portions of the embayment and the flood-dominant peripheral areas. Sediment will tend to be trapped along the edges of the embayment due to higher velocity flood currents, whereas deep channels will be maintained in the central portions of the embayment by higher velocity ebb flow. A relatively abrupt lateral transition from deep channel to intertidal flat, as characteristically seen in such systems, will tend to develop. In contrast, net sediment transport will be more-or-less landward throughout the inner embayment of systems with flood-dominant main channels. There is less tendency for abrupt discontinuities to develop between subtidal channels and intertidal flats in systems with flood-dominant main channels. Thus, depending on the velocity distortion of the main tidal channel, distinct embayment geometries may conceivably evolve along the lines of Dronkers' (1986b) Type 1 and 2 basins.

Velocity asymmetries produced by the zero-inertia equation also provide details of potential near-bed transport paths within the tidal channel cross-section. Across-channel patterns of velocity distortion reinforce bathymetry and are consistent with existing observations of morphology at tidal inlets (Hayes, 1975; 1980). Hayes observed that seaward transport occurs in main ebb channels, whereas landward transport occurs in marginal flood channels. The division of the channel cross-section into ebb- and flood-dominant sections also provides a mechanism for bedload transport in the direction opposite to that of cross-sectionally averaged velocity-dominance. For example, shallow systems having flood-dominant main channels tend to have cross-sectionally averaged currents near the tidal inlet which are ebb-dominant. Nonetheless, net landward bedload transport can still occur through the tidal inlets of such systems. If littoral transport supplies abundant sediment to the tidal inlet, more sediment transport may occur landward along the marginal flood channels than seaward through the main ebb channel.

This paper utilizes empirical relations between maximum spring tidal prism (or, interchangeably, maximum spring tidal discharge) and channel cross-sectional geometry and discusses the feedback between flow and morphology. For example, hydraulic engineers have long recognized the rapid filling of tidal inlets which often follows enlargement of inlets by dredging or reduction of prism by engineering works. Channel morphodynamics respond on much shorter time scales than changes in overall basin geometry due to sedimentation or changes in relative sea level. Gardner and Bohn (1980) presented geologic evidence from Holocene tidal channels which supports this causal interpretation. Thus pre-existing geologic control may dictate dominance type: Smaller inlet/bay systems that form with limited tidal prisms will tend to have shallower channels, whereas larger systems with relict morphology that dictates a larger tidal prism will have deeper channels. In basins forced by similar offshore tidal amplitudes, initially large or small tidal prisms will favor ebb- or flood-dominant main channels, respectively. Along a barrier-beach coast-line backed by a low-lying coastal plain, the occurrence of flood- or ebb-dominant embayments speculatively can be related to tidal range alone. Hayes (1979) and FitzGerald (1988) related the spacing of tidal inlets along barrier beach coastlines inversely to the tidal range. Therefore lower tidal amplitude may lead to wider spacing of inlets ($b_{high} >> b_{low}$), larger prisms, deeper channels ($h_{high} \approx h_{low}$), and ebb-dominant main channels. Higher tidal amplitude may lead to closer inlet spacing ($b_{high} \approx b_{low}$), smaller prisms, shallower channels ($h_{high} >> h_{low}$), and flood-dominant main channels.

These relationships are being explored as ongoing research.

15.6 Summary

1. Under conditions appropriate to many small (L <~20 km), well-mixed ($\zeta_o / \bar{h}$ > ~0.1) tidal embayments, friction is one-to-two orders of magnitude larger than acceleration in the cross-sectionally integrated momentum equation. Thus tidal flow in these systems may be described by a zero-inertia diffusive model. This representation captures the most important non-linearities which contribute to asymmetric peak tidal velocities (and, therefore, to net near-bed sediment transport), while greatly simplifying conceptual interpretation.

2. The linear analytic solution to the zero-inertia equation describes an exponentially decaying, partially progressive, partially standing wave governed by the frictional length scale L_f. L_f represents both the propagation speed ($c = L_f\, \omega$) and the rate of amplitude decay (e-folding length = L_f). Propagation speed decreases and amplitude decay increases as depth is reduced, friction is increased, or total embayment width is increased relative to the width of the channel. The zeroth-order decay and delay of the M_2 tide in four existing tidal embayments are consistent with behavior predicted by the linear solutions.

3. Non-dimensional parameters important to non-linear behavior are identified by considering variations in the magnitude of L_f during the tidal cycle. Since propagation speed is proportional to L_f, the tidal wave propagates most quickly during that portion of the tidal cycle having the greatest L_f. An extension of linear results indicates that the ratio of frictional lengths near high and low water is given by $L_{fhigh}/L_{flow} \approx (h_{high}/h_{low})^{10/9}(b_{high}/b_{low})^{-2/3}$. In four existing tidal embayments, non-linear behavior as indicated by M_2 and M_4 tidal components is consistent with conceptual predictions.

4. Non-linear numerical solutions to the zero-inertia equation recreate patterns of velocity distortion observed in the field: (a) Large changes in channel depth during the tidal cycle (h_{high}/h_{low} >> 1) are associated with flood-dominant main channels. (b) Large changes in total width during the tidal cycle (b_{high}/b_{low} >> 1) result in ebb-dominant main channels. (c) Systems with flood-dominant main channels tend to have ebb-dominant tidal inlets. (d) Systems with ebb-dominant main channels tend to have flood-dominant shallow margins and intertidal areas. (e) Tidal inlet cross-sections are often composed of separate deep ebb-dominant and shallow flood-dominant sections.

5. Both analytic and numerical solutions indicate that velocity asymmetry in these systems reinforces morphology and likely plays a central role in its creation and maintenance. Shallower areas tend to be flood-dominant and enhance landward near-bed transport; deeper areas tend to be ebb-dominant and enhance seaward near-bed transport. Changes in morphology are associated with reversals in velocity-dominance. The velocity patterns reinforce the morphologies: flood-dominance enhances landward near-bed transport, channel

shoaling, and further reductions in depth; ebb-dominance enhances seaward near-bed transport, channel erosion, and further increases in depth.

6. Geologic controls on tidal prism may ultimately dictate velocity distortion patterns and morphologic evolution of these shallow embayments. Large tidal prisms are associated with deep channels and wide intertidal areas (b_{high}/b_{low} >> 1), whereas small tidal prisms are associated with shallow channels and narrow intertidal areas (h_{high}/h_{low} >> 1). During the evolution of systems having large prisms, the morphologic transition between the ebb-dominant central portions of the embayment and the flood-dominant peripheral areas will be enhanced. In systems with small prisms, net sediment transport will be more-or-less landward throughout the inner embayment, leaving intertidal and channel areas morphologically less distinct.

Acknowledgments

Special thanks is extended to Dr. Ole S. Madsen who provided many helpful comments on the manuscript. This work was supported by the Office of Naval Research through the American Society for Engineering Education; by the National Science Foundation, under grant number CEE-8352226; by the Woods Hole Oceanographic Institution's Coastal Research Center; by the NOAA National Sea Grant College Program Office, Department of Commerce, under grant number NA86-AA-D-SG090, WHOI Sea Grant Project R/B-82; and by the A.W. Mellon Foundation. The U.S. Government is authorized to produce and distribute reprints for governmental purposes notwithstanding any copyright notation that may appear hereon.

15.7 References

Airy, G.B., 1845: Tides and Waves. Encyclopedia Metropolitan, London, Article 192.

Anderson, F.E., 1973: Observations of some sedimentary processes acting on a tidal flat. Marine Geology, 14, 101-116.

Aubrey, D.G., 1986: Hydrodynamic controls on sediment transport in well-mixed bays and estuaries. Physics of Shallow Estuaries and Bays, J. van de Kreeke, Ed., Springer-Verlag, 245-258.

Aubrey, D.G., and P.E. Speer, 1985: A study of non-linear tidal propagation in shallow inlet/estuarine systems. Part I: observations. Estuarine, Coastal and Shelf Science, 21, 185-205.

Boon, J.D. III, and R.J. Byrne, 1981: On basin hypsometry and the morphodynamic response of coastal inlet systems. Marine Geology, 40, 27-48.

Byrne, R.J., P. Bullock and D.G. Tyler, 1975: Response characteristics of a tidal inlet: a case study. Estuarine Research, Volume II: Geology and Engineering, L.E. Cronin, Ed., Academic Press, 201-216.

Chantler, A.G., 1974: The applicability of regime theory to tidal watercourses. Journal of Hydraulic Research, 12, 181-191.

Dronkers, J., 1986a: Tide-induced residual transport of fine sediment. Physics of Shallow Estuaries and Bays, J. van de Kreeke, Ed., Springer-Verlag, 228-244.

Dronkers, J., 1986b: Tidal asymmetry and estuarine morphology. Netherlands Journal of Sea Research, 20, 117-131.

Finley, R.J., 1976: Hydraulics and dynamics of North Inlet, South Carolina, 1974-75. US Army Coastal Engineering Research Center, G.I.T.I. Report 10, 188 pp.

FitzGerald, D.M., 1988: Shoreline erosional-depositional processes associated with tidal inlets. Hydrodynamics and Sediment Dynamics of Tidal Inlets, D.G. Aubrey and L. Weishar, Eds., Springer-Verlag, 186-225.

FitzGerald, D.M., and D. Nummedal, 1983: Response characteristics of an ebb-dominated tidal inlet channel. Journal of Sedimentary Petrology, 53, 833-845.

Friedrichs, C.T., and D.G. Aubrey, 1988: Non-linear tidal distortion in shallow well-mixed estuaries: a Synthesis. Estuarine, Coastal and Shelf Science, 27, 521-545.

Gardner, L.R., and M. Bohn, 1980: Geomorphic and hydraulic evolution of tidal creeks on a subsiding beach ridge plain, North Inlet, S.C. Marine Geology, 34, M91-M97.

Giese, B.S., and D.A. Jay, 1989: Modelling tidal energetics of the Columbia River Estuary. Estuarine, Coastal and Shelf Science, 29, 549-571.

Giese, G.S., D.G. Aubrey and J.T. Liu, 1989: Development, characteristics and effects of the New Chatham Harbor Inlet. Woods Hole Oceanographic Institution Technical Report, WHOI-89-19, 33 pp.

Hayes, M.O., 1975: Morphology of sand accumulations in estuaries. Estuarine Research, Volume II: Geology and Engineering, L.E. Cronin, Ed., Academic Press, 3-22.

Hayes, M.O., 1979: Barrier island morphology as a function of tidal and wave regime. Barrier Islands: From the Gulf of St. Lawrence to the Gulf of Mexico, S.P. Leatherman, Ed., Academic Press, 1-28.

Hayes, M.O., 1980: General morphology and sediment patterns in tidal inlets. Sedimentary Geology, 26, 139-156.

Henderson, F.M., 1966: Open Channel Flow, Macmillan, 522 pp.

Jarrett, J.T., 1976: Tidal prism-inlet area relationships. US Army Coastal Engineering Research Center, G.I.T.I. Report 3, 55 pp.

Kjerfve, B., 1978: Bathymetry as an indicator of net circulation in well mixed estuaries. Limnology and Oceanography, 23, 816-821.

Kjerfve, B., and J.A. Proehl, 1979: Velocity variability in a cross-section of a well-mixed estuary. Journal of Marine Research, 37, 409-418.

Kjerfve, B., L.H. Stevenson, J.A. Proehl, T.H. Chrzanowski and W.M. Kitchens, 1981: Estimation of material fluxes in an estuarine cross section: a critical analysis of spatial measurement density and errors. Limnology and Oceanography, 26, 325-335.

Lamb, H., 1932: Hydrodynamics. Cambridge University Press, 738 pp.

LeBlond, P.H., 1978: On tidal propagation in shallow rivers. Journal of Geophysical Research, 83, 4717-4721.

Lincoln, J.M., and D.M. FitzGerald, 1988: Tidal distortions and flood dominance at five small tidal inlets in southern Maine. Marine Geology, 82, 133-148.

Meyer-Peter, E., and R. Muller, 1948: Formulas for bedload transport. Proceedings of the Second Meeting, International Association of Hydraulic Research. Stockholm, Sweden, 39-64.

Münchow, A., and R.W. Garvine, 1991. Non-linear, barotropic tides and bores in shallow estuaries. Tellus, 43A, 246-256.

Nummedal, D., and S.M. Humphries, 1978: Hydraulics and dynamics of North Inlet, 1975-76. US Army Coastal Engineering Research Center, G.I.T.I. Report 16, 214 pp.

Park, M., 1985: Prediction of tidal hydraulics and sediment transport patterns in Stony Brook Harbor. M.S. thesis, State University of New York, Stony Brook Marine Environmental Science Program, 146 pp.

Parker, B.B., 1984: Frictional effects on the tidal dynamics of a shallow estuary. Ph.D. thesis, Johns Hopkins University, 292 pp.

Perry, F.C., W.C. Seabergh and E.F. Lane, 1978: Improvements for Murrells Inlet, South Carolina. US Army Engineer Waterways Experiment Station, Technical Report H-78-4, 339 pp.

Pingree, R.D., and L. Maddock, 1978: The M_4 tide in the English Channel derived from a non-linear numerical model of the M_2 tide. Deep-Sea Research, 25, 53-68.

Postma, H., 1961: Transport and accumulation of suspended matter in the Dutch Wadden Sea. Netherlands Journal of Sea Research, 1, 148-190.

Postma, H., 1967: Sediment transport and sedimentation in the marine environment. Estuaries, G.H. Lauff, Ed., American Association for the Advancement of Science, 158-186.

Robinson, I.S., L. Warren and J.F. Longbottom, 1983: Sea-level fluctuations in the Fleet, an English Tidal Lagoon. Estuarine, Coastal and Shelf Science, 16, 651-668.

Speer, P.E., 1984: Tidal distortion in shallow estuaries. Ph.D. thesis, Woods Hole Oceanographic Institution - Massachusetts Institute of Technology Joint Program in Oceanography, 210 pp.

Speer, P.E., and D.G. Aubrey, 1985: A study of non-linear tidal propagation in shallow inlet/estuarine systems. Part II: theory. Estuarine, Coastal and Shelf Science, 21, 207-224.

Speer, P.E., D.G. Aubrey and C.T. Friedrichs, 1991: Non-linear hydrodynamics of shallow tidal inlet/bay systems. Tidal Hydrodynamics, B.B. Parker, Ed., Wiley, in press.

Swift, M.R., and W.S. Brown, 1983: Distribution of bottom stress and tidal energy in a well-mixed estuary. Estuarine, Coastal and Shelf Science, 17, 297-317.

Uncles, R.J., and B. Kjerfve, 1986: Transverse structure of residual flow in North Inlet, South Carolina. Estuaries, 9, 39-42.

van de Kreeke, J., 1967: Water level fluctuations and flow in tidal inlets. ASCE Journal of the Waterways and Harbors Division, 43, 97-106.

Wallis, S.G., and D.W. Knight, 1984: Calibration studies concerning a one-dimensional numerical tidal model with particular reference to resistance coefficients. Estuarine, Coastal and Shelf Science, 19, 541-562.

Weisman, R.N., G.P. Lennon and F.E. Schuepfer, 1990: Resistance coefficient in a tidal channel. Estuarine and Coastal Modeling, M.L. Spaulding, Ed., American Society of Civil Engineers, 123-131.

Westerink, J.J., K.D. Stolzenbach and J.J. Connor, 1989: General spectral computations of the nonlinear shallow water tidal interactions within the Bight of Abaco. Journal of Physical Oceanography, 19, 1348-1371.

Appendix 1. Symbols defined in text

A total cross-sectional area of embayment, including intertidal flats
A_c cross-sectional area of momentum-transporting tidal channel

AA	advective acceleration term of momentum equation
b	total width of cross-section, including intertidal flats
$\underline{b}_c$	width of momentum-transporting portion of tidal channel
b	widths averaged over embayment length
b_c/b	embayment surface area (a_{bay}) at mean low water divided by a_{bay} at mean tide level
c	phase speed
c_d	bottom drag coefficient
D	non-linear diffusion coefficient
$\_\__{ebb}$	Mean value over the ebb for quantity __
F	friction term of momentum equation
$\_\__{flood}$	Mean value over the flood for quantity __
g	acceleration of gravity
$\underline{H}$	local total depth
h	depth at mean tide level, averaged over area of embayment
h	depth, averaged across momentum-transporting portion of channel cross-section
$\_\__{high}$	high tide value of quantity __
I	inertia term of momentum equation
$\_\__{inlet}$	value of quantity __ at inlet
L	system length
LA	local acceleration term of momentum equation
L_f	frictional length scale
$\_\__{low}$	low tide value of quantity __
n	Manning's coefficient
P	perimeter of entire embayment cross-section, including tidal flats
P_c	perimeter of momentum-transporting tidal channel
Q	volume transport
Q_m	maximum spring discharge
R_c	hydraulic radius of momentum-transporting tidal channel
t	time
u	depth-averaged velocity
u_o	characteristic scale of depth-averaged velocity
V_s/V_c	volume stored in intertidal areas/volume in channels
x	horizontal coordinate
α, β	constants determined by least-squares linear regression
$\delta\_\_$	characteristic variation in quantity __
Δt	numerical time step
ζ	surface elevation relative to a vertical datum
ζ_o	amplitude of sea-surface forcing
Θ	tidal lag
κ	"constant" diffusion coefficient
χ	dependent variable in least-squares linear regression
ψ	$2\ L/L_f$
ω	angular tidal frequency

16

Effects of Sea-Level Rise on Muddy Coastal Margins

R Kirby

Abstract

The cross-sectional shape, width and height of muddy shores is relevant to a wide range of engineering issues, but little is known of how these vary in nature. It has been appreciated for many years that both predominantly accretionary and predominantly erosional coasts occur in nature and more recently it was shown, for a large number of mud flats, that shore cross-sectional shape varies systematically, at least over a narrow band of tidal range.

Four coastal embayments in UK were selected for measurements of their height/area relationships. The sites were chosen on the basis of their sediment type, tidal range, depositional or erosional dominance and the availability of survey data. The area of shore at sequential heights was measured to permit area/height, (hypsographic), curves to be plotted. The data were non-dimensionalised to permit shores of widely varying size to be compared directly.

The study reveals an apparent systematic trend in hypsographic curve shape from microtidal through to hypertidal regimes. This is complicated by a second finding, that accretion-dominated shores appear to trend towards a high and convex-upwards cross-section, whilst erosion-dominated shores appear to trend towards a low and concave-upwards cross-section on coasts of the same tidal range. This may reflect the combined effects of relative wave exposure and sea level rise.

Arising from these findings it is suggested that changing tidal range, for example by constructing a tidal power barrage, will result in cross-sectional shape, width and height changing towards a new equilibrium. Furthermore, rising sea level, which might be expected to change the balance between accretion and erosion, is also likely to give rise to different changes. Further work is required, but if this confirms that shore cross-sectional shape does vary systematically, such empirical knowledge provides one approach to predicting long-term change.

16.1 Introduction

Our perception of muddy coastal margins has developed a great deal in less than 100 years. In the early days such areas were often termed "wastelands", a term carrying the implication that they were of no value other than for reclamation and perhaps wildfowling. Today we recognise a great variety of engineering and environmental significances, ranging from an initial defence against coast erosion to outlets for land drainage, stability of cables, pipes and other shore-attached structures, through to major nutrient sources in the coastal zone, nursery areas for fish, feeding areas for migratory birds and the habitat for invertebrate animals which provide the foodsource of these latter. In parallel with our relatively recent appreciation of the value of muddy coastal margins, our knowledge of them remains limited. Unlike sandy coasts, which have been extensively studied, relatively limited investigations of the dynamics and morphology of muddy coasts have been undertaken. We are further restricted by a general lack of good topographic surveys of intertidal zones and also by the absence of recognised techniques to evaluate change and to predict how such coastlines may evolve in the future.

Of those few investigations of muddy coasts which are available, the majority involve real-time monitoring of short-term changes, over periods ranging from hours to a year or more. Mathematical models which consider one or a few ebb and flood tides are available. In contrast, limited consideration has been given to longer term evolution, either in the past or as it is likely to occur in the future. Mathematical models are especially problematical in areas where long term prediction is required, where under-, normally or over-consolidated substrates are present and where mixed or layered substrates occur. It is these long term, real complexities which need to be assessed so that consequences of changes in water level on various timescales can be predicted.

The relative elevation of the land and sea can change suddenly, for example due to engineering works such as the construction of large dams for tidal power generation, or due to rapid earth movements. Relative changes can also occur on much longer timescales due to sea level rise or earth movements. Each of these will have major effects on a muddy coastline, which may take many years to adjust, even if the change itself occurs rapidly. Changes are likely to affect the height and width of the coastal zone, its cross-sectional shape and whether it is predominantly erosional or depositional.

Knowledge of the ambient height and shape of muddy coast-lines is often poor. These zones are at the junction between land or air-based and marine-based survey techniques. Other than incidentally in the course of hydrographic surveys for navigational purposes, muddy intertidal zones of a highly unstable and non-significant nature in navigational terms tend not to be surveyed with any degree of precision or regularity. Consequently, shore height, width and shape tends not to be well known. Only air photographic surveys flown at appropriate tidal times, from low level, with appropriate equipment can give comparable data and such surveys are never available on an incidental basis.

A preliminary evaluation of long term shore stability and shape and a tentative start in developing a methodology for predicting long-term evolution of muddy coastal zones has been made during initial feasibility studies into the effects of building a tidal power barrage in the Severn Estuary, UK. The results apply not only to such engineering works but also to natural changes in sea level, as well as those in prospect due to man-induced changes in climate. Having a foreknowledge of such changes is an important aspect of many engineering and environmental issues and must, in future, be an important aspect of coastal zone management.

16.2 Muddy Shore Evolution

Whilst intertidal topography is frequently surveyed incidentally and not necessarily with the precision required for a comprehensive plot of its shape to be derived, the sedimentary provenance of muddy shores is much more poorly studied and known. It is well established that some areas are long-term depositional sites, whereas others are erosional. Studying short-term change in muddy environments presents severe logistic problems, whereas for long-term change, where direct real-time monitoring cannot be undertaken, the problems are even more severe. There are, nevertheless, a wide range of physical, biological and chemical indicators of long-term change.

Evidence for change may come from old maps and charts, which in some parts of the world are available for the last several hundred years with steadily improving accuracy. More recently these have been supplemented by air photography and satellite imagery. Normally, and arising from the scales of changes, all such documents reveal areal change more readily than height variation.

Since the early 1950s and 1960s scientific investigations of muddy shores experiencing long term accretion have been available (van Straaten, 1950, Evans, 1965 etc.). In such regimes a coast-aligned sequence of mud and sand flats occurs across the shore, which is close to a long-term equilibrium with the hydrodynamic conditions shaping the shore. Physical properties of the sediment, such as sediment grain size and geotechnical properties are adjusted to the hydrodynamics of the overlying water mass. Similarly, biological characteristics, both floral and faunal, are in some form of dynamic equilibrium. Salt marshes rise by deposition and are generally transitional with the tidal flat with a seaward limit close to Mean High Water Neaps (MHWN). In the same way, hard-shelled, burrowing animals remain in position at the base of their burrows on death and are steadily entombed by the rising tidal flat surface. Finally, tidal flat sediments contain a wide range of anthropogenic chemicals, reflecting both deposition of sediment containing adsorbed contaminants and incorporation by animals. The sediments consequently reflect, and are in balance with, the chemistry of the overlying waters.

In contrast to this situation, tidal flats experiencing long-term erosion present many differences. The grain size and geotechnical properties of the sediment may not reflect the present hydrodynamic regime. The erosion may be detected in geological and archaeological indicators, such as the over-consolidated nature of the deposits, mud cracks, salt marsh or tidal flat cliffs and other indicators of erosion, peat beds, fossil forests and by a wide range of archaeological evidence, (Kirby, 1990, Kirby, in press, and SELRC, 1990). In addition to these natural indicators, measurements against man-made structures, especially of linear features crossing the shore such as pipelines and jetties, may be important, (Kirby, 1988).

In many tidal mud flat areas, biological indicators, in the form of scoured-out examples of burrowing or boring bivalves, such as *Mya arenaria*, *Scrobicularia plana*, and *Pholas dactylus*, indicate that an area is experiencing long-term erosion. Large scale, short-term episodic changes need to be distinguished from longer term, slower, evolution, but frequently there are a range of indicators of the timescale of change. Exposure of such bivalves has been used in a number of areas to indicate the sense but not the rate of long-term evolution, (Kirby, 1990, Kirby et al., in press, and Bradley, 1957). Frequently the progressive lowering of the mud flat surface which exposes the bivalves permits epifaunal species such as barnacles and algae to colonise the bivalve surfaces in bands. Band age decreases towards the sediment surface. Measurements of band thickness, reflected in yearly spatfall, provides a direct measurement of erosion rate, (Kirby, 1990).

Biological indicators may be complemented by chemical evidence, such as the absence of radionuclides from surficial sediments underlying contaminated waters, (Kirby and Parker, 1980) or the absence of other anthropogenic chemicals, (Little, 1989).

Despite these features permitting areas of long-term erosion and deposition to be distinguished, there is no systematic study of their distribution. Moreover, the reasons for different areas being dominated by long-term deposition or erosion remains obscure. No previous attempt has been made to use these sedimentary criteria to distinguish accretionary and erosional shorelines and use this as a basis for assessing whether such areas have characteristic and differing shapes and heights.

16.3 Muddy Shore Morphology

If shore shape and height were completely random, the prospects of finding any underlying similarity or control, which might in future serve as a basis for prediction, would be negligible. However, shore morphology, in general terms at least, is not completely random. In the 1960s and 70s Hayes carried out an assessment of the morphological features of soft sedimentary coasts around the world. He recognised that certain features occurred regularly and was able to use the criteria established by Davies (1964), Table 1, to demonstrate a link between coastal morphology and tidal range, (Hayes, 1975), Figure 1.

Figure 1 shows that, whereas river and tidal deltas, barrier islands and inlets are the dominant shore feature of microtidal coasts, linear sand ridges, tidal flats and salt marshes dominate macrotidal shores. The fact that this link between broad hydrodynamic conditions, specified in the tidal range, and coastal morphology is so general in nature gives some confidence that there is some link between the changing dominance of wave and current processes and the coastal morphology they give rise to.

Table 1: Mean tidal range divisions, following Davies (1964) and others

Name	Mean Tidal Range MTR (m)
Hypertidal	> 6.0
Macrotidal (High	5.0 - 6.0
Macrotidal (Low	4.0 - 5.0
Mesotidal (High	3.0 - 4.0
Mesotidal (Low	2.0 - 3.0
Microtidal	> 2.0

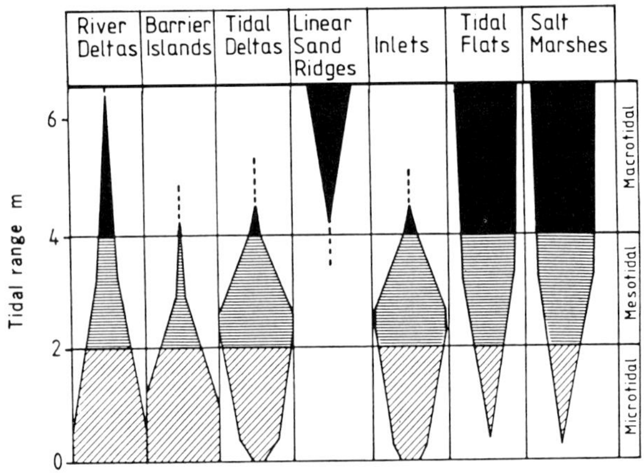

Figure 1: Variation in morphology of depositional shorelines compared to tidal range (Hayes 1975).

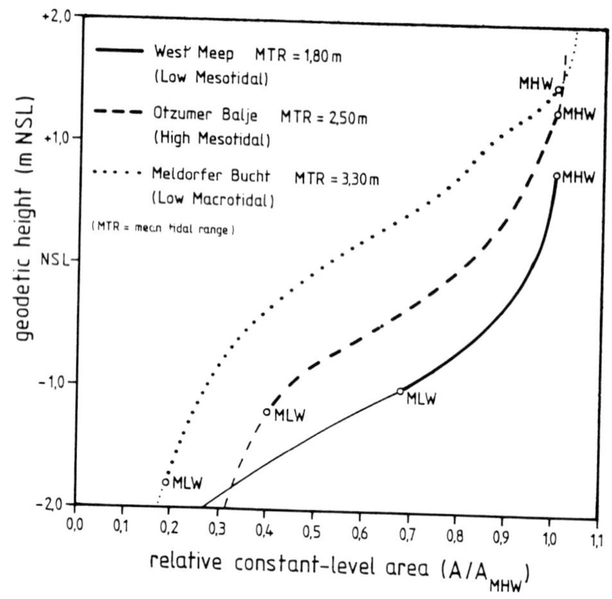

Figure 2: Relationship between shape and tidal range established for barrier-island tidal flats by Dieckmann et al (1987).

Table 2: Areas of tidal flats above MTL (from Dieckmann et al.)

High Mesotidal	62.5%
Low Mesotidal	33.0%
Microtidal	21.0%

Microtidal regimes are largely dominated by waves whereas, as tidal range increases, the relative importance of tidal currents must increase.

In addition, a further and apparently systematic characteristic of muddy shores was identified by Dieckmann et al. (1987). Dieckmann et al. measured the area/height (hypsographic) curves from a large number of inlets (37) ranging in size from 95-260km^2 and involving a total of 4,430km^2. The surveys used were also to a very high precision, often a few centimetres in height resolution. They used measurements of area at various heights on the shore to determine tidal flat shape. From these measurements non-dimensionalised hypsographic curves were produced. Non-dimensionalising the plots permits areas of differing size to be directly compared. For the narrow spread of tidal range variations, (microtidal to high mesotidal), along the Danish, German and Dutch coasts they showed that the area of tidal mud flat above Mean Tide Level (MTL) increases with increasing tidal range, whilst the shape of the area curve changes from concave-upwards in the case of microtidal to convex-upwards in the case of mesotidal, Figure 2, Table 2.

The very large number of measurements by Dieckmann et al., their high resolution and the extensive area they come from provide a cornerstone leading to confidence that the differences they report are real and typify the balance between wave and current forces in this barrier island coast of the North Sea. Dieckmann and Partensky (1985) have shown that the main factor governing morphological shape of the tidal flats is the tidal range. This confirms earlier work by Hayes (1979), who assessed the relationship between coastal morphology, tidal range and wave climate. The criteria of Hayes (1979) places the German Bight tidal flats clearly within the grouping of typical barrier island-type tidal flats, "tidally dominated mixed-energy coasts". The results produced by Dieckmann et al. thus led to the hypothesis that tidal mud flat shapes from higher or lower range areas might fit logically into the spectrum of types established for these relatively low tidal range coasts. The findings raised the possibility that hypsography might be applicable, in an empirical sense, to predict the consequences of changed tidal range.

Should a tidal power barrage be built in the Severn Estuary, mean tidal range inside the enclosed basin will change from the present 8.55m to about 4.5m. It remains to be established precisely what the revised hydrodynamic conditions will be within the enclosed basin, although a number of important attributes are well defined, as is their impact upon the shore. To provide comparative data on what the full spectrum of tidal flat shapes could potentially be in the post-closure regime behind the barrage, a careful appraisal was carried out to locate a small number of tidal mud flat localities of comparable mean tidal range to act as analogs. Sites representing exposed and sheltered, narrow and wide, erosional and depositional regimes were chosen. In each case a pre-requisite was that a good quality survey of the coastal zone in question was available. All four chosen sites are predominantly or entirely muddy, although in the Wash the classical, shore-parallel mud and sand flat environments of an accreting shore are developed. The Severn and Medway represent source-limited estuaries, in which fine sediment inputs are negligible or small at present. The Humber and Wash have more abundant inputs, especially from the sea. All four have a river wall forming their landward margin. The sedimentary regime of the Severn Estuary tidal flats has been specified by Kirby (1988), that of the Medway by Kirby (1990), of the Wash by Evans (1965) and of the Humber to some extent by Wilkinson et al. (1973). Some 34 sites from the Humber (MTR 4.8m at Immingham) and Medway (MTR 4.2m at Sheerness) estuaries and The Wash (MTR 4.8m at Hunstanton) coastal embayment were chosen. The Humber and Wash represent enclosed and exposed depositional areas respectively, whilst the Medway is an enclosed erosional site. It was considered that these sites would embrace the full range of possibilities for Severn estuary tidal flats at equilibrium post-closure. For comparison 9 sites within the Severn (MTR 8.55m at Cardiff, 9.4m at Avonmouth) were also measured, giving 43 in total. Sites chosen here ranged from enclosed (Cardiff Bay) to very exposed (Stert/Berrow) from both sides of the estuary and included localities which are both outside and inside the proposed barrage line. Unlike other localities, good quality surveys were not available from all sites in the Severn and specific aerial surveys of 7 sites had to be commissioned.

Almost incidentally, but consciously, the sites chosen in the four localities around the UK coast, when added to the German work, permitted almost a complete spectrum of hypsographic curves from microtidal through to hypertidal to be assembled for the first time.

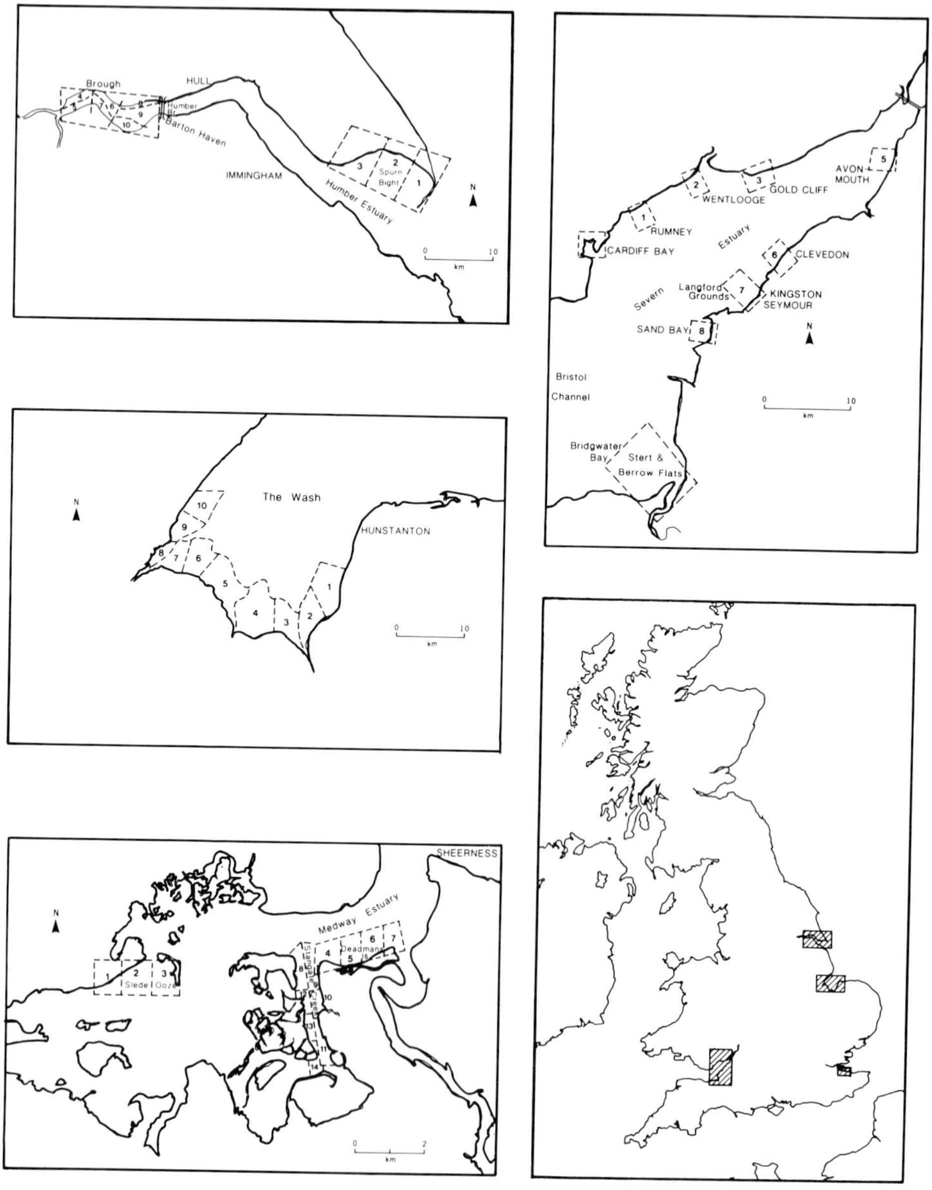

Figure 3: Locality map showing the four coastal embayments in which measurements were undertaken.Boundaries of numbered and named sub-areas measured are indicated. Note scales vary.

16.4 Methods

Input data from 2 sites, Humber and Medway, were conventional hydrographic surveys, whilst surveys from the Wash and Severn were mainly or entirely aerial surveys, Figure 3. In each case the surveys were first contoured at the narrowest contour interval possible (generally 0.25 - 0.5m) at a common datum. Areas at particular heights were then determined by planimeter out to the channel thalweg or a naturally occurring break of slope offshore. Areas were incremented cumulatively and non-dimensionalised to permit plotting of hypsographic curves, which were in every respect comparable to those obtained by Dieckmann et al., (1987).

Table 3: Total Areas Surveyed, Tidal Flat only

Area		km^2
Humber	(Spurn Bight	42.0
	(Brough	31.9
Wash		162.1
Medway	(Slede Ooze	1.4
	(Deadman's Island	1.0
	(Stangate Creek	0.9
Severn	(Stert and Berrow Flats	38.4
	(Cardiff Bay	1.9
	(Severn Tidal Power Group Areas 1-3, 5-8 (0.3 - 2.3)	9.0
German Bight (Dieckmann et al.) (95.0 - 260.0)		4430.0

The curves measured from the eight representative sub-areas presented here differ in a number of respects.

16.5 Results

A total of 43 sub-areas were calculated and both non-dimensional and dimensional plots were prepared. Sub-areas were amalgamated such that as large an area as could be justified has been calculated in order that results are as representative as possible. The 43 sub-areas were reduced to 15. The 10 Humber sites were amalgamated into 2 hypsographic curves from continuous areas, 14 sites in the Medway were compiled into 3 curves from continuous areas and 10 sites in the Wash were consolidated into 1 curve. The 9 sites measured in the Severn could not be consolidated further because all are from non-adjacent areas and all are different one from the other. Only 8 of the curves are reproduced here. In addition to individual plots, all Severn Estuary curves are individually compiled into a single plot. In addition, representative curves from this study are compiled together with the earlier Dieckmann et al data to produce a composite spanning the entire tidal range spectrum from 1.8 to 9.4m MTR. The areal coverage of the curves varies considerably from site to site, Table 3.

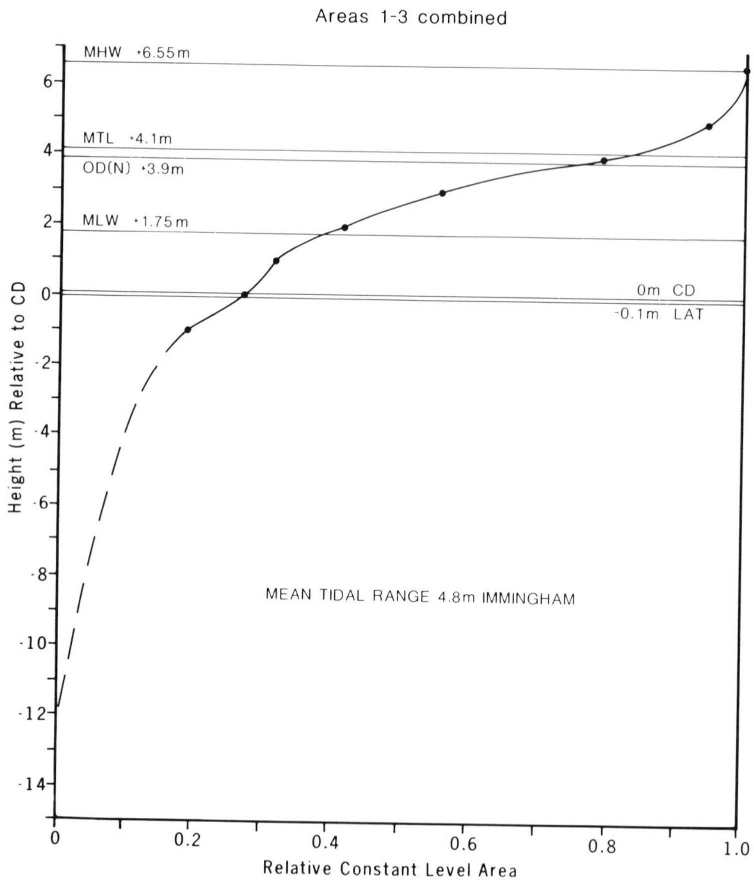

Figure 4: Hypsographic curve for areas 1-3 Spurn Bight, Humber Estuary.

16.5.1 Humber: Spurn Bight

The entire intertidal zone is a mud flat. The hypsographic curve, Figure 4, is linear to slightly convex upwards. The upper part of the tidal flat does not appear to rise as high as 6.0m above Chart Datum (CD), whereas Mean High Water (MHW) is at 6.55m. As a result the tidal flats must be covered right to the top on every tide (MHWN = 5.8m at Immingham).Calculation of the relative constant level area right out to the main channel axis has made it appear that the area above CD is relatively small. If only a narrow zone offshore had been considered the curve would have been virtually indistinguishable from that in the Wash. As plotted, the only differences are that the offshore subtidal component in the Wash is smaller, whilst the upper tidal flat rises above MHW. Spurn Bight does not have a salt marsh at its upper edge, whereas the Wash does. Possibly this, in part, explains the different levels of the upper shore. Also the higher mud flat in the Humber is poorly surveyed, in contrast to the Wash. The former was surveyed from a boat and the latter from the air.

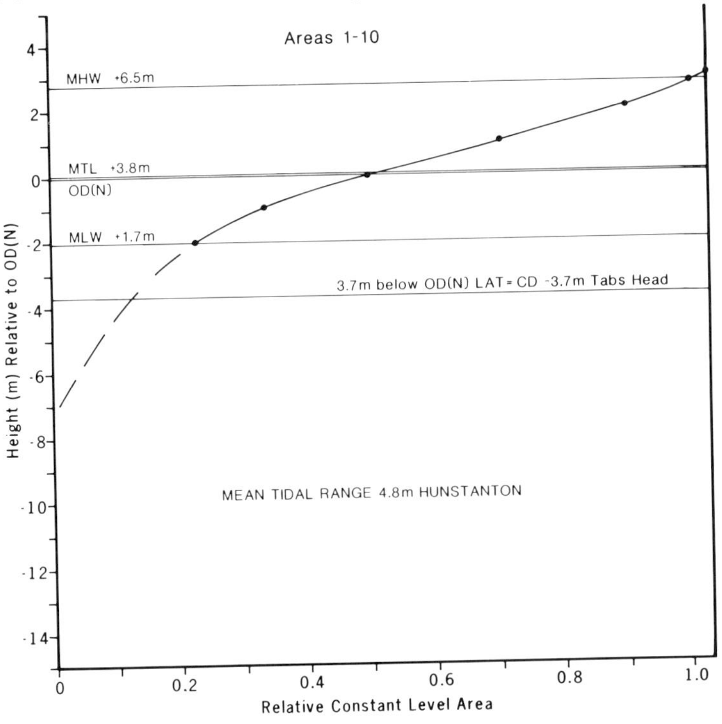

Figure 5: Hypsographic curve for areas 1-10 The Wash.

16.5.2 Wash

Only the marginal upper mud flat and upper sand flat region has been measured. The shore-attached finger banks towards the axis have been omitted. All 10 sub-areas into which the tidal flat was divided for convenience of measurement have a similar hypsographic curve. Arising from this it was justified to condense the 10 sub-areas onto a single composite curve (Figure 5). The Wash does not have a pronounced offshore channel, with the result that the offshore limit chosen was rather arbitrary. When the subtidal portion of the curve is considered together with the intertidal portion, the curve has a pronounced convex shape. Even neglecting the subtidal portion the curve is linear to slightly convex. The shape is most similar to Spurn Bight in the Humber and Cardiff Bay in the Severn both, like the Wash, areas of ongoing long-term net accretion.

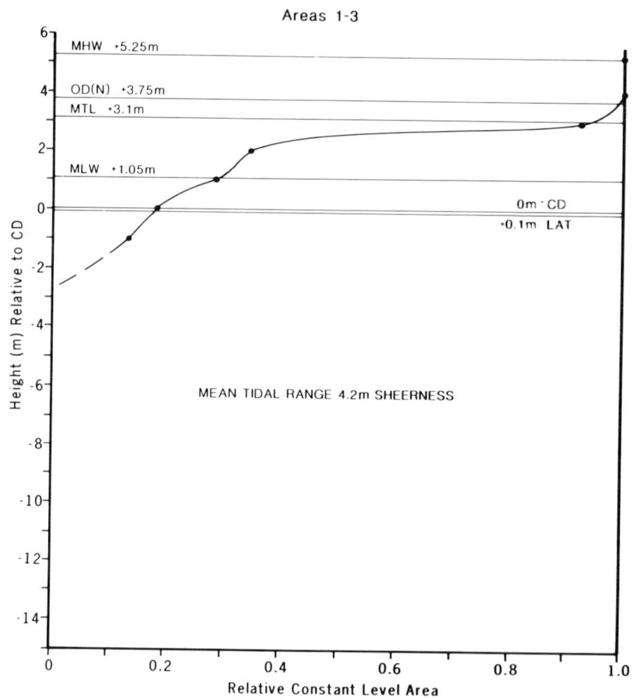

Figure 6: Hypsographic curve for areas 1-3 Slede Ooze, Medway Estuary.

16.5.3 Medway: Slede Ooze

This tidal flat is entirely muddy other than for a narrow zone with shellbanks. The hypsographic curve (Figure 6) has a pronounced sinuosity with a very broad flat zone just below MTL. The tidal flat is at a low elevation, only just reaching above MTL close to the coast. There is no upper shore, the tidal flat only rising to +3.80m (close to the national datum, Ordnance Datum (Newlyn), (OD{N})), and the mid-shore section is narrow and steep. The curve is most comparable with that from Deadman's Island in the Medway and the curves for sections 1,2,3, 5 and 6 in the Severn.

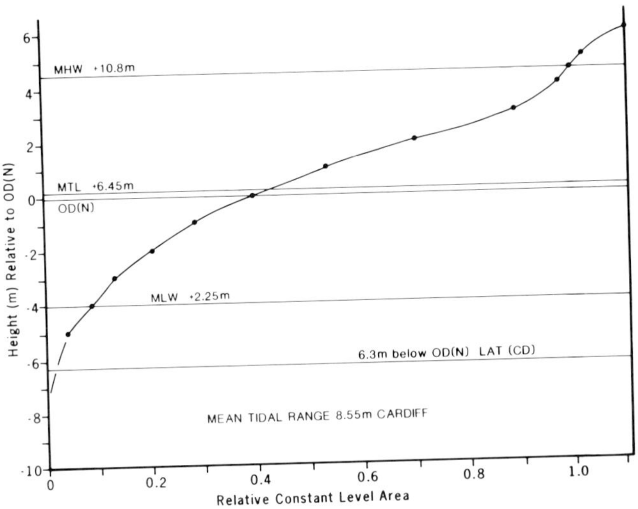

Figure 7: Hypsographic curve for Cardiff Bay, Severn Estuary.

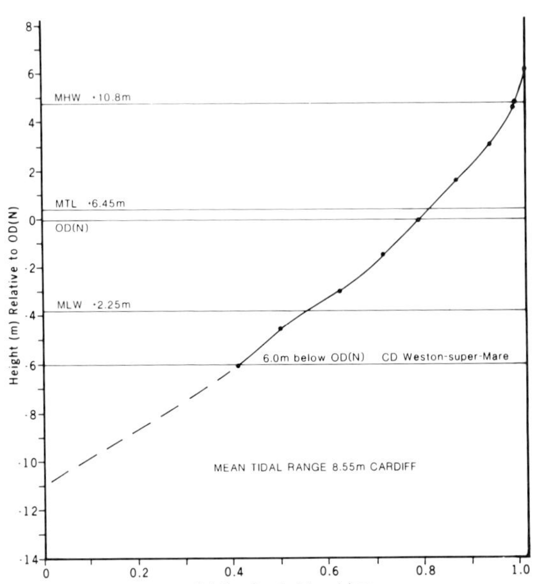

Figure 8: Hypsographic curve for Stert and Berrow Flats, Bridgwater Bay, Bristol Channel.

16.5.4 Severn: Cardiff Bay

The intertidal sediments are entirely muddy. The hypsographic curve has a distinctive pronounced "double-convexity" (Figure 7) both upper and lower shores are convex. The bay itself has a high proportion of its area in tidal flats. More than 65% of the whole area and 70% of the intertidal zone is above MTL, indicating a generally high level tidal flat. The near-shore zone is formed by a salt marsh.

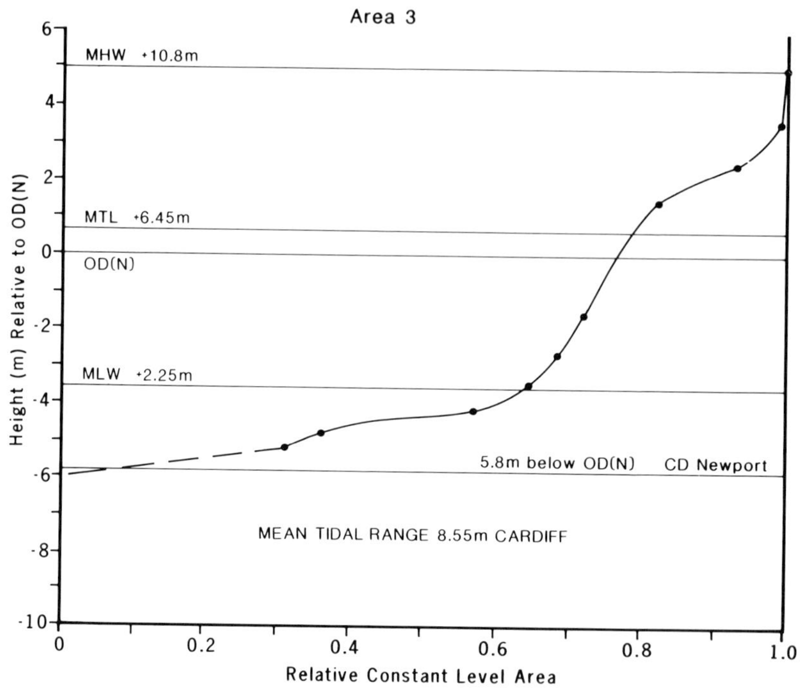

Figure 9: Hypsographic curve for Gold Cliff, Severn Estuary.

16.5.5 Severn: Stert and Berrow Flats

The inner intertidal zone, possibly 10-15% of the area, is a sand beach or has scattered sandy veneers. The remainder is mud flat. This is a geographically large area in Bridgwater Bay, which is possibly the most exposed locality for any mud flat in the estuary. Defining the subtidal limit of the shore profile presented a problem owing both to inadequate survey data as well as the difficulty of recognising any pronounced break of slope offshore.

The hypsographic curve (Figure 8) shows a pronounced and continuous concavity extending to the subtidal limit. The curve extends above both MTL and MHW but the inner section is narrow and in relative terms, steep. Only 20% of the profile is above MTL. There is no salt marsh at the top of the tidal flat for much of the coastal section and the salt marsh between Steart village and Hinkley occupies only 2% of the total intertidal zone. The section consequently provides a marked contrast with the hypsographic curve for Cardiff Bay, despite the latter being a similar distance down-estuary on the opposite bank.

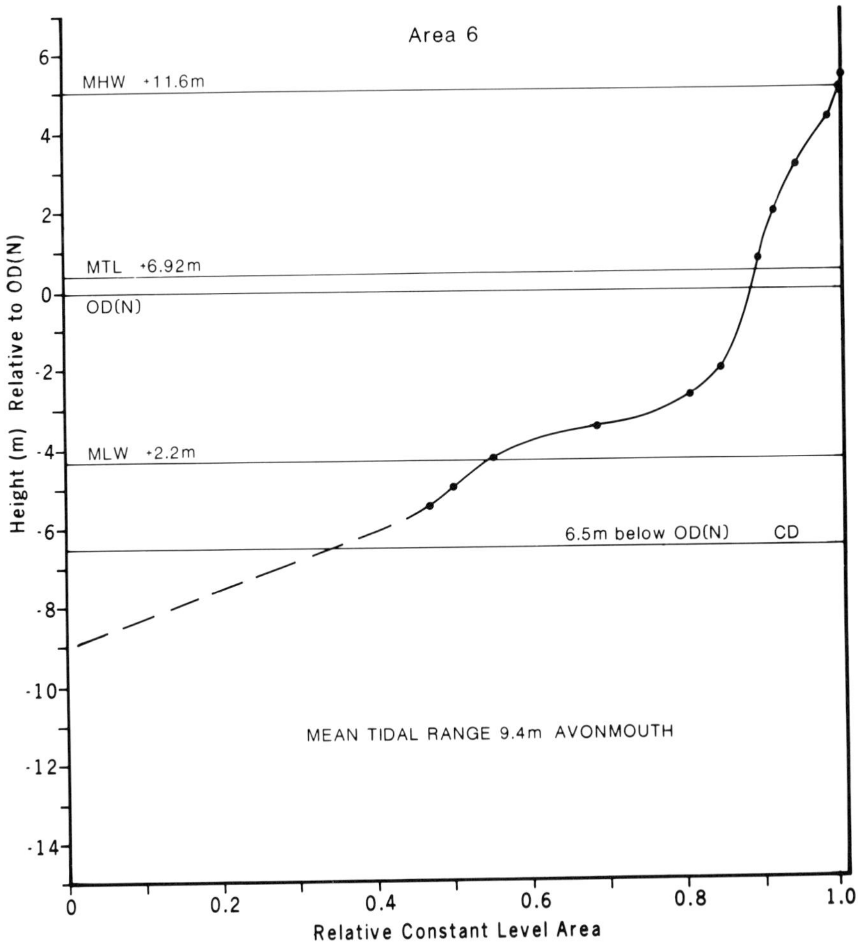

Figure 10: Hypsograpic curve for Clevedon, Severn Estuary.

16.5.6 Severn: Gold Cliff

The intertidal zone is entirely muddy, other than for small areas of exposed peat. The profile (Figure 9) has a complex shape with an overall concavity but with convex "shoulders" above MTL and below Mean Low Water (MLW). There are two gently inclined shore platforms, a relatively narrow upper one between +1.2m and +2.2m OD(N) and a broader, lower level platform starting at -3.5m and extending into the subtidal zone. Only 20% of the total curve is above MTL, although 70% is intertidal.

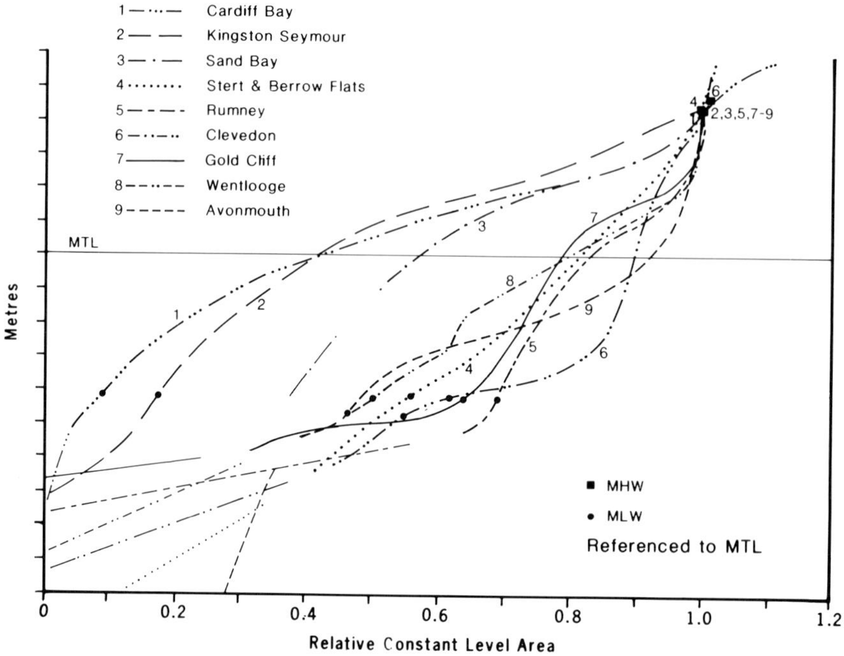

Figure 11: Superimposed Hypsographic curves for all nine Severn Estuary and Bristol Channel localities.

16.5.7 Severn: Clevedon

The shore is muddy, other than for an extensive narrow band of exposed rock at mid-shore level. The profile (Figure 10) has a similar shape to Gold Cliff with an overall concave-upward shape and two slight convexities. An upper shore platform is present between MHW and MTL but in this case could be, in part, an artefact arising from rock outcrops in some parts of this zone. The flatter, outer platform present in many other Severn Estuary curves is also present but here lies just above MLW, between -2.0m and -4.2m OD(N).

16.5.8 Severn: all 9 curves

Two distinctive suites of curves can be recognised (Figure 11). The larger group of 6 show closely comparable concave-upward curves. Most have high or low level shore platforms, or both, although these occur at different levels. The other group of 3 show higher, convex-upward shapes and apparently occur in zones which are shallower offshore. They also extend further across the graph.

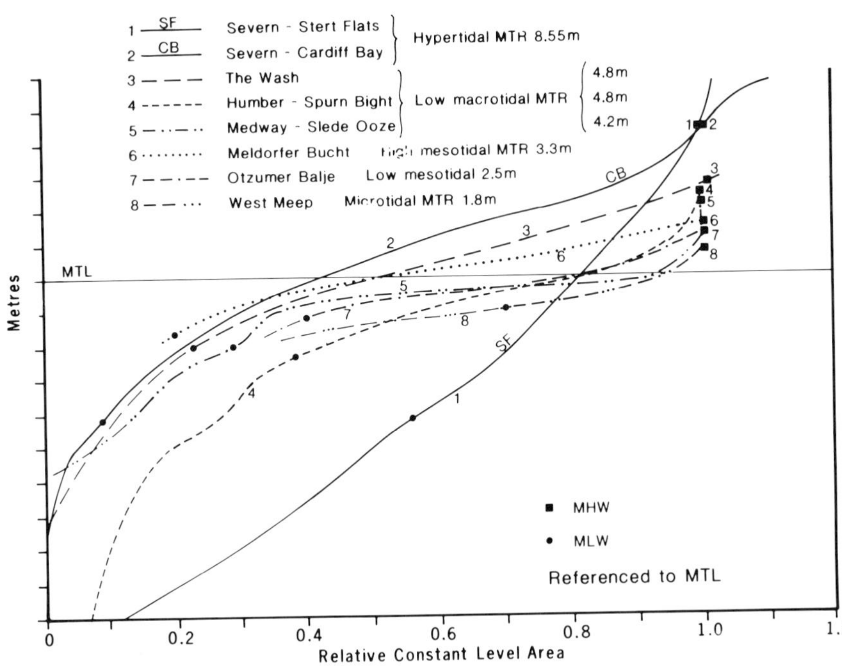

Figure 12: Superimposed selected Hypsographic curves representing hypertidal, low macrotidal, high mesotidal, low mesotidal and microtidal regimes in the UK and Germany classified according to Table 1.

16.5.9 Hypsographic curves from Micro- to Hypertidal regimes

It was initially postulated that if the trends apparent in the German suite of curves continued logically, then macrotidal and hypertidal hypsographic curves could be expected to be higher with respect to MTL, more convex in shape and shifted to the left on the graph, implying a greater proportion of the whole curve was intertidal. The comparison in Figure 12 is complicated by the fact that whilst the UK data is referenced to Mean Tide Level (MTL) the German data is referenced to Mean Sea Level (MSL). A strict intercomparison is not possible because MSL is not known for the UK curves.

The results show that in order of ascending tidal range, and providing that appropriate hypsographic curves are chosen, the Medway (Slede Ooze) MTR 4.2m, the Humber (Spurn Bight) MTR 4.8m, the Wash MTR 4.8m and the Severn (Cardiff Bay) MTR 8.55m do show the expected shift. However, superimposition of the curve for Stert and Berrow Flats in the Severn, or indeed certain other Medway or Severn main channel curves, shows that the full relationship is not quite so simple as the deliberately selected curves would make it appear.

16.6 Interpretation

It has been pointed out that muddy shorelines have been poorly surveyed and only very few determinations of shore shape are available. Arising from this any findings from this study must be treated with caution. Nevertheless the measurements of shore shape gain in reliability from the fact that the long-term sedimentary regime in the areas under scrutiny is so well established. The Humber, especially in the Spurn Bight region, is known to be experiencing infill or deposition over the longer term, Wilkinson et al (1973), as is the Wash, Evans (1965). Both these sites apparently have an adequate sediment input and both are gaining sediment from seaward as well as landward. In contrast, the Medway, at least in respect of its outer, main channel coastline, has been shown from a variety of evidence, including stake measurements, archaeological and map evidence, as well as biological evidence from *Mya* and other bivalves, to be erosional. This swing to an erosional regime has occurred sometime in the last 300 years, prior to which the Medway tidal flats were accretionary, Kirby (1990).

The Severn is a much larger system that the Medway or Humber, comparable with or bigger than the Wash and is not evolving identically throughout its entire intertidal zone.

Until 1905 Cardiff Bay was co-extensive with Rumney and Wentlooge and no doubt had comparable shaped flats. It was enclosed early in the century to create a large dock and has consequently shifted to become a zone of rapid long-term deposition with an abundant fine sediment supply. This site provides a marked contrast with the tidal flats of the main channel. These are experiencing rapid erosion over the longer term.

Although rates of shore retreat over the longer term are not possible to quantify, the tidal flats at Bridgwater Bay lie in the most exposed locality, whilst those at Avonmouth have been keyed-into a channel notched into rock for several thousand years, at a site where tidal range and hence currents are maximal. These tidal flat areas are interpreted as being the most severely degraded.

Tidal flats at Rumney, Wentlooge, Gold Cliff and Clevedon are known from geological and archaeological evidence to have experienced large scale retreat (a vertical erosion of 3.0m in 70 years at Rumney), although they are less exposed than Bridgwater Bay and Avonmouth. Kingston Seymour and Sand Bay represent sites hemmed in by headlands on one or both sides and hence protected to some degree from some components of the wave climate and the strongest tidal currents. Further protection is provided by an offshore sand bank. Sand Bay has a sandy beach at the top of the shore, but this represents 10% or less of the total shore area, the remainder being muddy. Kingston Seymour is entirely muddy.

This understanding of the long-term dynamics of the sedimentary regime may assist in interpreting the shapes of tidal flats which result. Firstly, tidal flat shapes at the Humber (Spurn Bight), Wash and Severn (Cardiff Bay) are all sites of long-term accretion and self-similar in shape (Curves 2, 3 and 4, Figure 12). All the curves have a convex profile and there is a possible tendency for a larger proportion to be shallower than MTR as tidal range increases. The Wash hypsographic curve is generally at a lower height than the Humber, although both have the same MTR. Possibly this is accounted for by the frequent reclamations in the Wash in the past. If so and the Humber is in equilibrium, the tidal flats in the Wash have still to rise to reach an equivalent height. The upper tidal flat in the Wash is, however, higher and has salt marshes.

The tidal flats on the main channel of the Medway are self-similar and most comparable to those of the main channel of the Severn. They have a pronounced sinuosity and a generally concave upward shape, often with two small convexities, one at mid and the other towards low tide level. In general the bulk of the curve lies at a low elevation compared to tidal flats in the Humber, Wash and Cardiff Bay and the inner tidal flat zone is narrow and steep. The low-level, pronounced concavity and sinuous shape may all be manifestations of the known predominantly long-term erosional regime of the main channel coasts of the Medway and Severn.

Kingston Seymour and possibly Sand Bay are known sites of long-term erosion, but have an overall convex-upward shape. If the hypothesis being proposed here is correct these are erosional but the least degraded of any erosional profiles in the main channel coast of the Severn. A possible explanation for this is their confined location between headlands and in the case of Kingston Seymour the protection afforded by the Langford Grounds Bank offshore.

By selecting particular hypsographic curves the suite of curves spanning the range microtidal through to hypertidal can be built up and seems to follow a logical progression. Curves do rise with respect to mid-tide level, become more convex upwards in shape and move to the left on the graph, implying a greater proportion of the curve is intertidal and a greater proportion at high level. The explanation for the shift presumably relates to the precise interplay of tidal current and wave forces on the tidal flat and remains to be confirmed with more data and explored in fundamental terms.

16.7 Conclusions

Despite some minor lithological variations between the various sites, three broad, though tentative, deductions can be drawn from the analyses carried out during the study, one independent and two linked one to the other.

The first is that a sequence of hypsographic curves representing a large proportion of the spectrum of tidal ranges encountered in nature has been assembled for the first time. In simplest terms these appear to form a suite of curves bearing a logical relation one to the other. If curves from elsewhere in the spectrum of tidal ranges fall within, or add to this spectrum then its validity will be enhanced. Possibly a closely related spectrum of curves does occur for tidal flat hypsographic curves at equilibrium? Should this prove to be the case such knowledge can be used to predict the revised shape of tidal flat profiles in circumstances where the tidal range changes in either the short or longer term.

Secondly, the understanding of the sedimentary regime in the sites studied and the hypsographic curves which have been measured suggest the possibility that curve shape may differ systematically depending upon whether tidal flats are accreting or eroding in the longer term. Equilibrium shapes for accretionary regimes, possibly indicated by the Humber, Wash and Cardiff Bay curves, appear consistently convex-upwards, whilst, in contrast, hypsographic curves for deeply denuded tidal flats in the main channel coasts of the Medway and Severn tend to be concave-upwards, implying a wide and low-lying outer shore and steep inner shore. Possibly convex shapes evolve through various intermediate curves into concave shapes when the sedimentary regime changes from accretionary to erosional?

Linked to this conclusion it is evident from the Medway, where over a 300 year period a long-term accretionary regime has swung to an erosional regime and from Cardiff Bay in the Severn, where an erosional regime has swung to accretionary over a 90 year time-span, that tidal flat shape changes through historic time. In the Medway, a convex-upward tidal flat shape has changed to a concave-upward shape, whilst in Cardiff Bay the reverse has occurred. In neither case is the old survey data adequate to producing accurate hypsographic curves to illustrate this feature.

The latter two conclusions have implications for sea level rise and a suggested narrowing of the coastline and enhanced tendency for erosion of muddy coasts which will materialise, (Goss-Custard et al. 1990), due to the Greenhouse Effect. They also have implications for tidal power schemes, for the relative proportions of intertidal to subtidal area, for coastal stability, fish and bird feeding times and the duration of water inundation required for invertebrates to reach maturity.

Acknowledgements

The author wishes to thank the Severn Tidal Power Group for permission to publish the results of contract work. Survey data was supplied by British Ports Association (Humber and Cardiff Bay), Binnie and Partners (Wash), Medway Ports Authority (Medway). Surveys for Severn Estuary main channel coasts were undertaken by contract aerial surveys. The survey of Bridgwater Bay was available in scientific literature. The early stages of the study benefitted from guidance provided by Dr-Ing Manuela Osterthun of the Franzius Institute in Hanover, Germany.

16.8 References

Bradley W.H., 1957 Physical and ecologic features of the Sagadahoc Bay tidal flat, Georgetown, Maine. Geol Soc America, Memoir 67, p.641-682.

Davies J.L., 1964 A morphogenic approach to world shorelines. Z. Geomorphol, 8, p.27-42.

Dieckmann R. and Partensky H-W., 1985 Quantitative geomorphology of tidal flat areas. Symposium on Coastal Morphology, Sedimentary Budgets, Coastal and River Hydraulics. Reykjavik, Iceland, p.52-62.

Dieckmann R., Osterthun M. and Partensky H-W., 1987 Influence of water-level elevation and tidal range on the sedimentation in a German tidal flat area. Progress in Oceanography, Vol 18, Nos 1-4, p.151-166.

Evans G., 1965 Intertidal flat sediments and their environments of deposition in the Wash. QJGS Lond, Vol 121, p.209-245.

Goss-Custard J.D., McGrorty S and Kirby R., 1990 Inshore birds of the soft coasts and sea-level rise. In: Expected effects of climatic change on marine coastal ecosystems, Editors: Beukema J.J., Wolff, W.J. and Brouns, J.J.W.M. Kluwer Academic Pub., p.189-193.

Hayes M.O., 1975 Morphology of sand accumulations in estuaries. In: Cronin L.E. (Ed) Estuarine Research, Vol 2, Geology and Engineering. Academic Press, NY, p. 3-22.

Kirby R. and Parker W.R., 1980 Settled mud deposits in Bridgwater Bay, Bristol Channel. IOS Report No. 107, 65pp (unpublished).

Kirby R., (in press) The evolution of the fine sediment regime of the Severn Estuary and Bristol Channel. Biological Journal of the Linnean Society. Proc. Symp. on evolution and change in the Bristol Channel and Severn Estuary.

Kirby R., 1990 The sediment budget of the erosional intertidal zone of the Medway Estuary, Kent. Proc. Geol. Assoc., Vol. 101, Pt 1, p.63-77.

Kirby, R., 1988 The stability of intertidal mud flats and their relationship to long term sources and sinks for fine sediment: with particular reference to the Severn Estuary and Bristol Channel. Report to STPG ref tasks 3.1(iii)b and 3.7(i)a, 103pp, (unpublished).

Kirby R., Bleakley R.J., Weatherup S.T.C., Raven P.J. and Donaldson N.D., (in press) Effect of episodic events on tidal mud flat stability, Ardmillan Bay, Strangford Lough, Northern Ireland. Workshop on nearshore and estuarine cohesive sediment transport. Springer Verlag.

Little D.I., 1989 The appraisal of contaminants in sediments of the Severn Estuary and Inner Bristol Channel, Report to STPG, 47pp, (unpublished).

Severn Estuary Levels Research Committee, 1988 The archaeological potential of the Severn Estuary. An initial assessment for STPG, 64pp, (unpublished).

Straaten L.M.J.U., van 1950 Environment of formation and facies of the Wadden Sea sediments. Tijdschr. Kon. Nederl. Aardr. Gen., LXVII, No. 3, p.354-368.

Wilkinson H.R., de Boer G. and Thunder A., 1973. A cartographic analysis of the changing bed of the Humber. A report to the Humber Estuary Research Committee, Dept. of Geography, Hull University, Misc. Series No. 14, 68pp, (unpublished).

17

Acoustic measurements of suspended sediment over sandwaves

P D Thorne, R L Soulsby and P J Hardcastle

Abstract

Bedforms are common features in our estuarine and coastal waters and the dynamic interaction between the bedforms and the flow is of considerable interest. In strong turbulent tidal currents the suspended load is the predominate form of transport and it is essential that the sedimentary processes of this component be understood. The work reported here is part of a study aimed at clarifying the role of suspended transport upon sandwave development and migration, and the feedback of these modifications on the suspended load. To discern the mechanisms of transport, measurements of the turbulent flow were taken at four heights above a sandwave and records of the response of the suspended sediment concentration were obtained using an Acoustic Backscatter Probe, ABP. In conjunction with the ABP the mean suspended sediment concentration was measured using conventional pump sampling. The ABP is increasingly becoming a prime technique for measuring suspended concentration fluctuations and consideration is given here to the assessment of the acoustic method for making such measurements.

17.1 Introduction

Over the past decade interest in sediment movement has increasingly focused on the detailed effects of turbulent currents and wave action on the transport of sedimentary material. Of particular interest is the sediment transport over bedforms in estuaries and coastal waters. Frequently suspended sediment and sandwaves occur together in these environments, however, there is a lack of knowledge of the mutual influences between the migrating bedforms, the suspended sediment and the flow. To develop our understanding of suspended sediment transport processes under these conditions requires detailed measurements of the turbulent current structure and the fluctuating suspended sediment concentration profiles.

For the flow measurements normally electromagnetic current meters, ECM, and acoustic travel time devices have been employed. For suspended load measurements miniature optical backscatter sensors, MOB, (Downing et al (1981)) and acoustic backscatter probes, ABP, (Hanes et al (1988)) are under development. The ABP is potentially the more powerful technique since a single instrument can be used to non-intrusively measure suspended sediment concentration profiles with a high spatial (~1cm) and temporal (~0.1s) resolution. However, when claiming this potential acknowledgement needs to be made of the statistical nature of the backscattered signal. The amplitude of the returned echo is incoherent and has a Rayliegh amplitude probability distribution. Therefore a degree of incoherent averaging has to be conducted on the pressure data before converting the signal to concentration levels. Also the interpretation of the backscattered echoes requires an understanding of the interaction of sound with suspensions and an accurate description of this process is still under development. To date the usual approach has been to conduct a series of laboratory calibration exercises on suspensions of various concentration and particle sizes and develop an algorithm for translating the acoustic observations to sediment parameters (Hanes et al). Such detailed exercises are unnecessary if the backscattering from marine suspensions can be described theoretically.

To examine the interaction of sound with marine suspensions a series of measurements have been taken at 3 MHz in the laboratory and in an estuarine environment. The laboratory observations are used to appraise a theoretical description of the acoustic scattering process and provide an estimate for the accuracy of the acoustic approach. The results of the laboratory work are applied to the field data to obtain acoustic estimates of the suspended load. These values are assessed using the pump sample measurements obtained in the field. The high spatial and temporal resolution facility of the ABP is then employed to look in detail at the suspended concentration field and to examine turbulent sediment fluxes over bedforms.

17.2 Acoustic analysis

17.2.1 Theory

The geometry for the scattering problem is shown in Fig 1. An acoustic transceiver transmits a short (of the order of 10μs duration), high frequency (typically a few MHz) acoustic pulse which propagates down through the water column. The range of these instruments is normally of the order of a couple of metres. If any material is in suspension then as the sound passes through the water a proportion of the sound will be backscattered and the returned echo is used to estimate the suspended sediment concentration profile.

The problem is approached by initially specifying the interaction of sound with a single particle. Employing the formulation generally applied (Neubauer(1974)) the backscattered pressure, P_s, from a particle in the beam of the transceiver is given by

$$P_s = a_s P_0 r_0 \frac{D^2}{2r^2} |f_m| e^{-2\alpha_w r} e^{i(2kr-\omega t)} \tag{17.1}$$

P_0 is the transmitted pressure at range r_0 (usually = 1m), a_s is the particle radius, r is the range to the particle, D is the transceiver directivity function, α_w is the attenuation due to the water and f_m is known as the form function and describes the scattering properties of the particle.

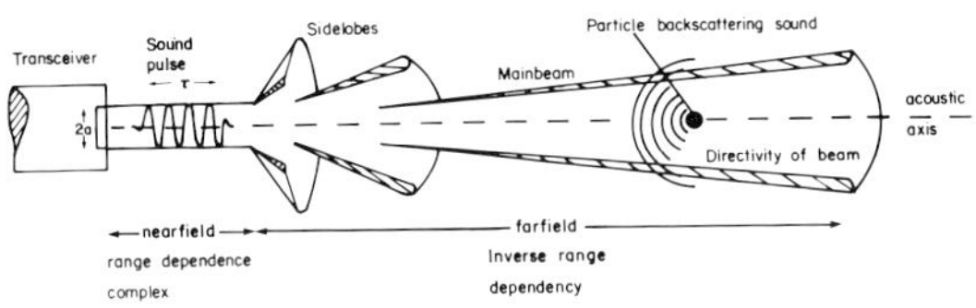

Figure 1: Schematic of the backscattering from a suspended particle.

For a cloud of scatterers equation (17.1) needs to be integrated over the volume insonified by the acoustic pulse as it propagates down through the water. When multiple scattering can be neglected the integral can be readily evaluated and the ensemble average root-mean-square backscattered pressure, $<P_{rms}>$, can be written (Sheng and Hay (1988) and Thorne et al (1990a)) as

$$<P_{rms}> = \frac{P_o r_o}{\psi r}|f_m|\left(\frac{3\tau c M}{16 a_s \rho_s}\right)^{1/2} e^{-2r(\alpha_w+\alpha_s)}\left[\int_o^{\pi}\left\{\frac{2J_1(ka_t Sin\theta)}{ka_t Sin\theta}\right\}^4 Sin\,\theta\, d\theta\right]^{1/2} \qquad (17.2)$$

M is the mass concentration, r is the range from the transceiver, α_s is the attenuation due to the suspension, ρ_s is the density of the sediment in suspension, c is the velocity of sound in the water, τ is the pulse duration and ψ is a term to modify the range dependence within the transceiver nearfield (Thorne et al (1990b)). k is the incident acoustic wave number, a_t is the radius of the transceiver, θ the angle subtended to the acoustic axis and J_1 is a first order Bessel function (this describes the beam pattern). Equation (17.2) specifies the forward problem; knowing the sediment parameters we can evaluate the backscattered signal. We need to know α_s, f_m, a_s, ρ_s and M(r). For laboratory work the latter three parameters can be readily measured, however, α_s and f_m are not generally known for irregularly shaped particles. To obtain estimates for these parameters it has recently been conjectured (Sheng and Hay) that relatively simple expressions for α_s and f_m may describe the backscattering behaviour of sedimentary material and the suggested appropriate expressions are

$$f_m = K\,(ka_s)^2/(1 + K(ka_s)^2)$$
$$K = \frac{2}{3}\,|\,\gamma_k - \gamma_p\,| \tag{17.3}$$
$$\gamma_k = (\kappa_s - \kappa_o)/\kappa_o \qquad \gamma_p = 3(\rho_s - \rho_o)/(2\rho_s + \rho_o)$$

Where κ and ρ are the compressibility and density respectively and subscripts s and o refer to the suspension and water. To evaluate α_s an integration over the concentration profile is required giving

$$\alpha_s = \frac{1}{r}\int_o^r \zeta(r)M(r)\,dr \tag{17.4}$$

The expression for ζ, the attenuation constant, has been investigated, by Sheng and Hay, using the data set of Flammers (1962) and resulted in

$$\zeta = (1/\rho a_s)\left\{\beta(ka_s)^4/[1+(ka_s)^2 + (\frac{4}{3})\beta(ka_s)^4]\right\}$$
$$\beta = (\gamma_k^2 + \gamma_p^2/3)/6 \tag{17.5}$$

The variation of f_m and ζ with ka_s are shown in Fig 2. For f_m there is a sharp decrease in the form function below $ka_s<0.5$, while above it there is a reduction in gradient leading to a uniform value of unity for $ka_s>5$. A similar response is seen for the attenuation constant. The prediction of the backscattered pressure, the forward problem, can therefore be expressed analytically and comparisons made with experimental data to assess its validity.

For marine measurements the inverse solution is required, using $<P_{rms}>$ the suspended concentration needs to be evaluated, a knowledge of α_s, f_m, a_s, ρ_s is again necessary. For a relatively well sorted bed a first approximation for a_s and ρ_s is to use values for the surficial layer of the bed (these are nominally known usually) and assume an invariant value for a_s in the water column above the bed. The parameters f_m and α_s are then estimated from equations (17.3)-(17.5). The concentration profile is then obtained by simply rearranging equation (17.2).

$$M(r) = \frac{<P^2>}{P_o^2 r_o^2}\,\frac{r^2\psi^2}{f_m^2}\,e^{4r(\alpha_w+\alpha_s)}\left\{\frac{16a_s\rho_s}{3\tau c}\right\}\Bigg/\left[\int_o^\pi\left(\frac{2J_1(ka_t\sin\theta)}{ka_t\sin\theta}\right)^4\sin\theta\,d\theta\right] \tag{17.6}$$

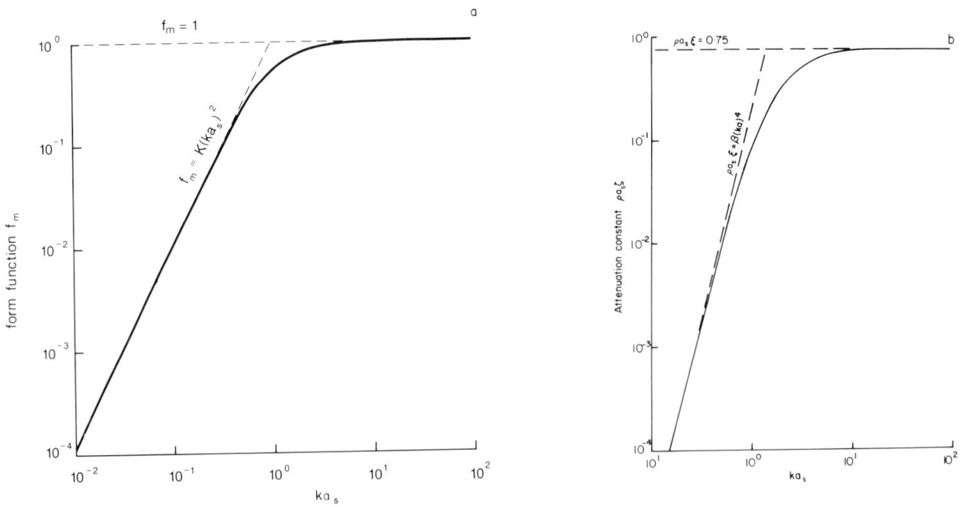

Figure 2: a) Variation of the form function with ka_s. b) The attenuation constant with ka_s.

It can be seen from equation (17.6) that to evaluate M(r) we need to know M(r) in equation (17.4). This therefore requires a computational procedure where sequentially M(r) is evaluated and a numerical integration conducted to evaluate α_s and this calculation repeated step-wise throughout the water column. This process has its problems; for example if M(r) is initially over estimated then the value of α_s is increased, this higher value is fed back into equation (17.6), M(r) is further over estimated at the following range and the effect can accumulate through the profile. There is a positive feedback process causing a divergent solution. This is particularly acute at high concentrations when there is a degree of uncertainty in α_s, which will invariably be the case since the definitive value for α_s for any particular sediment and conditions is not at present accurately known. However, the corollary to this is that for low values of α_s the problem is well defined. This is further considered when analyzing the experimental data.

17.2.2 Laboratory measurements

To examine the applicability of equation (17.2) a series of laboratory measurements were conducted in the vertical sediment tower shown in Fig 3. This consisted of a hopper and an impeller mounted on a perspex tower with a collection tray at the base. A homogeneous suspension of approximately 1.5m could be obtained with this arrangement. Absolute estimates of suspended concentration were taken using pumped sampling.

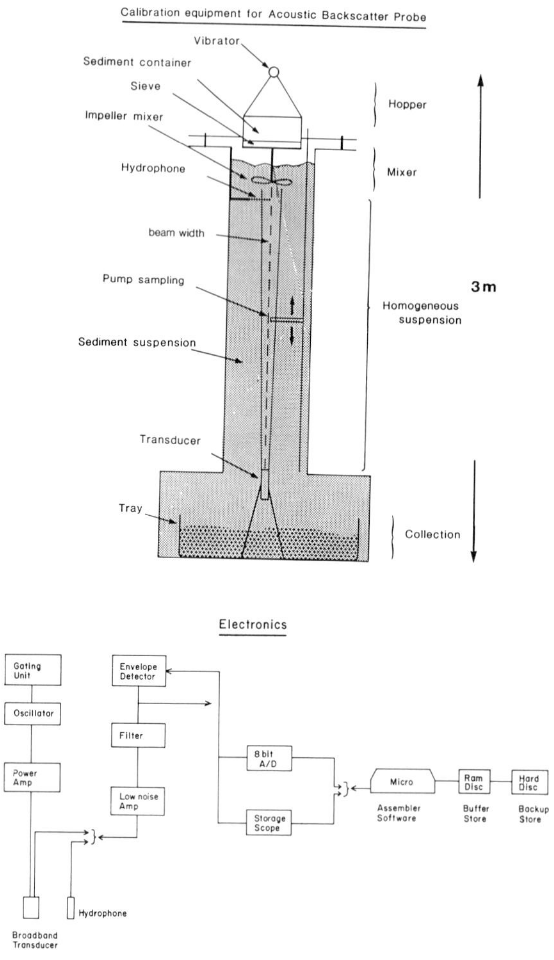

Figure 3: Experimental arrangement for laboratory measurements of backscattering.

The 3 MHz acoustic transceiver was mounted near the base of the tower and the envelope of the transceiver signal, backscattered from the sediment in suspension, was digitized and fed directly into a microcomputer. The system sampled the backscattered signal at 256 contiguous range bins to yield an acoustic profile of the returned echo. Typically 200 profiles were collected to mitigate the random fluctuations of the signal mentioned in the Introduction and the ensemble average rms backscattered pressure profile computed. These mean profiles were used for comparison with the theoretical developments. The material used for the suspension was material recovered from the estuarine site. This was clean sand with a suspended particle diameter of approximately d_{50}=160μm.

To establish experimentally the values for f_m and ζ a number of measurements at a selection of concentrations were carried out. Measurements of the attenuation for a range of concentrations were obtained by recording the signal level measured using a hydrophone placed at the top of the tower. The measurements gave ζ=0.8+0.15, this compares with a value obtained from equation (17.5) of ζ = 0.35. The form function was estimated from readings taken at low concentration when the effect of α_s was not significant. This was carried out by rearranging equation (17.2) and making f_m the dependent variable on the backscattered pressure.

$$|f_m| = \frac{<P_{rms}>\psi r}{P_o r_o}\left[\frac{16 a_s \rho_s}{3\tau c M}\right]^{1/2} e^{2r\alpha_w} \Bigg/ \left[\int_0^{\pi}\left\{\frac{2J_1(ka_t Sin\theta)}{ka_t Sin\theta}\right\}^4 Sin\theta d\theta\right]^{1/2} \qquad (17.7)$$

This resulted in a value of f_m=0.7+0.1, equation (17.3) gives f_m=0.5. The form function values are comparable while for attenuation the difference is significant. At present we are not in a position to resolve the latter discrepancy definitively although if a_s is being underestimated by 20% the readjusted calculated values would give f_m=0.58 and ζ=0.53 which would be roughly within the expected uncertainties. However, repeated measurements of a_s show a variability of the order of 10%. Also the analytical expression for ζ is a mean fit to the data of Flammer's which does include a scatter about a mean of 10%. However, these variations are not sufficient to account for the discrepancy in the attenuation constant and no convincing explanation is available at present to resolve the inconsistency. For the current laboratory analysis the measured value for ζ has been implemented.

The results of the variation in the backscattered signal with the range for a homogeneous suspension are shown in Fig 4a. There is seen to be a steady decrease in signal level with range which is due to a combination of spherical spreading from the source, ie the 1/r term in the denominator of equation (17.2) and the reduction due to attenuation in the argument of the exponent. It can be seen that for the higher concentration the signal level at about 1m is an order of magnitude below the lower concentration values due to the dominance of the sediment attenuation on the range dependence.

Computations for the rms backscattered signal using equation (17.2) are seen to be in excellent agreement with the observations over the order of magnitude change in concentration, a range variation of nearly two decades and a signal amplitude covering greater than two orders of magnitude.

Comparison of the variation with range of the ratio of the acoustic estimate of the concentration, calculated using equation (17.6), to the pump sample value are shown in Fig 4b. The results were computed by evaluating α_s using the measured concentration. Ideally the value should be unity and constant with range. The results are close to this with some scatter about unity, there is greater deviation within approximately 0.2m of the transceiver and this is considered to be due to the uncertainty in modelling backscattering in the transceiver nearfield. The general agreement is good though. A collection of all the observations are shown in Fig 5. This shows the results for M_A, averaged over the range 0.42-0.84m from the transceiver, against M_p for the concentration range covered.

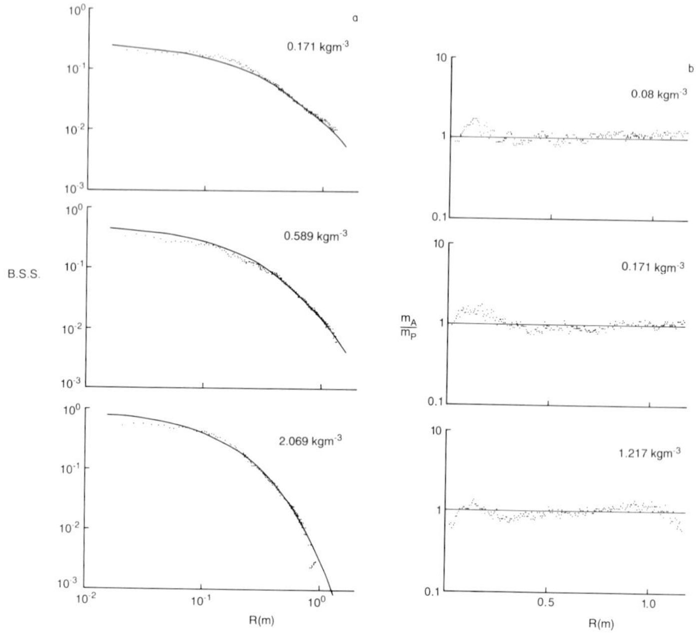

Figure 4: a) Range dependence of the backscattered signal. -Equation (17.2). b) Ratio of the acoustic estimate of concentration to the pumped sample measurement.

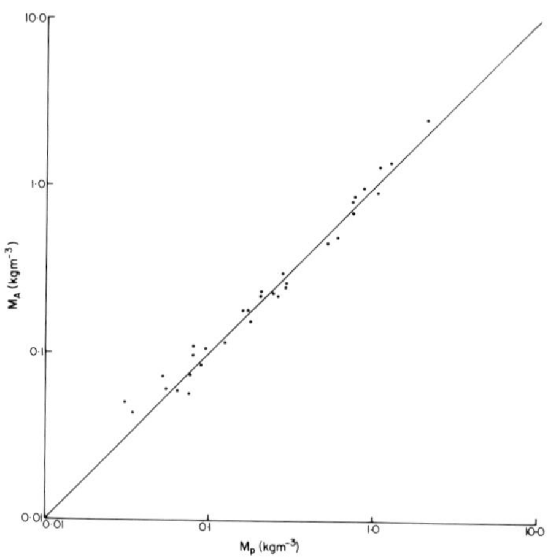

Figure 5: Comparison of acoustic measurement of suspended load M_A, with the pumped sample values M_p.

The variation about $M_A=M_p$ is of the order of 20%, some of which is assigned to uncertainties in M_p of 10%, and the rest to the acoustic measurements. The results, however, do show that the theoretical and measured concentrations are in excellent agreement and confirm the applicability of the acoustic technique for measuring suspended sediment concentration.

17.2.3 Marine observations

The estuarine measurements were carried out during a spring tidal period over floods alone, on a drying sandflat at Yelland Marsh in the river Taw estuary in North Devon. The estuary is sheltered and no wave activity was present for the duration of the experiment. During the period high concentrations of suspended sediment were observed. The bed consists of fine sand and the area is dominated by sandwaves which generally have a wavelength of 15-20m with a trough-to-crest height of approximately 0.8m. There is little suspended clay or silt in the water during the flood tide and temperature and salinity are both uniform with depth, due to strong mixing, so there is no thermohaline density gradient present. The site for the experiment is shown in Fig 6 with inserts displaying the experimental arrangement and the particle size distribution. At low water the site was dry and readily accessible for arranging the equipment.

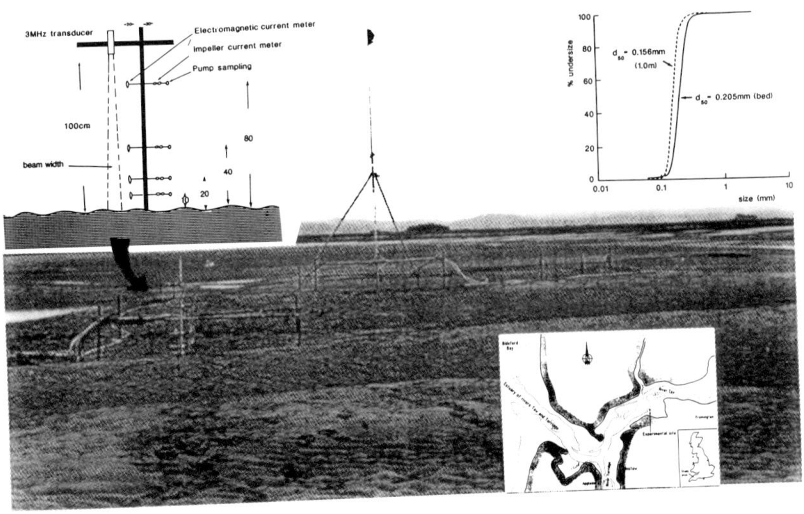

Figure 6: Field site, location, sediment distribution and experimental arrangement for estuarine measurements.

At high water the depth was approximately 5m. Measurements were taken above the bed at 0.1, 0.2, 0.4 and 0.8m using; 5cm diameter Electromagnetic Current Meters (ECM), Impeller Current Meters (ICM), the 3MHz ABP and pumped sampling. The ABP was also used to obtain continuous profiles down to the bed. Pumped samples were obtained at 5min intervals, therefore each height was sampled every 20min, the ICM's were recorded over a 4min count period each 5min. Analogue records of the ABP and ECM's were taken on an instrumentation tape recorder and were later digitised at 5Hz for analysis. The continuous ABP profiles were collected on a separate logging system consisting of 256 contiguous bins(ranges) with profiles being obtained at a rate of 3.6Hz. The profiles were digitised at 7us providing a spatial sampling interval of 0.525cm. The spatial resolution in this case was determined by the pulse length, this was 12us giving a resolution of 0.9cm.

Over the flood periods concentration profiles were obtained from the acoustic data using equation (17.6). To establish values for f_m and ζ for analyzing the field data a small number of pump sample measurements were employed. Using two of the pump sample observations at 0.8m height the value for f_m was calculated using equation (17.7). This is a first approximation with the assumption being made that for this short range, 0.2m, ignoring α_s would not introduce significant errors in the estimation of f_m. Employing this value of form function a single high concentration pump sample value at 10cm above the bed was noted and ζ adjusted so that the acoustic estimate of concentration, computed using equation (17.6), at the same range gave a corresponding value. This gave values for the form function and attenuation constant of $f_m = 0.83+0.15$ and $\zeta = 0.45+0.1$ respectively. The form function is similar in value to the laboratory observations although the attenuation is approximately half, which is closer to the theoretical estimate. An avenue considered for the discrepancy between the laboratory and field estimates for the attenuation constant was the assumption of uniform particle size distribution with height above the bed when estimating ζ. However, analysis of earlier observations showed no significant difference in particle size distribution at 0.25m and 1m above the bed. We have still yet to satisfactorily resolve this difference in ζ . For analyzing the field data the insitu observations of f_m and ζ were employed for all the acoustic estimates of concentration.

Estimates of concentration profiles, computed using equation (17.6), obtained from one minute average rms acoustic pressure data are shown in Fig 7a. The measurements near peak flow, around 1700 hours, show significant concentration levels of greater than a Kgm^{-3}. The suspension is well mixed through the water column with values close to 1 Kgm^{-3} being estimated at 1m above the bed.

Towards the end of the flood, 1900 hours, when concentrations are seen to be about two orders of magnitude down on the peak there is an almost uniform concentration profile. At the higher concentrations there is the possibility that uncertainties in the value of ζ can lead to significant errors in concentration estimates at the longer ranges, ie near the bed. As mentioned previously the step-wise sequential computation of M(r) is prone to accumulating errors, as M(r) is used to calculate α_s and is then used to continue evaluating M(r). An example of the variation which can occur is shown in Fig 7b. There is a steady divergence in predicted concentration as the bed is approached and therefore the confidence in concentration estimates is relatively low near the bed for high concentrations when the value for ζ is not precisely known. However, having taken cognisance of this, the results shown in Fig 8, where a comparison of the concentration estimates obtained using the ABP with the estuarine pump sample data collected over two consecutive floods show moderately good agreement.

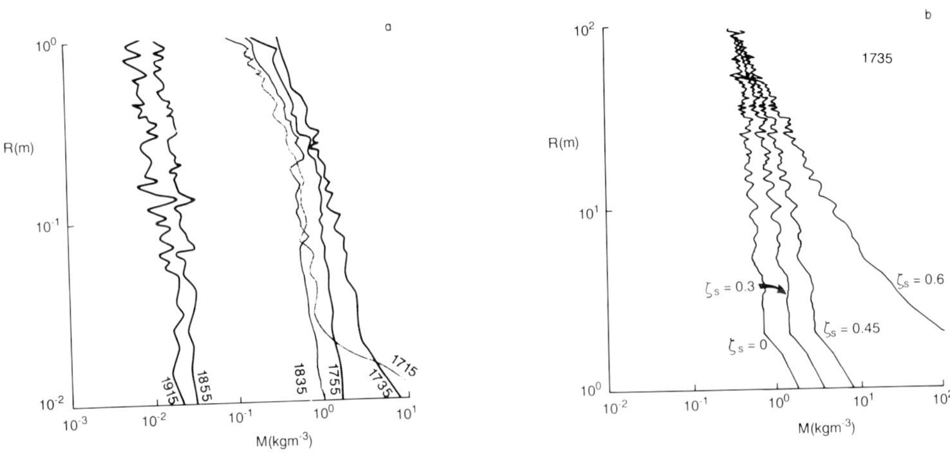

Figure 7: a) Acoustic measurements of suspension profiles at different times over a flood period. b) Effect of varying the attenuation constant.

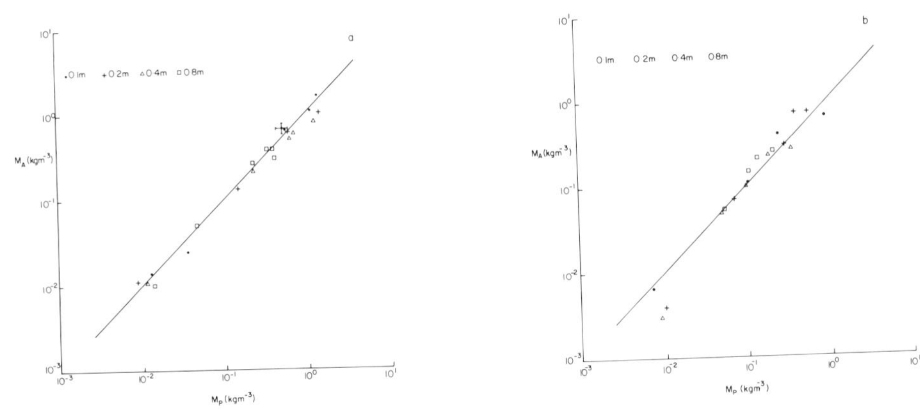

Figure 8: Comparison of the field measurements of the suspended concentration using acoustic, M_A, and pump sample, M_p, data on two separate days.

Figure 9: High temporal resolution acoustic measurements of suspended load. Values below 0.2kg m^{-3} have been set to zero for clarity.

The results are sufficiently similar to provide encouragement for progressing and examining the data to present its high temporal resolution characteristics. An example of the preliminary results presently being calculated are those shown in Fig 9. This covers one minute of data, equivalent to one profile in Fig 7 and shows the temporal and spatial variability of the concentration field. Surges in concentration are followed seconds later by relatively quiescent periods of similar duration, with the process being extremely variable.

This changeability in concentration is being studied to examine the processes of sediment transport subject to turbulent tidal current. However, as previously mentioned care must be adopted in the interpretation of such diagrams because of the statistical nature of the backscattered return. In Fig 9 an element of averaging has been carried out although the eye has been left to conduct a degree of the smoothing. Further work is ongoing on such representations.

17.3 Sediment observations

The analysis of the acoustic data to obtain concentration profiles which have synchronism with the turbulent data is still in progress. This is because for logistic reasons two independent systems were employed to record the acoustic and turbulence data. However, discreet measurements, sampled from the acoustic profiles, at the four ECM heights, was recorded simultaneously with the ECM data. Employing the temporal resolution provided by the acoustic technique the opportunity was taken to consider the longitudinal $-\overline{uc}$ and vertical $\overline{wc}$ diffusive fluxes at five locations over a sandwave (Atkins et al (1989)). Results from this analysis are presented in Fig 10.

The measurements were obtained over five flood periods, no data were collected over the ebb. To facilitate comparison of the observations at different stages of the flood the measurements were categorised into four velocity bands derived from the velocity at 0.4m above the bed on a reference mast approximately 50m upstream of the experimental site. The four bands were 0.1-0.3, 0.3-0.5, 0.5-0.7 and 0.7-0.9 ms^{-1} with band 1 being the lowest velocity range. Mean values of the flux were obtained by taking the product of the mean current and mean concentration over the period while the velocity was in the appropriate band. The diffusive flux was obtained from the product of the 5Hz velocity and concentration data over the same period of time.

Shown in Fig 10a is the longitudinal diffusive flux. This is positive along the flank of the sandwave indicating that the diffusion is occurring in the upstream direction. In the trough the flux is negative at low velocities, while as the velocity increases the depth of the downstream flux reduces. Values for the mean sediment flux, $\overline{U}\overline{C}$, have also been computed and this has allowed an estimate of $-\overline{uc}/\overline{U}\overline{C}$ to be calculated which, apart from the trough, gave a value generally less than 0.05. The longitudinal diffusive flux is therefore generally directed upstream and less than 5% of the mean sediment movement. Fig 10b shows the vertical diffusive flux which is the mechanism by which material is lifted into suspension. The negative values of vertical diffusive flux measured in and close to the trough implies that the diffusion is aiding the settling process. At all other observational positions the vertical diffusive flux is smaller than the calculated settling flux, $w_s\overline{C}$. Over most of the flank the ratio of $\overline{wc}/w_s\overline{C}$ is typically 0.2 at a height of 0.1m, increasing with height to about 0.7 at 0.8m. This is an indication of a net downward settling of sediment over the entire sandwave which is difficult to explain at present, however, previous observations have shown similar results. Analysis of the data is still in progress at this stage, although the potential of using the temporal resolution of the ABP to investigate turbulent fluxes has been demonstrated.

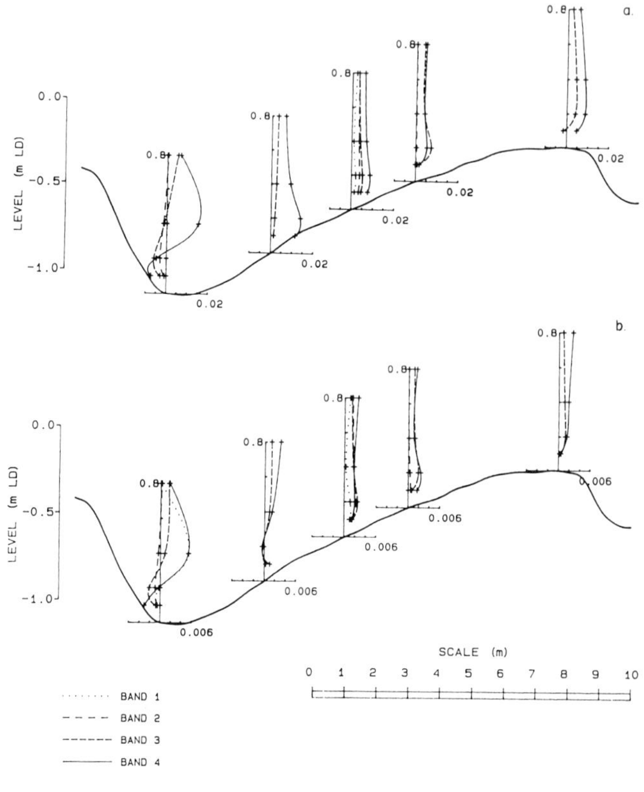

Figure 10: a) Longitudinal diffusive flux. b) Vertical diffusive flux. Scale kg $m^{-2}s^{-1}$.

17.4 Conclusions

An examination of the application of acoustic scattering for measuring suspended sediment concentration in an estuarine environment has been conducted. Laboratory, marine, and theoretical estimates for the acoustic parameters f_m and ζ are similar, although differences in ζ have been noted and discussed. Time averaged acoustic concentration predictions are in good agreement with pump sample data taken in both the estuary and the laboratory. The high temporal resolution profiles show the very variable nature of the transport process when turbulent currents are present. The sediment analysis is beginning to show the behaviour of sediment diffusive processes over sandwaves and work is underway to provide a better understanding of the connectivity between the bed, the suspension and the flow.

17.5 References

Atkins R., Soulsby R.L., Waters C.B., and Oliver N. (1989) Field measurements of sediment suspension above bedforms in a sandy estuary. Hydraulics Research Ltd., Wallingford, Oxford. Report SR 203.

Downing J.P., Sternberg R.W., and Lister C.B.R. (1981) New instrumentation for the investigation of sediment suspension in the shallow marine environment. Mar. Geol 42: 19-43.

Flammers G.H. (1962) Ultrasonic measurement of suspended sediments. Geo. Survey Bull. No. 1141-A (US GPO, Washington, DC)

Hanes D.M., Vincent C.E., Huntley D.A., and Clarke T.L. (1988) Acoustic measurements of suspended sand in the C^2S^2 experiment at Stanhope Lane, Prince Edward Island, Mar.Geol 81: 175-183.

Neubauer W.G., Vogt R.H. and Dragonett L.R. (1974) Acoustic reflection from elastic spheres 1 Steady-state signals. J. Acou. Soc. Amer. 55 6 1123-1129

Sheng J., and Hay A.E. (1988) An examination of the spherical scatter approximation in aqueous suspensions of sand. J. Acou. Soc. Am. 83 2 598-610

Thorne P.D., Hayhurst L., and Campbell S.C. (1990a) Measurements of scattering from a suspension of spherical scatters. Proceeding of the Institute of Acoustics. Acoustics'90. Held at Southampton University 28-30 March 1990.

Thorne P.D., Vincent C.E., Hardcastle P.J., Rehman S. and Pearson N. (1990b) Measuring suspended sediment concentration using acoustic backscatter devices. Mar.Geol. (In Press).

18

Some observations on fluid mud response to water waves

F Jiang and A J Mehta

Abstract

The behaviour of soft mud under progressive waves has been briefly examined in the field. The main objective was to demonstrate that under wave conditions that are too mild to cause significant particulate erosion, deep layers of fluid mud can undergo measurable oscillations due to wave loading. The field site was in the shallow littoral margin of Lake Okeechobee, Florida. In a test during which wind wave frequency was on the order of 0.4Hz and significant wave height around 10cm, wave-coherent mud motion was measured 20cm below the mud-water interface.

A shallow water wave model, which considers the water column to be inviscid and the mud layer to be a high viscosity fluid, has been used to aid in data interpretation. Given the wave energy spectrum, the model approximately simulates the water velocity and mud acceleration spectra; deviations between simulation and measurement are pronounced in the high frequency range wherein the shallow water assumption breaks down. A low frequency wave signal at about 0.04Hz, believed to be forced by the wind wave, was characteristic of measured spectra. This signal in the mud was enhanced relative to the forcing signal due to the dependence of wave attenuation on frequency, and led to horizontal mud displacements on the order of 2mm. It is possible that this signal is akin to surf beat which causes the mud to oscillate slowly, thereby contributing to its mobility. Such a motion may have implications for the rates of nutrient exchange between the bottom and the water column.

18.1 Introduction

It is generally well recognized that in shallow, episodic coastal or lacustrine environments with muddy beds, reworking of mud by waves causes the bottom to become fluidized, with the fluid state persisting as long as waves continue and thereafter, until the bottom material dewaters sufficiently to lead to hardening under calm conditions. Laboratory evidence points to fluidization under cyclic loading to be due to the breakup of the structural matrix of the bed held together by cohesive, inter-particle bonds (Ross and Mehta, 1990). This change of state from a bed to a fluid-supported slurry, and wave-induced oscillations within this slurry, have obvious implications for bottom boundary layer related phenomena including, for example, the availability of fluidized mud for transport by current, and possible changes in the effective bottom permeability or resistance to diffusion, leading to corresponding changes in the rate of exchange of nutrients or contaminants between the bottom and the water column. In this study the specific matter of fluid mud motion induced by progressive waves was examined from the following perspective.

18.2 A physical perspective

A simple physical perspective can be chosen to deal with a rather complex problem which is characterized by time-dependent changes in mud properties with continued wave action. Although such changes have been tracked to some extent in laboratory experiments, field evidence is scarce due to evident problems in deploying requisite transducers. Furthermore, the basis for any theoretical examination of the time-variability of such properties as mud shear strength is presently rudimentary. Therefore it must be borne in mind that in treating the problem these considerations impose certain operational limitations in data gathering and analytic constraints in analysis.

In the way of a general description of the problem consider Fig. 1, in which vertical profiles of sediment density (ρ) and the horizontal component of the wave-induced velocity amplitude (u_m) in the water column and bottom mud are depicted in a somewhat idealized manner. In the upper water column, in which pressure and inertia forces are typically dominant in governing water motion and the flow field is practically irrotational, sediment concentration usually tends to be quite low, so that the suspension density is close to that of water. The lower boundary of this upper column is characterized by a rather significant gradient in density, or lutocline, below which the density of the fluid mud is measurably higher.

Below the fluidized mud is the cohesive bed having yet higher densities (greater than ~1.2g/cm^3). Laboratory observations by Maa and Mehta (1987), and theoretical work by Foda (1989) for instance show that the wave orbits can penetrate the bed, thereby leading to deformations of the upper part of the otherwise stationary bed. Under continued wave loading such deformations can cause fluidization by elimination of the effective stress associated with the bed matrix. Due to the generally low rates of upward mass diffusion above the wave boundary layer however, the problem of mud motion by waves can be conveniently considered to be uncomplicated by the effects of particulate erosion due to stress-induced breakup and entrainment of the cohesive sediment aggregates at the mud-water interface (van Rijn, 1985). In fact, laboratory as well as field observations show that wave conditions required to generate measurable bottom motion can be quite moderate compared with conditions required to cause significant particulate erosion (Ross and Mehta, 1990; Mehta and Jiang, 1990). Accordingly, the following system is considered.

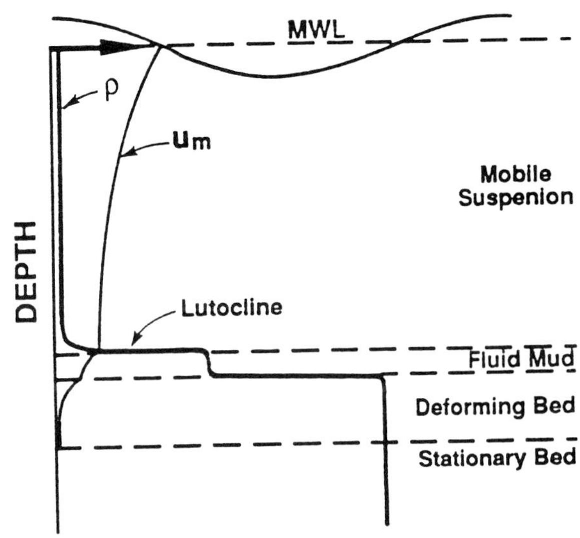

Figure 1: Schematic of mud bottom response to waves in terms of vertical profiles of sediment density and velocity.

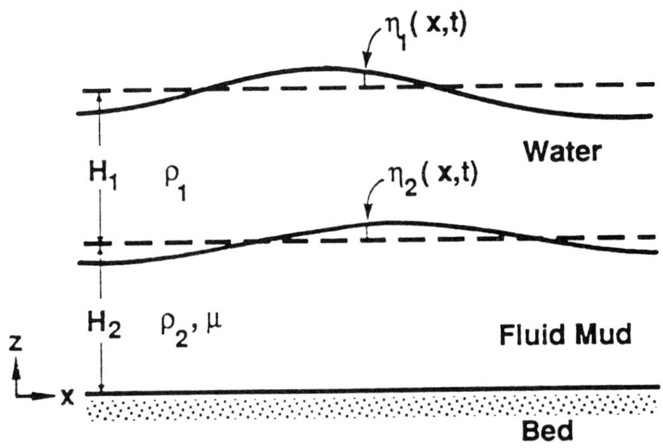

Figure 2: Two-layered water-fluid mud system subject to progressive wave action.

A two-layered, water-fluid mud system forced by a progressive, non-breaking surface wave of periodicity specified by frequency, σ, is depicted in Fig. 2. As far as wave dynamics is concerned we will restrict the problem to one of long waves, which would therefore be applicable to very shallow coastal or lacustrine water bodies, or to the margins of deeper ones where wave action often matters the most. In the case of a rigid bottom, the shallow water condition is satisfied when $H\sigma^2/g < 0.1$, where H is the water depth and g is the acceleration due to gravity. For a given σ, this relationship specifies H such that, for shallow water condition to hold, the actual depth must be equal to or less than that value of H. When the bottom is non-rigid the maximum water depth to which shallow water condition is satisfied will be somewhat larger, inasmuch as the wave length is greater than in the rigid bottom case.

The upper water layer of thickness H_1 and density ρ_1 is considered to be inviscid, which is not unreasonable in comparison with the highly viscous lower, homogeneous layer of fluidized mud having a thickness H_2, density ρ_2 and dynamic viscosity μ. However, the implications of this assumption of inviscid upper layer relative to the absence of the bottom boundary layer in the water column and the associated interfacial shear stress are noted later. Beneath the lower mud layer is the bed, which is assumed to be rigid. The surface and interface variations about their respective mean values are $\eta_1(x,t)$ and $\eta_2(x,t)$. The amplitude of the simple harmonic surface wave is assumed to be small enough to conform to linear theory, as also the response of the mud layer. Accordingly, the relevant governing equations of motion and continuity can be written as (Gade, 1958):

Upper layer:

$$\frac{\partial u_1}{\partial t} + g\frac{\partial \eta_1}{\partial x} = 0 \tag{18.1}$$

$$\frac{\partial}{\partial t}(\eta_1 - \eta_2) + H_1\frac{\partial u_1}{\partial x} = 0 \tag{18.2}$$

Lower layer:

$$\frac{\partial u_2}{\partial t} + rg\frac{\partial \eta_2}{\partial x} + (1-r)g\frac{\partial \eta_1}{\partial x} = \nu\frac{\partial^2 u_2}{\partial z^2} \tag{18.3}$$

$$\int_0^h \frac{\partial u_2}{\partial x}dz + \frac{\partial \eta_2}{\partial t} = 0 \tag{18.4}$$

where $u_1(x,t)$ and $u_2(x,z,t)$ are the wave-induced velocities, $h = H_2 + \eta_2$, $r = (\rho_2 - \rho_1)/\rho_2$ is the normalized density jump, and $\upsilon = \mu/\rho_2$ is the kinematic viscosity of mud. Considering the fact that the fluid domain is bounded between z = 0 and $H_1 + H_2$, is infinite in extent in the ±x direction, the lower layer is viscous, and the solution sought is harmonic, the following boundary conditions are imposed:

$$\eta_1\,(0,t) = a_o \cos\sigma t \tag{18.5a}$$

$$u_1\,(\infty,t),\ u_2(\infty,z,t),\ \eta_1(\infty,t) \text{ and } \eta_2(\infty,t) \to 0 \tag{18.5b}$$

$$u_2\,(x,0,t) = 0 \tag{18.5c}$$

$$\partial u_2\,(x,H_2,t)\,/\,\partial z = 0 \tag{18.5d}$$

where a_0 is the surface wave amplitude at x = 0. Equation 18.5a specifies the surface wave form, Equation 18.5b represents the fact that, due to viscous dissipation, all motion must cease at infinite distance, Equation 18.5c is the no-slip bottom boundary condition, and Equation 18.5d states that because the upper layer fluid is inviscid, there can be no stress at the interface. Note that by virtue of the assumption of rotationality in the lower layer only, and the shallow water condition, u_2 can vary with z, but not u_1.

In order to generalize the solution of the above equations and the boundary conditions, the following convenient dimensionless quantities are introduced: $u_1' = u_1/\sigma H_1$, $u_2' = u_2/\sigma H_1$, $\eta_1' = \eta_1/H_1$, $\eta_2' = \eta_2/H_1$, $H_2' = H_2/H_1$, $h' = H_2' + \eta_2'$, $t' = \sigma t$, $k' = kH_1$ (where k is the wave number), $x' = x/H_1$, and $z' = z/H_1$. Thus Equations 18.1 through 18.4 become:

Upper layer:

$$\frac{\partial u_1'}{\partial t'} + \frac{1}{F_r^2}\frac{\partial \eta_1'}{\partial x'} = 0 \tag{18.6}$$

$$\frac{\partial}{\partial t'}(\eta_1' - \eta_2') + \frac{\partial u_1'}{\partial x'} = 0 \tag{18.7}$$

Lower layer:

$$\frac{\partial u_2'}{\partial t'} \frac{r}{F_r^2} \frac{\partial \eta_2'}{\partial x'} + \frac{1-r}{F_r^2} \frac{\partial \eta_1'}{\partial x'} = \frac{1}{Re} \frac{\partial^2 u_2'}{\partial z^{2\prime}} \tag{18.8}$$

$$\int_0^{h'} \frac{\partial U_2'}{\partial x'} dz' + \frac{\partial \eta_2'}{\partial t'} = 0 \tag{18.9}$$

where $F_r = \sigma(H_1/g)^{1/2}$ is the Froude number and $Re = \sigma H_1^2/\upsilon$ is the Reynolds number. Note that σH_1 is the characteristic wave velocity. Note further that the dimensionless surface slope term in Equation 18.6, as well as the interfacial and surface slope terms in Equation 18.8 are scaled by $1/F_r^2$, while the dissipation term in Equation 18.8 is scaled by 1/Re. For typical wave-forced situations it can be easily shown that the dissipation term is far less significant than the surface and interfacial gradient terms, particularly the former. Yet, of course, dissipation plays a critical role in the problem in terms of wave damping and boundary layer effect within the mud. Note also that at typically low values of the wave Reynolds number, fluid mud motion is wholly laminar (Maa and Mehta, 1987).

The normalized boundary conditions are expressed as:

$$\eta_1'(0,t') = A \cos t' \tag{18.10a}$$

$$u_1'(\infty,t'),\ u_2'(\infty,z,t'),\ \eta_1'(\infty,t') \text{ and } \eta_2'(\infty,t') \rightarrow 0 \tag{18.10b}$$

$$u_2'(x',0,t') = 0 \tag{18.10c}$$

$$\partial u_2'(x,H_2',t')/\partial z' = 0 \tag{18.10d}$$

where $A = a_0/H_1$. The solutions of Equations 18.6 through 18.9 with these boundary conditions are relatively straightforward. Omitting details which may be referenced elsewhere (Mehta and Jiang, 1990), we state the following solutions for the elevations and the velocities:

$$\eta_1' = A \exp[i(k'x' - t')] \tag{18.11}$$

$$\eta_2' = A\left[1 - \left(\frac{k'}{F_r}\right)^2\right] \exp[i(k'x' - t')] \tag{18.12}$$

$$u_1' = A\frac{k'}{F_r^2}\exp[i(k'x' - t')] \tag{18.13}$$

$$u_2' = A\frac{k'}{F_I^2}\left[1 - r\left(\frac{k'}{F_I}\right)^2\right][1 - \cosh(mz') + \tanh(mH_2') \sinh(mz')]\exp[i(k'x' - t')] \tag{18.14}$$

where $m = \left(\frac{R_e}{i}\right)^{1/2}$

As observed η_2' is damped relative to η_1' by the multiplier $1-(k'/F_r)^2$. Likewise, u_2' is damped relative to u_1' by the two multipliers of Ak'/F_r^2 (i.e. amplitude of u_1').

Next we require the dispersion relationship for wave celerity based on the real part, k_r, of the wavenumber k, and an expression for the wave attenuation coefficient k_1, which is equal to the imaginary part of k. Note that the surface wave amplitude is specified as $a_x = a_0\exp(-k_1 x)$, where a_x is the amplitude at a distance x. Equations 18.11 through 18.14 satisfy the boundary conditions 18.10b, as the amplitudes and velocities vanish at infinite distance. We have, by definition,

$$k' = k_r' + ik_i' \tag{18.15}$$

where

$$k_r' = \left\{\frac{1}{2} F_I^2 [(Y_R^2 + Y_I^2)^{1/2} + Y_R]\right\}^{1/2} \tag{18.16}$$

which is the dispersion relationship, and

$$k_i' = \left\{\frac{1}{2} F_I^2\left[\left(Y_R^2 + Y_I^2\right)^{1/2} - Y_R\right]\right\}^{1/2} \tag{18.17}$$

which is the expression for the wave attenuation coefficient. Here

$$Y_R = \frac{1}{2r(R^2+I^2)}\{R[1+R-\cos\frac{\theta}{2}(p^2+q^2)^{1/4}] + I[I-\sin\frac{\theta}{2}(p^2+q^2)^{1/4}]\} \tag{18.18a}$$

$$Y_I = \frac{1}{2r(R^2+I^2)}\{R[I-\sin\frac{\theta}{2}(p^2+q^2)^{1/4}] - I[1+R-\cos\frac{\theta}{2}(p^2+q^2)^{1/4}]\} \tag{18.18b}$$

where

$$p = (1 + R)^2 - 4rR - I^2 \tag{18.19a}$$

$$q = 2I(1 + R - 2r) \tag{18.19b}$$

$$\theta = \tan^{-1}\left(\frac{q}{p}\right) \tag{18.19c}$$

$$R = H_2'\left[\frac{\exp(4\chi) - 1 + 2\sin(2\chi)\cdot\exp(2\chi)}{2\chi[\exp(4\chi) + 1 + 2\cos(2\chi)\cdot\exp(2\chi)}\right] \tag{18.19d}$$

$$I = H_2'\left[1 - \frac{\exp(4\chi) - 1 - 2\sin(2\chi)\cdot\exp(2\chi)}{2\chi\,[\exp(4\chi) + 1 + 2\cos(2\chi)\cdot\exp(2\chi)}\right] \tag{18.19e}$$

$$\chi = H_2'\left(\frac{Re}{2}\right)^{1/4} = H_2\left(\frac{\sigma}{2\nu}\right)^{1/4} \tag{18.19f}$$

It can be readily shown that k_r'/F_r depends on H_2', r and χ. Note that χ is the mud layer thickness normalized by $(2\upsilon/\sigma)^{1/2}$, which is representative of the mud boundary layer thickness. For the selected value of r = 0.15 (e.g. corresponding to field condition), this dependence is shown in Fig. 3 for H_2' ranging from 0.1 to 1. It is observed that for a given H_2', K_r'/F_r decreases from 1 at $\chi = 0$, becoming practically constant above a certain χ. This trend can be easily examined, for example, for a given system (H_1, H_2 fixed) subject to a wave of a given frequency. At $\chi = 0$, the bottom is rigid, hence $k_r' = F_r$ means $\sigma/k = C_o = (gH_1)^{1/2}$, which is the shallow water, rigid bottom wave celerity. Increasing χ implies decreasing viscosity υ, hence decreasing μ, since r and, therefore, ρ_2 are held constant in this case. Note that $k_r'/F_r = C_o/C$, where C is the celerity at any $\chi > 0$. Initially therefore, as the bottom becomes soft the wave speed increases over that due to rigid bottom. As $\chi \rightarrow \infty$, the lower layer becomes inviscid, and the wave speed equals $[gH_1(1 + H_2')]^{1/2}$. For practical purposes, the lower layer becomes "watery" for values of χ exceeding 2 to 3, and the celerity does not change too rapidly with further increase in χ as seen from Fig. 3.

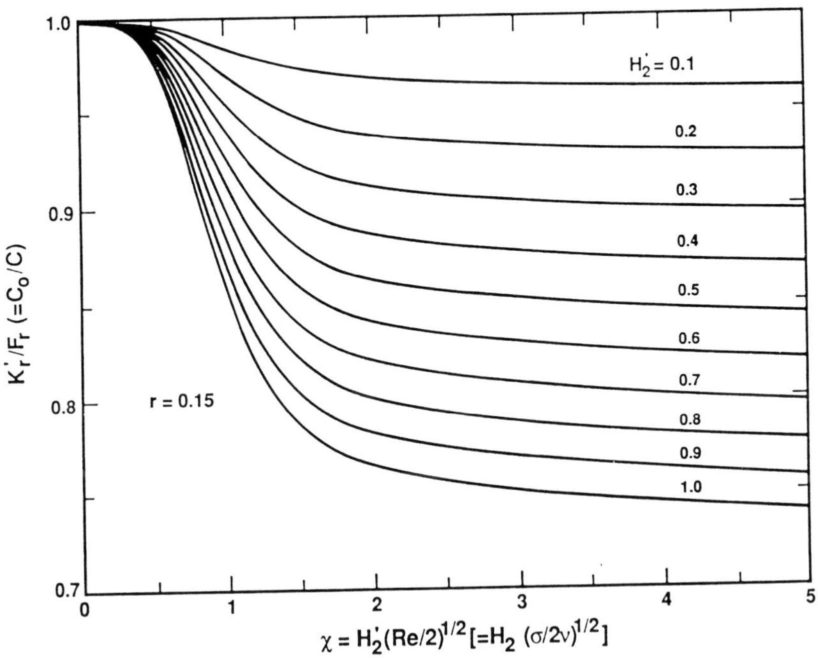

Figure 3: Dispersion relationship based on the inviscid-viscid model

In Fig. 4 the normalized attenuation coefficient k_i'/F_r, which also depends on H_2', r and χ, is plotted against χ for values of H_2' ranging from 0.1 to 1. We note that since $k_i'/F_r = k_iC_o/\sigma$, the plot can be considered to represent the manner in which k_i changes with increasing χ, starting with $\chi = 0$ at which $k_i = 0$ (rigid bottom case). An interesting feature of the observed variation in k_i is the occurrence of a peak in k_i as χ approaches unity. In other words, wave damping is greatest when the mud layer thickness (H_2) is of the order of the boundary layer thickness. As χ increases beyond this value, k_i decreases with decreasing thickness of the boundary layer and approaches zero as $\chi \to \infty$, i.e. the lower layer also becomes inviscid. As χ decreases below one, although the entire mud thickness constitutes the boundary layer, the rate of energy dissipation reduces relative to that at $\chi = 1$, hence k_i reduces as well.

In Fig. 5, profiles of the velocity amplitude, u_m, are plotted using typical test parameters, and χ ranging from 0.1 to 1.5. Notice the heavy damping of velocity with decreasing χ in the fluid mud layer. By virtue of the model assumptions, no boundary layer is found in the water layer, even though the velocity is damped with decreasing χ, by virtue of the influence of χ on the wave number (Equation 18.13). Besides the absence of boundary layer effect particularly just above the interface, the lack of velocity equality at the interface is another manifestation (limitation) of the assumed inviscid-viscid behaviour. The outcome is the suggestion of steeper gradients and hence rotationality at the interface than in reality, although laboratory measurements (e.g. Maa and Mehta, 1987) do indicate a rather drastic reduction in the mud velocity relative to that in water across a thin interfacial layer.

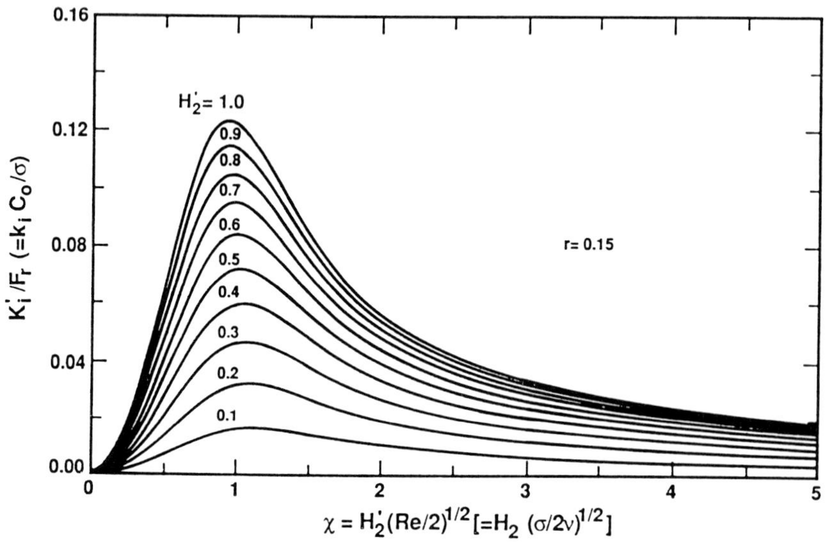

Figure 4: Wave attenuation relationship based on the inviscid-viscid model

18.3 Test Site and Experiment

Fig. 6 shows the test site located about 1km offshore of the eastern edge of Lake Okeechobee, Florida, in the proximity of the Okeechobee Waterway. A large part (528 sq. km) of this shallow, hard bottom lake is overlain with mud having a thickness of up to 80cm (Kirby et al., 1989). During the field test, wave conditions may be characterized as having been mild. While such waves did generate horizontal motions within mud, particulate erosion was minimal because the wave-induced bottom stresses were below 0.5 Pa, the critical shear stress for particulate erosion of mud (Hwang, 1989). When storm waves do occur, the top ~5cm thick layer of the mud tends to dilate to ~10cm due to sediment diffusion. Above this dilated layer, upward sediment mass transport tends to be comparatively very small, but the presence of the dilated layer does tend to complicate the near-bed processes as far as sediment motion is concerned (Hwang, 1989).

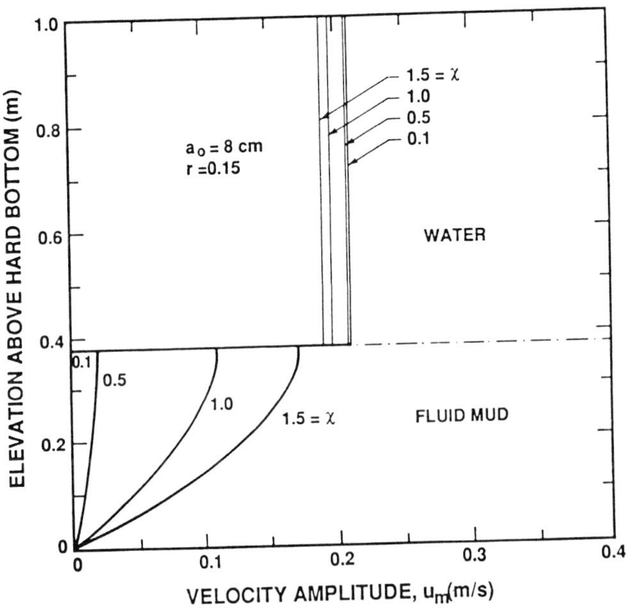

Figure 5: Simulated profiles of velocity amplitude, u_m for different values of χ using typical field test parameters

Two features of the bottom environment are noteworthy. Firstly, the mud throughout includes about 40% (by weight) material that is essentially organic and, secondly, the top 5 to 20cm of the mud has negligible vane shear strength (less than 0.39kN/m^2), and persists in the fluid state even under calm conditions (Kirby et al., 1989, Hwang, 1989). This state is brought about partly by the occasionally significant wave action, but it is believed that an important factor is the presence of the organic material which, presumably by virtue of having an open and comparatively strong, fibrous structure of peaty origin, prevents rapid dewatering of the top layer of the deposit.

In consonance with the nature of the problem and the two-layered flow formulation, the bulk density and the dynamic viscosity are the two important parameters characterizing the fluid mud. Vertical density profiles in the mud were obtained by a simple bottom coring procedure that yielded approximate variation of density with depth. A mud bulk density profile is shown in Fig. 7. The substrate underneath the mud layer may be considered as "hard"; the transition to hardness being here defined as the level at which the instrumented tower assembly (noted later) rested on its own account.

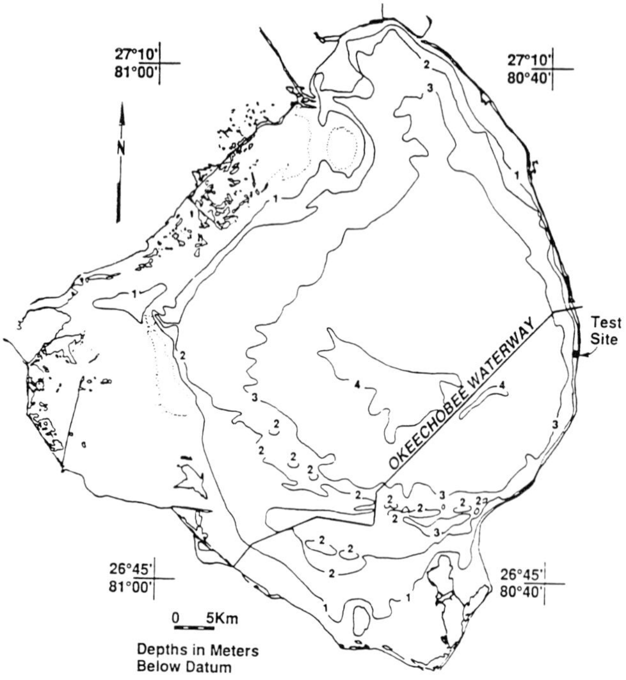

Figure 6: Test site in Lake Okeechobee. Depths are relative to a datum which is 3.81 m above msl

Mud viscosity was measured in a Brookfield viscometer (model LVT) with miniature vanes, at different mud bulk densities. Fig. 8a shows the stress-shear rate relationship obtained at a density of 1.05g/cm^3. The behaviour of the mud is observed to be pseudoplastic at rates less than about 3s^{-1}, and Newtonian at higher rates. Since the wave-induced shear rates in the mud during the study period were on the order of 0.1-0.2s^{-1}, it was decided to calculate a representative viscosity, μ, as the slope of the tangent to the curve passing through the origin as shown by the dashed line. The stress-shear rate relationships at all other densities were found to be qualitatively similar, and were used to obtain the corresponding values of viscosity. The resulting relationship between the viscosity (relative to water, μ_W) and density is shown in Fig. 8b. Due to limitations in the apparatus, mud denser than 1.12g/cm^3 could not be tested. It was therefore assumed that at higher (up to 1.18g/cm^3) densities, the viscosity could be obtained by extrapolation of the log-linear curve at higher densities in Fig. 8b, i.e. assuming a power law.

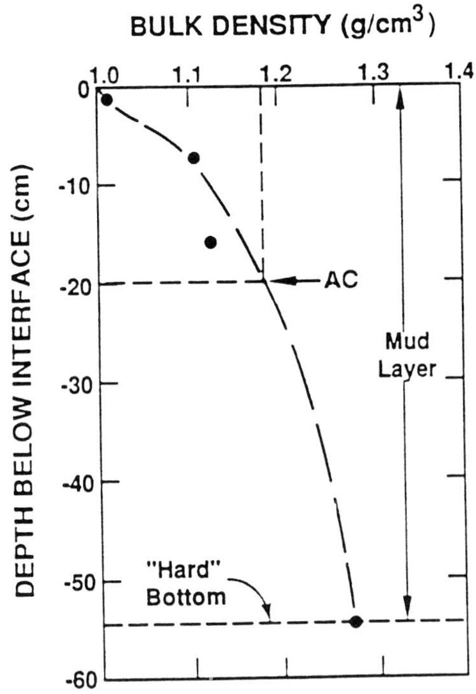

Figure 7: Mud density profile at the test site

Any interpretation of viscometer data obtained using vanes, in terms of a relationship between viscosity and the structure of the sediment matrix, must be treated with circumspection since vanes, by their very presence, break up the matrix at the cylindrical surface defined by the vane dimensions. Furthermore, viscosity is not characterized by density in a unique sense, insofar as viscosity and density are independent physical properties of fluids. Nevertheless the observed trend in Fig. 8b does seem to suggest a rather significant influence of sediment packing, as reflected by the bulk density, on viscosity. Apparently, at densities less than about 1.02g/cm^3, the aggregated matrix rapidly became tightly packed with increasing density. However, further increase in density did not seem to have drastically altered the compact arrangement attained at 1.02g/cm^3. Since the "granular" density of the organics-rich sediment was 2.14g/cm^3, the sediment volume fraction corresponding to 1.02g/cm^3 would be 0.018. It is conceivable that this volume fraction is equal to the so-called critical volume fraction above which the rigidity of the matrix (as reflected by the shear modulus of elasticity), increases rapidly (Williams and Williams, 1989).

During the experiment, time-series of water level, horizontal water velocity and mud motion were obtained. Water level was measured with a subsurface mounted pressure gage

(Transmetrics, Model P21LA). Water motion was measured by an electromagnetic (EM) current meter (Marsch-McBirney, Model 521), and a biaxial accelerometer (Entran, Model EGA2-C-5DY) was used to record mud motion. A 4Hz low pass filter was used to remove high frequency noise from the accelerometer output. Compensative correction to the output had to be made to account for a very slow linear drift. Finally, a running-mean correction was made to the output to eliminate low frequency fluctuations of the order of 10^{-3}Hz.

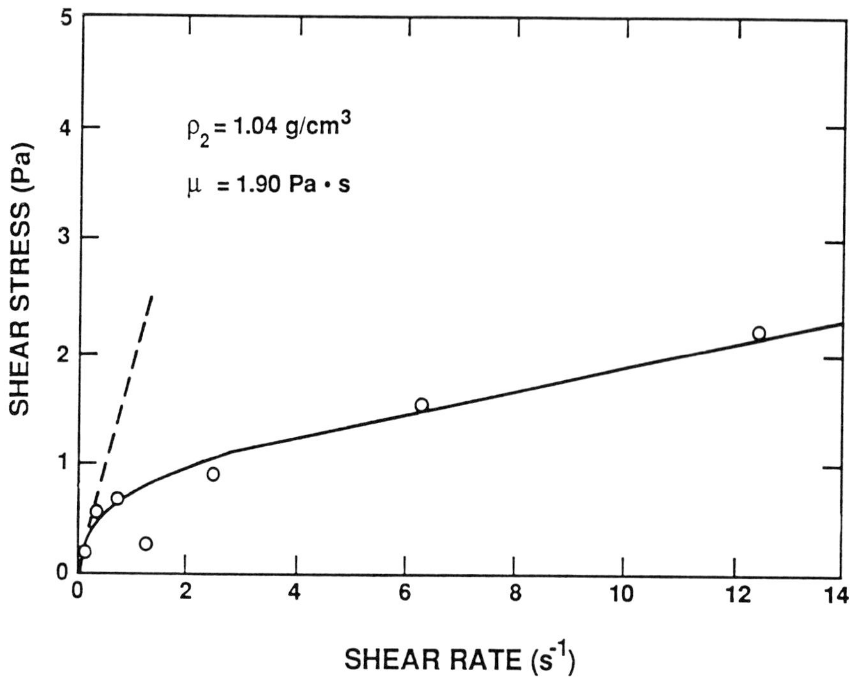

Figure 8a: Stress-shear rate relationship for Lake Okeechobee mud at a bulk density of 1.04 g/cm^3

The transducers were mounted appropriately on an 2.45m tall aluminum tower assembly with a triangular base for bottom stability. The depth of water at the test site was 1.43m, and mud thickness was 0.55m. The EM current meter was mounted 0.87m below the still water level in such a manner as to allow measurement of the two horizontal components of velocity, in order to enable the determination of the velocity in the direction of wave propagation. The accelerometer was mounted 0.20m below the mud surface for measuring a horizontal and the vertical acceleration components. The tower was designed to allow positioning of the device in such a way as to direct the horizontal axis in the dominant direction of wave motion.

The sampling rate for all the transducers was 4Hz for 10min every hour. This rate may be considered to be minimally adequate based on previous studies (Mehta and Dyer, 1990).

The transducers were connected to a data acquisition system (Tattletale, Model 6) mounted on a wooden base at the top of the tower. The test was conducted for a total of seven hours, during which the wave direction was consistently westerly, corresponding to a fetch of about 50km.

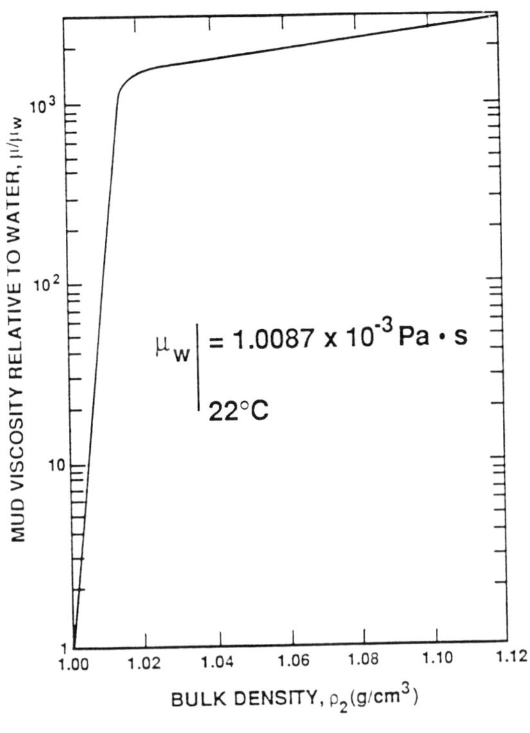

Figure 8b: Selected relationship between dynamic viscosity and density for Okeechobee mud

18.4 Water and mud motions

The time-series were analyzed in terms of their spectral properties and central tendencies via standard procedures. Fig. 9 shows the variation of the significant wave height, $H_{1/3}$, over test duration. Note that each hourly data point is a 10min average over the record length. $H_{1/3}$ is observed to have been rather small, peaking to 10cm at 3hr.

In Fig. 10 the variation of the dominant (modal) surface wave frequency, f_m, defined as the frequency at the peak of the wave energy density spectrum, is shown. An example of the spectrum itself is shown in Fig. 11 The modal frequency variation is compared in Fig. 10 with the same determined from the water velocity spectra, an example of which (corresponding to Fig. 11) is shown in Fig. 12.

As observed from Fig. 10, the dominant wave frequency was comparatively constant, with a mean of 0.42Hz (obtained from f_m values for the wave and the velocity spectra). This value can be shown to correspond to a wind of 20km/hr (moderate breeze), using shallow water forecasting curves for wave generation over a rigid bottom (Coastal Engineering Research Center, 1984). This value is believed to be a reasonable estimate of the wind speed at the time of study; however, at this speed the forecasted wave height in the rigid bottom case can be shown to be about 40cm, which would be four times greater than what was measured. It is surmised that the discrepancy is due to significant wave damping over the mud bottom.

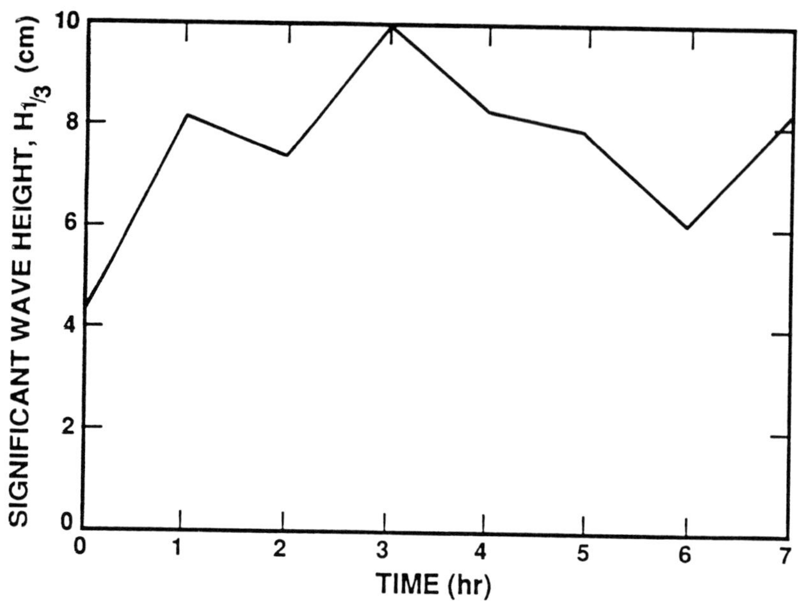

Figure 9: Variation of significant wave height during test

A mud acceleration spectrum (corresponding to the wave energy spectrum in Fig. 11 and velocity spectrum in Fig. 12) is shown in Fig. 13. The corresponding modal frequency (Fig. 10) is seen to be roughly in agreement with the wave- and velocity-based modal frequencies in Fig. 10, thus implying that the accelerations were wave-coherent. The slightly higher values (mean 0.51Hz) in the case of acceleration in comparison with those for the waves and the current are believed to be attributable to anomalies associated with the in situ mount used for the accelerometer. The acceleration spectrum shows a marked peak at a very low frequency corresponding to a long period oscillation. Commensurate but comparatively smaller peaks also appear in the wave and water velocity spectra of Figs. 11 and 12. This low frequency is distinct from the wind wave-induced frequency, and is discussed further later.

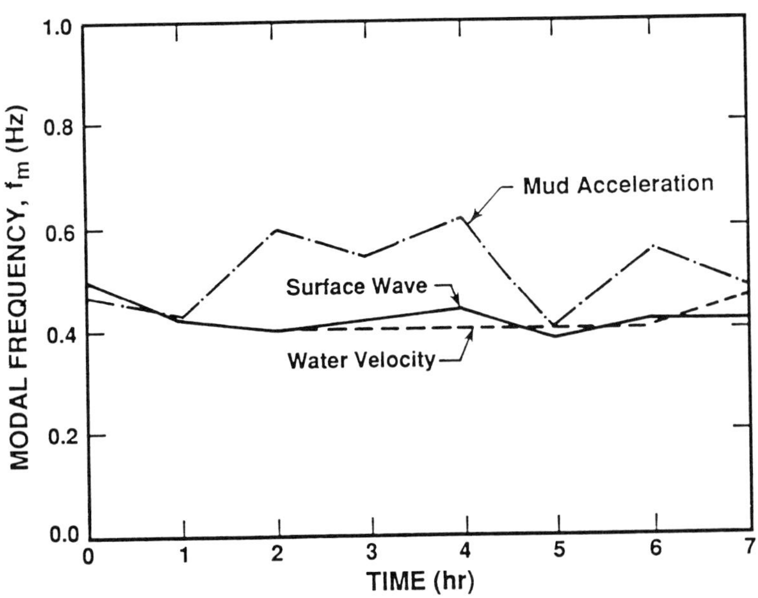

Figure 10: Variation of modal wave frequency during test

Vertical acceleration measurements did not exhibit long wave signatures for evident reasons but, on the other hand, the magnitude of the variance obtained from the time-series of vertical acceleration showed this variance to range from almost nil to ~0.004cm^2/s^4; the latter value being as high as the variance of the horizontal acceleration. This behaviour indicates that, in general, the waves at the site did not quite conform to the shallow water assumption. In fact, it can be easily shown that this condition was valid for waves having frequencies less than around 0.2Hz. A subsequent experiment at the same site, but with the current meter set to measure a horizontal and the vertical water velocities, confirmed this observation (Mehta and Jiang, 1990).

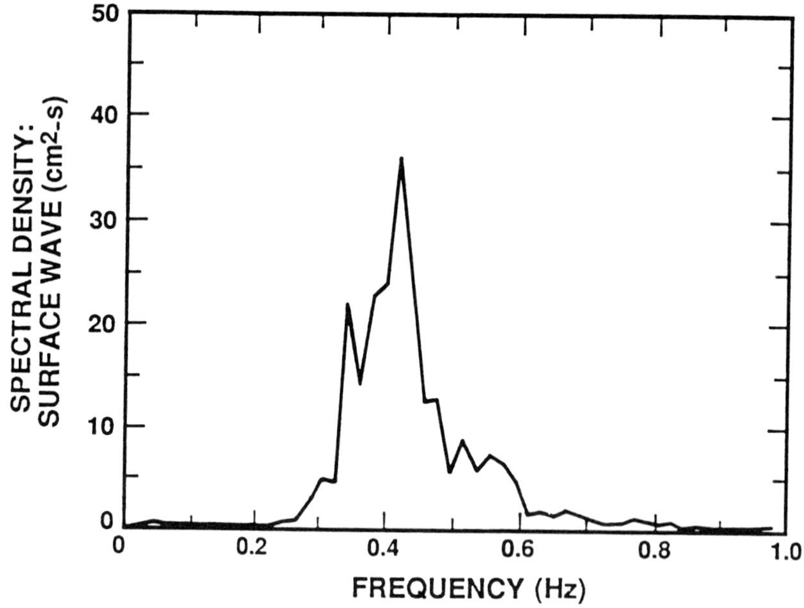

Figure 11: Wave energy spectrum at 1 hr

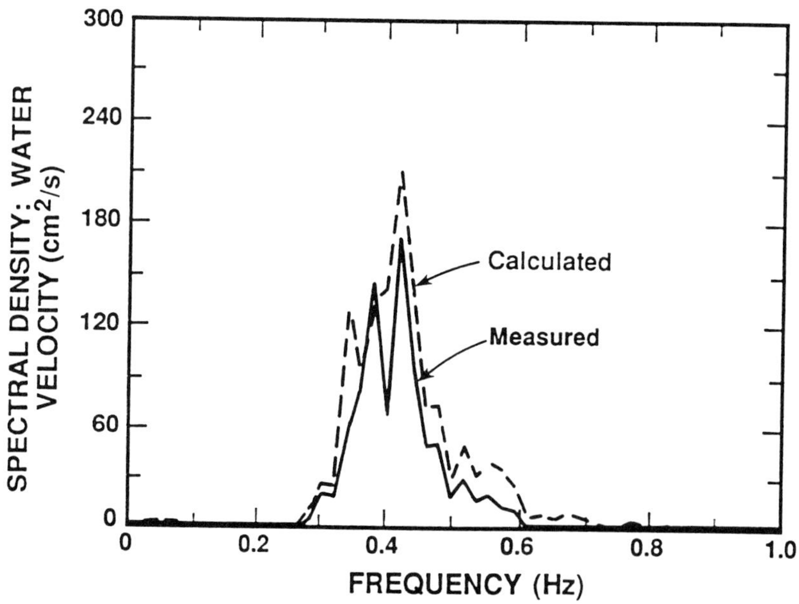

Figure 12: Measured and calculated water velocity spectra at 1 hr

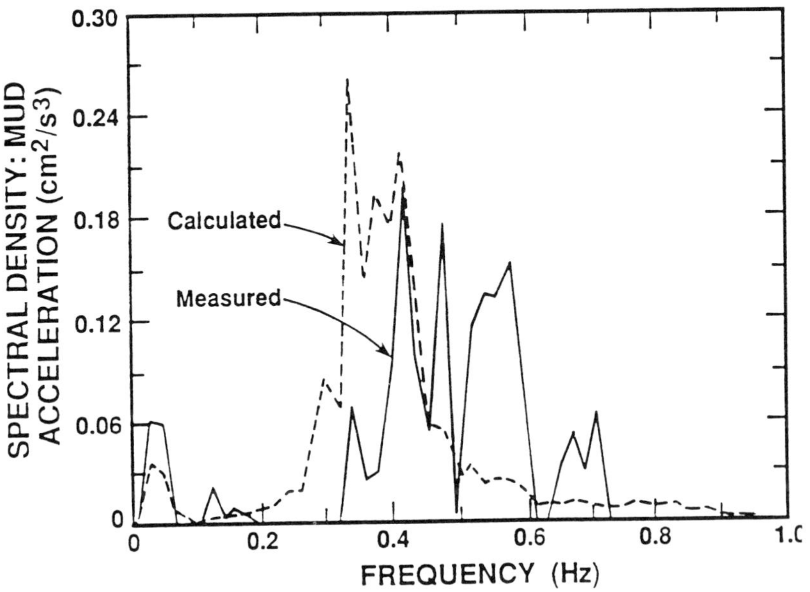

Figure 13: Measured and calculated mud acceleration spectra at 1 hr

18.5 Response of Linearized Fluid Mud-Water System

Given the wave energy spectrum, which essentially characterizes the measured wave amplitude (A) variation with wave frequency, Eq. 18.13 can be used via the corresponding spectral transfer functions (Mehta and Jiang, 1990) to calculate the water velocity spectrum, and the time-derivative of Eq. 18.14 to calculate the mud acceleration spectrum.

An evident difficulty in adapting the model to measurement involves the selection of a representative mud density since, as seen from Fig. 7, the density characteristically increased quite rapidly with depth below the interface in the top ~10cm layer. The matter of selecting density of course bears critically on the rate of energy dissipation, via the dependence of the viscosity on density (Fig. 8b). In order to account for this problem, Maa and Mehta (1987) and others (e.g. Shibayama et al., 1989) have developed layered bed models in which mud properties can vary arbitrarily with depth. It was felt that the use of such models for the present case would amount to an over-specification of the physical system, whose understanding was constrained by the rather limited data collection effort. On the other hand, the assumption of an invariant mud density, especially in view of its evident depth-dependence (Fig. 7), is a limitation of model application which must be borne in mind while considering model results.

The choice of appropriate mud thickness, H_2, was difficult in the same sense as choosing a representative density. With regard to the latter quantity, a value of 1.18g/cm^3 was selected, corresponding to the position (AC) of the accelerometer shown in Fig. 7. The computations were found to be particularly sensitive to the selected value of H_2 by virtue of the effect of χ on the kinematics, which in turn signifies the strong inter-relationship between the mud boundary layer and energy dissipation. It should be further pointed out that, as a rule mud becomes rather immobile at densities exceeding around 1.2g/cm^3. Therefore, referring to fig. 7,it would be unreasonable to select the entire mud thickness of 0.55M for computational purposes. It was decided to select H_2 = 0.283m. For the cases examined, this selection seemed to give the best results by way of agreement with the data. Unfortunately however, considering the limitations associated with model assumption of constant density, no independent arguments can be offered with regard to any justification for the choice of H_2 on physical grounds.

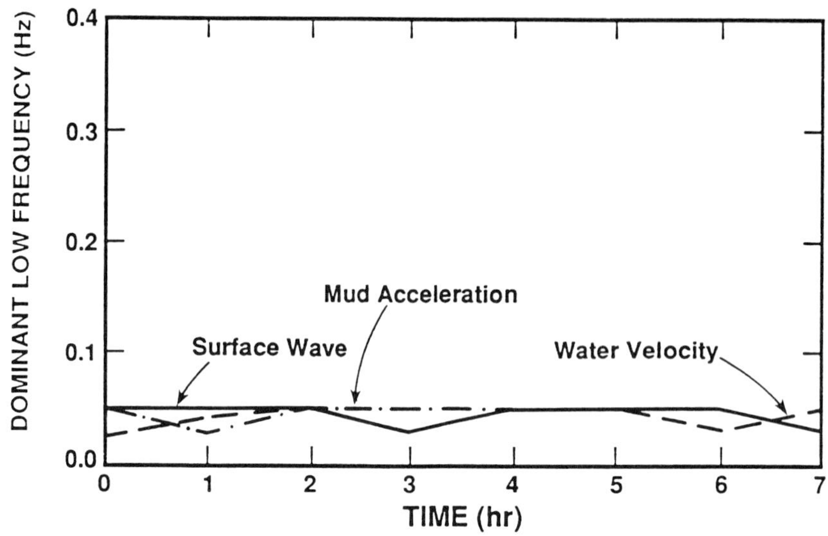

Figure 14: Dominant low frequency spectral signature during the test

Using the surface wave energy spectrum in Fig. 11, the calculated water velocity and mud acceleration spectra are shown in Figs. 12 and 13. The velocity measurement was 0.56m above the interface, at an elevation that was likely to have been above the wave boundary layer, since the latter would be on the order of a centimetre only. Therefore, the model assumption of inviscid upper layer, when comparing calculated values with data essentially in the upper portion of the water column, may not be overly limiting. Data and calculation show approximately similar trends. Note that by virtue of the assumptions of linearity and inviscid upper layer, the calculated velocity spectrum is nearly self-similar to the wave energy spectrum.

Comparing the calculated and measured mud accelerations in Fig. 13 indicates that while model results generally agree with the data, the data exhibit accelerations at high frequencies which cannot be accounted for by the shallow water model. These high frequency waves were proportionately less damped most probably due to these being outside the shallow water domain, in intermediate water depth. Note that the measured velocity spectrum of Fig. 12 indicates greater damping at high frequencies relative to the acceleration spectrum in Fig. 13, possibly due to transducer limitations. In general however, selective damping of water wave components has been well documented for example by Wells and Kemp (1986), in their study on wave propagation over a large mud flat off Surinam.

18.6 Low Frequency Signature

The dominant frequency of the low frequency wave signature is plotted in fig. 14. It is observed that all the spectra yield rather consistent values with a mean of 0.043Hz. It is also noteworthy that by integrating the acceleration twice, it can be shown that a maximum horizontal displacement of mud on the order of 2mm occurred at ~0.03Hz.

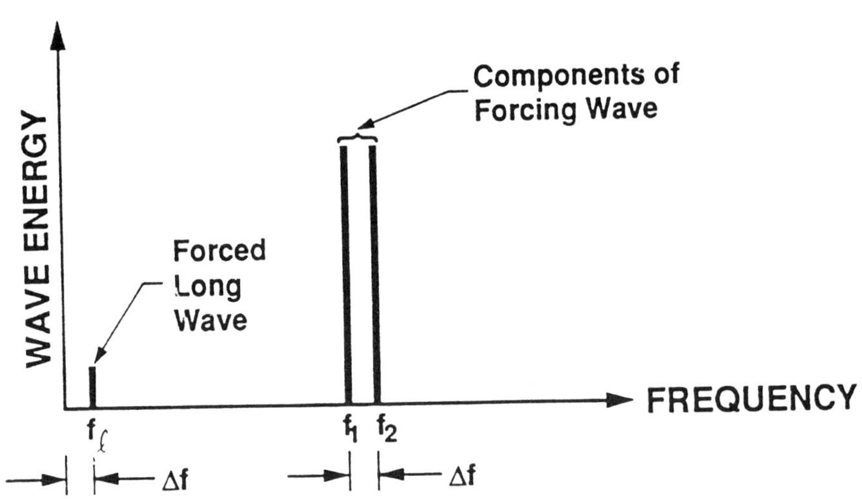

Figure 15: Wave energy spectrum showing two forcing wave components and forced long wave

It is seen from Figs. 11, 12 and 13 that the low frequency wave peak is accentuated in mud relative to that in water, when this peak is compared with the corresponding peak of the forcing wave. This relative enhancement is due to frequency dependent shallow water wave damping as can be gleaned from the attenuation coefficient characterized in Fig. 4. Thus, for example, selecting 0.4Hz and 0.04Hz as representative frequencies for the forcing wave and the low frequency wave, respectively, the corresponding damping coefficients (assuming shallow water condition) would be $0.0034m^{-1}$ and $0.0013m^{-1}$, which indicates relatively much less damping at the lower frequency.

Some comments on the causative mechanism for the low frequency signature are in order. In that context it must be noted that the dominant seiching frequency in Lake Okeechobee is considerably smaller, on the order of 10^{-4}Hz. It seems plausible that the long wave is in fact a second order effect resulting from wind wave forcing leading to surf beat. This effect, which commonly occurs in the frequency range of $3x10^{-3}$ to $8x10^{-3}$Hz at open coasts, may be modified by a compliant bottom as in the present case, but it would serve the purpose to examine this mechanism by assuming a rigid bottom, at least for arriving at a possible qualitative explanation for the occurrence of the low frequency wave.

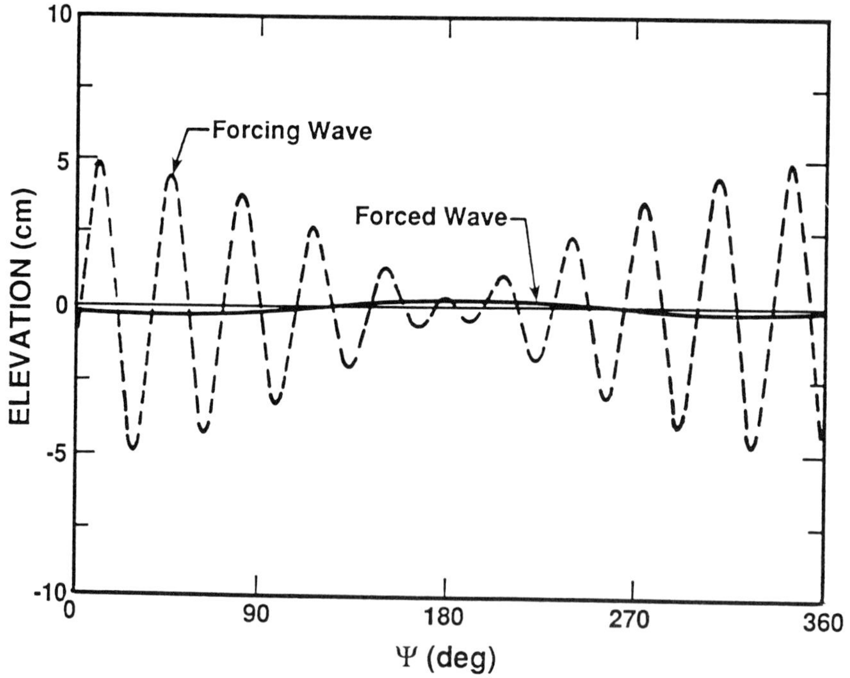

Figure 16: Forcing wave group and low frequency wave derived from test data at 1 hr

We consider two waves, $\eta_1(0,t)$ and $\eta_2(0,t)$, of respective frequencies σ_1 and $\sigma_2 = \sigma_1 + \Delta\sigma$, where $\Delta\sigma$ is a small difference:

$$\eta_1 = a_1\cos(\sigma_t - \epsilon_1) \tag{18.20a}$$

$$\eta_2 = a_2\cos(\sigma_2 t - \epsilon_2) \tag{18.20b}$$

Here, ε_1 and ε_2 are the phase lags. Assuming further that the amplitude $a = a_1 = a_2$, the resultant (forcing) wave, $\eta_r = \eta_1 + \eta_2$, is

$$\eta_r = 2a\cos(\sigma t - \epsilon)\cdot\cos\left(\frac{\Delta\sigma t - \Delta\epsilon}{2}\right) \tag{18.21}$$

where $\sigma = (\sigma_1 + \sigma_2)/2$, $\varepsilon = (\varepsilon_1 + \varepsilon_2)/2$, and $\Delta\varepsilon = \varepsilon_1 - \varepsilon_2$. It is thus seen that the wave form, $2a\cos(\sigma t - \varepsilon)$, is modulated by $\cos[(\Delta\sigma \bullet t - \Delta\varepsilon)/2]$, which causes the beat effect due to wave groupiness.

The fact that real waves have finite amplitudes means that second order effects arising from changes in water surface elevation and associated kinematics cannot always be ignored. An effect germane to the present case is the setting up of a forced long wave which modulates the wave given by Eq. 18.21. This long wave follows from the inclusion of the kinetic head term in the dynamic free surface boundary condition, and is obtained by calculating the mean (relative to the wave period $2\pi/\sigma$) water surface profile by carrying out the computations to second order (Longuet-Higgins and Stewart, 1962; Dean and Dalrymple, 1984). Omitting details which are given elsewhere (Mehta and Jiang, 1990), and without restricting the basic formulation to shallow water, the second order wave can be shown to be (the overbar representing short period, i.e. $2\pi/\sigma$, averaging):

$$\bar{\eta} = -\frac{a^2 k}{\sinh kH}\,[1 + \cos(\Delta\sigma\cdot t - \epsilon)] \tag{18.22}$$

where $k = (k_1 + k_2)/2$. The first term on the right hand side represents a steady set down and the second is the forced long wave. The energy spectrum thus obtained is shown in Fig. 20, which shows that the long wave, corresponding to two forcing waves at frequencies $f_1 = \sigma_1/2\pi$ and $f_2 = \sigma_2/2\pi$ and amplitude a, has a frequency $f_\ell = \Delta f = f_2 - f_1$ and amplitude $a^2k/\sinh kH$. It is important to note that, comparing Eq. 18.22 with Eq. 18.21 indicates that the long wave is π radians (180°) out of phase with the envelope of the short period wave group.

When forcing is represented by a continuous spectrum, the treatment for determining the forced long wave spectrum becomes involved (Sharma and Dean, 1979). Here a very approximate approach is selected to interpret the present data. Consider for example the measured, double peaked velocity spectrum corresponding to the forcing wave in Fig. 12.

Assuming the two peak frequencies to be the primary contributors to the corresponding long wave peak, we have the following parameters (obtained from the Fourier series of the corresponding surface wave record): a_1 = 2.4cm, a_2 = 2.6cm, long wave amplitude a_ℓ = 0.30cm, f_1 = 0.38Hz, f_2 = 0.42Hz, f_ℓ = 0.049Hz, ε_1 = 191°, ε_2 = 171°, and ε_ℓ = -65°. Note that Δf = 0.040Hz, which is reasonably close to f_ℓ. The forcing wave (η_r) and the forced wave (η_ℓ) are plotted against $\psi = \Delta\sigma.t-\Delta\varepsilon$ in Fig. 16. It is evident that the forced wave is almost π radians out of phase with the short period wave envelope as theory would require. If we assume the shallow water condition, the forced wave amplitude (from Eq. 18.22) would be a^2/H. Given a = 0.5(a_1 + a_2) = 2.50cm and H = 143cm, a^2/H = 0.044cm. Thus the theory under-predicts the amplitude significantly, which might be due to inherent theoretical limitations as well as the rather gross assumptions made in applying the theory to data.

18.7 Concluding Remarks

Notwithstanding the limited nature of the data obtained, the complexities in modelling mud motion, and the rather obvious constraints in simulation arising from the shallow water assumption as well as others (e.g. inviscid water layer, linearized response, homogeneous mud layer), it is demonstrated that measurable mud accelerations can occur tens of centimetre below the mud-water interface, under wave action that is mild enough to preclude measurable particulate erosion of the interface. Although bioturbation does not seem to be a significant factor in Lake Okeechobee (Kirby et al., 1989), the effect of persistent mud oscillation, even though very small, can be germane to likely changes in the rates of exchange of phosphorus and other constituents responsible for the highly trophic state of this lake. Similarly, there may be an effect on the formation and upward transport of gas bubbles which occur abundantly in the muddy area of the lake (Kirby et al, 1989).

In this study, we considered mud up to a density of 1.18g/cm^3 to be a highly viscous fluid. In general, however, careful measurements (e.g. Sills and Elder, 1981) indicate that, at least under quiescent conditions, at densities exceeding about 1.1g/cm^3, clayey mud typically exists as a porous solid rather than a fluid. Vane shear measurements (Hwang, 1989) suggest that the organics-rich Lake Okeechobee mud is prone to occur in a fluidized state at densities at least up to about 1.065g/cm^3 in quiescent conditions, although viscosity data presented here seem to suggest the occurrence of a structured phase for densities greater than 1.02g/cm^3.

It is unclear if at 1.18g/cm^3 density this mud is normally (i.e. in the absence of wave action) fluidized, but it is worth examining this issue by calculating values of χ. Thus given H_2 = 0.283m, υ = 1.76x10^{-2} m^2/s (selected value for model calculations) and σ = 5.03rad/s (corresponding to the dominant forcing frequency of 0.4Hz) yields χ = 2.4, which would mean, as noted previously, that the mud would respond like a "watery" material. On the other hand, at the low frequency of 0.04Hz, χ = 0.76, which would mean that the mud would respond as a "harder" material at this frequency. Hence the chosen model description may be approximate in this latter respect. Nonetheless, since in general the model calculated mud accelerations compare favourably with the measured ones, at least in order of magnitude, the description of mud as a fluid seems acceptable, although a better description, which recognizes the transition from the fluid phase to the solid phase, would constitute a worthwhile improvement.

Acknowledgement

Support provided by the South Florida Water Management District, West Palm Beach (Lake Okeechobee Phosphorus Dynamics Study), and the U.S. Army Engineer Waterways Experiment Station, Vicksburg, MS (Contract DACW39-89-M-4639), is acknowledged.

18.8 References

Coastal Engineering Research Center (1984). Shore Protection Manual, Vol. 1, U.S. Army Corps of Engineers, Washington, D.C. 608p.

Dean R.G. and Dalrymple R.A. (1984). Water Wave Mechanics for Engineers and Scientists. Prentice-Hall, Englewood Cliffs, NJ, 364 p.

Foda M.A. (1989). Sideband damping of water waves over a soft bed. Journal of Fluid Mechanics, 201, 189-201.

Gade H.G. (1958). Effects of non-rigid, impermeable bottom on plane surface waves in shallow water. Journal of Marine Research, 16(2), 61-82.

Hwang K.-N. (1989). Erodibility of fine sediment in wave-dominated environments. Rept. UFL/COEL-89/017, Coastal and Oceanographic Engineering Dept., Univ. of Florida, Gainesville, FL, 77 p.

Kirby R.R., Hobbs C.H. and Mehta A.J. (1989). Fine sediment regime of Lake Okeechobee, Florida. Rept. UFL/COEL-89/009, Coastal and Oceanographic Engineering Dept., Univ. of Florida, Gainesville, FL, 77 p.

Longuet-Higgins M.S. and Stewart R.W. (1962). Radiation stress and mass transport in gravity waves, with application to 'surf beat'. Journal of Fluid Mechanics, 13, 481-504.

Maa P.-Y. and Mehta A.J. (1987). Mud erosion by waves: a laboratory study. Continental Shelf Research, 7(11/12), 1269-1284.

Mehta A.J. and Dyer K.R. (1990). Cohesive sediment transport in estuarine and coastal waters, In: The Sea: Ocean Engineering Science, B. LeMehaute and D.M. Hanes eds., Vol. 9, Part B, Wiley-Interscience, New York, 815-839.

Mehta A.J. and Jiang F. (1990). Some field observations on bottom mud motion due to waves. Rept. UFL/COEL-90/008, Coastal and Oceanographic Engineering Dept., Univ. of Florida, Gainesville, FL, 85p.

Ross M.R. and Mehta A.J. (1990). Fluidization of soft estuarine mud by waves. In: The Microstructure of Fine-grained Sediment: From Mud to Shale. Ch. 19, R.H. Bennett ed., Springer-Verlag, New York, 185-191.

Sharma J.N. and Dean R.G. (1979). Development and evaluation of a procedure for simulating a random directional second order sea surface and associated wave forces. Ocean Engrg. Rept. No. 20, Dept. of Civil Engrg., Univ. of Delaware, Newark, DE, 139 p.

Shibayama T., Aoki T. and Sato S. (1989). Mud mass transport due to waves: a visco-elastic model. Proceedings of the 23rd Congress of I.A.H.,R., Ottawa, Canada, B567-B574.

Sills G.C. and Elder D. McG. (1986). The transition from sediment suspension to settling bed. In: Estuarine Cohesive Sediment Dynamics, A.J. Mehta ed., Springer-Verlag, Berlin, 192-205.

van Rijn L.C. (1985). The effect of waves on kaolinite/sand beds. Rept. M2060, Delft Hydraulics, Delft, The Netherlands.

Wells J.T. and Kemp G.P. (1986). Interaction of surface waves and cohesive sediments: field observations and geologic significance. In: Estuarine Cohesive Sediment Dynamics, a.J. Mehta ed., Speinger-Verlag, Berlin, 43-65.

Williams D.J.A. and Williams P.R. (1989). Rheology of concentrated cohesive sediments. Journal of Coastal Research, Special Issue No. 5, 165-173.

19

The reflection of waves off beaches

J Darbyshire

Abstract

The reflection of waves off beaches has been given considerable attention recently as it is important in civil engineering applications and also in academic problems such as the production of microseisms. A numerical method, based on Lamb's formula, has been developed to calculate the reflection coefficient for a given wave frequency off a beach of any profile. The method has been tested in wave tanks with artificial beaches having a wide range of topography and the results have always been found to agree well with the observations.

It was decided to apply the method to real beaches and use was made of many observations of beach profiles and the corresponding wave records taken by the author over several years at Hell's Mouth Bay in the Llyn Peninsula, North Wales.

The waves were registered by three recorders placed at low tide level and situated at the corners of a right-angled triangle, such that the second recorder was in most cases 10 m away from the first, measured along a line parallel to the beach and the other recorder was also 10m from the first but along a line normal to the beach. From Fourier and correlation analysis of three wave records taken at or near high tide, it was possible to calculate and plot the phase differences against the frequency for all three combinations. It was also possible to calculate the wave direction. If the waves were progressive, at this depth (where $c^2 = gh$, independent of frequency), the variation of phase difference with frequency should be linear. In actual fact, however, the trend line has a "wavy" character due to the effect of reflection.

It is difficult to calculate the reflection coefficient directly from these variations as it is an inverse problem. It is possible, however, to apply the numerical method to the observed beach profile and calculate the resulting variations in phase difference. Ten such examples were taken. Reasonable agreement was obtained between observed and calculated variations. The values of the reflection coefficients for various frequencies so obtained turn out to be quite high, often being of the order of 10-60%.

The method can be extended to two dimensions.

19.1 Introduction

The reflection of waves off beaches has been given considerable attention recently as it is important in civil engineering applications and also in academic problems such as the generation of microseisms. A good deal of theoretical work has been done by Bremmer (1951), Yoshida (1959), Kujura (1961), Wait (1962), and Meyer (1979). Their work is summarized by Mei (1982) and its implications will be discussed in the next section. The present paper has a more practical approach and deals with measurements in a wave tank and on an actual beach. It is very difficult to measure the reflection coefficient directly on a beach as can be done with a tank as it is difficult to get controlled conditions. A possible solution is to calculate the reflection coefficient for each frequency from the beach contours by a numerical method based on Lamb's formula. It is then possible to work out the effect of the reflection on the waves and compare this with what is observed.

19.2 The reflection program

This is based on the formula given by Lamb (1932). Art 176. He showed that when a wave of amplitude a_1 in a channel of width b_1 and depth h_1 enters a channel of width b_2 and depth h_2, then the reflection coefficient is given by

$$(b_1c_1 - b_2c_2)/(b_1c_1 + b_2c_2) \qquad (19.1)$$

and the transmission coefficient is

$$2b_1c_1/(b_1c_1 + b_2c_2) \qquad (19.2)$$

where

$$c_{1,2} = (gh_{1,2})^{1/2} \qquad (19.3)$$

Lamb had some reservations about the accuracy of the formula because of the rapid variations in depth and width at the point of discontinuity.

If the width is constant, the reflection formula reduces to :

$$(c_1 - c_2)/(c_1 + c_2) \qquad (19.4)$$

This formula was derived for long waves (where $c = \sqrt{gh}$), by equating the flux on both sides of the boundary and equating the displacement. By this means the same formula can be extended to any gravity wave. However, in the long wave case, the rate of passage of energy across the boundary is also conserved whereas this is not so in the general case as the group velocity varies with depth in a different manner from the phase velocity. Even so, as the results will show, the formula applies over cases where the ratio of depth to wavelength approaches the deep water condition.

In this paper, the formula is applied one-dimensionally over a succession of very small intervals over a variable profile. This gets over to some extent Lamb's objections as there are no sudden changes.

If the origin is referred to the measuring position, x^0 refers to the distance of the water-line from the origin, and x^1 refers to any point between them, then the reflected wave at x^1 will be contributed to by all positions from it to the water-line x^0. The phase of the contribution will depend on the time taken by the wave to travel from x^1 to x and back. Expressed mathematically and using the formula, then the reflected wave at x^1 will be given by:

$$\int_{x^1}^{x^0} \left(\frac{1}{2c}\right)\cdot\left(\frac{dc}{dx}\right).\exp\left(i \int_{x^1}^{x^0} 2kdx\right) \qquad (19.5)$$

where c and k, phase velocity and wave number, are functions of the depth which is in turn a function of x.

The integration can be carried out numerically for each value of x^1 between 0 and x^0 and thus gives the reflection coefficient at all points between 0 and x^0. A computer program was prepared to do this.

In the work described by Mei (1982), the mild slope condition is assumed which impies that:

$$(\partial h/\partial x)/kh << 1 \qquad (19.6)$$

and the bottom profile is denoted by:

$$h=\mu x \text{ where } \mu \text{ depends on x and is } <<1 \qquad (19.7)$$

Under these conditions, by a perturbation method, the relevant equations can be solved to a first order in μ and an expression is found for the reflection coefficient.

$$R = -\int_{-\infty}^{\infty} dx\left[(d/dx)\log(kh)^{1/2}\right]\exp 2iS/\mu \qquad (19.8)$$

For the shallow depth condition, the integrand is identical with that in equation (19.5), the only difference being that the term $\exp 2iS/\mu$ is based on an average value over the range of integration, whereas in (19.5), this replaced by a second integration over the same limits.

A further approximation is made in assuming that the depth varies linearly with x with a very small gradient although with the shallow depth conditions (19.6) and (19.7) seem sufficient. They then obtain by using the delta function, a value of R for a discontinuous change in depth which turns out to be the same as equation (19.4) with $c_1 \approx c_2$. For the case of a linear slope over a distance L, having a mean depth h and a change Δ between the ends, they found

$$R = -\Delta \sin kL / (4hkL) \qquad (19.9)$$

where k is the wave number corresponding to the depth h. This expression can be derived directly from equation (19.5) for the conditions assumed.

In deriving the equations, conditions (19.6) and (19.7) have been assumed. Equation (19.5) is derived from more elementary considerations and no condition is immediately apparent besides the shallow depth one. However, these conditions will be borne in mind when discussing the results. As equation (19.8) is correct to a first order in μ, a value =of μ less than 0.33 should give an accuracy of 10% or more.

19.3 Application to tank experiments

The reflection program was tested with artificial analytical profiles inside wave tanks in the laboratory. The first example shown in Figure 1a refers to data given by Davies and Heathershaw (1983) and deals with a sinusoidal bottom consisting of 10 waves of wavelength 1 metre and amplitude 5cms. The frequency of the water waves was adjusted by the wave-maker so that for the depth used, the water wavelength was twice the bottom ripple wavelength. b is the bottom ripple amplitude (5cms in these cases), and d is the depth for each experiment, and the reflection coefficient has been obtained experimentally for the seven cases shown and are indicated by the points. The curve gives the results obtained by the program and shows very close agreement. This is the case even when the ratio of d/λ approaches the deep water condition. Thus the two cases with the lowest value of b/d, 0.08 and 0.10, give good agreement although the wave periods were 1.17 and 1.20 secs. respectively corresponding to depths of 62.5cms and 50cms and giving values of d/λ of 0.306 and 0.245 which approach the deep water condition which is often taken to be reached when d/λ is 0.25. The data used in Figures 1b, c and d, were obtained by Drs. M. Belzons and E. Guazzelli of the Dept. of Physics, Centre St. Jerome, University of Provence, Marseille and have not been published.

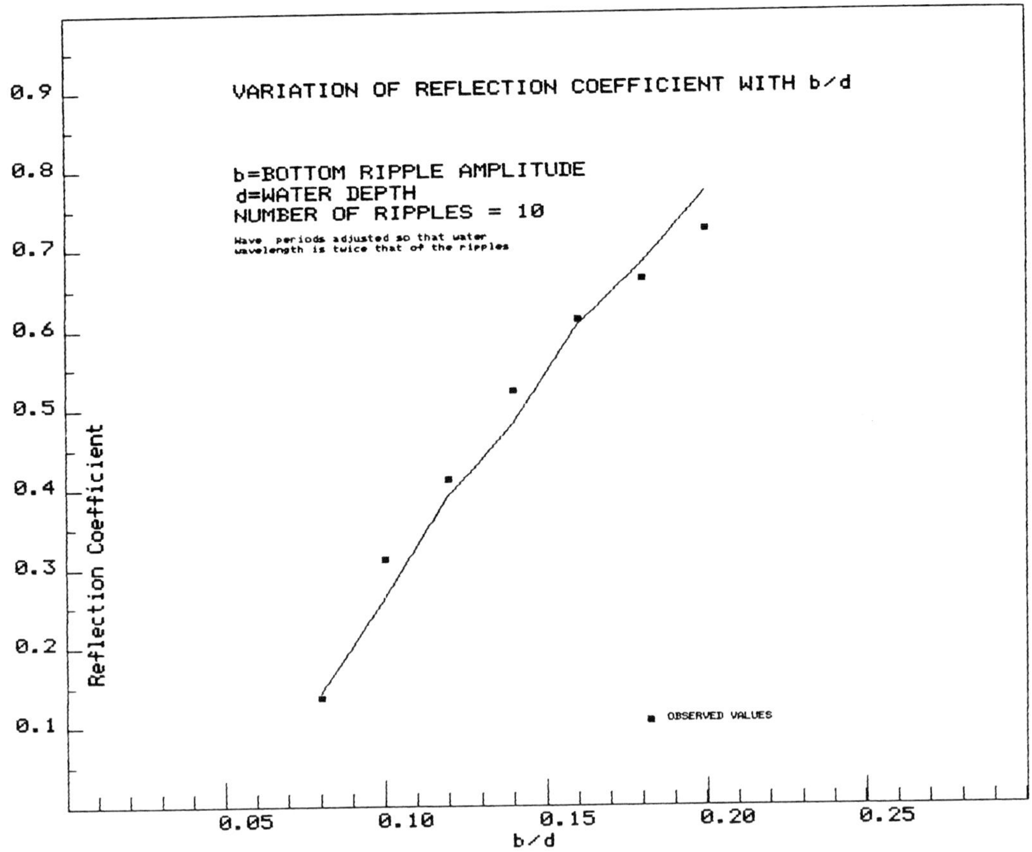

Figure 1a: Comparison of observed and calculated reflection coefficients for a sinusoidal bottom profile.

Here the bottom ripples consist of a mixture of two sine waves as shown on the diagrams. In these cases the measured reflection coefficients are shown by the points whilst the curves are given by the theory. The agreeement is particularly good for Figure 1b. Figure 1c shows not so good agreement for the 2.5cm depth. The agreement is not so good for Figure 1d but here, probably, a smaller integration interval would have improved matters. In Figure 1b, there is good agreement for 3.25 Hz. and 0.04m depth corresponding to a wavelength of 0.14m and $h/\lambda = 0.286$, close to the deep water condition.

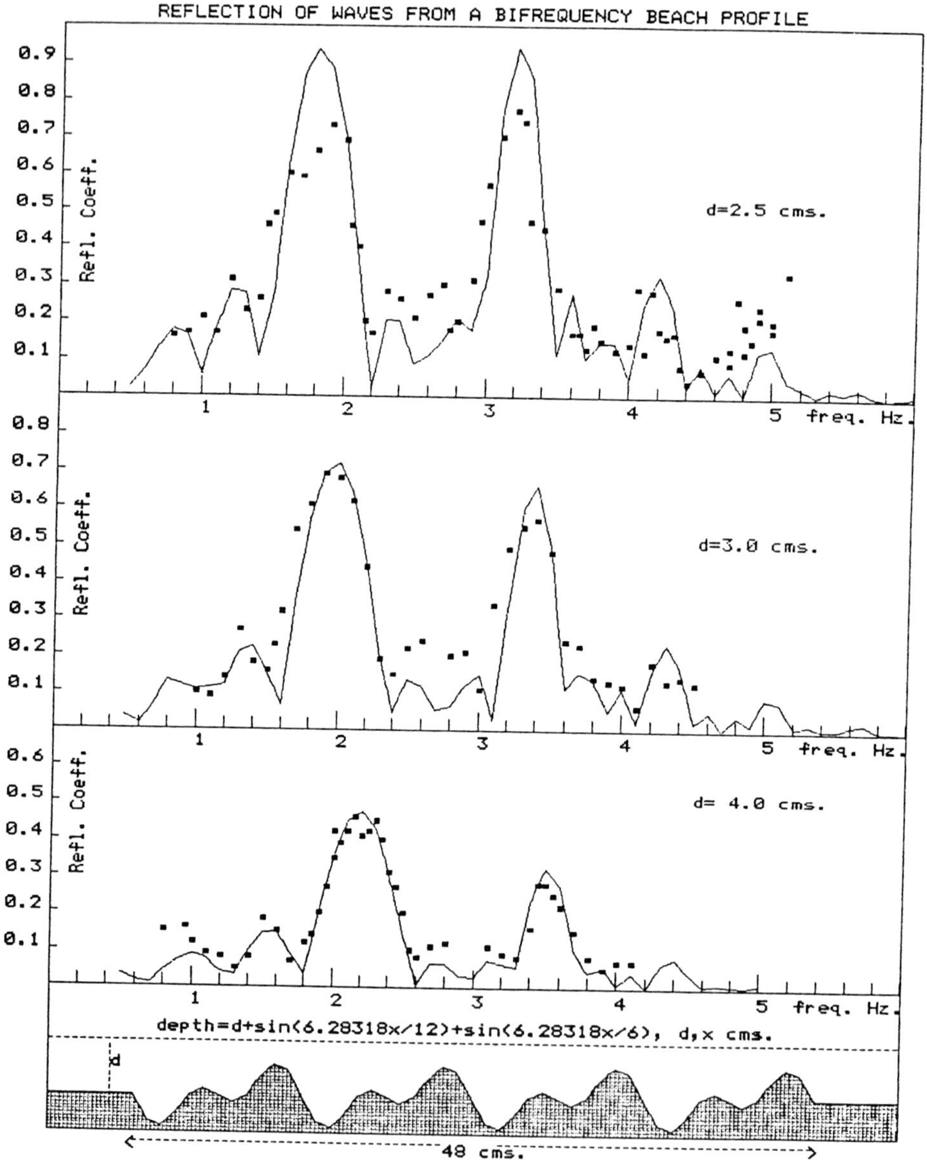

Figure 1b: Comparison of observed and calculated reflection coefficients for the bifrequency bottom profile shown.

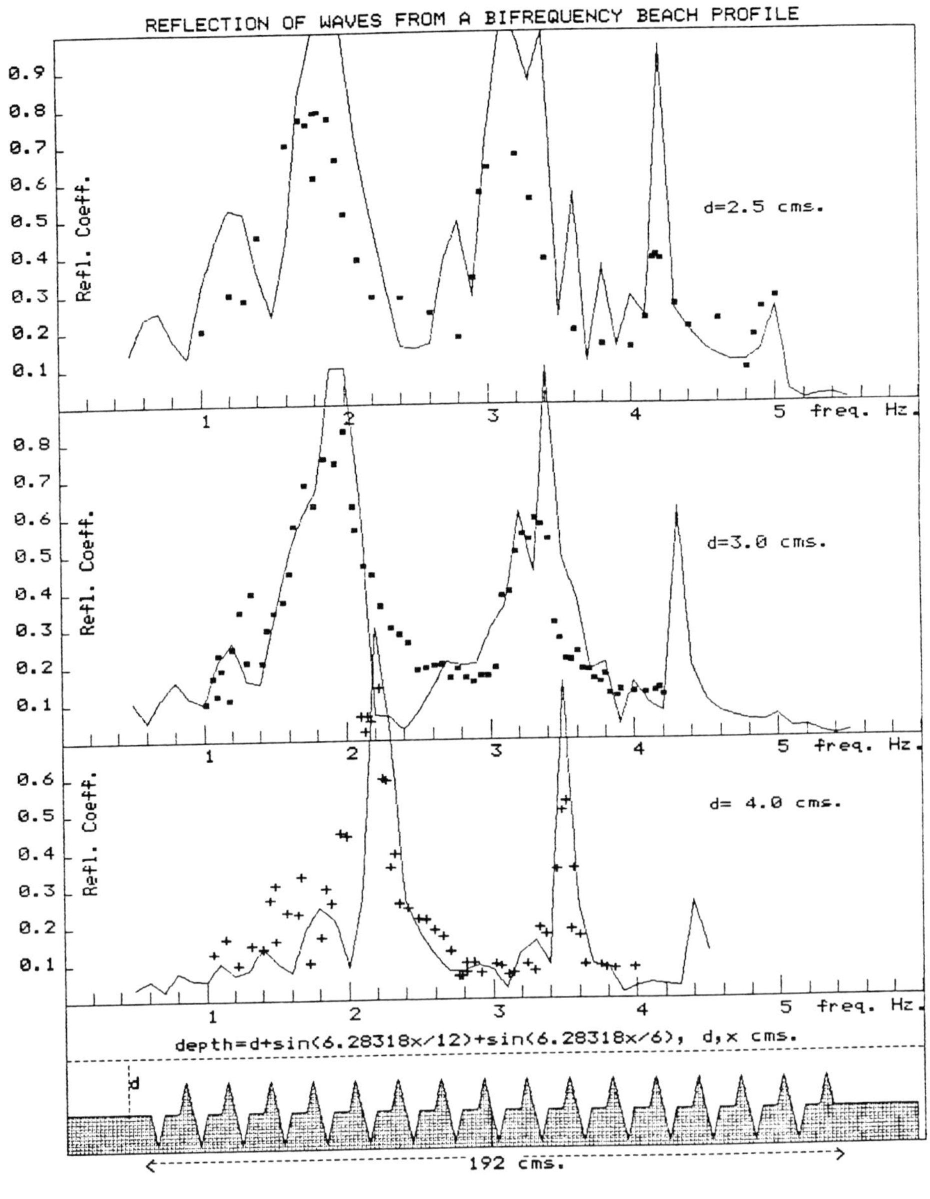

Figure 1c: Comparison of observed and calculated reflection coefficients for the bifrequency bottom profile shown.

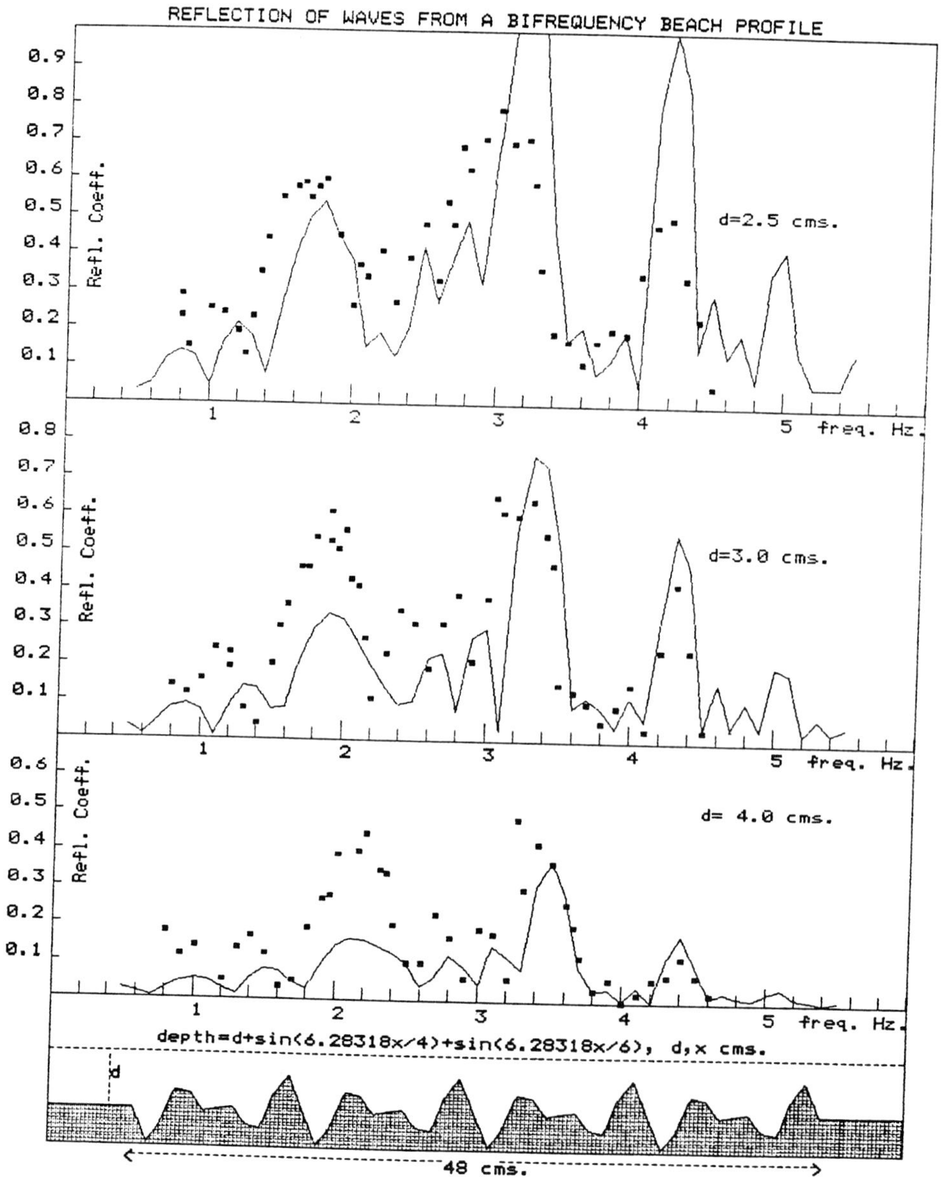

Figure 1d: Comparison of observed and calculated reflection coefficients for the bifrequency bottom profile shown.

Table 1: Gradient and gradient/(kh) for tank experiments.

h = mean depth (m), G = r.m.s. gradient
f = frequency Hz., k = wave number (m^{-1})

	h	G	f	k	G/kh	f	k	G/kh	f	k	G/kh
Fig. 1a	0.42	0.22	0.81	3.14	0.28						
	0.25	0.22	0.70	3.14	0.55						
	0.63	0.22	0.85	3.14	0.11						
Figs 1b and 1c	0.025	0.83	1.0	12.9	2.56	2.5	35.5	0.93	4.0	68.7	0.48
	0.030	0.83	1.0	11.8	2.33	2.5	33.1	0.83	4.0	66.7	0.41
	0.04	0.83	1.0	10.3	2.00	2.5	30.1	0.69	4.0	64.4	0.32
Fig. 1d	0.025	1.33	1.0	12.9	4.10	2.5	35.5	1.49	4.0	68.7	0.78
	0.030	1.33	1.0	11.8	3.76	2.5	33.1	1.34	4.0	66.7	0.67
	0.040	1.33	1.0	10.3	3.22	2.5	30.1	1.10	4.0	64.4	0.52

Table 1 shows the values of the r.m.s. gradient and $(\partial h/\partial x)/kh$ for these experiments. Those shown in Figure 1a just obey conditions (19.6) and (19.7). In the case of Figures 1b, 1c and 1d, the r.m.s. gradients satisfy condition (19.7) but in no case is the ratio less than unity by an order of magnitude. The ratio is only less than unity for all three diagrams for 1 Hz. frequency. It is also less than unity for 2 Hz. in Figures 1b and 1c. The close agreement between the calculated and observed values, particularly those in Figure 1b, suggests that condition (19.6) is too stringent.

Taken as a whole, the results are satisfactory and one can use the program with some confidence with real conditions as found on a beach.

19.4 Application to beach conditions

With waves on beaches there is an important difference compared to the tank experiments which have been discussed as the depth becomes zero at the end. A direct application of the formula over a region containing a zero depth would give an infinite answer so the method has to be altered to allow for the breaking of waves in shallow depths.

In the breaking zone, it is usually assumed that the wave height is proportional to the depth. If we take the point of measurement as origin, the x_b gives the position of the breaking point, x_0 that of the still water line and x that of any other point. The depth d between x_b and x_0 will be assumed to be:-

$$d = d_b(x_0 - x)^n / (x_0 - x_b)^n \tag{19.10}$$

where d_b is the depth at the breaking point.

n can have any real value and determines the shape of the profile between x_0 and x_b.

The wave height in this region then becomes proportional to d and so the reflection formula for this region becomes, ignoring the phase term:

$$\int_{x_b}^{x_0} (1/2c)\ (x_0 - x)^n dx / (x_0 - x_b)^n \tag{19.11}$$

$$c = \sqrt{gd} = \left(gd_b\,(x_0 - x)^n / (x_0 - x_b)^n\right)^{1/2} \tag{19.12}$$

$$dc/dx = -\ 0.5(gd_b)^{1/2}(x_0 - x)^{n/2-1} / (x_0 - x_b)^{n/2} \tag{19.13}$$

$$(1/2c)\, dc/dx = -n/4(x_0 - x) \tag{19.14}$$

Thus the integral becomes

$$-n/4 \int_{x_b}^{x_0} (x_0 - x)^{n-1} dx / (x_0 - x_b)^n = 1/4 \tag{19.15}$$

The value 1/4 is thus independent of n i.e. of the shape of the profile in the breaking zone.

If the zone covers several integration steps, then the value 1/4 has to be divided between them and multiplied by the appropriate phase factors determined in the usual way. In our case, however, the standard program with n^0 steps is taken and extended to cover n^0+1 steps, the (n^0+1)th. step being given a modulus of 0.25 and this is multiplied by the phase factor for this distance. The value x_b is taken to be the n^0 th. value and all the breaking is assumed to take place between the n^0th. and n^0+1 th. steps.

19.5 Site of wave measurements

The shore observations were carried out at Hell's Mouth Bay in the Llyn Peninsula. Figure 2. This bay faces south-west and receives the full onslaught of the winter gales. The beach has not been touched by man and all the changes that have and are taking place are entirely due to the effect of the sea on the coast. There is extensive beach erosion, the recession of the coastline averaging about 0.3 metres per year.

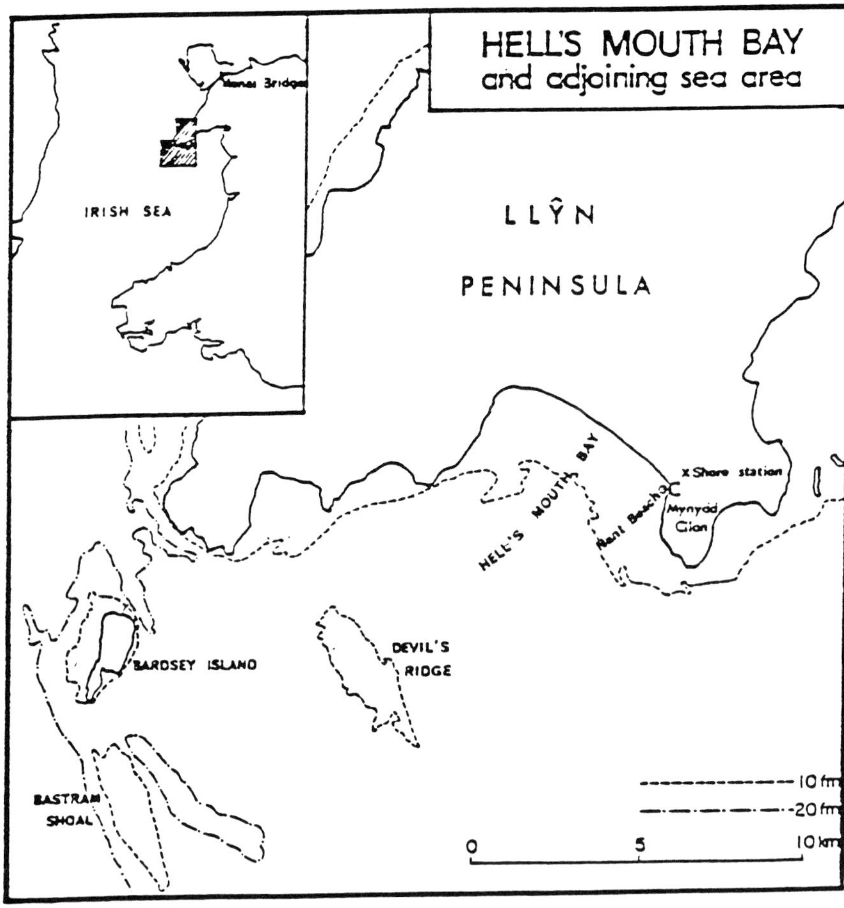

Figure 2: Map showing site of wave and beach measurements.

The name is found on the earliest maps and reflects the fact that the old sailing ships were driven there by sea and wind and were often stranded. The beach has been regularly surveyed since 1972 and for a great deal of that time, fortnightly surveys were carried out at low spring tides. For most of the period 1975-86, the waves were monitored by a wave recorder and during 1977-1981 there were three recorders which enabled the direction as well as the height and period, to be assessed. The site of the wave recorders is shown by the point C on the map but the scale is too small to show the individual recorders.

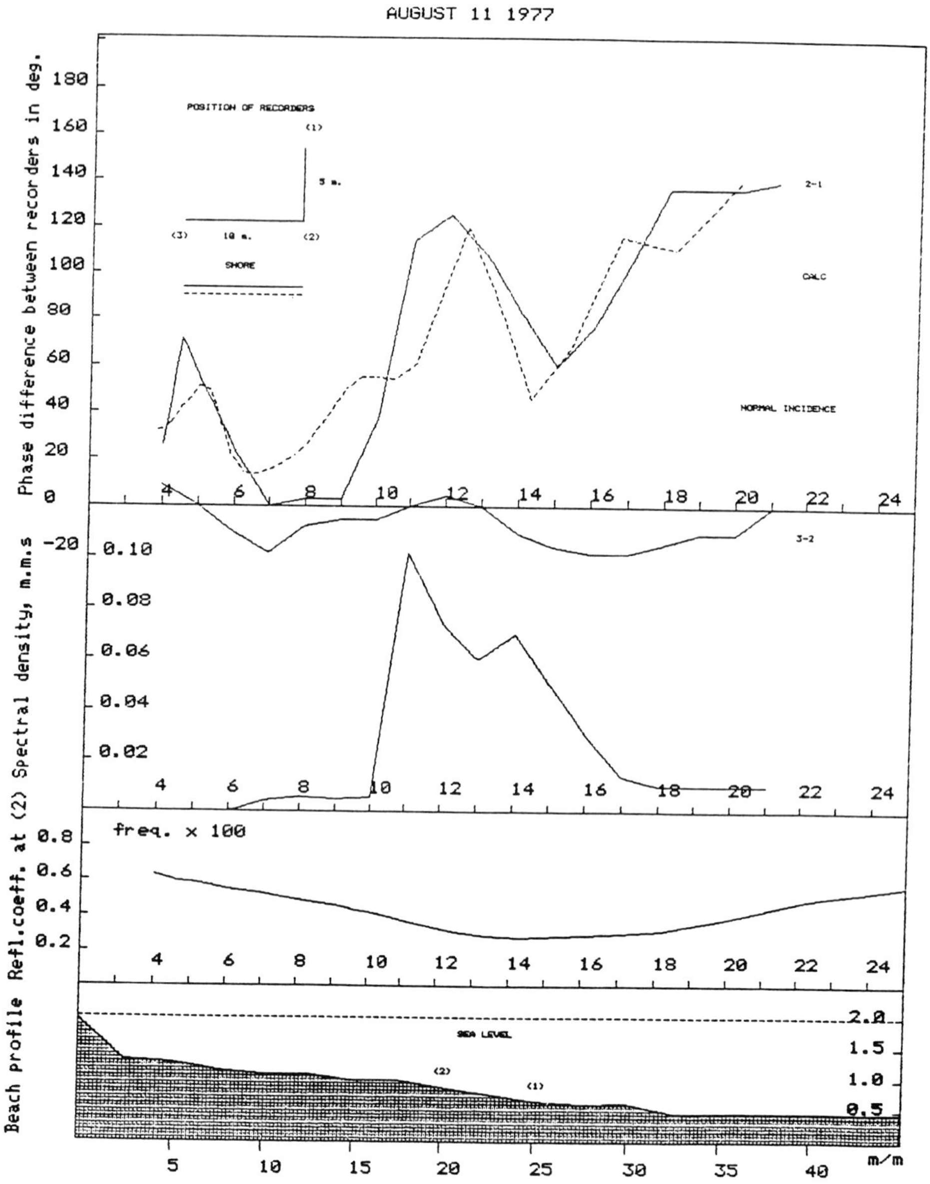

Figure 3a: Variation of observed and calculated phase differences between waves on (2) and (1), and (2) and (3) and the calculated reflection coefficients against the frequency for August 11 1977. The wave power spectrum and the beach profile are shown.

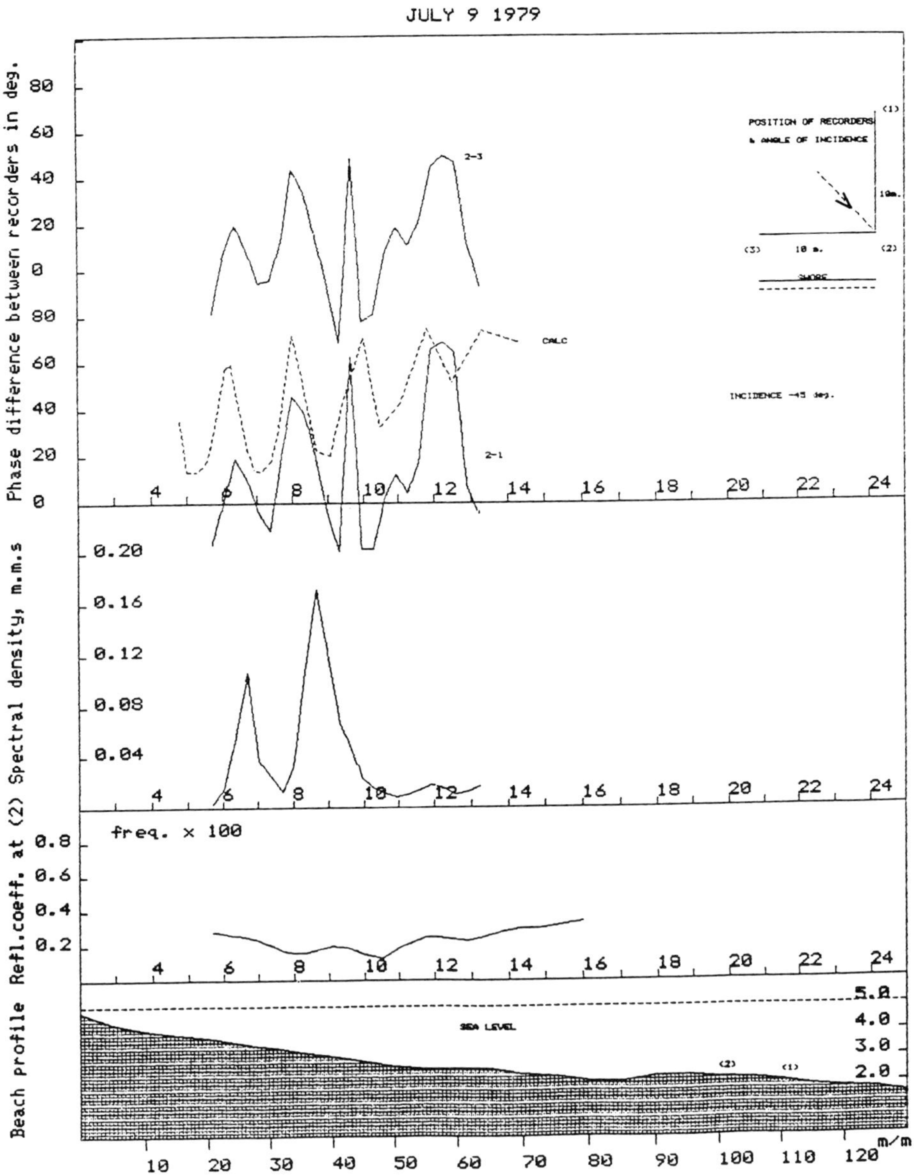

Figure 3b: Variation of observed and calculated phase differences between waves on (2) and (1), and (2) and (3) and the calculated reflection coefficients against the frequency for July 9 1979. The wave power spectrum and the beach profile are shown.

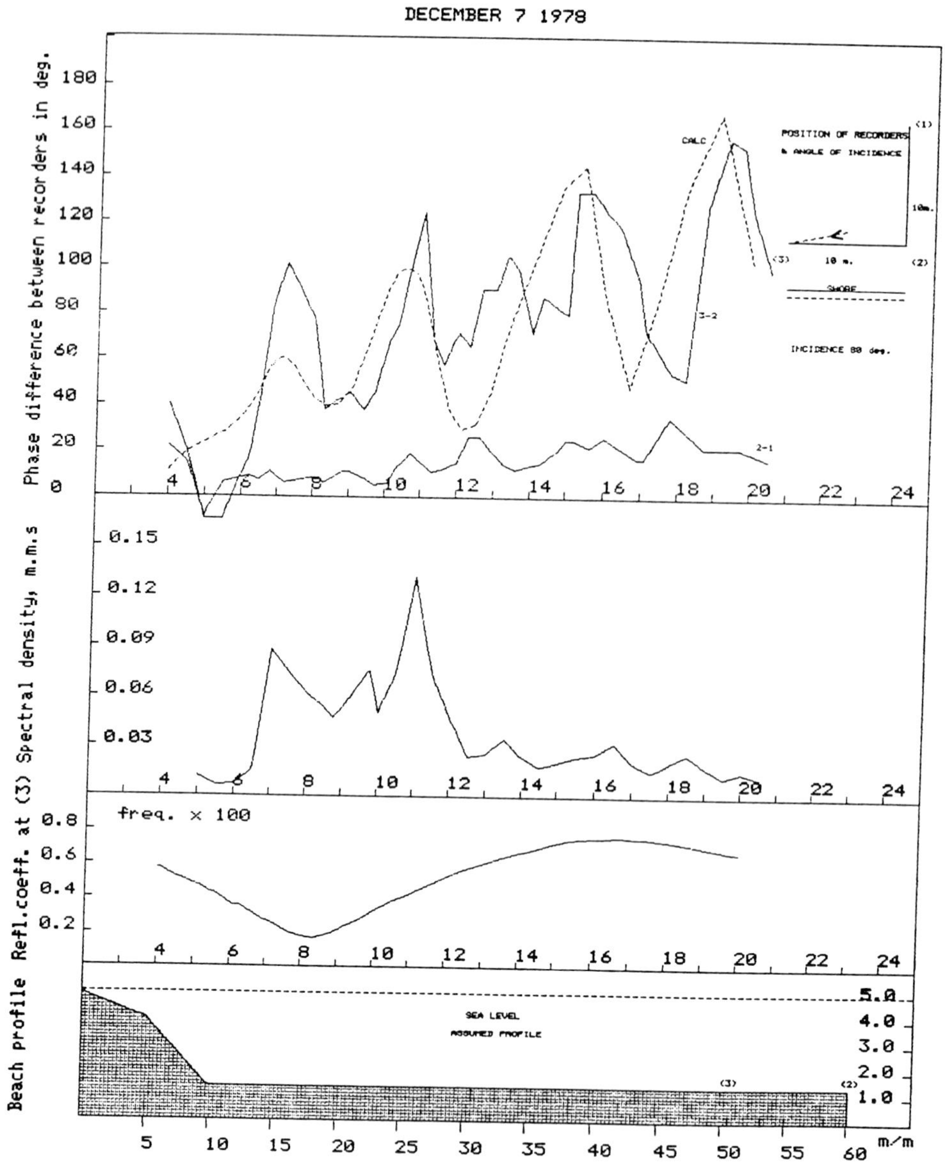

Figure 3e: Variation of observed and calculated phase differences between waves on (2) and (1), and (2) and (3) and the calculated reflection coefficients against frequency for December 7 1978. The wave power spectrum and the beach profile are shown.

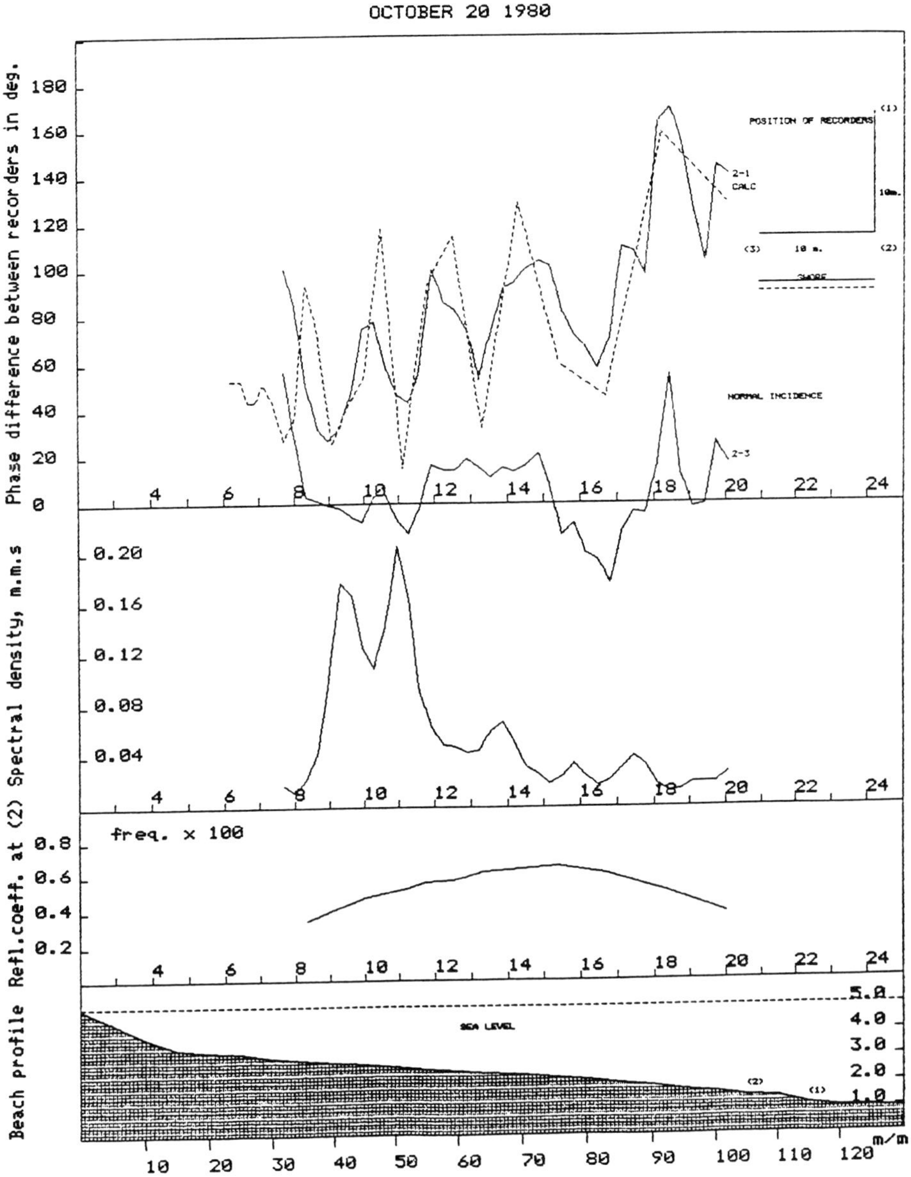

Figure 3f: Variation of observed and calculated phase differences between waves on (2) and (1), and (2) and (3) and the calculated reflection coefficients against the frequency for October 20 1980. The wave power spectrum and the beach profile are shown.

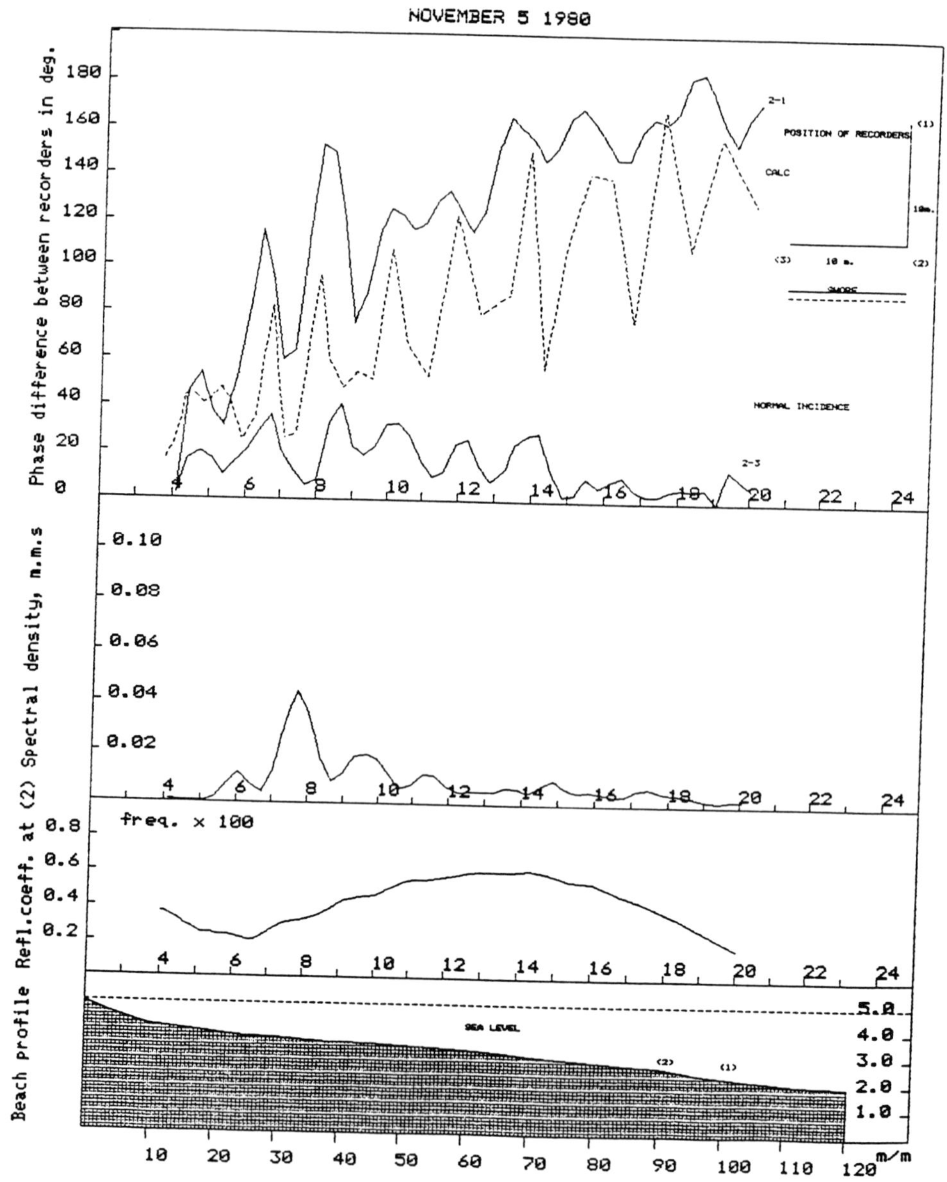

Figure 3g: Variation of observed and calculated phase differences between waves on (2) and (1), and (2) and (3) and the calculated reflection coefficients against the frequency for November 5 1980. The wave power spectrum and the beach profile are shown.

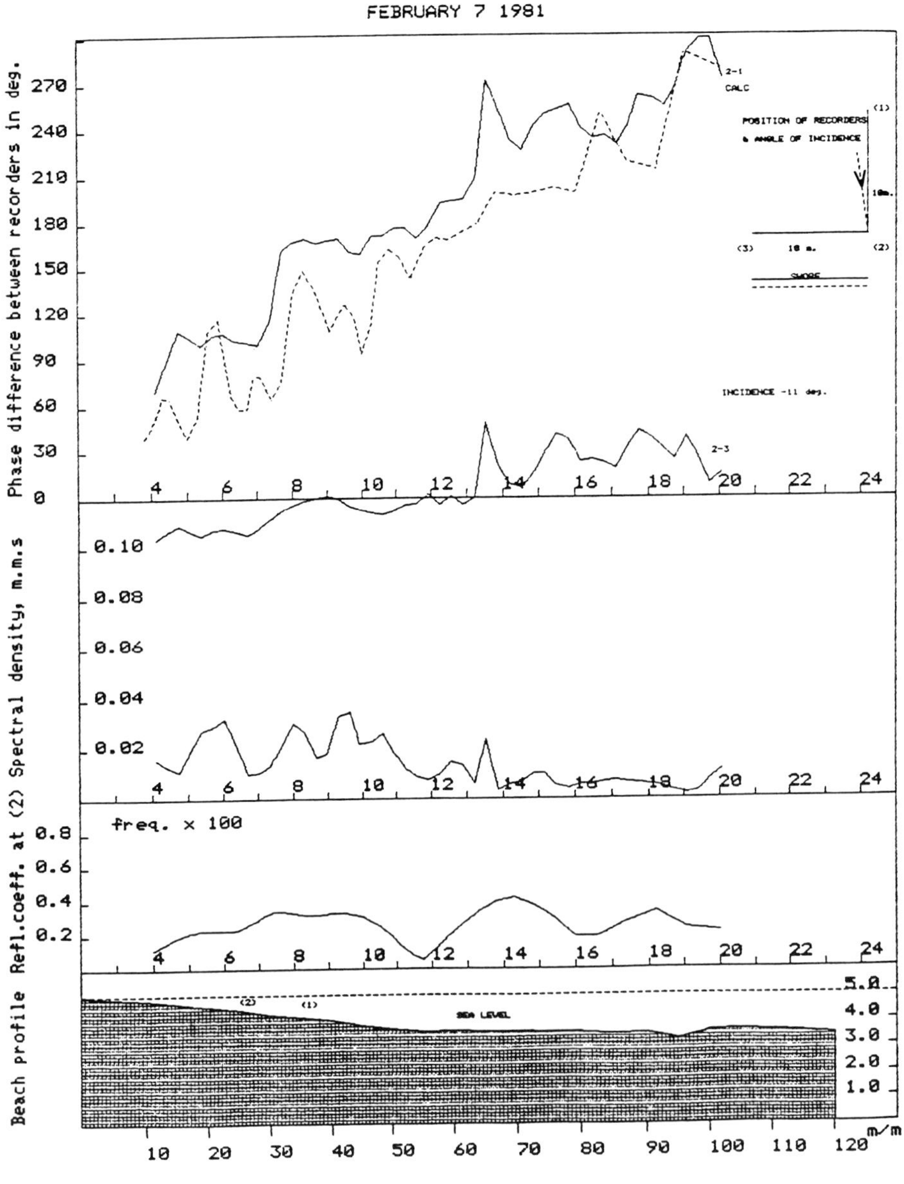

Figure 3h: Variation of observed and calculated phase differences between waves on (2) and (1), and (2) and (3) and the calculated reflection coefficients against the frequency for February 7 1981. The wave power spectrum and the beach profile are shown.

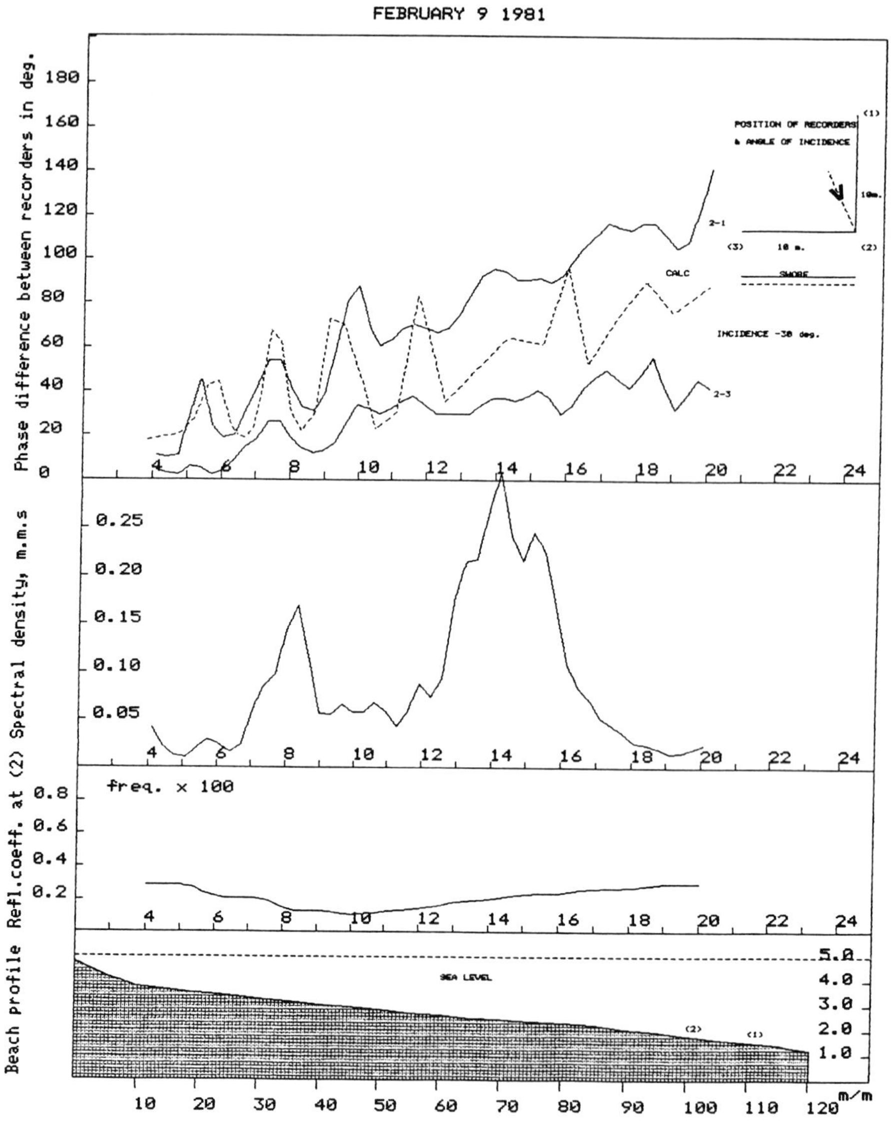

Figure 3i: Variation of observed and calculated phase differences between waves on (2) and (1), and (2) and (3) and the calculated reflection coefficients against the frequency for February 9 1981. The wave power spectrum and the beach profile are shown.

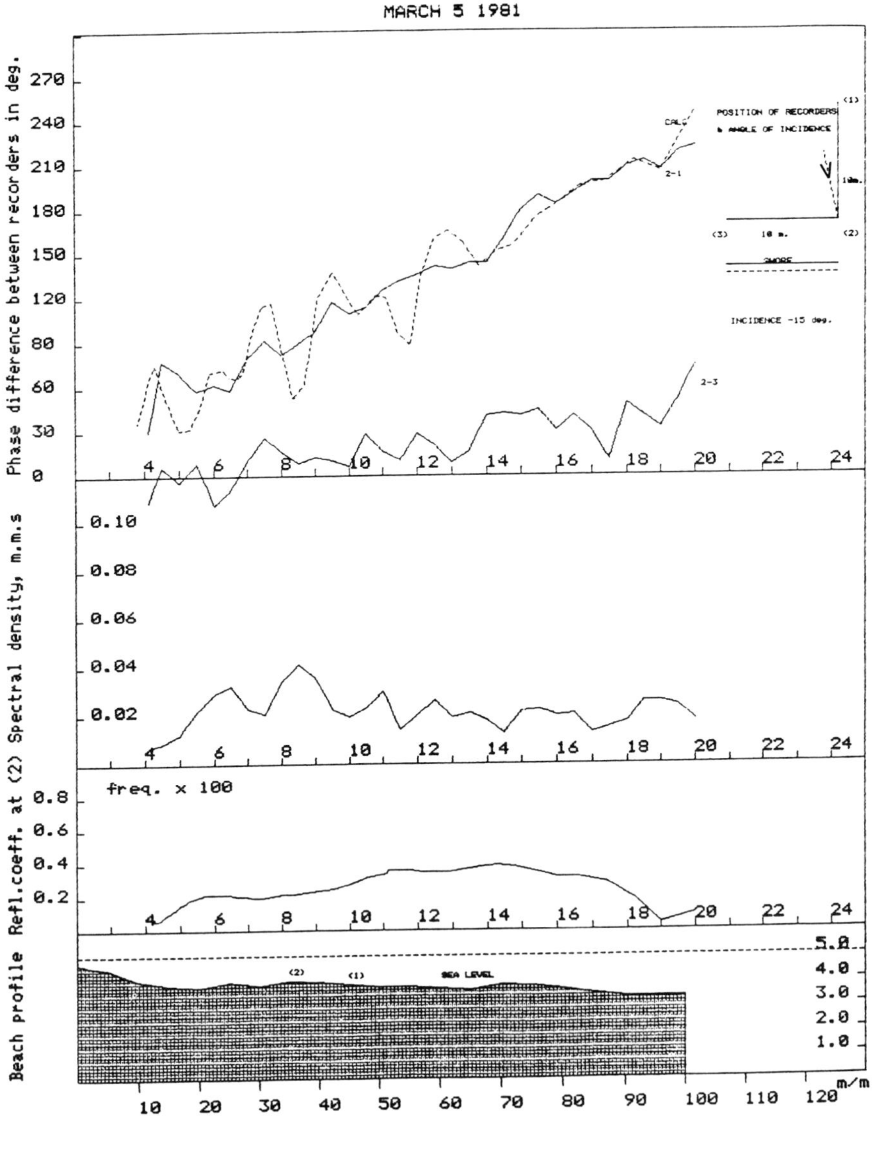

Figure 3j: Variation of observed and calculated phase differences between waves on (2) and (1), and (2) and (3) and the calculated reflection coefficients against the frequency for March 5 1981. The wave power spectrum and the beach profile are shown.

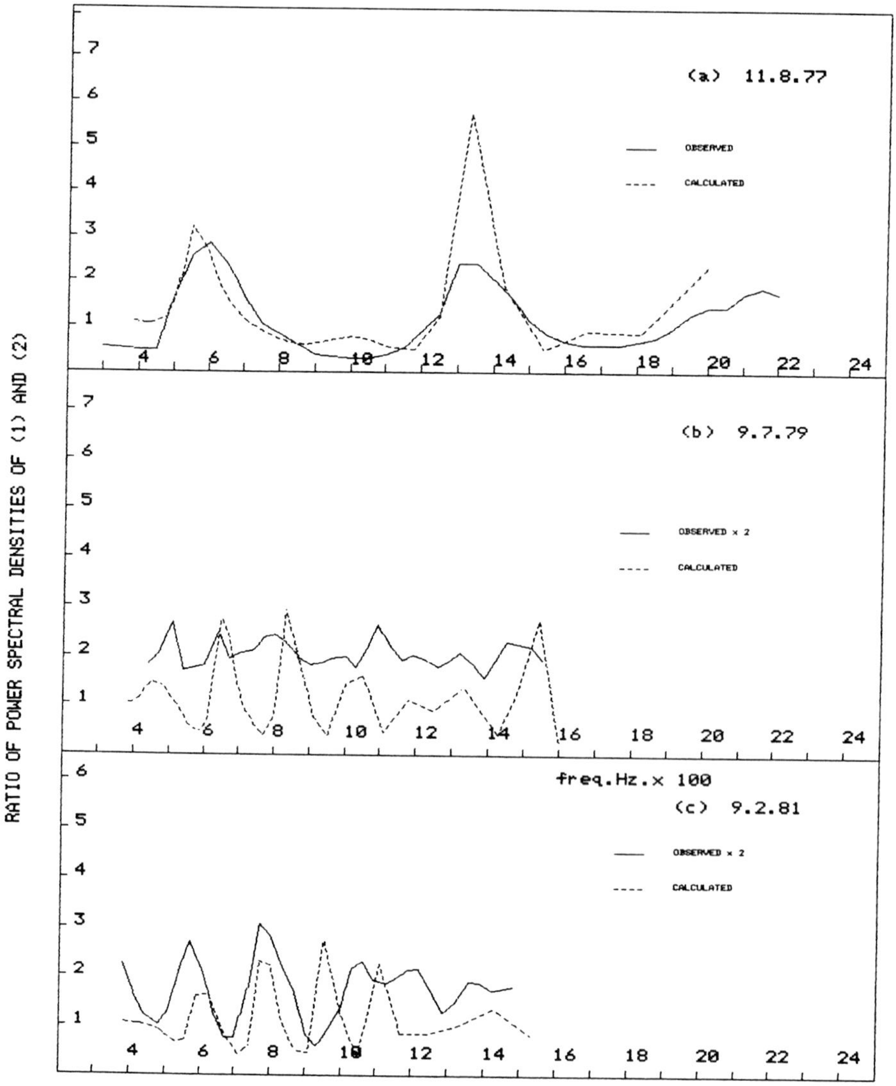

Figure 4: Comparison of observed and calculated variation of the ratio of power spectral densities of waves on (1) to those on (2), against the frequency. (a) For the same data as Figure 3a. (b) For the same data as Figure 3b. (c) For the same data as Figure 3i.

19.6 The instruments and their lay-out

The three recorders were aligned in the form of a right-angled isosceles triangle, Figures 3, so that the line joining the recorder 2 to the outermost recorder 1 was perpendicular to the beach line. The line joining recorder 2 to recorder 3 was parallel to the beach line. In all cases but one, (Figure 3a), the lengths 1-2 and 2-3 were 10 metres. The recorders were situated about 120 metres seaward of the cliff edge and were exposed at low tide. The wave recorders were pressure transducers. The responses were digitized and sent by a cable to a shore station about 200 metres beyond the cliff edge. The maintenance of the recorders was very difficult as they had to be left unattended for long periods. On one or two occasions, high seas carried the whole system up to the shore and the system had to be relaid. A more common hazard, however, was for the recorders to get buried under about 2 metres of sand. In a way this was the least of the evils as the units were then quite safe but obviously their sensitivity was affected. In analysing the results, therefore, the amplitudes were considered not to be so reliable and the work was concentrated more on the phase difference between the recorders.

The digitized output of the three recorders were fed simultaneously into the shore receiver and recorded on magnetic tape (5 hole paper tape in the case of the earlier records). The results were read by a micro-computer and then analysed by a Fast Fourier program which gave the power spectra for the three outputs for a period of 512 or 1024 seconds. The corresponding co- and quadrature spectra were also obtained between each pair of units and these were reduced to values of coherence ratio and phase difference for each frequency interval. The unit frequency interval over which the spectral estimates were calculated was 0.02 s^{-1} frequency in the case of 512 secs and half this for 1024 secs. With the method of analysis used, this corresponded to about 20 degrees of freedom, allowing a statistical range of about 3 to 1 on the 5% limits. The variation between consecutive values is nevertheless much less than would be expected for a random scatter.

Referring to Figures 3, if l_{ij} is the distance between recorders i and j and L is the local wavelength, and Θ is the angle between the wave direction and the perpendicular to the beach line then the phase difference between recorders i and j is proportional to the projection of line l_{ij} on a line parallel to the wave direction so if i=1 and j=2 then

$$\phi_{12} = 360 l_{12} \cos\theta / L \qquad (19.16)$$

and if i=2 and j=3:

$$\phi_{23} = 360 l_{23} \sin\theta / L \qquad (19.17)$$

In shallow water the velocity becomes $\sqrt{gh}$ where h is the depth. If f is the frequency then

$$L = (\sqrt{gh})/f \tag{19.18}$$

and the formulae for the phase differences become:

$$\phi_{12} = 360 1_{12} f cos\theta/\sqrt{gh} \quad \text{and} \quad \phi_{23} = 360 1_{23} f sin\theta/\sqrt{gh} \tag{19.19}$$

Thus the phase difference becomes directly proportional to the frequency. The direction can be found from the two phase differences

$$\phi_{23}/\phi_{12} = (1_{23}/1_{12})\tan\theta \tag{19.20}$$

Thus knowing the ls and the ϕs, can be found and then $\sqrt{}$gh obtained from both the phase difference formulae.

In the straightforward case where the wave is progressive, these formulae completely solve the problem for any frequency. When there is reflection, however, we do not get a straight line but patterns as shown by the full lines on Figures 3, which show a wavy trend superimposed on a linear one. In the extreme case, where there is total reflection and the wave is completely stationary, formed possibly by reflection off some eminence near the water-line, there would be nodes and antinodes set up whose position would change with the frequency. Thus at one frequency two recorders would be on the same side of a node and have zero phase difference whereas at another frequency they would be on opposite sides and have a phase difference of 180°. The phase-frequency diagram would thus show a series of jumps between 0° and 180°. In the real examples in Figures 3, there is no perfect reflection but we have a system approaching the stationary wave pattern.

In practice, however, it is easier to deal with the problem by calculating the phase difference between two recorders, for a progressive wave, using the values of $\sqrt{}$gh and Θ found from the mean trends. The observed beach profile is then adjusted to give the right depth at the recorder positions. The amplitude and phase of the reflected wave at both recorder positions are then calculated from the reflection program. The reflected waves are then combined with the straight progressive waves to give a resultant wave at each positon. The phase difference between the two should correspond to the observed phase change. As the reflection program assumes normal incidence, if it is not zero the integration interval must be divided by cos.Θ. This assumes some uniformity along the beach but this may not always be so.

19.7 Discussion of results

The results are shown in Figures 3a.....j. In each case the top part of the diagram gives the observed variations in phase difference against frequency in full lines. The predicted phase differences are given by the broken line. The second part gives the power spectral densities. The third part gives the predicted reflection coefficient plotted against the frequency. The bottom part gives the observed profile, referred to the sea level at the time of recording and the positions of recorders 1 and 2.

Figure 3a is the most striking example. This example was different from the others in that the wave recorders were installed specially only 25-30 metres from a sharp rise in the beach. This was the closest approach to a laboratory tank experiment. The results show a classical case of a wavy phase difference frequency variation. In this case l_{12} was 5 metres and l_{23} was 10 metres. The reflection coefficient varies from 25% to 60%. In all the other cases, both l_{12} and l_{23} are 10 metres. In these cases conditions were not so idealistic as in Figure 3a as the recorders had in many cases been installed some time before. The beach surveys were carried out at low spring tide so the wave observations could not be carried out at exactly the same time and there was usually a difference of one or two tidal cycles, so that there could be some changes in the beach profile. This could explain the constant phase shift found in Figures 3d and 3g. Nevertheless in most of the nine cases the results are satisfactory. In Figures 3c, 3f, 3g, the waves are moving normal to the beach. The agreement is good for Figure 3c and Figure 3f. In Figure 3g, the actual variations in the predicted values agree with those of the observed values apart from the constant phase shift. Figures 3b and 3d show an angle of approach of nearly 45° to the shore line. The variations of the predicted values for Figure 3d, like those of Figure 3g agree with those of the observed ones but again there is a constant phase shift. The agreement for Figure 3d is not unsatisfactory. Figures 3i and 3j show examples where the wave direction is about 20° off the normal. Agreement with both is satisfactory. A most remarkable case is shown in Figure 3e. Here the wave direction is at 70° to the shore line. In this case no observed profile was available but a constant depth was assumed in the direction of l_{23} and then a sudden rise about 40 metres beyond position 3, (on the other side to position 2). On this basis, some agreement can be obtained with the observed results.

In most of the cases, the reflection coefficients vary appreciably with the frequency, from 10% to 60%. The 60% value is reached in Figures 3a, 3e, 3f and 3g.

Table 2 shows the values of the mean gradient and $(\partial h/\partial x)/kh$ for Figures 3a, 3b, 3c, 3d, 3f, 3g, 3h, 3i and 3j. In all these cases conditions (6) and (7) are met.

The reflection effect should also cause a variation in the wave amplitude at the three recorders. The ratio of the spectral power densities for each frquency for the waves recorded at (1) to those at (2) are shown in Figure 4. The reflection program can calculate this ratio and the calculated results are also shown for three examples in Figure 4. These examples were chosen because at these times, the sensitivity of recorders (1) and (2) were nearly equal. The first example is for 11.8.77 where there is good agreement between the observed and calculated frequencies corresponding to maximum and minimum ratios. The other examples are not quite so good but are reasonable. The calculated ratios are higher than the observed ones. This can be attributed to some extent to the coherence between waves at (1) and (2) not being 100% but usually 75-95%.

Table 2: Gradient and gradient/(kh) for beach observations.

h = mean depth (m), G = mean gradient,
f = frequency Hz., k = wave number (m^{-1})

	h	G	f	k	G/kh	f	k	G/kh
Fig. 3a	1.0	0.047	0.083	0.17	0.28	0.17	0.34	0.14
Fig. 3b	2.5	0.025	0.083	0.11	0.09	0.17	0.22	0.045
Fig. 3c	1.5	0.017	0.083	0.14	0.08	0.17	0.28	0.040
Fig. 3d	3.0	0.033	0.083	0.10	0.11	0.17	0.21	0.054
Fig. 3f	2.5	0.033	0.083	0.11	0.12	0.17	0.22	0.060
Fig. 3g	1.5	0.025	0.083	0.14	0.12	0.17	0.28	0.059
Fig. 3h	1.5	0.017	0.083	0.14	0.08	0.17	0.28	0.040
Fig. 3i	2.5	0.033	0.083	0.11	0.12	0.17	0.22	0.060
Fig. 3j	1.0	0.015	0.083	0.17	0.09	0.17	0.34	0.045

19.8 Application to two dimensions

It is not the purpose of this paper to deal with two-dimensional cases but some suggestions can be made how to extend the process. The mild slope equation (Williams et al., 1980), when solved over an area does include the effects of reflection and indeed a one dimensional form of it has been applied to the case of Figure 1a, and given similar results. With a two dimensional case, however, its solution presents some difficulties. A progressive wave is usually input at the open end but if there is reflection, this input is incorrect and leads to errors. A plot of the phases and to a lesser extent the amplitudes across a longtitudinal section from the open to the closed end would show 'wavy' variations like those shown in Figures 3. If these were smoothed out, the results for a purely progressive wave would remain. If, furthermore, the progressive wave values found by this method for the closed end were reinput at that end, the final result would be reasonably accurate and show reflection at the open end.

Another possible method of attack is to extend the reflection program to cover the width b as well as the velocity so that bc rather than c becomes the variable parameter. A separate wave refraction program is constructed which stores the track of an initially parallel and equidistant set of rays over a series of points and giving and storing the velocity at each point as well as b the ratio of the perpendicular distance between two adjacent rays and the original separation. The reflection program can be modified to read these values and to carry out the numerical integration along the ray paths, using the appropriate values of b and c. The program also uses a similar procedure with the transmission coefficient based on Lamb's value. To a first approximation, ignoring diffraction, there can be no transfer of energy across the ray paths which can be regarded as solid boundaries. As the wavelength will usually be comparable to the other lengths involved and much greater than the depth, the problem becomes identical with that of the response of bays and inlets to long waves and the ray paths correspond to the "talweg". The main difference is that in the response of bays, there is assumed to be complete reflection giving rise to a stationary wave whereas in our case there is only partial reflection. The value of b will vary from quite large values to zero and then large negative values. The effect of very small and very large values is largely offset by breaking and diffraction and the underlying assumptions regarding the ray paths no longer hold so in the program the signs are ignored but the absolute values are kept between limits which are usually taken to be 0.5 and 2. Apart from these limitations, the procedure is exactly the same as for waves moving up a channel of variable depth and width. When wave ray channels intersect, then the wave activity is summed up over the point of intersection, allowing for amplitude, phase and direction. On this basis, it is possible to calculate the reflection coefficients at each point as well as the resultant amplitude and phase.

Such a program has actually been used with some success but at the moment there are only very few sets of simultaneous beach topographical and wave observations available to check the program.

There is an interesting consequence of this method. The reflected waves would retrace the path of the incident waves and so in theory a straight incident wave front, although greatly modified by refraction in shallow water, would still ultimately give rise to a reflected wave in the open sea travelling in an opposite direction to the incident wave, although this would be relatively small.

Such a result might provide an explanation for the formation of microseisms. According to Longuet-Higgins (1950), microseisms are caused by the interaction of waves travelling in opposite directions. This could happen either in mid-ocean where waves generated by opposing winds in a single storm or in two storms can meet or by reflection off coasts. An exhaustive survey of microseisms generated during the winter of 1987-88, Darbyshire (1990), showed that there were 20 instances of large microseisms being formed but these could all be attributed to coastal reflection and during this period the mid-ocean conditions were never met. One must therefore conclude that the formation of microseisms in mid-ocean is a rather rare event and that coastal reflection is the main cause. Most of the coasts involved, however, are very indented like the coast of Norway and it is difficult to see at first sight how there can be much reflection.

19.9 Conclusions

A method of evaluating the reflection coefficient for any wave frequency for a one-dimensional profile is developed. It has been applied successfully with bottom profiles in tanks. The method can be applied to real beaches by using a triangular arrangement of wave recorders. In the cases dealt with in this paper, some very interesting results were obtained, indicating that for some frequencies, the reflection coefficient in some cases can be as high as 60%.

The one-dimensional approach can be extended to two dimensions.

Acknowledgments

The author wishes to thank Mrs. Eileen Pritchard and the late Mr. W.R. Davis for their invaluable assistance throughout the years in obtaining the beach profiles and maintaining the pressure recorders. My thanks are also given to Drs. Belzons and Guazzelli for the use of their data.

19.10 References

Belzons, M. and E. Guazzelli, 1989: (personal communication).

Bremmer, H., 1951: The WKB approximation as the first term of a geometric optical series. Comm. Pure Appl. Math., 4, 105-115.

Darbyshire, J., 1990: Analysis of twenty microseism storms during the winter of 1987-88 and comparison with wave hindcasts. Physics of the Earth and Planetary Interiors, 63, 181-195.

Davies, A.G. and A.D. Heathershaw, 1984: Surface wave propagation over sinusoidally varying topography. Institute of Oceanographic Sciences, J. Fl. Mech., 144, 419-443.

Kajiura, K., 1961: On the partial reflection of water waves passing over a bottom of variable depth. Proc. Tsunami Meetings 10th Pacific Science Congress, I.U.G.G. Monograph, 24, 206-234.

Lamb, H., 1932: Hydrodynamics. Cambridge University Press, 738 pp.

Longuet-Higgins, M.S., 1950: A theory of the origin of microseisms. Phil. Trans. Roy. Soc. Lond., A243, 1-35.

Mei, C.C., 1982: The Applied Dynamics of Ocean Waves. Wiley Interscience, 740 pp.

Meyer, R.E., 1979: Surface-wave reflection by underwater ridges. J. Phys. Oceanogr., 9, 150-157.

Ogawa, K. and K. Yoshida, 1959: A practical method for a determination of long gravitational waves. Records Oceanographic Works Jap., 5, 38-50.

Wait, J., 1962: Electromagnetic Waves in Stratified Media. Pergamon, New York.

Williams, R.G., J. Darbyshire, and P. Holmes, 1980: Wave refraction and diffraction in a caustic region. Proc. Inst. Civ. Engrs., Part 2, 69, 635-649.

Part V

Modelling (Sediment)

20

Dispersion in tidally-averaged transport equation

Ralph T Cheng and Vincenzo Casulli

Abstract

A general governing inter-tidal transport equation for conservative solutes has been derived without invoking the weakly nonlinear approximation. The governing inter-tidal transport equation is a convection-dispersion equation in which the convective velocity is a mean Lagrangian residual current, and the inter-tidal dispersion coefficient is defined by a dispersion patch. When the weakly nonlinear condition is violated, the physical significance of the Stokes' drift, as used in tidal dynamics, becomes questionable. For nonlinear problems, analytical solutions for the mean Lagrangian residual current and for the inter-tidal dispersion coefficient do not exist, they must be determined numerically. A rectangular tidal inlet with a constriction is used in the first example. The solutions of the residual currents and the computed properties of the inter-tidal dispersion coefficient are used to illuminate the mechanisms of the inter-tidal transport processes. Then, the present formulation is tested in a geometrically complex tidal estuary -- San Francisco Bay, California. The computed inter-tidal dispersion coefficients are in the range between 5×10^4 and 5×10^6 $cm^2/sec.$, which are consistent with the values reported in the literature.

20.1 Introduction

One of the underpinning motivations for marine science research is to understand and to quantify processes that affect the balances in marine ecosystems. In tidal estuaries, bays, and shallow coastal waters, the transport of dissolved and suspended substances is determined by the principle of conservation. The governing conservation equation accounts for the balances between advection, local turbulent mixing, and the sources and sinks in a control volume. The relative importance of these transport fluxes determines the mechanisms that control the long-term consequences of the transport process. As most biochemical processes take place over a time interval of several tidal periods or longer, the basic issues concerning transport processes must address solutions of the conservation equation over an extended period of time.

Studies of conservative solute transport are fundamental; the sources and sinks are considered known or non-existing. The transport of non-conservative solutes is more complex. Although the ultimate goal is to understand the transport of both conservative and non-conservative substances over an inter-tidal time span, without the basic and thorough understanding of the inter-tidal transport of conservative solutes, solutions to the transport of non-conservative matters are at best qualitative. Thus, this research focuses on improving the present knowledge of long-term transport of conservative solutes, such as salt, in a tidal estuary. Future research will be expanded to include the transport of fine suspended sediments, nutrients, and ultimately the processes concerning non-conservative matters such as planktons or reactive chemicals.

Without considering the sources or sinks, the transport of a dilute solute in a tidal system becomes the balance between the convective flux due to tidal currents and the dispersive flux due to local turbulent mixing. Using the characteristic dimensions of San Francisco Bay, California, an order of magnitude estimate suggests that the turbulent dispersive fluxes are at least two orders of magnitude smaller than the tidal current convective flux. This assessment is probably valid for a large number of tidal estuaries, bays, and coastal seas. The results of this estimate imply that the intra-tidal transport processes are usually convection dominated. Thus, the use of a Lagrangian method in analysis of such processes is advantageous. Using a perturbation technique and assuming the tidal system is weakly nonlinear, Feng et al. (1986) have shown that the tidally-averaged transport of solute, at the first order approximation, is a pure convection process in which the convective velocity is the mass transport velocity (Longuet-Higgins, 1969) or the first order Lagrangian residual current as used by Cheng et al. (1986). In the second order theory, the Lagrangian residual currents form an ellipse in the hodograph plane (Cheng et al., 1986). These theoretical developments have pointed out some properties of the Lagrangian residual current and its role in the long-term transport processes. Similar investigations have been reported in atmospheric literature. Dunkerton (1980) gave a good review of the related research, and Middleton and Loder (1989) have reached similar conclusions in weakly nonlinear analysis for small amplitude waves. Within the limits of weakly nonlinear approximation and for shallow water tidal systems, the weakly nonlinear analyses have been extended to three-dimensional and stratified flows, and their results have provided additional insights to residual currents and long-term transport processes (Feng et al., 1990; Hamrick, 1986, 1990).

In realistic tidal embayments, the basin geometry and water depth distribution (bathymetry) are usually quite complex. The distributions of tidal current speed and direction are rectified by the basin geometry and basin bathymetry resulting in a complex structure of transient tidal flow. The nonlinearities in the flow are embedded in the complex structure of velocity shears which give rise to macro-scale mixing (tidal pumping and tidal trapping). The variations in velocity shear can be large and variable, the weakly nonlinear approximation may not be valid for part or all of the embayment. As the flows change from weakly nonlinear to nonlinear or strongly nonlinear, the practical use of the results derived from weakly nonlinear analysis becomes increasingly unsatisfactory. An extension of the properties of Lagrangian residual currents and long-term transport to general nonlinear tidal flows is highly desirable.

In this paper, a governing equation is presented for the inter-tidal transport processes without invoking the weakly nonlinear assumption. The governing equation turns out to be again a convection-dispersion equation, except the convective velocity is the mean Lagrangian residual current. The tidal-cycle averaged dispersion is shown to have resulted from nonlinear interactions of tides and tidal currents with basin geometry and from complex structure of tidal current shear. Neither the analytical solutions of the mean Lagrangian residual current nor the tidally-averaged dispersion coefficient are known to exist, they will be determined numerically. In the examples, the computed inter-tidal dispersion coefficient is shown to be small and negligible in areas where weakly nonlinear conditions prevail. In areas of complex bathymetry and complex shear structures, the computed inter-tidal dispersion coefficients are consistent with the values reported in the literature (Eg. Okubu, 1971; Fischer et al., 1979; Ridderinkoff and Zimmermann, 1990).

20.2 Tidally-averaged transport equation

20.2.1 Weakly nonlinear analysis

There are two possible pathways that can lead to solutions of inter-tidal transport of a conservative solute over a time span of several tidal periods or longer. A straightforward approach is the direct integration of the governing intra-tidal transport equation over a long period of time. The tidally-averaged solute concentration is derived from time-averaging the time dependent solution, or by applying a low-pass filter to remove tidal fluctuations. When the numerical solution for a long period of time is sought, a slight numerical error per time step of integration may render the final solution meaningless. To provide an accurate solution for long-term transport, the time integration must be carried out with sufficiently fine spatial and temporal resolutions, and the numerical integration must be both robust and efficient.

An alternative approach is to pose the question from a different view point by asking if it is possible and advantageous to derive another governing equation which gives directly the tidally-averaged solute concentration. This approach has been attempted by applying a time-averaging operator (a low-pass filter) to the intra-tidal governing equation. A form of the intra-tidal transport equation for a conservative solute is

$$\frac{\partial C}{\partial t} + \vec{u}_e \cdot \nabla C = D \nabla^2 C \tag{20.1}$$

where C(t) denotes the concentration of solute, t is time, $\vec{u}_e$ is the Eulerian tidal velocity, and D is the turbulent dispersion coefficient. Denote both the tidal current and the solute concentration as the sum of an Eulerian mean and a time dependent complementary part. A tidally-averaged Eq.(20.1) becomes

$$\frac{\partial<C>}{\partial t} + \vec{u}_{er}\cdot\nabla<C> + <\vec{u}_e'\cdot\nabla C'> = 0 \tag{20.2}$$

where < > is a tidal cycle averaging operator, $\vec{u}_{er}$ is the Eulerian residual current, and the 'primed' variables are the respective time-dependent components (Eg., Fischer et. al, 1979). In Eq.(20.2), the turbulent dispersion has been neglected when compared to convection. Strictly, Eq.(20.2) is a consequence of mathematical formality. Eq.(20.2) is not a well posed problem without resolving the closure problem associated with the correlation term $<\vec{u}_e'\cdot\nabla C'>$. A common approach is to introduce additional assumptions or ad hoc approximations relating the correlation term $<\vec{u}_e'\cdot\nabla C'>$ to the mean variable $<C>$.

For a class of weakly nonlinear tidal dynamic problems, using a small perturbation technique, Feng et al. (1986a,b), Cheng et al. (1986) reported some interesting properties of Lagrangian residual current and inter-tidal transport processes. Subsequently, numerous extensions of the weakly nonlinear analyses of tidal dynamics have been reported [For example, see Feng and his co-workers (1990) and also see Hamrick (1987, 1990)]. Severe limitations exist in using the results of weakly nonlinear analysis in realistic applications. Before extending the analysis to general nonlinear flows, some of the key results from the weakly nonlinear analyses are reviewed and summarized.

Following the dynamics of a labelled water parcel in Lagrangian mechanics, an accepted definition of the Lagrangian residual current, $\vec{u}_{lr}$, is given as

$$\vec{u}_{lr}(t_0) = \frac{\vec{X}(T+t_0) - \vec{X}(t_0)}{T} \tag{20.3}$$

where $\vec{X}(T+t_0)$ and $\vec{X}(t_0)$ are the ending and beginning positions of a labelled water parcel released at the time t_0 (referring to a tidal phase), and T is the tidal period, (Zimmerman, 1979; Cheng and Casulli, 1982). Cheng and Casulli (1982) pointed out that the Lagrangian residual current is a function of the tidal phase t_0. Under the weakly nonlinear approximation, Feng et al. (1986a) have shown that the Lagrangian residual current can be given in a small parameter expansion as

$$\vec{u}_{lr}(t_0) = \vec{u}_{er} + \vec{u}_{sd} + \varepsilon\,\vec{u}_{ld}(t_0) + O(\varepsilon^2) \tag{20.4}$$

in which $\vec{u}_{sd}$ is the Stokes' drift, ε is a small parameter on the order of the ratio of residual current over tidal current. The time dependency of the Lagrangian residual current appears in the Lagrangian drift, $\vec{u}_{ld}(t_0)$, [see Cheng et al (1986)]. The Lagrangian drift inscribes an ellipse on the hodograph plane in a tidal cycle. Within the weakly nonlinear analysis, Feng et al. (1986b) continued to show that the first order tidally-averaged transport equation for a conservative solute is a pure convection equation,

$$\frac{\partial <C>}{\partial t} + (\vec{u}_{er} + \vec{u}_{sd})\cdot\nabla <C> = 0 \quad , \tag{20.5}$$

where $(\vec{u}_{er} + \vec{u}_{sd})$ is the first order Lagrangian residual current, Eq.(20.4). Note that $(\vec{u}_{er} + \vec{u}_{sd}$ is the same as the mass transport velocity given in Longuet-Higgins (1969). Similar conclusions have been reported by Middleton and Loder (1989). It is interesting to point out further that a closure relation for Eq.(20.2) has been obtained without invoking additional ad hoc assumptions. The closure condition is, of course,

$$<\vec{u}_e{}'\cdot\nabla C'> = \vec{u}_{sd}\cdot\nabla <C> \quad , \tag{20.6}$$

in which the correlation term has been shown to be a convective transport with the Stokes' drift as the convective velocity. These results are instructive from an analytical point of view. Solutions of Eq.(20.5) are not necessarily more accurate than an ad hoc solution of Eq.(20.2) in practical applications because the weakly nonlinear condition is a fairly stringent requirement for most realistic embayments. Eq.(20.5) correctly states that the tidally-averaged transport is principally convection due to a mass transport velocity (or the first order Lagrangian residual current), all other mechanisms are of secondary importance. When the weakly nonlinear condition breaks down, the claims of Eq.(20.5) become invalid.

Confusions may arise in tidal dynamics of horizontal flows regarding the definition of the Stokes' drift because the Stoke's drift is a concept "borrowed" from surface waves. To eliminate this confusion in this and future tidal dynamic analyses, the Stokes' drift is defined to be the tidally-averaged difference between the Lagrangian tidal velocity and the Eulerian tidal velocity truncated at the first order of Taylor's series expansion, (Longuet-Higgins, 1969). The Lagrangian and Eulerian tidal velocity difference at (x_0, y_0) can be approximated by a Taylor's series of $\vec{u}_e(t)$ and truncated at the first order as

$$\Delta\vec{u} = \vec{u}_1(t, t_0) - \vec{u}_e(t) = \left[\int_{t_0}^{t} \vec{u}_e dt'\right]\cdot\nabla\vec{u}_e + \text{Higher Order Terms,} \tag{20.7}$$

where $\vec{u}_1(t, t_0)$ and $\vec{u}_e(t)$ are the Lagrangian and Eulerian tidal velocities, respectively. Specifically, the Stokes' drift is defined as

$$\vec{u}_{sd} = <\Delta\vec{u}> = \frac{1}{T}\left\{\int_{t_0}^{t+T}\left[\int_{t_0}^{t} \vec{u}_e dt'\right]\cdot\nabla\vec{u}_e dt\right\} \quad , \tag{20.8}$$

with all variables defined at (x_0, y_0). Following this definition, the sum of the Eulerian residual current and the Stokes' drift equals the mass-transport velocity, or the first order Lagrangian residual current.

In fact Eq.(20.7) can be used to set the limits for the weakly nonlinear approximation to be considered valid. The expression in the [ ] in Eq.(20.7) is the displacement of a water parcel at time t released from t_0. If at any time in a tidal cycle the value of the [ ] is of the same order of magnitude as the characteristic length of the velocity gradient, then the truncated Taylor series in Eq.(20.7) is insufficient to approximate the difference between the Eulerian and Lagrangian tidal velocities (Zimmerman, 1979). When the weakly nonlinear approximation becomes invalid, it further implies that the physical significances of the Stokes' drift (as defined above) and the mass transport velocity are also questionable, (Zimmerman, 1979).

20.2.2 Generalized Lagrangian dynamics

Without invoking the weakly nonlinear approximation, a tidally-averaged transport equation can be derived by following the Lagrangian particle dynamics. First, consider the physical process of solute transport in a tidal embayment for which the local turbulent mixing can be assumed to be small and neglected. Under the stated conditions, the solute is transported by tidal current convection alone. The macroscopic mixing is generated by tidal velocity shears which are produced from spatial variations in tidal current amplitude and phase resulting from topographical rectification of tidal currents in tidal basins of complex geometry and variable bathymetry. These phenomena of macroscopic mixing are commonly known as tidal pumping and tidal trapping.

The mathematical definition of the tidally-averaged solute concentration at a fixed position, $\langle C(x_0, y_0)\rangle$, is equivalent to measuring solute (salinity, for example) at a fixed station and averaging the results over a tidal cycle. Thus, the tidally-averaged concentration of an Eulerian variable is the averaged values of solute concentration sampled from water parcels that travelled through the sensor at (x_0, y_0), in sequence, over a period of a tidal cycle. Specifically, the solute concentrations recorded by the sensor are NOT determined from the same water mass. When the local turbulent mixing is neglected, the time-averaged solute concentration at (x_0, y_0) would be the same as an ensemble average of solute concentrations taken from a subset of water parcels that move through (x_0, y_0) in a tidal cycle. A common property that defines the water parcel subset is that each water parcel in the subset must move through (x_0, y_0) at least once in a tidal period. Once this subset of water parcels can be determined, an ensemble-averaged value of solute concentration can be computed from the solute concentrations sampled from these water parcels.

In order to compute an ensemble-averaged concentration, the question is shifted to the identification and the determination of this subset of water parcels. When a labelled water parcel is released in a tidal system from (x_0, y_0), the Lagrangian residual current is defined at the end of a tidal period. In repeating tides, by continuously releasing labelled water parcels over a tidal cycle, the ending locations of these water parcels form a closed locus enclosing a small area in space. An inverse question can be posed: What is the distribution of the starting positions of a subset of water parcels satisfying the requirement that each labelled water parcel must move through (x_0, y_0) at least once in a tidal period? Assuming that the starting positions of this subset of water parcels can be defined, they enclose a small region which will be referred to as a dispersion patch. By definition then, an equivalence of an ensemble-averaged solute concentration and the time-averaged concentration at (x_0, y_0) can be established. To help visualize this complex situation, consider again a weakly nonlinear tidal system. The dispersion patch in this case can be shown to be a mirror image of the Lagrangian residual ellipse (Cheng et al., 1986) because the governing equations are Hamiltonian. In the general nonlinear case, a dispersion patch exists but it is not necessarily a mirror image of the region occupied by the Lagrangian residual currents.

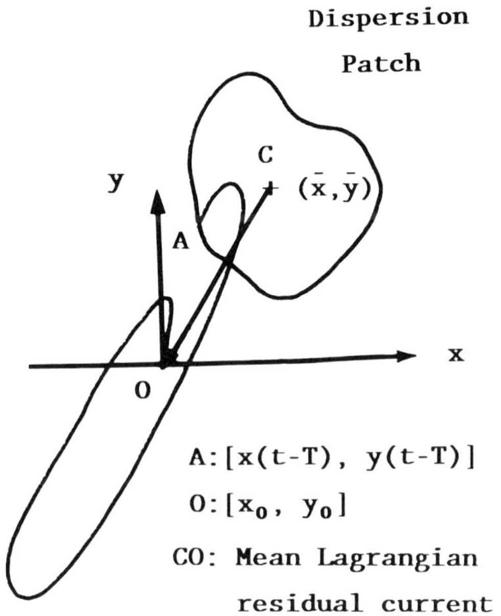

Figure 1: A schematic diagram showing the dispersion patch, mean Lagrangian residual current. Point A is a general point on the edge of the dispersion patch.

Depicted in Figure 1 is a schematic illustration of a dispersion patch. When a labelled water parcel is released at time (t-T) from a general point A, [x(t-T),y(t-T)], on the edge of the dispersion patch, this water parcel will arrive at point (x_0, y_0) at time t where T is the tidal period. For repeating tides, but not necessarily weakly nonlinear systems, the water parcels defining the dispersion patch form a closed locus in space. Thus, the governing tidally-averaged transport equation is a mathematical expression stating the equivalence between the tidally-averaged solute concentration at (x_0, y_0) and an ensemble-averaged concentration on the edge of the dispersion patch. Mathematically, the time-averaged C at (x_0, y_0) over a tidal period is

$$<C(x_0,y_0,t_0)> = \frac{1}{T}\int_{t_0-T/2}^{t_0+T/2} C(x_0,y_0,t)dt \tag{20.9}$$

On the other hand, when the turbulent dispersion is neglected, Eq.(20.1) implies that the solute concentration is constant along each streakline. In particular, $C(x_0,y_0,t) = C[x(t-T),y(t-T),t-T]$, where $[x(t-T), y(t-T)]$ is a general point on the edge of the dispersion patch. Thus, Eq.(20.9) can be written as

$$<C(x_0,y_0,t_0)> = \frac{1}{T}\int_{t_0-T/2}^{t_0+T/2} C[x(t-T),y(t-T),t-T]dt \quad , \tag{20.10}$$

where the right hand-side of Eq.(20.10) is an ensemble average. Because of the periodic tide, the integral on the right hand side of Eq.(20.10) can be written in terms of tidally-averaged concentrations. Thus, the equivalence of a time-averaged and an ensemble-averaged solute concentration can be established as

$$<C(x_0,y_0,t_0)> = \frac{1}{T}\int_{t_0-T/2}^{t_0+T/2} <C[x(t),y(t),t_0-T]>dt \tag{20.11}$$

Since the dispersion patch is defined by $[x(t), y(t)]$, the time integration on the right hand side of Eq.(20.11) is an average of the solute concentrations taken from the water mass on the edge of the dispersion patch. Denote the centroid of the dispersion patch by $(\bar{x},\bar{y})$, Figure 1, so that $\bar{x}$ and $\bar{y}$ are defined by

$$\bar{x} = \frac{1}{T}\int_{t_0-T/2}^{t_0+T/2} x(t)dt \quad \text{, and} \quad \bar{y} = \frac{1}{T}\int_{t_0-T/2}^{t_0+T/2} y(t)dt \ . \tag{20.12}$$

By subtracting $<C(\bar{x},\bar{y},t_0-T)>$ from either side of Eq.(20.11) and dividing Eq.(20.11) by T, Eq.(20.11) becomes

$$\frac{<C(x_0,y_0,t_0)> - <C(\bar{x},\bar{y},t_0-T)>}{T} = \frac{\frac{1}{T}\int_{t_0-T/2}^{t_0+T/2} <C[x(t),y(t),t_0-T)]>dt - <C[\bar{x},\bar{y},t_0-T]>}{T} \tag{20.13}$$

The truncated Taylor series expansion of $<C(x_0,y_0,t_0)>$ and $<C[x(t),y(t),t_0-T]>$ about $(\bar{x},\bar{y},t_0-T)$ can be written as

$$<C(x_0,y_0,t_0)> = <C(\bar{x},\bar{y},t_0-T)> + T\frac{\partial<C>}{\partial t} + (x_0-\bar{x})\frac{\partial<C>}{\partial x} + (y_0-\bar{y})\frac{\partial<C>}{\partial y} + \text{HOT} \tag{20.14}$$

$$<C[x(t),y(t),t_0-T]> = <C(\bar{x},\bar{y},t_0-T)> + [x(t)-\bar{x}]\frac{\partial <C>}{\partial x} + [y(t)-\bar{y}]\frac{\partial <C>}{\partial y}$$
$$+\frac{1}{2}[x(t)-\bar{x}]^2\frac{\partial^2 <C>}{\partial x^2} + [x(t)-\bar{x}][y(t)-\bar{y}]\frac{\partial^2 <C>}{\partial x \partial y} + \frac{1}{2}[y(t)-\bar{y}]^2\frac{\partial^2 <C>}{\partial y^2} + \mathrm{HOT} \tag{20.15}$$

where all the derivatives are evaluated at $(\bar{x},\bar{y},t_0\text{-}T)$ and HOT stands for higher order terms. Substitution of these Taylor series expansions into Eq.(20.13), after neglecting the higher order terms, yields the following tidally-averaged transport equation,

$$\frac{\partial <C>}{\partial t} + <\vec{u}_{lr}>\cdot\nabla<C> = \nu_{xx}\frac{\partial^2 <C>}{\partial x^2} + 2\nu_{xy}\frac{\partial^2 <C>}{\partial x \partial y} + \nu_{yy}\frac{\partial^2 <C>}{\partial y^2}\,. \tag{20.16}$$

In Eq.(20.16),

$$<\vec{u}_{lr}> = \frac{1}{T}\begin{pmatrix} x_0-\bar{x} \\ y_0-\bar{y} \end{pmatrix}$$

is the mean Lagrangian residual current following the definition given in Eq.(20.3). The dispersion coefficients are given by

$$\nu_{xx} = \frac{1}{2T^2}\int_{t_0-T/2}^{t_0+T/2}[x(t)-\bar{x}]^2\,dt \tag{20.17a}$$

$$\nu_{xy} = \frac{1}{2T^2}\int_{t_0-T/2}^{t_0+T/2}[x(t)-\bar{x}][y(t)-\bar{y}]\,dt \tag{20.17b}$$

$$\nu_{yy} = \frac{1}{2T^2}\int_{t_0-T/2}^{t_0+T/2}[y(t)-\bar{y}]^2\,dt \tag{20.17c}$$

The tidally-averaged transport equation, Eq.(20.16), is valid up to the limit of the Taylor series expansions used in Eqs.(20.14) and (20.15). The exact condition for the convergence of the Taylor series is difficult to establish, however, the characteristic length scale for the dispersion patch is expected to be an order of magnitude smaller than the length scales of tidal velocity gradients used in the definition of the Stokes' drift, Eq.(20.7). Thus, the validity of this equation has been extended to general nonlinear flows without the limitation of weakly nonlinear approximation. In strongly nonlinear flows, the dispersion patches can be highly contorted. The Taylor series expansion given in Eqs.(20.14) and (20.15) may not converge, and the validity of Eq.(20.16) becomes questionable.

Nevertheless, the present results, Eq.(20.16), reveal some important properties of tidally-averaged transport and mixing processes. The governing transport equation remains in a form of convection-dispersion equation. In the transport equation, the mean Lagrangian residual current is the convective velocity. The tidally-averaged dispersion is related to the spatial spread of the Lagrangian residual currents in a tidal cycle. These results are plausible and not at all surprising from a physical point of view. In regions where tidal current distributions are complex and highly variable, the residual dispersion patches are expected to occupy bigger and irregular areas than in regions of smooth or weak velocity shears. Intensified mixing and high values of tidally-averaged dispersion can be anticipated in regions where intense tidal velocity shears are expected. In the limiting case of a weakly nonlinear approximation, the dispersion patch becomes negligibly small, and the mean Lagrangian residual current is reduced to the first order Lagrangian residual current or the mass transport velocity. The area of the dispersion patch is proportional to ε^2 which is small and negligible when $\varepsilon \sim o(1)$. Obviously, in this limiting case, Eq.(20.16) is reduced to Eq.(20.5), a result that was reported earlier by Feng et al. (1986b).

20.3 A model for the tidally-averaged dispersion coefficient

Strictly, Eq.(20.13) or (20.16) are equivalent forms of an inter-tidal transport equation, either one can be used for solution of <C>. Eq.(20.13) is a discrete form of the transport equation; the ensemble average of solute concentrations representing dispersion can be evaluated in the course of numerical integration. In highly nonlinear cases, the local mixing approximation may not apply, the use of Eq.(20.13) may be preferred. In contrast, when the governing inter-tidal transport equation is expressed in Eq.(20.16), properties of Eq.(20.16) can be related to the conventional formulation of convection and dispersion.

Although the shapes of the dispersion patch are generally irregular, at this order of the approximation the dispersion tensor is a symmetric second rank tensor, Eqs.(20.17). To better understand the physical processes described by the inter-tidal dispersion, a simplified isotropic dispersion model is suggested to ascertain the present approach gives at least a correct order of magnitude of the inter-tidal dispersion coefficient. General tensorial properties of the inter-tidal dispersion coefficients will be further pursued if the simple dispersion model proves to be useful in practical applications. A simplified model is proposed by re-writing Eq.(20.16) as

$$\frac{\partial <C>}{\partial t} + <\vec{u}_{lr}> \cdot \nabla <C> = \nu_t \nabla^2 <C> , \tag{20.18}$$

where $<\vec{u}_{lr}>$ is the mean Lagrangian residual current, υ_t is a bulk inter-tidal dispersion coefficient. Here, υ_t is defined to be the mean distance between the edge of centroid of the dispersion patch squared divided by a tidal period. In the following examples and in practical applications, both the mean Lagrangian residual current and the inter-tidal dispersion coefficient will be evaluated by means of a numerical hydrodynamic model.

20.4 Computations of the tidally-averaged dispersion coefficient

Assuming that a robust numerical hydrodynamic model is available, the transient tides and tidal currents can be adequately described by the model [for example the models described in Casulli (1990) or in Cheng and Casulli (1991)]. Specifically, the present interest is in using the numerical hydrodynamic model to determine the mean Lagrangian residual current and the dispersion patch. In order to define the dispersion patch, the governing equations must be integrated in a reversed time sequence. Because the shallow water equations are nonlinear, the tidal dynamics is not a reversible process. An integration of the governing nonlinear equations with a negative increment of time does not produce the time-history of tidal dynamics in a reversed time sequence. Nonetheless, in a discrete numerical domain, computations to generate a "reversed time history" of the tidal dynamics are possible.

The numerical solutions of the shallow water equations are solved by integrating the governing equations forward in time until the solution has reached a dynamic steady state. The computed tidal velocity distributions are saved at every integration time step by writing the numerical solution to a time-history file for a complete tidal period. If the numerical model must be integrated with a relatively small Δt due to reasons of numerical stability, then the numerical solutions can be saved at appropriate multiple time intervals. The sequence of records in this file contains the complete time-history of the tidal circulation in the embayment. A 'reversed time-history file' is created by switching the orders of the sequential records from the 'forward time-history file' to a reversed order without altering the contents of the records. The first record in the reversed time-history file is the tidal current distribution saved at the last time step. The information in the 'reversed time-history file' is used to perform the reversed time integration of the Lagrangian movements of water parcels. It is important to note that in the reversed time-integration the nonlinearities of the tidal characteristics are preserved. When a labelled water parcel is released from a fixed location at t_0, by integrating for one tidal cycle along the streakline backward in time, the origin of the labelled water parcel can be determined. During the reversed time integration of the Lagrangian water parcel movements, the local tidal velocities at previous times are found by interpolation from the numerical solutions saved in the reversed time-history file. If the starting position for an arbitrary water parcel can be determined, then the determination of a dispersion patch is simply a systematic release of labelled water parcels in a reversed time sequence over an entire tidal period. Because the tide is assumed to be repeating and cyclic, the starting locations of this subset of labelled water parcels form a closed locus that defines the dispersion patch. Finally, the values of the inter-tidal dispersion coefficient are evaluated by computing the mean distance between the edge of centroid of the dispersion patch squared and divided by the tidal period for every point.

20.5 Examples

Results from two examples are given to illustrate the properties and the spatial distribution of tidally-averaged dispersion coefficient. In both cases, the numerical solutions of the shallow water equations are solved by a semi-implicit, Eulerian-Lagrangian, finite-difference method (Casulli, 1990). The details of the *T*idal, *R*esidual, and *I*nter-tidal *M*ud-flat model (TRIM) are reported in Cheng and Casulli (1991), and they are not repeated here.

20.5.1. Tidal inlet with a constriction

The first example is designed to gain additional insights. This problem represents an idealized tidal inlet whose properties including the transient tide-induced eddies, tidal and residual flows have been investigated by Imasato (1983). The test basin used in this investigation is similar to the test problem of Imasato (1983) consisting of a rectangular basin (55 km x 46 km) with a constriction that divides the inner and outer basins. The inner and outer basins are connected through an opening 8 km in width (Figure 2a). The water depth is assumed to be 10 m throughout the basin, and a uniform M_2 tide of 1.0 m amplitude is specified across the open boundary. The dimensions of this idealized basin are chosen to represent a typical coastal tidal embayment.

The TRIM model is used in the computations of tidal and residual currents with 0.5 km x 0.5 km finite-difference square meshes. The numerical method used in the TRIM model, under linearized analysis, has been shown to be unconditionally stable (Casulli, 1990). The numerical integration is carried out with Δt = 15 min. without any sign of instability (the grid Courant number is on the order of 20). Thirteen drogues are followed for five days, the starting positions of the drogues are shown in Figure 2a. The Lagrangian trajectories of these drogues, Figure 2b, show a complicated pattern in the headland regions. The drogue trajectories exhibit regular patterns away from the center of the basin. The complex trajectory pattern is also an indication that the spreading of the Lagrangian residual current can be quite large. This pattern shows a high degree of nonlinear interactions between the tide and the current in regions of high velocity shear. In this example, the spatial variations of the velocity field (tidal current shears) are induced by the headland constriction. A typical tidal current distribution is depicted in Figure 2c showing a pair of tidal vortices located near the inlet constriction (headland) in the inner basin during flood. A pair of vortices are developed near the headland in the outer basin during ebb, Figure 2d. A slight antisymmetry can be observed in the flow field, which is caused by the Coriolis acceleration. In regions where the vortex quadripoles developed, the tidal velocity shears are higher than the rest of the region. The sizes of the velocity eddies are of the same order of magnitude as the characteristic length of the basin. Therefore, the weakly nonlinear approximation is not valid for the headland region covered by the vortex quadripoles. Figure 2e and 2f are the computed Eulerian and

Lagrangian residual current distributions.

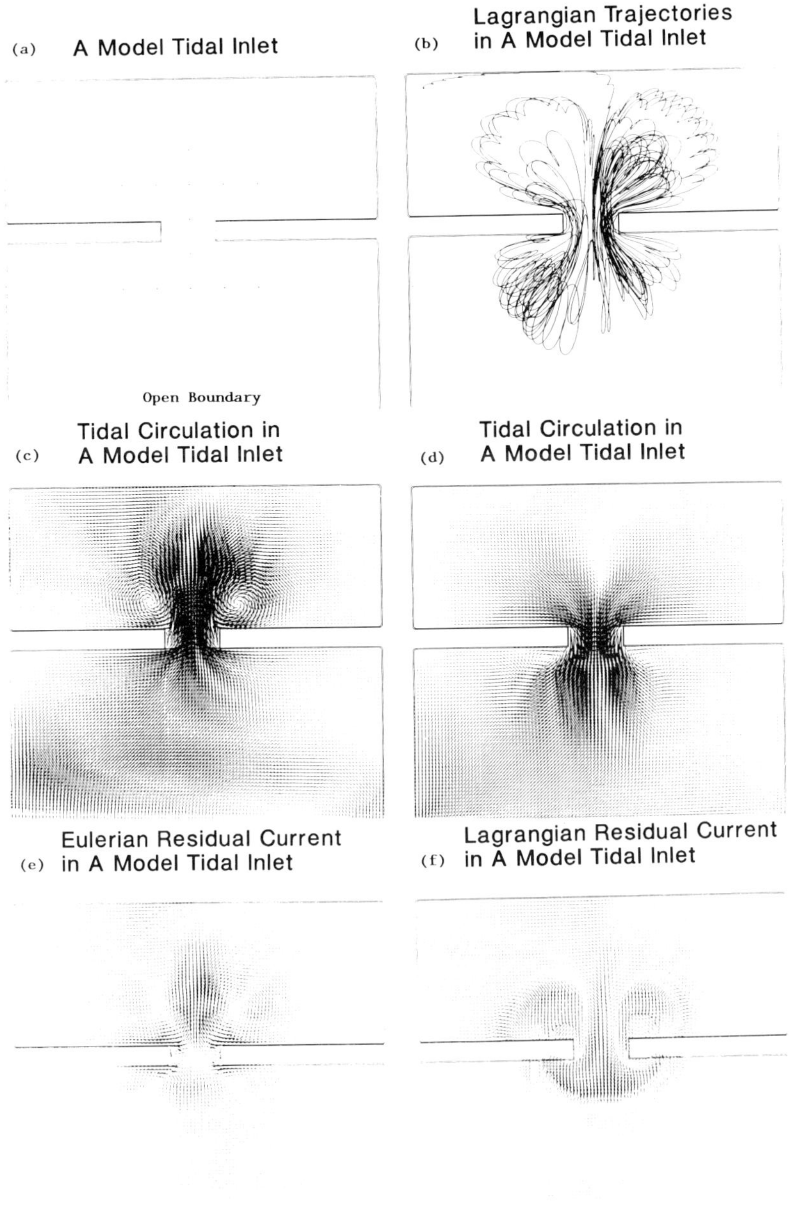

Figure 2: a) A model tidal inlet, x's are the starting positions of drogues. b) The trajectories of thirteen drogues covering a period of five days. c) and d) are the computed tidal current distributions during flood and ebb, respectively. The tidal vortices on the two sides of the constriction have developed. e) and f) The computed Eulerian and Lagrangian residual current distributions, respectively.

Following the procedures described above, the distribution of the tidally-averaged dispersion coefficient, υ_t, is computed and depicted in Figure 3. The computed υ_t ranges between 5×10^4 and 5×10^6 cm^2/sec. The region in which the values of υ_t is greater than 1×10^5 cm^2/s nearly coincides with the residual quadripoles. The mean Lagrangian residual current pattern is similar to the Lagrangian residual current distribution at a given time. The sizes and shapes of the dispersion patches are measures of the variance of the Lagrangian residual currents. In regions near the constriction headland, the dispersion patches are large and irregular and the nonlinear interactions are significant. In this region, the weakly nonlinear approximation is violated, and the physical meaning of the Eulerian residual current is questionable, Imasato (1983) . Likewise, the usage of the Stokes' drift and the mass transport velocity become questionable in regions of strong nonlinearity (Zimmerman, 1979). Away from the constriction, the nonlinearity is expected to be weak. The computed dispersion patches show neatly closed loci. In some cases the dispersion patches can be approximated by ellipses consistent with the results of weakly nonlinear analysis.

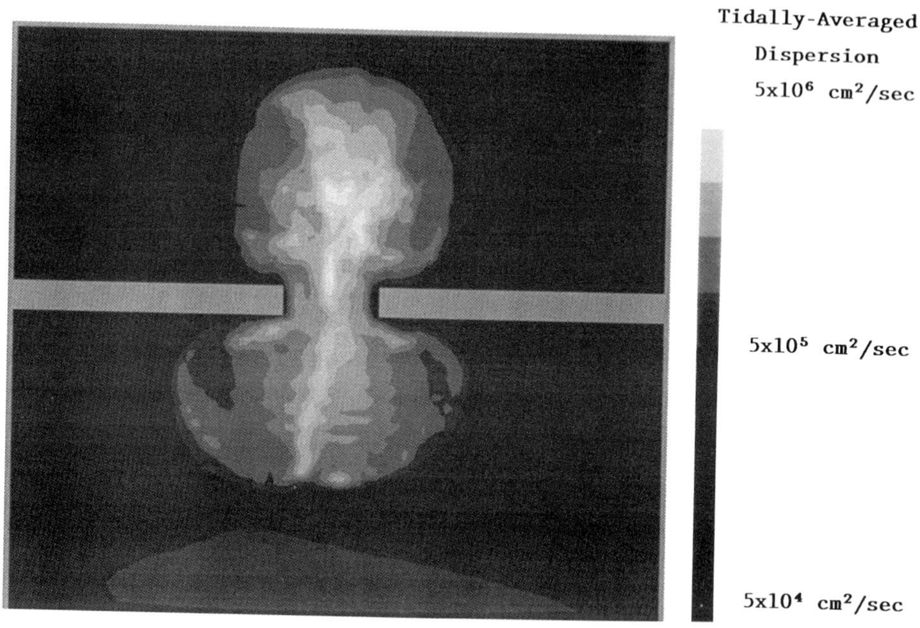

Figure 3: The distribution of tidally-averaged dispersion coefficient in a rectangular tidal inlet with a constriction.

20.5.2 San Francisco Bay, California

To evaluate the practical value of the proposed model, this approach is tested in San Francisco Bay, California. San Francisco Bay is a complex estuary consisting of interconnected embayments, sloughs, marshes and channels, Figure 4. Among the many factors that affect the flow properties in the Bay, the water depth distribution is one of the most important factors controlling the spatial variations of both the magnitude and the direction of tidal currents (Cheng and Gartner, 1985; Walters et at., 1985). The tides entering the Bay through a narrow opening at the Golden Gate have a range of roughly 2 m. composed primarily of the M_2 and K_1 tides. Within Central Bay the tides bifurcate and propagate concurrently into South Bay and the northern reach which include San Pablo Bay, Carquinez Strait, and Suisun Bay. Central Bay is not only geometrically complex, the variations in the depth distribution are also quite large. The deepest region in Central Bay is located near Golden Gate, where a depth close to 100 m. is found. At the eastern boundary of Central Bay is a broad region of inter-tidal mud-flat. The geometry of Central Bay is further complicated by the presence of several islands. Without further elaborations, Central Bay is simply a very complex tidal basin in which large variability of tidal current can be anticipated.

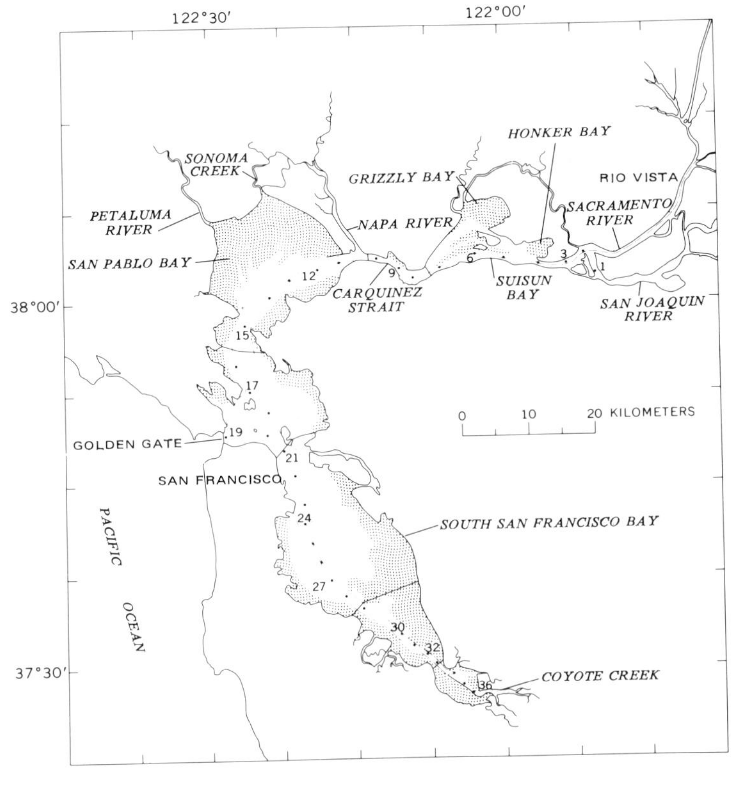

Figure 4: San Francisco Bay, a bathymetrically complex tidal estuary.

Both San Pablo and South Bays can be characterized by a deep channel which is surrounded by broad shoals. Tidal current distributions in these basins are less complex. High intensity tidal currents are distributed within deep channels, lower magnitude tidal velocities are observed over shoal regions due to increased friction. The tidal velocity shears in regions of shoals and in the middle of deep channels are relatively weak. However, in the transition zones between the channel and the shoals, the tidal velocity shears become large, and the nonlinearities are relatively intense. The computed Stokes' drift, using Eq.(20.8), shows an irregular pattern and a high degree of uncertainty in the transition zone irrespective of the model grid resolution. This observation further ascertains that the weakly nonlinear approximation is invalid in the transition zones. Carquinez Strait is a deep channel consisting of relatively simple geometry that connects San Pablo Bay to its west and Suisun Bay to its east. Although the intensities of the tidal currents are high, the spatial variability of the tidal currents is relatively low. Therefore, the weakly nonlinear approximation might be marginally valid in Carquinez Strait. Suisun Bay is a shallow and complex basin, it consists of braided channels and shoals. The complex bathymetry structure suggests that the characteristic lengths of the bathymetric features are relatively short. The flows in Suisun Bay is expected to be highly nonlinear. Additional discussion of the tides and tidal currents in San Francisco Bay is referred to Walters et al. (1985).

The inter-tidal dispersion coefficients in San Francisco Bay are computed using the same numerical hydrodynamic model, Cheng and Casulli (1991). The Bay is discretized by a 500 m x 500 m rectangular finite-difference mesh giving approximately 18,000 total grid points. Of all the model grid points, about 30% are active. The computing time required by the TRIM model is roughly proportional to the active computational points. For the purpose of computing the tidally-averaged dispersion coefficients, only an M_2 tide is imposed at the open boundaries, except the tidal amplitude is increased to represent realistic tidal range values. The numerical solutions are achieved using an integration time step Δt = 15 min. In this application, the horizontal tidal velocity shears are, in a large part, the direct consequence of the spatial variations of the basin bathymetry. The computed tidal current distribution in the Bay is quite complex. Cheng and Gartner (1985) have shown that the amplitudes of tidal current in San Francisco Bay are linearly proportional to the local water depth. It is not surprising to identify a fairly strong correlation between the spatial variabilities of the water depth and the tidal currents. Without discussing the complex structures of the Eulerian and Lagrangian residual currents and the Stokes' drift, the computed values of the inter-tidal dispersion coefficient is depicted in Figure 5. The values of the dispersion coefficients are in the range of 3.0×10^4 to 3.0×10^6 cm^2/sec which are consistent with the patterns and the values reported in the literature (e.g. Ridderinhkof and Zimmerman, 1990).

The inter-tidal dispersion is resulted from the time dependency of the Lagrangian residual currents, whose distribution is principally a function of the tidal current shears. While the tidal velocity shears are strongly dependent upon the spatial variations of the basin depth distribution, the values of inter-tidal dispersion coefficients are also closely correlated to the variability of the basin bathymetry. High values of inter-tidal dispersion coefficient are found in regions of high tidal velocity shears. Or high values of the inter-tidal dispersion coefficients can be anticipated in regions with complex geometry and large water depth variations. Central Bay is just such a region, the dispersion coefficient in Central Bay is on the order of ~10^6 cm^2/sec. The velocity shears in the middle of deep channels in both South and San Pablo Bays, and in Carquinez Strait are not necessarily high, the computed values of the dispersion coefficient are small to moderate (~ $5x10^5$ cm^2/sec). Similarly, over the regions of broad shoals where the weakly nonlinear approximation may be valid, the computed dispersion coefficients are less than 10^5 cm^2/sec. However, at the edge of the deep channels, the dispersion coefficients become locally large because the tidal velocity shear is high in the transition zones, Figure 5. The computed dispersion coefficients in the western Suisun Bay are greater than 10^6 cm^2/sec. The complex basin geometry induces high velocity shear, and thus high tidal dispersion for that part of Suisun Bay

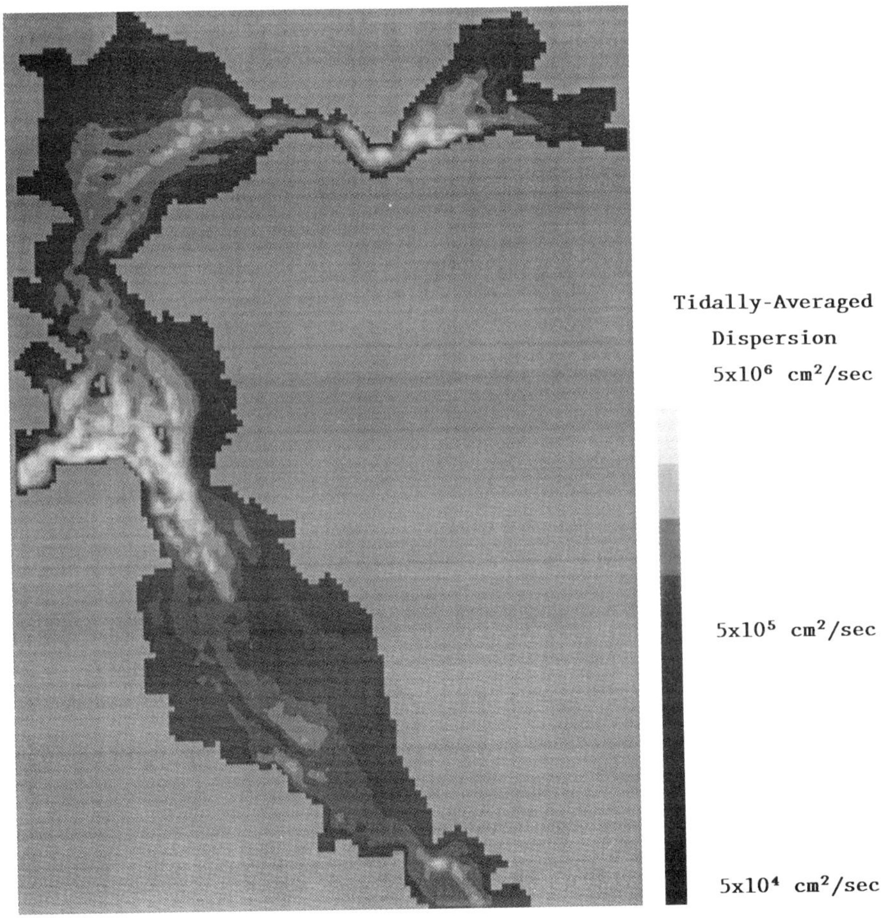

Figure 5: The computed tidally-averaged dispersion coefficient distribution in San Francisco Bay.

20.6 Discussion and Conclusion

Of the two pathways to solutions of long-term transport problems, the direct integration of the intra-tidal conservation equation is applicable only when an accurate, robust, and efficient numerical method of solution is available. In practical applications, this approach gives only marginally satisfactory answers. Alternatively, the solution of tidally-averaged solute concentration can be attempted using an inter-tidal transport equation, if such a solution can be shown to give practical answers. Earlier analysis under a weakly nonlinear approximation has shed some interesting light in this approach, the resulting inter-tidal transport equation gives a lucid explanation of the inter-tidal transport processes.

In realistic applications, the weakly nonlinear analysis is often a poor approximation in a tidal estuary. Because the Lagrangian residual current is a function of time (tidal phase), a residual dispersion patch can be introduced. Without invoking a weakly nonlinear approximation, the dispersion patch is defined by a subset of water parcels that move through a fixed point (x_0, y_0) at least once every tidal cycle. An ensemble-averaged solute concentration is defined as the averaged solute concentration sampled from water masses on the edge of the dispersion patch. By examining the intra-tidal Lagrangian movements of the water parcels, an inter-tidal transport equation has been established by the equivalence of time-averaged and ensemble averaged solute concentrations. The transport of a dilute conservative solute over inter-tidal time spans is again governed by the combination of convection and dispersion. The convection is due to a mean Lagrangian residual current, and the dispersion is represented by an ensemble-average along the edge of the dispersion patch.

To gain further insights, a simplified isotropic dispersion model is proposed and tested in two numerical examples. Through numerical solutions of the shallow water equations, the distributions of inter-tidal dispersion coefficients are computed. The distribution of the inter-tidal dispersion can be reasonably explained by relating the properties of dispersion to the distributions of tidal velocity shear and nonlinearity in the basin. In regions where the nonlinearity is high, the values of the computed inter-tidal dispersion coefficient are also high. In areas where weakly nonlinear approximation is valid, numerical solutions reproduce results of the analytical solutions. The present formulation is further tested in a complex realistic tidal basin -- San Francisco Bay. The distribution of the computed inter-tidal dispersion coefficient is quite complex. It is encouraging that the properties of the inter-tidal dispersion are well correlated with the anticipated distributions of nonlinearity in the Bay. The computed values of the dispersion coefficient are consistent with values reported in the literature.

The preliminary results obtained in this study of long-term, inter-tidal transport of conservative solute are quite encouraging. Future investigations should consider the anisotropic dispersion properties. The inter-tidal transport equation will be solved using the computed mean Lagrangian residual current to cover a seasonal time-span. The numerical solutions will be compared with solutions of the intra-tidal transport equation and with field data.

20.7 References

Casulli, V., 1990, Semi-implicit finite difference methods for the two-dimensional shallow water equations, J. Comp. Phys., Vol.86, No.1, p.56-74.

Cheng, R. T., and Casulli, V., 1991, A Tidal, Residual, Inter-tidal Mud-flat (TRIM) model using semi-implicit, Eulerian-Lagrangian Method, to appear.

Cheng, R. T. and V. Casulli, 1982, On Lagrangian residual currents with applications in South San Francisco Bay, California, Water Resour. Res., Vol. 18, p. 1652-1662.

Cheng, R. T. and J.W. Gartner, 1985, Harmonic analysis of tides and tidal currents in South San Francisco Bay, California, Estuarine, Coastal and Shelf Sci., Vol. 21, p. 57-74.

Cheng, R. T., S. Feng, and P. Xi, 1986, On Lagrangian residual ellipse, in Physics of Shallow Estuaries and Bays, Ed. van de Kreeke, Lecture Notes on Coastal and Estuarine Studies, Vol. 16, p. 102-113., Springer-Verlag.

Dunkerton, T., 1980, A Lagrangian mean theory of wave, mean-flow interaction with applications to nonacceleration and its breakdown, Reviews of Geophysics and Space Physics, Vol. 18, No.2, P. 387-400.

Feng, S., R. T. Cheng, and P. Xi, 1986a, On tide-induced residual current and residual transport, 1. Lagrangian residual current, Water Resour Research, Vol. 22., No. 12, p. 1623-1634.

Feng, S., R. T. Cheng, and P. Xi, 1986b, On tide-induced residual current and residual transport, 2. Residual transport with application in South San Francisco Bay, California, Water Resour Research, Vol. 22., No. 12, p.1635-1646.

Feng, S., R.T. Cheng, Sun Wenxin, Xi Pangen and Song Lina, 1990, Lagrangian residual current and long-term transport processes in a weakly nonlinear baroclinic system, in Physics of Shallow Seas, Eds., Wang Huatong et at., China Ocean Press, p. 1-20.

Fischer, H.B., E.J. List, R.C.Y. Koh, and N. H. Brooks, 1979, Mixing in inland and coastal waters, Academic Press.

Hamrick, J.M., 1986, Long-term dispersion in unsteady skewed free surface flow, Estuarine, Coastal and Shelf Science, Vol. 23, p. 807-845.

Hamrick, J.M., 1990, The dynamics of long-term mass transport in estuaries, in Residual Currents and Long-Term Transport, Ed., Ralph T. Cheng, Spring-Verlag, p. 17-33.

Imasato, N., 1983, What is tide-induced residual current?, J. of Phys. Ocean., Vol.13, p. 1307-1317.

Longuet-Higgins, M. S., 1969, On the transport of mass by time-varying ocean currents, Deep Sea Research, Vol.16, p. 431-447.

Middleton, J. F., and J. W. Loder, 1989, Skew fluxes in polarized wave fields, J. of Physical Oceano., Vol. 19, No.1, p. 68-76.

Okubu, A., 1971, Oceanic diffusion diagram, Deep-Sea Research, Vol.18, p.789-802.

Ridderinkhof, H. and Zimmerman, J.T.F., 1990, Mixing processes in a numerical model of the western Dutch Wadden Sea, in Residual Currents and Long-Term Transport, Ed., Ralph T. Cheng, Spring-Verlag, p. 194-209.

Walters, R.A., R.T. Cheng, T.J. Conomos, 1985, Time scales of circulation and mixing processes of San Francisco Bay waters, Hydrobiologia, Vol. 129, p. 13-36.

Zimmerman, J.T.F., 1979, On the Euler-Lagrangian transformation and the Stokes drift in the presence of oscillatory and residual currents, Deep Sea Research, 26A, p.505-520.

21

Effect of bends on dilution rates

R Smith

Abstract

Centrifugal effects as the flow goes around bends contribute to secondary flow in the estuary cross section. For particles which are not neutrally buoyant the vertical sampling of the secondary flow is non-uniform: buoyant material tends to be carried to the outside of the bend and dense material to the inside of the curve. This paper focuses attention on the sensitivity of the horizontal dilution rates to the secondary flow and to the vertical rise (or sinking) velocity of the particles.

21.1 Introduction

The work described in this paper was motivated by my being invited to watch some spectacular large scale experiments conducted by Guymer, Brockie & Allen (1990) in the SERC flood channel facility at Hydraulics Research Wallingford. The main concern of the experiments was to investigate the lateral spreading of neutrally buoyant particles in a meandering channel. However, some flow visualisation experiments indicated that the direction of travel and rate of dilution could be markedly different for particles which are not neutrally buoyant. Indeed, it is intended that over the next 3 years a number of research groups will utilize the SERC flood channel facility to investigate a variety of questions involving sediments and buoyant particles. This paper is a theoretician's attempt to predict quantitative differences in the transport and dilution of both rising and sinking particles.

21.2 Vertical concentration profile

For a shallow water flow the vertical distribution $\gamma(z)$ of rising (or sinking) particles is determined by a balance between vertical mixing κ_3 and the upwards drift velocity w:

$$\partial_z(\kappa_3\partial_z\gamma) - \partial_z(W\gamma) = 0 \tag{21.1}$$

with

$$w\gamma - \kappa_3\partial_z\gamma = 0 \quad on \quad z = -h,0 \tag{21.2}$$

The boundary condition (21.2) expresses the zero flux of tracer into the bed $z = -h$ and out of the free surface $z = 0$. The solution of these equations (21.1,2) is given by

$$\gamma(z) = \exp\left(\int_{z0}^{z}\frac{w}{\kappa_3}dz'\right) \tag{21.3}$$

with

$$\| \gamma \| = \frac{1}{h}\int_{-h}^{0} \gamma dz = 1 \tag{21.4}$$

where the reference level z_0 is chosen so that the depth averaged value $\| \gamma \| = 1$ is correctly reproduced.

For simplicity we shall restrict attention in this paper to the idealised case of constant vertical diffusivity κ_3 and constant drift velocity w. So, the equilibrium vertical concentration $\gamma(z)$ is normalised exponential.

$$\gamma = \frac{P\exp(P\xi)}{1 - \exp(-P)}, \quad with \quad P = \frac{wh}{\kappa_3}, \quad \xi = \frac{z}{h} \tag{21.5}$$

The vertical Péclet number P is positive for a rising substance and negative for a sinking substance. Figure 1 shows the profiles for $P = -1,1$.

For air bubbles of radius 5×10^{-5} m (.05 mm) with a dirty surface the rise velocity is 5×10^{-3} ms^{-1} (5 mms^{-1}). For spherical particles of sand of the same size and with a density twice that of water, the fall velocity would have the same value. If we specify the water depth and vertical diffusivity

$$h = 5m, \quad \kappa_3 = 0.025m^2s^{-1}, \tag{21.6}$$

then the Péclet numbers for the bubbles is $P = 1$, and that for the sand is $P = -1$. For bubbles or particles of other sizes the Péclet number scales as the square of the radius.

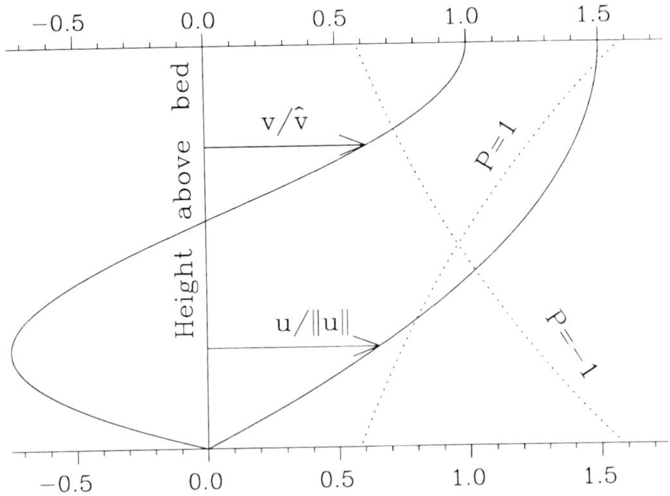

Figure 1: Velocity profiles of the longitudinal and cross-stream flow. The dotted curves show the vertical concentration profiles $\gamma(z)$ for rising ($P = 1$) and for sinking ($P = -1$) particles.

21.3 Longitudinal and transverse velocity profiles for flow around a bend

Locally the flow can be viewed as being circular. Instead of using the r, θ polar coordinate notation, we shall interpret the x coordinate as being directed along the curving depth-averaged flow and the y coordinate as being radially outwards. For shallow water flow, the dominant terms in the horizontal momentum equations are

$$\frac{R}{R+y}u\partial_x u + v\partial_y u + \frac{uv}{R+y} = -\frac{r}{R+y}\partial_x \|p\| + \partial_z(\nu\partial_z u), \tag{21.7}$$

$$\frac{R}{R+Y}u\partial_x v + v\partial_y v - \frac{u^2}{R+y} = -\partial_y \|p\| + \partial_z(\nu\partial_z v) \tag{21.8}$$

with

$$u = v = 0 \quad at \quad z = -h \tag{21.9}$$

and

$$\nu \partial_z u = \nu \partial_z v = 0 \quad at \quad z = 0 \tag{21.10}$$

Here R is the radius of curvature of the flow line $y = 0$, $\| p \|$ is the vertically averaged dynamic pressure, and ν is the vertical eddy viscosity. At the bed there is a no-slip zero velocity boundary condition (21.9), while the upper surface $z = 0$ is assumed to be stress free. (The omitted terms in equations (21.7,8) involve lateral viscosity and lateral gradients of velocities).

Again, for simplicity we shall assume that there is only gradual x, y-dependence of the flow and we shall model the viscosity as being constant. If the transverse flow v is comparatively weak, then the longitudinal velocity can be approximated (see figure 1)

$$u = \| u \| \frac{3}{2}(1 - \xi^2) \tag{21.11}$$

with

$$\| u \| = \frac{-Rh^2}{3(R + y)\nu} \partial_x \| p \| \tag{21.12}$$

For a given pressure gradient the longitudinal velocity is faster at the inside of the bend. By hypothesis the transverse velocity v has zero vertical value. The centrifugal term $u^2/(R + y)$ in equation (21.8) is associated with the secondary circulation (see figure 1)

$$v = \frac{\hat{V}}{5}(5 - 33\xi^2 + 35\xi^4 - 7\xi^6) \tag{21.13}$$

with

$$\hat{V} = \frac{3h^2 \| u \|^2}{56\nu \, (R + y)}, \quad \partial_y \| p \| = \frac{54 \| u \|^2}{35(R + y)} \tag{21.14}$$

Here $\hat{V}$ is the maximum outwards velocity, at the free surface. Equation (21.4) qualifies the slightly higher pressure at the outside of the bend caused by the curvature of the channel.

21.4 Weighted-average velocities

The vertical sampling of the flow by the particles leads to a weighted average horizontal velocity

$$\| \gamma u \| = \| u \| U(P), \quad \| \gamma v \| = \hat{v} V(P) \tag{21.15}$$

The explicit formulae (21.5, 21.11, 21.13) for γ, u, v, enable us to evaluate the dimensionless proportionality factors U,V:

$$U = \frac{3}{2(1-\exp(-P))}\left\{1 - \frac{2}{P^2} + 2\exp(-P)\left[\frac{1}{P} + \frac{1}{P^2}\right]\right\} \tag{21.16}$$

$$V = \frac{\hat{v}}{5(1-\exp(-P))}\left\{5 - \frac{66}{P^2} + \frac{840}{P^4} - \frac{5040}{P^6}\right.$$

$$\left. + \exp(-P)\left[-\frac{32}{P} - \frac{144}{P^2} + \frac{1680}{P^4} + \frac{5040}{P^5} + \frac{5040}{P^6}\right]\right\} \tag{21.17}$$

At $P = 0$ there are no singularities:

$$U = 1, \quad V = 0 \tag{21.18}$$

(i.e. by definition the depth averaged flow is in the x-direction)

Figure 2 shows the velocity ratios U and V as functions of the vertical Peclét number. The curves have been plotted in a way which gives emphasis to the qualitative similarity to the velocity profiles $U/\| u \|$, $V/\hat{v}$ shown in figure 1.

21.5 Shear dispersion

Fischer (1969) showed experimentally and theoretically that for neutrally buoyant tracers the effective cross-stream mixing rate in a curved channel could be up to 10 times what one would expect in a straight channel. The secondary flow causes the augmented cross-stream mixing by the shear dispersion mechanism first explained by G.I. Taylor (1953). Recently, the author (Smith 1991) has derived expressions for the effects of rise (or sinking) velocities on the horizontal shear dispersion tensor

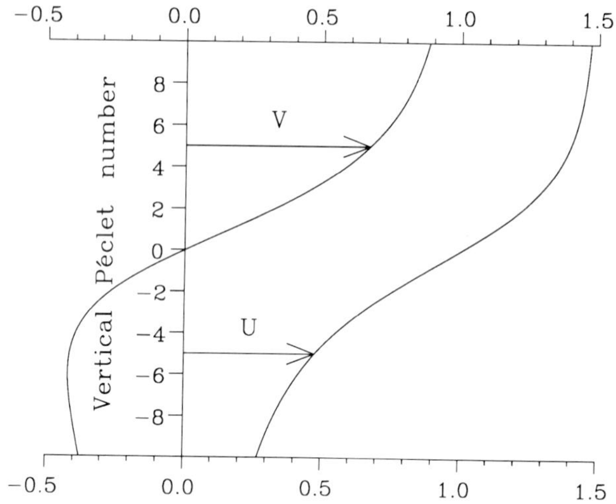

Figure 2: The dependence upon the dimensionless rise velocity $P = wh/\kappa_3$ of the dimensionless longitudinal and transverse velocities U,V.

$$D_{11} = \frac{1}{h}\int_{-h}^{0} \frac{1}{\gamma\kappa_3}\left[\int_{-h}^{z} \gamma(u - U \| u \|)dz'\right]^2 dz, \tag{21.19}$$

$$D_{22} = \frac{1}{h}\int_{-h}^{0} \frac{1}{\gamma\kappa_3}\left[\int_{-h}^{z} \gamma(v - V\,\hat{v})dz'\right]^2 dz \tag{21.20}$$

$$D_{12} = D_{21} = \frac{1}{h}\int_{-h}^{0} \frac{1}{\gamma\kappa_3}\left[\int_{-h}^{z} \gamma(u - U \| u \|)dz'\right]\left[\int_{-h}^{z} \gamma(v - V\,\hat{v})dz'\right] dz \tag{21.21}$$

Here D_{12}, D_{21} are mixed terms.

It is comparatively easy to evaluate the inner integrals:

$$\int_{-h}^{z} \gamma(u - U \| u \|)dz' = H \| U \| I(\xi) \tag{21.22}$$

$$\int_{-H}^{Z} \gamma(V - \nu\hat{V})\, DZ' = H\hat{V}\hat{I}(\xi) \tag{21.23}$$

where

$$I(\xi) = \frac{3}{2(-\exp(-P))}\left\{\exp(P\xi)\left[1 - \xi^2 + 2\frac{\xi}{P} - \frac{2}{P^2}\right] + \exp(-p\left[\frac{2}{P} + \frac{2}{P^2}\right.\right.$$

$$\left. - \frac{3(\exp(P\xi) - \exp(-P))}{2(1 - \exp(-P))^2}\left\{1 - \frac{2}{P^2} + \exp(-P)\left[\frac{2}{P} + \frac{2}{P^2}\right]\right\}\right\} \tag{21.24}$$

$$\hat{I}(\xi) = \frac{1}{5(1-\exp(-P))}\left\{\exp(P\xi)\left[5 - 33\xi^2 + 35\xi^4 - 7\xi^6 + \frac{66\xi - 140\xi^3 + 42\xi^5}{P}\right.\right.$$

$$+ \frac{420\xi^2 - 210\xi^4 - 66}{P^2} - \frac{840}{P^3}(\xi^3 - \xi) + \frac{840 - 2520\xi^2}{P^4}$$

$$\left. + \frac{5040\xi}{P^5} - \frac{5040}{P^6}\right] - \exp(-P)\left[\frac{32}{P} + \frac{144}{P^2} + \frac{1680}{P^4} - \frac{5040}{P^5} - \frac{5040}{P^6}\right]\Bigg\}$$

$$- \frac{\exp(P\xi) - \exp(-P)}{5(1 - \exp(-P))^2}\left\{5 - \frac{66}{P^2} + \frac{840}{P^4} - \frac{5040}{P^6} - \exp(-P)\left[\frac{32}{P} + \frac{144}{P^2} - \frac{1680}{P^4} - \frac{5040}{P^5} - \frac{5040}{P^6}\right]\right\} \tag{21.25}$$

In terms of the two functions I, $\hat{I}$ we define the bulk flow, curvature and mixed coefficients

$$a_B(P) = \frac{1 - \exp(-P)}{P} \int_{-1}^{0} \exp(-p\xi) I^2 d\xi \qquad (21.26)$$

$$a_R(P) = \frac{1 - \exp(-P)}{P} \int_{-1}^{0} \exp(-P\xi) \hat{I}^2 \, d\xi \qquad (21.27)$$

$$a_M(P) = \frac{1 - \exp(-P)}{P} \int_{-1}^{0} \exp(-P\xi) I\hat{I} \, d\xi \qquad (21.28)$$

In principle these integrals can be evaluated explicitly. However, for our present purposes numerical values suffice (see figure 3). In the neutrally buoyant case ($P = 0$) the explicit values of the coefficients are

$$a_B = \frac{2}{105}, \quad a_R = \frac{37888}{1126125}, \quad a_M = \frac{832}{34650} \qquad (21.29)$$

With the dependence upon vertical drift accounted for in the a_B, a_R, a_m coefficients, the formulae (21.19, 20, 21) for the shear dispersion tensor take the simple form

$$D_{11} = \frac{h^2 \, \| u \|^2}{\kappa_3} a_B, \quad D_{22} = \frac{h^2 \hat{v}^2}{\kappa_3} a_R, \quad D_{12} = D_{21} = \frac{h^2 \, | u | \, \hat{v}}{\kappa_3} a_m \qquad (21.30)$$

From figure 3 it is noteworthy that sinking particles have greater longitudinal but smaller transverse shear dispersion than rising particles. This can be explained in terms of the velocity profiles. The longitudinal velocity has zero shear at the free surface but the transverse velocity also has zero shear near the bed.

21.6 Gradual separation between plumes of neutral and of rising particles

In practice the effective longitudinal velocity $U \, | u |$ greatly exceeds the effective transverse velocity $V \hat{v}$ of particles. This leads to greatly elongated particle distributions. Consequently, the strongest gradients and strongest shear dispersion is in the cross-stream direction. The corresponding simplified version of the two-dimensional shear dispersion equation is

$$h\left(1 + \frac{y}{R}\right) \partial_t c + \partial_x(hU \parallel u \mid c) + \partial_y(h\left(1 + \frac{y}{R}\right) V\hat{v}c)$$

$$-\partial_y \left(h(1 + \frac{y}{R}\right) [\mid \gamma\kappa_2 \mid + D_{22}]\partial_y c) = 0$$

(21.31)

Here c is the vertically averaged concentration and $\mid \gamma\kappa_2 \mid$ is the appropriate weighted average of the transverse turbulent eddy diffusivity.

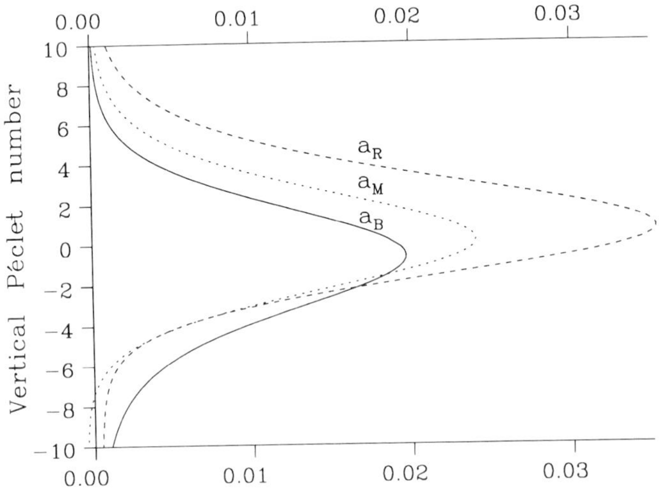

Figure 3: The dependence upon the dimensionless rise velocity $P = wh/\kappa_3$ of the numerical coefficients a_B——, a_R-----, a_M ..... in the formulae (5.5a,b,c) for the longitudinal, transverse and mixed shear dispersion coefficients.

For a steady discharge the local (constant coefficient) solution of equation (21.1) is a Gaussian

$$c = \frac{Q}{(2\pi)^{\frac{1}{2}} \sigma H u \mid U \mid} \exp\left(- \frac{1}{2\sigma^2} \left(y - y_o - \frac{V\hat{v}}{U \mid u \parallel} (x - x_0)\right)^2 \right)$$

(21.32)

with

$$\sigma^2 = 2[\| \gamma\kappa_2 \| + D_{22}] \frac{x - x_0}{U} \| u \| \tag{21.33}$$

To derive this solution we regard all the flow properties h, U, $\| u \| V \hat{v}$, $\| \gamma\kappa_2$, D_{22} as being constant and we neglect y/R.

If a neutrally buoyant (V = 0) and rising (or sinking) particles are released together then eventually the plumes will become distinct. If we denote the velocities and dispersion coefficients for the two species by superscripts $^{(0)}$, $^{(1)}$ then we can estimate the necessary downstream distance for separation

$$\sigma^{(0)^2} + \sigma^{(1)^2} = \left[\frac{V^{(1)}\hat{v}}{U^{(1)} \| u \|} (x - x_0) \right]^2 \tag{21.34}$$

$$i.e. \quad x - x_0 = 2 \frac{U^{(1)} \| u \|}{(V^{(1)}\hat{v})^2} \left[\| \kappa_2 \| + \| \gamma\kappa_2 \|^{(1)} + D_{22}^{(0)} + D_{22}^{(1)} \right] \tag{21.35}$$

This formula equates the square of the centroid separation to the sum of the variances of the individual plumes. So, the spreading and separation are of comparable size.

The formula (21.30) for D_{22} allows us to assess how the distance (21.35) for plume separation varies with the strength $\hat{v}$ of the secondary flow. The functional form is

$$x - x_0 = 2 [\| \kappa_2 \| + \| \gamma\kappa_2 \|^{(1)}] \frac{U^{(1)}}{V^{(1)^2}} \frac{\| u \|}{\hat{v}^2} + \frac{h^2 \| u \|}{\kappa_3} X_\infty \tag{21.36}$$

$$where \quad X_\infty(P) = 2 \frac{U^{(1)}}{V^{(1)^2}} \left[\frac{37888}{1126125} + a_R(P) \right] \tag{21.37}$$

Thus, the distance for plume separation decreases rapidly as the strength of the secondary flow increases. The work of Fischer (1969), as updated by Guymer, Brockie & Allen (1990), reveals that for neutrally buoyant particles D_{22} can be up to 10 times $\| \kappa_2 \|$ in meandering channels. For neutral and rising particles released together, the distance (21.36) for plume separation can be within 10% of the asymptote (i.e. retaining just the X_∞ term).

Figure 4 shows how the dimensionless separation distance X_∞ varies as a function of the vertical Péclet number P. It is in accord with common observation that strongly buoyant flotsam can rapidly become separate from a neutrally buoyant plume. Figure 4 reveals that the separation is just as pronounced for sinking particles.

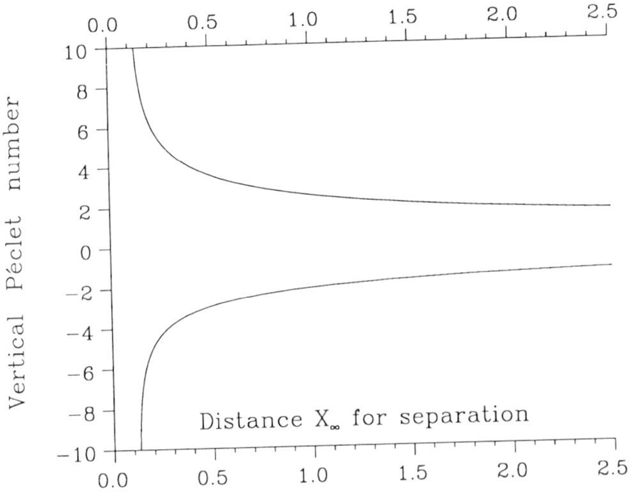

Figure 4: The dependence upon the dimensionless rise velocity $P = wh\ \kappa_3$ of the dimensionless longitudinal distance X_∞ required for a plume of rising (or sinking) particles to become separate from a plume of a neutrally buoyant particles.

Acknowledgements

I am most grateful to Ian Guymer, Nick Brockie & Cath Allen for letting me see what really goes on in turbulent flow round bends. The typescript was kindly prepared by the Department of Electronic & Electrical Engineering. When this work was commenced the author held a post-doctoral fellowship funded by the Royal Society.

21.7 References

Fischer, H.B. 1969: The effects of bends on dispersion in streams. Water Resources Res. 5, 496-506.

Guymer, I . Brockie, N.J.W. & Allen, C.M. 1990: Towards random walk models in a large scale laboratory facility. Proc 5th Int Conf on Mixing and Dispersion, MIT (to appear).

Smith, R. 1991: Wind Augmented transport and dilution. J Fluid Mech., 228, 549-560.

Taylor, G.I. 1953: Dispersion of soluble matter in solvent flowing slowly through a tube. Proc Roy Soc Lond, A219, 186-203.

22

Modelling the vertical distribution of suspended sediment in combined wave-current flow

A G Davies

Abstract

A numerical model of the vertical distribution of suspended sand is described. This is an extension of an existing "one-equation" K-model of the rough-turbulent bottom boundary layer, which has been used to predict the vertical current structure in steady, oscillatory and combined wave-current flows. This hydrodynamical model is used to supply the vertical eddy viscosity distribution (as a function of phase angle in the case of oscillatory flows) for the sediment suspension model. It is assumed that the sediment diffusivity is equal to the eddy viscosity and also that the sediment concentration is sufficiently low that it does not affect the flow. Model results are presented showing cycle-averaged distributions of suspended sediment concentration for combined wave-current flow, and these results are compared with those for both the waves and current alone. The influence of waves on the suspended sediment flux is examined, and the sensitivity of the results to the form of the (time-varying) sediment reference concentration is assessed. It is concluded that the "wave-related" contribution to the flux is likely to be significant for grains which are confined to the oscillatory boundary layer.

22.1 Introduction

There have been several recent studies of the combined effects of waves and currents in the bottom boundary layer of shelf seas. These studies have been concerned with the influence of wave-current interaction on the rate of energy dissipation at the seabed, on vertical mixing processes and, hence, on such practically important matters as storm-surge prediction, scour around offshore structures, and sediment resuspension.

Oscillatory boundary layers are qualitatively similar whether they are induced by waves or tides. However, whereas turbulent wave boundary layers are thin [O(10-20cm)] and confined to the near-bed layer, tidal boundary layers are thick [O(20-80m)] and may occupy the entire water depth over much of the continental shelf. In the combined wave-current case, the oscillatory boundary layer thickness is significantly greater than that for waves in isolation, and the amount of near-bed turbulence is greatly increased. On an erodible bed, these factors are known to lead to increased entrainment and transport of sediment (e.g. Grass,1981). As a result of wave action, substantial loads may be carried by weak currents which, in isolation, would be unable to move sediment. Sediment transport rates are likely to be greatest during storm events when, in addition to bottom-generated turbulence, turbulent energy is introduced into the water column at the free-surface by wave breaking. However, it has been pointed out by Soulsby(1987) that the largest contributions to long term transport are due to fairly large, frequently occurring waves, rather than to extreme events (e.g. major storms during spring tides).

Various models have been developed to predict the vertical structure of wave-current bottom boundary layer flow, and to assess the nonlinear enhancement of the bed shear stress and the (oscillatory) boundary layer thickness. The combined flow was modelled originally by Bijker(1967) and, subsequently, by Smith(1977), Grant and Madsen(1979), Bakker and van Doorn(1979), Fredsoe(1984) and others. The present model of rough-turbulent boundary-layer flow, on which the sediment transport calculations are based, was first used to study the boundary layers beneath waves and tidal currents separately (King, Davies and Soulsby, 1985; Davies, 1986), and more recently to study the combined case (Davies, Soulsby and King, 1988; Davies, Barber and Soulsby, 1990; Davies, 1991). The essential difference between the present numerical model and previous models is that it utilises a general turbulence closure scheme which is applicable to waves, currents and any combined case (e.g. small waves on a strong current or large waves on a weak current). The turbulence scheme involves the determination of the eddy viscosity as a function of height above the bed and phase angle in the wave cycle, via the turbulent energy equation and various turbulence scaling laws. Thus the model solution provides information not only about the vertical distribution of velocity, but also the turbulent energy.

The hydrodynamical model is used here to provide the sediment diffusivity required for the determination of vertical profiles of suspended sediment concentration. In particular, the diffusivity is set equal to the derived eddy viscosity. It is assumed also, for simplicity, that the suspended load does not damp the turbulence and, hence, modify the eddy viscosity. At the bed, a "reference concentration" boundary condition is adopted, the bottom concentration being prescribed as function of the bed shear stress. On this basis, calculations are made of the vertical distribution of suspended sediment in the cases of waves and currents alone, and also in the case of combined wave-current flow and, hence, estimates are made of the enhancement of the suspended load by wave action.

22.2 Formulation

Only the essential details of the formulation are described here; for a complete account the reader is referred to Davies, Soulsby and King(1988). The principal features of the model are shown in outline in Figure 1.

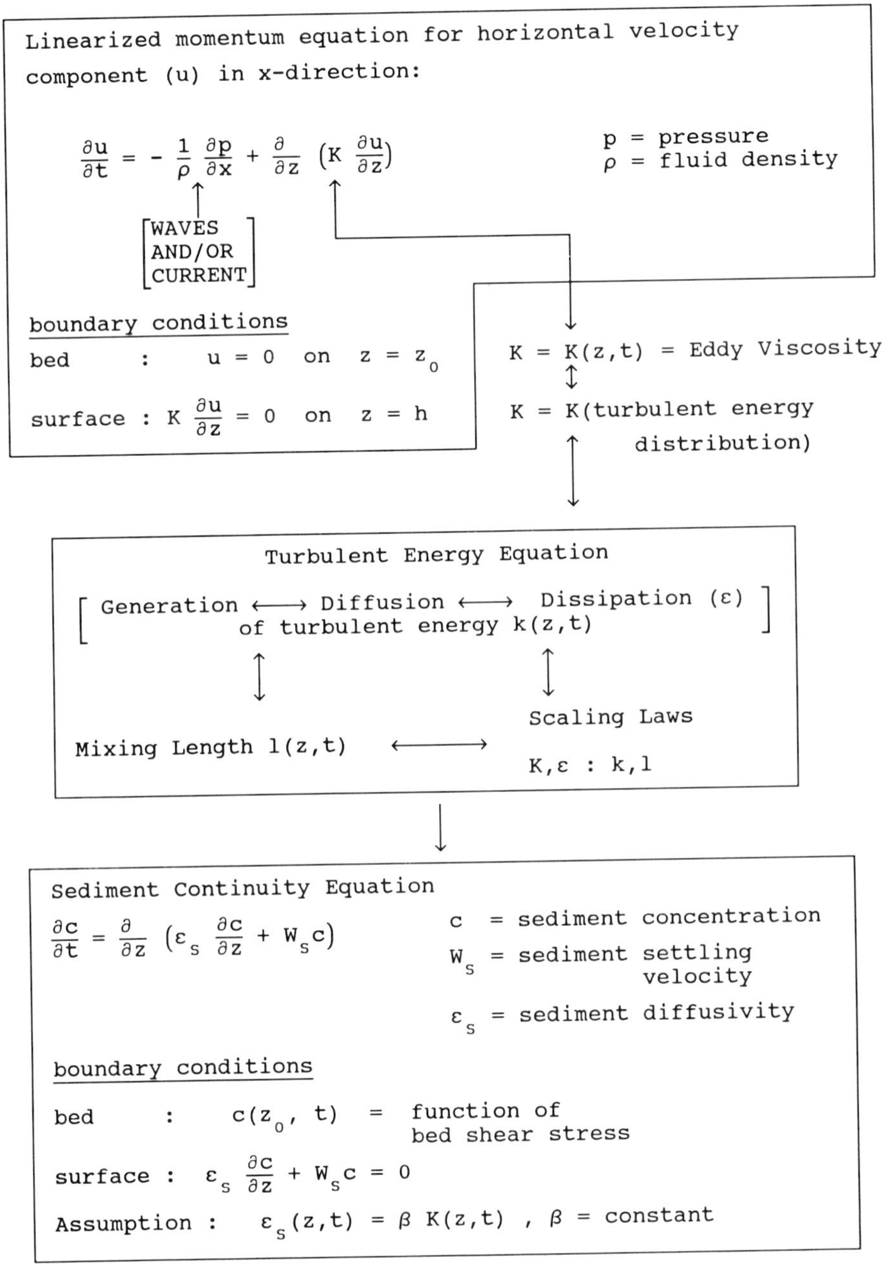

Figure 1. Model equations

Initially, a steady current is generated in the x-direction, driven by a (constant) pressure gradient corresponding to a chosen value of the depth-averaged current for water of depth (h) and a bed roughness of lengthscale (z_0). The boundary layer is assumed to be rough-turbulent. It is assumed also that the surface wind stress is zero, and that the effects of thermal stratification and of the rotation of the Earth are negligible.

Once the current is established, waves are superimposed upon it co-linearly. After the transients have died away, the combined motion consists of a thin oscillatory (wave+current) boundary layer of thickess O(1m) embedded in a thicker steady (current) boundary layer which may occupy the entire water column. The problem is solved in such a way that the depth-averaged current is maintained at its initial chosen value. On account of the increased bottom friction due to the presence of the waves, this requires an increased (steady) pressure gradient, which is determined by the model. The wave motion comprises a uniform horizontal oscillation, driven by a uniform oscillatory pressure gradient in the x-direction. Essentially, the model can be viewed as having a "rigid lid", such that there is a uniform horizontal flow between the fixed free-surface ($z=h$) and the flat rough bed ($z=z_0$). The model solution predicts the vertical structure of this flow.

The problem which arises in solving the momentum equation is the familiar one of what to assume for the eddy viscosity K(z,t). As compared with previous models in which the eddy viscosity has simply been prescribed as a function of height above the bed and has been allowed no time-dependence, here we adopt a more general formulation in which K(z,t) is taken as a function of the instantaneous turbulent kinetic energy (t.k.e.) distribution. The turbulence closure scheme (see Figure 1) is based on (i) the turbulent energy equation, which models the local balance between the generation, diffusion and dissipation of t.k.e. (k); (ii) a mixing length (l) which depends upon the distribution of t.k.e.; and (iii) standard scaling laws relating both the eddy viscosity (K) and the turbulence dissipation rate (ε) to k and l. The formulation differs from that adopted by Davies et al(1988a,b) only by the inclusion of a "surface damping function" in the expression for the mixing length. [The choice of damping function has been discussed by Davies(1990)]. Nezu and Rodi(1986) have demonstrated the need for such a function in connection with laboratory experiments on steady flow . Thus the eddy viscosity (K) is determined as a function of height (z) above the bed and phase angle in the wave cycle. The distribution of K differs from the functional forms adopted by previous workers (e.g. Grant and Madsen(1979)) and, hence, gives rise to rather different predictions for the peak bed shear stress in the wave cycle and other quantities of interest.

The vertical distribution of suspended sediment has been modelled on the basis of the sediment continuity equation (see Figure 1). Solution of this equation requires a knowledge of the sediment settling velocity (W_s) and physically plausible assumptions for both the sediment diffusivity (ε_s) and the bottom boundary condition. The results in Section III are based on three assumptions. Firstly, it has been assumed that the presence of suspended sediment does not affect the distribution of t.k.e in the flow; while this might be reasonable for low sediment concentrations, it is not a generally valid assumption (Soulsby and Wainwright(1987)). Secondly, the sediment diffusivity (ε_s) has been assumed to be equal to the eddy viscosity (K). Again, this is probably an oversimplification (see Dyer and Soulsby(1988)). Thirdly, and most importantly, a knowledge of the "reference concentration" at the bed has been assumed. This form for the bottom boundary condition has been adopted mainly for simplicity. In particular, it has been assumed that the reference concentration at level $z=z_0$ can be prescribed as a function of the instantaneous bed shear stress in the wave-current cycle. At the free water surface, the condition of zero sediment flux is applied.

The system of equations in Figure 1 has been solved in two stages, by means of numerically conservative, finite difference methods on a log-linear vertical grid. Initially, the required hydrodynamical solution has been obtained for wave, current and combined wave-current flows. The respective velocity and eddy viscosity distributions, determined as functions of height above the bed and phase angle in the wave cycle, have then been used in the solution of the sediment continuity equation. The results presented below relate to a case study carried out with the following parameter settings. The bed roughness lengthscale has been taken as z_0 = 0.5cm, corresponding to sand ripples having the approximate dimensions: height ≈5cm, wavelength ≈50cm. The water depth has been taken as 10m.

Initially, a steady current has been generated with depth-averaged velocity 50cm/s. Waves having velocity amplitude 50cm/s and period 10s have then been generated with the same values of z_0 and h. On the basis of linear wave theory, the surface wavelength corresponding to this period is 92m, which is considerably greater than the water depth and justifies the assumption of horizontal uniformity in the formulation. The flow in the oscillatory boundary layer is rough turbulent. Finally, the waves have been superimposed co-linearly on the steady current. This superimposition has been carried out such that the depth-averaged velocity has been maintained at its initial value of 50cm/s. The sediment continuity equation has been solved for the current, wave and combined wave-current flows, with various values for the sediment settling velocity W_s, for example W_s=0.1, 0.25 and 1.0cm/s corresponding to coarse silt (nominal grain diameter D≈0.04mm), very fine sand (D≈0.07mm) and fine sand (D≈0.14mm). The use of a constant value of z_0 of 0.5cm implies the availability of sediment from a rippled bed comprising a mixture of grain sizes, this mixture being consistent with the existence of ripples of the size indicated above.

22.3 Model results

22.3.1 Flow solutions

(i) Steady current
As expected, the vertical profile of horizontal (mean) velocity (U(z)) is logarithmic in the lower part of the flow (Fig. 2i), and the (mean) shear stress increases linearly from zero at the free-surface to its maximum value of $\tau(z_0)$=8.72dyn/cm^2 (u_*=2.95cm/s) at the bed (Fig. 2ii).

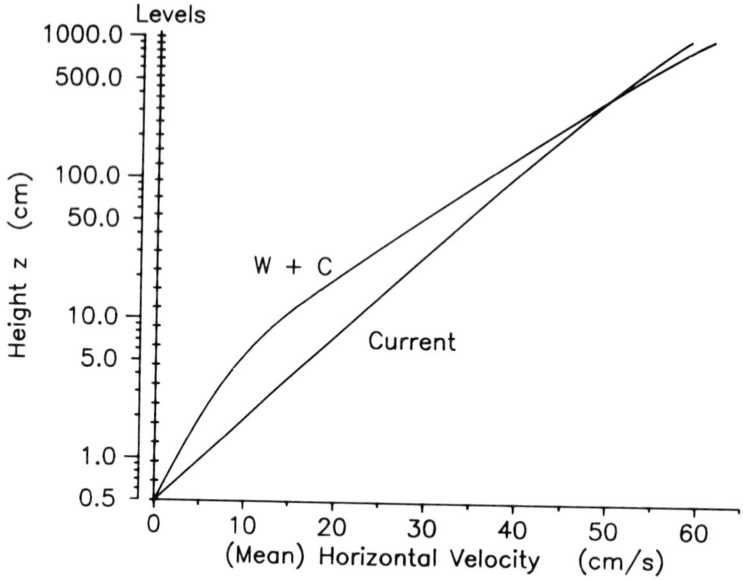

Figure 2i: Vertical profiles of horizontal velocity U(z) for the steady current and <u(z,t)> for the combined wave-current (W+C) flow. (Angle brackets <> denote time-averaging.) The model grid is shown on the inner vertical axis.

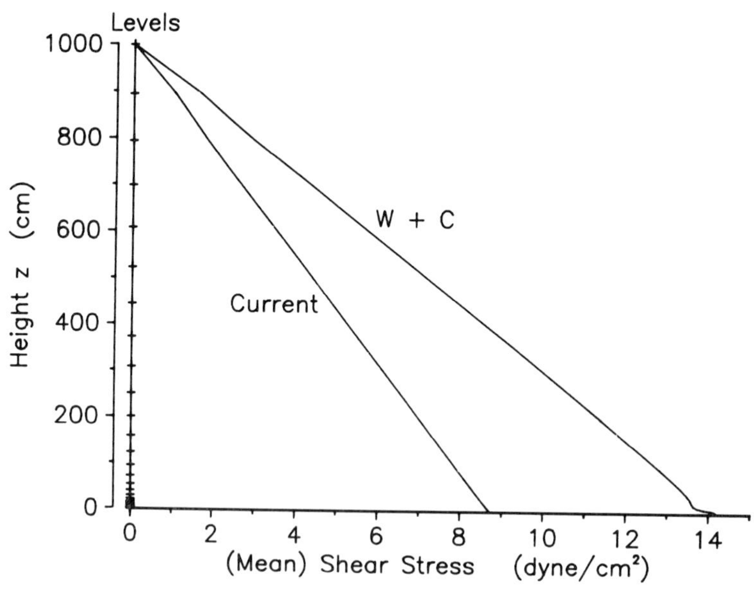

Figure 2ii: Vertical profiles of shear stress $\tau(z)$ for the current and $<\tau(z,t)>$ for the combined wave-current flow.

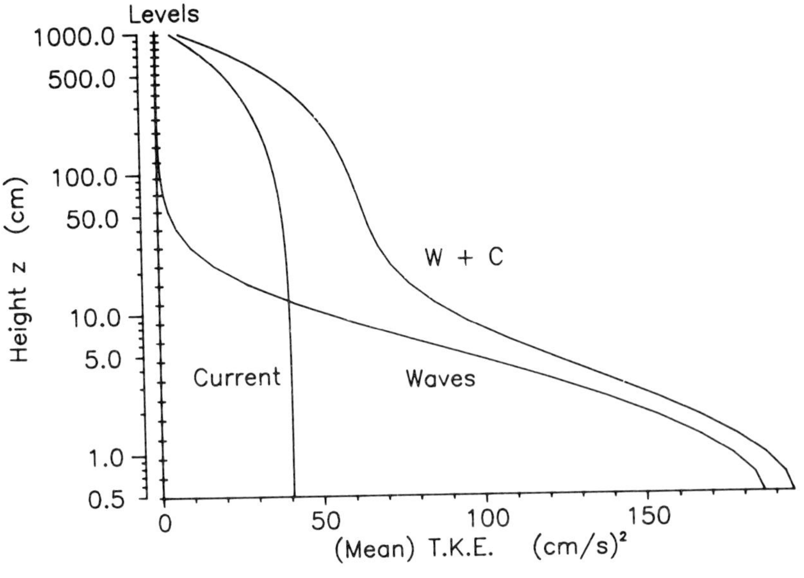

Figure 2iii: Vertical profiles of (cycle-averaged) turbulent kinetic energy for the current, waves alone and the combined wave-current flow.

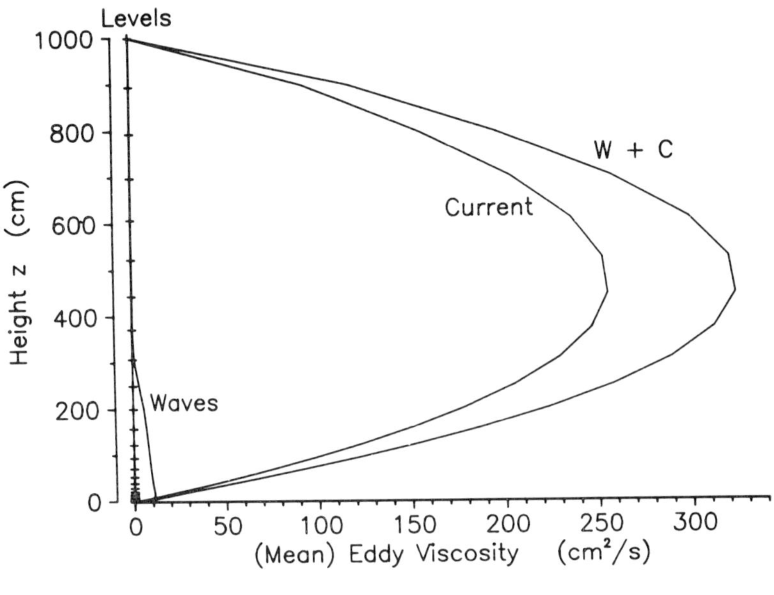

Figure 2iv: Vertical profiles of (cycle-averaged) eddy viscosity for the current, waves alone and the combined wave-current flow.

The associated distributions of t.k.e. (k) (Fig. 2iii) and mixing length (not shown) give rise to an essentially parabolic distribution of eddy viscosity (K) (Fig. 2iv).

(ii) Waves
The vertical profile of the amplitude of wave velocity (Fig. 2v) has the familiar structure of a free-stream region ($z \geq 20$cm), an "overshooting" region ($5 \leq z \leq 20$cm), and a logarithmic near-bed region. The associated cycle-averaged distributions of t.k.e. and eddy viscosity for the waves are shown in Figures 2iii and 2iv. The solution converged after 60 cycles (10 min real time).

(iii) Combined waves and current
The results for wave-current flow relate to the co-linear superimposition of the waves (of Section 22.3 1(ii)) and the current (of Section 22.3 1(i)). Cycle-averaged profiles of velocity, t.k.e., eddy viscosity and shear stress are shown in Figures 2i to 2iv. The effect of the waves is to greatly enhance near-bed t.k.e. compared with the current alone, and hence to decrease the velocity in the near-bed layer. Since the depth-averaged velocity is required to remain at its value for the current alone of 50cm/s, this necessitates an increase in the velocity in the upper part of the flow, which is brought about by an increase in the mean component of the pressure gradient. After all the transients have died away (after about 80 cycles) the final converged, cycle-averaged shear stress profile is linear with a value of mean bed shear stress of 14.0dyn/cm^2 (as compared with 8.72dyn/cm^2 for the current alone).

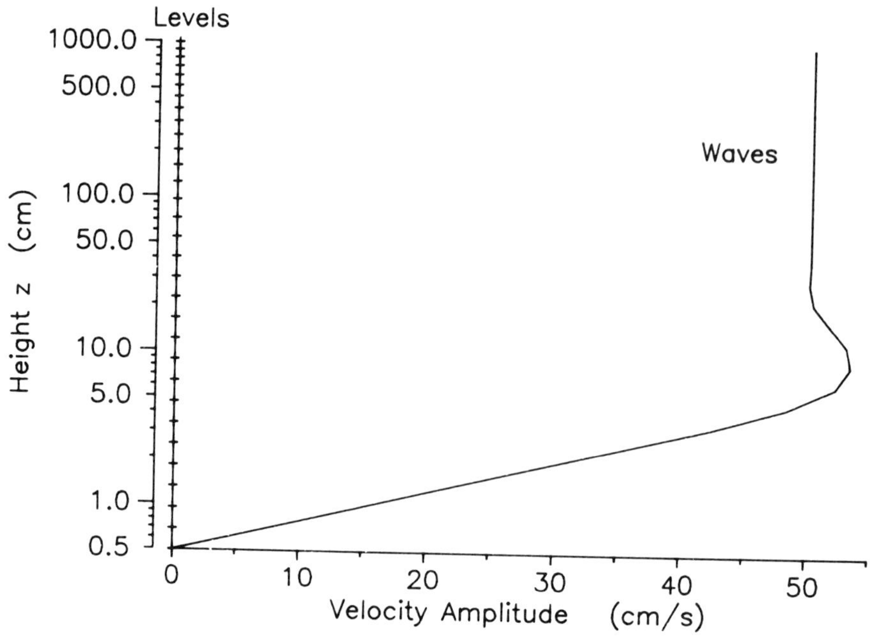

Figure 2v: Vertical profile of the amplitude of horizontal velocity for the waves alone.

Extrapolation of the straight-line, logarithmic segment of the wave-current profile ($15 \le z \le 200$cm) in Figure 2i indicates an "apparent roughness" for the wave-current boundary layer of 2.5cm. This is 5 times the physical roughness lengthscale (z_0). In addition, the slope of the straight-line segment suggests an "apparent mean u_*" value of 3.97cm/s. This is somewhat larger than the true value of 3.74cm/s (i.e. the value calculated from the mean stress at the bed level). This discrepancy is explained by the non-logarithmic form of the mean velocity profile in the upper part of the flow.

22.3.2 Solutions for suspended sediment concentration

(i) Steady current

Vertical profiles of suspended sediment concentration (c) are shown in Figure 3i for three grain sizes having the settling velocities W_s = 0.1, 0.25 and 1.0cm/s. Each profile has been normalized by the reference concentration $c=C_{0c}$ at level $z=z_0$. No attempt is made to evaluate C_{0c} in what follows; C_{0c} is simply taken as the value of reference concentration appropriate to u_* = 2.95cm/s for the respective grain sizes. Thus the problem is reduced to the prediction of the vertical distribution of concentration relative to C_{0c}.

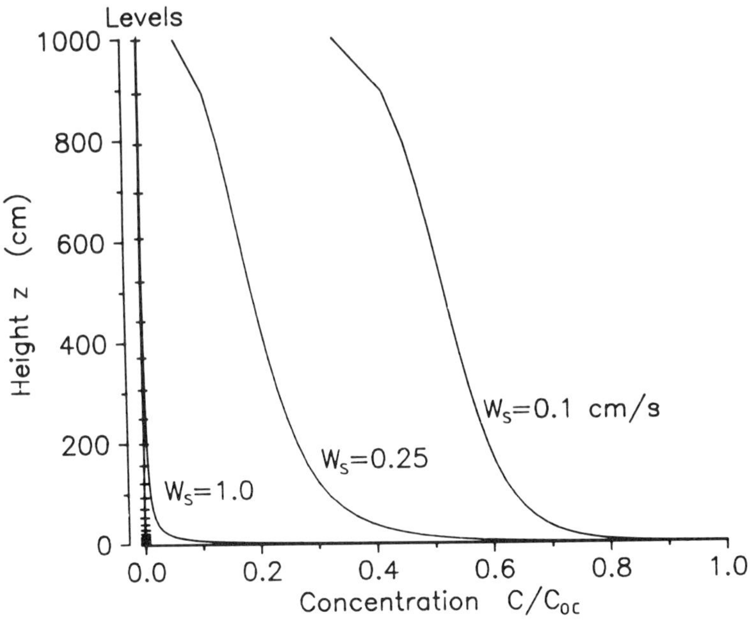

Figure 3i: Vertical profiles of suspended sediment concentration for the steady current, for three values of settling velocity W_s. Each profile has been normalized by the appropriate reference concentration C_{0c}.

As expected, the profiles indicate a relatively higher concentration of sediment in the mid-depths as W_s decreases. [Strictly, each profile should tend to zero concentration at the free surface on account of the mixing length damping function. In fact, due to the rather coarse vertical grid interval adopted near the surface, a non-zero value is plotted. This value has been forced, through a modified surface boundary condition, to be consistent in the upper part of the flow with the sediment distribution associated with the familiar Rouse profile.]

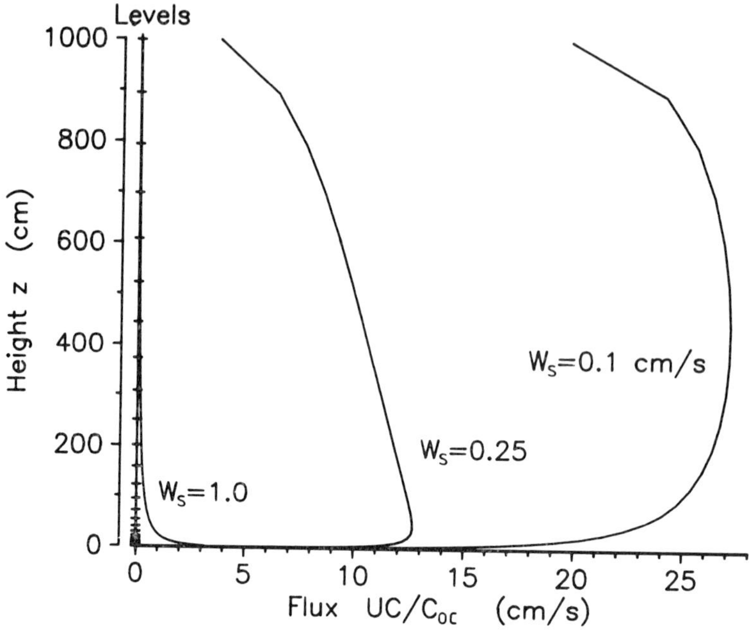

Figure 3ii. Vertical profiles of suspended sediment flux for the steady current, for three values of settling velocity W_s.

Assuming that the horizontal velocity component of the sediment grains is equal to that of the fluid, the sediment flux at each level (normalized by the reference concentration C_{0c}) is $U(z)C(z)/C_{0c}$. Vertical profiles of this quantity are shown in Figure 3ii for the three settling velocities W_s = 0.1, 0.25 and 1.0cm/s. The levels of maximum sediment flux are 445, 41 and 1.8cm in the respective cases. Values for the depth-averaged concentration and suspended sediment transport rate for steady flow are discussed in Section 22.4.

(ii) Waves Alone

From the instantaneous profiles of horizontal velocity u(z,t) and eddy viscosity K(z,t) generated by the model, the bed shear stress $\tau_0(t)=\tau(z_0,t)$ has been determined as a time series through the wave cycle. The values of $\tau_0(t)$ have then been used to generate a time series of reference concentration $c(z_0,t)$ such that

$$c(z_0,t) = C_{0c}|\tau_0(t) \;/\; \tau_{0c}|^N \tag{22.1}$$

where τ_{0c} = 8.72dyn/cm^2 is the value of bed shear stress for the steady current. It should be noted, firstly, that $c(z_0,t)$ has been taken as a power of the magnitude of the normalized instantaneous bed shear stress and, secondly, that no sediment threshold motion criterion has been included in the formulation. Various expressions for $c(z_0,t)$ have been suggested by previous workers, incorporating different power law behaviours, for example N = 1/2 (Shi, 1985) and N = 1 (Hagatun and Eidsvik, 1986). Since higher values than these in the range N = 3/2 to 3 may be inferred from results for the cycle-averaged reference concentration (see Dyer and Soulsby(1988) and Nielsen(1988)), this remains an area of considerable uncertainty. The sensitivity of the present model results to changes in N is discussed in Section 22.5. The omission of a threshold shear stress from the expression for $c(z_0,t)$ takes some account both of the persistent nature of near-bed sediment motion in oscillatory flow and of the effect of settling grains.

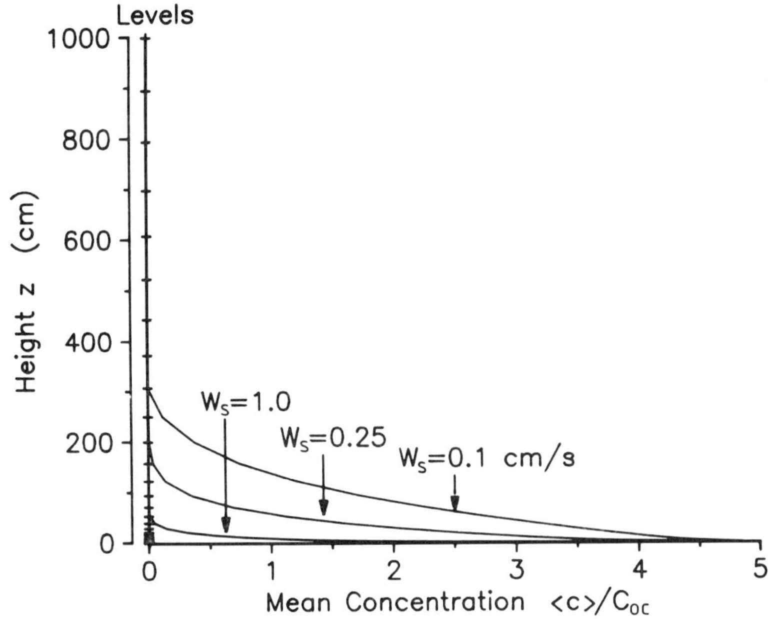

Figure 4. Vertical profiles of cycle-averaged sediment concentration for the waves alone, for three values of settling velocity W_s. Each profile is normalized by the reference concentration C_{0c} for the steady current.

On the basis of the above formulation for $c(z_0,t)$ solutions of the sediment continuity equation have been obtained with N=1 and ß=1 (i.e. the sediment diffusivity is equal to the eddy viscosity, see Figure 1). In each case, the converged hydrodynamical solution for waves alone has been used repeatedly to provide the sediment diffusivity as a function of height and phase angle in the wave cycle. Thus the determination of the sediment concentration profile has been tackled as an initial value problem, with zero concentration assumed initially at each level. The cycle-averaged concentration profiles for the three sediment settling velocities are shown in Figure 4. At the bed, each profile takes the value $\langle c\rangle/C_{0c} = 4.83$ as dictated by the cycle-averaged value $\langle |\tau_0(t)/\tau_{0c}|\rangle$. In each case, the sediment remains confined to the near-bed layer with the settling velocity strongly influencing the mean profile. Converged solutions were obtained after 400, 150 and 25 cycles (67, 25 and 4 min. real time) for the cases W_s = 0.1,0.25 and 1.0cm/s, respectively.

(iii) Combined Wave-Current Flow

As for the case of waves alone, the eddy viscosity generated by the hydrodynamical model has been used in the solution of the sediment continuity equation. The reference concentration has been calculated in the way described earlier and an initial value problem has been solved for c(z,t). Converged solutions were obtained after 1400, 800 and 200 cycles (233, 133 and 33 minutes real time) in the cases W_s = 0.1, 0.25 and 1.0 cm/s, respectively. Each of the cycle-averaged profiles in Fig 5i takes the value $\langle c\rangle/C_{0c} = 5.02$ at the bed level $z=z_0$ (as compared with the value 4.83 for waves alone). The concentration distributions for the case of combined wave-current flow may be compared directly with the results for the current alone in Figure 3i. An enhancement in the suspended sediment concentration is evident for each value of settling velocity considered. This arises both on account of the increase in mean bottom concentration, and also on account of the enhanced diffusivity.

The profile of suspended sediment flux has been calculated in two ways. Firstly, the flux has been taken as the product of the cycle-averaged values of velocity and concentration at each level $\langle u\rangle\langle c\rangle$. Although this is a commonly used approach, it is potentially inaccurate on account of its omission of the "wave-related" contribution to the flux. If the horizontal velocity (u) and concentration (c) are written

$$u = U + u_w + u_t \text{ and } c = C + c_w + c_t \tag{22.2}$$

where U and C are the time-invariant components of u and c, and the subscripts w and t denote the wave (i.e. periodic) and turbulent (i.e. random) components of each quantity, the cycle-averaged flux at a given level is

$$\langle uc\rangle = UC + \langle u_w c_w\rangle + \langle u_t c_t\rangle \tag{22.3}$$

where angle brackets < > denote averaging over a large (integral) number of wave periods. Dyer and Soulsby(1988) have suggested that the turbulent contribution to this expression is likely to be small (<1%) and that

$$\langle uc\rangle \approx UC + \langle u_w c_w\rangle \tag{22.4}$$

The first term on the right hand side of this expression is simply the mean component of flux <u><c> since <u>=U and <c>=C.

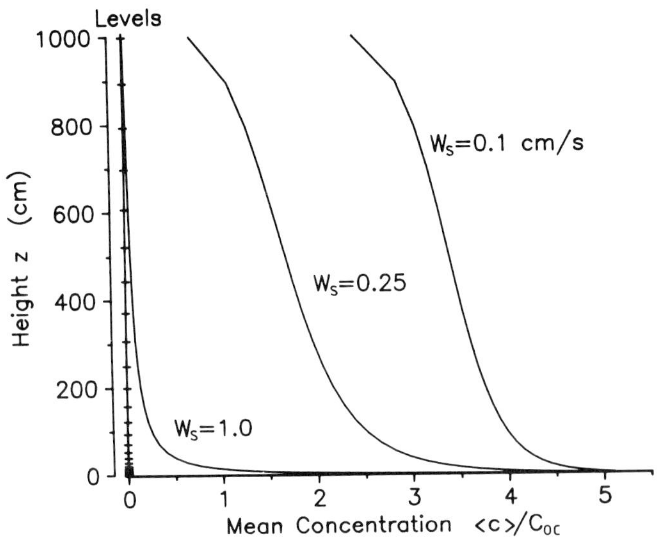

Figure 5i: Vertical profiles of cycle-averaged suspended sediment concentration for wave-current flow, for three values of settling velocity W_s. Each profile is normalized by the reference concentration C_{0c} for the steady current.

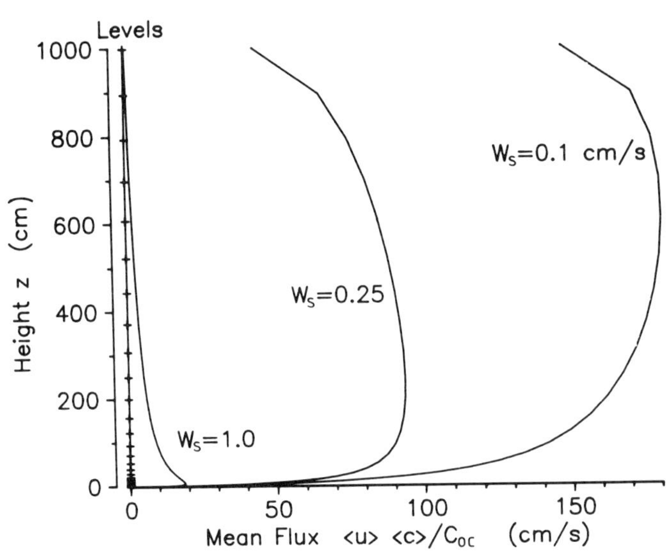

Figure 5ii: Vertical profiles of cycle-averaged sediment flux <u><c> for the wave-current flow, for three values of settling velocity W_s. Each profile is expressed in terms of the reference concentration C_{0c} for the steady flow.

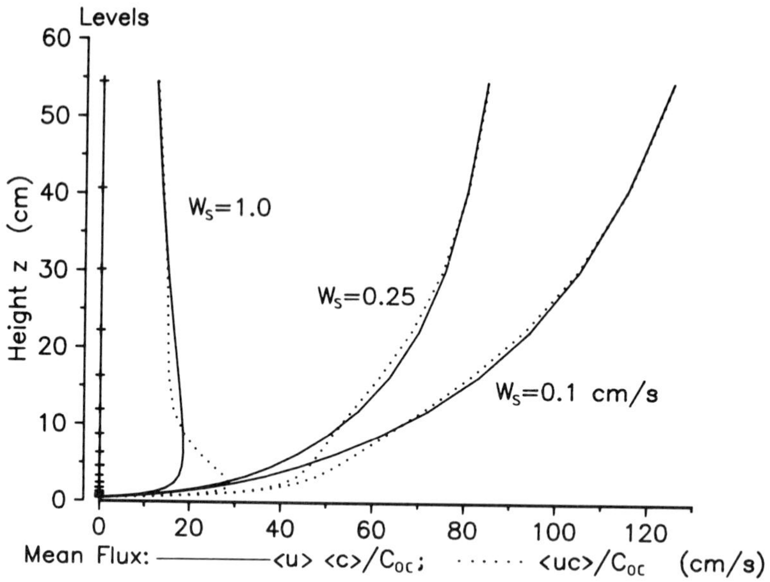

Figure 5iii: Vertical profiles of cycle-averaged sediment flux for the lower part of the wave-current flow, for three values of settling velocity W_s. Profiles based upon <u><c> (full line) and <uc> (dotted line) are shown in each case.

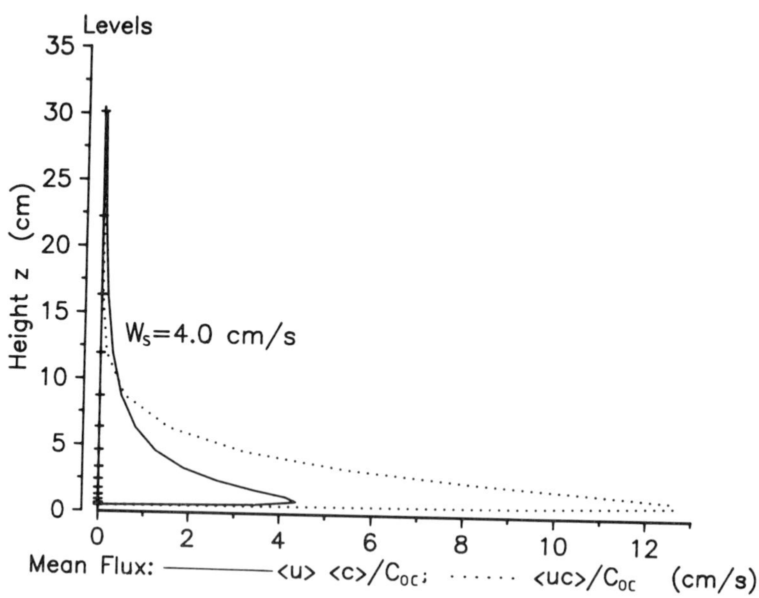

Figure 5iv: Vertical profiles of cycle-averaged sediment flux as in Fig. 5iii, but for a larger value of W_s.

The profiles of flux calculated in these two ways are identical above the wave boundary layer, as shown in Figures 5ii-iv. As for the current alone, each profile of flux displays a maximum at a height which increases as W_s decreases. Figure 5ii shows profiles of UC over the entire depth. The levels of maximum sediment flux are 608, 226 and 6.4cm for W_s = 0.1, 0.25 and 1.0 cm/s, respectively. Figure 5iii shows profiles of both <u><c> and <uc> in the bottom part of the flow. Above the wave boundary layer (z>35cm) the curves for given W_s are identical. However, within the wave boundary layer, substantial differences arise, the present formulation (with N=1) suggesting that the wave-related contribution to the flux increases the net transport. This is due to the fact that very close to the bed peak instantaneous concentrations occur at about the time in the wave cycle of maximum forward velocity, giving rise to a nonlinear enhancement in the flux. Evidently, this effect is greatest for coarse sediments which are confined to the oscillatory boundary layer. Figure 5iv illustrates this for a fourth, larger value of settling velocity (W_s=4cm/s).

22.4 Depth-averaged concentration and sediment flux

22.4.1 Notation

In order to allow a general assessment of the results, the profiles of concentration and flux have been averaged over the depth. These results are denoted by an overbar notation as follows. For the steady current,

$$\textbf{Average concentration} \quad \bar{C} = \frac{1}{h-z_0}\int_{z_0}^{h} C(z)dz ,$$

$$\textbf{Suspended sediment flux} \quad q_c = \int_{z_0}^{h} C(z)U(z)dz . \tag{22.5}$$

Following Dyer and Soulsby(1988) and Soulsby(1988) we express the flux in terms of a non-dimensional "shape function" I_c defined as follows:

$$q_c = C_{0c}\bar{U}(h-z_0)I_c \tag{22.6}$$

where $\bar{U}$ is the depth-averaged velocity. Here the flux is divided into two parts: (i) a "stirring term" given by the reference concentration C_{0c}, which is assumed to depend only on the bed shear stress and properties of the grains; and (ii) a "transporting term" comprising the product of the volumetric flow rate per unit width $\bar{U}(h-z_0)$ and the shape function I_c. The later calculations concern only the transporting term.

For wave-current flow, depth-averages are taken after cycle-averaging as follows:

$$\text{Average concentration } \overline{<c>} = \frac{1}{h-z_0}\int_{z_0}^{h} <c(z,t)>dz \tag{22.7}$$

The suspended sediment flux is calculated on the basis of the two approaches discussed above from the expressions:

$$q_{+1} = \int_{z_0}^{h} <u><c>dz = \int_{z_0}^{h} U(z)C(z)dz \tag{22.8}$$

$$q_{+2} = \int_{z_0}^{h} <uc>dz = \int_{z_0}^{h} \{U(z)C(z) + <u_w c_w>\}dz \tag{22.9}$$

Following Soulsby(1988) we again express these fluxes in terms of non-dimensional "shape functions" I_{+1} and I_{+2}, defined for wave-current flow in the respective cases by:

$$q_{+i} = C_{0m}\bar{V}(h-z_0)I_{+i}\,, \quad (i=1,2), \tag{22.10}$$

where $C_{0m} = <c(z_0,t)>$ is the mean over an integral number of wave periods of the reference concentration at $z=z_0$. [It should be noted that C_{0m} and C_{0c} will, in general, be different.] The quantity $\bar{V}$ is the cycle-averaged, depth-averaged velocity in wave-current flow, such that the product $\bar{V}(h-z_0)$ is the volumetric flow rate per width. Since the waves are superimposed on the current co-linearly, $\bar{V}$ has the same direction as $\bar{U}$.

22.4.2 Results

Calculations of the above depth-averaged quantities have been made for a range of values of settling velocity W_s. The results tabulated below for the current, wave and wave-current cases are based upon the flow solutions discussed earlier.

(i) Steady current

The results for depth-averaged concentration $\bar{C}$ and transport rate per unit width q_c (both normalized by the reference concentration C_{0c}), together with the shape function I_c, are listed in Table 1. As expected, each quantity decreases rapidly with increasing W_s.

Further model runs have been performed with the roughness to depth ratio $z_0/h=10^{-4}$. In Figure 6 the results obtained for I_c are plotted against the parameter $S_c = C_D\bar{U}/3W_s$, where the drag coefficient $C_D = \tau(z_0)/\rho\bar{U}^2$, on the graph of Dyer and Soulsby(1988). The results relate to a single settling velocity W_s, and a single roughness lengthscale z_0 (i.e. the simultaneous presence of different bedform types is not permitted). The various curves plotted in the inset are for different functional forms for the eddy viscosity, and hence sediment diffusivity (with ß=1). Evidently, I_c is fairly insensitive to the choice of eddy viscosity, and the present model results conform closely to the general trend. Although all the plotted results are derived for low concentrations involving no damping of the turbulence by the sediment grains, it has been suggested by Dyer and Soulsby(1988) that the results may also be applicable in cases in which drag reduction by sediment in suspension does occur.

Table 1: Depth-averaged results for the steady current obtained with $\beta = 1$. Flow parameters: $z_0 = 0.5$ cm, h = 1000 cm, $\bar{U} = 50$ cm/s, $u_* = 2.95$ cm/s.

W_s (cm/s)	$\bar{c}/C_{0c}$	q_c/C_{0c} (cm^2/s)	I_c
0·1	$5\cdot22\times10^{-1}$	$2\cdot54\times10^{4}$	$5\cdot09\times10^{-1}$
0·25	$2\cdot06\times10^{-1}$	$9\cdot55\times10^{3}$	$1\cdot91\times10^{-1}$
0·5	$5\cdot12\times10^{-2}$	$2\cdot14\times10^{3}$	$4\cdot29\times10^{-2}$
0·75	$1\cdot54\times10^{-2}$	$5\cdot57\times10^{2}$	$1\cdot12\times10^{-2}$
1·0	$5\cdot73\times10^{-3}$	$1\cdot69\times10^{2}$	$3\cdot38\times10^{-3}$
1·5	$1\cdot47\times10^{-3}$	$2\cdot47\times10^{1}$	$4\cdot95\times10^{-4}$
2·0	$6\cdot77\times10^{-4}$	$6\cdot39\times10^{0}$	$1\cdot28\times10^{-4}$
3·0	$3\cdot03\times10^{-4}$	$1\cdot26\times10^{0}$	$2\cdot52\times10^{-5}$
4·0	$1\cdot91\times10^{-4}$	$4\cdot46\times10^{-1}$	$8\cdot94\times10^{-5}$
6·0	$1\cdot07\times10^{-4}$	$7\cdot39\times10^{-2}$	$1\cdot48\times10^{-6}$

(ii) Waves only and combined wave-current flow

As noted earlier, the cycle-averaged concentration at the bed level $z=z_0$ for the case of waves alone is $\langle c\rangle/C_{0c} = 4.83$, and for the combined wave-current case is $\langle c\rangle/C_{0c} = 5.02$. Table 2 shows results for the mean, depth-averaged concentration $\overline{\langle c\rangle}/C_{0c}$ calculated for a range of settling velocities W_s. If the results for waves alone are compared with those for the current alone in Table 1, it is seen that the depth-averaged concentration for given W_s is generally greater for the waves. Only for the two smallest values of W_s, for which the grains are strongly diffused into the outer flow, is the opposite result predicted.

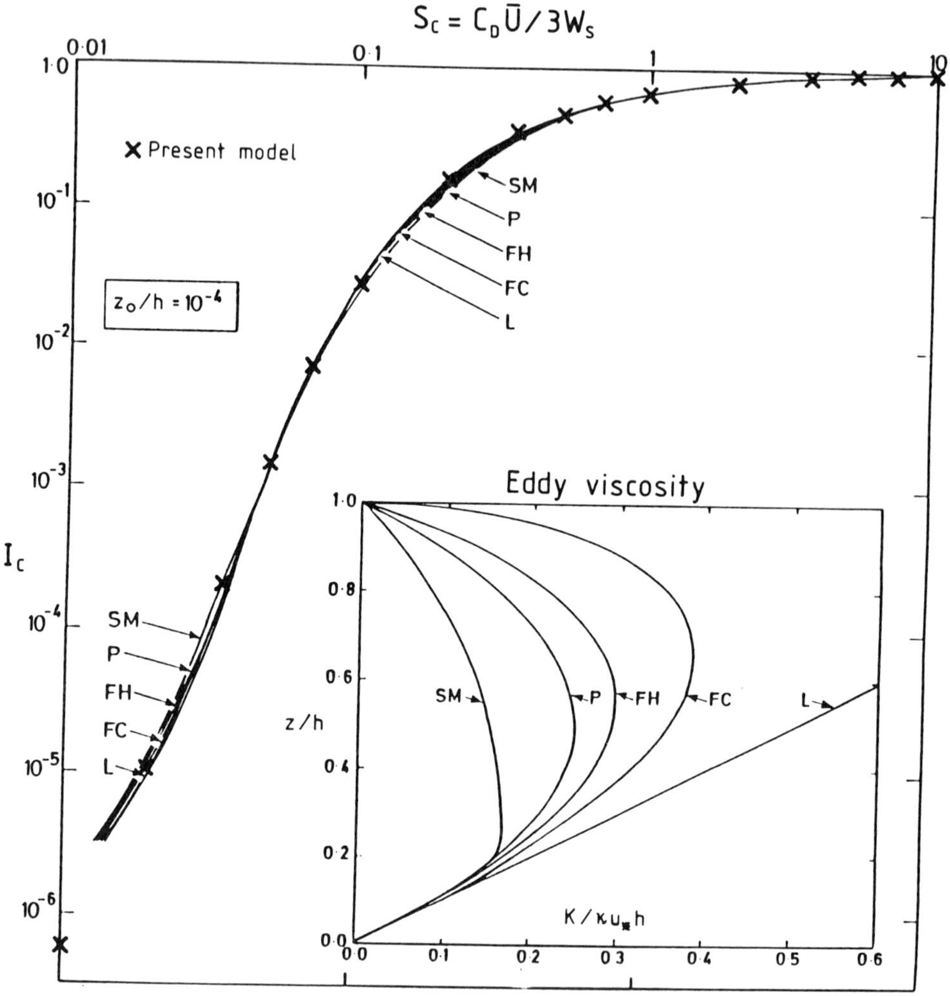

Figure 6: Shape function I_c versus S_c for steady flow with $z_0/h=10^{-4}$. The labelled curves correspond to the vertical profiles of eddy viscosity shown in the inset, as explained in detail by Dyer and Soulsby(1988) and Soulsby(1988) (κ = Von Karman's constant). The present model results are shown by the crosses.

Table 2: Cycle-averaged, depth-averaged concentration results obtained with $\beta = 1$ and N = 1 for both waves only and combined wave-current (W+C) flow. Flow parameters: $z_0 = 0.5$cm, h = 1000 cm; mean velocity $\overline{V} = \overline{\langle u \rangle} = 50$ cm/s; wave velocity amplitude $\hat{u}_w = 50$ cm/s, period = 10s.

W_s (cm/s)	$\overline{\langle c \rangle}/C_{0c}$		$\overline{\langle c \rangle}$ ratios	
	Waves	W + C	$\frac{W+C}{Waves}$	$\frac{W+C}{Current}$
0·1	$3{\cdot}97\times10^{-1}$	$3{\cdot}40\times10^{0}$	8·6	6·5
0·25	$1{\cdot}65\times10^{-1}$	$1{\cdot}76\times10^{0}$	10·7	8·5
0·5	$7{\cdot}33\times10^{-2}$	$6{\cdot}38\times10^{-1}$	8·7	12·5
0·75	$4{\cdot}27\times10^{-2}$	$2{\cdot}63\times10^{-1}$	6·2	17·0
1·0	$2{\cdot}82\times10^{-2}$	$1{\cdot}23\times10^{-1}$	4·4	21·5
1·5	$1{\cdot}50\times10^{-2}$	$3{\cdot}90\times10^{-2}$	2·6	26·6
2·0	$9{\cdot}43\times10^{-3}$	$1{\cdot}68\times10^{-2}$	1·8	24·8
3·0	$4{\cdot}86\times10^{-3}$	$6{\cdot}63\times10^{-3}$	1·4	21·8
4·0	$3{\cdot}07\times10^{-3}$	$3{\cdot}84\times10^{-3}$	1·3	20·1
6·0	$1{\cdot}66\times10^{-3}$	$1{\cdot}95\times10^{-3}$	1·2	18·3
8·0	$1{\cdot}10\times10^{-3}$	$1{\cdot}26\times10^{-3}$	1·2	-
12·0	$6{\cdot}37\times10^{-4}$	$7{\cdot}19\times10^{-4}$	1·1	-
16·0	$4{\cdot}65\times10^{-4}$	$5{\cdot}18\times10^{-4}$	1·1	-

The results for wave-current flow in Table 2 are compared with those for both the current alone and the waves alone. As compared with the values for the current, there is a fairly uniform order of magnitude increase in $\overline{\langle c \rangle}/C_{0c}$. In contrast to this, comparison with the results for waves alone suggests rather similar results for the larger values of W_s, but an order of magnitude difference for the smaller values. Results for the sediment transport rate (q_+) per unit width in combined wave-current flow, and also for the shape function (I_+), are shown in Table 3 for various settling velocities W_s. The results have been calculated both excluding (subscript 1) and including (subscript 2) the wave-related contribution to the sediment flux, as discussed earlier.

Comparison with the results in Table 1 for the current alone reveals a substantial enhancement in the flux (q_+) as a result of wave action. For the values of W_s considered, the flux increases by between one and two orders of magnitude. Both q_+ and I_+ decrease rapidly as W_s increases. However, I_{+1} decreases much more rapidly than I_{+2}, indicating the importance of the wave-related contribution to the flux for the larger grains.

Table 3: Mean sediment transport rate (q_+ per unit width) and shape factor (I_+) for combined wave-current flow. Flow parameters are as for Table 2.

W_s (cm/s)	q_{+1}/C_{0c} (cm^2/s)	q_{+1}/q_c	I_{+1}	q_{+2}/C_{0c} (cm^2/s)	q_{+2}/q_c	I_{+2}	I_{+2}/I_{+1}
0·1	$1\cdot65\times10^{5}$	6·5	$6\cdot59\times10^{-1}$	$1\cdot65\times10^{5}$	6·5	$6\cdot59\times10^{-1}$	1·00
0·25	$8\cdot15\times10^{4}$	8·5	$3\cdot26\times10^{-1}$	$8\cdot16\times10^{4}$	8·5	$3\cdot26\times10^{-1}$	1·00
0·5	$2\cdot70\cdot10^{4}$	12·6	$1\cdot08\times10^{-1}$	$2\cdot70\times10^{4}$	12·6	$1\cdot08\times10^{-1}$	1·00
0·75	$9\cdot91\times10^{3}$	17·8	$3\cdot96\times10^{-2}$	$9\cdot94\times10^{3}$	17·8	$3\cdot97\times10^{-2}$	1·00
1·0	$4\cdot01\times10^{3}$	23·8	$1\cdot60\times10^{-2}$	$4\cdot04\times10^{3}$	23·9	$1\cdot61\times10^{-2}$	1·01
1·5	$9\cdot13\times10^{2}$	36·9	$3\cdot65\times10^{-3}$	$9\cdot42\times10^{2}$	38·1	$3\cdot76\times10^{-3}$	1·03
2·0	$2\cdot45\times10^{2}$	38·3	$9\cdot77\times10^{-4}$	$2\cdot73\times10^{2}$	42·7	$1\cdot09\cdot10^{-3}$	1·12
3·0	$4\cdot72\times10^{1}$	37·5	$1\cdot88\times10^{-4}$	$7\cdot44\times10^{1}$	59·1	$2\cdot97\times10^{-4}$	1·58
4·0	$1\cdot74\times10^{1}$	39·1	$6\cdot96\times10^{-5}$	$4\cdot04\times10^{1}$	90·5	$1\cdot61\times10^{-4}$	2·32
6·0	$4\cdot87\times10^{0}$	65·9	$1\cdot94\times10^{-5}$	$1\cdot74\times10^{1}$	235	$6\cdot94\times10^{-5}$	3·57
8·0	$2\cdot01\times10^{0}$	-	$8\cdot03\times10^{-6}$	$8\cdot35\times10^{0}$	-	$3\cdot33\times10^{-5}$	4·15
12·0	$5\cdot23\times10^{-1}$	-	$2\cdot09\times10^{-6}$	$2\cdot51\times10^{0}$	-	$1\cdot00\times10^{-5}$	4·81
16·0	$1\cdot71\times10^{-1}$	-	$6\cdot83\times10^{-7}$	$8\cdot90\times10^{-1}$	-	$3\cdot55\times10^{-6}$	5·20

As for the current alone, further calculations of shape function have been made with $z_0/h=10^{-4}$, and in Figure 7 these results have been included on the graph of I_+ versus S_+ presented by Soulsby(1988). Here the parameter S_+ is equal to $(C_{D+}\bar{V}/3W_s)$, where the drag coefficient $C_{D+} = |<\tau_0(t)>|/\rho\bar{V}^2$. As explained in detail by Soulsby, the results shown are for the values (i) $\alpha_+=5$ where α_+ is the ratio of the maximum (instantaneous) to mean (cycle-averaged) value of u_* in the wave-current cycle; and (ii) $\beta_+=100$ where β_+z_0 is the thickness of the oscillatory boundary layer. The various curves included in Figure 7 have been obtained from different assumed functional forms (shown in the inset) for the time-invariant eddy viscosity in wave-current flow. As compared with the results in Figure 6, the flux does now depend significantly upon the vertical profile of eddy viscosity. As far as the present model results are concerned, a consistent comparison can be made only between I_{+1} and the other curves plotted, since the wave-related contribution has not been included in Soulsby's calculations. However, values for I_{+2} are also included in Figure 7. These indicate a marked difference, by a factor of about 6, between I_{+1} and I_{+2} for small S_+, for the reasons discussed earlier. Despite this difference, the present values all fall well within the wide range of uncertainty displayed by the other model results.

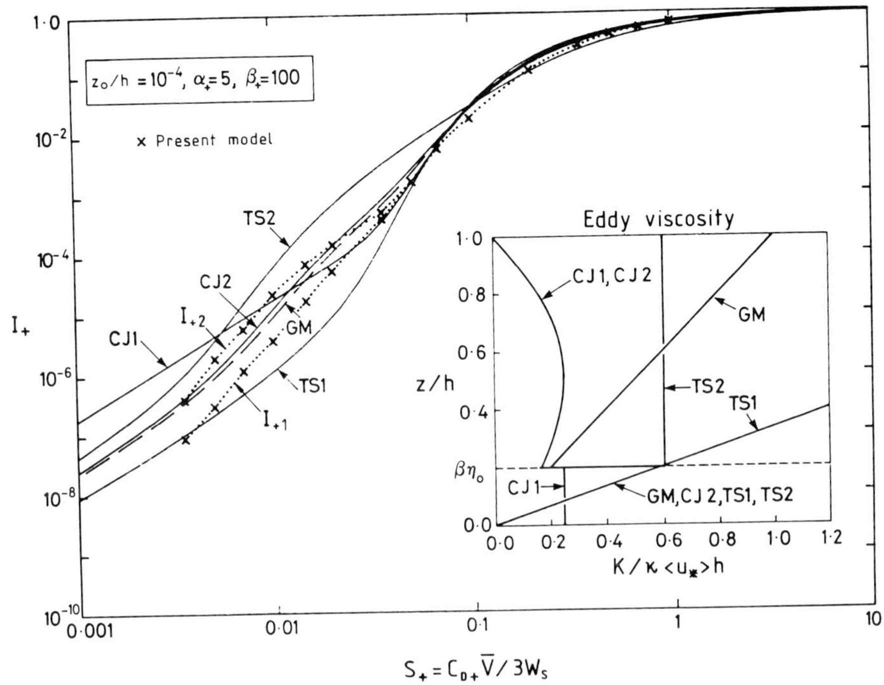

Figure 7: Shape function I_+ versus S_+ for wave-current flow with $z_0/h=10^{-4}$, $\alpha_+=5$ and $\beta_+=100$. The labelled curves correspond to the (time-invariant) vertical profiles of eddy viscosity shown in the inset, as explained in detail by Soulsby(1988) (κ = Von Karman's constant). The present model results for I_{+1} and I_{+2} are shown by the crosses.

22.5 Discussion

The results in Sections 22.3 and 22.4 were obtained with the sediment diffusivity set equal to the eddy viscosity (ß=1) and with N=1 in the formula for the reference concentration. These assumptions are now discussed.

22.5.1 Effect of changing ß

For steady flow the sediment continuity equation (see Figure 1) reduces to

$$\varepsilon_s \frac{dC}{dz} + W_s C = 0 \tag{22.11}$$

It follows that if $\varepsilon_s = \beta K$, where β is independent of z,

$$K \frac{dC}{dz} + (W_s/\beta) C = 0 \tag{22.12}$$

For unsteady flow, β cannot be absorbed into the settling term in this way on account of the time rate of change term $\partial c/\partial t$. In order to assess the importance of this term in the context of the earlier case studies, model runs have been performed in which both β and W_s have been varied at a fixed value of W_s/β.

For the waves alone, halving the value of β from 2 to 1 at W_s/β = 0.5cm/s, and from 1 to 0.5 at W_s/β = 2cm/s, produces in each case an increase of 5% in the cycle-averaged, depth-averaged concentration. This increase, which would not occur with a time-invariant eddy viscosity, is attributable to the combined effect of unsteadiness in the eddy viscosity and sediment concentration distributions. For the wave-current case, the results arising from the same changes in β differ depending upon the value of W_s/β. The effect of halving β from 2 to 1 with W_s/β = 0.5cm/s is to increase $\overline{\langle c \rangle}/C_{0c}$ by 28%, and to increase both q_{+1} and q_{+2}, and hence I_{+1} and I_{+2}, by 32%. The effect of halving β from 1 to 0.5 with W_s/β = 2cm/s is to increase both $\overline{\langle c \rangle}/C_{0c}$ and also q_{+1} and I_{+1} by 6%, but to leave q_{+2} and I_{+2} virtually unchanged. This indicates that the effect of changing β is rather less for coarse grains confined to the near-bed region than for finer grains which are diffused into the outer part of the flow.

It would appear from these results that the effect of the time rate of change term $\partial c/\partial t$ is significant, and that the value of β should not, in general, be absorbed into W_s in the way indicated above. It should be added that β may, in reality, be a function of z, and that our understanding of its behaviour is presently incomplete (Dyer and Soulsby,1988).

22.5.2 Effect of changing the power N

Previous investigators have suggested different values for the power N in their formulae defining the reference concentration. In order to assess the sensitivity of the case study results to changes in N, calculations for waves alone and for the combined wave-current case have been repeated with N=1/2 and 3/2 for the values W_s = 4.0, 1.0 and 0.25cm/s. For W_s = 4cm/s the sediment grains are confined to the near-bed layer, and for 0.25cm/s they are well diffused into the outer part of the flow. At the bed level $z=z_0$ the effect of changing N is to alter significantly the value of $\langle c(z_0,t) \rangle$. For waves alone $\langle c \rangle/C_{0c}$ takes the values 2.079, 4.829 and 11.87 for N=1/2, 1 and 3/2, respectively. For the wave-current case the values are 2.098, 5.019 and 12.89, respectively. These variations at the bed exert the primary control on the depth-averaged values $\overline{\langle c \rangle}/C_{0c}$. In particular, the values obtained for $\overline{\langle c \rangle}/C_{0c}$ with N=1/2 and 3/2 fall within 5% of the value obtained by scaling the model result for N=1 by the appropriate bottom value <c(z_0,t)>. Similar comments can be made about the results for transport rate q_{+i} (i=1,2) listed in Table 4, though for the case of W_s=4cm/s the variations are more pronounced than those obtained by a simple scaling based on <c(z_0,t)>. This is apparent in the results for I_{+i} (i=1,2). If changes in I_+ with N are expressed in terms of the value for N=1 in each case then, for W_s=0.25cm/s, both I_{+1} and I_{+2} change by -4% with N=1/2 and +3% with N=3/2. For W_s=1.0cm/s, the respective changes in both I_{+1} and I_{+2} are -6% and +5%. For W_s=4.0cm/s the values for I_{+1} are -10% and +8%, and for I_{+2} are -25% and +23%, indicating again the important role of the wave-related contribution to the flux.

It may be concluded that the choice of power law behaviour has a strong influence on the results as far as the average bottom concentration <c(z_0,t)> is concerned, but that its effect on the shape factor I_+ is relatively small, except for grains confined to the oscillatory boundary layer.

Table 4: Mean sediment transport rate (q_+ per unit width) and shape factor (I_+) for combined wave-current flow. Values for power N in the reference concentration formula are as indicated. Flow parameters are as for Table 2.

W_s (cm/s)	q_{+1}/c_{0c} (cm^2/s)			q_{+2}/c_{0c} (cm^2/s)		
	N=1/2	N=1	N=3/2	N=1/2	N=1	N=3/2
0·25	$3\cdot28\times10^{4}$	$8\cdot15\times10^{4}$	$2\cdot16\times10^{5}$	$3\cdot28\times10^{4}$	$8\cdot16\times10^{4}$	$2\cdot17\times10^{5}$
1·0	$1\cdot59\times10^{3}$	$4\cdot01\times10^{3}$	$1\cdot08\times10^{4}$	$1\cdot59\times10^{3}$	$4\cdot04\times10^{3}$	$1\cdot09\times10^{4}$
4·0	$6\cdot56\times10^{0}$	$1\cdot74\times10^{1}$	$4\cdot85\times10^{1}$	$1\cdot27\times10^{1}$	$4\cdot04\times10^{1}$	$1\cdot28\times10^{2}$
	I_{+1}			I_{+2}		
	N=1/2	N=1	N=3/2	N=1/2	N=1	N=3/2
0·25	$3\cdot13\times10^{-1}$	$3\cdot26\times10^{-1}$	$3\cdot37\times10^{-1}$	$3\cdot13\times10^{-1}$	$3\cdot26\times10^{-1}$	$3\cdot37\times10^{-1}$
1·0	$1\cdot51\times10^{-2}$	$1\cdot60\times10^{-2}$	$1\cdot68\cdot10^{-2}$	$1\cdot52\times10^{-2}$	$1\cdot61\times10^{-2}$	$1\cdot70\times10^{-2}$
4·0	$6\cdot26\times10^{-5}$	$6\cdot96\times10^{-5}$	$7\cdot54\times10^{-5}$	$1\cdot21\times10^{-4}$	$1\cdot61\times10^{-4}$	$1\cdot99\times10^{-4}$

22.5.3 Phase of c(z_0,t) in relation to the bed shear stress τ(z_0,t)

The formula for the reference concentration which has been used throughout is based on the assumption of an instantaneous response of c(z_0,t) to changes in the bed shear stress. This assumption may be oversimplistic, since grains in close proximity to the bed may take time to respond to changes in τ(z_0,t). In other words, a phase lag may be appropriate in the specification of the reference concentration, as follows:

$$c(z_0,t) = C_{0c}|\tau(z_0,t-\Delta t)/\tau_{0c}|^N \tag{22.13}$$

Such an approach has been taken, in effect, by Bailard(1981) who neglected the phase difference between the bed shear stress and the near-bed fluid velocity and assumed the time-varying sediment transport rate to be in phase with the free-stream velocity.

A number of runs have been conducted to test the sensitivity of the model results to variations in Δt. For a lag Δt corresponding to 36° of phase (such that the reference concentration is roughly in phase with the outer free-stream velocity), the depth-averaged concentration for waves alone increases by 12% for W_s=1cm/s and by 3% for W_s=4cm/s. These results were obtained with N=1; percentage changes were slightly less with N=1/2, and slightly greater with N=3/2. For the case of wave-current flow with W_s=4cm/s and N=1, the same phase lag gives rise to a 5% increase in $\overline{\langle c\rangle}/C_{0c}$, a 12% increase in I_{+1} and an 11% decrease in I_{+2}. A more complete account of these comparisons will be given elsewhere. This is a topic which warrants careful investigation, not least in connection with vortex formation and shedding above rippled beds, which is known to be important in the process of sediment entrainment (Nielsen, 1988).

22.6 Conclusions

A substantial enhancement in the suspended sediment transport rate may occur when waves are superimposed on a current. This arises from the combined effect of an increase in the time-averaged, bottom reference concentration, and an increase in the vertical mixing of sediment in the water column. The results of a case study indicate enhancements in the transport rate of between one and two orders of magnitude, depending upon the sediment settling velocity.

For sediment grains which are well diffused into the outer part of a wave-current flow, a good approximation to the net sediment flux may be obtained from the product of the cycle-averaged profiles of horizontal velocity $\langle u\rangle$ and sediment concentration $\langle c\rangle$. However, for larger grains confined to the oscillatory boundary layer, the "wave-related" contribution to the flux (incorporated in the term $\langle uc\rangle$) may dominate the net transport.

The model results have been expressed in terms of the "shape factor" of Dyer and Soulsby(1988) and Soulsby(1988), which represents a non-dimensionalisation of the sediment flux by the product of the volumetric flow rate (per width of the flow) and the bottom reference concentration (cycle-averaged in the case of wave-current flows). For currents alone, the present model is in close agreement with all previous models, confirming that the shape function is insensitive to the form of the vertical profile of eddy viscosity. However, for wave-current flows, the shape function is strongly dependent upon the eddy viscosity profile, particularly for grains confined to the oscillatory boundary layer. The predictions of the present model for such cases fall well within the broad band of uncertainty associated with previous models.

The present results are dependent upon the assumed relationship between the (instantaneous) bed shear stress and the (instantaneous) bottom reference concentration. The power law relationship between these quantities exerts a significant influence on the transport rate. This arises primarily through changes in the reference concentration, and only to a small extent through the shape function. The phase relationship between the two quantities also affects the results, not least in respect of the "wave-related" contribution to the flux. This is a matter which warrants further investigation.

Acknowledgements

Richard Soulsby (Hydraulics Research Ltd., Wallingford) is thanked for his helpful comments, and also for allowing use to be made of Figures 6 and 7. Bob Everitt and Angela Williams are thanked for their help with the figures and the text. Financial support for the work was provided by the U.K. Ministry of Agriculture, Fisheries and Food.

22.7 References

Bailard, J.A., 1981: An energetics total load sediment transport model for a plane sloping beach. Journal of Geophysical Research, 86, C11, 10938-10954.

Bakker, W.T., and T. van Doorn, 1979: Near-bottom velocities in waves with a current. Proceedings of the 16th International Conference on Coastal Engineering, 1978, American Society of Civil Engineers, 1394-1413.

Bijker, E.W., 1967: The increase of bed shear in a current due to wave motion. Proceedings of the 10th International Conference on Coastal Engineering, 1966, American Society of Civil Engineers, 746-765.

Davies, A.G., 1986: A numerical model of the wave boundary layer. Continental Shelf Research, 6, 6, 715-739.

Davies, A.G., 1990: A model of the vertical structure of the wave and current bottom boundary layer. pp.263-297 in Modeling Marine Systems, Vol.2, ed A.M.Davies, CRC Press.

Davies, A.G., 1991: Transient effects in wave-current boundary layer flow. Ocean Engineering, 18, 1/2, 75-100.

Davies, A.G., Soulsby, R.L., and H.L.King, 1988: A numerical model of the combined wave and current bottom boundary layer. Journal of Geophysical Research, 93, C1, 491-508.

Davies, A.G., Barber, M.E., and R.L.Soulsby, 1990: Vertical current structure and bedload sediment transport in a combined wave-current boundary layer flow. pp.52-67 in Physics of Shallow Seas, eds. Wang H.,Wang J. and Dai H., China Ocean Press.

Dyer, K.R., and R.L.Soulsby, 1988: Sand transport on the continental shelf. Annual Review of Fluid Mechanics, 20, 295-324.

Fredsoe, J., 1984: Sediment transport in current and waves. Institute of Hydrodynamics and Hydraulic Engineering, Series Paper 35, Technical University of Denmark, Lyngby, 37pp.

Grant, W.D., and Madsen, O.S., 1979: Combined wave and current interaction with a rough bottom. Journal of Geophysical Research, 84, C4, 1797-1808.

Grass, A.J., 1981: Sediment transport by waves and currents. Report No. FL29, SERC London Centre for Marine Technology, 26pp.

Hagatun, K., and K.J. Eidsvik, 1986: Oscillating turbulent boundary layer with suspended sediments. Journal of Geophysical Research, 91, C11, 13045-13055.

King, H.L., Davies, A.G., and R.L. Soulsby, 1985 : A numerical model of the turbulent boundary layer beneath surface waves and tides. Institute of Oceanographic Sciences Report No. 196, 90pp.

Nezu, I., and W.Rodi, 1986: Open-channel flow measurements with a laser Doppler anemometer. Journal of Hydraulic Engineering, 112, 5, 335-355.

Nielsen, P., 1988: Three simple models of wave sediment transport. Coastal Engineering. 12, 43-62.

Shi, N.C., Larsen, L.H., and J.P.Downing, 1985: Predicting suspended sediment concentration on continental shelves. Marine Geology, 62, 255-275.

Smith, J.D., 1977: Modeling of sediment transport on continental shelves, pp. 539-577 in The Sea, Vol. 6, eds. Goldberg, E.D., McCave, I.N., O'Brien, J.J., and Steele, J.H., Interscience, New York, 1977.

Soulsby, R.L., 1987: The relative contributions of waves and tidal currents to marine sediment transport. Hydraulics Research Ltd., Wallingford, Report SR 125.

Soulsby, R.L., 1988: The structure of suspended sediment transport formulae for unidirectional and wave-plus-current flows. Proceedings of the Symposium on Mathematical Modelling of Sediment transport in the Coastal Zone. Danish Hydraulic Institute, May 1988, 68-78.

Soulsby, R.L., and B.L.S.A. Wainwright, 1987: A criterion for the effect of suspended sediment on near-bottom velocity profiles. Journal of Hydraulic Research, 25, 3, 341-356.

23

Some considerations on mathematical modelling of morphological processes in tidal regions

Z B Wang

Abstract

The construction and application of a mathematical model for morphological development in tidal regions is considered. The model consists of a flow module, a sediment transport module and a bed level change module. The structure of the model and the choice of the flow module and sediment transport module are discussed. The application aspects of the model are considered with a case study for the Yantze estuary.

23.1 Introduction

The morphological development of estuarine and coastal regions is a long term process. The time scale of morphological processes may range from weeks to decades or even centuries. Morphological development changes the geometry and bathymetry of a region and thereby it influences the flow pattern and all other physical processes in the region. Therefore knowledge about the morphological development is important for hydraulic engineering as well as for studying long term variability of other physical processes in estuarine and coastal regions.

Morphological development of estuarine and coastal regions is a very complicated process. Many factors are involved in this process: river flow, tide, density flow, wind and waves, sediment properties, etc. Although considerable knowledge exists on all these individual elements, very little is known how to include them into a mathematical model for morphological development. Mathematical modelling of this process is still in an early stage.

At Delft Hydraulics the morphological model DELMOR is under development. At this moment it is not wise to try to develop a general applicable model by including all the significant factors one can imagine. This will lead to a very complicated model. Little is known about the behaviour of such a complicated model. Therefore the step by step strategy is followed for the development of DELMOR. In the first step the model will only include river flow and tidal motion as driving forces for sediment transport. Density flow, wind and waves are not taken into account for the time being. Further it is restricted to non-cohesive sediment. The fundamental aspects of such a simplified model were studied by Wang (1989), who developed the research model ESMOR. Some considerations on the construction and application of such a relative simple morphological model are made in this paper. In the future when the behaviour of such a simpler model is sufficiently understood the model will be extended step by step by taking more factors into account.

In the following section the structure of the model is considered. The model consists of three sub-models viz. the flow model, the sediment transport model and the bed level model. The three individual sub-models are discussed in section 23.3. Section 23.4 considers the application aspects of the model. Attention is paid to the dominating condition and the choices of the major parameters controlling the computation procedure. In this section the case study for the Yantze estuary is used as an example. In section 23.5 some numerical aspects are considered. Attention is paid to the modelling of the source term and the boundary conditions. Finally the major conclusions are summarised in section 23.6.

23.2 Structure of the model

Despite of the impressive advance in computer technology, a major problem for the mathematical modelling of morphological processes is still the computational cost. Because of the complexity and the long running time of the model, an improperly designed model will cost too much computation time making it useless in practice.

The morphological development involves the variations of flow, sediment transport and bed level. These three subprocesses all interact with each other. This means that the equations describing the flow field variables (h, u, v,...), the sediment transport variables (c, s_x, s_y...) and the bed level (z_b) are coupled with each other. If these coupled equations are solved together the time step used should be much smaller than the smallest significant time scale, i.e. the tidal period. Obviously such a computation for a long period (decades or even centuries) is impossible. In other words, simplifications in morphological models are virtually inevitable.

In order to simplify the problem two important assumptions are made. The first one is the so called low concentration assumption: the influence of the sediment transport on the flow field is negligible. This assumption is justified in most estuarine and coastal regions since the sediment concentration by volume mostly is of the order of magnitude of 10^{-3}. This assumption makes it possible to separate the flow module and the sediment transport module into uncoupled sub-models. The second assumption is the so called quasi-steadiness assumption: the large difference between the morphological time scale and the time scale of the flow makes it possible to separate the flow computation and the bed level computation. This assumption is justified for small and moderate Froude numbers (de Vries, 1959, 1965), common in most estuarine and coastal regions. This assumption makes it possible to distinguish between the long term variation processes (i.e. the bed level change) and the short term variation processes (i.e. the flow and the sediment transport). It is then possible to apply two different time steps in the model, one for the bed level change and one for the flow and sediment transport.

The two assumptions make it possible to divide the morphological model into three sub-models, viz. the flow model, the sediment transport model and the bed level model. In one morphological time step these three sub-models have to be run successively. It is further assumed that the flow field and the sediment transport field remain periodic during several tidal periods. In other words the morphological time step can be as large as a number of tidal periods. Furthermore, the bed level change after one morphological time step is often so small that the continuity correction method can be applied for calculating the flow field during the next few steps (Hauguel, 1978, de Vriend, 1985). In this method the water level and the discharge are assumed to be the same as in the previous step. The water depth and the velocity field change only due to the bed level change. All these considerations together lead to a structure of the model as shown in fig.1.

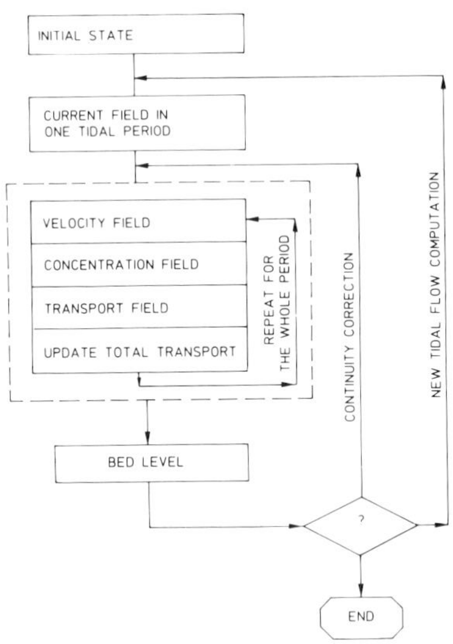

Figure 1: Structure of the model

The morphological time step and the number of times of continuity correction after each complete tidal flow computation are two of the major parameters in the model and the choice of them will be the subject of a discussion in section 23.4.

23.3 The three sub-models

As explained in the previous section, the morphological model under consideration can be divided into three sub-models, viz. the flow model, the sediment transport model and the bed level model. Much work was done on the individual sub-models by many researchers. Different combinations of all types of the sub-models are possible. However, the construction of the morphological model as a whole is less elaborated. Therefore the choice of each sub-model is paid attention to in this section. These choices should be based on the validity (accuracy) and applicability (cost) considerations. The details of the individual sub-models self are not considered here.

The Flow Model: In the case of estuarine and coastal problems one can choose from two possibilities for the flow model, viz. the two-dimensional (horizontal) and the three-dimensional model. At present, computer capacity restricts affordable long term morphological computations based on a fully 3D flow model. Therefore the quasi-3D approach is chosen in ESMOR as well as in DELMOR. The depth averaged flow is computed with a 2DH tidal flow model. A profile model is applied to describe the vertical distribution of the velocity in the longitudinal direction and a secondary flow model is applied in order to take into account the velocity component in the normal direction. In DELMOR, the depth averaged flow is computed with TRISULA. The secondary flow model takes into account the influence of the curvature of the streamlines of the depth averaged flow and the geostrophic acceleration (de Vriend, 1981, Kalkwijk and Booij, 1986). It should be mentioned that TRISULA is a multi-dimensional flow model developed at Delft Hydraulics. which is able to perform fully 3D computations. This makes it easier to extend DELMOR to a full 3D model in the future if necessary.

The Sediment Transport Model: The sediment transport is divided into bed load transport and suspended transport. The bed load transport is assumed to adjust instantaneously to the local transport capacity which is related to the local flow conditions via a transport formula. For the suspended transport one can choose from various possibilities varying from a simple transport formula to a sophisticated fully 3D model (van Rijn, 1987, Wang, 1989). The available computer capacity restricts the possibility to apply the full 3D transport model to be applied in the long term morphological computations. On the other hand the simple transport formula is usually insufficient to model the suspended sediment transport in a varying flow field as in an estuary. Therefore the quasi-3D approach, originally developed by Galappatti (1983, see also Galappatti and Vreugdenhil 1985) and later extended by Wang (1989), is chosen.

This quasi-3d model is derived by solving the full 3D sediment concentration equation asymptotically. From the asymptotic solution of the 3D sediment concentration equation a 2DH equation is derived for the depth averaged sediment concentration.

$$\frac{\partial c}{\partial t} + \gamma u \frac{\partial c}{\partial x} + \gamma v \frac{\partial c}{\partial y} - \frac{\partial}{\partial x}(D_x \frac{\partial c}{\partial x}) - \frac{\partial}{\partial y}(D_y \frac{\partial c}{\partial y}) = \frac{c_e - c}{T_a} \tag{23.1}$$

in which

c = depth averaged sediment concentration;
u, v = depth averaged velocity components;
T_a = adaptation time;
γ = coefficients;
c_e = depth averaged equilibrium sediment concentration.

The term on the right hand side of this equation is a source term. When the actual concentration c is smaller than the equilibrium value c_e erosion occurs causing increase of c. When c is larger than c_e sedimentation occurs causing decrease of c. The coefficient γ represents the influence of the non-uniform distribution of the sediment concentration and the flow velocity over the depth on the advection.

After solving this mean concentration equation the sediment transport rates can be calculated. Detailed information about this model is given in Galappatti (1983) and Wang (1989).

Remark: Equation (23.1) applies for the case of sand. In case of cohesive sediment the right hand side of the equation should be replaced by the sedimentation and erosion rate (Krone, 1962, Mehta and Partheniades, 1982).

Extensive analyses, theoretically as well as experimentally, on the validity of this model was carried out (Wang and Ribberink, 1986, Wang, 1989, 1990). Fortunately, the analyses show that this quasi-3D model is sufficiently accurate to replace the full 3D model in most practical problems in estuarine and coastal regions.

The Bed Level Model: The bed level model is the most simple one of the three sub-models. The bed level change is described by the mass-balance equation for sediment.

$$\frac{\partial z_b}{\partial t} + \frac{\partial s_x}{\partial x} + \frac{\partial s_y}{\partial y} + \frac{\partial (hc)}{\partial t} = 0 \tag{23.2}$$

This equation is exact and has to be solved with the large time scale so it should be integrated over the tidal period.

$$\frac{\Delta z_b}{NT} + \frac{\partial T_x}{\partial x} + \frac{\partial T_y}{\partial y} = 0 \tag{23.3}$$

In this equation

Δz_b = bed level change in N tidal periods;
T = tidal period;
NT = morphological time step;

T_x, T_y= total sediment transport per unit of width during a tidal period.

23.4 The dominating condition and major parameters

Figure 1 shows implicitly that a typical tide is used to describe the dominating condition and that the major parameters characterizing this computation procedure are the morphological time step and the number of continuity corrections after a new tidal flow computation. The dominating condition as well as the major model parameters are discussed in the following for a practical case study of the morphological process in the Yantze Estuary in China (fig.2). The results presented are computed with ESMOR. The conclusions drawn from this case cannot directly be applied to other cases but it may provide useful insight into this kind of problems.

The Yantze river with a length of 6380 km and a catchment area of 1.8 million km^2 is the largest river in China. It flows into the East China Sea north of Shanghai. Its estuary covers an area of about 100x100 km^2. Unfortunately during the case study it was impossible to take the whole estuary in the model, because of the restricted computer capacity as well as lack of field data. However, the most interesting part of the estuary, i.e. the south branch, where also the most field measurements have taken place, is modelled. The model covers an area of about 80x20 km^2 (see fig.2). With a space step of 1 km in both horizontal directions the model contains about 1000 active computational grid points.

The flow conditions as well as the sediment transport conditions in the estuary are varying constantly. The variations even contain certain stochastic components. Nevertheless two systematic variation patterns can be recognized, viz. the spring-neap tidal variation and the seasonal variation. These two components of variations are dominating in many other estuaries as well.

Both the spring-neap tidal variation and the seasonal variation are considered for defining the dominating condition. In order to examine the spring-neap tidal variation, a one week measuring period (from 28 Aug. to 4 Sept. 1984) is simulated. During this survey the velocity and sediment transport in 15 stations distributed in 3 cross sections in the model area were measured during 3 days (the spring tide, averaged tide and the neap tide). The time step for the flow computation is 5 minutes and the bed level was adjusted each day. The calculated total sediment transport field and the sedimentation-erosion pattern during spring tide, averaged tide and neap tide are shown in fig.3 and fig.4. The transport patterns as well as bed level change patterns are similar during spring tide, averaged tide and neap tide. Therefore it is concluded that it is perfectly possible to carry out morphological computations with a properly chosen standard tide.

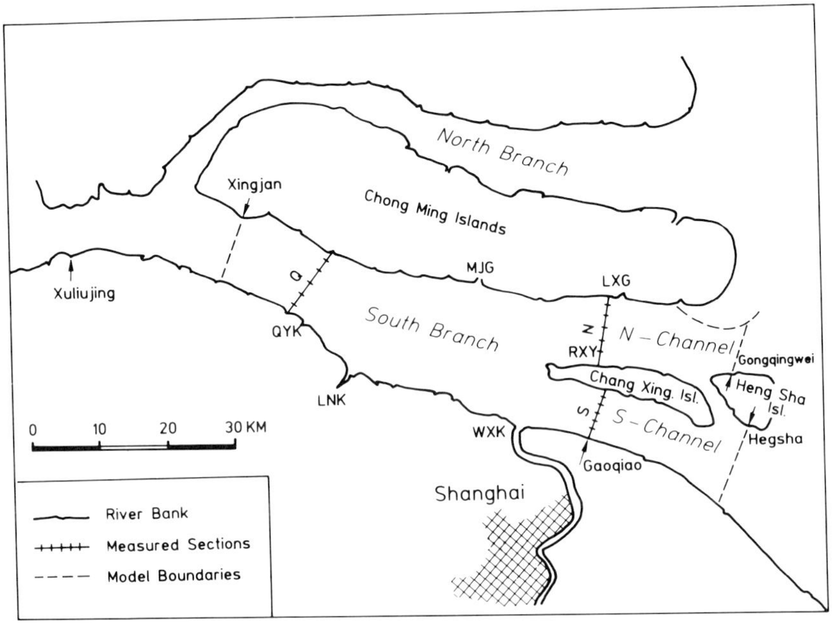

Figure 2: The Yantze Estuary

The choice of the standard tide should be based on the net sediment transport during the tidal period. The net transport in every cross section has to be about the averaged net transport in the spring-neap tide cycle. In fig.5 the computed total ebb- and flood-transport through the three measured cross sections during the survey are depicted together with the available measurements. The figure shows that the averaged total ebb- as well as flood- transport are represented by a tide between the averaged tide and the spring tide. However, concerning the net transport, i.e. the difference between the ebb transport and the flood transport, the averaged value is represented by the averaged tide. Therefore the averaged tide is chosen as the standard tide in the long term morphological computations.

The seasonal variation would have to be studied in a similar way but the available data did not allow this. Therefore this variation is taken into account simply by adjusting the upstream boundary conditions and the coefficient in the bed boundary condition in such a way that the averaged annual sediment transport through the estuary is correctly reproduced.

The morphological time step Δt_m and the number of continuity corrections M_r are model parameters only. They do not have any physical meaning and they have to be chosen such that the computational effort required is as small as possible while sufficient accuracy of the computations is ensured. For this purpose a sensitivity analysis was carried out. In this analysis a number of runs have been carried out with different values of Δt_m and M_r. The computed bed level changes after 6 months from these runs are compared with each other. From the comparison it is concluded that Δt_m can be as large as 40 times the tidal period (i.e. about 20 days), and that M_r can be as large as 3.

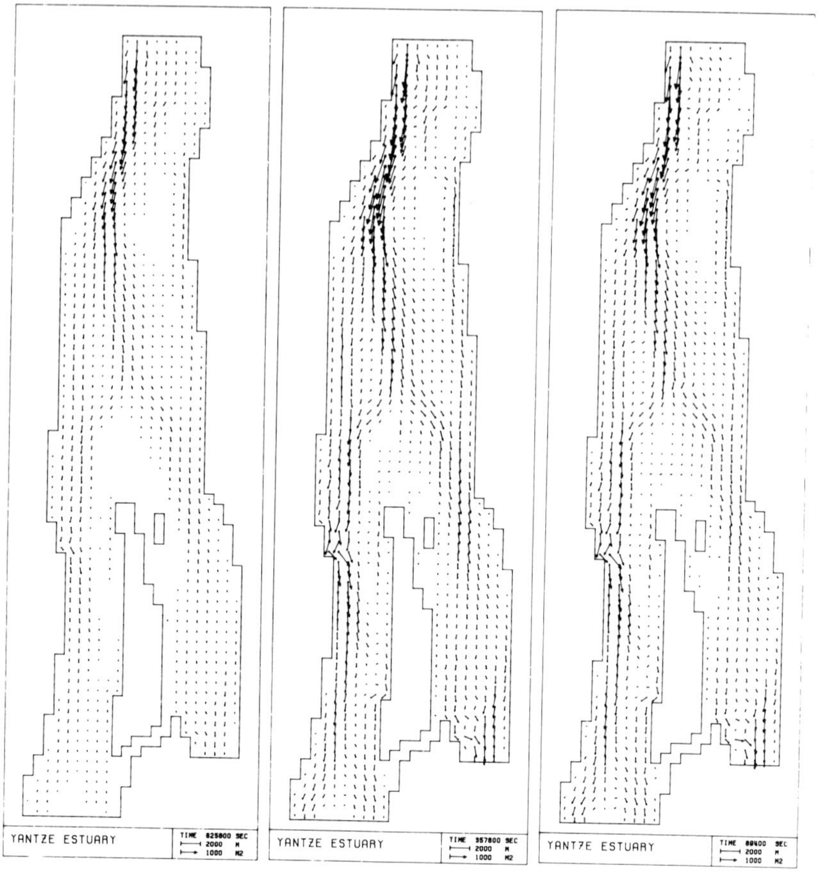

Figure 3: Net sediment transport during spring tide (top), averaged tide (middle) and neap tide (bottom)

It should be pointed out that if the simple transport formula mentioned in section 3 is applied instead of the quasi-3D model for the sediment transport, a time step much smaller than 40T would be required because of the CFL stability criterion. Because of the complicated character of the problem the bed level equation (23.3) can only be solved explicitly. A Courant number larger than 1 is only possible if relaxation effects of the suspended sediment transport are taken into account. This means that the more sophisticated quasi-3D model is not necessarily more expensive than the simpler equilibrium transport model in a long term morphological simulation.

Figure 4: Bed level change during spring tide (top), averaged tide (middle) and neap tide (bottom)

An interesting result was that with M_r=3 the computational effort used for the flow computation and that for the sediment transport computation are about equal. This is remarkable as in the sediment transport model one differential equation (i.e. the concentration equation) has to be solved while in the flow computation 3 differential equations have to be solved. For a totally different case, (steady flow in a meandering river section (Waal in the Netherlands) with only bed load transport), it also appeared that the most efficient computation is obtained when the effort for flow computation and that for the sediment transport computation are about equal. It seems to be that in a well constructed morphological model the computational effort for flow and for the sediment transport should be of the same order of magnitude.

After the determination of the model parameters some sensitivity analyses on the physical parameters, e.g. the fall velocity and the bed boundary condition etc., were carried out. Next, the model was calibrated and a 5 year simulation was carried out. Detailed information about this simulation is given in Wang (1989). Here only some results of the 5 year simulation are illustrated in fig.6 and fig.7.

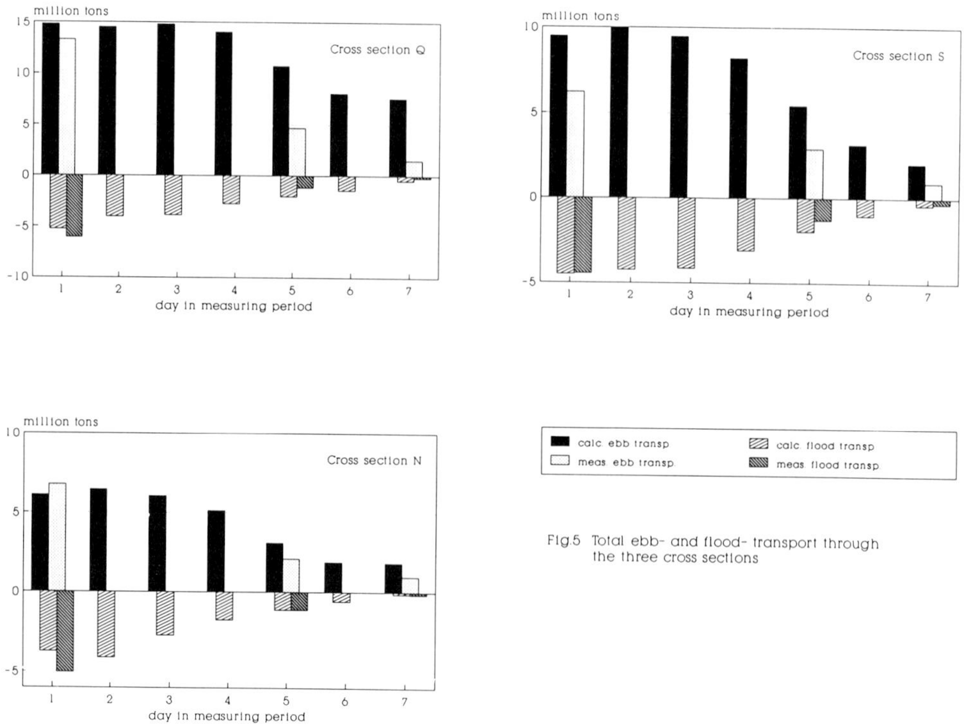

Figure 5: Total ebb- and flood transport through the three cross sections

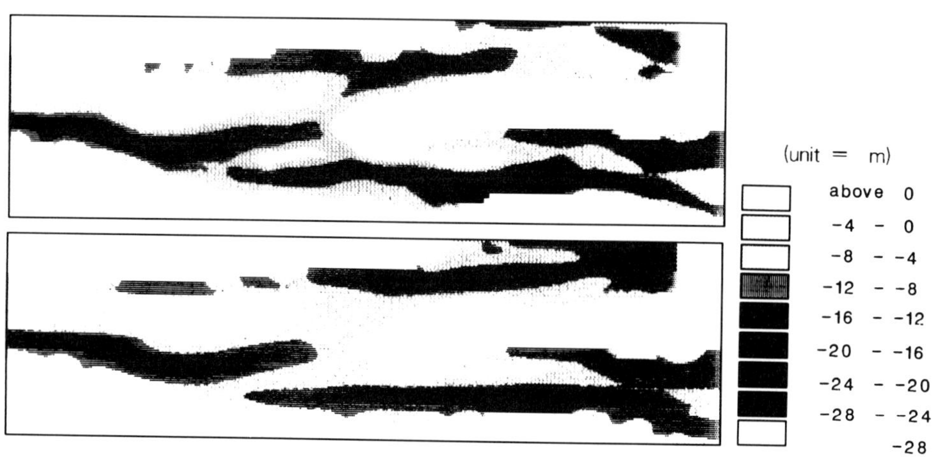

Figure 6: The initial bed level (top) and the calculated bed level after 5 year

23.5 Numerical aspects

It is not the intention of this paper to give a detailed numerical analysis for the model. Only two problems are discussed. The first one is about the discretization of the source term in the mean sediment concentration equation (23.1) and the second one is about the upstream boundary condition for this equation.

The source term (right hand side) in the mean concentration equation (23.1) is a linear term. It is strongly recommended to discretise this term implicitly, even when an explicit scheme is applied to the rest of the equation. This is possible because this term is a linear term and it only depends on the local sediment concentration. In this case the total scheme will remain effectively explicit. If an explicit discretization is applied to the source term, the stability of the computation can be strongly affected. This is especially important when shallow regions are present in the model area. In the shallow regions the adaptation time T_a (~ h/w_s) is small and the factor $\Delta t/T_a$ can be extremely large. For an explicit scheme an absolute limit for this parameter is $\Delta t/T_a<2$. Especially when an existing solver for the convection-diffusion equation is used, one tends to build the model simply by adding the source term to it with an explicit discretization. From the point of view of robustness of the model this should certainly be avoided.

In the case of the Yantze Estuary, data were available on the sediment transport neither at the upstream nor at the downstream model boundary. This is a common rather than exceptional problem of morphological modelling. Therefore some attention is paid to this problem here.

During inflow at the upstream boundary the sediment concentration or the sediment transport has to be prescribed. A wrongly chosen upstream boundary condition disturbs the computation at the boundary region. The disturbance propagates into the model area as the computation proceeds. Its celerity is the small propagation velocity of the bed forms, if the equilibrium transport model is applied. In morphological models for rivers this problem is sometimes solved by applying the boundary condition far away from the region of interest so that it is ensured that during the simulation the disturbance has no chance to enter the region of interest. However, in the quasi-3D model the disturbance will penetrate as far as the adaptation length of the suspended sediment transport ($\approx uT_a$~ $uh/w_s \approx 1000h$) after the very first morphological time step. In the case of the Yantze estuary model an erroneous boundary condition will affect the computation in the whole model area within a few morphological time steps. Moreover it is much more difficult to extend a 2DH model in the upstream direction of an estuary than in the case of a one-dimensional model for a river. Therefore another solution to this problem is needed.

In the Yantze Estuary model this problem is solved as follows. At the upstream boundary the dynamic equilibrium concentration is applied. The dynamic equilibrium concentration is the solution of the following equation:

$$\frac{\partial c}{\partial t} = \frac{c_e - c}{T_a} \qquad (23.4)$$

This condition is sufficient for solving the sediment concentration equation but it is not sufficient for the whole morphological model. This becomes clear if the steady case is considered. For the steady case equation (23.4) becomes

$$c = c_e \qquad (23.5)$$

which means in fact that the equilibrium transport rate at the upstream boundary is applied. In this case the bed level at the upstream boundary has also to be prescribed in order to run the whole morphological model. Fortunately the bed level change from 1984 to 1985 in the area is available, so for the short term runs the measured bed level change can be applied in combination with equation (23.4). For runs longer than 1 year the problem has been solved by recording the sediment transport rate at the upstream boundary during a whole tidal period in a short term simulation and using it as boundary condition in the long term simulation. This implies that the sediment transport through the upstream boundary will not change much during the simulation. This seems to be reasonable as the total sediment transport through the estuary is mainly determined by the river inflow in the case of the Yantze estuary.

The solution presented here is certainly not ideal. Without any doubt it will introduce some error in the computations. However, this method ensures that the bed level change varies smoothly in the boundary region. No significant disturbance due to the boundary condition can be observed in the computational results. With the available data it seems to be the best solution, if it is not the only solution.

23.6 Conclusions

Morphological development in estuarine and coastal regions is an important but also a very complicated process. Mathematical modelling of this process is only possible with some simplifications. In this paper the construction and the application of mathematical morphological models are considered.

The structure of the specific model discussed here is based on two important assumptions, viz. the low concentration assumption and the quasi-steadiness assumption. These two assumptions make it possible to distinguish three sub-models, i.e. the flow model, the sediment transport model, and the bed level model. The assumptions also make it possible to apply two different time steps in the model, one for the flow and the sediment transport model, the other for the bed level model. Further the continuity correction method has been applied in order to save computational costs.

Many possibilities exist for the flow model and the sediment transport model. Based on the validity (accuracy) and the applicability (cost) considerations the quasi-3D approach has been chosen for both sub-models.

In order to apply the model, a dominating condition in terms of a standard tide has to be defined. For determining this condition the spring-neap tidal cycle as well as the seasonal variation are considered. For a practical case it has been shown that it is possible to define a standard tide for the long term morphological simulation. In the case of the Yantze estuary it has been shown that the averaged tide can be used as standard tide. However, it should be pointed out that this can vary from case to case. This strongly depends on the upstream river inflow, the tidal force from the sea as well as the formulation used for the bed boundary condition.

The choice of the morphological time step and the number of times of continuity corrections were investigated with a sensitivity analysis. The conclusion is that a large morphological time step is allowed because of using the quasi-3D model for the sediment transport. It is also concluded that the continuity correction method is a useful tool for saving computational cost.

Further, the numerical schematization of the source term in the sediment concentration equation and the upstream boundary condition have been discussed. It is recommended to discretise the source term implicitly. For the case of lacking information at the upstream boundary a solution for the boundary condition is presented.

23.7 Acknowledgement

The main part of the work described in this paper is carried out during the preparation of the doctoral thesis at the Delft University of Technology. The writer wishes to express his appreciation of the advice and encouragement given by the supervisors prof. dr. M. de Vries and prof. dr. C.B. Vreugdenhil. Thanks are due to Mr. J.C. Winterwerp for his encouragement and constructive comments during the preparation of this paper.

23.8 References

Galappatti, R. (1983), A depth-integrated model for suspended transport, Report no. 83-7, Delft University of Technology, Dept. of Civil Eng.

Galappatti, R. and Vreugdenhil, C.B. (1985), A depth-integrated model for suspended sediment transport, J. of Hydr. Res., Vol. 23, no. 4.

Hauguel, A. (1978), Utilization des modeles mathematiques pour L' etude du transport solide sous l' action des courants de Maree, Report E42/78.41, EDF, Direction des Etudes et Recherches.

Kalkwijk, J.P.Th. and Booij, R. (1986), Adaptation of secondary flow in nearly horizontal flow, J. of Hydr. Res., Vol. 24, no. 1.

Krone, R.B. (1962), Flume studies of transport of sediment in estuarial environments, Final report, University of California, Hydraulic Engineering and Sanitary Engineering Laboratory, Berkeley.

Mehta, A.J. and Partheniades, E. (1982), Resuspension of deposited cohesive sediment beds, Proceedings of the XVIII Coastal Engineering Conference, ASCE, Vol. 2, Cape Town, South Africa, pp 1569_1588.

Rijn, L.C. van (1987), Mathematical modelling of morphological processes in the case of suspended sediment transport, Doctoral Thesis, Delft University of Technology

Vriend, H.J. de (1981), Steady flow in shallow channel bends, Doctoral thesis, Delft University of Technology, Delft, The Netherlands.

Vriend, H.J. de (1985), Flow formulation in mathematical models for 2Dh morphological Changes, Report no. R1747-5, Delft Hydr. Lab., Delft, The Netherlands.

Vries, M. de (1959), Transients in bed-load transport (Basic considerations), Report no. R3, Delft Hydr. Lab., Delft, The Netherlands.

Vries, M. de (1965), Considerations about non-steady bed load transport in open channels, Publ. no. 36, Delft Hydr. Lab., Delft, The Netherlands.

Wang, Z.B. (1984), The validity of a depth-integrated model for suspended sediment transport and the extension of this model to tidal rivers, Report no. 10-84, Delft University of Technology, Dept. of Civil Eng.

Wang, Z.B. and Ribberink, J.S. (1986), The validity of a depth integrated model for suspended sediment transport, J. of Hydr. Res., Vol. 25, no.1.

Wang, Z.B. (1989), Mathematical modelling of morphological processes in estuaries, Doctoral thesis, Delft University of Technology.

Wang, Z.B. (1990), Theoretical analysis on depth-integrated modelling of suspended sediment transport, paper submitted to J. of Hydr. Res.

24

A three-dimensional transport model for dissolved and suspended matter in estuaries and coastal seas

G C van Dam and R A Louwersheimer

Abstract

This paper describes a model which simulates transport of dissolved or suspended matter in three dimensions by sets of discrete particles being displaced in three dimensions at each time step. The transport model is uncoupled from the models providing the required three-dimensional deterministic velocity fields. The deterministic fields are supplemented with movements that account for vertical exchange and for sub-grid structures in the fields of horizontal velocities. These movements have a random character but in the horizontal directions they are organized (scaled) as to account for the spectral structure of the field in nature.

At present, the 3D deterministic fields are generated by an analytical extension in vertical direction of the velocity field computed by a horizontally two-dimensional numerical flow model. Distortion of vertical velocity distributions by wind or density gradients can also be taken into account. In all cases the analytical functions are given the simplest possible shape while care is taken that the vertically averaged velocities are not affected. The random walk in vertical direction corresponds with an exchange coefficient K_z that does not depend on vertical position. In the case of sediments, the particles are given a fall velocity and criteria for deposition and erosion at the bottom. The horizontal displacements by the depth dependent velocities are supplemented with scale dependent random displacements that take care of the "sub-grid" velocities not modelled by the numerical flow model. Computational results based upon the above principles are compared with data from tracer experiments for calibration. For various cases, computed concentration distributions at small and large scales are presented and discussed.

24.1 Introduction

The transport and dispersion of matter in natural bodies of water is governed by physical mechanisms in three dimensions. The horizontal spreading of material can only be properly simulated when processes in the vertical are explicitly taken into account.

In the present three-dimensional transport model, the necessary 3D deterministic flow field is derived by extending a computed vertically integrated velocity field into three dimensions by means of simple analytical functions. The analytical velocity profiles are steady and apart from additional distortions by wind or density gradients they are unidirectional, so they do not account for phase variations and skewed surface and bottom Ekman layers (Hamrick 1986). This provisional simplification seems to be justified by certain results such as those of section 23.3.1 and it also makes possible to investigate the sensitivity of the dispersion processes to various parameters. Nevertheless it is intended to evaluate the improvements to be obtained by quasi-3D or "2½D" approaches (Prandle 1982, Van der Giessen en Jansen 1988, Davies 1990) for obtaining the required 3D velocity fields. Finally, the ideal way to derive these fields would be by complete three-dimensional hydrodynamic numerical modelling which has become feasible in recent years. These various alternatives for obtaining better 3D velocity structures would leave the structure of the transport module unchanged: discrete particle approach, displacement in three directions at each time step including supplementary displacements which account for vertical exchange and for spectrally structured horizontal "sub-grid" velocities.

The particle approach and the scaled or spectrally structured sub-grid modelling described below has essentially no lower bound to spatial resolution. Resolving finer structures is just a matter of particle numbers and sufficiently small timesteps (and therefore a matter of economy as well). This sub-grid modelling differs essentially from sub-grid parameterization in the sense of accounting for the effect of sub-grid processes on phenomena at larger scales.

24.2 The model

24.2.1. The flow model and the use of the velocity fields

The two-dimensional hydrodynamic numerical model used for the computation of the 2D horizontal velocity field, in all applications so far, is a recent version of the shallow water flow simulation model WAQUA, which originates from the well-known concept of Leendertse (1967). The present version includes the improvements in the numerical schemes developed by Stelling (1983). All applications with the present transport model have so far been made with implementations of the flow model on a rectangular grid. The applications were made with the North Sea model GENO (mesh $\Delta x = \Delta y = 8000$ m), the southern North Sea model ZUNOWAQ (mesh = 3200 m) and an Ems estuary model with a mesh of 300 m. Two-dimensional predecessors of the particle transport model (Van Dam 1985[a]) have been used (or are still in use) with WAQUA-models of Western Scheldt (mesh = 400 m) (Van Dam 1985[c], 1986[b]) and IJsselmeer (mesh = 2000 m). The latter is a non-tidal fresh water lake. In the present article the special characteristics of non-tidal applications will not be discussed.

In the present applications, flow computation and transport computation are uncoupled. The tidal flow computations were made with cyclic boundary conditions (containing several Fourier modes). After the flow in the computation has become cyclic, all velocities and water levels for the last cycle at a certain number of time levels (between 10 and 30 minutes apart) are stored. In the transport computation, this set of velocity fields is used repeatedly, as many times as required for the length of the simulation period. This may amount to several hundred cycles. The time steps of the transport computation are usually somewhat smaller than the interval between two successive time levels of the stored velocities. The velocities used, are obtained by linear interpolation between time levels; further there is a bilinear interpolation in space among grid points. In some applications the velocity fields were Fourier analyzed (in time) in order to save on computer memory.

In general, the amount of velocity data for one cycle is too large for the working store of the computer so that only two successive levels are present at a given time. Updating the velocity fields for successive intervals takes a considerable amount of the total processing time. Depending upon the type of computer it is of the same order as the necessary computing time (CPU time) or larger.

The separation of flow computation and transport computation makes it possible to perform great numbers of transport computations (various locations and types of sources, various durations, transport parameter values and vertical velocity distributions) on the basis of just one flow computation. In practice the uncoupled approach can only be used in case of strictly periodic flow, short non-periodic flow histories and stationary fields. Non-periodic transport computations for long periods have to be performed in a coupled mode of flow model and transport model.

24.2.2. Supplemental horizontal velocities

Because of its finite mesh, the numerical flow model cannot reproduce details of the 2D horizontal velocity field below a certain length scale. This length is usually larger than the mesh of the grid (Van Dam, 1988, 1989) but for convenience the terms sub-grid scales and sub-grid velocities will be used in this connection.

The particle technique makes it possible to resolve details of concentration distributions on much smaller scales than the grid mesh. This can only be done realistically if the computed flow field is supplemented in one way or another. The supplementary mechanism must be of a stochastic character as long as the given flow model is not refined. If a deterministic representation of smaller details is for some reason required, the usual approach is the nesting of a finer grid flow model within the given one.

Two methods have been applied for supplementing the horizontal velocity fields: 1. The addition of an artificial eddy velocity field, composed of harmonic functions of space, covering the spectral range missing in the given flow field, down to some lower bound of the range of wave lengths, corresponding with the required resolution. 2. The addition of a scaled random walk process, i.e. using a random step depending upon the size of the particle (sub)cloud, which is usually related to its age.

Both methods have a (gradual) cut-off at the upper end of the spectral gap that has to be filled. From a physical point of view, the method of the additional flow field is preferable, but it is more time consuming and gives some technical problems at boundaries. The second method is more economic but has physical drawbacks. It assumes that particles have a memory or at least "know" the average distance to the other particles of their group. Further it produces rather smooth distributions while those from the other method have a more realistic aspect. In certain cases the unsatisfactory physical basis may cause unacceptable deviations in concentration values (Van Dam 1988[b]) but for most practical purposes it is good enough. The idea of a "scaling" of a random walk was first proposed by Maier-Reimer (1973); the principle of a gradual cut-off at the end of the spectral gap was added by Van Dam (1982, 1985[a]).

The numerical experiments with the method of the additional flow fields have played an important part in gaining insight and in the interpretation of field data and results of computations. Some details will be given in section 24.3.

24.2.3. Extension in vertical direction

The 3D extension of the model exists of

- a decomposition of vertically integrated velocities into vertical velocity distributions in analytic form;
- the attribution of a vertical position (z) to each particle (z has been chosen positive upwards, zero at the bottom);
- the introduction of a vertical exchange mechanism by means of an ordinary random walk in vertical direction corresponding with a depth independent exchange coefficient K_z.

The particles are of course displaced in horizontal directions by the horizontal velocity corresponding with their vertical position.

The analytical expressions for the vertical velocity distributions were chosen as simple as possible i.e. Van Veen distributions (Van Veen 1936, 1938) for undistorted verticals (neglecting phase differences):

$$\vec{v}\,(z) = \vec{v}\,(\alpha + 1)\left(\frac{z}{H}\right)^{\alpha} \tag{24.1}$$

($\vec{v}$ = given vertically averaged velocity; H = water depth; α = shape parameter) and the sum of two expressions of the same form to account for possible distortions v_d (z) by wind or density currents:

$$\vec{v}_d(z)_d = \vec{v}_{d_0}\left\{\frac{1+\beta}{\beta-\gamma}\left(\frac{z}{H}\right)^{\beta} + \frac{1+\gamma}{\gamma-\beta}\left(\frac{z}{H}\right)^{\gamma}\right\} \qquad (24.2)$$

($\vec{v}_{d_0}$ = value at the surface; β and γ shape parameters).

Both distributions do not effect the given 2D horizontal velocity field. The vertical average of (24.1) is equal to the given velocity v; the vertical average of (24.2) is zero. Both distributions lie in vertical planes, but (24.2) can make any angle with (24.1) and a sum of (24.1) and (24.2) is generally not in a plane. The distortion v_d (z) is taken neutral (vertical average is zero); the effect of the cause of the distortion upon the vertically averaged flow field is taken into account in the numerical flow computation.

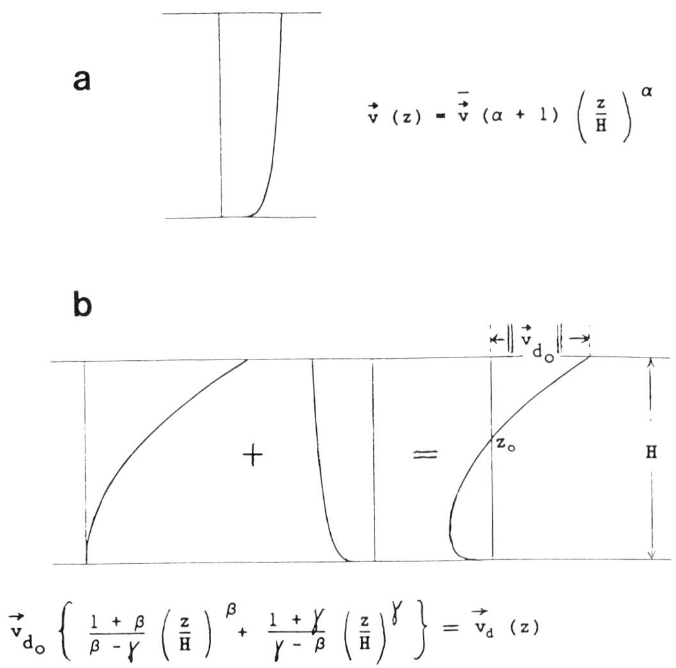

Figure 1: Construction of "distortion" v_d (z) of vertical velocity distribution by wind or density current.

The simple power expressions have been chosen for reasons of simplicity. The numerical difference between a logarithmic distribution and a power distribution (with proper exponent α) is extremely small. The effect of this difference upon dispersion patterns would not be noticeable.

The value of α for neutral buoyancy is << 1; a common choice for shallow waters is α = 1/6. It will tend to be even smaller for deeper water (upper parts of the velocity distributions become more uniform). To approximate vertical velocity distributions in stratified situations, much larger values of α can be appropriate, even α = 1 or even larger. Of course, in this case the approximation remains always poor because of the usual jumps that correspond with density discontinuities. This can only be improved by using explicit 3D flow computations.

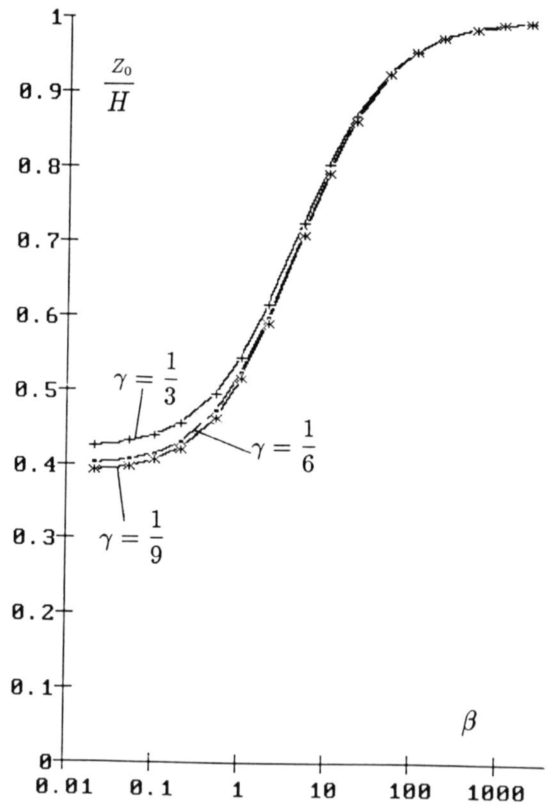

Figure 2: z_0/H (see fig. 1) as a function of β, for γ = 1/9, 1/6 and 1/3.

If the vertical modelling is to be refined, the assumption of height independent K_z would be the first one to be modified since it is in fact inconsistent with almost any non-uniform vertical velocity distribution. A possible refinement should be preceded by sensitivity tests as to the effects upon dispersion. The matter is probably rather insignificant for dissolved matter dispersion but it could be of interest in the case of sediments. Another refinement could be to account for the time variation of K_z in a tidal flow.

In (24.2), one of the exponents, at least in practice, has to be > 1, say $\beta > 1$, while the other is < 1. The term with $\beta > 1$ can be considered as the primary influence of the wind, down from the surface (fig. 1b). A higher value of β means a smaller relative penetration depth, so that β could be higher for deeper water. The term with γ (normally $\gamma < 1$) can be seen as the compensation current that neutralizes the effect of the other term upon the depth averaged velocity. Therefore it seems appropriate to take γ roughly equal to α. The height z = z_o where v_d (z) = 0 lies closer to the bottom as β is smaller (relative penetration depth greater). Figure 2 gives z_o/H as a function of β for fixed values of γ.

24.2.4 Two dispersion mechanisms

The model contains two (horizontal) dispersion mechanisms. When the results of the simulations are compared with field data, as will be shown below, it may be concluded that the main mechanisms in nature are indeed contained in the mechanisms of the model.

The two mechanisms are (i) the horizontal dispersion by vertical current shear and the accompanying vertical mixing and (ii) the spreading caused by the inhomogeneity and time dependence of the field of horizontal velocities. In the second mechanism horizontal shearing of velocities plays a similar role as the vertical shear in the first mechanism, but in the second case the effect occurs at numerous different length scales and in all directions.

In the model, the (vertical) shear dispersion (i) is dealt with by the vertical extension of velocity distribution and accompanying vertical mixing. The spreading (ii) caused by the structure of the field of horizontal velocities can be modelled by an additional field or a (scaled) random walk in the "sub-grid" range of the flow model (section 24.2.2); the larger scales are taken care of by the structure of the computed flow field itself.

The two scale ranges do not act entirely independent. The large scale structure of a velocity field alone does not induce the same patterns as its combination with smaller scale structures. Smaller scale velocity structures have a smoothing effect upon dispersion patterns as would be caused by the large scale structure alone.

Much insight in the effects of the horizontal structures has been obtained from numerical experiments with synthetic velocity fields in two dimensions (Van Dam 1987, 1990[a,c]). It was found that the spatial structure is in fact determining the spreading. Growth rate of patches is virtually independent of the life time of the eddies of various sizes. Only if all spectral components are completely stationary, special effects such as trapping of particles may occur. A minor dependence on time is sufficient to avoid this. This means that from the viewpoint of growth rate or dilution of patches of dissolved matter the effect of persistent structures related to topography is the same as the effect of stochastic contributions of the same kinetic energy at the same length scales. Therefore, from the viewpoint of growth rate of single patches the 2D horizontal velocity structure of an entire region such as the North Sea can be fully described by a spectrum of kinetic energy of spatial velocity differences. From numerical experiments combining such spectra with the effects of (vertical) shear dispersion and comparing these with field data, it was concluded that for the North Sea the present empirical data agree fairly well (figs. 3 and 4) with a kinetic energy spectrum of horizontal velocities of the simple form

$$E_k = c_k\, k^{-m} \tag{24.3}$$

with E_k = energy per unit of mass and unit of wave number k; c_k and m positive constants. On the basis of comparison between numerical simulation and experiments the values used for the North Sea at present are $c_k = 2.87 \times 10^{-4}\ m^2\ s^{-2}$, m = 1 (Van Dam 1987). A spectrum of the simple form (24.3) in two dimensions and in absence of other effects (in particular 3D mechanisms) corresponds with a behaviour of the maximum concentration of a single patch (instantaneous point source)

$$C^{(1)}_{(2D)\max} = c_o\, t^{-\beta} \qquad \text{with } \beta = 4/(3-m) \tag{24.4}$$

The superscript (1) refers to the release of one unit; the 2D character implies a dimension $[L]^{-2}$ (units per unit of area); c_o and β are positive constants. For m = 1, $\beta = 2$ the value $c_k = 2.87 \times 10^{-4}\ m^2\ s^{-2}$ (in combination with m = 1, $\beta = 2$) corresponds with $c_o = 7.09 \times 10^3\ m^{-2}\ s^2$.

If we choose a scaled random walk technique for simulation of the effects of the horizontal velocity structure, the (average) size of the random step in horizontal direction for the case $C_{max} \sim t^{-2}$ will grow as $t^{0.5}$. In general, an increasing random step $\sim t^{\alpha_o}$ corresponds to $C_{max} \sim t^{-(2\alpha_o+1)}$ (Van Dam 1985[b]). It should be remembered that the correspondence between (24.4) and (24.3) and the relationship $\beta = 4/(3-m)$ only hold as long as the kinetic energy of horizontal velocities according to (24.3) is the only mechanism of dispersion. Somewhat similar, a scaled random walk with a random step $\sim t^{\alpha_o}$ will only cause a $C_{max} \sim t^{-\beta}$ (with $\beta = 2\alpha_o + 1$) if the random walk is the only mechanism of horizontal dispersion in the simulation. Of course, the physical cause of the scale effect is not the time but the increasing size of the patch, allowing larger eddies to affect it. Therefore, if other mechanisms are present in the simulation, such as (vertical) shear dispersion, a correct scaled random walk, corresponding with the assumed spectrum of horizontal velocities, has to be established by relating the random step size to the size of the patch, in such a way that the step size used with a certain patch size has to be the same as it would be for the same patch size if the horizontal velocity structure were the only mechanism.

If there is a structured given (computed) velocity field present, the random step has to be levelled off (gradually) at the scale where the given velocity field begins to provide the correct structures.

24.3 Model studies for further investigation of dispersion mechanisms

24.3.1 Relative importance of the two mechanisms of horizontal dispersion

For investigating the relative importance and the combined effects of the two mechanisms, it is sufficient and convenient to use a simplified model with a uniform velocity field and a uniform water depth. For a rather wide range of scales, up to the resolution of the computed velocity field concerned, the results with the uniform field and with the computed field (in a region of uniform depth) will be identical. For the larger scales, the non-uniformity of the computed field is taken into account in the simpler model by continuing the scaled random walk beyond the scale at which otherwise the computed field would provide the proper velocity structures and their dispersive effect.

On the basis of these considerations, computations have been made in a model with a uniform velocity field of a simple harmonic time dependence (tidal ellipse shown in the inset of figures 3a and b) and a period of 44700 s. The uniform depth H implies that vertical exchange coefficient K_z and vertical mixing coefficient T_z, related by

$$K_z = \frac{H^2}{2\,T_z} \qquad \text{or} \qquad T_z = \frac{H^2}{2\,K_z} \tag{24.5}$$

become equivalent parameters. By working with T_z, the value of H can remain undefined. The factor 1/2 in (24.5) corresponds to a definition of T_z as the time needed to reach a $\sigma_z = H$ after an instantaneous point release in absence of boundaries (water surface and bottom).

Since in the computation in question both mechanisms act simultaneously, it is necessary to relate the random step size to patch size and not to time as explained in the preceding section. The shape of the patches is not isotropic; patch size is defined by

$$r = \sqrt{2\,\sigma_\xi\,\sigma_\eta} \tag{24.6}$$

if ξ and η are the directions of the principal axes of the patch. The use of r instead of t implies that σ_ξ and σ_η have to be computed from the particle distribution quite frequently.

In figure 3, the result of a number of simulations is presented in the form of a 2D maximum concentration as a function of time after an instantaneous point release of one unit at the surface, for various values of T_z and a fixed kinetic energy distribution for the horizontal velocity variations. The chosen energy distribution is the spectrum assumed for the North Sea as described in the preceding section. This choice is particularly interesting because of the large number of available field data for the North Sea (Joseph et al 1964; Barrett et al 1969; Talbot and Talbot 1974; Van Dam 1982; Suijlen et al 1990[a,b]; Weidemann 1973) (fig. 4). The assumed tidal ellipses are pictured in the inset. Of course there is a physical relation between tidal energy and spectral energy, the tides being the main source of kinetic energy. Therefore it is not meaningful to take just any amplitude for the tidal velocities in these simulations.

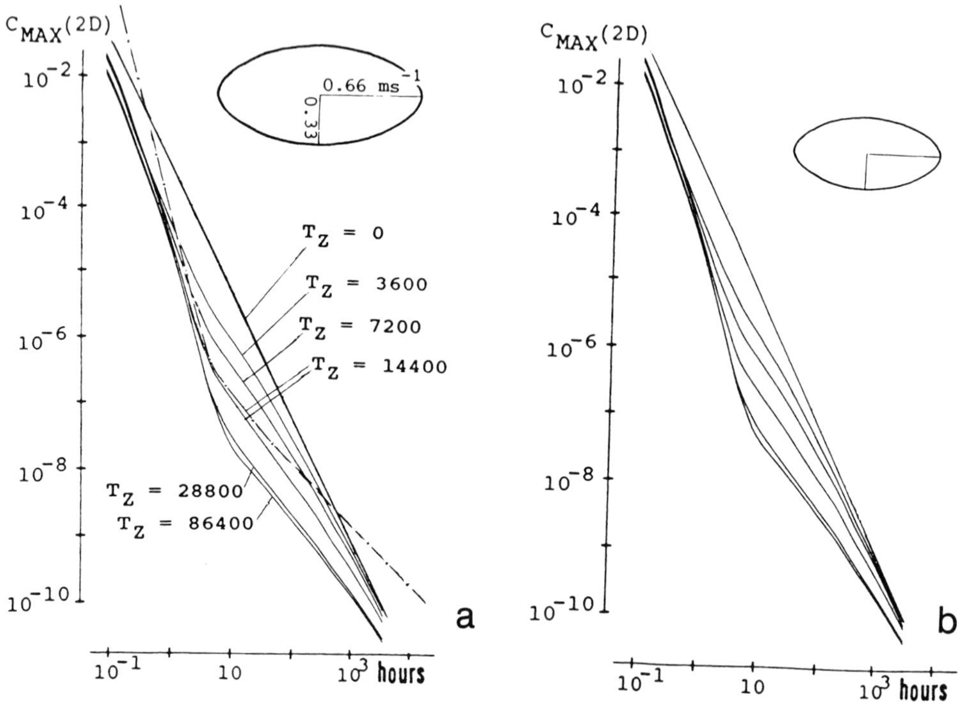

Figure 3: Maximum vertically integrated concentration $C^{(1)}_{max}(2D)$ after an instantaneous point release of one unit of dissolved matter at the surface, computed with uniform depth and a simple tidal ellipse (inset). Kinetic energy of tide in case b is half of that in a.

It is interesting to compare figure 3 with the compiled data from tracer experiments in the North Sea (fig. 4). The comparison supports the chosen approximation of the energy spectrum and it strongly suggests that the natural variations of T_z (the higher values are realistic in stratified areas) are the main cause of the spread among the various experiments, rather than regional differences in kinetic energy content or its spectral structure. The correspondence between the band of observations and the band of simulations is quite satisfactory. Note that the band of simulations has a lower bound due to the fact that from a certain value of T_z the increase of dispersion rates with increasing T_z comes to an end and even changes into a decrease at still larger T_z values.

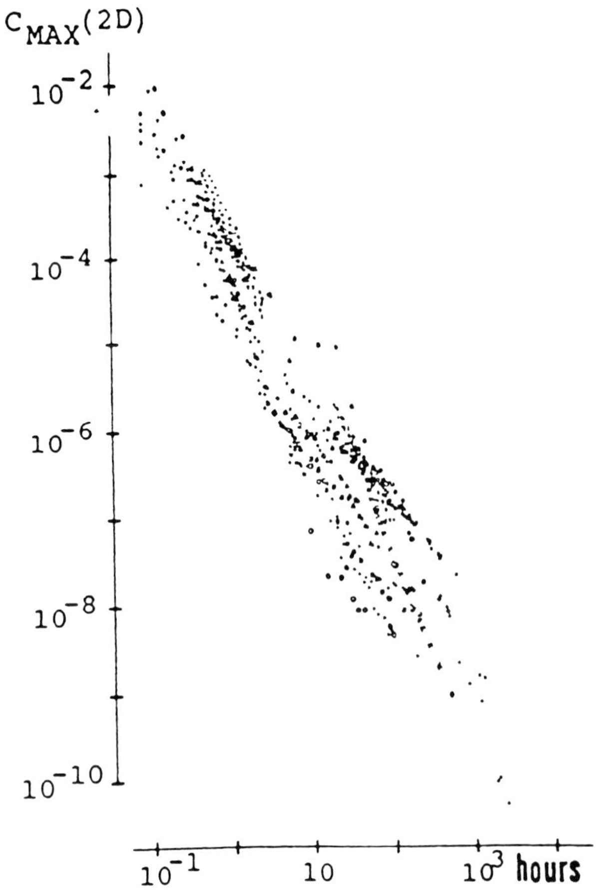

Figure 4: Compiled North Sea data of instantaneous point releases of tracer (see refs.).

The latter effect has been studied separately in a vertically two-dimensional (2DV) model (essentially the same model was presented in 1982 by Allen). A result of these studies is presented in figure 5. The dispersion coefficients plotted in the vertical direction are the constant values (no tide, net current only) or the effective tidal averages, both reached after a certain initial period related to the vertical mixing time.

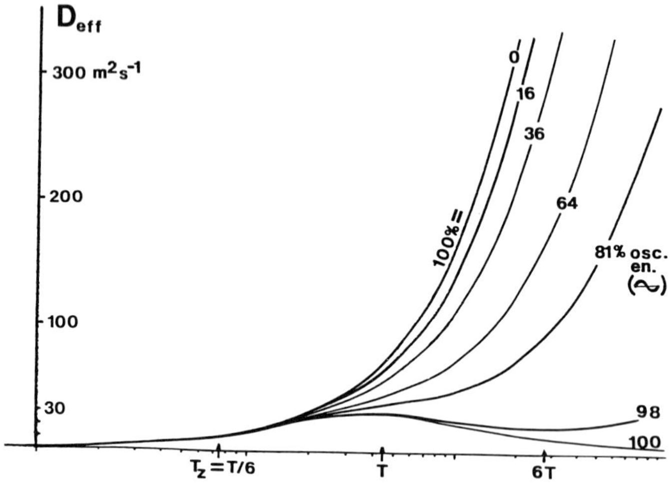

Figure 5. Effective (longitudinal) dispersion coefficient (net values per period) as a function of vertical mixing time for various mixtures of tidal and net current, computed in a 2DV particle model. For all curves the total tidally averaged kinetic energy is the same ($\overline{v^2}$ = ½ m^2 s^{-2}).

Figure 6 assists in explaining the role of T_z in (vertical) shear dispersion. It depicts the behaviour of a vertical line source during a tidal period, with vertical velocity distribution and vertical exchange as only mechanisms, for a T_z << tidal period T, a $T_z \approx$ T and a T_z >> T.

Figure 3 shows that with the relatively low energy level of the horizontal velocity variations found for the North Sea, there is a fairly long period in which the mechanism of (vertical) shear dispersion plays an important or even dominant role, the more so when T_z is larger (in the range $T_z \leq$ T); the field data show that all T_z values of this range do occur (and possibly beyond; this is certainly so in strongly stratified regions). After the period that the relative importance of the mechanism itself is significant, there is a period in which the obtained "lead" in dispersion still has an impact on concentrations ("memory" effect).

In a well mixed estuary, the situation is different. Vertical mixing times are short, so that the (vertical) shear dispersion becomes less important. On the other hand, because of the much more pronounced topography and the usually higher tidal energies, the energy level of the velocity variations on length scales up to the tidal excursion length is considerably higher.

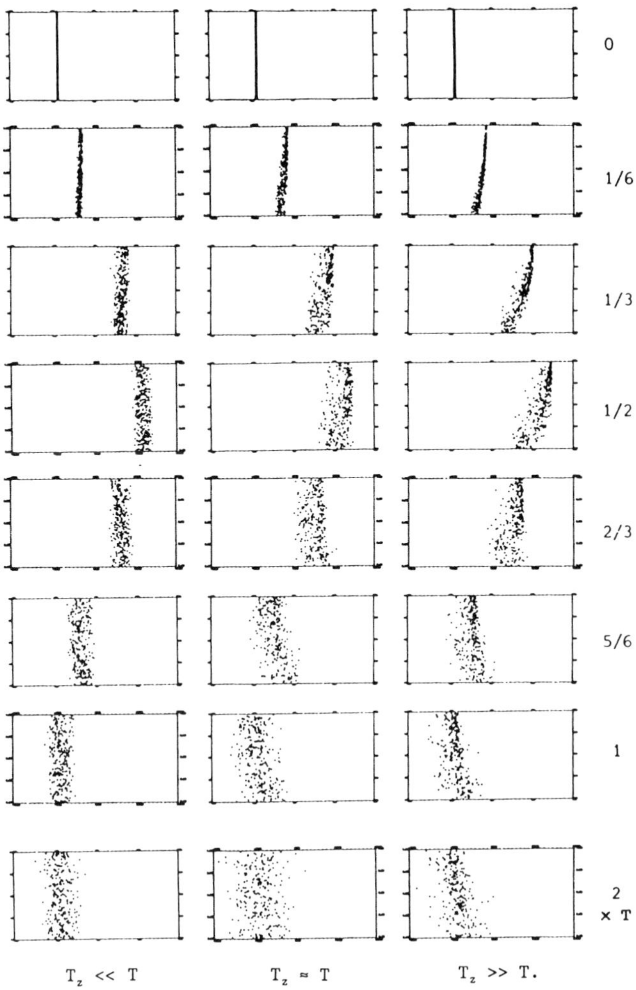

Figure 6. Illustration of (vertical) shear dispersion of an instantaneous vertical line source for 100% tidal current (compare fig. 5), showing how the longitudinal dispersion decreases compared with the case $T_z \approx T$, when T_z becomes (much) smaller or (much) larger.

24.3.2 Chaotic advection

If a fine mesh numerical flow model of an estuary is available, from a certain scale onwards the dispersive effect of the spatial velocity structure can be demonstrated by using the computed velocity field without any supplement. An example is given in figure 7, based upon a 300 m mesh rectangular grid (WAQUA-) model of the Ems estuary. It shows the evolution during one tidal period of a contour which is initially a circle of 1500 m radius, using the computed flow velocities only. On this scale, the relative part of the horizontal velocity structure in the dispersion is dominant and the main part of its energy is reproduced in the flow computation. Similar results have been published by Ridderinkhof (1990). In figure 7c the distribution of a dense set of particles, filling the circle, is given at the end of the same tidal period, including the effects of "sub-grid" scale velocity variations and vertical shear.

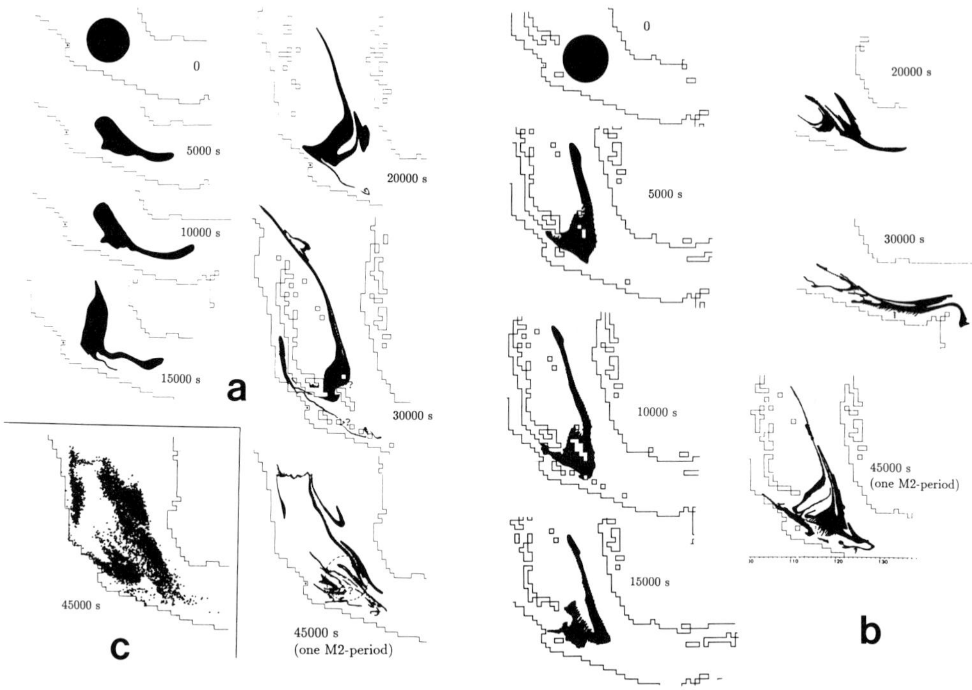

Figure 7. Deformation of a circular contour during one tidal period in Ems Estuary model (mesh = 300 m) by computed velocity field only. Difference in initial phase between a and b is one half of tidal period (22500 s). Extra dispersion by "sub-grid" effects (vertical shear with vertical mixing; "sub-grid" horizontal velocity variations) for case a shown in inset (c) (situation at end of period only).

The advection by the complex time dependent velocity field is typically a chaotic process. The pattern after one tidal period strongly depends upon the tidal phase at the beginning (figures 7a and 7b). In other words, the displacement of a particle during a tidal period and therefore its tidally averaged Lagrangian velocity is very much dependent upon tidal phase (Van Dam, 1984,1986[a]; Ridderinkhof 1990). This limits the physical meaning and practical utility of "Lagrangian residual velocities".

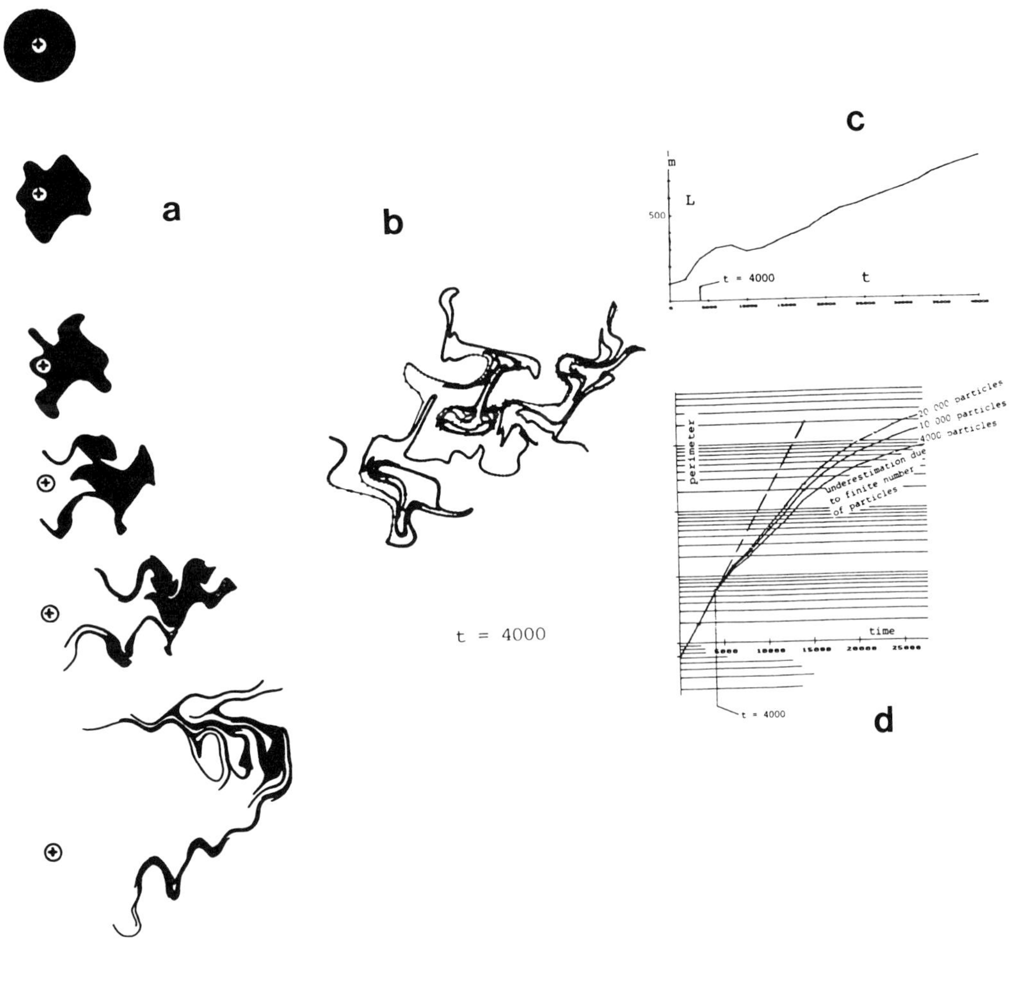

Figure 8: Effects similar to figure 7, obtained by simulation with 2D synthetic eddy fields (8 pairs of harmonic components) in an unbounded region.
a. Stationary field
b. Eddy life time 1.5 revolutions.
In case b, only the contour of the fluid blob has been drawn; graphs (c and d) show evolution of overall size and contour length in time.

Similar patterns were obtained in the separate investigations of synthetic velocity fields. Figures 8a and 8b were obtained with 8 "eddy components" (8 pairs of harmonic functions), figure 8a in a stationary mode, fig. 8b with finite life times (1.5 revolutions per eddy). Units are essentially arbitrary, but fig. 8b can be interpreted as a patch of some 250 m diameter and an age of 4000 s. If the exponential behaviour of the contour length is extrapolated, which is justified for reasons of similarity, one arrives at a length of some 1000 000 000 000 m when the diameter has become 1 km (Van Dam 1989[a]).

The numerical experiments with synthetic fields are a quick means to interpret dispersion patterns in terms of energy spectra (section 24.2.4).

24.4 Practical applications

24.4.1 Dissolved matter

Applications may vary from time scales of a few hours (or less; there is no lower bound to spatial resolution) to time scales of the order of a year (or longer). In practice, at this moment, simulation times of the order of a year are only possible with merely periodic velocity fields.

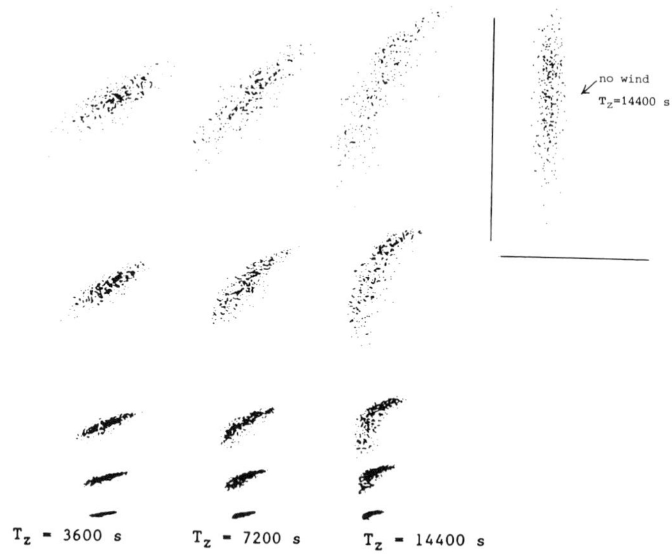

Figure 9: Bird's-eye view of particle sets in a tidal current (North going phase) with moderate wind from the West, computed for three different vertical mixing times T_z.
Point release at the surface at t = 0; distributions drawn after 40, 80, 120, 200 and 360 minutes. Patch without wind drawn for t = 360 minutes only.

The simulation of periods up to a few days has been operational with preceding 2D versions of the model for accidental spills at sea (Delft Hydraulics Laboratory 1989) or in the Western Scheldt (Van Dam 1985[c], 1986[b]; Lefèvre and Van der Male 1988). For the same type of applications a 3D version of the model is now being implemented for the North Sea (Diependaal 1990; Hendriks 1990[a,b]) and will be adopted for other regions in the near future.

An illustration of an application on a short time scale is given in figure 9, presenting a bird's-eye view only (particles at all depth levels). The chosen number of particles (300) is smaller than in cases where iso-concentration lines are required. The picture gives an impression of wind influence (moderate cross wind) and its dependence upon T_z.

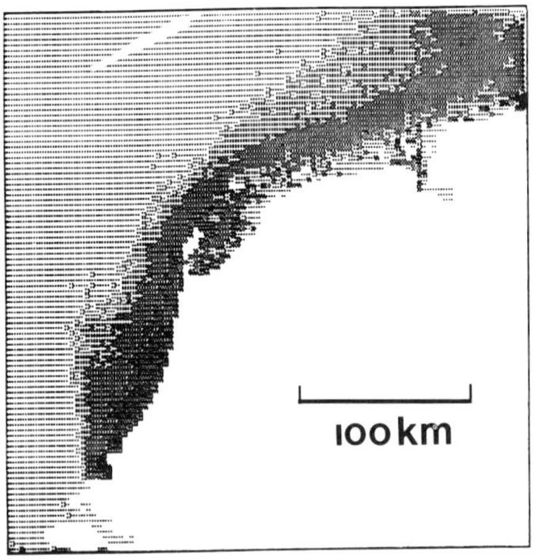

Figure 10a

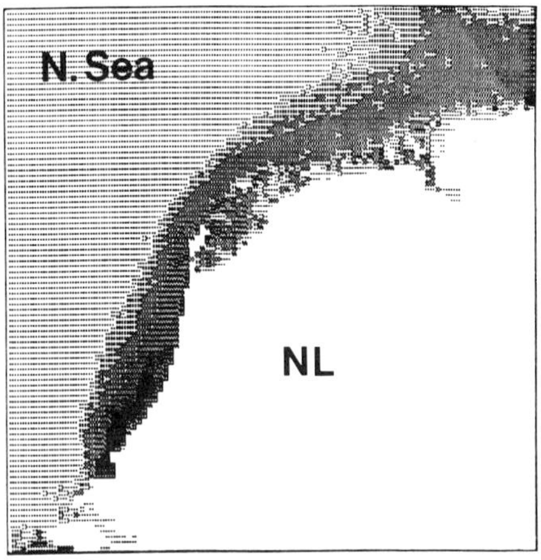

Figure 10b

Figure 10. Simulation with 3D mode of transport model of a continuous release (near the mouth of rivers Rhine and Meuse) in the North Sea. Periodic flow; (periodic) steady state concentration distribution.
a. Computed with non-uniform account of density current perpendicular to the coast and suppressed vertical mixing, decreasing with distance from river mouth.
b. No density current and mixing uniform over the whole area.
White = land (figures 10, 11, 12). Shades in figures 10 through 15 represent concentration ranges of a factor 4.

Figures 10, 11 and 12 give some typical long term applications, which were only feasible by the use of a strictly periodic velocity field. Of course, the periodicity over such a long time (order of one year) limits the realistic meaning of the result. The periodicity does not only enable the separation of the (relatively short) flow computation and the transport computation but it further enables the use of superposition (Van Dam 1985a), which saves a great amount of computer time. The results have been obtained by following only 300 particles from the source, amounting to a total of some 100 000 particles in the final distribution by superposition of almost 400 successive states (for larger times, parts of the groups of 300 particles have left the model area at the open boundary), each representing a period of 90000 s (= the cycle time of the flow computation). The figure represents some general results obtained by simulations over a period of 405 days (in order to obtain a (periodic) steady state concentration distribution in the entire region). Shades represent vertically averaged (3D) concentrations. The underlying cyclic 2D velocity field was obtained by one particular run of the WAQUA-model "ZUNOWAQ" (mesh 3200 m). One should realize that different results are obtained when other boundary conditions are applied.

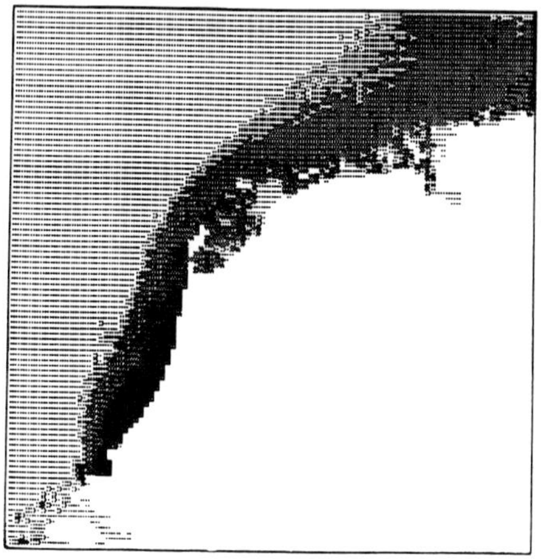

Figure 11a

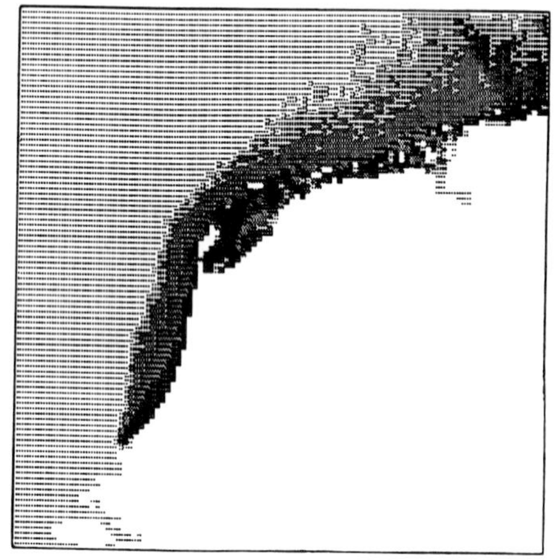

Figure 11b

Figure 11: a. Simulation with 2D mode of transport model of a continuous release from same source as figure 10. No extra dispersion added to parameterize absence of dispersion by vertical shear.
b. Same as a, but tidally averaged velocities (Eulerian residual transport velocities) used instead of complete time dependent velocity field.

Figures 11a and 12a were computed in a 2D mode of the particle model. In figure 10b the vertical shear effects are taken into account in a uniform way (corresponding to just one of the lower curves of figure 3a). Figure 10a accounts also for the local stratification near the main river mouth, vanishing at larger distances, and for the density currents perpendicular to the coast, decreasing at larger distances from the main source of fresh water (compare De Ruijter et al 1990 and Visser 1990). Only the approach leading to figure 10a produces realistic river water concentrations near the river mouth if the source is identified with the fresh water outflow of the local river mouth. At large distances from the source the differences between the results of figures 10a, b and 11a become insignificant.

Figure 11b and 12b as compared to 11a and 12a are to illustrate the effects due to differences in advection (Van Dam 1984, 1986[a], 1988[a]) that arise if Eulerian averages of tidal velocities are used in order to save computer time. It is seen that the effects are more pronounced in one area than in another. However, there is little reason to use the faster but less accurate approach (figures 11b and 12b) now the complete time dependent simulation has become feasible.

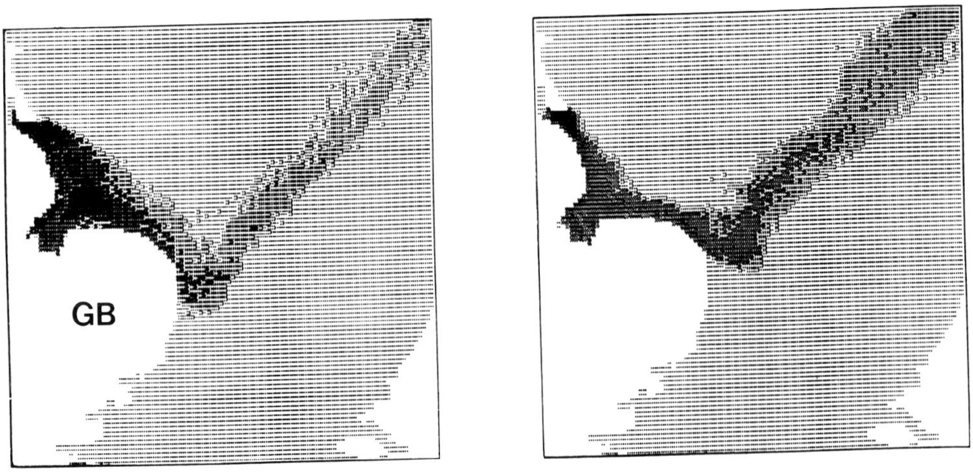

Figure 12a Figure 12b

Figure 12: As figure 11, except for location of source.

It should be pointed out that the (very) long term simulations have to be regarded as preliminary. Some refinements of the model for this type of applications are possible and may be desirable. One aspect is the uniformity of various parameters in the present programme, especially K_z (or alternatively T_z). For events in restricted areas this is usually not a problem. On the other hand it is not easy to indicate how K_z will vary over the whole area. The global dependency on location in figure 10a, has been acquired by a "trick"; the computation of the fate of the basic group of 300 particles was interrupted at certain stages and restarted with other values of some parameters (T_z; shape parameter of Van Veen vertical; density current intensity) dependent on the changing position of the particle cloud. In recent long term simulations, particularly those with sediments (next sections) a uniform K_z was preferred to a uniform T_z in order to avoid unrealistic slow mixing in shallow areas. In the near future it may become possible to make K_z height dependent in a meaningful way. One of the agents for K_z is the energy of the current. This also means that in fact K_z varies during a tidal period. Also this effect may mainly be of interest for sediment applications.

24.4.2 Extension to suspended matter

The extension of the model into a version for simulating the transport of (settling) suspended materials is quite recent and the provisions made for this purpose will most likely need some refinements in order to obtain meaningful results, hopefully of some predictive value.

A fall velocity was introduced as an input variable. Fractions with different fall velocities have to be dealt with separately. For each fraction two critical velocities can be chosen (which can be related to shear stresses at the bottom) on a height that can be specified. The smaller velocity corresponds with sedimentation, the larger with erosion. No time lags are used so far. This means that a particle is immediately reactivated when the erosion velocity (on a specified height) is exceeded.

An important improvement compared to models in two dimensions, is the fact that a particle will only settle on the bottom, when it hits the bottom and at the same time the velocity (on the prescribed height) is below the critical value. Reversely, when the particle is re-activated at a later time, its initial position is very close to the bottom.

The above provisions include the case that all material remains in suspension, even with high fall velocities; one only needs to specify a threshold velocity equal to zero. In a uniform region of depth H, vertical exchange coefficient K_z and M kg suspended material in a water column (unit area), the vertical distribution will then become (Van Dam 1982):

$$C(z) = C(0)\, e^{-\frac{|w_p|\, z}{K_z}} \tag{24.7}$$

where

$$C(0) = \frac{|w_p|\, M}{K_z\left(1 - e^{-\frac{H|w_p|}{K_z}}\right)} \tag{24.8}$$

($|w_p|$ = absolute value of fall velocity; with z positive upwards, $w_p < 0$) or, with (24.5) for mixing time T_z and $T_v = H/|w_p|$ for falling time and $z^* = z/H$ for relative vertical position:

$$C(z^*) = C(0)\, e^{-2\frac{T_z}{T_v} z^*} \tag{24.9}$$

For $T_z \ll T_v$, the vertical distribution is practically uniform (very fine sediments). For $T_z \gg T_v$, most of the material is concentrated in a thin layer near the bottom. Maybe the latter case can be used as a first approximation of a "fluid mud" layer.

Figures 13, 14 and 15 give some examples of typical results that can be obtained with the present provisions. They pertain to specific fractions with a fixed fall velocity. The behaviour of mixtures can be studied by separate computations for different fractions. At this moment it is not yet possible to evaluate the results. It may be rather difficult to find proper field data for verification when the simulations acquire sufficient realistic value.

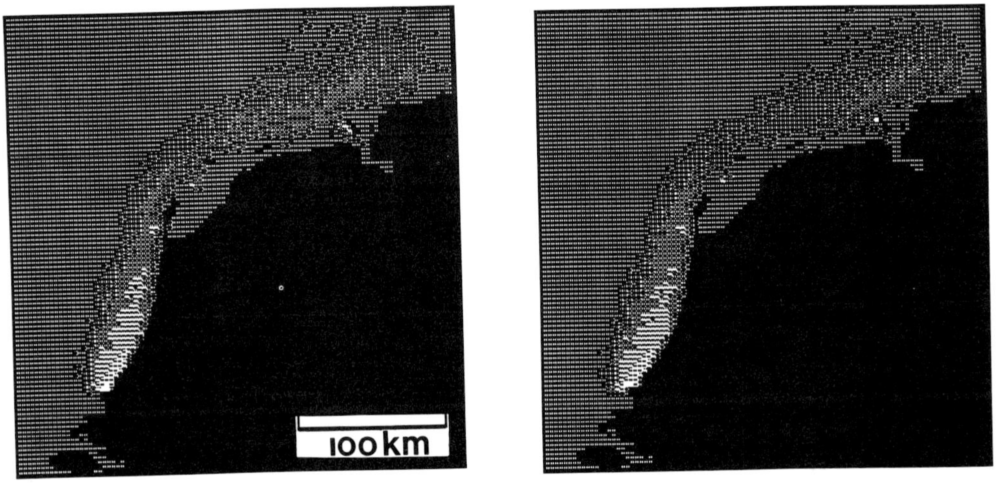

Figure 13: Computed distribution along the Netherlands' coast of suspended part of a certain sediment fraction from a continuous source (same location as fig. 10 & 11), for two different tidal phases (¼ M2-period apart).

Black = land (figures 13, 14, 15)

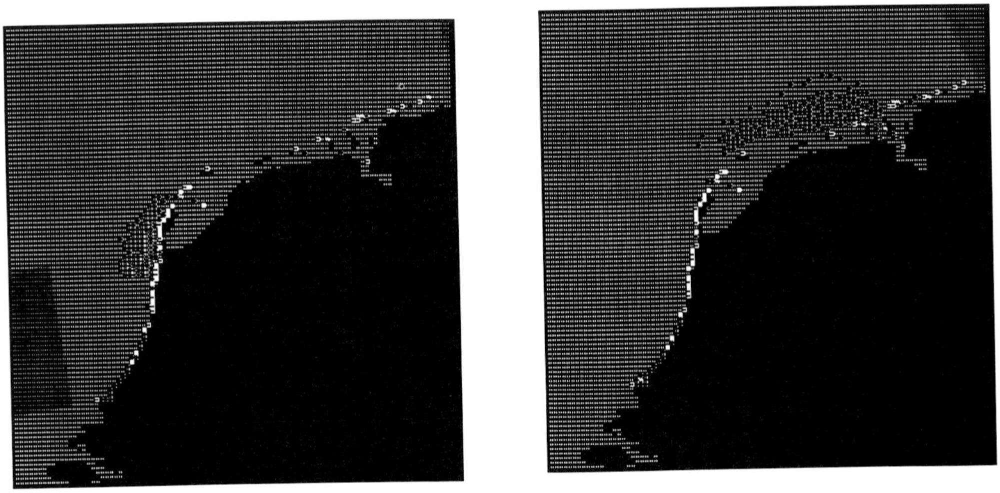

Figure 14: Settled part of the sediment of figure 13 at same tidal phases. Sedimentation zone moves along the coast with the tidal wave and corresponds with zones of lowered concentration in figure 13. The non changing spots represent permanently settled material (under the rigidly constant conditions of the simulation).

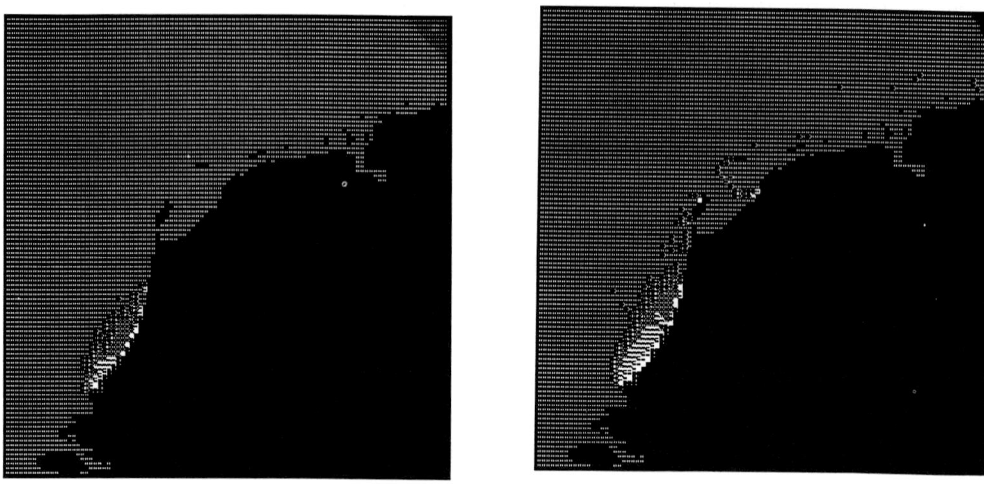

Figure 15: Simulation with strong density current (or similar circulation in vertical plane) perpendicular to the coast. Source of sediment as in preceding figures.

a. Fraction with same characteristics as preceding figures.

b. Critical velocities for sedimentation and erosion reduced by a factor ⅔ (fall velocity unchanged, 0.0003 m s^{-1}). Note great impact of this relatively small alteration.

Another difficulty will be the behaviour of the flow model near the closed boundaries. In figures 14 and 15 it is visible that the present flow field of ZUNOWAQ creates narrow zones of low velocities along the Netherlands' coast, which act as sedimentation zones in the simulation. This effect will be very sensitive to the rules applied in the flow model at the closed boundaries: "slip", "no-slip" or some intermediate prescription.

In the list of possible refinements, the introduction of a scour lag is probably the most urgent. Depth and flow dependence of K_z may come next. Flocculation and disintegration are important but for the near future seem to be beyond the reach of meaningful modelling.

24.5 Conclusions

The described 3D transport model takes explicitly into account the mechanism of horizontal dispersion due to vertical velocity shear and the mechanism of horizontal dispersion by horizontal eddies and it reproduces very well the shape and the width of the bundle of curves obtained from the numerous dispersion experiments with dissolvable tracers in the North Sea (figures 3 and 4). In the simulation, the upper bound of the set of curves can be adjusted by calibration of the energy spectrum associated with the second of the two mechanisms.

From this result it can be concluded that all substantial dispersion mechanisms are modelled and this thus forms the basis for a wider application (non-dissolvable matter, dispersion in estuaries etc.). The rather simple way in which the hydrodynamically computed 2D (vertically averaged) velocity field is extended into three dimensions is apparently good enough for several applications. Nevertheless it is intended to evaluate the improvements to be obtained by a more explicit calculation of the 3D velocity field, either by a "2½ D" approach or by a fully three-dimensional hydrodynamic computation of the flow.

24.6 References

Allen, C.M., 1982. Numerical simulation of contaminant dispersion in estuary flows. Proc. Roy. Soc. Lond. A 381, 179-194.

Barrett, N.J., D. Munro and A.R. Agg, 1969. Radiotracer dispersion studies in the vicinity of a sea outfall. Proc. 4th Int. Conf. on Water Pollution Research, Pergamon Press, New York, 863-882.

Davies, A.M., 1990. On extracting tidal current profiles from vertically integrated two-dimensional hydrodynamical models. Jour. of Geophysical Res. 95, No.C10, 18317-18342.

Delft Hydraulics Laboratory, 1989. Gebruikershandleiding MARS (Marine Spills). Rept. T 470, 47p.

Diependaal, R.J., 1990. MARS 2.0, Processen en Transport. Delft Hydraulics Laboratory, T 607.

Hamrick, J.M., 1986. Long-term dispersion in unsteady skewed free surface flow. Est., Coast. and Shelf Sc. 23, 807-845.

Hendriks, A., 1990[a] MARS 2.0, Gebruikershandleiding. Delft Hydraulics Laboratory, T 607.

Hendriks, A., 1990[b]. MARS 2.0, De constructie van een hydrodynamische database. Delft Hydraul. Lab., T 607.

Joseph, J., H. Sendner and H. Weidemann, 1964. Untersuchungen über die horizontale Diffusion in der Nordsee. Dtsch. Hydrogr. Z., 2, 57-75.

Leendertse, J.J., 1967. Aspects of a computational model for long-period waterwave propagation. The Rand Corporation, Santa Monica, Calif., rpt. RM-5294-PR / Ph.D. Thesis, Technical University Delft 1967, 165p.

Lefèvre, F.O.B. and C. van der Male, 1988. Gebruikersdokumentatie programma pakket CALM, Rijkswaterstaat (Neth.), Tidal Waters Division, rpt. GWWS-88.407, 62p.

Maier-Reimer, E., 1973. Hydrodynamisch-numerische Untersuchungen zu horizontalen Ausbreitungsvorgängen in der Nordsee. Dissertation Univ. Hamburg/Mitteilungen Inst. für Meereskunde der Univ. Hamburg, XXXI, 1-56.

Prandle, D. 1982. The vertical structure of tidal currents and other oscillatory flows. Cont. Shelf Res. 1, No. 2, 191-207.

Ridderinkhof, H., 1990. Residual currents and mixing in the Wadden Sea. Ph.D. Thesis, University Utrecht, Neth., 91p.

De Ruijter, W.P.M., and A. van der Giessen, 1990. Current and density structure in the Netherlands' coastal zone. Proc. 5th Int. Bienn. Conf. Physics of Estuaries and Coastal Seas, Gregynog, Wales.

Suijlen, J.M., J.S. Sydow, C. Heins and P.C. Beukenkamp, 1990[a]. Measurement of turbulent diffusion and residual displacements by dye experiments in the southern North Sea in 1971, 1973, 1978 and 1979. Data Report. Rijkswaterstaat (Neth.), Tidal Waters Div. GWAO-90.022.

Suijlen, J.M., 1990[b]. Measurement of turbulent diffusion and residual displacements by dye experiments in the southern North Sea in 1982. Data Report. Rijkswaterstaat (Neth.), Tidal Waters Div. GWAO-90.023.

Stelling, G.S., 1983. On the construction of computational methods for shallow water flow problems, Ph.D. Thesis, Technical University Delft, Neth.,232 p.

Talbot, J.W. and G.H. Talbot, 1974. Diffusion in shallow seas and in English coastal and estuarine waters. Rapp., P.V. Cons. Int. Explor. Mer., 167, 93-110.

Van Dam, G.C., 1982. Models of Dispersion. In: Pollutant transfer and transport in the Sea, G. Kullenberg ed. CRC Press Inc., Boca Raton, Florida, 91-160.

Van Dam, G.C., 1984. Reststromen en resttransporten in modelberekeningen. Rijkswaterstaat (Neth.), Dir. W & W, Physics Div., rpt. FA 8402, 33p. (English version: Van Dam, 1988[a]).

Van Dam, G.C.,1985[a]. Deeltjesmodellen, 2DH (Particle models 2DH). Rijkswaterstaat (Neth.), Dir. W & W, Physics Div., rpt. FA 8507, 44p.

Van Dam, G.C., 1985[b]. Konstante en aangroeiende randomstap in deeltjesmodellen. Rijkswaterstaat (Neth.), Dir. W & W, Physics Div., note 85-FA-219, 5p.

Van Dam, G.C., 1985[c]. A particle model applied to the Western Scheldt. Liverpool Conference on mixing and dispersion in estuaries, 4-25 Sept. 1985/ Rijkswaterstaat (Neth.), Dir. W&W, Physics Div., 85-FA-226

Van Dam, G.C., 1986[a]. Residual currents and transport in two-dimensional model computations. Neth. J. Sea Res. Publ. Series, 13, 66-67.

Van Dam, G.C., 1986[b]. Particle modelling of 2DH transport phemomena in the Western Scheldt Estuary. Abstracts International Symposium on Physical Processes in Estuaries, Noordwijkerhout, The Neth.: I.1.2.

Van Dam, G.C., 1987. Eddy simulation of two-dimensional dispersion. Rijkswaterstaat (Neth.), Tidal Waters Division, rept. GWAO-87.022, 26p.

Van Dam, G.C., 1988[a]. Residual currents and transport in connection with two-dimensional model computations. Rijkswaterstaat (Neth.), Tidal Waters Division, GWAO-88.042, 20p.

Van Dam, G.C., 1988[b]. The mathematical modelling of transport phenomena in the North Sea in relation to experiments. Rijkswaterstaat (Neth.), Tidal Waters Division, rept. GWAO-88.043, 27p.

Van Dam, G.C., 1989[a]. Eddy simulation of dispersion. Workshop on applying non-linear dynamics and chaos in fluid mechanics, June 1989, Woudschoten, Zeist, The Netherlands/Rijkswaterstaat (Neth.) Tidal Waters Division, GWAO-89.313.

Van Dam, G.C.,1989[b]. Dispersion of dissolved matter in the North Sea. ICES, Hydrogr. Committee, C.M. 1989/C:13, 21p.

Van Dam, G.C.,1990[a]. Two- and three-dimensional modelling of dispersion and mixing in shallow waters. IUTAM Symp. on Fluid Mechanics of Stirring and Mixing, La Jolla, 1990.

Van Dam, G.C., 1990[b]. Two- and three-dimensional models of dispersion based upon measurements in the North Sea. Proc. IAWPRC/EWPCA Int. Conf. on North Sea Pollution, Amsterdam, 14p.

Van Dam, G.C., 1990[c]. Numerical simulation of dispersion by eddy velocity fields in two dimensions. Submitted for publication.

Van der Giessen, A. and R.R. Jansen, 1988. Het 2,5 D-model. Een nieuwe methode ter bepaling van de vertikale stromingsstruktuur in de Nederlandse kustzône. Rijkswaterstaat (Neth.) Tidal Waters Division, GWAO-88.013.

Van Veen, J,, 1936. Onderzoekingen in de Hoofden in verband met de gesteldheid der Nederlandse kust. Alg. Landsdrukkerij, Den Haag, Neth., 258p.

Van Veen, J., 1938. Water movements in the Straits of Dover., J. Cons. Int. Explor. Mer. 13, 7-36.

Visser, D., 1990. A non-linear diffusion model for spreading of river discharge in a shallow sea. 5th Int. Bienn. Conf. Physics of Estuaries and Coastal Seas, Gregynog, Wales.

Weidemann, H. (editor), 1973. The ICES diffusion experiment RHENO 1965. Rapp. P.V. Réun. Cons. Int. Explor. Mer. 163, 111p.

25

An estuarine and coastal sand transport model

B A O'Connor and J Nicholson

Abstract

This paper contains a description of an estuarine and coastal sand transport model which is currently under development. Also described are the results of tests to choose a suitable mathematical analogue for the various processes which control the transport of sediment. The resulting model and its manner of operation were then checked against a field situation and shown to give promising results.

25.1 Introduction

Engineering works in estuaries and on coastlines are increasing in both number and scope. As these works can cause marked changes in the local morphology, whether intentional or unintentional, there is a need for sediment transport models which can predict such changes.

A depth-averaged model of sand transport, which can be applied to both estuaries and coasts, is currently under development. The proposed scheme is modular in form, each module representing one or more of the processes governing the transport of sediment. Described herein are the model itself, including the results of tests which have been undertaken in order to establish a suitable method of simulating the various transport processes, and the application of the model to a field situation.

25.2 Model description

25.2.1 General

Fine sand is commonly found in estuarine and coastal areas, the suspended component of which, unlike the bed component, is unlikely to attain equilibrium with the unsteady and non-uniform ambient flow conditions. For this reason, therefore, the model treats the suspended and bed components of the sediment load separately. In addition, a cross-shore transport module is included in order to reproduce the development of beach profiles.

The model computes the bed level changes in two stages. Firstly, the depth-averaged version of the unsteady-state diffusion-advection equation is solved for the suspended sediment concentrations. This equation takes the form:

$$\frac{\partial}{\partial t}(h_F\bar{C}) - \underbrace{\frac{\partial}{\partial x}\left(h_F D_x \frac{\partial \bar{C}}{\partial x}\right) - \frac{\partial}{\partial y}\left(h_F D_y \frac{\partial \bar{C}}{\partial y}\right)}_{DISPERSION}$$

$$+ \underbrace{\frac{\partial}{\partial x}(h_F\overline{U}\overline{C}) + \frac{\partial}{\partial y}(h_F\overline{V}\overline{C})}_{ADVECTION} + \underbrace{R_D - R_E}_{SOURCE\ /SINK} = 0 \tag{25.1}$$

where:

$\bar{C}$ = depth-averaged suspended sediment concentration (kg m^{-3});
D_x = longitudinal dispersion coefficient (m^2 s^{-1});
D_y = lateral dispersion coefficient (m^2 s^{-1});
h_F = flow depth (m);
R_D = sediment deposition rate (kg m^{-2} s^{-1});
R_E = sediment entrainment rate (kg m^{-2} s^{-1});
t = time (s);
$\bar{U}$ = depth-averaged longitudinal velocity (m s^{-1});
$\bar{V}$ = depth-averaged lateral velocity (m s^{-1});
x = longitudinal co-ordinate (m);
y = lateral co-ordinate (m).

In Eq. 25.1, details of the flow field (h_F,$\bar{U}$,$\bar{V}$) are supplied by a separate model, for example that of Yoo and O'Connor (1988). The deposition rate (R_D), which is a function of the unknown variable ($\bar{C}$), is described below along with the entrainment rate (R_E), which is computed independently. An alternating direction implicit (ADI) finite difference scheme is used to solve the overall equation. The results of two tests which simulated the progress of a slug of suspended sediment subject to advection and advection combined with dispersion, are shown in Figs. 1a and 1b respectively. The flow velocity was 1.4 ms^{-1}, the dispersion coefficient was 0.1 m^2s^{-1} and the time interval was 0.5 s. As is indicated by the maintenance of the shape of the sediment slug in Fig. 1a and the lack of any disturbance in the wake of the slug in both Figs. 1a and 1b, the effects of numerical diffusion and disperson were minimal. In addition, a check on the quantity of sediment present during the progress of the slug showed that mass was conserved.

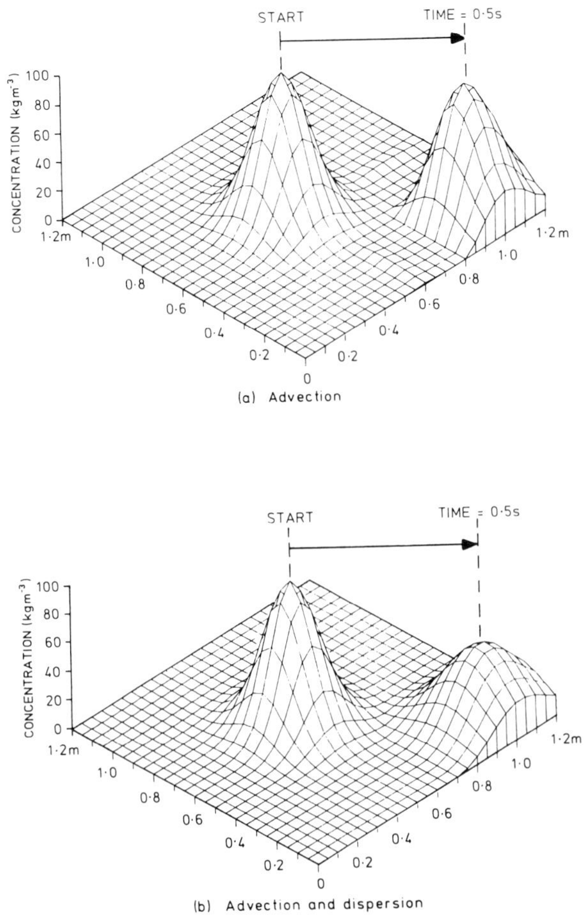

Figure 1: Model test - transport of suspended sediment

The second stage of the computation of the bed level changes involved solving the bed mass continuity equation. Bed mass continuity is described by the expression:

$$(1-p)\rho_s\frac{\partial z_b}{\partial t} + \frac{\partial}{\partial x}(Q_{BX} + Q_{CX}) + \frac{\partial}{\partial y}(Q_{BY} + Q_{CY}) - R_D + R_E = 0 \qquad (25.2)$$

where:

p = porosity of the bed material ($\approx$ 0.4);
Q_{BX} = longitudinal bedload transport rate due to tidal and wave-induced littoral currents (kg m^{-1} s^{-1});
Q_{BY} = lateral equivalent of Q_{BX} (kg m^{-1} s^{-1});
Q_{CX} = longitudinal cross-shore transport rate due to undertow, wave asymmetry and local bed slope (kg m^{-1} s^{-1});
Q_{CY} = lateral equivalent of Q_{CX} (kg m^{-1} s^{-1});
z_B = elevation of the bed relative to an arbitrary datum (m);
ρ_s = density of the bed material ($\approx$ 2650 kg m^{-3}).

The only unknown in Eq. 25.2 is the elevation of the bed (z_B), details of the computation of the independent variables (Q_{BX}, Q_{BY}, Q_{CX}, Q_{CY}, R_D, R_E) being set out below. As the deposition rate (R_D) is a function of the depth-averaged concentration ($\bar{C}$), this variable is dependent on the solution of Eq. 25.1. Eq. 25.2 is solved by means of an explicit finite difference scheme.

The structure of the model set up to solve Eqs. 25.1 and 25.2 is contained in Fig. 2. A description of the constituent parts of the model follows.

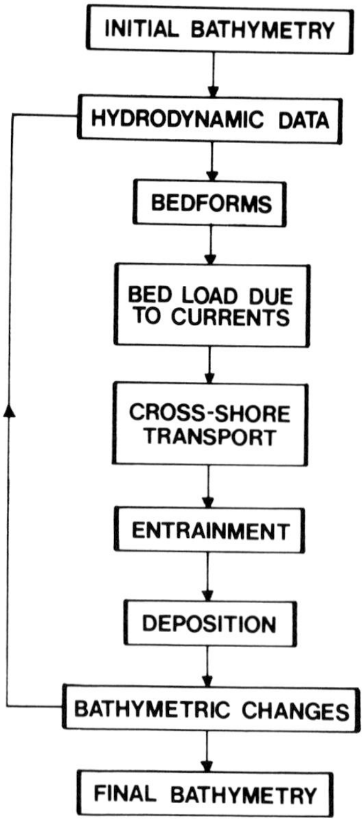

Figure 2: Model structure

25.2.2 Bedforms

The height of the bedforms at any point in plan is needed to fix the thickness of the mobile bed layer. Both current and wave action may be present at any given location but the authors are unaware of any quantitative information in the literature about bedforms generated under such conditions. This problem was overcome, therefore, by determining the bedform height produced by currents and waves separately and then taking the larger of the two to be the representative value. The determination of the characteristics of the current-generated ripples is based on the work of Yalin (1977) and that of wave-generated ripples is based on the work of Nielsen (1985) or Swart (1976). The present version of the model makes no provision for the occurrence of marine dunes.

The results of a comparison of the Nielsen and Swart methods of deriving the wave-generated ripple heights is shown in Fig. 3. Clearly, these two approaches yield values which differ markedly in trend. However, recent work by Ribberink and Al-Salem (1990) has confirmed that the Nielsen method gives the more qualitatively correct results, a sand bed tending to become plane under intense wave activity.

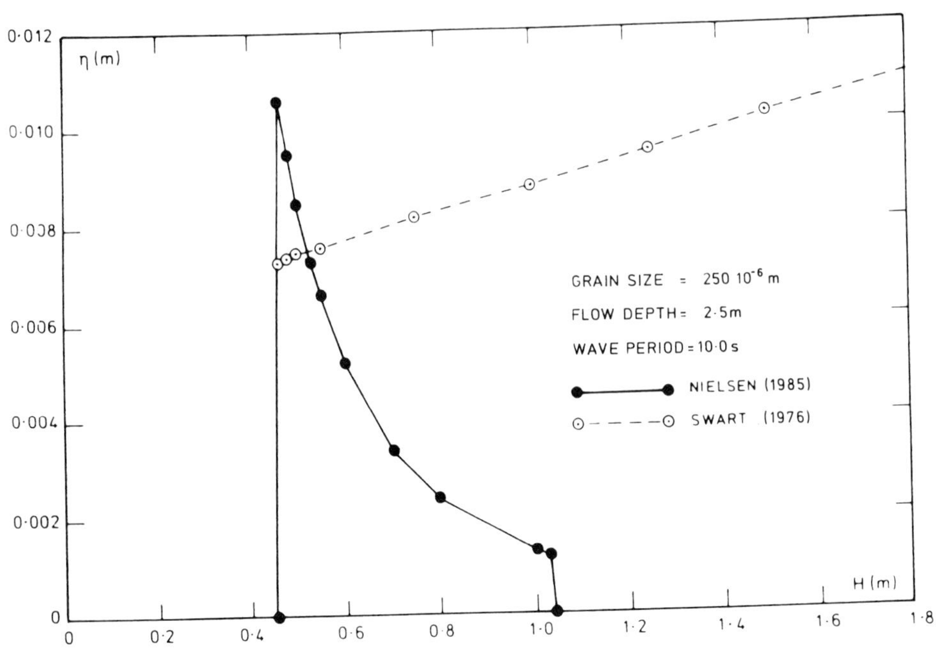

Figure 3. Bedform height (η) versus wave height (H)

25.2.3 Bedload Transport

Bedload transport due to currents, with or without the presence of waves, is computed in three stages. Firstly, the steady-state suspended load transport rate is calculated using an empirical expression based on laboratory and field data (unpublished). In its simplest form, this expression applies to unidirectional flows but allowance is also made for the effects of wave action and reversing (tidal) flows. Wave action increases the transport rate but the effects of tidal action are more complex. The finite duration of a tidal cycle may cause the water column to be undersaturated and inertia may have the same effect during accelerating flows but the opposite effect during decelerating flows. The present form of the empirical expression does not differentiate between accelerating and decelerating flows.

Secondly, the ratio of the steady-state suspended load transport rate to the bed load transport rate is determined using the Bijker (1968) version of the method originally proposed by Einstein (1950). The Bijker-Einstein ratio, however, has been slightly altered for use in the present model in that the thickness of the mobile bed layer is set equal to half the bedform height rather than the total bedform height adopted by Bijker. This results in an expression of the form:

$$\frac{Q_S}{Q_B} = 2.47\left(\ln\left(\frac{33h_F}{r_B}\right)I_1 + I_2\right) \tag{25.3}$$

where:

I_1 = first Einstein integral;
I_2 = second Einstein integral;
Q_B = bedload transport rate (kg m^{-1} s^{-1});
Q_S = steady-state suspended load transport rate (kg m^{-1} s^{-1});
r_B = bed roughness height (m).

The latter variable, in turn, is derived using the following relationship, which was proposed by van Rijn (1982):

$$r_B = 3D_{90} + 1.1\eta\left(1 - \exp\left(-25\frac{\eta}{\lambda}\right)\right) \tag{25.4}$$

where:

D_{90} = ninetieth percentile-by-weight grain size (m);
η = bedform height (m);
λ = bedform length (m).

Finally, the known steady-state suspended load transport rate (Q_S) is combined with Eq. 25.3 to yield the bedload transport rate (Q_B). This method of determining the latter quantity has been adopted because it ensures that the bed load is compatible with the steady-state suspended load and the latter is computed using an expression based on a wide range of laboratory and field conditions. Typical values produced for Q_S/Q_B and Q_B are given in Fig. 4.

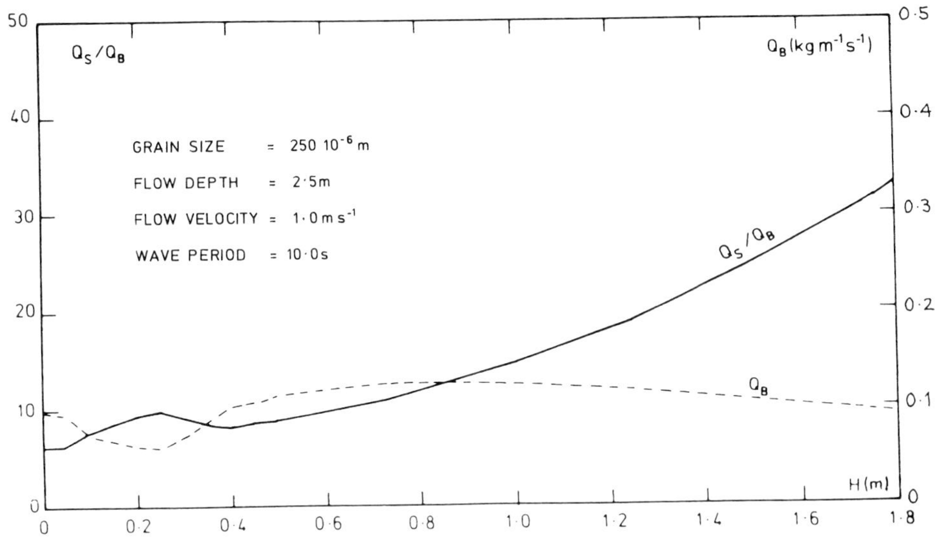

Figure 4: Q_S/Q_B and Q_B versus wave height (H)

25.2.4 Cross-Shore Transport

A comparison was made of the following cross-shore transport sub-models in order to select a suitable scheme for inclusion in the main model:

Bailard (1982)
Nishimura and Sunamura (1986)
Watanabe (1985)
Shibayama and Horikawa (1985).

This comparison entailed running the four schemes for various deep-water wave characteristics (H_O, L_O), grain sizes (D_{50}) and initial plane beach slopes (i_o); hydrodynamic data, needed to drive the sediment models, were generated by means of the Yoo and O'Connor (1988) wave model. The results, expressed in the form of the occurrence of accretion, erosion or no change (stability) at the intersection point between the mean water level and the beach profile, were then checked against field data collated by Sunamura and Horikawa (1974). The latter have shown that the overall shape of a beach profile is characterised by the movement of the mean water level intersection point. Based on this approach, the Watanabe method gave the best results, a summary of the latter being given in Fig. 5. Even so, the Watanabe method produced eroding beaches when they should have been stable and stable beaches when they should have been accreting. Clearly, this is a subject requiring future improvement.

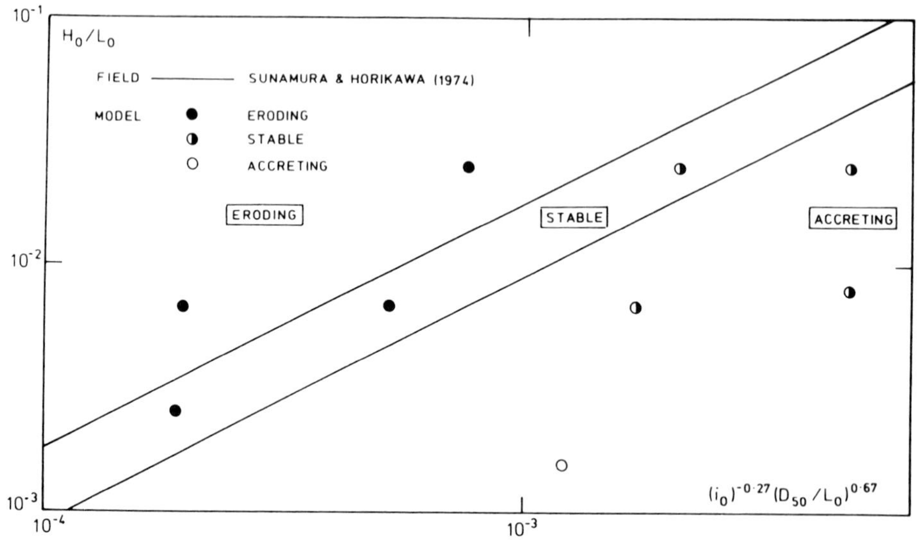

Figure 5: Beach type as function of wave and sediment characteristics

25.2.5 Sediment Entrainment

Nine possible methods of deriving the sediment entrainment rate were considered. These various methods fall into two categories, namely those which give the entrainment rate directly and those in which the entrainment rate is derived from the steady-state near-bed suspended sediment concentration. In the latter case, the entrainment rate is set equal to the steady-state deposition rate so that:

$$R_E - W_S C_B^{\infty} \tag{25.5}$$

where

C_B^{∞} = steady-state near-bed suspended sediment concentration (kg m^{-3});
W_S = sediment fall velocity (m s^{-1}).

The relevant entrainment theories, and the conditions for which they were derived, are as follows:

Direct approach:

Van Rijn and Tan (1985) - unidirectional flow and waves;
Van Rijn (1985) - unidirectional flow;
Nakagawa and Tsujimoto (1980) - unidirectional flow;
Fernandez Luque (1974) - unidirectional flow;
De Ruiter (1982) - unidirectional flow.

Indirect approach:

Nielsen (1986) - waves;
Engelund and Fredsoe (1976) - unidirectional flow;
Fernandez Luque and van Beek (1976) - unidirectional flow;
O'Connor (O'Connor, Nicholson and Yoo, 1989) - unidirectional flow and waves.

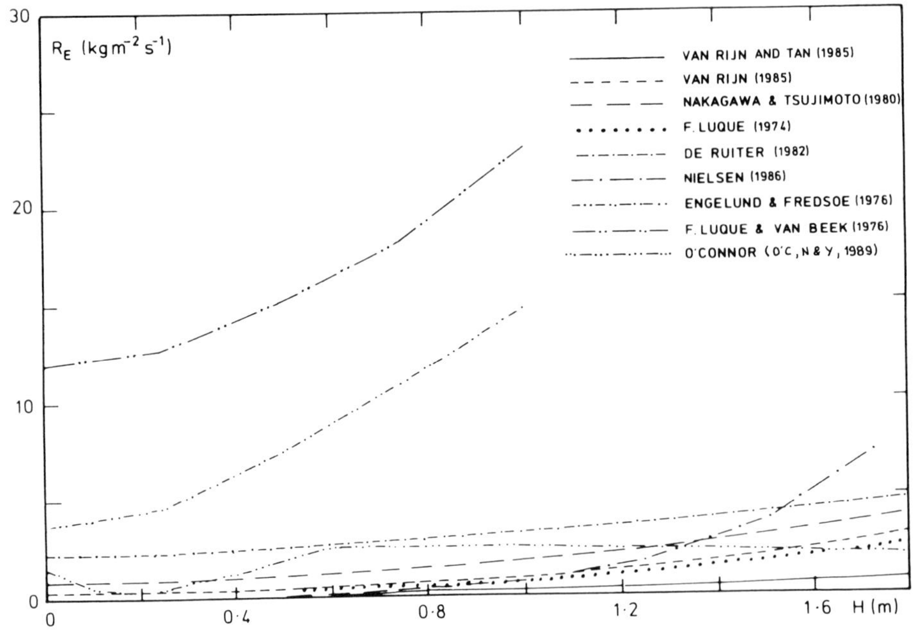

Figure 6: Entrainment rate (R_E) versus wave height (H)

An example of the results produced by the various theories is shown in Fig. 6. When producing the data contained in this figure, it was assumed that theories derived for unidirectional flow conditions could also be applied to situations involving wave action, provided that the appropriate value was used for the bed shear stress; the latter was calculated using the O'Connor and Yoo (1988) bed friction model. The contents of Fig. 6 indicate that all the results are in general quantitative agreement, except for the very large entrainment rates given by the Engelund and Fredsoe and the Fernandez Luque and van Beek methods. This discrepancy can be explained by the fact that a value of one or two grain diameters was adopted for the thickness of the mobile bed layer in the latter two methods, regardless of the presence of bedforms. However, if it is assumed that the mobile bed thickness is directly related to the bedform height, then the near-bed concentrations, and also the entrainment rates, are reduced accordingly.

25.2.6 Sediment Deposition

The sediment deposition rate is defined by the relationship:

$$R_D = W_S C_B \tag{25.6}$$

where:

C_B = instantaneous near-bed suspended sediment concentration (kg m^{-3}). However, the solution of the depth-averaged diffusion- advection equation (Eq. 25.1) yields the instantaneous depth-averaged concentration $\bar{C}$. Hence, if the ratio between these two quantities is written as:

$$\beta = \frac{C_B}{\bar{C}} \tag{25.7}$$

combining Eqs. 25.6 and 25.7 yields:

$$R_D = \beta\, W_S C_B \tag{25.8}$$

The only unknown quantity in Eq. 25.8 is β. A value is established for this variable by assuming that the form of the vertical concentration profile can be represented by the Rouse (1937) distribution. The reference level contained in the Rouse distribution is equated to the mobile bed thickness; the latter, in turn, is set equal to half the roughness height given by Eq. 25.4. This approach yields the following expression for the concentration ratio:

$$\beta = 0.22\,\frac{\left(1 - \dfrac{h_B}{h_F}\right)}{\left(\dfrac{h_B}{h_F} I_1\right)} \tag{25.9}$$

where:

h_B = mobile bed layer thickness (m).

Values of β, as a function of flow velocity ($\bar{U}$) for a fixed median grain size (D_{50}), flow depth (h_F), wave height (H) and wave period (T), are shown in Fig. 7; both the Nielsen and the Swart methods have been used to compute the wave-generated bedform characteristics. The unrealistic change in the value of β at a flow velocity of approximately 0.4 m s^{-1} reflects the separation of unidirectional flow effects from wave effects in the determination of the bedform characteristics (see Sect. 25.2.2).

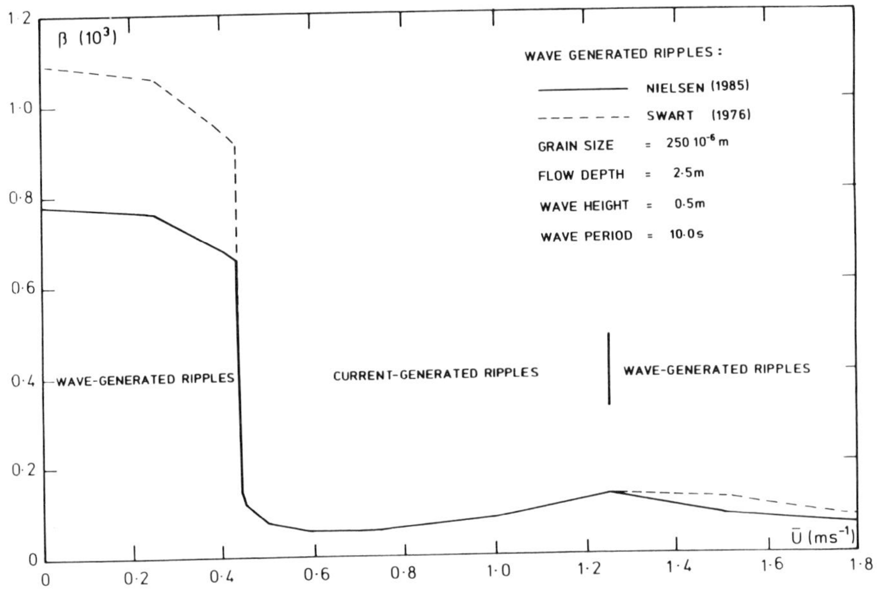

Figure 7: β-function versus flow velocity ($\bar{U}$)

25.3 Model application

25.3.1 Field Data

The model was tested by applying it to a field situation, details of which had been monitored by the Danish Hydraulic Institute (1984). The field situation consisted of a trench approximately 40m wide by 6m deep which had been dredged across the inshore zone of the Danish North Sea Coast in order to bring ashore oil and gas pipelines. The trench, the layout of which is given in Fig. 8, had been surveyed at regular intervals and continuous measurements made of the wave climate. Details were known, therefore, of the siltation in the trench and the associated wave conditions.

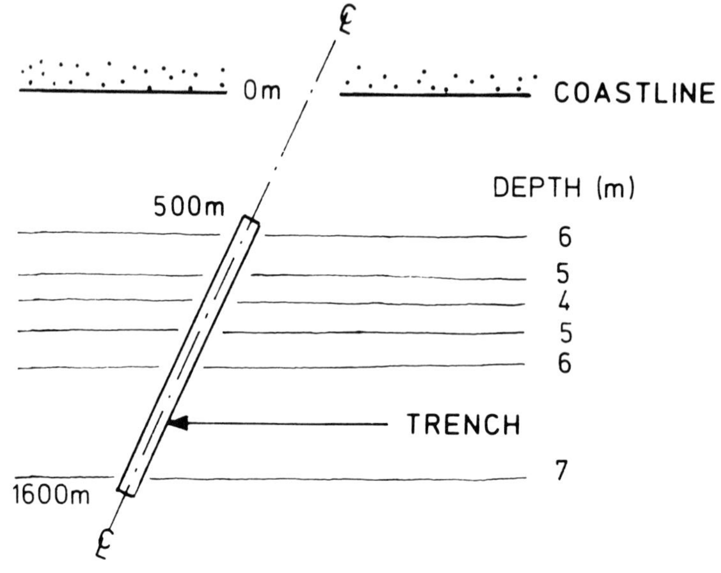

Figure 8: Layout of the dredged trench

A severe storm occurred between two successive surveys and, as the wave records indicated that virtually all the sediment movement associated with the monitoring period could be ascribed to this one storm, only the latter was reproduced in the model. In addition, it was assumed that a single wave height could be used to represent the storm due to the abrupt growth and decay of the wave intensity. The resulting representative conditions were as follows:

Wave period	= 5.5 s
Breaker depth	= 4.8 m
Breaker height	= 3.0 m
Breaker approach angle	= 7.5 °
Storm duration	= 4.3 x 10^5 s (5 days)
Median grain size	= 200 μm.

The hydrodynamic data needed to drive the sediment model consisted of a cross-shore profile of the longshore current, which had been produced by Deigaard et al (1986), and a wave field based on a mixture of available information and information derived from a plane beach configuration. Tidal effects were considered to be negligible, the mean tidal range being less than one metre.

25.3.2 Model Results

Before the start of a model run, the system was operated with fixed bathymetry until the suspended sediment concentrations had attained a state of equilibrium. The resulting concentrations were then used as the initial values for the model run. The latter took place in a series of time steps, at the end of each of which the flow field was adjusted to allow for the effects of depth changes. This direct coupling between the hydrodynamic "model" and the sediment model caused the output from the latter to be dependent on the time step.

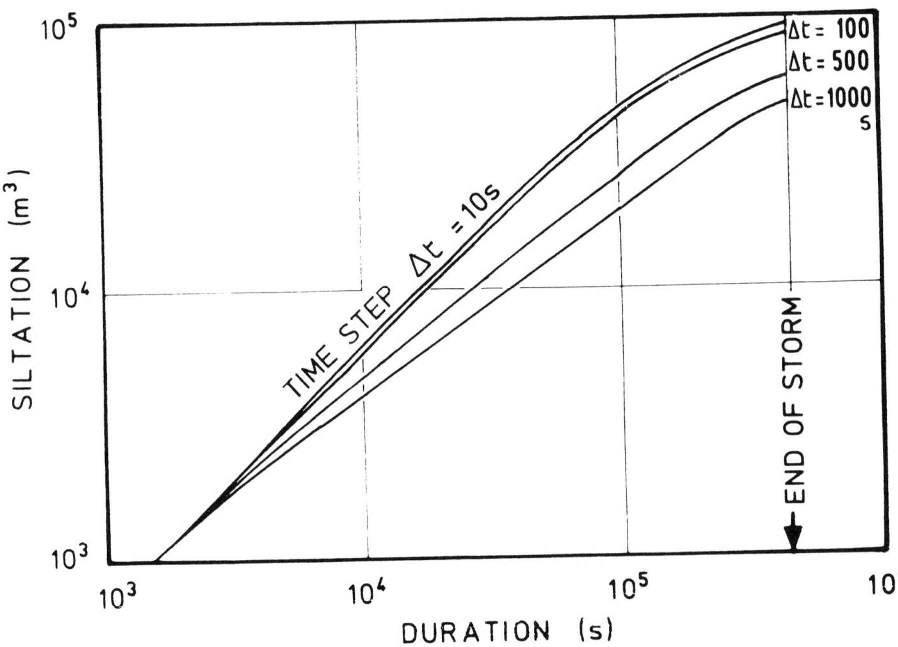

Figure 9. Time step effects

The effect of the time step on the model output is illustrated in Fig. 9, which shows the volume of siltation in the trench as a function of time. The particular version of the model associated with Fig. 9 incorporated the Swart bedform sub-model and the van Rijn entrainment sub-model. From the contents of the figure, it can be seen that the amount of siltation varied inversely with the time step but that this trend became negligible if the time step was less than 100 s. It was concluded, therefore, that a time step of 100 s yielded the best compromise between operating requirements and accuracy.

The contents of Fig. 9 also indicate that the rate of infilling decreases with the passage of time. This factor can cause the model output to be misinterpreted if short-term results are extrapolated to give long-term values. For example, running the model for 10^5 s real time with a time step of 100 s leads to an extrapolated infilling volume of 184 000 m^3 after 4.3 10^5s. This compares with a fully computed value of only 86 000 m^3.

The total measured siltation in the dredged trench was then compared with the output given by versions of the sediment model which incorporated various combinations of the bedform and entrainment sub-models. Only a limited number of these tests could be completed, however, due to instability problems. The latter were caused by very large deposition rates which so altered the bathymetry during the course of a time step that the water depths were no longer compatible with the flow field. Although it may have been possible to complete a failed test by employing a reduced time step, the run times under such circumstances would have been unacceptably large. Nevertheless, the fact that a test failed was a strong indication that the particular version of the model being examined yielded unrealistic values. The results of the successful tests are contained in Fig. 10, which shows that the version of the model containing the Swart bedform sub-model and the de Ruiter entrainment sub-model yielded a total siltation volume of 114 000 m^3. This quantity compares favourably with the measured value of 150 000 m^3, an error of 24 per cent being well within the accepted level of engineering accuracy.

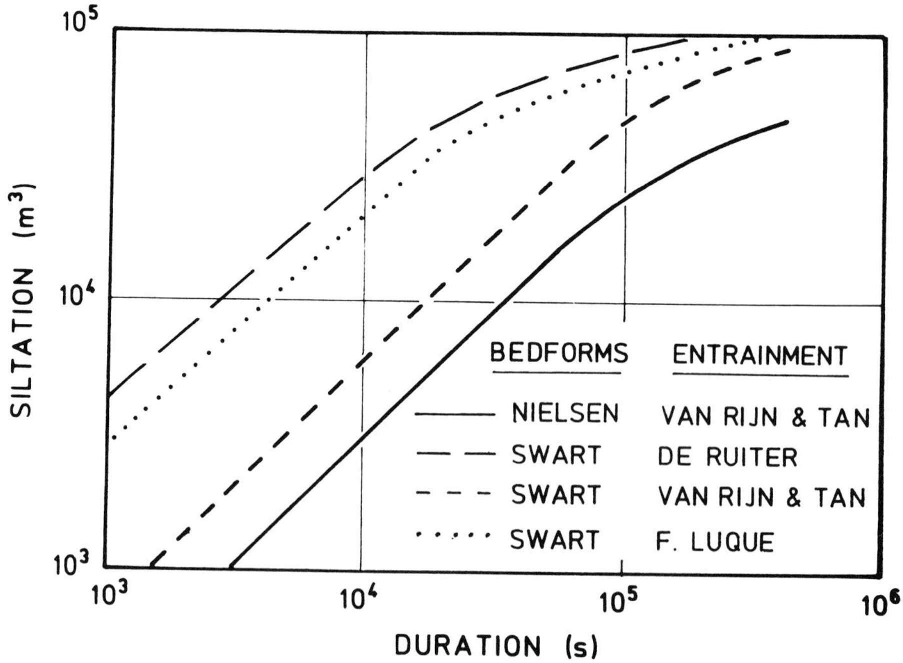

Figure 10: Comparison of computed infilling rates

A more detailed comparison between the model output and the field data, based on the distribution of infilling along the length of the trench, is shown in Fig. 11. There are three distributions in the figure, namely the output given by the present model (incorporating Swart bedforms and de Ruiter entrainment), the output given by a longshore transport model developed by Deigaard et al (1986) and the measured values. Again, there is a reasonable measure of agreement between the output from the present model and the field data, especially when allowance is made for the approximate nature of the hydrodynamic data used to drive the former.

Finally, the computed and measured distributions of infilling in two trench cross-sections are compared in Fig. 12. The agreement between the computed and measured values is very poor in the inshore cross-section (chainage 700), as is to be expected from the comparison of the bulk siltation quantities shown in Fig. 11. A probable reason for this discrepancy is the omission of any allowance for the effect of breaking waves on sediment entrainment inside the surf zone. On the other hand, the agreement between the computed and measured distributions of infilling in the offshore cross-section (chainage 1350) is good. Again, this reflects the comparison contained in Fig. 11.

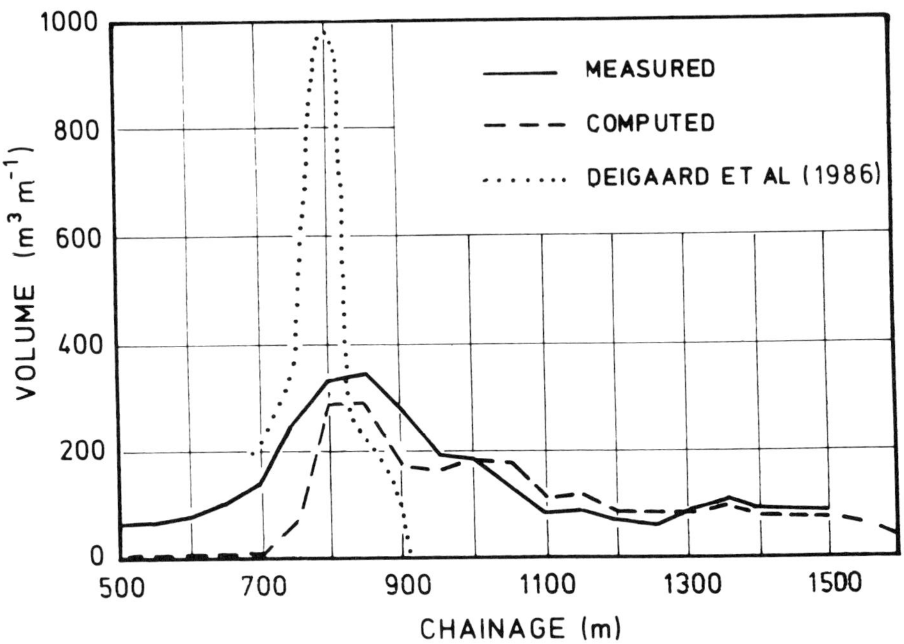

Figure 11: Comparison between measured and computed longitudinal infilling

25.4 Conclusions

A depth-averaged estuarine and coastal sediment transport model has been established which incorporates the effects of waves and currents. The model is of modular construction so that it can be updated in the future as understanding of sediment processes is improved through further research. Tests on the various sediment process modules have indicated that bedload transport and sediment entrainment are adequately represented, bedforms and cross-shore transport are rather crudely represented and may need improvement while sediment deposition appears to be unrealistic under certain flow conditions and may also need improvement.

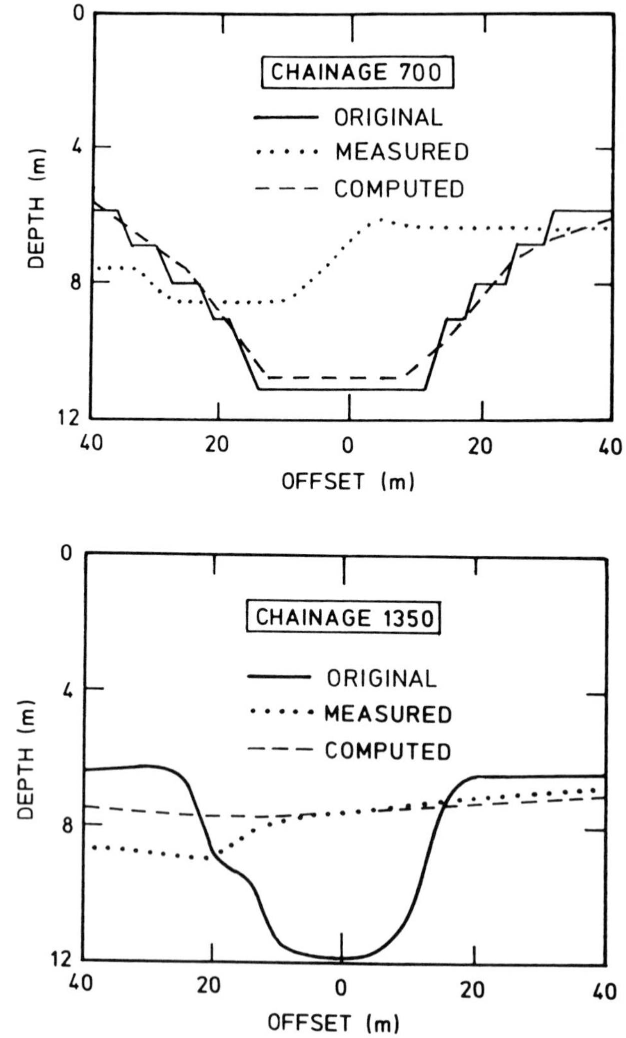

Figure 12: Comparison between measured and computed lateral infilling

Application of the model to a simplified field situation involving the siltation of a dredged trench in a coastal environment yielded realistic answers, which were well within engineering accuracy. The field application also showed the importance of including "feedback" between morphological changes and the hydrodynamic flow field, the need to choose a suitable model time step and the need to operate the model long enough to ensure that any "scaling-up" of short-term runs, which reduces model operation costs, is representative of the final conditions. Finally, a detailed examination of the longitudinal siltation pattern in the dredged trench showed that the model underestimated the siltation in the inshore section of the trench, which suggests that additional field testing is required in order to provide further checks on the accuracy of the scheme.

Acknowledgement

This work was undertaken as part of the MAST G6 Coastal Morphodynamics research programme. It was funded by the Commission of the European Communities, Directorate General for Science, Research and Development, under MAST contract no. 0035.

Symbols

$\bar{C}$ = depth-averaged suspended sediment concentration (kg m^{-3});
C_B = instantaneous near-bed suspended sediment concentration (kg m^{-3});
C_B^{∞} = steady-state near-bed suspended sediment concentration (kg m^{-3});
D_{50} = median grain size (m);
D_{90} = ninety per cent finer-by-weight grain size (m);
D_x = longitudinal dispersion coefficient (m^2 s^{-1});
D_y = lateral dispersion coefficient (m^2 s^{-1});
h_B = mobile bed layer thickness (m);
h_F = flow depth (m);
H = wave height (m);
H_o = deep water wave height (m);
i_o = initial beach slope;
I_1 = first Einstein integral;
I_2 = second Einstein integral;
L_o = deep water wave length (m);
p = bed layer porosity;
Q_{BX} = longitudinal bed load transport rate due to tidal and wave-induced littoral currents (kg m^{-1} s^{-1});
Q_{BY} = lateral equivalent of Q_{BX} (kg m^{-1} s^{-1});

Q_{CX} = longitudinal cross-shore transport rate due to undertow, wave asymmetry and local bed slope (kg m^{-1} s^{-1});
Q_{CY} = lateral equivalent of Q_{CX} (kg m^{-1} s^{-1});
Q_S = suspended load transport rate (kg m^{-1} s^{-1});
r_B = bed roughness height (m);
R_D = sediment deposition rate (kg m^{-2} s^{-1});
R_E = sediment entrainment rate (kg m^{-2} s^{-1});
t = time (s);
T = wave period (s);
$\bar{U}$ = depth-averaged longitudinal velocity (m s^{-1});
$\bar{V}$ = depth-averaged lateral velocity (m s^{-1});
x = longitudinal co-ordinate (m);
y = lateral co-ordinate (m);
W_S = sediment fall velocity (m s^{-1});
z_B = elevation of the bed relative to an arbitrary datum (m);
β = $C_B/\bar{C}$;
Δt = model time step (s);
η = ripple height (m);
λ = ripple length (m);
ρ_s = sediment density (kg m^{-3}).

25.5 References

Bailard, J.A., 1982: Modeling on-offshore sediment transport in the surf zone. Proc. Eighteenth Coastal Engineering Conf., ASCE, 1419-1438.

Bijker, E.W., 1968: Littoral drift as function of waves and current. Proc. Eleventh Coastal Engineering Conf., ASCE, 415-435.

Danish Hydraulic Institute, 1984: Monitoring of sedimentation in a dredged trench. Report, Horsholm, Denmark.

Deigaard, R., Fredsoe, J. and Hedegaard, I.B., 1986: Mathematical model for littoral drift. Jour. Waterway, Port, Coastal and Ocean Engineering, ASCE, Vol. 112, No. 3, 351-369.

De Ruiter, J.C.C., 1982: The mechanism of sediment transport on bed forms. Mechanics of Sediment Transport, Euromech 156, 137-142.

Einstein, H.A., 1950: The bed-load function for sediment transportation in open channel flows. Technical Bulletin No. 1026, United States Department of Agriculture, Soil Conservation Service, Washington D.C.

Engelund, F. and Fredsoe, J., 1976: A sediment transport model for straight alluvial channels. Nordic Hydrology, Vol. 7, 293-307.

Fernandez Luque, R., 1974: Erosion and transport of bed-load sediment. Dissertation, KRIPS Repro BV, Meppel, Netherlands.

Fernandez Luque, R. and van Beek, R., 1976: Erosion and transport of bed-load sediment. Jour. Hyd. Res., Vol. 14, No. 2, 127-144.

Nakagawa, H. and Tsujimoto, T., 1980: Sand bed instability due to bed load motion. Proc. ASCE, Jour. Hyd. Division, Vol. 106, No. HY12, 2029-2051.

Nielsen, P., 1985: A short manual of coastal bottom boundary layers and sediment transport. Technical Memo. 85/1, Public Works Dept., Sydney, Australia.

Nielsen, P., 1986: Suspended sediment concentrations under waves. Coastal Engineering, Vol. 10, 23-31.

Nishimura, H. and Sunamura, T., 1986: Numerical simulation of beach profile changes. Proc. Twentieth Coastal Engineering Conf., ASCE, 1444-1455.

O'Connor, B.A., Nicholson, J. and Yoo, D.H., 1989: Coastal Impact Modelling. Volume 3. Report, University of Liverpool.

O'Connor, B.A. and Yoo, D.H., 1988: Mean bed friction of combined wave-current flow. Coastal Engineering, Vol. 12, 1-21.

Ribberink, J.S. and Al-Salem, A., 1990: Bedforms, sediment concentrations and sediment transport in simulated wave conditions. Abstrs. Twenty-Second Coastal Engineering Conf., ASCE, 501-502.

Rouse, H., 1937: Modern conceptions of the mechanics of fluid turbulence. Trans. ASCE, Vol. 102, 463-543.

Shibayama, T. and Horikawa, K., 1985: A numerical model for two-dimensional beach transformation. Proc. JSCE, No. 357/II-3, 167-176.

Sunamura, T. and Horikawa, K., 1974: Two-dimensional beach transformation due to waves. Proc. Fourteenth Coastal Engineering Conf., ASCE, 920-938.

Swart, D.H., 1976: Coastal sediment transport. Computation of longshore transport. Report 968-I, Delft Hydraulics Lab., Netherlands.

Van Rijn, L.C., 1982: Equivalent roughness of alluvial bed. Proc. ASCE, Vol. 108, No. HY10, 1215-1218.

Van Rijn, L.C., 1985: Mathematical models for sediment concentration profiles in steady flow. Publ. No. 354, Delft Hydraulics Lab., Netherlands.

Van Rijn, L.C. and Tan, G.L., 1985: Two-dimensional vertical mathematical model for sedimentation in dredged channels and trenches by currents and waves. Rijkswaterstaat Communications, No. 41.

Watanabe, A., 1985: Three-dimensional predictive model of beach evolution around a structure. Proc. Water Wave Research Conf., Hannover, 123-141.

Yalin, M.S., 1977: Mechanics of sediment transport. Pergammon Press, Oxford, England.

Yoo, D.H. and O'Connor, B.A., 1988: Numerical modelling of waves and wave-induced currents on groyned beach. Proc. IAHR Symp. on Mathematical Modelling of Sediment Transport in the Coastal Zone, Copenhagen, 127-136.

Part VI

Applied Studies

26

Current and density structure in the Netherlands coastal zone

W P M de Ruijter, A van der Giessen and F C Groenendijk

Abstract

Results are described of a measurement programme in the outflow zone of the Rhine river that has been going on over the past years with varying intensity. Its aim is to describe the three-dimensional hydrographic structure in the area at different time and space scales.

The density distribution in the Dutch coastal zone is controlled by the salinity field. Here the largest salinity gradients of the North Sea are observed. This is due to the relatively high river discharge and the fact that its northward spreading is mostly confined to a coastal band within 20km from the shore.

Varying conditions of both wind and fresh water discharge lead to a highly variable density and current structure. Especially in the upper layer and up to 80km from the river outlet almost all individual salinity patterns deviate significantly from the mean. In this region the water column also often is strongly stratified.

Currents were measured continuously over a one year period in a cross-shore transect located approximately 40km north of the main outflow port. As expected, the mean southwesterly winds mainly drive a longshore northward current, with average values in the upper layer between 7 and 11cm/s. The large cross-shore density gradient is related both to a longshore geostrophic flow component with large vertical shear and to a cross-shore estuarine circulation component. Thus, this longshore flow is enhanced near the surface and reduced near the bottom. Almost permanently the observed near bed residual current was directed on-shore, with a one year mean value of 3cm/s.

Under extreme discharge conditions combined with strong winds on-shore near bed currents of ca. 10cm/s were observed. During this event stratification was very strong throughout the coastal zone. In the upper layer longshore current speeds appeared of up to 50cm/s, whereas in the lower layer these were only a few cm/s.

Vertical current profiles are very sensitive to wind direction. Wind- and density driven components may counteract or reinforce each other. Consequently, during easterlies a large veering of the current direction with depth was observed. During southwesterlies this is only minor, but the amplitude decreases strongly with depth. In case of northerlies only a small vertical variation of residual current speed was observed.

26.1 Introduction

The input of nutrients and contaminants into the Netherlands coastal zone (Fig. 1) by the Rhine river outflow has a strong impact on its water quality. In the Southern Bight of the North Sea the discharge of the Rhine/Meuse estuary constitutes the major source of fresh water. Consequently, stratification occurs in a large part of the Dutch coastal region (Van Alphen et al, 1988). The three-dimensional character of the density field is certainly not limited to the immediate vicinity of the main river water outlets, the Rotterdam Waterway and the Haringvliet sluice complex (see Fig. 2 for locations).

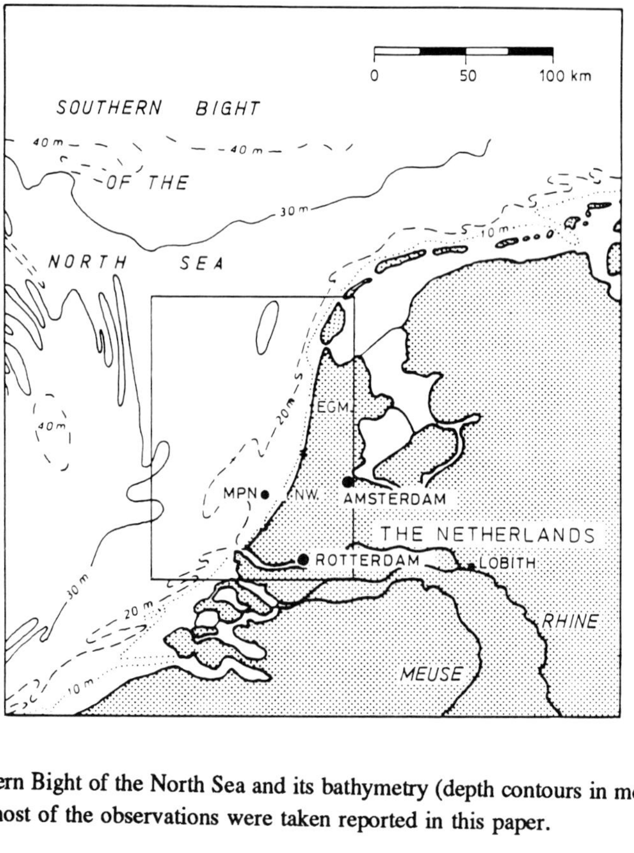

Figure 1: The Southern Bight of the North Sea and its bathymetry (depth contours in meters). The area is indicated where most of the observations were taken reported in this paper.

Over a large distance after its debouchment into the North Sea river water spreading causes spatial inhomogeneities in the salinity field. Associated with it is a marked three-dimensional structure both of currents and suspended matter (Van der Giessen et al, 1990, Visser et al, 1991). Information about this vertical distribution is essential to develop appropriate models to support the management of the coastal environment. The extent of the stratified area is largely controlled by wind and river discharge conditions, both largely variable. Consequently, the salinity pattern is highly variable as well.

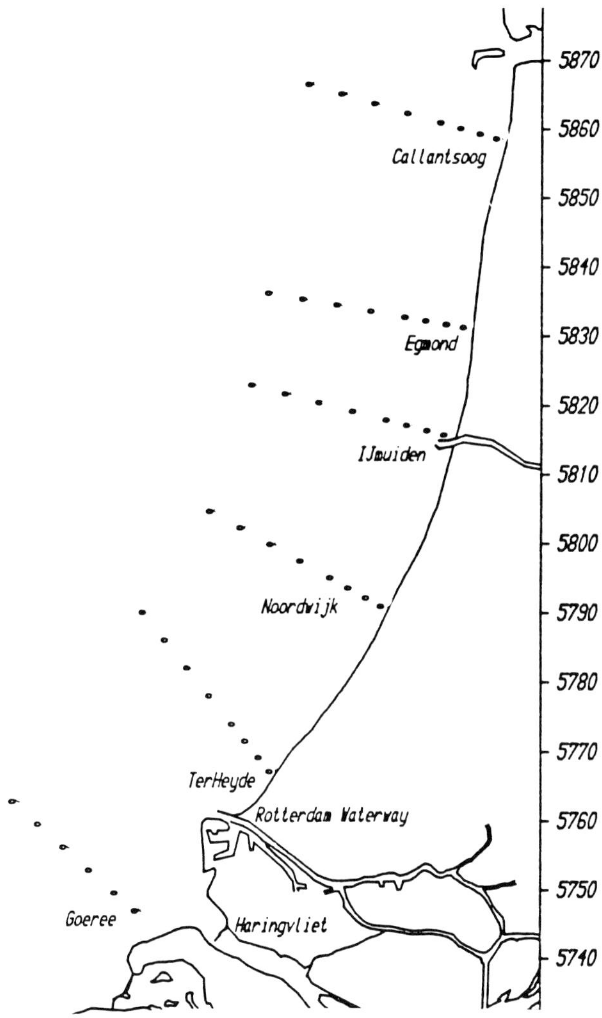

Figure 2: Location of cross-shore transects where repeated CTD measurements were taken. Coordinates are in km (UTM-system)

In this paper the mean density field and its variability are presented and analyzed (section 26.3), as determined from repeated cruises in 1986 and 1987 in the area between the Dutch coast and 30km off-shore (Fig. 2). Before that, in section 26.2, a general description of the hydrography of the area is given. Section 26.4 reviews the residual current structure measured for a one-year period from a series of current meter moorings in a cross-shore transect near Noordwijk, located 40km north of the main river outlet. Finally, in section 26.5 the results are shortly summarized and discussed.

26.2 General Characteristics of the Dutch Coastal Zone

The southern part of the North Sea is characterized by its shallowness (Fig. 1) and strong tidal currents, with maximum current speeds ranging between 0.7 and 1.1m/s and directed mainly parallel to the coast. Consequently, over most of the area the water column is well mixed throughout the year. Major exceptions are the outflow regions of the rivers where strong haline stratification exists throughout the year. Current variations over the spring-neap tidal cycle are substantial (Van der Giessen et al., 1990. The discharge of the Rhine river varies strongly. It ranges between 600 and 13000m^3/s with a long term average of 2200m^3/s (Fig. 3). Generally, discharges are low in late summer and fall, higher discharges are observed in winter and spring. Shorter individual peaks frequently occur, mainly due to heavy rain events over the catchment area.

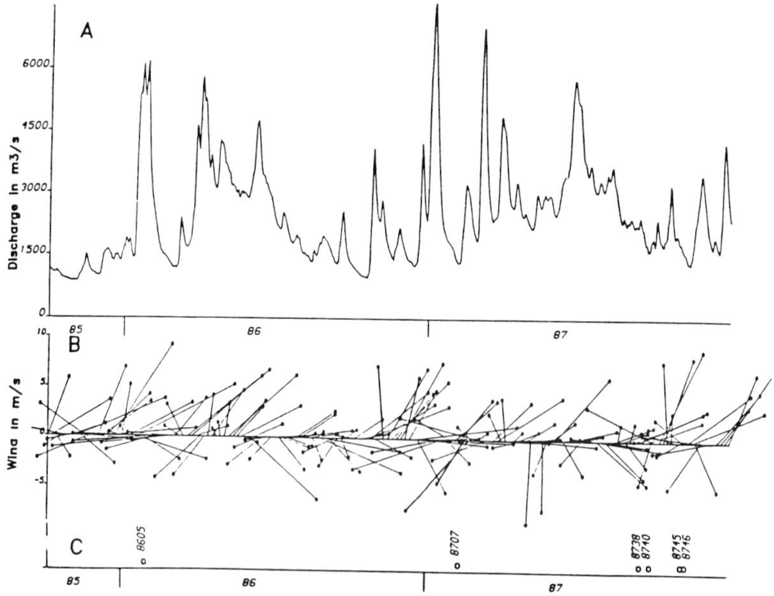

Figure 3: Discharge and wind conditions during the observation periods. A: Rhine discharge. B: wind vector. C: period numbers indicating individual cruises in the coastal zone

Evidently, the other key factor controlling the density and sub-tidal flow fields is the wind. It is also highly variable, with south westerlies characterizing the climatological mean (Fig. 3).

The bulk of the Rhine water combines with the Meuse river discharge (averaging $300m^3/s$) and debouches into the North Sea through both the Rotterdam Waterway and the Haringvliet sluices (Fig. 2). The latter are manipulated with the aim to guarantee a discharge of $1500m^3/s$ through the Waterway as long as possible to prevent salt wedge intrusion. Above a fresh water discharge of 1500 m^3/s the remainder enters the coastal zone via the Haringvliet. Only at discharges over $6000m^3/s$ an extra transport through the Rotterdam Waterway takes place. In front of the Dutch coast the river water mixes with open sea water that has entered the southern North Sea through the Dover Strait. This influx of Channel water varies considerably with wind speed and direction (Prandle, 1978).

Important length scales that characterize the near field area of the outflowing river water are the internal Rossby radius of deformation and the inertial deformation radius. The former defines the horizontal scale over which the Coriolis force and the pressure gradient force related to the horizontal salinity gradient become of equal order of magnitude. The latter estimates the scale after which inertial and Coriolis forces may balance. Both length scales are close to 10km in the Rhine outflow (Van Alphen et al, 1988) which is the same order of magnitude as the tidal excursion length. At larger distances from the mouth the influence of earth rotation becomes more significant. Observations near the mouth of the Rotterdam Waterway reveal formation of a sharp surface front when the convergence is strongest between buoyant river outflow and maximum southward ebb-tidal currents. At this phase of the tide river outflow is maximal and driven southward by the longshore tidal current. The front is clearly visible by an abrupt change in colour and collection of foam and floating material at the convergence line. The very strong surface convergence is accompanied by significant downwelling and internal front formation at the northern edge of the outflow (Van Alphen et al, 1988). Near field observations have shown that plume dimensions vary considerably depending on discharge and wind field.

26.3 The Observed Density Field

Over the years 1986 and 1987 repeated cross-shore CTD-sections were taken in a zone between the coast and 30km offshore (Fig. 2). Measurements were performed by the research vessel 'Holland' using a Guildline CTD probe. The observational period comprises a wide variety of meteorological and river discharge conditions (Fig. 3). At every station the vertical temperature and salinity structure was observed at depth intervals of ca. 1m. Underway, near surface salinities were taken continuously.

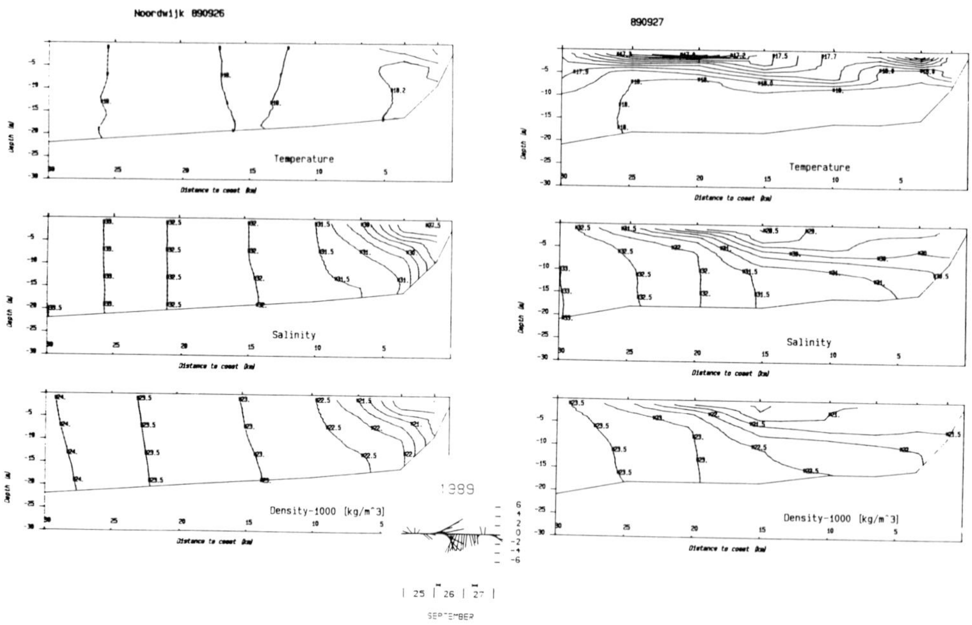

Figure 4: Temperature, Salinity and sigma-t-fields cross-shore near Noordwijk (40km north of the river mouth). The observations were taken on two consecutive late September days when the wind veered from southwest to north-northeast (wind shown on the bottom of the figure). It shows that in this area the salinity field determines the density structure. The temperature then follows and adjust locally to atmospheric conditions. This response appears to be strongly affected by the local depth of the wind-mixed layer.

In this region the structure of the density field is determined by the salinity field due to the dominant influence of the river outflow. As an example, Fig. 4 shows temperature, salinity and density fields in the Noordwijk transect measured on two consecutive late September days when the wind veered from southwest to north-northeast. The response to the change in wind direction is remarkable, due to the associated Ekman transport in the upper layer changing from onshore to offshore. Most striking is the change of the temperature field. Temperature is clearly non-conservative, in particular in the upper layer where it adjusts locally to the atmospheric conditions. In this example (Fig. 4) the relatively cool northerlies lead to lower temperatures in the upper layer, the effect being largest around 4 and 20km offshore, because there the mixed layer depth is minimal and thus cooling maximal. The temperatures in the lower layer still reflect the conditions of the prior day.

As a consequence of the above, temperature is not a useful tracer to derive e.g. advective properties of the flow field, nor is an analysis using TS-diagrams very useful in this region.

In the sequel the analysis will therefore be limited.

Expectedly, the average salinity field (Fig. 5,6) shows a gradual weakening of vertical stratification with increasing distance from the main river water outlet. Averages were taken over 8 cruises (see Fig. 3). In the Ter Heijde transect (Fig. 5a) 10km north of the mouth, stratification is strongest. The relatively freshwater layer is confined mainly to the upper 5-8m. Variations in wind and discharge are reflected mainly in large variability of the upper layer characteristics. This close to the river mouth the large horizontal salinity gradient is associated with a persistent estuarine-type cross shore circulation component. It shows in a large slope of the isopycnals, far inshore penetration of dense, highly saline, open sea water and relatively small variability in the lower layer (also due to tidal mixing). A relative minimum of variability (in terms of standard deviation) around 4km off-shore may indicate a preferred upwelling area. Northerlies and offshore winds may further the occurrence of upwelling, related to offshore Ekman transport. A similar phenomenon is observed off Noordwijk (Fig. 5b), 30km further northward.

However, the structure of average and standard deviation of the salinity field in the Ter Heijde transect is also strongly affected by near field processes as it is only one internal Rossby and inertial radius away from the source. Both analytical (Garvine, 1987) and laboratory models (McClimans, 1988) of river plumes and fronts indicate that near the mouth there is an inertial turning region. Streamlines of the outflow separate from the coast on both sides of the source and start turning to the right under the influence of earth rotation. Within several radii of deformation the plume reattaches to the coast and forms a coastal current.

Although these models ignore wind and tidal effects the structure of mean salinities and their standard deviation at Ter Heijde and Noordwijk seem to reflect the effect of the dynamical processes incorporated. The mean position of the turning plume is situated approximately 10km offshore from Ter Heijde (Fig. 5a); reattachment takes place between Ter Heijde and Noordwijk and the position of a coastal front around 7km from the shore is suggested by the crowding isolines of the standard deviation (Fig. 5b) at Noordwijk.

Spreading and mixing processes, involving considerable shear dispersion, act to weaken stratification and cross-shore gradient. Consequently, moving northward salinity variability decreases, in particular in the upper layer, but the width of the area affected by river water becomes larger. This is a result both of vertical exchange in combination with highly sheared currents and reduced on-shore penetration of the saline wedge, related to a weakening of the cross-shore estuarine flow component. On moving further away from the river water source the water column becomes more and more vertically homogeneous (Fig. 5c, d), reflected in steeper isopycnals and isolines of variability. Nevertheless, in the Callantsoog transect, located ca. 100km north of the river mouth, even the mean salinity field is still weakly stratified.

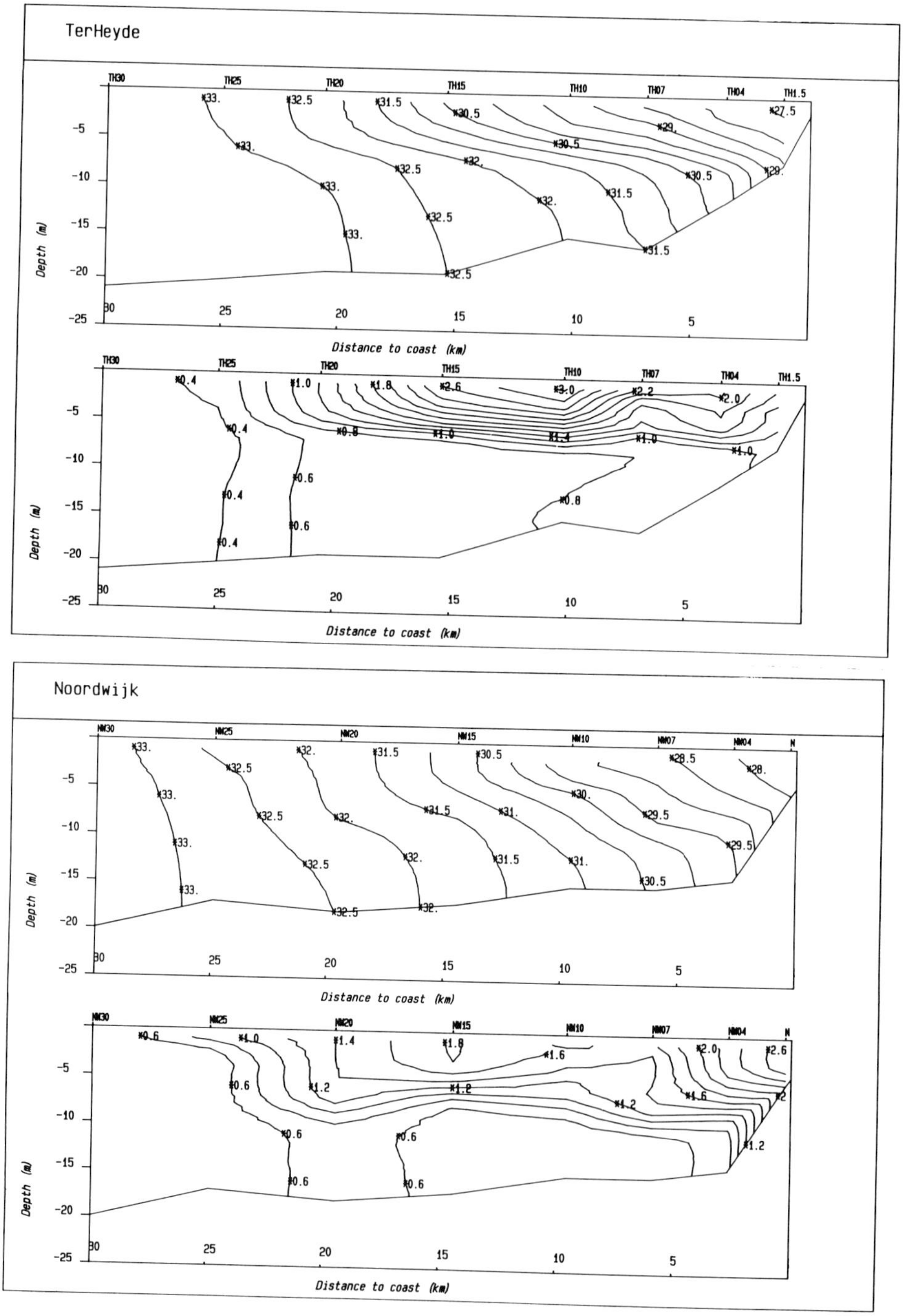

Figure 5a,b: Average salinity distribution (in 0/00) and standard deviation over 8 cruises in 1986 and 1987; a: Ter Heijde, b: Noordwijk

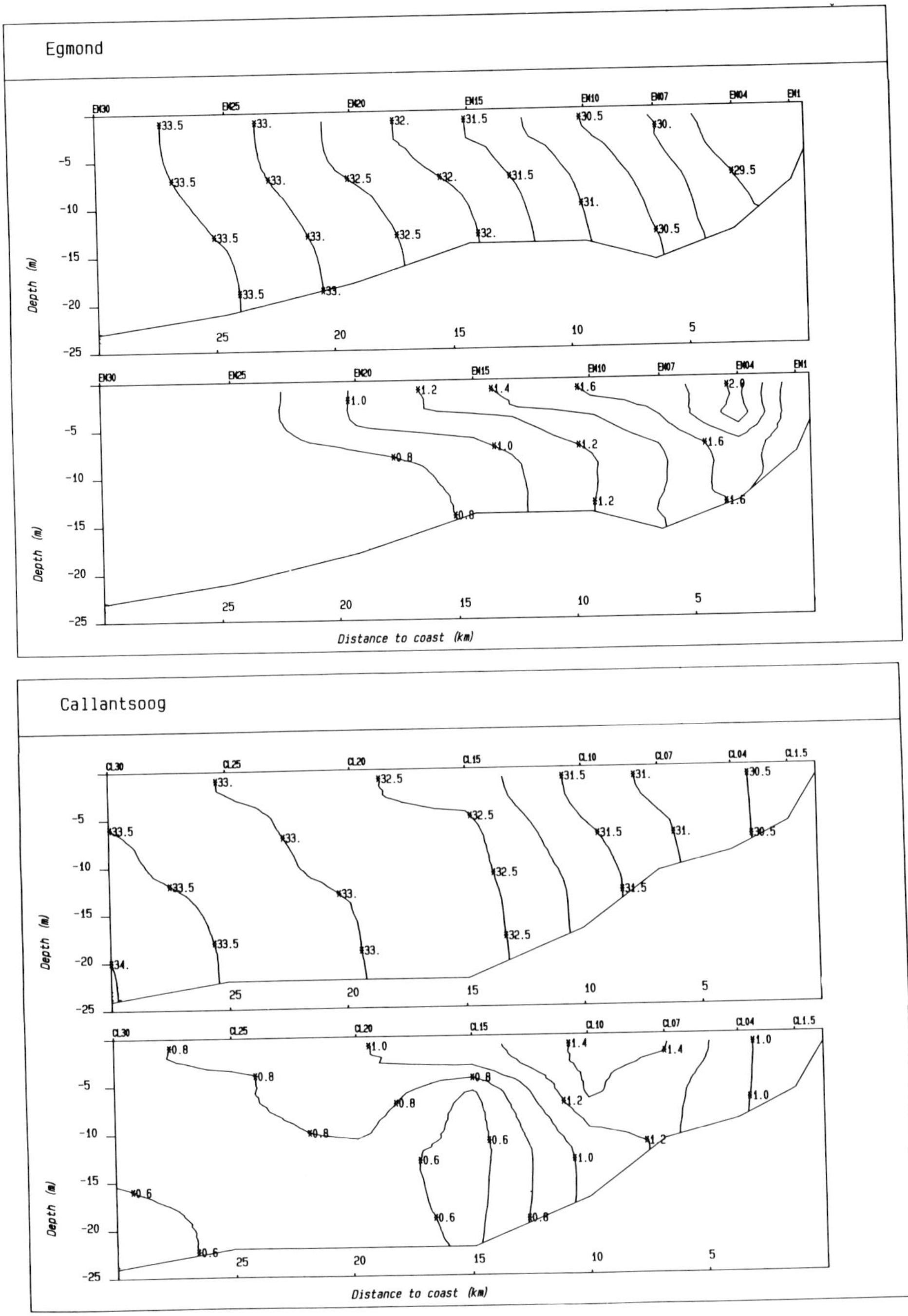

Figure 5c,d: Average salinity distribution (in 0/00) and standard deviation over 8 cruises in 1986 and 1987; c: Egmond, d: Callantsoog, e: Goeree (see Figure 2 for locations of transects

It is interesting to note the increased gradient of salinity variability between 10 and 15km offshore from Callantsoog (fig. 5d). This is also the area of relatively maximum stratification and cross shore density gradient, suggesting that here the position of the frontal zone separating open sea waters from the region of major river influence is relatively stable. This may be a consequence of the rather steep bottom slope inhibiting (by rotational constraints) cross-shore displacement of water columns and steering the flow along it. However, so far the number of cross-shore sections taken at this location is too small to draw firm conclusions. In this area also a minimum of suspended matter has often been observed, mostly during winter cruises (Visser et al, 1991).

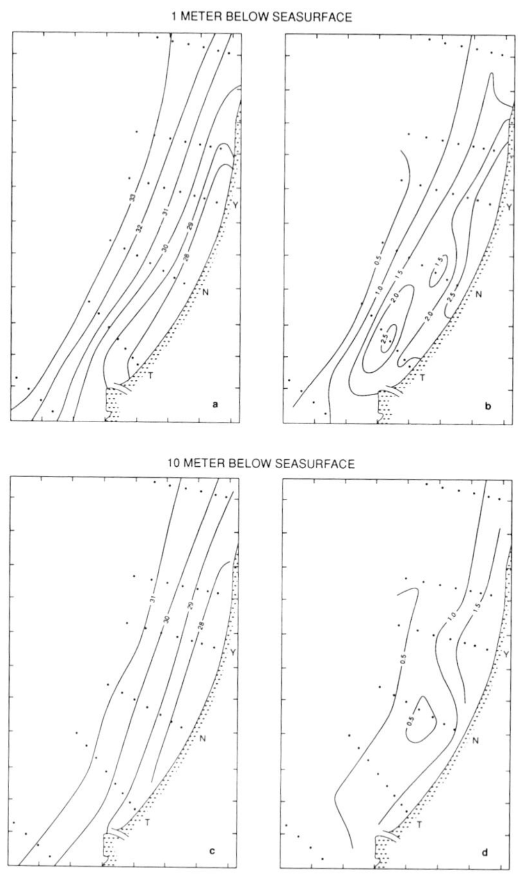

Figure 6: Average salinity fields (over 8 cruises) and their standard deviation (in o/oo) at 1m and 10m below the sea surface. Clearly, variability in the upper layer is much larger than in the lower parts of the water column.

The above three-dimensional picture of the salinity field is further confirmed by its horizontal distribution at a few selected depths (fig. 6). However, these mean fields don't appear exciting as many of the interesting features have been averaged out (though the fields of standard deviation indicate that there's more going on).

A much more revealing picture appears if the average salinity fields of upper and lower layers are subtracted. What emerges is a figure showing the horizontal distribution of stratification in the coastal zone (Fig. 7). The buoyant plume spreading from the river mouth stands out clearly. North of Noordwijk (i.e. 40km north of the source) the plume axis seems to split into two branches. This may indicate two favourite spreading modes. In the first one the plume reattaches and flows northward as a narrow coastal jet within 10km from the shore. In the other mode it either doesn't reattach or, while on the way northward, it separates from the coast, due to winds favouring offshore upper layer transport.

It is tempting to try to interpret the whole figure in terms of horizontal advection and spreading. However, as described earlier, cross-shore circulation in the vertical plane may play an equally important role. The band of relatively large stratification around 15km from the shore may also indicate the main position of a front, separating the region of major river influence from the open sea.

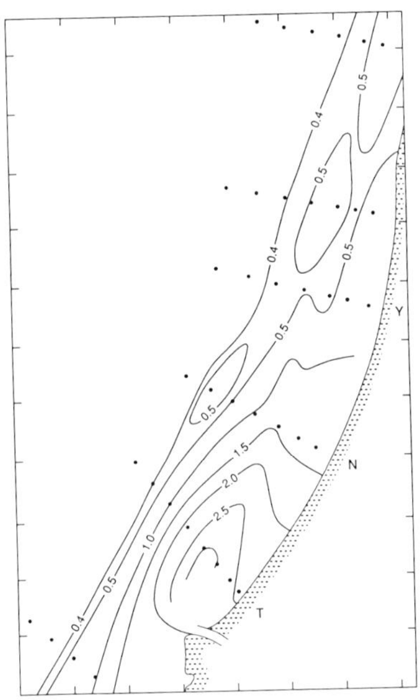

Figure 7: Horizontal distribution of mean salinity stratification, defined as the difference between the average salinities at 1m and 10m below the sea surface. It suggests that river spreading takes place both along the coast and offshore in a band centred between 10 and 15km from the coast. In the northern part of the region this band of relatively large stratification may also be related to the frequent existence of an offshore front (compare Fig. 5d).

After its debouchment into the North Sea some river water is transported southward, particularly in case of northerlies. Consequently, in the Goeree-section some 20km south of the Rotterdam Waterway (Fig. 2) the influence of the outflowing river is still significant up to 15km offshore. Variability is of the same order of magnitude as in the Noordwijk-transect. Vertical density inhomogeneities are smaller due to the shallowness of the coastal zone near Goeree. Under persistent winds from the north Rhine water may even reach the mouth of the ecologically vulnerable Eastern Scheldt. In such an extreme situation combined with a high river discharge a river water fraction of about 25% was observed at this location, which is situated nearly 40km south of the Haringvliet (De Ruijter et al., 1991).

In view of the large variability of the density structure it is useful to consider a few characteristic individual situations. Some distinct patterns, representing periods of high, mean and low river discharge, respectively, are presented for the transects near Ter Heijde and Noordwijk. Figs. 8-10 show the salinity fields during such situations.

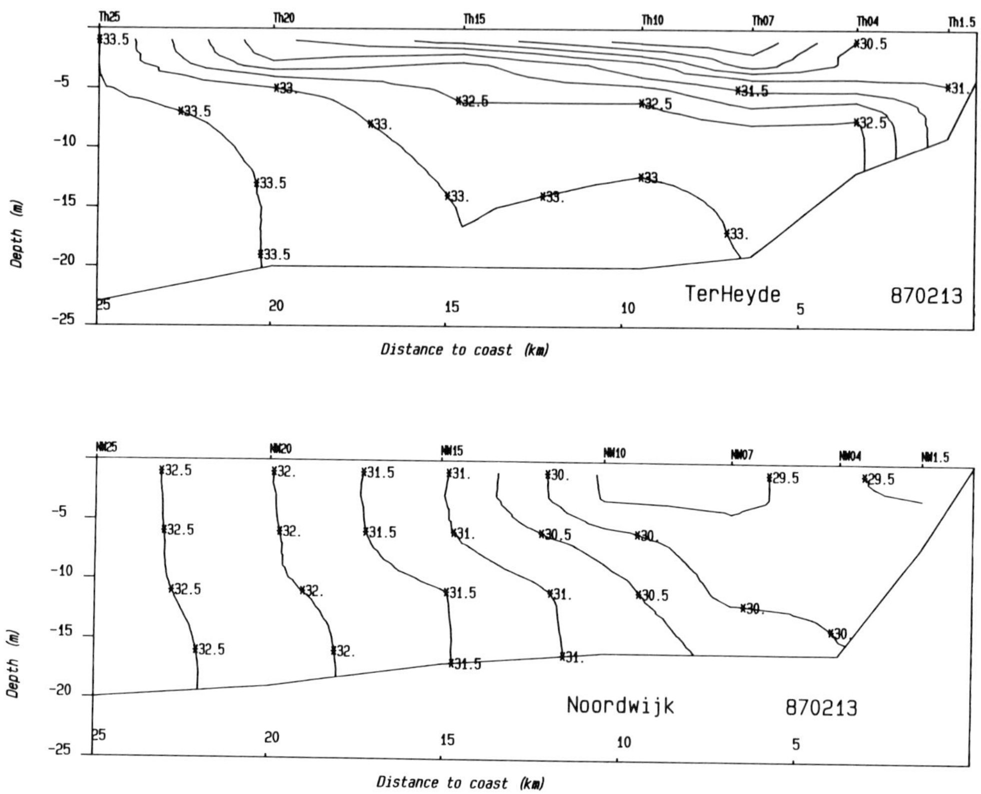

Figure 8: Salinity distributions in the Ter Heijde (a) and Noordwijk (b) transects at average river discharge (2300 m^3/s) and winds with an easterly component. Date of survey: February 13, 1987

In the situation with discharge close to average in the Ter Heijde-transect up to 20km offshore stratification is strong. Mixing processes have substantially reduced the stratification in the Noordwijk-section, where it appears mainly in a zone between 5 and 15km offshore. Fig. 8 also shows some earlier mentioned phenomena observed in these transects. For instance, near Ter Heijde the pattern of isohalines indicates an estuarine-like circulation, with less saline water driven off-shore and open sea water transported on-shore. Ca. 4km from shore a front near the bottom (Ter Heijde) can be observed. Northerlies and off-shore winds may stimulate the occurrence of these phenomena.

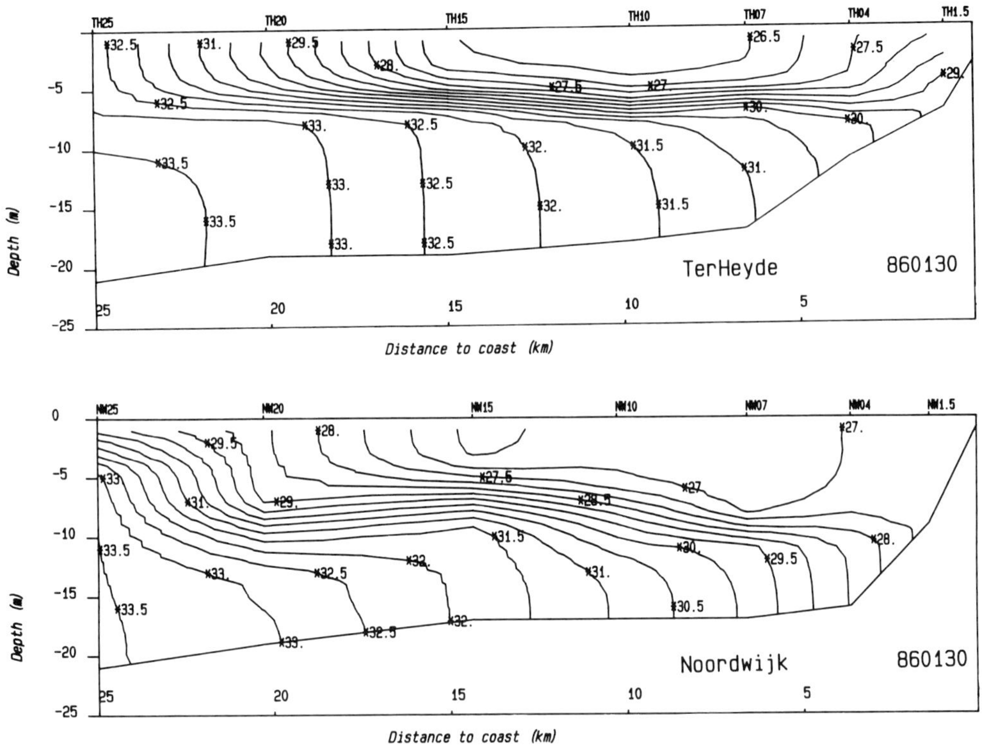

Figure 9: As in Fig. 8, now at above normal discharge conditions (6000m^3/s). Date of survey January 30, 1986

Extreme river discharge conditions lead to substantial deviations from the average salinity pattern. A salinity distribution during a peak discharge (±6000m^3/s) is depicted in Fig. 9. Stratification is extreme now, throughout the coastal zone. Compared to the Ter Heijde transect one may notice the deepening of the upper wind mixed layer in the stratified system near Noordwijk, 30km further northward. It shows the impact of strong winds (5 Bf) occurring during this specific period. However, there are indications that also at average wind conditions the buoyant spreading and mixing of the river water plume is affected strongly by wind-induced stirring of the upper layer.

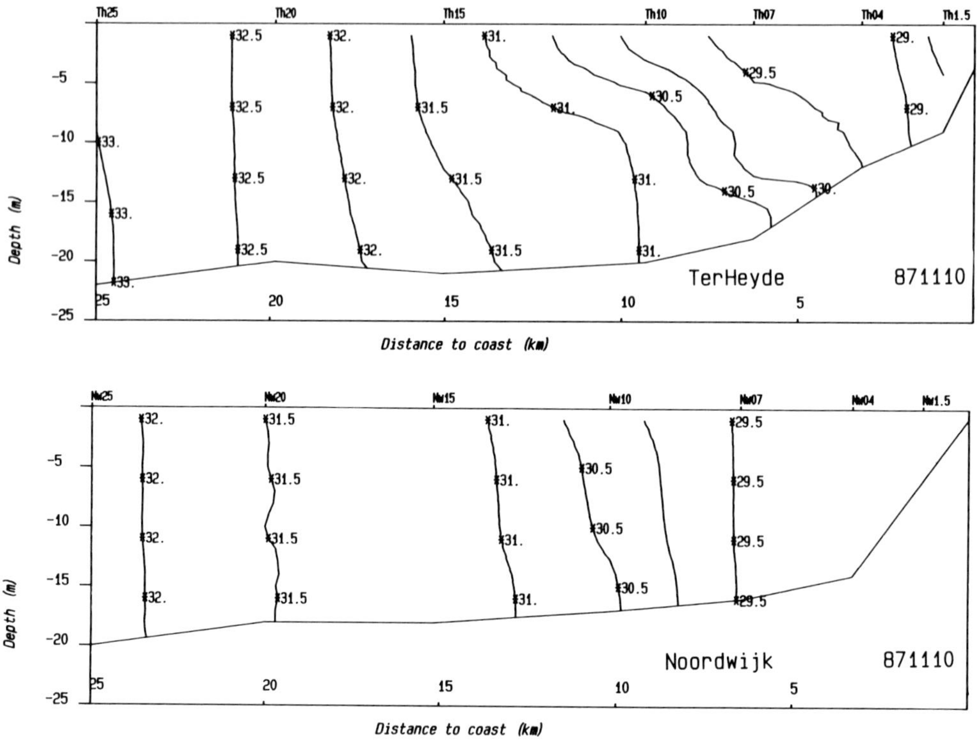

Figure 10: As in Fig. 8, now at relatively low Rhine river discharge (1400m^3/s). Date of survey November 10, 1987

During periods of low fresh water discharge, stratification is almost absent (Fig. 10) except in a small area of the Ter Heijde-section, close to the major river outlet. Nevertheless, significant horizontal density gradients still exist.

In the long-term average situation these lateral density gradients decrease with increasing distance from shore. However, in a specific case the actual meteorological and hydraulic conditions may disturb this pattern considerably. Therefore any individual density pattern may deviate significantly from the average situation.

26.4 The Observed Residual Current Field

Currents were measured continuously in the transect near Noordwijk from August 1985 through September 1986. (see Fig. 11 for location of moorings and positioning of current meters in the water column.) Some remarks about the tidal currents in this region were given in section 26.2. Here we will focus on the residual currents.

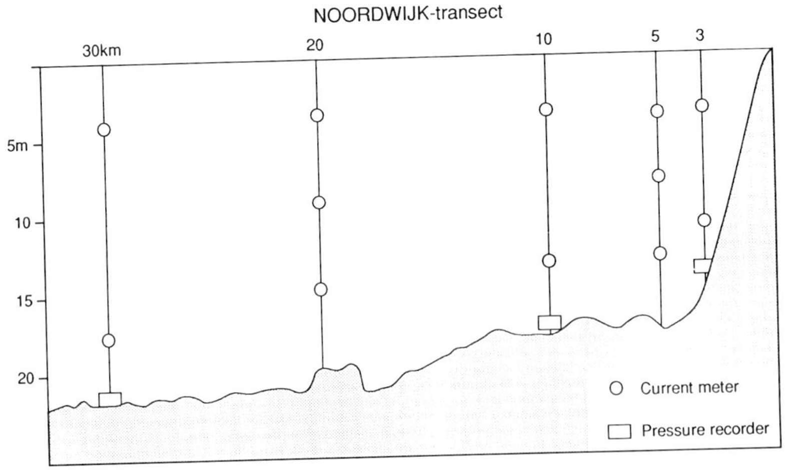

Figure 11: Location of current meters during the one-year observational period in the cross-shore transect near Noordwijk. Current meters were robust self-recording Flachsees. Every 10 minutes they registered 10 minute-averaged velocities and instantaneous current directions

Residual currents are induced by several physical mechanisms. A basic contribution is due to nonlinear tide-topography interactions (e.g. Zimmerman, 1978). Another dominant component is wind-induced. In case of winds with a predominant coast-parallel component, within several tens of kilometres from the shore, the flow mainly follows the coastline as a geostrophic jet in downwind direction at every depth (decreasing in magnitude with depth due to friction). Besides alongshore flow, winds also induce a cross-shore circulation component, either by a combination of friction and Coriolis force (as an Ekman spiral) or as a drift related to wind setup. In coastal waters density driven residual cross-shore currents can be important. For instance, for the Irish Sea Heaps (1972) has shown that the observed average on-shore flow near the seabed may result from the buoyancy effect due to the lateral variation in the density field with relatively fresh water inshore and higher salinities offshore. The same applies to the Dutch coastal zone where the horizontal density contrast is larger due to the higher mean river discharge coming from Meuse and Rhine.

Such an estuarine-type circulation component is constant in direction and therefore may contribute significantly to the long-term distribution and transport of water properties.

The long term average currents near the bottom vary from 2.5 to 3.5cm/s (Fig. 12). The maximum of $\partial\rho/\partial x$ ($\approx 2.10^{-4}$kg/m^4, Fig. 4b) between 5 and 10km off-shore from Noordwijk corresponds with a maximum in the on-shore near bed flow and a relatively large longshore mean current near the surface, indicating that both the estuarine and geostrophic current components are important here. This may be the heart of the coastal current (see section 26.3).

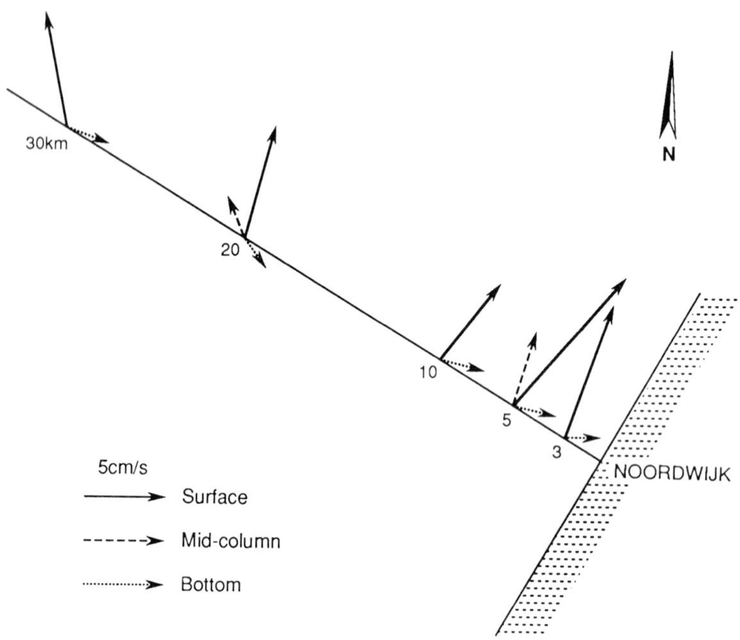

Figure 12: Mean currents (close to the surface, mid-column and near the bottom) in the Noordwijk transect. The average was taken over the full period of observation, August 1985-September 1986. Relatively large longshore mean currents at 5 and 3km offshore are indicative of frequency occurrence of a narrow coastal jet.

Large variability was observed of the three-dimensional current structure on different time scales, mainly induced by variations in wind conditions (Van der Geissen et al., 1990). As an illustration periods characterizing four different persistent wind situations were selected from the one-year data set, otherwise with comparable hydrological conditions such as average Rhine discharge and equal magnitude of wind speed (of 4 à 5 Bf, Fig. 13).

To a large degree, the three-dimensional current structure can be qualitatively understood by combining the earlier mentioned physical mechanisms: a persistent cross-shore estuarine circulation component, longshore geostrophic flow (related to the river induced salinity gradient) northward in the upper layer with significant vertical shear, and varying wind-induced currents (e.g. Csanady, 1982). Southwesterlies reinforce the longshore jet by adding a barotropic component to it (average surface velocity of ±14cm/s), northerlies do the reverse to the longshore flow and reinforce the cross-shore circulation cell. The effect of the Ekman spirals driven by cross-shore winds is also clear from the observations (Fig. 13).

Consequently, during easterlies the current direction veers significantly with depth; for southwesterlies this is only minor but then the amplitude decreases strongly with depth. When the wind has a component blowing from the north a small vertical variation of residual current speed was observed.

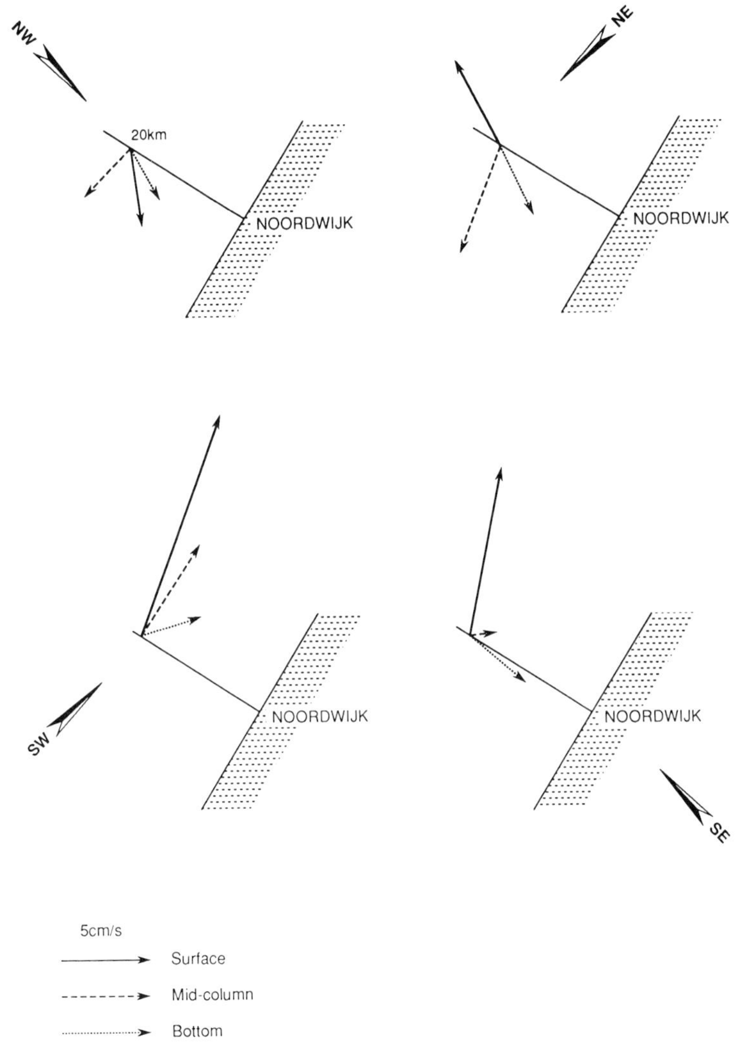

Figure 13: Residual currents 20km offshore from Noordwijk under persistent winds blowing from four distinct sectors

These qualitative current profiles can be rather easily reproduced using a diagnostic (i.e. given the observed density field) linear model involving the above given processes. However, correct current velocities can only be obtained by imposing a longshore sea surface slope of a few cm per 100km, (Groenendijk, 1988). Apparently, due to the wedge shaped geometry of the Southern North Sea, with only a small gap at the Dover Strait, the momentum input by the wind can be partly balanced by wind set-up against a southern boundary (see also De Ruijter et al., 1991).

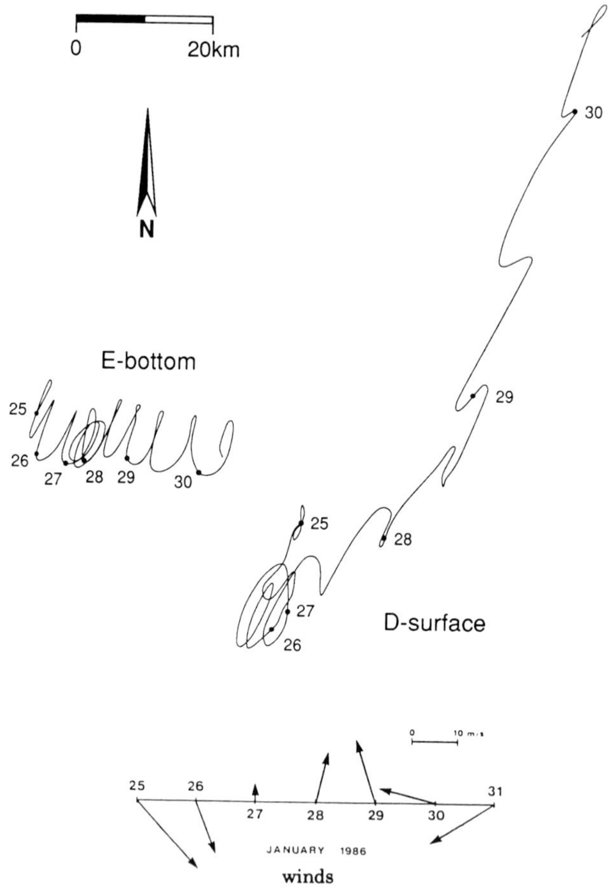

Figure 14: Progressive velocity vector diagrams, near the surface at 10km (D) and near the bottom (E) at 20km offshore from Noordwijk. Period of observation: January 25 - January 30, when the Rhine discharge was large (±6000m^3/s) and winds strong and variable (shown in the lower part of the Figure).

Southwesterly winds are most common over the Dutch coastal zone. These conditions prohibit cross-shore mixing of river with sea water as the wind- and density induced cross-shore flow components counteract. The Rhine outflow zone is therefore rather narrow, with the major transport of river water mostly taking place within 15-20km offshore. The strong current fluctuations induced both by varying river discharge and wind are illustrated clearly from the high discharge event of January 1986. When the wind turned from northwesterly to southwesterly, longshore wind- and density driven current speeds close to 0.5m/s resulted (Fig. 14). Furthermore during this period and the month following it, when there was still much fresh water in the coastal zone, all five near bottom current meters registered an on-shore residual current around 10cm/s (Fig. 15).

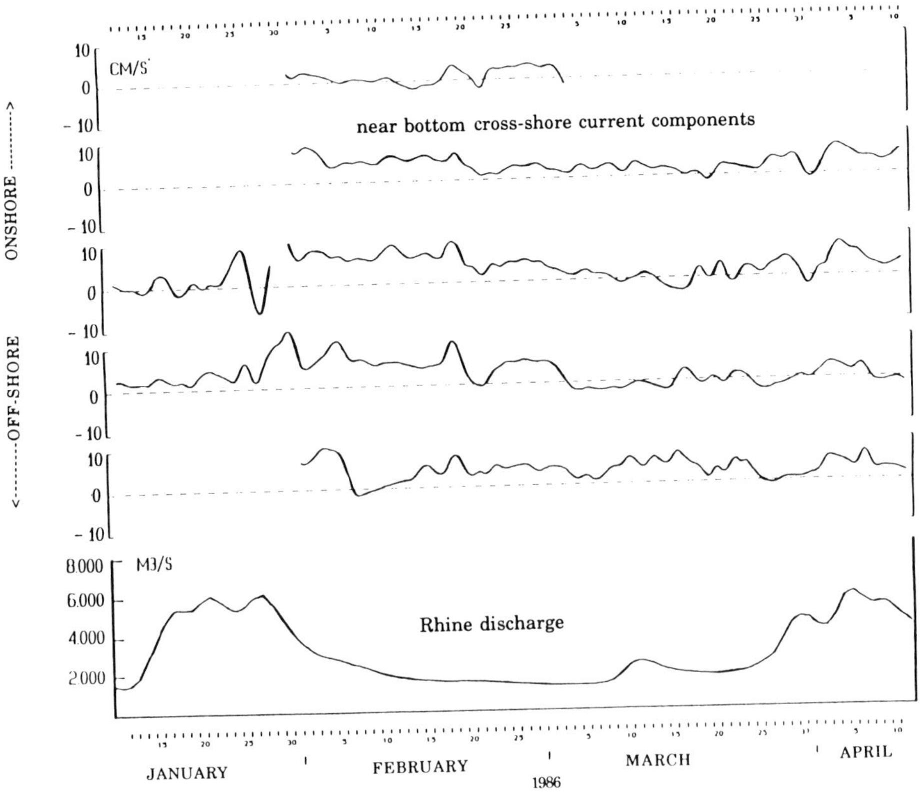

Figure 15: Low-pass filtered cross-shore current components at five near bottom current meters off-shore from Noordwijk. Persistent on-shore currents exist after a period of high Rhine discharge. Fluctuations on time scales of several days are mainly wind-induced.

26.5 Summary and Discussion

The average salinity distribution in the Netherlands coastal zone shows the outflow of Rhine river water to take place mainly northward within an area with offshore width shrinking from 20-25km near the mouth to 10-15km some 100km northward. Mostly the less saline layer is confined to the upper 4-8m of the water column. Up to 100km north of the Rotterdam Waterway stratification is observed frequently in the outflow zone.

The horizontal distribution of mean stratification (Fig. 7) suggests two favourite modes of river plume spreading. In both cases the outflow jet separates from the river mouth and turns to the right influenced by earth rotation. In this first mode the plume reattaches to the coast (probably between Ter Heijde and Noordwijk), to form a narrow (±10km) coastal jet. In the other mode reattachment doesn't take place, or while moving northward, separation from the coast is induced by winds favouring offshore upper layer transport (particularly winds between North and East). In the mean field a band of relatively strong stratification is observed between 10 and 15km offshore. This band may also be related to the frequent existence of an offshore front.

The salinity field shows a large variability, especially in the upper part of the water column in an area between Goeree and Noordwijk. Most actual density distributions occurring in this region deviate significantly from the mean. Wind strongly affects the spreading characteristics of the river plume both by the wind-driven current field and vertical turbulent mixing. Obviously, the time-varying discharge is another crucial parameter.

The large observed variability of the current field is mainly wind-induced. If strong winds and a high river discharge coincide then a distinct two-layer system can develop. Non-tidal current speeds of over 40cm/s were measured during such periods. Averaged over a full year of observations the residual currents in a cross-shore transect near Noordwijk ranged between 7 and 11cm/s in the upper layer. An estuarine circulation component showed dominantly near the bottom, where the sub-tidal current was almost permanently directed on-shore, with speeds up to 10cm/s under favourable density and wind conditions. One year average values are between 2.5 and 3.5cm/s.

The large variability necessitates an elaborate analysis of individual density patterns (fronts, stratification), characterizing a wide variety of wind and river discharges conditions.

Extensive field studies, including repeated hydrographic sections and long term observations from moored arrays, are indispensable for the development and validation of 3-D hydrodynamic and dispersion models. The results reported here indicate that emphasis should be on a 25km wide zone between Goeree and Noordwijk (i.e. some 40km north of the outflow port). Longshore resolution of the observational network has to be much finer than was used for this study. (Recently, as part of the PROFILE-project (Processes in Regions of Freshwater Influence) within the European Community MAST-program a collaborative effort has started to fill in this gap).

All of the above shows that for most applications a treatment of the Dutch coastal zone using long-term mean hydrographic parameters can lead to totally erroneous results. The same holds if vertically integrated instead of three-dimensional currents (and models) are applied.

Acknowledgement

We gratefully acknowledge the many people who contributed to the completion of this study, in particular the crew from the RV "Holland", the personnel from the North Sea Directorate, The Tidal Waters division and from the former division Hellevoetsluis of Rijkswaterstaat who organised and assisted in the data collection and analysis.

26.6 References

Csanady, G.T., 1984: Circulation in the coastal ocean. D. Reidel Publishing Company, Dordrecht/ Boston/Lancaster, 279pp.

De Ruijter, W.P.M., J.C. Borst and J.S.L.J. van Alphen, 1991: Spreading of Rhine river water at high discharge and strong winds. Subumitted.

Garvine, R.W., 1987: Estuary plumes and fronts in shelf waters. J. Phys. Oceanog., 17, 1877-1896.

Groenendijk, F.C., 1988: Residual current structure in the Netherlands coastal zone. Masters Thesis (in Dutch). Univ. of Utrecht, Inst. of Meteorology and Oceanography. Report IMOU V-88-6, 94pp.

Heaps, N.S., 1972: Estimation of density currents in the Liverpool Bay area of the Irish Sea. Geophys. J.R. Astron. Soc., 30, 415-432.

McClimans, T.A., 1988: Estuarine fronts and river plumes. In: Physical processes, In: Physical processes in estuaries, J. Dronkers and W. van Leusden (eds.), Springer Verlag, 55-70.

Prandle, D. 1978: Monthly-mean residual flows through the Dover Strait, 1949-1972. J.Mar.Biol.Ass.UK, 58, 956-973.

Van Alphen, J.S.L.J., W.P.M. de Ruijter and J.C. Borst, 1988: Outflow and thre-dimensional spreading of Rhine river water in the Netherlands coastal zone. In: Physical processes in estuaries, J. Dronkers and W. van Leussen (eds.), Springer Verlag, 70-92.

Van der Giessen, A., W.P.M. de Ruijter and J.C. Borst, 1990: Three-dimensional current structure in the Dutch coastal zone. Neth.J.Sea Res., 25, (1.2), 45-55.

Visser, M., W.P.M. de Ruijter and L. Postma, 1991: The distribution of suspended matter in the Dutch coastal zone. Submitted.

Zimmerman, J.T.F., Topographic generation of residual circulation by oscillatory (tidal) currents. Geophys. Astrophys. Fluid Dyn., 11, 35-47.

27

On the distribution of suspended matter and the density driven circulation in the Dutch coastal area

M Visser

Abstract

Especially in winter, observations of the near-surface suspended matter distribution in the Dutch coastal zone show a suspended matter minimum located between salinity maximum and the coast. This minimum is often present in an area with relatively strong (cross-shore) density changes, related to Rhine river outflow.

A simple model of cross-shore density driven circulation shows the possible influence of a density front on the transport and sedimentation of suspended matter. Particles with a settling rate comparable to the vertical velocity scale of the circulation are forced to move off-shore until they reach an area where they can sink to the bottom layer. In the bottom layer they are transported onshore again. Wind conditions are important in modifying the strength and position of a suspended matter minimum.

27.1 Introduction

The Southern Bight of the North Sea is exposed to high inputs of contaminants, especially through rivers and dumping activities. Part of those contaminants are dissolved and transported with the (residual) circulation, others interact with suspended particles (silt, clay). Therefore the distribution and transport mechanisms of suspended matter may give more insight into the fate of pollutants.

From 1975 to 1983 a water quality monitoring program has taken place in the Dutch coastal area (Fig. 1). During this program salinity, temperature, suspended matter and a range of other water quality parameters were measured. The suspended matter patterns show a clear seasonal variability (Visser et al., 1991). In winter a local concentration minimum is present between salinity maximum and the Dutch coast. A similar minimum was found by Lee &

Folkard (1969) and Postma (1981). Postma (1981) assumed its position to coincide with the seaward end of a density driven circulation.

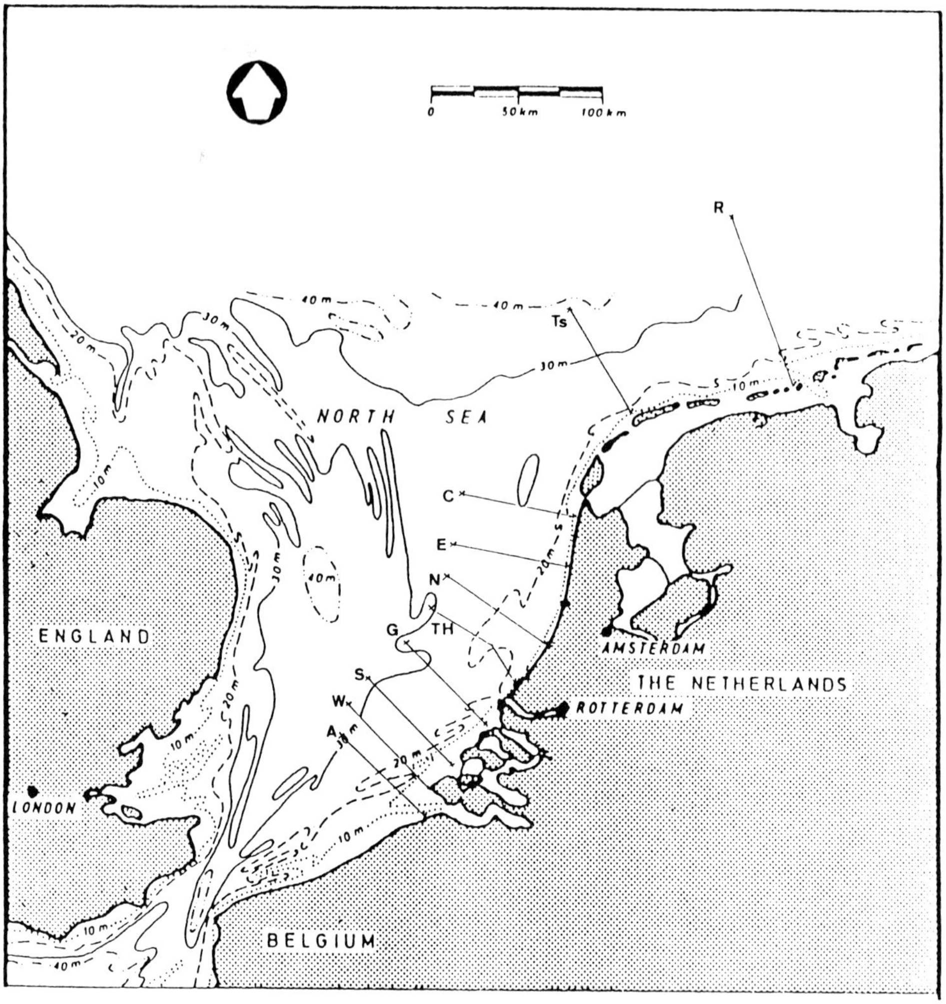

Figure 1: Southern Bight of the North Sea with isobaths and investigated area. The 76 measuring locations are distributed over ten cross-transects along the Dutch coast. The locations are indicated by the first letter(s) of the transect and their distance from the shore (km). From south to north the transects are: Appelzak (A), Walcheren (W), Schouwen (S), Goeree (G), Ter Heijde (TH), Noordwijk (N), Egmond (E), Callantsoog (C), Terschelling (Ts) and Rottumerplaat (R).

In this paper we study the possible influence of the (cross-shore) density driven circulation on the suspended matter distribution using a simple analytical model. In section 27.2 a description of the Dutch coastal zone is given. Section 27.3 shows observed salinity and suspended matter distributions. A simple model of the density driven circulation and the associated (advective) transport of suspended matter is presented in section 27.4. In section 27.5 we summarize and discuss the results of this study.

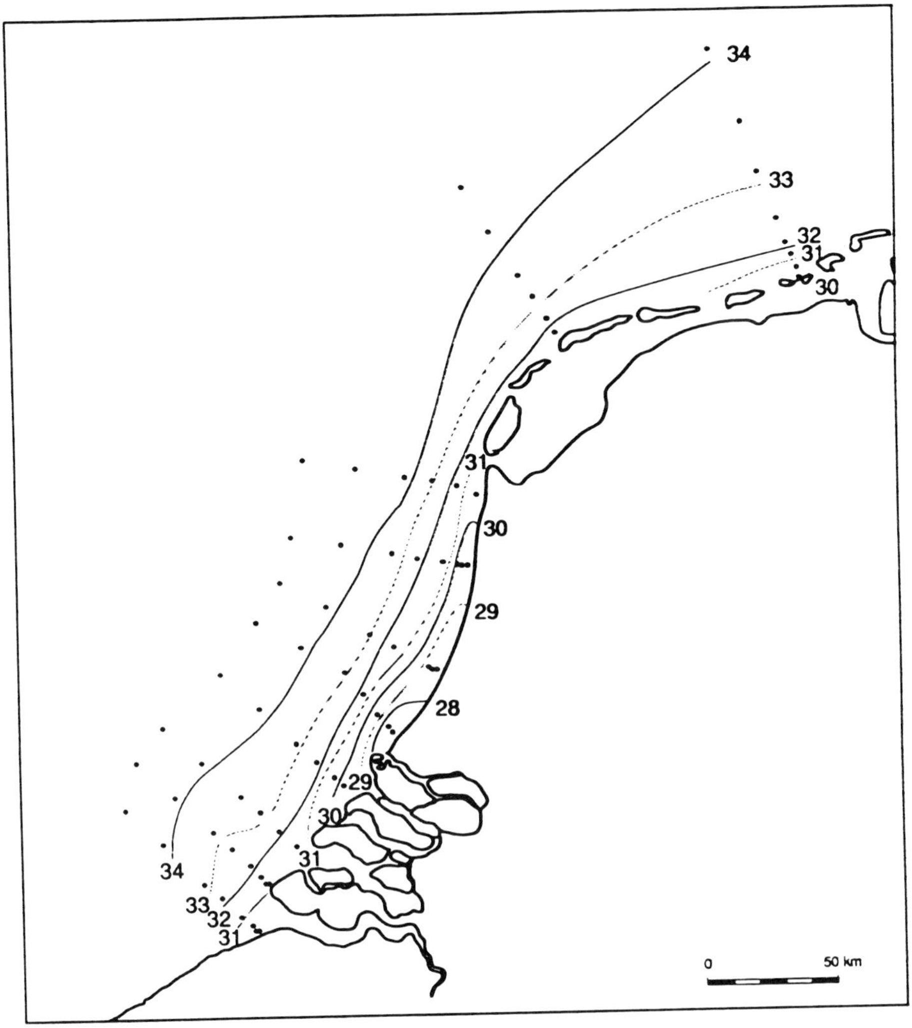

Figure 2: Average salinity distribution (0/00) over the full period (1975 - 1983). The fresh river water is spread north-eastward along the Dutch coast.

27.2 Dutch Coastal Zone

The Dutch coastal zone is located in a shallow part of the southern North Sea with water depths less than 30 m (Fig. 1). The greater part of the water mass originates from the English Channel. A small part enters the coastal zone through the continental rivers, in particular the Rhine.

Spatial density variations are mainly due to variations in salinity; temperature variations are of less importance. Over a large portion of the area the water column is stratified. Strong stratification occurs in the immediate vicinity of the outflow port. The three-dimensional density structure in the Ter Heijde (TH) transect (ca. 10 km north of the main outlet of the Rhine-Meuse estuary) is still to a great extent determined by the initial mixing and outflow processes of river water. Going to the north stratification diminishes, but even in the Callantsoog (C) transect, located some 100 km north of the river mouth, weak stratification is often observed (De Ruijter et al., this volume). Cross-shore the density gradient is still very strong.

Tidal currents are strong and nearly parallel to the shore. On time scales of days to weeks the sub-tidal circulation is predominantly wind-driven. The dominant south-westerlies (Visser, 1987) drive long-shore currents in northeastward direction with magnitudes of the order of 10^{-1} ms^{-1} in the Dutch coastal zone. Changing wind conditions cause considerable deviations from the average circulation (Van der Giessen et al., 1990).

The average Rhine-Meuse discharge is ± 2500 m^3s^{-1}. It varies strongly with an associated time varying density distribution in the coastal zone. Remarkably, the position of the salinity maximum is almost constant at the axis of Channel water inflow some 75 km off-shore (Visser, personal communication). The eight year average surface salinity distribution shows the river water spreading northeastward along the Dutch coast (Fig 2).

27.3 Observations

From May 1975 to February 1983 every two weeks a survey was made by the R.V. "Holland". During each cruise near-surface water samples were taken at 76 locations, distributed over ten cross-transects (70 km wide) along the Dutch coast (Fig. 1). The locations are indicated by the first letter(s) of the transect and their distance from the shore (km). From south to north the transects are: Appelzak (A: 1, 2, 4, 10, 20, 30, 50 and 70 km off-shore), Walcheren (W: 1, 2, 4, 10, 20, 30, 50 and 70 km off-shore), Schouwen (S: 1, 4, 10, 20, 30, 50 and 70 km off-shore), Goeree (G: 6, 10, 20, 30, 50 and 70 km off-shore), Ter Heijde (TH: 1, 2, 4, 10, 20, 30, 50 and 70 km off-shore), Noordwijk (N: 1, 2, 4, 10, 20, 30, 50 and 70 km off-shore), Egmond (E: 1, 2, 4, 10, 20, 30, 50 and 70 km off-shore), Callantsoog (C: 1, 2, 4, 10, 20, 30, 50 and 70 km off-shore), Terschelling (Ts: 4, 10, 20, 30, 50 and 70 km off-shore) and Rottumerplaat (R: 3, 5, 10, 15, 20, 30, 50, 70 and 100 km off-shore). At every location salinity, temperature, suspended matter concentration and a range of water quality parameters were measured. The suspended matter concentration was determined from a water sample of 1 litre that was filtered with a 0.45 μm cellulose-acetate filter. The residue was washed with demineralized water to remove the salt. After it was dried at a temperature of 105°C, the residue was cooled and weighed. No distinction was made between organic and inorganic suspended matter.

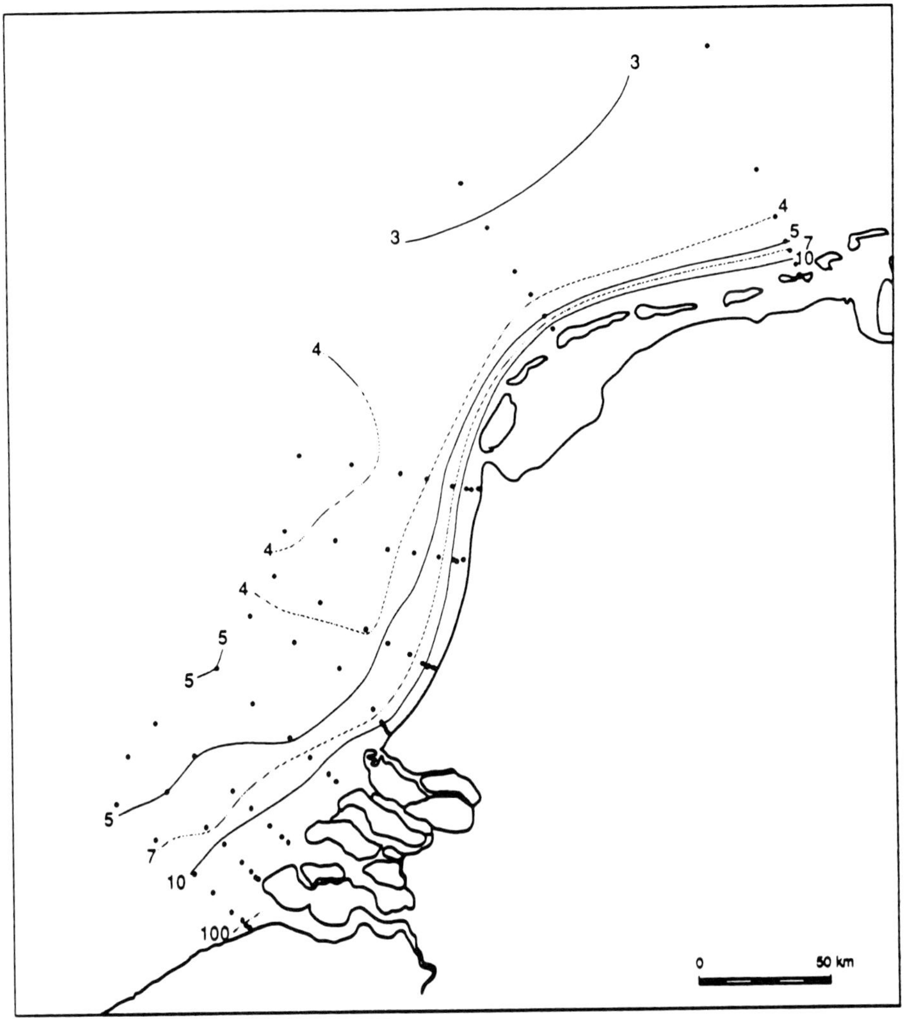

Figure 3: Average suspended matter distribution (mgl^{-1}) over the full period (1975 - 1983). High concentrations are measured in front of the coast. Further off-shore the turbidity is much smaller and more horizontally homogeneous. However a weak local minimum is present at some 25 km off-shore north of Noordwijk.

The average (over the full period) suspended matter distribution (Fig. 3) shows a very steep concentration gradient in a narrow strip along the coast, say within 5 - 10 km. In this study our emphasis will be on the area off-shore from this zone, where the water is more saline. There the turbidity is much smaller and more horizontally homogeneous. However, even the average suspended matter pattern shows a weak minimum north of Noordwijk between 20 and 30 km off-shore.

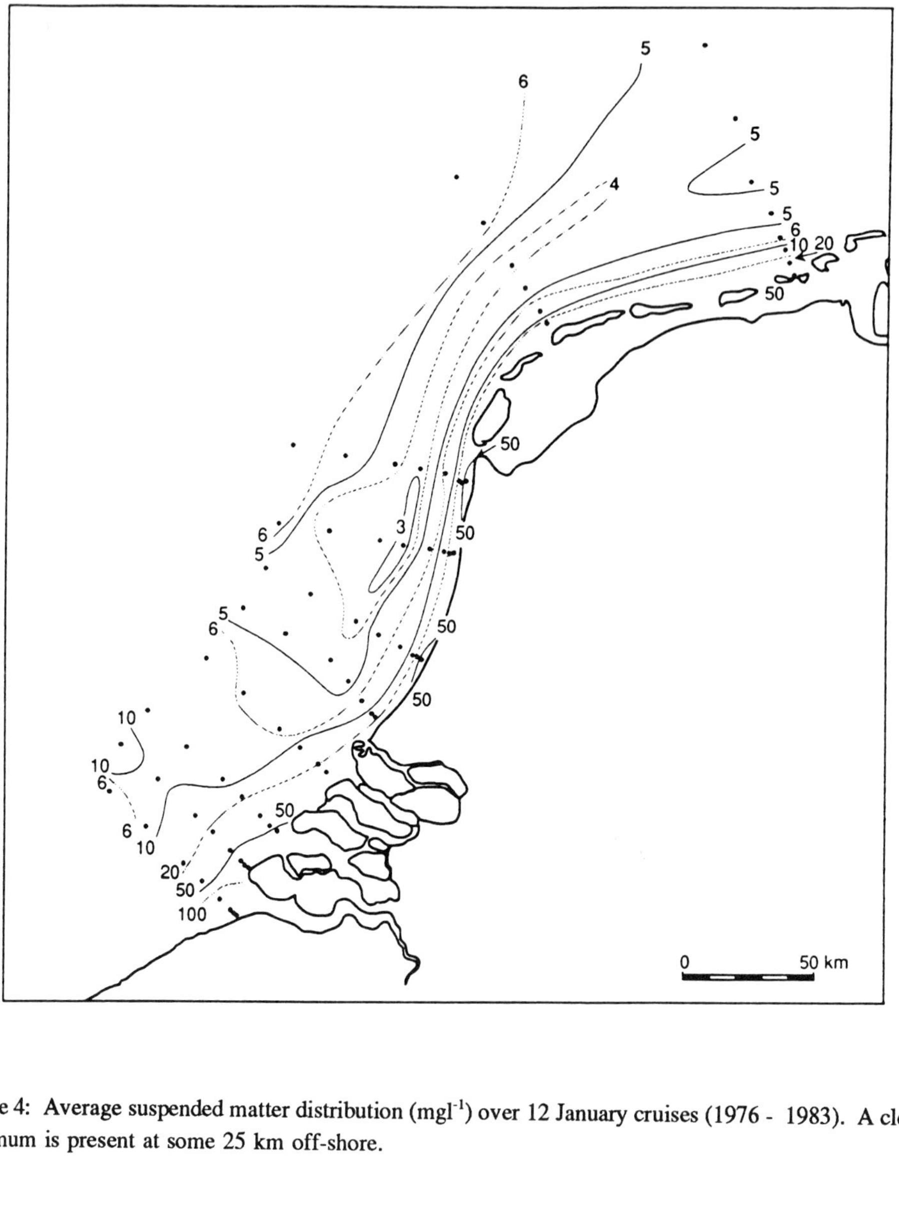

Figure 4: Average suspended matter distribution (mgl^{-1}) over 12 January cruises (1976 - 1983). A clear minimum is present at some 25 km off-shore.

Especially in winter a pronounced suspended matter minimum is present. The average January pattern (Fig. 4) shows a clear minimum at some 25 km off-shore. Individual measurements often show a local minimum in which abrupt concentration changes occur. These frontal changes often coincide with a large density gradient.

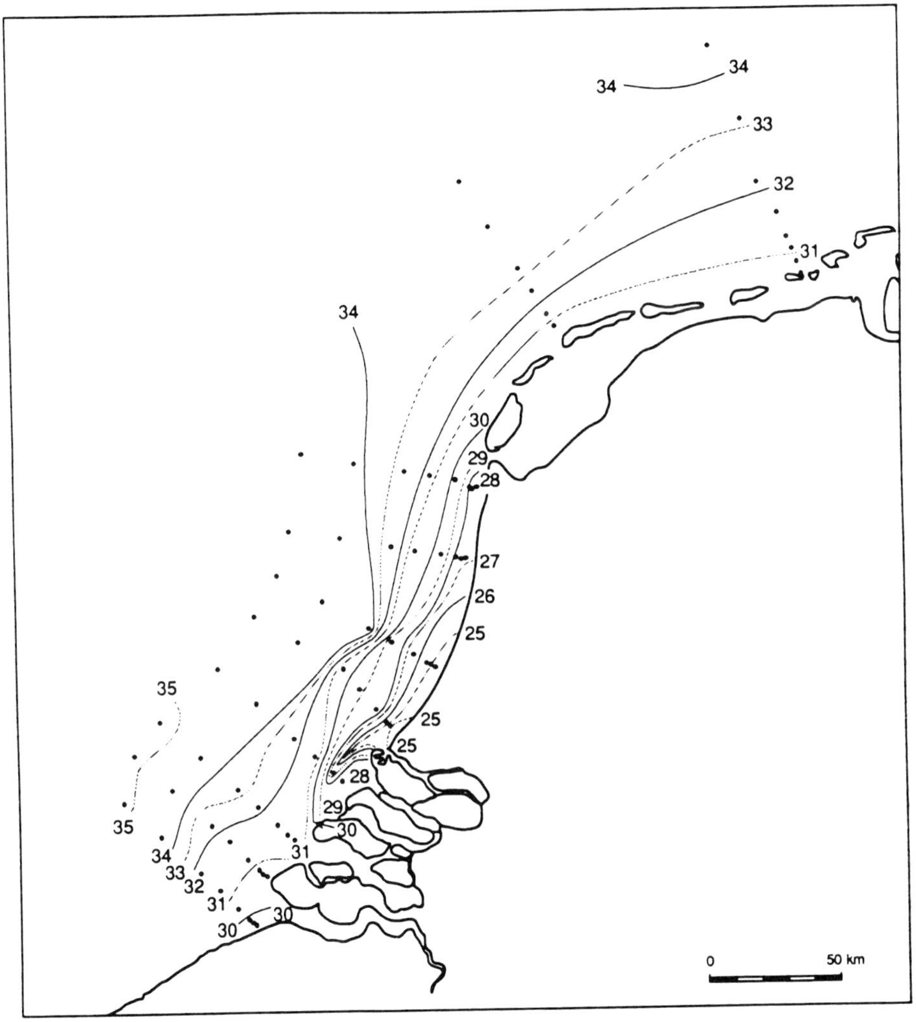

Figure 5a: Salinity in o/oo (a) and suspended matter in mgl^{-1} (b) measured during the survey of 5 - 8 March 1979. A salinity front is coinciding with a suspended matter minimum.

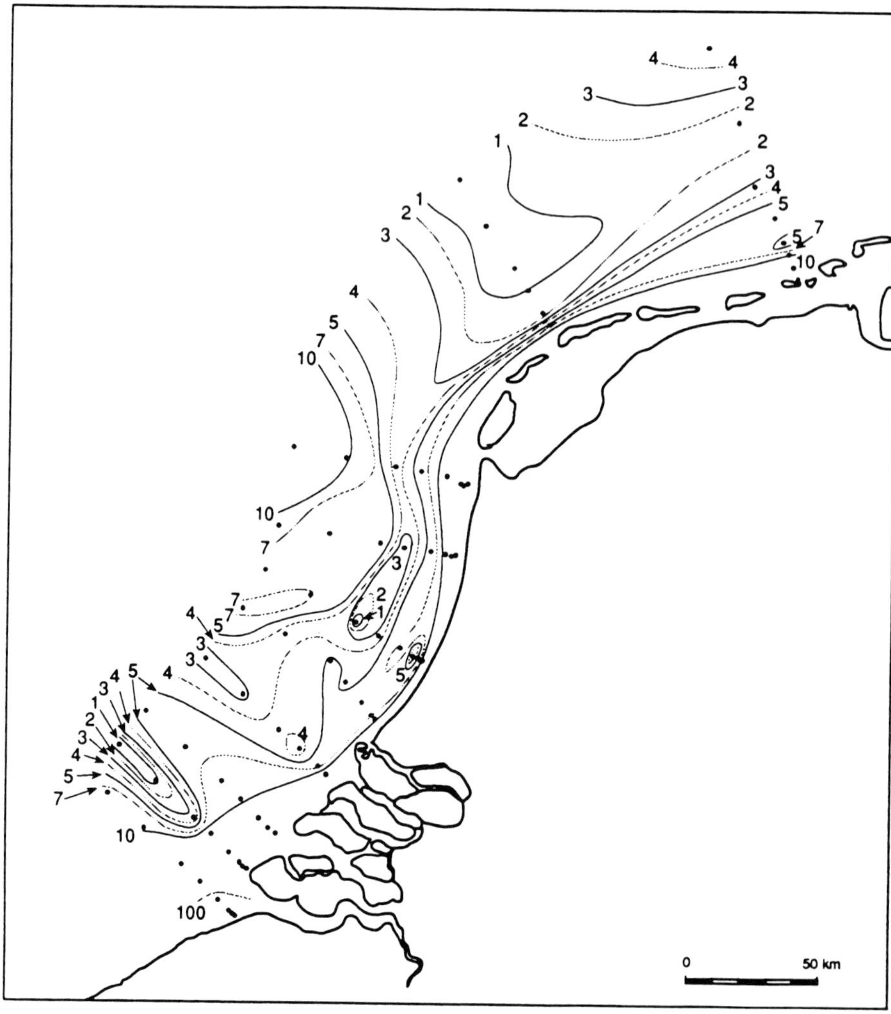

Figure 5b: Salinity in o/oo (a) and suspended matter in mgl^{-1} (b) measured during the survey of 5 - 8 March 1979. A salinity front is coinciding with a suspended matter minimum.

Fig. 5 shows the salinity and suspended matter distribution, observed during a cruise from 5 to 8 March 1979. In this characteristic winter pattern a narrow strip of low suspended matter concentration is present parallel to the Dutch coast some 25 km off-shore (minimum concentration less than 1.0 mgl^{-1} at 30 km off-shore in the Noordwijk transect, Fig. 5b). High suspended matter concentrations are found in the southern transects, indicating the influence of the Flemish Banks. The salinity distribution (Fig. 5a) shows a density front near the suspended matter minimum (Noordwijk transect, salinity at 20 km off-shore: 28.3 o/oo, at 30 km: 34.1 o/oo).

In the absence of large sources in the Dutch coastal zone the global distribution of suspended matter (high concentrations in winter; low concentrations in summer) is expected to be largely determined by the boundary. Particularly seasonally varying fluxes through the English Channel are important (Visser et al., 1991). Locally for the existence of a minimum in suspended matter concentration (between salinity maximum and shore) a large enough river discharge is necessary. In combination with a favourable wind (long-shore, southwest) a salinity front can develop parallel to the coast. If the associated density driven circulation is strong enough, the cross-shore component may cause a depletion of suspended matter in the neighbourhood of this front.

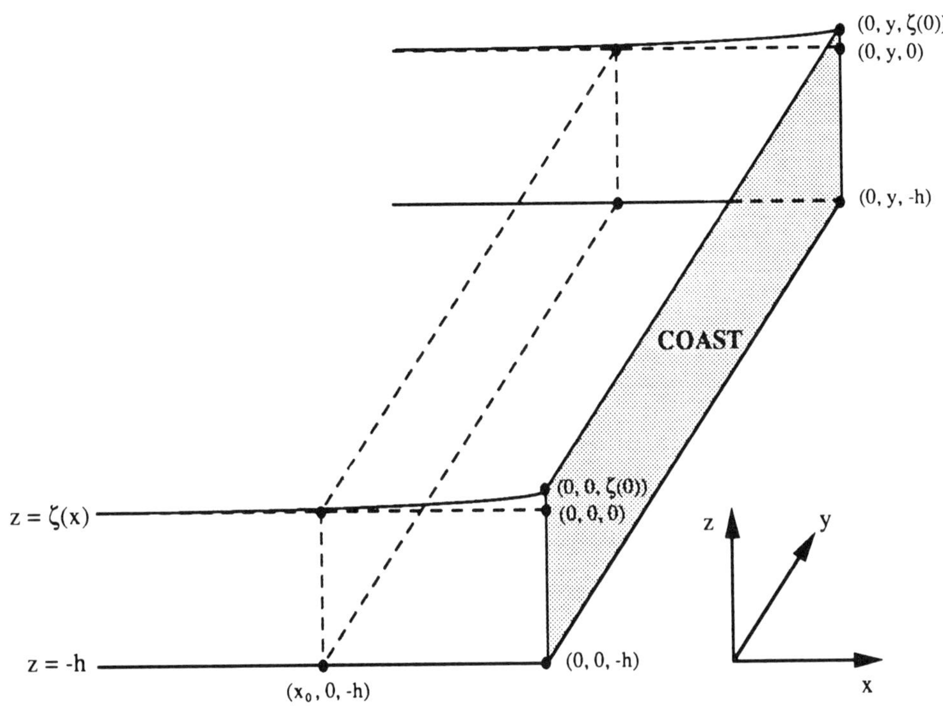

Figure 6: Model set up: x-axis cross-shore positive eastward; y-axis long-shore positive northward; z-axis vertical positive upward; surface at $z = \zeta(x)$; bottom at $z = -h$ (uniform water depth); coast at $x = 0$.

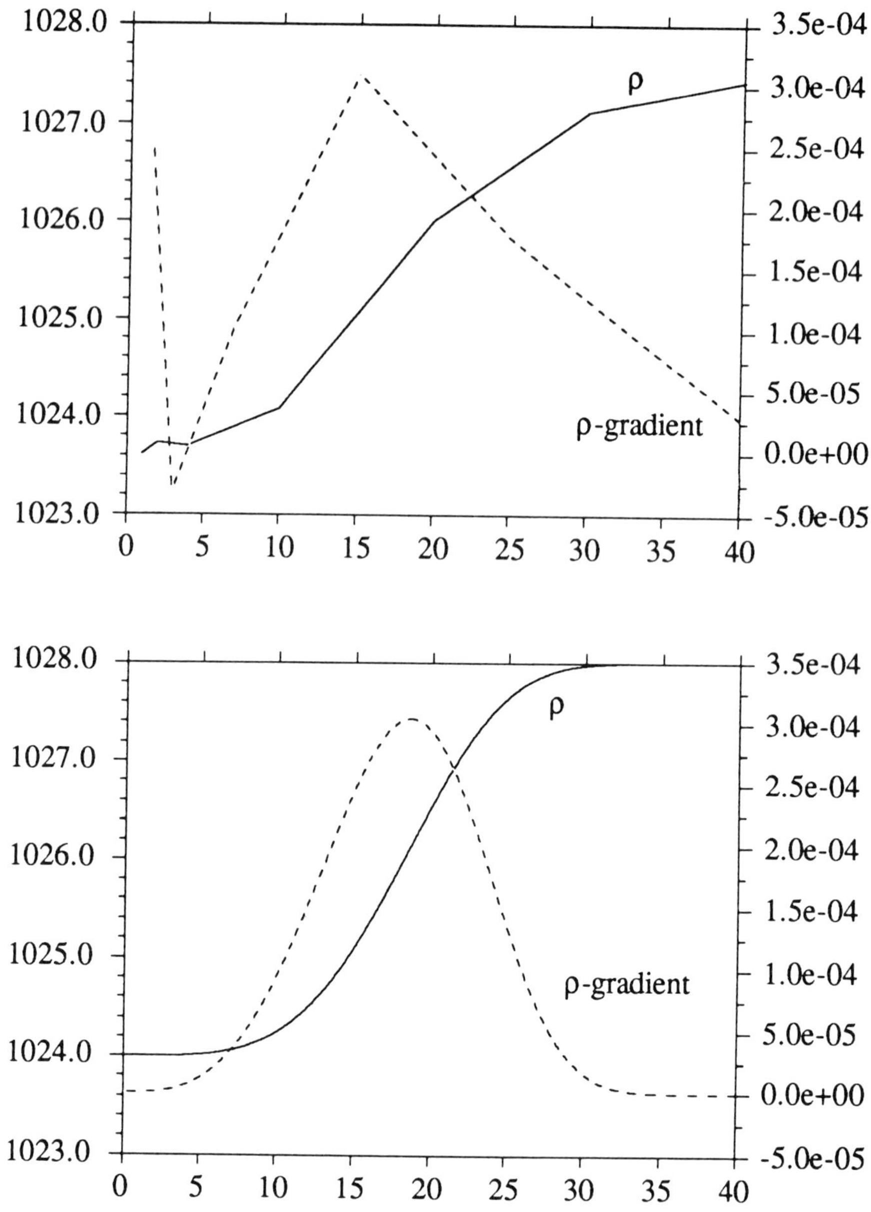

Figure 7: (a): Average density in kgm^{-3} (----) and density gradient (- - -) over 12 January cruises in the Callantsoog transect in kgm^{-3} (along horizontal axis distance from the shore in km). (b): Prescribed density field (-----), $\rho(x) = 1028.0 - 4.0 \exp\{-(x/-x_0)^4\}$, with $x_0 = 2\ 10^4$ m and density gradient (- - -).

27.4 Modelling

To illustrate the possible effect of a density front we use a linear, stationary model, with a coordinate system as shown in Fig. 6. In the model we describe the advective transport due to the density driven circulation. In the northern part of the Dutch coastal zone (north of Noordwijk) long-shore and vertical density changes are inferior to the still very large cross-shore gradients. In this part of the area a local suspended matter minimum often occurs. Therefore long-shore density changes are not taken into account in the model.

Further we assume the density distribution to be stationary. The fresh water inflow from the Rhine-Meuse estuary is supposed to maintain a constant cross-shore density gradient. This implies that the velocity field is only depending on the cross-shore x-coordinate and on the vertical z-coordinate. To simulate a relatively strong density gradient at $x = -x_0$ we choose:

$$\rho(x) = \rho_0 - \rho_d \exp\left\{ - \left(x/-x_0\right)^4 \right\} \tag{27.1}$$

in which $\rho_0 = 1028$ kgm^{-3} the standard density of sea water; ρ_d the density difference between coastal and sea water; and x_0 a cross-shore length scale, approximately the position of the largest density gradient. This density field is in reasonable agreement with measurements of the density profile in the Callantsoog transect during January (Fig. 7).

The pressure field p(x,z) simply follows from the hydrostatic balance:

$$\frac{\partial p}{\partial z} = -\rho(x)\, g \tag{27.2}$$

The horizontal velocities u(x,z) and v(x,z) satisfy the momentum equations:

$$\frac{1}{\rho_0}\frac{\partial p}{\partial x} - f\,v = k\frac{\partial^2 u}{\partial z^2} \tag{27.3}$$

$$f\,u = k\frac{\partial^2 v}{\partial z^2} \tag{27.4}$$

where $f = 10^{-4}$ s^{-1} is the Coriolis parameter; $g = 9.8$ ms^{-2} the gravitational acceleration and k an eddy viscosity. The vertical velocity w(x,z) follows from the continuity equation, which reduces to:

$$\frac{\partial u}{\partial x} + \frac{\partial w}{\partial z} = 0 \tag{27.5}$$

due to the fact that the long-shore velocity component v is independent of the y-coordinate.

This provided that we can introduce a 'stream' function $\psi(x,z)$ defined by:

$$u = \frac{-\partial\psi}{\partial z}, \quad w = \frac{\partial\psi}{\partial x} \tag{27.6}$$

The 'stream' function describes the circulation in the x-z plane. The boundary conditions we have to use are (partly) determined by the area to which we want to apply the model. For (27.2) we have to use continuity of pressure at the sea surface $z = \zeta$:

$$p(x,\zeta) = 0 \tag{27.7}$$

where $\zeta = \zeta(x)$ is the surface elevation. Further we have the kinematic condition at the surface:

$$u\frac{\partial\zeta}{\partial x} = w \tag{27.8}$$

At the bottom $z = -h$ the kinematic condition is:

$$w(x,-h) = 0 \tag{27.9}$$

At the coastal boundary $x = 0$ we have the condition of no net transport through the coast:

$$\int_{-h}^{\zeta} u(x,z)\, dz = 0 \tag{27.10}$$

Properly we have to require that the velocity component normal to the coast vanishes. But in reality the bottom $z = -h(x)$ depends on the cross-shore coordinate. Approaching the coast the depth diminishes and is zero at the coast. In that case we have to use the boundary condition that the velocity component normal to the bottom vanishes and that $u(0) = 0$. In this simple model, in which we use an uniform depth h, we have to use a transport condition on the coastline $x = 0$. This implies that we allow that stream lines intersect the coast. We assume that there exists a 'small' boundary layer in which the depth diminishes to zero. The coastal boundary condition provides that there is no net transport into or out of this boundary layer.

Depth integration of the continuity equation (27.5) and applying boundary conditions (27.8), (27.9) and (27.10) gives that there is no net transport in cross-shore direction over the whole area. Depth integrating (27.4) shows that bottom and surface stresses in long-shore direction have to be equal:

$$\left[\frac{\partial v}{\partial z}\right]_{z=\zeta} = \left[\frac{\partial v}{\partial z}\right]_{z=-h} \tag{27.11}$$

This is a direct consequence of the fact that we neglect the (small) density variations in long-shore direction. Cross-shore bottom and surface stresses do not have to be equal.

At the bottom we use a stress condition:

$$\left[\frac{\partial u}{\partial z}\right]_{z=-h} = \frac{\tau_b^x}{\rho_0 k}, \quad \left[\frac{\partial v}{\partial z}\right]_{z=-h} = \frac{\tau_b^y}{\rho_0 k} \tag{27.12}$$

with the bottom stress proportional to the bottom velocity (Csanady, 1982):

$$\tau_b^x = \rho_0 L\, u(x,-h), \quad \tau_b^y = \rho_0 L\, v(x,-h) \tag{27.13}$$

Further we apply a wind stress condition at the surface:

$$\left[\frac{\partial u}{\partial z}\right]_{z=\zeta} = \frac{\tau_w^x}{\rho_0 k}, \quad \left[\frac{\partial v}{\partial z}\right]_{z=\zeta} = \frac{\tau_w^y}{\rho_0 k} \tag{27.14}$$

The wind stress is given by

$$\tau_w^x = \rho_{air} c_d |W| W^x, \quad \tau_w^y = \rho_{air} c_d |W| W^y \tag{27.15}$$

with W = (W^x, W^y) the wind vector.

For ψ we choose:

$$\psi(x,-h) = 0 \tag{27.16}$$

Representative parameter values are $k = 0.01\ m^2s^{-1}$ (Bowden, 1983), $x_0 = 2\ 10^4$ m, h = 20 m, $L = 10^{-3}\ ms^{-1}$ (Csanady, 1982) and $\rho(x)$ prescribed by (1) with $\rho_d = 4.0\ kgm^{-3}$. With these boundary conditions we can determine the horizontal velocities u and v, the 'stream' function ψ and the surface slope ζ (the formulae are given in Appendix A).

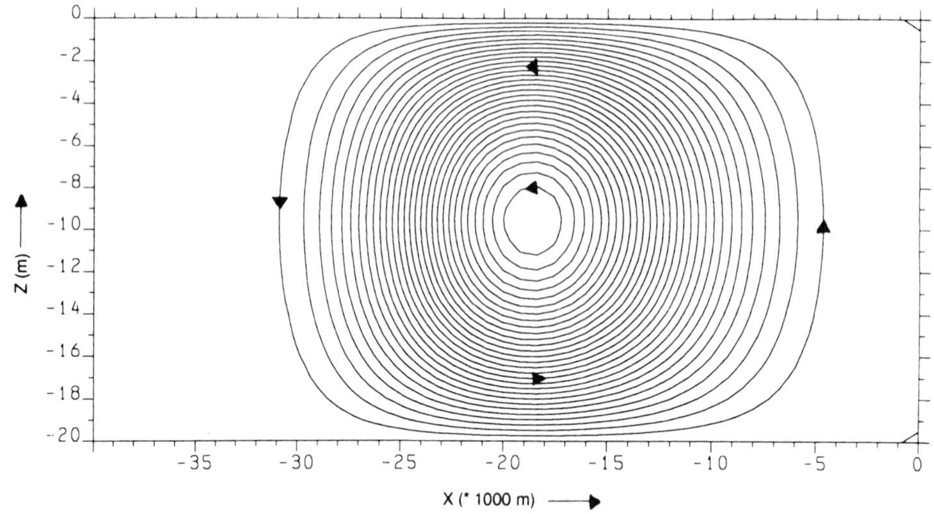

Figure 8: Stream function $\psi(x,z)$ in absence of wind: off-shore motion in the surface layer, onshore motion near the bottom. The circulation is restricted to the area where the density gradient differs from zero. This is a projection on the x-z plane. The long-shore velocity component $v(x,z)$ is not included in this picture. The axes are defined in Fig. 6.

In absence of wind the associated 'stream' function is depicted in Fig. 8. As expected there is off-shore motion in the upper layer and onshore motion near the bottom. The circulation is restricted to the area where the density gradient differs from zero. To illustrate the possible effect on the sedimentation of suspended matter an equation for the sediment concentration is added:

$$u(x,z)\frac{\partial C}{\partial x} + v(x,z)\frac{\partial C}{\partial y} + \left[w(x,z) + W_s\right]\frac{\partial C}{\partial z} = 0 \qquad (27.17)$$

describing advective transport of suspended matter with constant settling rate W_s. Because of the fact that we only look at advective transport we can determine the characteristic curves along which the suspended matter is transported. The set of characteristic curves is given by the intersection of two families of surfaces (see e.g. Kevorkian, 1990):

$$\psi(x,z) + W_s x = \text{constant} \tag{27.18a}$$

$$\int_{-h}^{z} v(x,\tilde{z})\, d\tilde{z} - y\left[w(x,z) + W_s\right] = \text{constant} \tag{27.18b}$$

In absence of wind, the projection of the trajectories on the x-z plane is given in Fig. 9. For different settling velocities the patterns show considerable differences. Obviously, if the vertical velocity component w is much smaller than the settling velocity all the particles will immediately sink to the bottom (Fig. 9a). If the vertical velocity component is much larger than the settling rate the particles will follow the residual circulation and hardly settle down (Fig. 9c).

If settling velocity and vertical velocity component are of the same order a complex suspended matter circulation pattern exists (Fig. 9b). There is an area in which the trajectories do not intersect the bottom, in this area suspended matter particles will stay in suspension. This area is comparable to the 'region of retention' as described by Stommel (1949) in a study about small particles sinking through convection cells. Suspended matter particles present in the surface layer (outside the 'region of retention') are forced to move in off-shore direction until they reach the area where the vertical velocity can not compensate the settling rate anymore. This happens at the seaward side of the density front, where the particles sink to the bottom layer in which they are transported in the onshore direction. The settling rate of a particle depends on its size among other things. Most suspended particles are finer than 30μm (Eisma et al., 1980). Settling velocities of such fine grained particles are of the order of 10^{-5} ms^{-1} (Meade, 1972). Murray (1970) has shown that the average fall velocity determined by experiment in various turbulent fields is reduced by 30% below the corresponding still-water terminal fall velocity.

Only particles with a settling rate comparable to the magnitude of the vertical velocity contribute to the forming of a suspended matter minimum. To investigate the influence of wind direction and velocity we choose in the following $W_s = -10^{-5}$ ms^{-1}. The patterns for different wind directions are given in Fig. 10. With a northerly (long-shore) wind the associated Ekman transport is off-shore directed. The suspended matter transport paths are stretched in the off-shore direction. Particles present in the surface layer are forced to move further off-shore before they can sink, than in the case of no wind (Fig. 9). Southerly (long-shore) winds lead to an onshore directed Ekman transport. Material present far off-shore in the surface layer is forced to move in the onshore direction due to the wind, while material present in the surface layer of the coastal area is forced to move off-shore due to the density gradient.

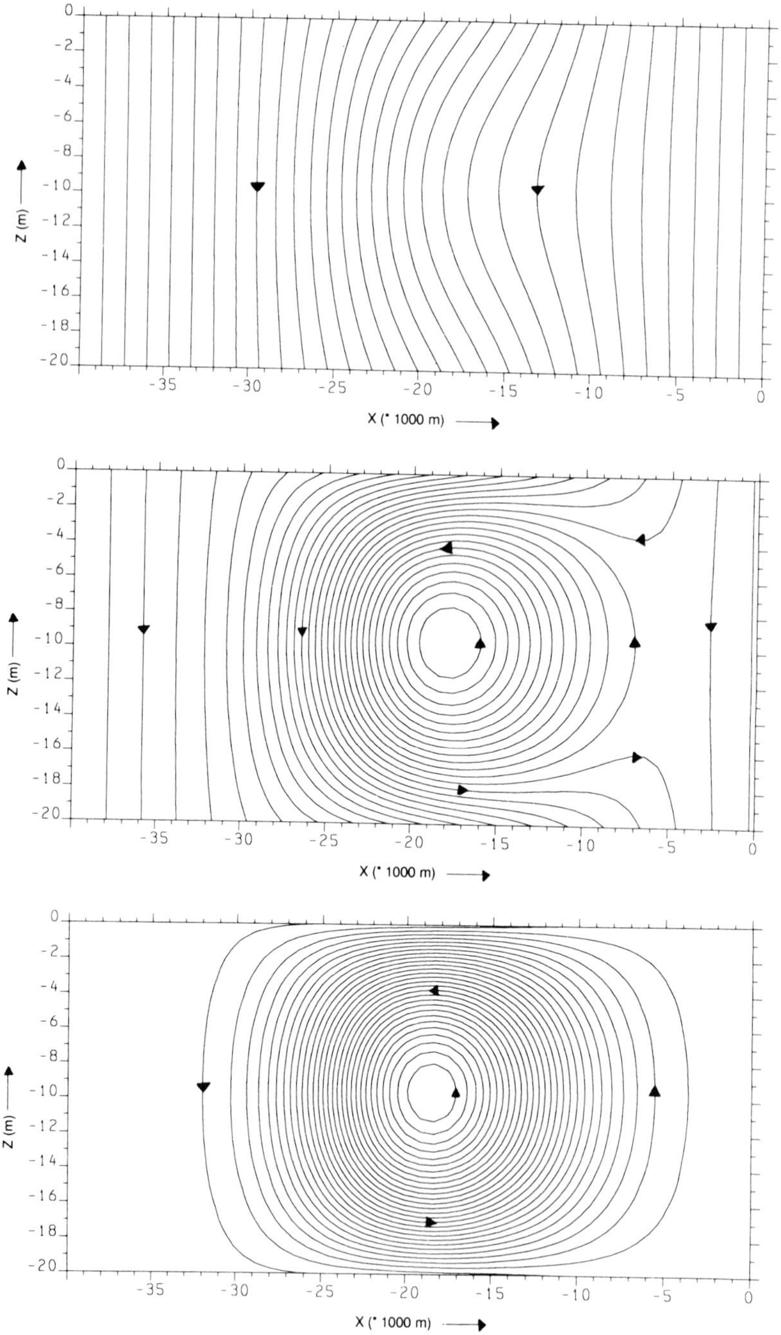

Fig.9a

Fig.9b

Fig.9c

Figure 9: Characteristic curves of suspended matter (projection on the x-z plane) with different settling velocities (a) $W_s = -10^{-4}$ ms^{-1}; (b) $W_s = -10^{-5}$ ms^{-1}; (c) $W_s = -10^{-6}$ ms^{-1}.

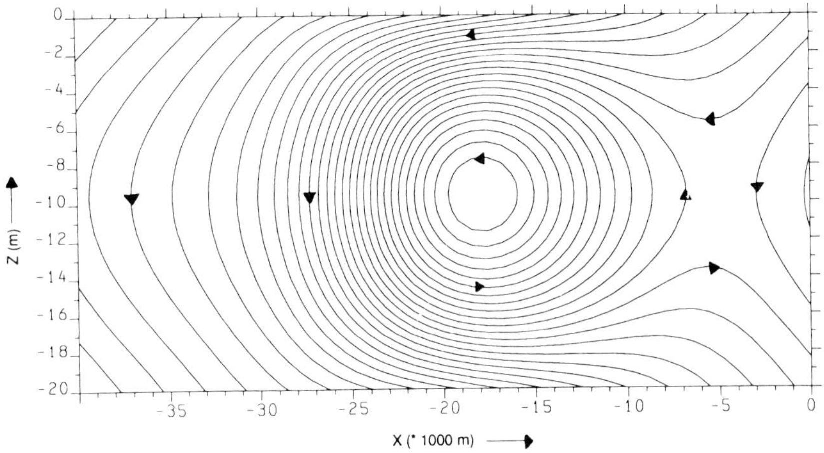

Fig. 10 a

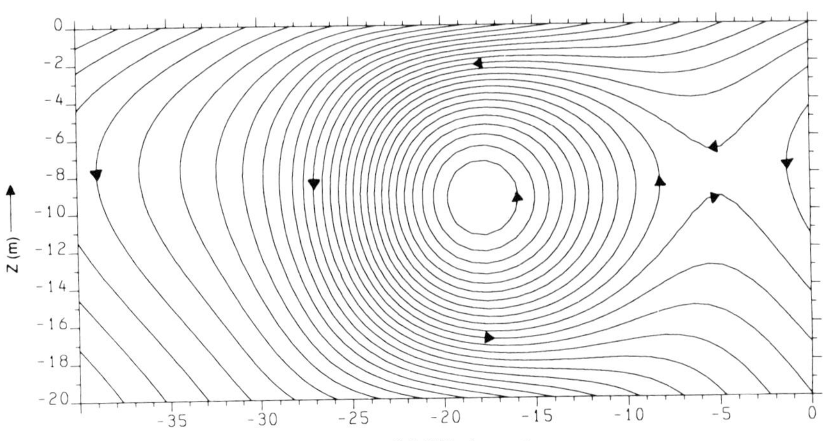

Fig.10 b

Figure 10: Characteristic curves of suspended matter with different wind directions and a wind speed of 5 ms^{-1} (projection on the x-z plane). a: northerly (long-shore) wind, b: easterly (off-shore) wind, c: southerly (long-shore) wind, d: westerly (onshore) wind.

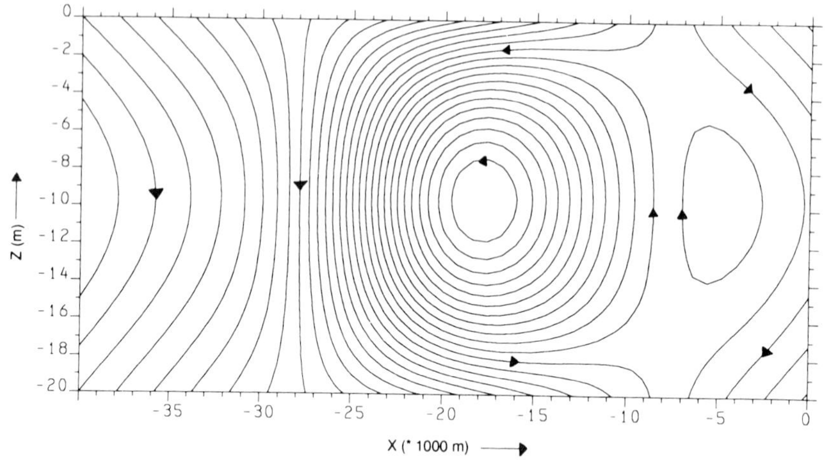

Fig.10c

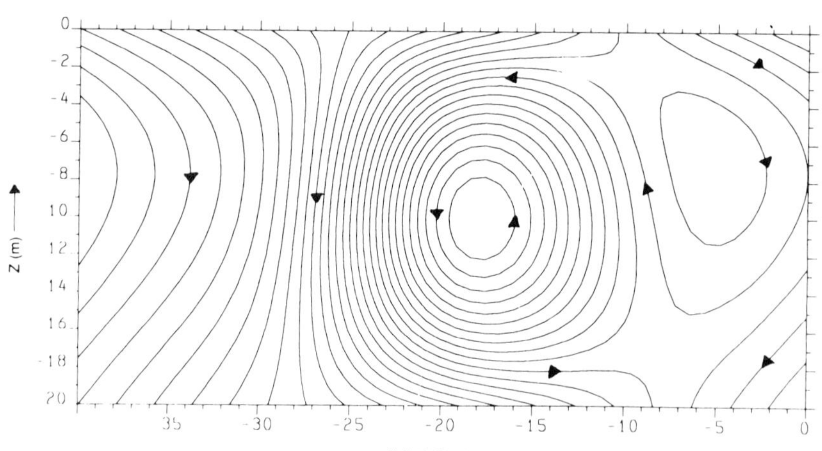

Fig.10d

Figure 10: Characteristic curves of suspended matter with different wind directions and a wind speed of 5 ms^{-1} (projection on the x-z plane). a: northerly (long-shore) wind, b: easterly (off-shore) wind, c: southerly (long-shore) wind, d: westerly (onshore) wind.

The vertical structure of the wind-driven circulation shows an Ekman spiral with surface currents in a 45° direction to the right of the wind. Therefore easterly (off-shore) winds give rise to an off-shore surface velocity component and in this way the transport pattern is comparable to the pattern with a northerly wind. In the other way, westerly (onshore) winds give rise to an onshore surface velocity component and the transport pattern of suspended matter is comparable to the case of southerly wind. The wind-driven circulation is added to the density driven circulation, therefore the suspended matter minimum is more pronounced in circumstances with westerly or southerly winds.

Not only the wind direction is important but also its magnitude. Fig. 11 shows the pattern with a southerly wind with a magnitude of 10 ms^{-1}. In the case of a wind velocity of 5 ms^{-1} (Fig. 10c) the suspended matter minimum is located at some 28 km off-shore, while in the case of a magnitude of 10 ms^{-1} the minimum is located closer by the coast (at some 24 km off-shore). This is due to the fact that the circulation in the surface layer is predominantly forced by the wind. The influence of the density driven circulation is smaller in the case of a stronger wind. Therefore the particles present far off-shore in the surface layer can move further to the coast before they sink out of the surface layer.

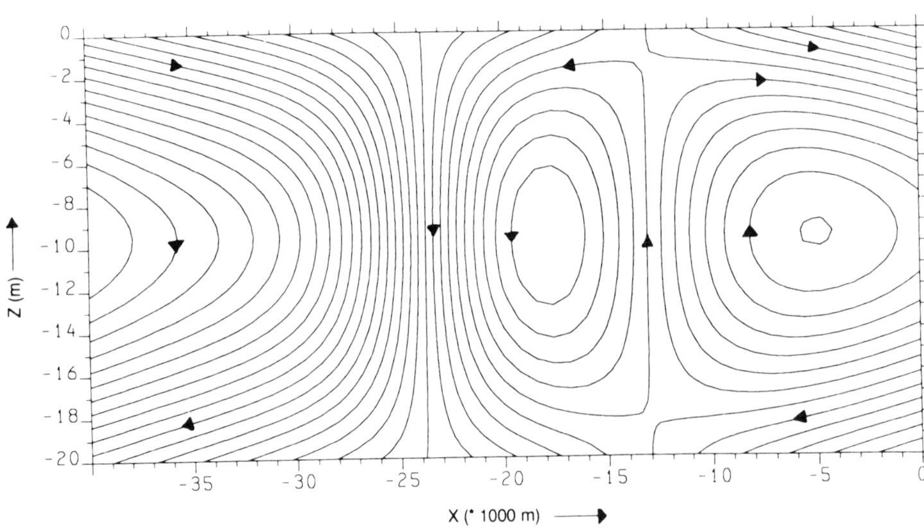

Figure 11: Characteristic curves of suspended matter with a southerly (long-shore) wind of 10 ms^{-1} (projection on the x-z plane). Comparing to Fig. 10c (southerly wind 5 ms^{-1}) the position of the suspended matter minimum is switched to the coast at some 24 km off-shore (against 28 km in Fig. 10c).

27.5 Discussion and Conclusions

Locally for the existence of a minimum in suspended matter concentration between salinity maximum and the shore a large enough river discharge is necessary. In combination with a favourable long-shore southerly wind a salinity front can develop parallel to the coast. If the associated density driven circulation is strong enough, the cross-shore component may cause a depletion of suspended matter in the surface layer.

For particles with a settling rate comparable to the vertical velocity scale there exists an area in which the circulation is strong enough to keep the particles in suspension. Particles present in the surface layer are forced to move off-shore before they can sink to the bottom layer. The depletion of suspended matter in the surface layer takes place at the seaward side of the density front. During winter observations of the near-surface suspended matter distribution a suspended matter minimum is often located in an area with strong density changes. Southerly and westerly winds can strengthen a minimum. During the cruise of 5-8 March 1979 there was a long-shore southerly wind. In that case the minimum was located some 25 km off-shore.

In the model we describe the advective transport of suspended particles only, turbulent mixing is not taken into account for reasons of simplicity. Partly the effect of turbulent mixing can be simulated by taking an "effective" settling velocity smaller than the actual settling rate for particles with that size (Murray, 1970). Stommel (1949) argues that the 'region of retention' must gradually lose its particles due to turbulent exchange through the boundaries. In our model the region of retention is embedded in an asymmetric circulation pattern. Therefore it is possible that the exchange of material takes place at a favourable side. The region of retention is separated from the area of trajectories intersecting the bottom by a separatrix, which contains a saddle node. In a study on diffusion across characteristic boundaries with critical points Matkowsky et al. (1983) have shown that the distribution of exit points is peaked at the critical points. In our case this would imply that the exchange of material between the region of retention and the surrounding area takes place most probably at the coastal side. In this way the density front may act like a kind of barrier through which exchange between coastal and off-shore waters is hardly possible (as proposed by Postma, 1981).

Variation of the eddy viscosity k has its influence on the vertical structure of the velocity field. The suspended matter circulation patterns will look slightly different for other values of k. Another uncertainty is the magnitude of the settling velocity W_s, although the settling rate of the particles in suspension is determined to a great extent by the magnitude of the vertical velocity component. In a numerical study on turbidity maxima in partially mixed estuaries Festa & Hansen (1978) concluded that individual estuaries will select from the full range of particles supplied to them, a small range for storage in the water column. This is due to the estuarine dynamics. The same situation occurs in our model: only particles with a settling rate comparable to the magnitude of the vertical velocity contribute to the formation of a suspended matter minimum; larger particles will settle down on the bottom while smaller particles remain in suspension over the whole area. The range size distribution of suspended matter varies as the dynamics of the system change. Stommel (1949) describes a comparable sorting process. Flocculation is not an important process in this area: salinity differences are "small" and suspended matter concentrations are relatively low.

The model presented in this paper is a strong simplification of reality. The fact that the pressure (and velocity) field is independent of the long-shore y-coordinate is a strong assumption. It provides that bottom and surface stresses in long-shore direction are equal. Further it provides that we can use a stream function to describe the circulation. Including an y dependence will make the model more realistic, but also more complicated; an analytical solution can not be found easily. Suspended matter sources (and sinks) are also not included in the model. This is done because the major sources (English Channel and Flemish Banks) are located outside the investigated area. Because of the y dependence of the concentration in the sediment equation (27.17) it is possible to include an inflow of suspended matter over a southern boundary (e.g. at $y = 0$). Then a concentration field $C(x,y,z)$ can be determined with the characteristic curves. The wind-driven circulation has an important effect on the suspended matter distribution, not only by affecting the position and strength of a local minimum, but also by affecting the total suspended matter amount. It affects the activity of sources and the large scale (wind-driven) circulation, which are important factors in determining the seasonal variations in the suspended matter distribution (Visser et al., 1991), although they are not included in the model presented here. Therefore in this form the model is not directly applicable to the Dutch coastal zone. But it shows clearly that strong density changes can have a large influence on the formation and position of a local suspended matter minimum and therefore it is useful.

Acknowledgements

I would like to thank W.P.M. de Ruijter, H.E. de Swart and D. Visser for discussions and information. Further I acknowledge the North Sea Directorate of Rijkswaterstaat (Rijswijk, NL) for putting the data at my disposal. The investigations were supported by the Working Group on Meteorology and Physical Oceanography (MFO) with financial aid of the Netherlands Organization for the Advancement of Research (NWO).

27.6 References

Bowden, K.F., 1983, Physical Oceanography of Coastal Waters, Ellis Horwood Ltd., Chichester, pp.302

Csanady, G.T., 1982, Circulation in the Coastal Ocean, D. Reidel Publishing Company, Dordrecht, pp. 279

De Ruijter, W.P.M., Van der Giessen, A., & Groenendijk, F.C., 1991, Current and density structure in the Netherlands coastal zone, this volume.

Eisma, D., Kalf, J., & Veenhuis, M., 1980, The formation of small particles and aggregates in the Rhine estuary, Neth. J. Sea Res., 14, 172-191

Festa, J.F. & Hansen, D.V., 1978, Turbidity maxima in partially mixed estuaries: A two-dimensional numerical model, Estuarine and Coastal Marine Science, 7, 347-359

Kevorkian, J., 1990, Partial Differential Equations Analytical Solution Techniques, Wadsworth & Brooks/Cole, Pacific Grove, California, pp. 547

Lee, A.J. & Folkard, A.R., 1969, Factors affecting turbidity in the southerly North Sea, J. Cons. int. Explor. Mer, 32, 291-302

Matkowsky, B.J., Schuss, Z., & Trier, C., 1983, Diffusion across characteristic boundaries with critical points, SIAM J. Appl. Math., 43, 673-695

Meade, R.H., 1972, Transport and deposition of sediments in estuaries, In: Nelson, B.W., (ed.), Environmental framework of coastal plane estuaries, Geol. Soc. Am. Memoir., 133, Boulder, 91-120

Murray, S.P., 1970, Settling velocities and vertical diffusion of particles in turbulent water, J. Geophys. Res., 75, 1647-1654

Postma, H., 1981, Exchange of materials between the North Sea and the Wadden Sea, Mar. Geol., 40, 199-213

Stommel, H., 1949, Trajectories of small bodies sinking slowly through convection cells, J. Mar. Res., 8, 24-29

Van der Giessen, A., De Ruijter, W.P.M., & Borst, J.C., 1990, Three dimensional current structure in the Dutch coastal zone, Neth. J. Sea Res., 25, 45-55

Visser, D., 1987, Analysis of wind measurements at lightvessels Noord-Hinder and Texel over the period 1951-1981, Report 87-3, Inst. Meteorology and Oceanography, University of Utrecht (NL), pp. 31

Visser, M., De Ruijter, W.P.M., & Postma, L., 1991, The distribution of suspended matter in the Dutch coastal zone, Neth. J. Sea Res., 27, 127-143

Appendix A

In this appendix the formulae for the pressure field, the horizontal velocities, the stream function and the surface elevation are given.

Integrating (27.2) and applying boundary condition (27.7) gives the pressure field:

$$p(x,z) = g\,\rho(x)\,[\zeta(x) - z] \tag{27.A0}$$

Substitution of (27.3) and (27.A0) in (27.4) and applying (27.12) - (27.14) leads to the horizontal velocity components u and v:

$$
\begin{aligned}
u = {} & \frac{\delta g}{2\rho_0 f C}\frac{\partial \rho}{\partial x}\left[\begin{array}{l}
\sin(\eta - 2d)\cosh(\eta) - \cos(\eta - 2d)\sinh(\eta) - \\
- \cosh(\eta - 2d)\sin(\eta) + \sinh(\eta - 2d)\cos(\eta) + \\
\sin(\eta + d)\cosh(\eta - d) - \cos(\eta + d)\sinh(\eta - d) - \\
- \sin(\eta - d)\cosh(\eta + d) + \cos(\eta - d)\sinh(\eta + d)
\end{array}\right] \\
& + \frac{\delta}{2\rho_0 k C}\tau_w^x\left[\begin{array}{l}
- \sin(\eta + d)\cosh(\eta - d) - \cos(\eta + d)\sinh(\eta - d) + \\
\sin(\eta - d)\cosh(\eta + d) + \cos(\eta - d)\sinh(\eta + d)
\end{array}\right] \\
& + \frac{\delta}{2\rho_0 k C}\tau_b^x\left[\begin{array}{l}
- \sin(\eta - 2d)\cosh(\eta) - \cos(\eta - 2d)\sinh(\eta) + \\
\sin(\eta)\cosh(\eta - 2d) + \cos(\eta)\sinh(\eta - 2d)
\end{array}\right] \\
& + \frac{\delta}{2\rho_0 k C}\tau_w^y\left[\begin{array}{l}
\sin(\eta + d)\cosh(\eta - d) - \cos(\eta + d)\sinh(\eta - d) - \\
- \sin(\eta - d)\cosh(\eta + d) + \cos(\eta - d)\sinh(\eta + d)
\end{array}\right] \\
& + \frac{\delta}{2\rho_0 k C}\tau_b^y\left[\begin{array}{l}
\sin(\eta - 2d)\cosh(\eta) - \cos(\eta - 2d)\sinh(\eta) - \\
- \sin(\eta)\cosh(\eta - 2d) + \cos(\eta)\sinh(\eta - 2d)
\end{array}\right]
\end{aligned}
\tag{27.A1}
$$

$$
\begin{aligned}
v = \frac{\delta g}{2\rho_0 f C}\frac{\partial \rho}{\partial x}&\begin{bmatrix} -\sin(\eta-2d)\cosh(\eta) - \cos(\eta-2d)\sinh(\eta) + \\ \cosh(\eta-2d)\sin(\eta) + \sinh(\eta-2d)\cos(\eta) - \\ -\sin(\eta+d)\cosh(\eta-d) - \cos(\eta+d)\sinh(\eta-d) + \\ \sin(\eta-d)\cosh(\eta+d) + \cos(\eta-d)\sinh(\eta+d) \end{bmatrix} \\
&+ \frac{\delta}{2\rho_0 kC}\tau_w^x \begin{bmatrix} -\sin(\eta+d)\cosh(\eta-d) + \cos(\eta+d)\sinh(\eta-d) + \\ \sin(\eta-d)\cosh(\eta+d) - \cos(\eta-d)\sinh(\eta+d) \end{bmatrix} \\
&+ \frac{\delta}{2\rho_0 kC}\tau_b^x \begin{bmatrix} -\sin(\eta-2d)\cosh(\eta) + \cos(\eta-2d)\sinh(\eta) + \\ \sin(\eta)\cosh(\eta-2d) - \cos(\eta)\sinh(\eta-2d) \end{bmatrix} \\
&+ \frac{\delta}{2\rho_0 kC}\tau_w^y \begin{bmatrix} -\sin(\eta+d)\cosh(\eta-d) - \cos(\eta+d)\sinh(\eta-d) + \\ \sin(\eta-d)\cosh(\eta+d) + \cos(\eta-d)\sinh(\eta+d) \end{bmatrix} \\
&+ \frac{\delta}{2\rho_0 kC}\tau_b^y \begin{bmatrix} -\sin(\eta-2d)\cosh(\eta) - \cos(\eta-2d)\sinh(\eta) + \\ \sin(\eta)\cosh(\eta-2d) + \cos(\eta)\sinh(\eta-2d) \end{bmatrix} \\
&+ \frac{\delta g}{\rho_0 f}\frac{\partial}{\partial x}[\rho(d-\eta)]
\end{aligned}
\tag{27.A2}
$$

Integrating (27.6) and applying boundary condition (27.16) lead to the 'stream' function ψ:

$$
\begin{aligned}
\psi = \frac{\delta^2 g}{2\rho_0 f C}\frac{\partial \rho}{\partial x}&\begin{bmatrix} \cos(\eta-2d)\cosh(\eta) - \cosh(\eta-2d)\cos(\eta) + C + \\ \cos(\eta+d)\cosh(\eta-d) - \cos(\eta-d)\cosh(\eta+d) \end{bmatrix} \\
&- \frac{\delta^2}{2\rho_0 kC}\tau_w^x[\sin(\eta-d)\sinh(\eta+d) - \sin(\eta+d)\sinh(\eta-d)] \\
&- \frac{\delta^2}{2\rho_0 kC}\tau_b^x[\sin(\eta)\sinh(\eta-2d) - \sin(\eta-2d)\sinh(\eta)] \\
&+ \frac{\delta^2}{2\rho_0 kC}\tau_w^y[\cos(\eta+d)\cosh(\eta-d) - \cos(\eta-d)\cosh(\eta+d)] \\
&+ \frac{\delta^2}{2\rho_0 kC}\tau_b^y[\cos(\eta-2d)\cosh(\eta) - \cos(\eta)\cosh(\eta-2d) + C]
\end{aligned}
\tag{27.A3}
$$

where:

$$\delta = \left[\frac{2k}{f}\right]^{1/2} \quad \text{Ekman depth ;}$$

$$\eta = \frac{z+h}{\delta} \; ; \; d = \frac{\zeta + h}{\delta}$$

$$C = \cosh(2d) - \cos(2d)$$

(27.A4)

From (27.11) - (27.14) it follows that:

$$\tau_w^y = \rho_0 L \, v_{|\eta=0} \tag{27.A5}$$

Substituting (27.A2) in (27.A5) gives an expression for the surface gradient:

$$\begin{aligned}
\rho \frac{\partial d}{\partial x} &= - \frac{\partial \rho}{\partial x} \left\{ d + C^{-1} \left[\begin{matrix} \sin(d)\cos(d) - \sinh(d)\cosh(d) \\ -\sin(d)\cosh(d) + \cos(d)\sinh(d) \end{matrix} \right] \right\} \\
&+ \tau_w^y \left\{ \frac{f}{\delta g L} - \frac{f}{gkC} \left[\begin{matrix} \sin(d)\cos(d) - \sinh(d)\cosh(d) \\ -\sin(d)\cosh(d) + \cos(d)\sinh(d) \end{matrix} \right] \right\} \\
&+ \frac{f}{gkC} \tau_w^x \left[\sin(d)\cosh(d) + \cos(d)\sinh(d)\right] \\
&- \frac{f}{gkC} \tau_b^x \left[\sin(d)\cos(d) + \cosh(d)\sinh(d)\right]
\end{aligned}$$

(27.A6)

From (27.13) and (27.A1) an expression for the cross-shore bottom stress is derived:

$$\begin{aligned}
\tau_b^x &= \frac{\delta L}{CQ} \left[\frac{g}{f} \frac{\partial \rho}{\partial x} + \frac{\tau_w^y}{k} \right] \left[\begin{matrix} \sin(d)\cosh(d) + \sinh(d)\cos(d) \\ -\sinh(d)\cosh(d) - \cos(d)\sin(d) \end{matrix} \right] \\
&+ \frac{\delta L}{kQC} \tau_w^x \left[\sinh(d)\cos(d) - \cosh(d)\sin(d)\right]
\end{aligned}$$

(27.A7)

where:

$$Q = 1 + \frac{\delta L}{2kC} [\sinh(2d) - \sin(2d)] \tag{27.A8}$$

Now our system is completely determined if we prescribe wind and density field.

28

Coastal dynamics along a rugged coastline

B King and E Wolanski

Abstract

A two-dimensional (depth-averaged) numerical model is proposed to describe the hydrodynamics of the Central Great Barrier Reef and is used to investigate the dynamics of its coastal waters. The model incorporates the dominant hydrological and meteorological forces and possesses sufficient resolution to examine the small-scale dynamics which are influenced by both the reefs and the complex coastal topography. Field measurements collected over the shelf confirm the broad-scale features of the model, thus yielding support to predictions of fine-scale processes. Model simulations suggest that the presence of the reef matrix enhances the velocity shear within the coastal zone, in both the long-shore and cross-shore directions. Tracking simulations suggest that the shear acts as a steering mechanism for particulates within the water column. This shear will hinder the exchange of water across the shelf, thus coastal trapping can occur. Further implications for effective regional ecological management are that little terrestrially-derived matter will reach the reef system.

28.1 Introduction

The Central Zone of the Great Barrier Reef Marine Park (CGBR) encompasses most of the coastal and continental shelf waters off Townsville in North-east Australia. The shelf in this region is about 120 km wide and its bathymetry gently slopes from the coast to a depth of about 100m at the shelf-break (figure 1) with the isobaths approximately running parallel to the shoreline. Thus the along-shore direction of this part of the shelf is defined to be roughly aligned with the north-westward, south-eastward axis.

The CGBR is characterized by: a complex series of headlands, capes and shallow embayments at its western boundary; an extensive reef matrix towards its eastern boundary; and a relatively topographic-free region in between, known as the 'lagoon'. The cross-shelf width of the lagoon is comparable to that of the offshore reef system and the two domains are generally separated by the 40m isobath.

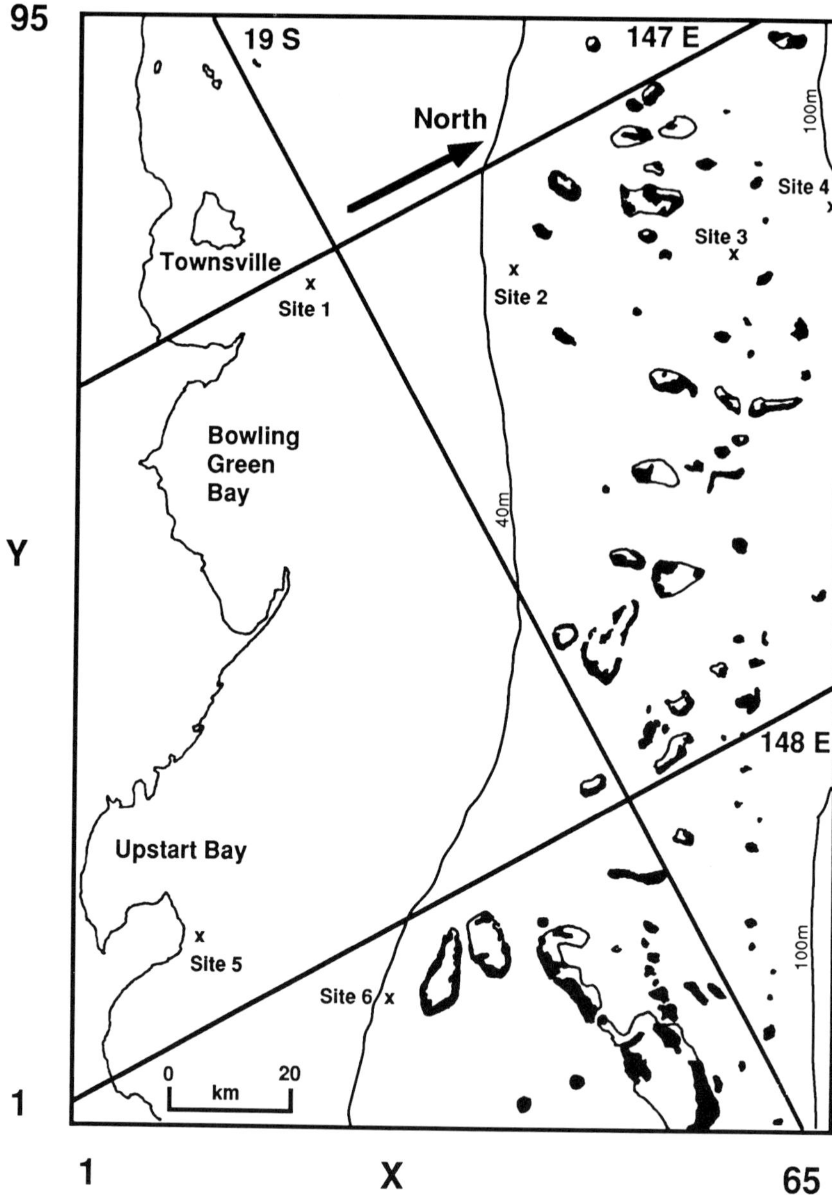

Figure 1: Grid boundaries of the study area within the central Great Barrier Reef. The current data collected from within the region are from moorings located at sites 1-6.

The shelf waters of this area are usually vertically well-mixed in temperature and salinity during the austral winter (April - September) while some stratification occurs following periods of calm weather in the austral summer (Wolanski and Bennett, 1983). The inner shelf waters also receive freshwater pulses from river runoff; but runoff occurs in events, which usually lasts only a few days at a time and occurs principally in the austral summer.

The currents in the CGBR are predominantly barotropic and are influenced by three phenomena (Wolanski and Pickard, 1985): diurnal and semi-diurnal tidal forcing, meso-scale wind events and a poleward ocean current known as the East Australian Current (EAC).

28.1.1 Tidal currents

Analytical models of the currents in the CGBR predict tidal ellipses with the major axis predominantly in the across-shelf direction (Church *et al.* 1985). Further, numerical simulations by Andrews and Bode (1988) show that the tidally-induced component of the currents in this region is generally weak (peak tidal currents $\leq$ 0.2 ms^{-1}) and field measurements indicate that the tidal currents are highly modulated and dominated by the sub-tidal flows (Wolanski and Pickard, 1985). Work undertaken by Andrews and Furnas (1986) suggest that the tidal flows would not contribute significantly to the large-scale advective processes of this region.

28.1.2 Wind forcing

During the austral winter, the wind field is dominated by the south-east trades and is highly coherent over distances greater than 1000 km (Wolanski, 1982). The wind speed generally ranges from 0 to 15 ms^{-1} and blows predominantly north-westward (i.e. along-shelf) in the CGBR with typical wind events lasting for a period of 1-2 weeks. Non-tidal, along-shore currents have been found to be highly coherent with the local wind forcing during this time (Wolanski and Bennett, 1983; Burrage *et al.* 1991a).

Due to the narrowing of the continental shelf north of the CGBR and the presence of the dense reef matrix in this northern region, there exists a natural barrier to the equatorward, along-shore wind-driven currents. This results in a pile-up of water and establishes a longshore pressure gradient on the shelf, set up in response to the uniform along-shore wind stress.

28.1.3 East Australian Current

The EAC flows poleward along the outer shelf and shelf edge in this region and is part of the south-west Pacific Ocean circulation (Church, 1987). It is a boundary current which is associated with an offshore along-shelf slope. Measurements of the currents offshore of the CGBR show speeds typically of the order of 0.3 ms^{-1} in the top 200m of the outer shelf and shelf break waters (Church and Boland, 1983). In the absence of local wind forcing, the effect of the EAC propagates across the entire shelf, through the generation of a longshore pressure gradient on the shelf. This gradient drives a quasi-steady southward flow in the shallow coastal waters as well as through the reef matrix (Burrage *et al.* 1991a; Wolanski *et al.* 1989).

In summary, Godfrey (1973 a,b) shows that the two pressure gradients set up on the shelf as a result of the wind forcing and EAC combine and the resulting pressure gradient on the shelf drive the low frequency currents which have been shown to play the major advective exchange role in this region. Wolanski and Pickard (1985) and Burrage *et al.* (1991b) also show that at low frequencies, the along-shore currents are significantly larger than the cross-shelf currents. Thus given the above assumptions, we can simplify our work by only considering the role that the dominant, non-tidal phenomena contributes to the coastal dynamics of the CGBR. This was also the procedure used in the study of oil dispersion in the CGBR by Andrew *et al.* (1983).

Finally, given the above knowledge of the CGBR, we implemented a numerical model of this region as an extension of the work of Wolanski and Bennett (1983) and Burrage *et al.* (1991 a,b). These authors have shown that spatial variability of the sub-tidal currents exists, in both the cross-shelf and the along-shelf direction; and it is the aim of this paper to use numerical techniques to: (a) resolve the nature and structure of this spatial variability in much finer detail and (b) investigate the influence of the unusual topographical features of the CGBR on the advective characteristics of its coastal waters. This is important in order to determine the fate of land-derived organic and inorganic compounds and various pollutants and to understand what their influence would be on the reef system and the coastal zone.

The model utilizes hydrological and meteorological data of the region. Currently, such data sets are available only for the austral winter.

28.2 Available data

During 1985, current, wind and water elevation data were collected in the CGBR (see Burrage *et al.* 1991b for details). In particular, four current meter moorings were located across the shelf (shown as sites 1-4 in figure 1). Each mooring comprised of multiple current meters situated at various depths throughout the water column. Since Burrage *et al.* (1991b) reported only very weak vertical shear in the currents measured on the shelf, only the mid-depth current meter data were used for model verification, as this was assumed to be representative of the depth-averaged current velocity. Additional current meter data were also collected in 1980 at sites 5 and 6 (see Wolanski and Bennett, 1983).

The mid-depth current meter data that were collected at the four locations across the shelf were analysed to examine only the sub-tidal currents and their response to wind forcing. During the SE trade wind season, high correlation between the sub-tidal along-shelf currents (V), lines (a) - (d) in figure 2 and the along-shelf wind velocity (W_y), line (e) in figure 2, was observed at all inshore sites.

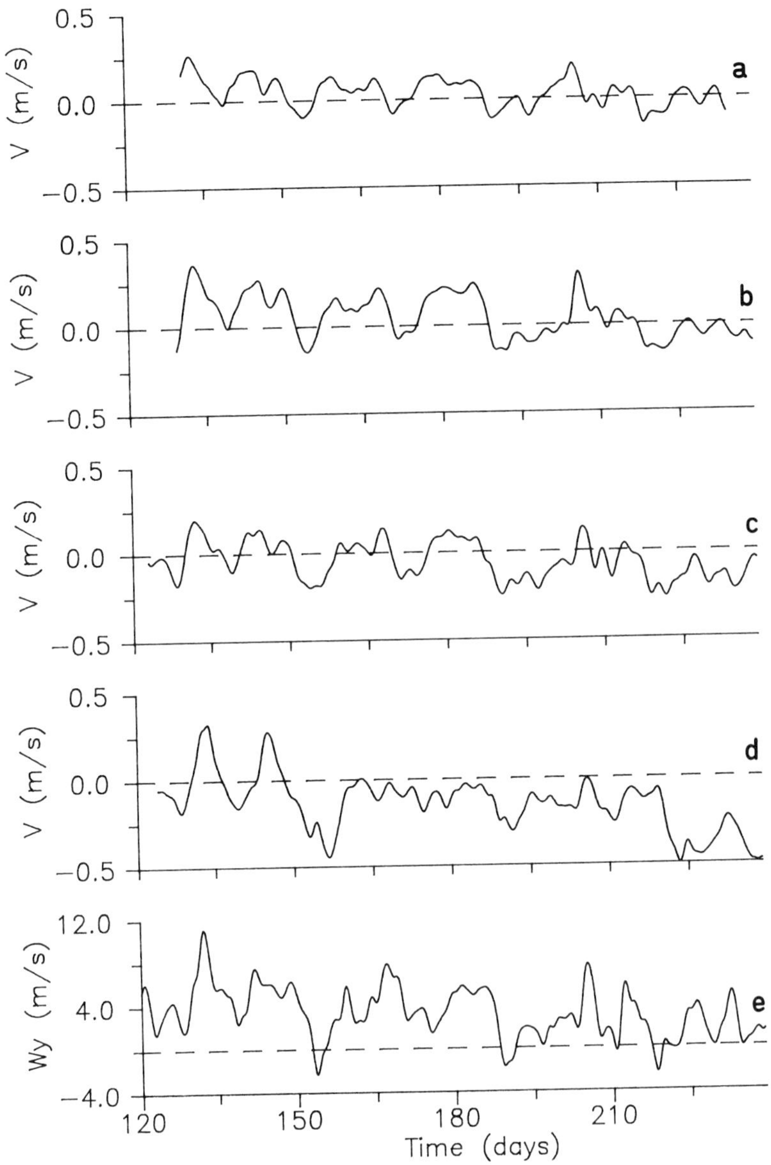

Figure 2: Time-series plots of the hydrological and meteorological data used for model calibration and forcing. The sub-tidal along-shelf currents (V), from sites 1 - 4, are depicted in graphs (a) - (d) respectively. The low-pass filtered, along-shelf wind forcing (W_y), as measured in Mackay, is shown in graph (e). The time is given as the day number in 1985 and covers the austral winter season.

Regression analysis, using the Methods of Least Squares, was employed to describe a general linear relationship between V and W_y at each site, which is independent of time. The resulting regression equations are given in table 1. The confidence levels of the correlations between longshore currents and winds are particularly high in the shelf waters but lost significance at the shelf-break (site 4). These simple functions give a measure of the mid-depth current response to local wind forcing and are used for model verification.

Table 1: Linear regression equations giving the along-shelf, mid-depth current (V) response to local along-shelf wind velocity (W_y) for each mooring site during the austral winter; with the corresponding R^2 and the time (in hours) that V lags W_y, calculated from the analysis. * denotes R^2 values which are significant at the 95% confidence interval.

Site	Regression Eqtn.	R^2	Lag (HR)
4	$V_4 = 4.92 \times 10^{-2}\, W_y - 0.312$	0.463	24
3	$V_3 = 4.00 \times 10^{-2}\, W_y - 0.182$	0.737*	15
2	$V_2 = 3.70 \times 10^{-2}\, W_y - 0.0973$	0.738*	15
1	$V_1 = 2.94 \times 10^{-2}\, W_y - 0.0619$	0.740*	9

28.3 A numerical model

To understand and study the sub-tidal hydrodynamic characteristics of the coastal flows in the CGBR, a detailed numerical model of the entire shelf was set-up in order to encompass the small-scale and shelf-scale influences in this region. The numerical model used to achieve this is based on the work of Leendertse (1970) and Falconer *et al.* (1986) and solves, using an implicit scheme, the following depth-averaged two-dimensional hydrodynamic equations:

$$\frac{\partial UH}{\partial t} + \beta\left(\frac{\partial U^2H}{\partial x} + \frac{\partial UVH}{\partial y}\right) - fVH + gH\frac{\partial \eta}{\partial x} + \frac{gU|U+V|}{C^2} - W_x - \frac{1}{\rho}\frac{\partial HT_{xy}}{\partial y} = 0 \tag{28.1}$$

$$\frac{\partial VH}{\partial t} + \beta\left(\frac{\partial VUH}{\partial x} + \frac{\partial V^2H}{\partial y}\right) + fUH + gH\frac{\partial \eta}{\partial y} + \frac{gV|U+V|}{C^2} - W_y - \frac{1}{\rho}\frac{\partial HT_{yx}}{\partial x} = 0 \tag{28.2}$$

$$\frac{\partial \eta}{\partial t} + \frac{\partial UH}{\partial x} + \frac{\partial VH}{\partial y} = 0 \qquad (28.3)$$

where t is the time; x and y the horizontal axes, orientated across-shelf and along-shelf respectively; H the total depth; U and V the depth-averaged horizontal velocities; β is a correction factor for the non-uniformity of the vertical velocity profile (=1.016); f the Coriolis parameter; g the gravitational acceleration (=9.81); η the free surface elevation; ρ is the water density; T is the depth-integrated Reynolds stress; C is the Chezy coefficient ($=H^{1/6}/n$, where n is the Manning Coefficient); and W is the wind forcing.

The depth-averaged equations are valid for this application due to the predominantly horizontal flow which occurs in shallow shelf waters. Three dimensional currents have been observed around reefs and in wakes behind islands and headlands but these presumably will contribute little to the large-scale flows and are most likely to be sub-grid scale phenomena in this model (Wolanski and Hamner, 1989; Deleersnijder, *et al.* 1989). Baroclinic effects and atmospheric pressure differences during the austral winter are negligible.

Further, this numerical scheme has been used extensively in other hydrodynamic studies, including regions within the CGBR. This presents us with the advantage that the formulations and parameters used in equations (28.1) - (28.3) have already been previously tested, calibrated and verified with numerous field data sets (e.g. Wolanski *et al.* 1989; Falconer *et al.* 1986; Wolanski and King, 1990) thus allowing us to utilize our limited data for verification of the present scheme. The work of Burrage *et al.* (1991b) also provides realistic estimates of boundary conditions and model parameters which we have employed in establishing a representative shelf-model of the CGBR.

The study domain was represented by a mesh with 65 points in the across-shelf direction (X axis, positive eastward) and 95 points in the along-shelf direction (Y axis, positive northward) as shown in figure 1. The resulting unit mesh size was 2031.2 meters square, which was a suitable size in order to encompass the influences of both the large-scale shelf dynamics and that of the topographically-complex, small-scale reefs and coastal topography.

The coastline closed the western boundary of the model domain while the remaining three boundaries were open and were forced by sea-level (η) fluctuations.

The eastern boundary occurred at the shelf-break and approximately coincided with the 100m isobath. This boundary is used to control the along-shelf gradient in the water elevation (Godfrey, 1973b), which is due to the combined presence of large-scale wind forcing and the EAC as detailed earlier in this text. At this stage, the absolute value for this along-shelf gradient is unknown, therefore it is simply determined by trial-and-error. That is, the slope is calibrated by adjusting the water elevation along this boundary; such that the model predicted current at site 4, match those measured at site 4, for any given wind forcing. Thus, having calibrated this open boundary, the remaining inshore sites (1-3) are used to verify the model's response to this forcing across the shelf. These comparisons are listed in table 2.

Table 2: Comparison of the numerically predicted along-shelf component of the current velocity (V_m; ms^{-1}), with the observed (V_p; ms^{-1}) values extrapolated from the corresponding regression equation, listed in table 1, for the 6 given wind conditions.

Run #	1		2		3		4		5		6		7	
Site #	V_m	V_p	V_m	V_p	V_m	V_p	V_m	V_p	V_m	V_p	V_m	V_p	V_m	V_p
4	-0.31	-0.31	-0.21	-0.21	-0.12	-0.12	-0.02	-0.02	0.08	0.08	0.18	0.20	-0.31	-0.25
3	-0.18	-0.18	-0.10	-0.09	-0.02	-0.01	0.06	0.09	0.14	0.12	0.22	0.17	-0.18	-0.18
2	-0.10	-0.10	-0.02	-0.03	0.05	0.06	0.13	0.11	0.19	0.20	0.27	0.26	-0.10	-0.08
1	-0.06	-0.06	0.00	-0.01	0.06	0.05	0.12	0.08	0.17	0.20	0.23	0.22	-0.06	-0.03

For the northern and southern across-shelf open boundaries, Wolanski and Bennett (1983) and Burrage *et al.* (1991a) suggest the across-shelf slope is in geostrophic balance with the sub-tidal along-shelf flows. Therefore, both these boundaries are controlled by:

$$gH\frac{\partial \eta}{\partial x} = fVH \tag{28.4}$$

where V and H were locally determined from the previous time-step.

Simulations were performed for six different wind conditions which dominate the wind field during the austral winter (Wolanski, 1982). These conditions assume south-easterly trade wind with velocities ranging from 0 to 10 ms^{-1} at 2 ms^{-1} increments, and are identified as run #1 to #6 respectively. A simulation was also performed identically to run #1, but without the reef matrix (run #7) in order to assess the influence of the reef on the near inshore waters.

The model was started from rest and allowed to run for 120 hours in order to reach a steady state. Tests were then performed, utilizing the along-shelf slope along the outer boundary under no wind conditions, using various timesteps, to examine the effects of numerical diffusion on the simulation. The predicted flows for each test revealed no significant dependency on timestep. The timestep was set at 600 seconds.

Finally, the numerically-predicted currents of the first 6 runs were used to drive a simple advection model. The advection model utilised the shelf-model velocities to determine drogue trajectories. For each run, these simulated drogues, representing any passive particle, were released; one at each grid cell, in an across-shelf transect within the coastal and lagoon region. This technique better enabled us to examine the effects of the across-shelf shear by taking advantage of the greater resolution of the shelf-model. Thus we were able to assess the transport dynamics of the coastal region for a variety of weather scenarios which are likely to occur during the austral winter.

28.4 Results

The results from each of the first 6 simulations were compared with data from sites 1 to 3 for model verification after the eastern boundary sea-level gradient was adjusted to match the flows observed at site 4. They were generally close to those predicted by the regression equations (see table 2). This is not surprising considering the previous work with this numerical scheme within this region. However it is encouraging to see that the sub-tidal currents, as reported by Burrage *et al.* (1991a) and described by the regression equations of table 1, are reproduced. Thus we were now confident that the finer detail given by the shelf-model of the shear in the sub-tidal currents, is a reasonable representation of reality.

Analysis of the momentum balance for run #1 reveals that, under no wind conditions, the coastal waters are predominantly influenced by the offshore sea level gradient and flow southward along-shelf. Preliminary results for run #1 (reef, no wind) and #7 (no reef, no wind) were compared with data from sites 1 to 6. This exercise showed the inshore currents predicted by run #7 under-estimated those observed in the field. However, run #1 yielded predicted currents which compared well to the observed ones. From table 2, we can see that velocity shear exists within the lagoon due to across-shelf variations in the frictional influence of this sloping shelf in both these cases. However, while shear will exist across any shelf under these conditions, it can be noted from the data in table 2, that the presence of the reef matrix strengthens the inshore water velocity (comparing run #1 with run #7). This suggests that the reef matrix plays an important role in the hydrodynamics of the coastal waters, probably by channelling waters from the reef complex to the inner shelf. This channelling is enhanced by the presence of the large southern reef systems in the domain and this can be seen in figure 3a.

Data from Wolanski and Bennett (1983) suggested that the along-shelf currents at site 5 and 6 would be approximately 0.2 m/s and 0.4 m/s respectively, to the south in the absence of wind forcing. This agrees with the predicted velocities given by run #1. Given that the frictional influence at site 1 is similar to 5 and site 2 is similar to 6, then the along-shore variation in velocity strength can also be attributed to the influence of the reef matrix and coastal topography. Topographic steering by the coastal headlands and reef clusters promotes the channelling of water through the lagoon and this will contribute to the along-shore variation in the advection mechanisms. This can be seen in detail in figures 3a and 3b.

Flows within embayments also demonstrate some cross-shelf shear; mostly due to the increased frictional influence of the shallow waters. Stagnation zones also occur within the embayments as a result of upstream blocking by headlands and capes.

The influence of this across-shelf shear on the transport of neutrally-buoyant matter is demonstrated by simulating drogue trajectories with the advection model, given the flow pattern shown in figure 3a. Snapshots of the drogue positions can be seen in figure 4. Velocity shear in the embayments and near the coast prevent the inshore drogues from moving offshore. Thus, inshore waters appear to remain coastally-trapped for long periods and are not readily flushed under these conditions, that is, by the advective exchange processes present in these coastal waters.

The drogues released in deeper waters are subjected to stronger and less obstructed flows and are advected further downstream than those inshore. The velocity shear across the lagoon tends to align these drogues into a slick formation (see figure 4f). Thus the advective process here, under the influence of velocity shear, aggregate the drogues which may oppose any diffusive processes operating.

In the presence of light SE winds (1-3 m/s; run #2), decoupling of the shallow coastal currents from the offshore shelf currents occurs in the lagoon (see figure 5). The inshore currents are dominated by the wind forcing and move northward, while the offshore currents remain under the influence of the along-shelf slope of the EAC and flow southward. There is little current in the zone between. This decoupling results in a separation of coastal and offshore shelf waters and can be expected to reduce the exchange of waters from the coastal zone to the reef matrix and vice versa. The offshore extent of the resulting 'separation front' is found to be dependent on the wind strength and water depth.

The 'separation front' runs continuously through the lagoon of the CGBR under these light wind conditions, except for one break offshore from Cape Bowling Green (see figure 5). Local coastal topographic steering of the inshore currents, by the headland in this particular case, have caused the front to become segmented. This may reduce the along-shore progression of shallow water, but would locally enhance the cross-shelf exchange of water.

Figure 6 shows the synoptic views of the drogue locations from run #2. The embayment waters are very slow to flush due to the weak velocities in the coastal zone. Again the drogues deployed in the deeper region of the lagoon align themselves into a slick. Two slicks are present in the later stages of the simulation, emphasising the role of the separation front. No evidence of water exchange from the coastal zone to the reefal system is present from these trajectories.

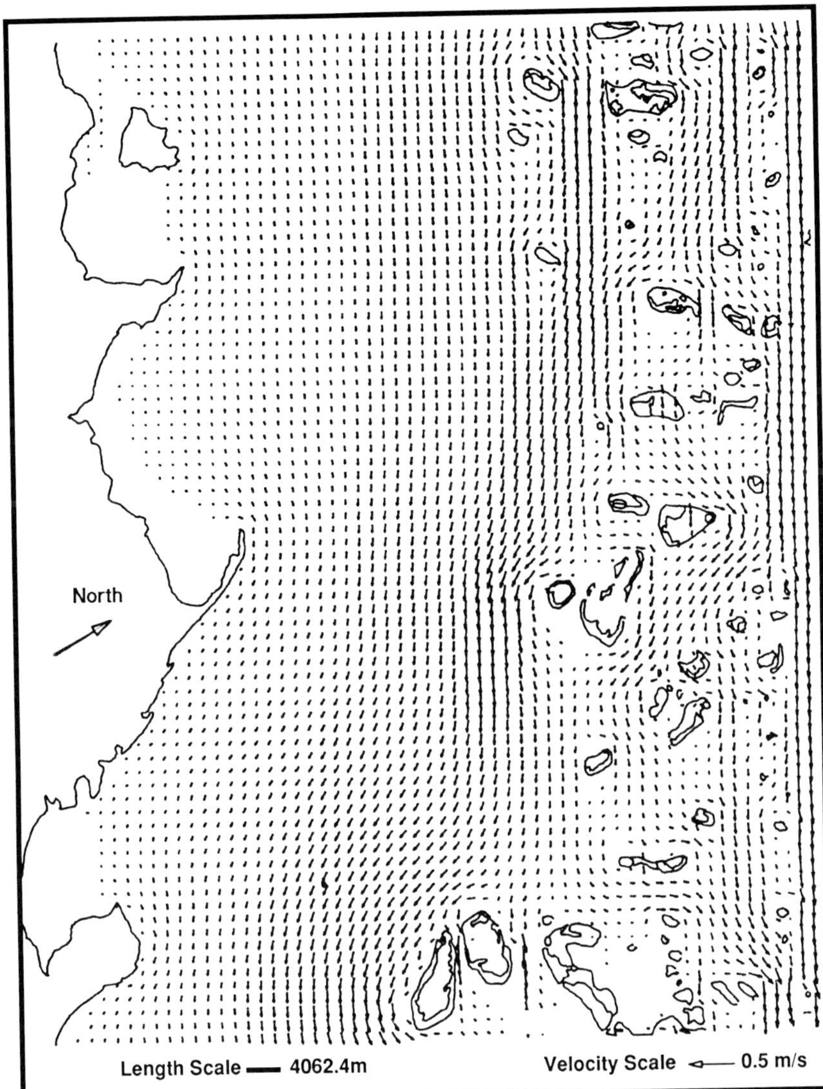

Figure 3a: The predicted current profiles of the model domain, forced by an along-shelf slope only (run #1). The length of the arrows represents the relative current velocity strength and its orientation indicating the corresponding direction of flow. Part (a) shows the predicted shelf-scale flows while (b) gives a subset of the domain focusing on the predicted flows within the coastal zone of the CGBR lagoon and its embayments. Note the velocity shear across the lagoon and the variation in current strength along the lagoon.

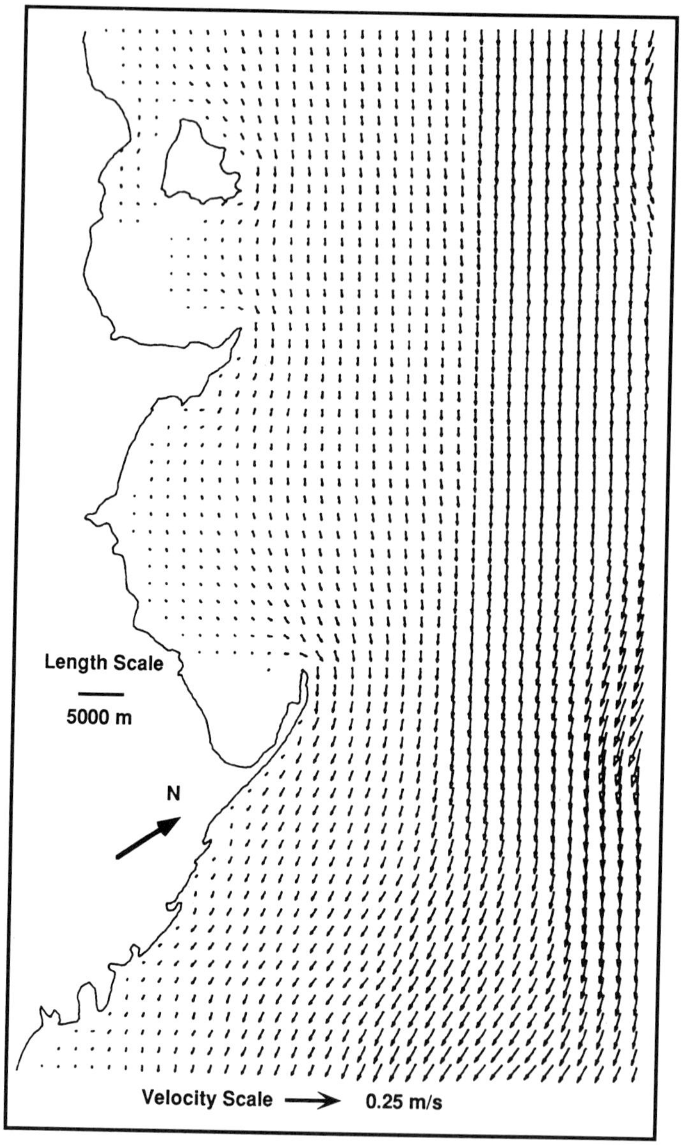

Figure 3b: The predicted current profiles of the model domain, forced by an along-shelf slope only (run #1). The length of the arrows represents the relative current velocity strength and its orientation indicating the corresponding direction of flow. Part (a) shows the predicted shelf-scale flows while (b) gives a subset of the domain focusing on the predicted flows within the coastal zone of the CGBR lagoon and its embayments. Note the velocity shear across the lagoon and the variation in current strength along the lagoon.

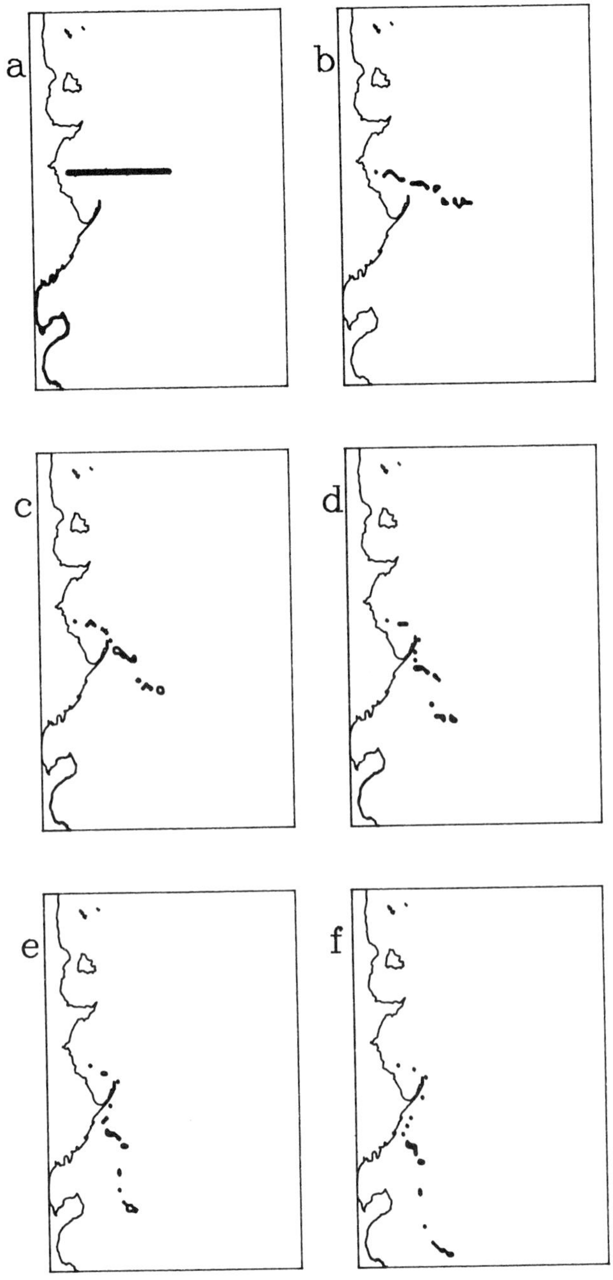

Figure 4: Snap shots of drogue distribution driven by an advection model using the output of run #1. Plot (a) shows the initial distribution of 26 drogues along the transect Y=55 for $10 \leq x \leq 35$ at time = 0 days. The remaining plots give the distribution after (b) 2 days, (c) 4 days, (d) 6 days, (e) 8 days and (f) 10 days.

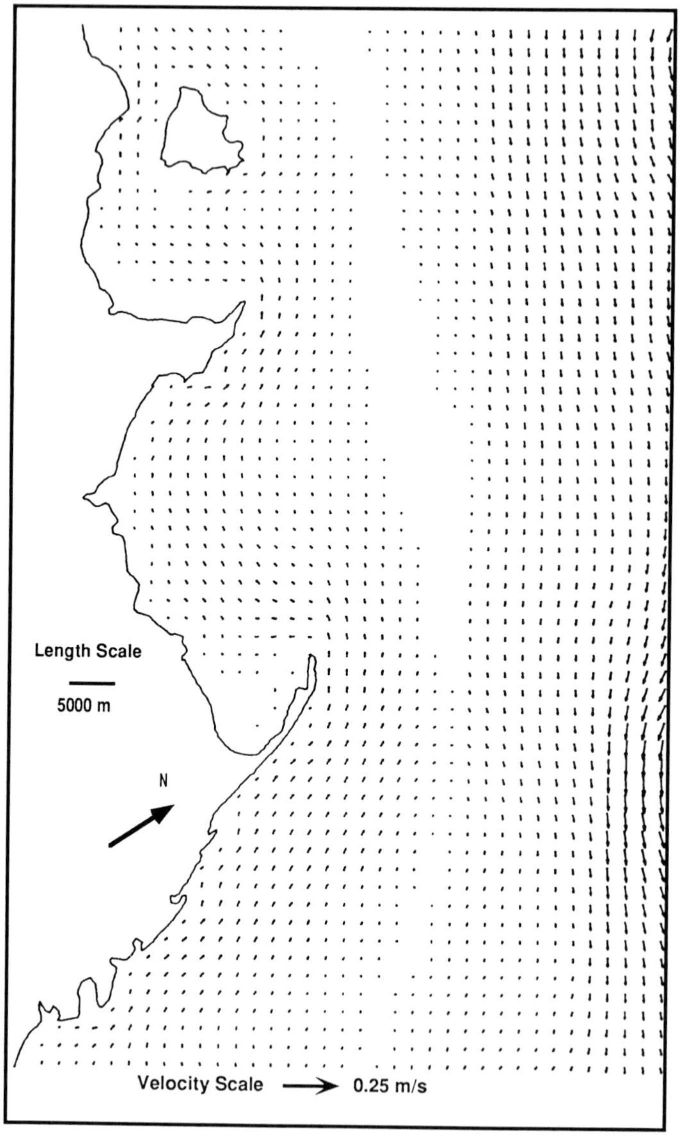

Figure 5: The flow patterns which exist in the coastal region from simulation #2. The decoupling of the currents across the shelf that results is highlighted by a separation zone and strong shear patterns. The wind forcing is directed up the page.

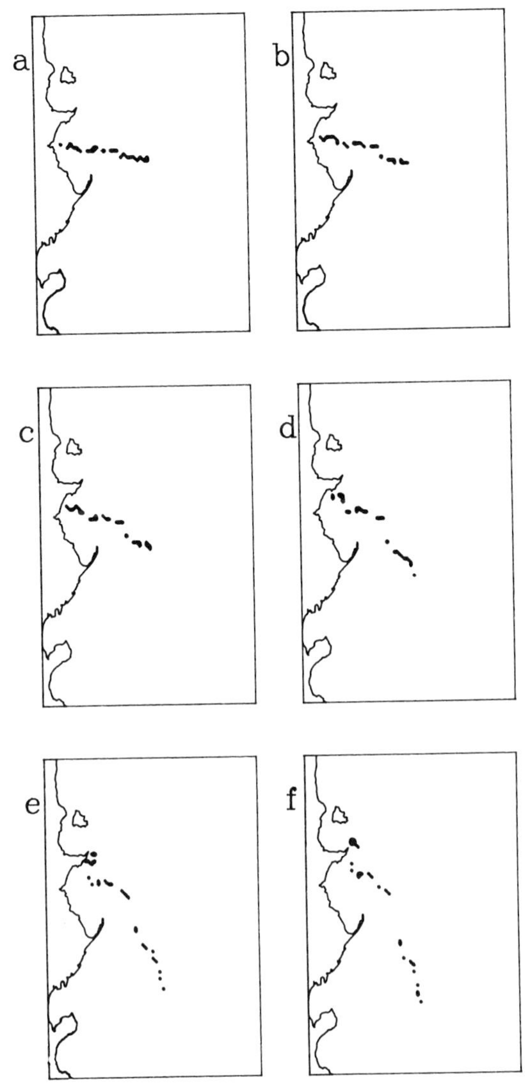

Figure 6: Snap shots of drogue position driven by an advection model using the output of run #2, given the same initial distribution shown in figure 4a. These plots give the distribution after (a) 2 days, (b) 4 days, (c) 6 days, (d) 10 days, (e) 15 days and (f) 20 days.

As the wind velocity strengthens from the south-east, the local wind forcing dominates and flow reversal occurs across the entire shelf. Under these conditions, velocity shear still occurs near the coast and is particularly evident in the embayments. See figure 7 for example. The existence of an eddy in Bowling Green Bay, in the lee of the cape, can also be observed. This will result in some trapping of the water in this area.

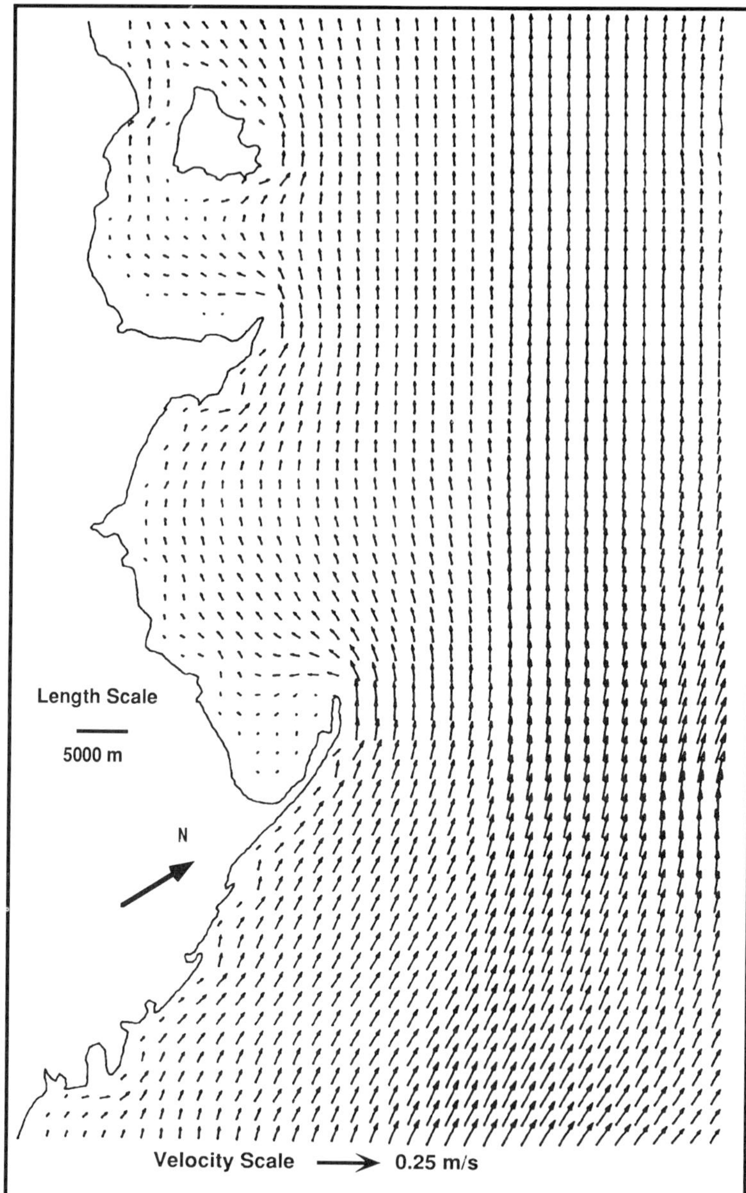

Figure 7: Increasing wind strength (run #3) causes flow reversal across the entire lagoon and shelf. This leads to strong shear zones across the mouths of the embayments and an eddy develops in the lee of Cape Bowling Green. The wind forcing is directed up the page.

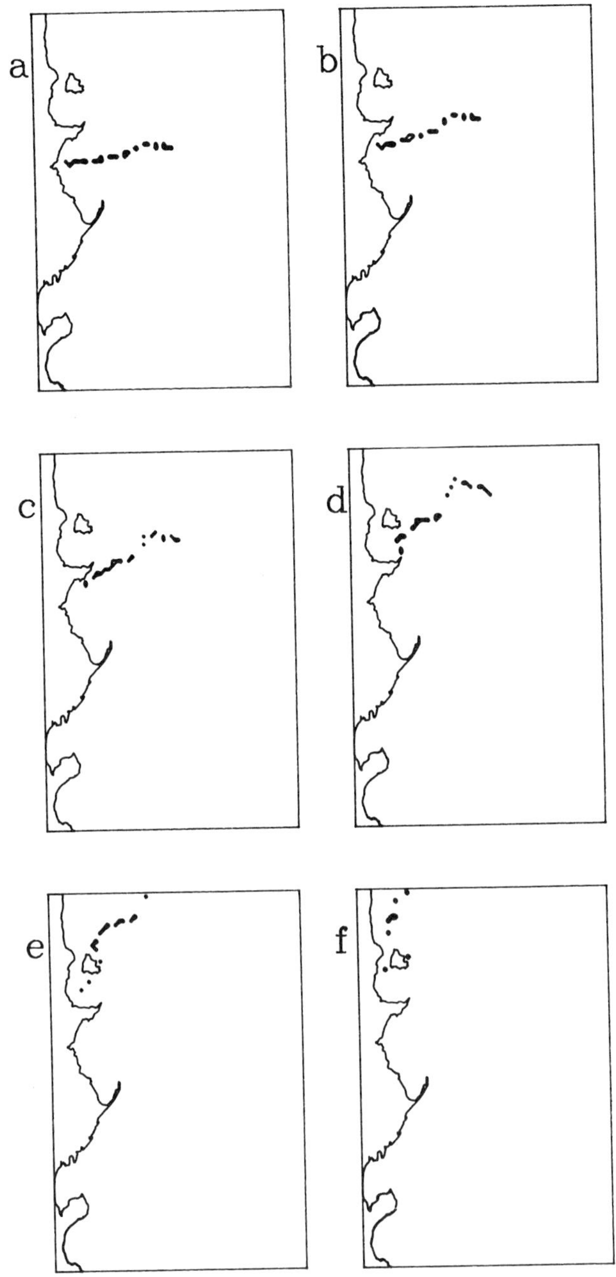

Figure 8: Snap shots of drogue position driven by an advection model using the output of run #3 given the same initial distribution shown in figure 4a. These plots give the distribution after (a) 2 days, (b) 4 days, (c) 6 days, (d) 10 days, (e) 15 days and (f) 20 days.

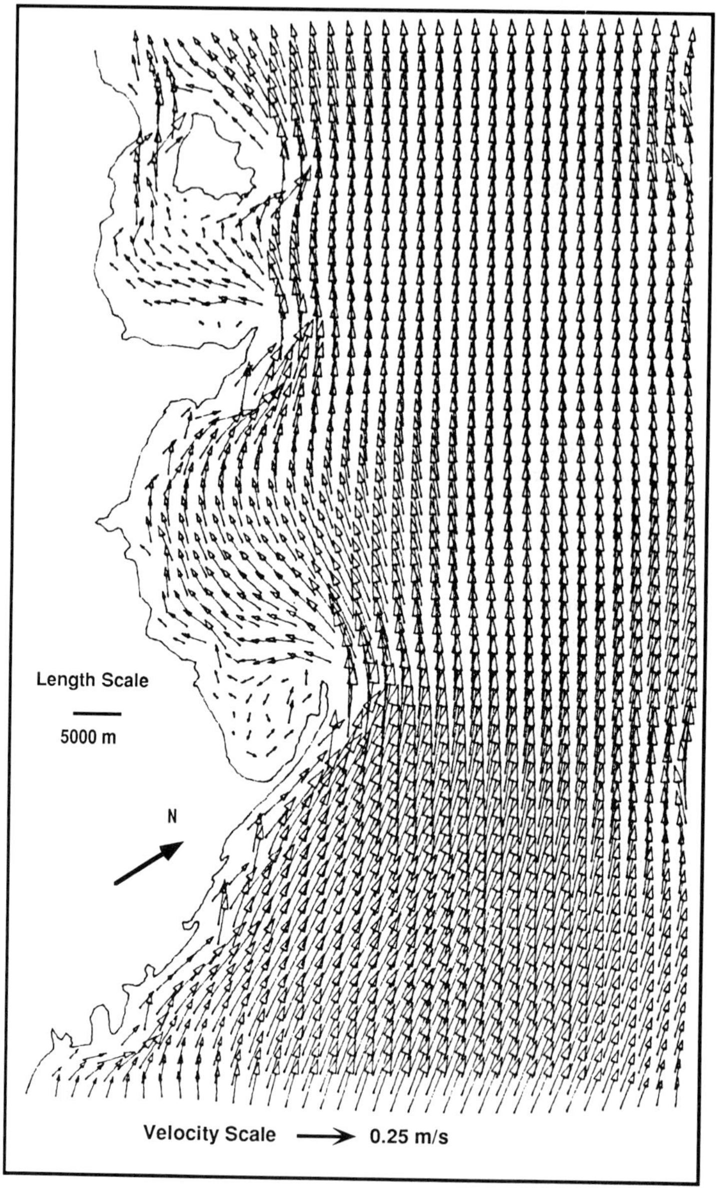

Figure 9: Strong wind (run #6) drives strong inshore currents. This leads to strong shear zones across the mouths of the embayments and an eddy develops in the lee of Cape Bowling Green. The wind forcing is directed up the page.

The inshore flow increases correspondingly with wind strength as shown in table 2. Under the milder conditions of run #3, the drogues remain resident within Bowling Green Bay for up to 10 days (figure 8) while run #6 predicts rapid flushing of the embayment by the second day (not shown).

However the observed trajectories of the drogues for runs #3 to #6 all show a similar distribution to that depicted in figure 9. This is due to the similar velocity shear patterns which exist under northward-moving coastal flows. All simulations show slick aggregation of the off-shore drogues.

28.5 Discussion

The behaviour of inshore water advective processes for the CGBR have been described for a variety of hydrological and meteorological scenarios, commonly occurring during the austral winter period. This was achieved by utilizing existing field data and a numerical model of the shelf of the CGBR to resolve the detailed flow regime of the region. This analysis highlighted the role that the reef matrix plays in modifying the dynamics of the coastal waters, by topographically-steering and channelling which enhances the inshore flow. This process also leads to increased velocity shear within the lagoon region of the shelf.

The channelling effects produced by the irregular topography of the coast and reef matrix in the CGBR also explain the along-shelf variation in current velocities within the lagoon (comparing sites 1 with 5 and 2 with 6). This demonstrates the importance of our approach here in designing a fine-scale shelf model in order to study the coastal dynamics of the region. Conversely one must proceed with caution when using simpler models for predicting coastal and shelf flows in such a complex environment.

From our understanding of the dominant processes controlling long-term water transportation in the CGBR; numerical simulations based on these processes were analysed, in order to determine the likely coastal dynamics and cross-shelf exchange processes present. These are made apparent by calculations of the fate of neutrally-buoyant particles. These simulations reveal that under all wind conditions, strong velocity shear exist across the lagoon which inhibit water exchange from the coastal zone to the reef system. Thus inshore waters remain coastally trapped. Flushing is apparently reduced by the velocity shear across the openings of embayments and the generation of either stagnation zones upstream and eddies in the lee of headlands which promote local trapping. In the embayments, flushing of coastal waters is further inhibited by trapping in the fringing mangrove swamps (Wolanski and Ridd, 1990).

This study has important implications for the sound ecological management of the coastal and reef systems. That is, any land-derived matter (such as dissolved nutrients, pesticides and other pollutants and suspended sediment) entering the coastal system are not readily flushed and remain coastally trapped under most weathers conditions, in the austral winter. Depending on the nature of the injected matter, accumulation may pollute the coastal waters in this region.

However, since little across-shelf exchange processes are evident, the reef system would appear to be isolated from this threat. Indeed, observations of low levels of agricultural pesticides found in reef organisms in the region by Olafson (1978) support our claim. Olafson postulated that the CGBR was not under threat from land-derived pesticides from adjacent river runoff due to the nature of the oceanographic features of the region as observed from terrestrial sediment distributions on the seabed.

Finally, while the across-shelf tidal excursion does not significantly contribute to the across-shelf exchange of water, the role of tidal diffusion in mixing the waters is yet-to-be determined. Although, from an ecological perspective, this process would help dilute any injected contaminants, further research into its effects is warranted, in order to give some measure of the mixing rates in the coastal zone.

28.6 References

Andrews, J.C. and Bode, L., 1988, "The tides of the central Great Barrier Reef". Continental Shelf Research., Vol. 8, 9, 1057-1085.

Andrews, J.C. and Furnas, M.J., 1986, "Subsurface intrusions of Coral Sea water into the central Great Barrier Reef - I. Structures and shelf-scale dynamics". Continental Shelf Research., Vol. 6, 4, 491-514.

Andrews, J.C., Mitchell, A.W. and Bellamy, N.F., 1983, "Field studies of currents and simulations of oil dispersion in the Central Great Barrier Reef". Sixth Australian Conference on Coastal and Ocean Engineering. Gold Coast, Australia, 13-15th July, 1983.

Burrage, D.M., Steinberg, C.R. and Church, J.A., 1991a, "Current variability and shear across the continental shelf and slope of the Central Great Barrier Reef". (submitted to Journal of Physical Oceanography).

Burrage, D.M., Church, J.A. and Steinberg, C.R., 1991b, "Linear systems analysis of momentum on the continental shelf and slope of the Central Great Barrier Reef". (submitted to Journal of Geophysical Research).

Church, J.A., 1987, "East Australian current adjacent to the Great Barrier Reef". Australian Journal of Marine and Freshwater Research, 38, 671-683.

Church, J.A., Andrews, J.C. and Boland, F.M., 1985, "Tidal currents in the central Great Barrier Reef". Continental Shelf Research, 5, 515-531.

Church, J.A. and Boland, F.M., 1983, "A permanent under current adjacent to the Great Barrier Reef". Journal of Physical Oceanography, 13, 1747-1749.

Deleersnijder, E., Wolanski, E. and Norro, A., 1989, "Numerical simulation of the three-dimensional tidal circulation in an island's wake". Proceedings of the 4th International Conference on Computational Methods and Experimental Measurements, Capri, Italy, 23-26 May, 1989.

Falconer, R.A., Wolanski, E. and Mardapitta-Hadjipandeli, L., 1986, "Modelling tidal circulation in an island's wake". Journal of Waterway, Port, Coastal and Ocean Engineering, American Society of Civil Engineering, 112, 234-254.

Godfrey, J.S., 1973a, "On the dynamics of the western boundary current in Bryan and Cox's (1968) numerical model ocean". Deep-sea Research, 20, 1043-1058.

Godfrey, J.S., 1973b, "Comparison of the East Australian Current with the western boudary flow in Bryan and Cox's (1968) numerical model ocean". Deep-sea Research, 20, 1059-1076.

Leendertse, J.J., 1970, "A water-quality simulation model for well-mixed estuaries and coastal seas: Volume 1, Principles of computation". Report No. RM-6230-RC. The Rand Corp., 77.

Olafson, R.W., 1978, "Effect of agricultural activity on levels of organochlorine pesticides in hard corals, fish and molluscs from the Great Barrier Reef". Marine Environmental Research, Vol. 1, 2, 87-107.

Wolanski, E., 1982, "Low-level trade winds over the western Coral Sea". Journal of Applied Meteorology, 21, 881-882.

Wolanski, E. and Bennett, A.F., 1983, "Continental shelf waves and their influence on the circulation around the Great Barrier Reef". Australian Journal of Marine and Freshwater Research, 34, 23-47.

Wolanski, E., Burrage, D.M. and King, B., 1989, "Trapping and dispersion of coral eggs around Bowden Reef, Great Barrier Reef, following mass coral spawning". Continental Shelf Research, Vol. 9, 5, 479-496.

Wolanski, E. and Hamner, W.M., 1988, "Topographically controlled fronts in the ocean and their biological influence". Science, 241, 177-181.

Wolanski, E. and Jones, M., 1981, "Physical properties of Great Barrier Reef lagoon waters near Townsville. 1. Effects of Burdekin River floods". Australian Journal of Marine and Freshwater Research, 32, 305-319.

Wolanski, E., Jones, M. and Williams, W.T., 1981, "Physical properties of Great Barrier Reef lagoon waters near Townsville. 2. Seasonal variation". Australian Journal of Marine and Freshwater Research, 32, 321-334.

Wolanski, E. and King, B., 1990, "Flushing of Bowden Reef lagoon, Great Barrier Reef". Estuarine, Coastal and Shelf Science, 31, 789-804.

Wolanski, E. and Pickard, G.L., 1983, "Currents and flushing of Britomart Reef lagoon, Great Barrier Reef". Coral Reefs, 2, 1-8.

Wolanski, E. and Pickard, G.L., 1985, "Long-term observations of currents on the central Great Barrier Reef continental shelf". Coral Reefs, 4, 47-57.

Wolanski, E. and Ridd, P., 1990, "Mixing and trapping in Australian tropical coastal waters". In "Residual Currents and Long-term Transport", R.T. Cheng (Ed). Springer-Verlag, New York.

29

Transport of hypoxic waters: an estuary-subestuary exchange

A Y Kuo and K Park

Abstract

Hypoxic or anoxic conditions in the subpycnocline water of Chesapeake Bay persist throughout the summer. The effect on the dissolved oxygen concentration in the deep basin of the lower Rappahannock River, a subestuary on the western side of the bay, was studied with an observational program. The data indicate that in the lower portion of the water column the subtidal (or residual) current was directed into the subestuary most of the time. The mass fluxes of salt and dissolved oxygen into the subestuary through a point near the estuarine bottom at the river mouth were calculated for tidal and subtidal components respectively. From the analyses, we conclude that the mass exchange owing to the tidal component is at least an order of magnitude smaller than that resulting from the subtidal component. Characteristic differences in the properties of water transported into the subestuary were observed. On five occasions, each lasting about two days, during the one month period of field measurements, the imported water was characterized by low dissolved oxygen and high salinity, typically 3 to 4 mg l^{-1} lower dissolved oxygen and 2 to 3 psu higher salinity than at other times. The low dissolved oxygen, high salinity water masses were all accompanied by strong subtidal current and southwest wind. During periods of strong wind from the southwest quadrant, the density-driven current near the bottom was enhanced by the wind-driven circulation. Furthermore, the surface set-up favors the transport of the water from the deep portion of the bay resulting from (1) tilting of the pycnocline in the bay and (2) shoreline and bathymetric configurations around the estuary-subestuary junction.

29.1 Introduction

Problems of hypoxia and anoxia in estuarine and coastal waters have received increased attention in recent years (Schroeder, 1985). The depletion of dissolved oxygen (DO) generally results from a combination of biological and physical factors. It is most frequently observed in subpycnocline waters during summer when vertical mixing is weak and water temperature is high. In Chesapeake Bay, for example, summer anoxia has been observed in the deep bottom waters since the 1930s (Newcombe and Horne, 1938), and has become more widespread and of longer duration during recent times (Flemer et al., 1983).

All major subestuaries on the western side of Chesapeake Bay have a deep basin near their mouths (Fig. 1). Hypoxia has been observed frequently in the deep basins in all these subestuaries except in the James River, the southernmost subestuary. Kuo and Neilson (1987) made a comparative investigation of the bottom DO in the three Virginia subestuaries. They reported that hypoxia occurred most frequently in the deep water of the Rappahannock River, less frequently in the York River, and rarely in the James River, even though it received the heaviest wastewater loadings among the three subestuaries. This was attributed to the relatively strong gravitational circulation in the James River.

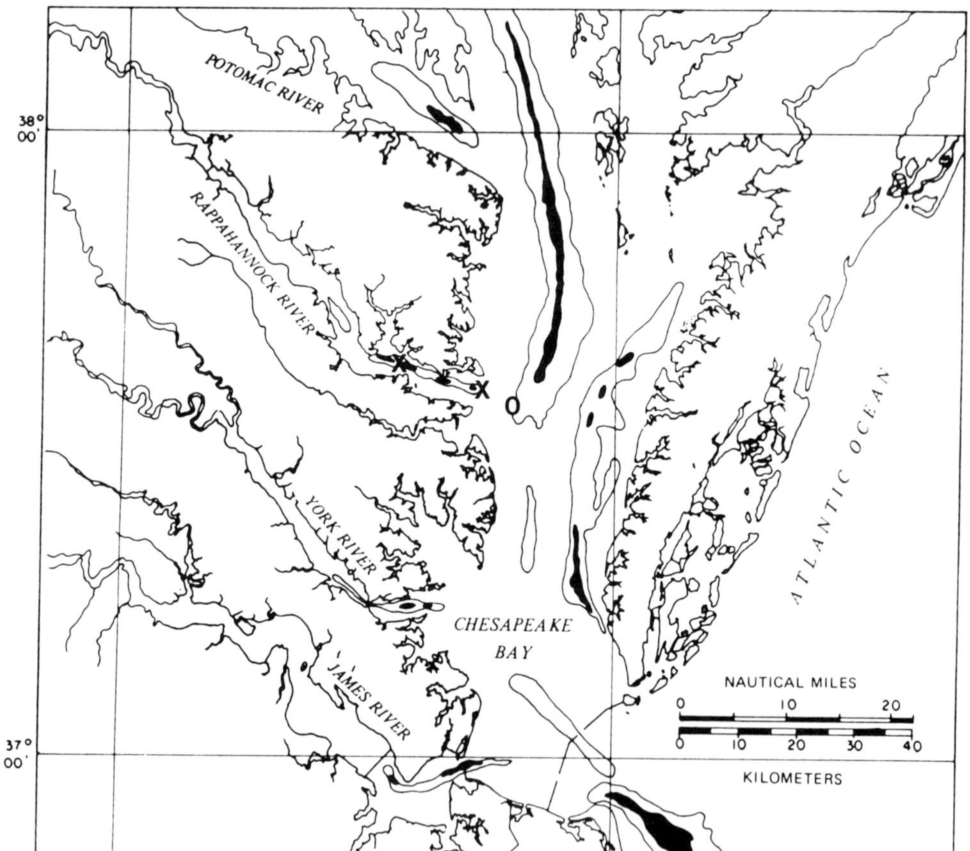

Figure 1:. Lower Chesapeake Bay with its four major subestuaries (The shaded areas are deeper than 20 m, the depth contours are 13 m; **X** marks the station with time series measurements and **O** marks the slackwater survey station in the bay).

Since the variability in hypoxic conditions among the three Virginia subestuaries follows a similar trend of decreasing bottom DO from south to north in the main stem of Chesapeake Bay, it is possible that the variability may be due, at least in part, to the quality of the waters imported into the mouths of the subestuaries from the bay. In this paper, the effect of the hypoxic or anoxic conditions in the bay on water quality in the lower Rappahannock River is investigated by examining the mass transport near the bottom at the river mouth. Time series data from simultaneous measurements of DO, temperature, salinity and velocity at the bottom of the river mouth were analyzed to demonstrate the primary factors dominating the estuary-subestuary exchange.

29.2 Field observation

Field data for this study were collected during summer 1987. Data collection can be divided into two major groups.

29.2.1 Slackwater surveys

A total of 13 slackwater surveys were conducted from June to September at slackwater before ebb. During each survey, temperature, conductivity and DO were measured at 11 stations along the river axis and one station off the river mouth in the main stem of the bay. The river stations extended from the river mouth to 57.8 km upriver beyond the deep basin in the lower subestuary.

Temperature and conductivity were measured with an Applied Micro System conductivity-temperature-depth probe (CTD). Continuous vertical profiles, top to bottom, for these variables were obtained at each station. DO was measured using an oxygen probe (Yellow Springs Instruments) every meter from the surface to 15 m depth, then every 2 m until the bottom was reached. Conductivity and temperature measurements were converted to salinity as psu (UNESCO, 1983). Salinity, temperature and DO data for the river stations were displayed as isopleths in the vertical-longitudinal plane for each survey.

Two examples of spatial distributions of DO and salinity are presented in Fig. 2. In general, the surface DO concentrations were always high, with little variability along this stretch of the subestuary. The DO concentrations in the bottom water were low in all surveys, with a characteristic longitudinal pattern. The bottom DO concentrations decreased in the upriver direction from the river mouth, reached a minimum and then increased as water depth decreased. There was a striking difference in vertical salinity and DO distributions at the river mouth between the two dates. The water column at the river mouth was moderately stratified on August 10 and became nearly homogeneous on August 17 (Fig. 2a and 2b). Fig. 2b also shows an apparent intrusion of high salinity water at mid-depth, thus implying probability of reversed three layered circulation (Elliott, 1978). This created an unstable density stratification in the lower portion of water column, and enhanced the vertical mixing.

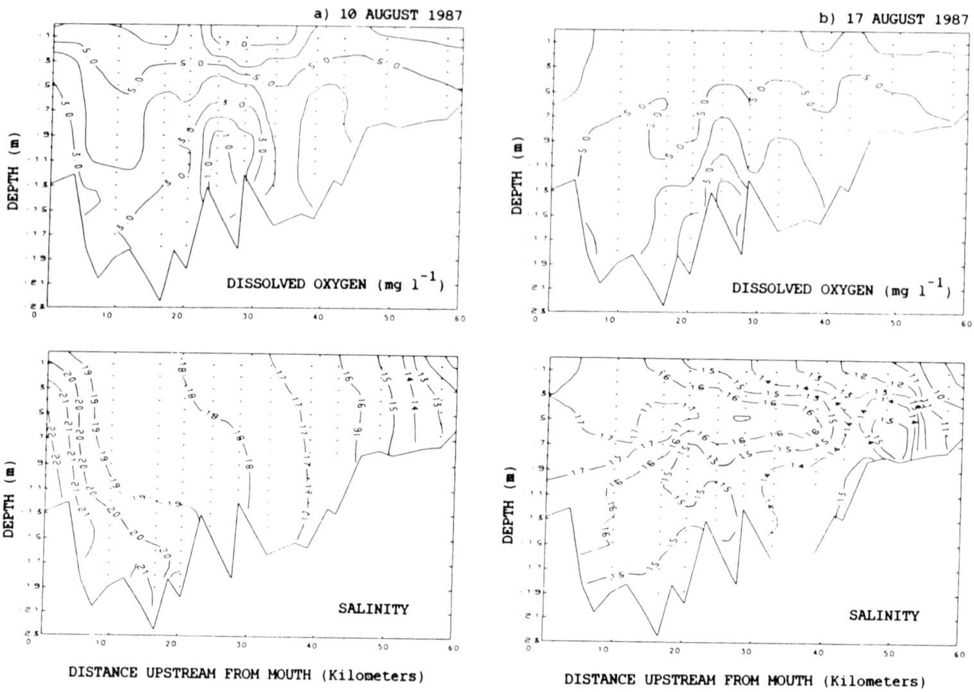

Figure 2: Distribution of dissolved oxygen (mg l^{-1}) and salinity (psu) in the lower Rappahannock River. a) August 10, 1987 and b) August 17, 1987 (• marks the measuring location).

US EPA (Environmental Protection Agency) has a Chesapeake Bay monitoring program measuring water quality parameters throughout the bay at semi-monthly intervals. As indicated by the August 3-4, 1987 sampling, the bay was highly stratified and DO in the bottom water was severely depressed (Fig. 3). To demonstrate the time history of the vertical distribution of DO and salinity in the bay, the EPA program data and our slackwater survey data, both at the station off the Rappahannock River mouth (Fig. 1), were combined to construct constant value contours in the depth-time plane (Fig. 4). It shows that the bay was stratified throughout the period. Surface DO was always at or near saturation level while hypoxic conditions in the bottom water started to develop in May and persisted throughout the summer.

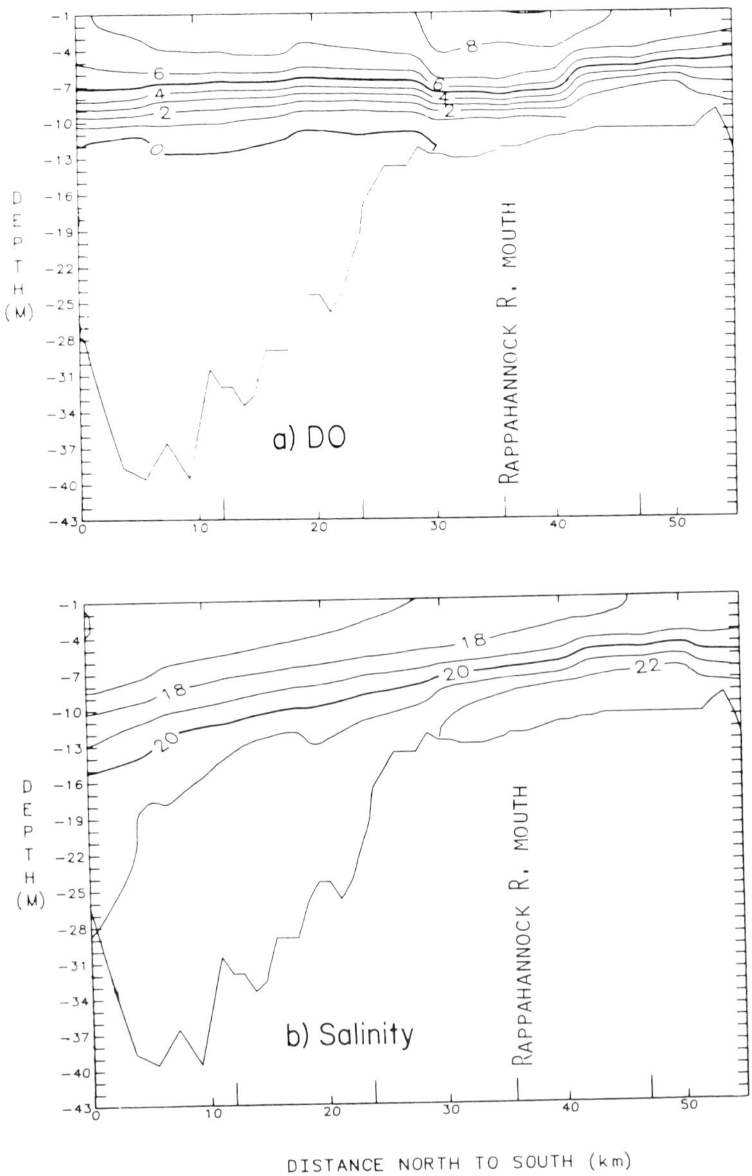

Figure 3: Dissolved oxygen (mg l^{-1}) and salinity (psu) in the main stem of Chesapeake Bay on August 3-4, 1987 (The larger tick marks on abscissa indicate measuring stations).

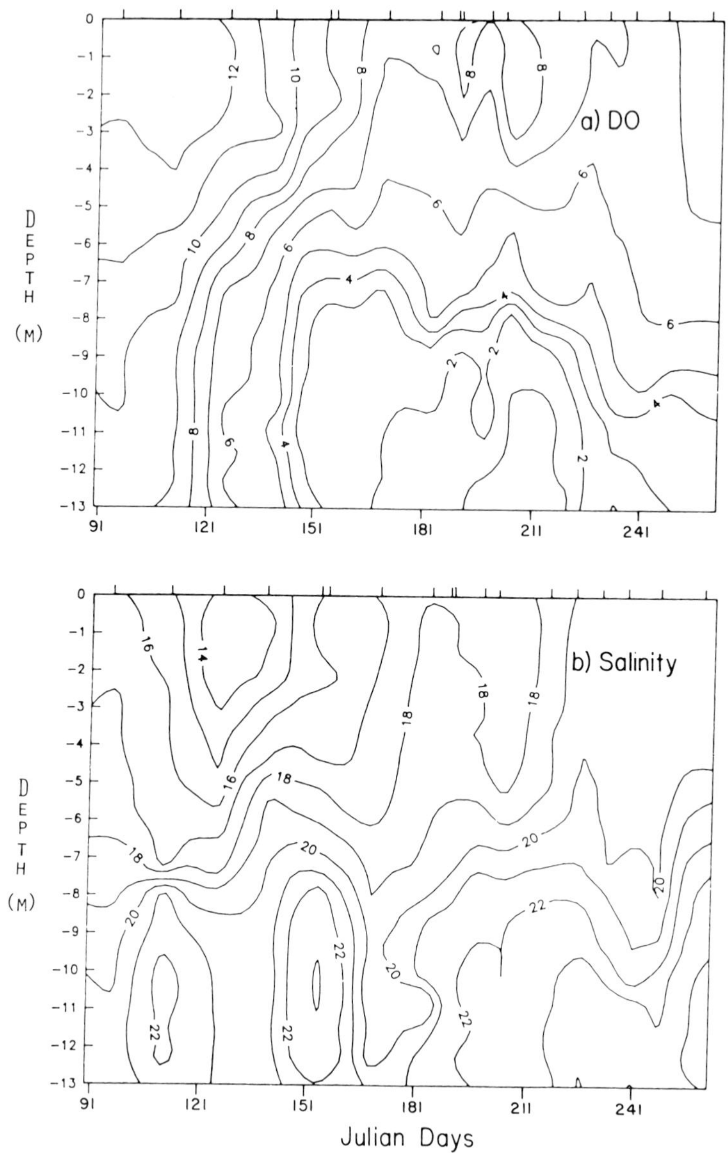

Figure 4: Dissolved oxygen and salinity isopleths in depth-time plane (The tick marks at the top of the figure indicate sampling dates).

29.2.2 Time series measurements

Time series measurements of temperature, conductivity and dissolved oxygen were taken at two stations, one at the river mouth and the other at 16.6 km upriver in the deep basin (Fig. 1). The data were measured and recorded every half hour with Hydrolab DataSounde 1. One sensor package was deployed at the river mouth near the bottom at 11.8 m depth, while at the upriver station two were deployed, one at mid-depth (7.5 m) and one at the bottom (16.2 m). The meter at the river mouth station was deployed from 5 August to 2 September. During this period, it was repeatedly serviced, calibrated and redeployed for periods ranging from 7 to 10 days. Battery life and fouling restricted the deployment period, thus creating gaps in the time series data. The meters at the upriver station were not deployed until 25 August because of availability.

Current velocities were measured with in-situ, self recording meters that were deployed with taut wire moorings at the river mouth station and the station 16.6 km upriver. Two types of current meters were used; Inter-Ocean S4 meters and modified Braincon Histogram meters. The S4 meter is an electromagnetic type current meter with solid state memory, some of which are also equipped with temperature, conductivity and pressure sensors. The meters were set to record the average values of all variables every 30 min. The Braincon meter measures current magnitude by a savonius rotor, with a vane attached for direction measurement. All three Braincon meters deployed functioned properly for a period of only a few days and then succumbed to electronic problems or fouling. This leaves available data for two depths at the river mouth station, and for three depths at the upriver station.

The velocity data from each current meter were analyzed to determine the principal axis of the flow at the meter location. The direction of the principal axis is defined as the direction along which the sum of the absolute values of velocity components from all data points is maximum. This direction was taken as the longitudinal axis of a coordinate system for further current analyses.

Three tide gauges were installed during this study at locations spanning the length of the subestuary. Water surface elevations were recorded every 6 min. These data were then converted to hourly surface elevations.

Wind data from Norfolk and Richmond airports were examined. Norfolk is located 65 km to the south and Richmond is located 80 km to the west of the Rappahannock River mouth. Both sets of wind data exhibit similar wind direction and magnitude except during periods of weak wind with variable direction. The wind data in Norfolk is presented in Fig. 5a and used for later analyses. Fig. 5 also displays the time series data of water surface elevation, longitudinal velocity components, salinity and DO at the river mouth station.

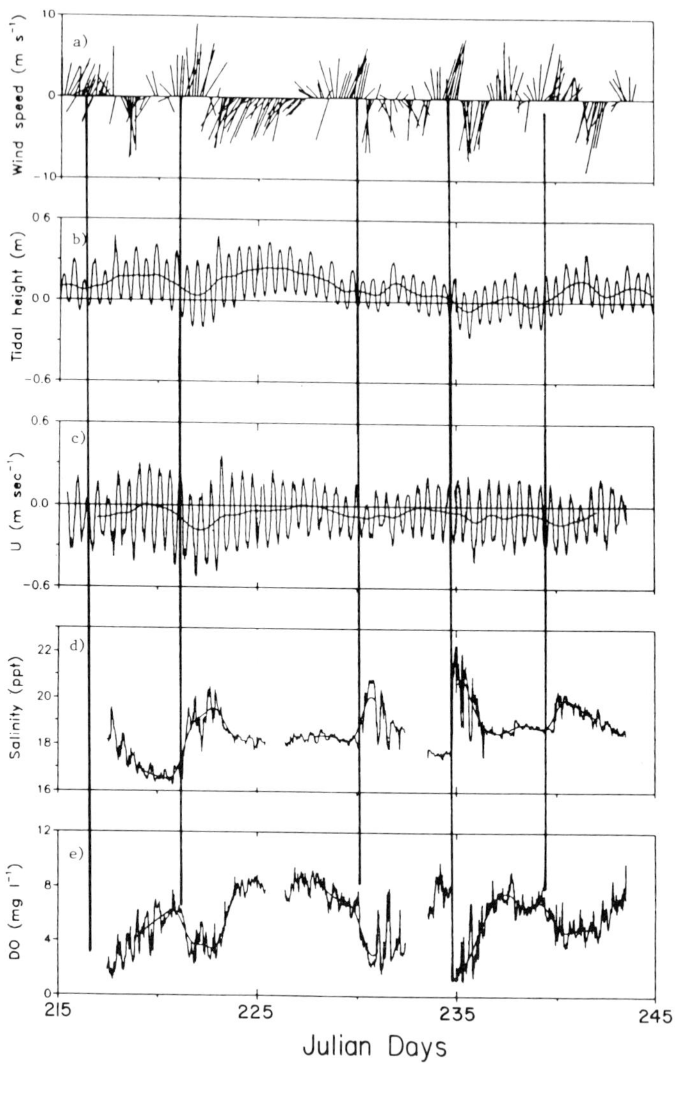

Figure 5: Time series data. a) Wind data at Norfolk Airport (y-axis points to the north), b) Filtered and unfiltered surface elevation at river mouth and c-e) Filtered and unfiltered time series data near the bottom at the river mouth (positive velocity is in ebb direction). The vertical lines indicate the events of high salinity, low DO water intrusion.

29.3 Data analyses and mass flux calculation

All time series data were subjected to a low-pass filter with a cut-off frequency of $(36\ \text{hr})^{-1}$. The filter was a modification of the low-pass filter designed by Godin (1972). The filtering process essentially eliminates the diurnal and semi-diurnal tidal constituents and fluctuations of higher frequencies. The resulting filtered series were considered as subtidal components. The difference between the measured time series and a subtidal component is considered as the tidal component for the following discussion. That is

$$q(t) = \langle q \rangle (t) + Q(t)$$

where

q = total component, or the measured time series data,
$\langle q \rangle$ = subtidal component, or the filtered time series data,
Q = tidal component,
t = time.

Fig. 5 also presents the subtidal components of surface elevation, longitudinal current, salinity and DO. The subtidal current (or Eulerian residual current) was directed into the subestuary along the bottom at all times (Fig. 5c). The average value over the period of measurement is -6.3 cm s^{-1}. The average values of the longitudinal velocity components at the upriver station were 2.9, -3.0 and -3.7 cm s^{-1} at depths of 1.2, 10.0 and 18.7 m, respectively. This agrees with the classical estuarine circulation of seaward flow in the upper layer and landward flow in the lower layer.

Since the hypoxic conditions exist only at the lower portion of the water column, both in the bay and subestuary, the influence of the bay on the subestuary may be investigated by considering the mass transport through the lower portion of the water volume at the subestuary mouth. Quantitative estimates of mass transport should be obtained by spatial integration of the product of velocity and concentration over the portion of the cross-section through which the transport is landward.

$$M(t) = -\iint_A u(y,z,t)\, c(y,z,t)\, dy\, dz$$

where

M = mass transport rate into the subestuary,
u = longitudinal velocity component (positive is out of subestuary),
c = concentration of dissolved substance,
A = the portion of the cross-section in which the transport is into the subestuary,
y, z = the spatial coordinates in the cross-section.

In this study, there are measurements of velocity, salinity and DO at only one point in the lower layer of the transect at the river mouth. No information on the spatial variability of mass transport may be extracted. However, the temporal variability may be investigated, at least qualitatively. The mass flux, or the mass transport per unit area, at the current meter location may be calculated as:

$$m(t) = -u(t)c(t)$$

where the concentration c may be salinity or DO (or dissolved oxygen deficit). If it is assumed that spatial distributions of u and c over the area A do not vary significantly with time, temporal variability of m should be the same as that of M.

The net flux of mass may be decomposed into two components, i.e.,

$$\overline{m} = -\overline{\langle u\rangle\langle c\rangle} - \overline{UC}$$

where the overbars designate the temporal average over the study period. The first term on the right hand side is the net transport by the subtidal component and the second term is the net transport by the tidal component. The data collected near the bottom of the river mouth were used to calculate the net mass fluxes of salt and DO, contributed by total, tidal and subtidal components respectively. The results are listed in Table 1, together with the average values of each component of longitudinal velocity, salinity and DO. The root mean square values of tidal current, and tidal components of salinity and DO fluctuation are also presented in Table 1. It shows that, even though the tidal current is much stronger than the average current, the salinity and DO fluctuations at tidal time scales are much smaller than their respective average values. Furthermore, Fig. 5 indicates that the salinity and DO fluctuations at time scales longer than tidal are much larger than those at tidal or shorter time scales.

Table 1. Average values of measured quantities and mass fluxes

	Total Component	Subtidal Component	Tidal Component	Root Mean Square of Tidal Component
(a) Average over 596 hours				
u (cm s^{-1})	-6.31	-6.28	-0.03	16.6
s (psu)	18.5	18.5	0.00	0.26
- u × s	119	118	1	-
(b) Average over 333 hours				
u (cm s^{-1})	-7.93	-7.90	0.03	17.4
DO (mg l^{-1})	5.63	5.61	0.02	0.72
- u × DO	37.9	41.1	-3.19	-

The net mass fluxes of salt and DO by the tidal component amount to only 1% and 8% respectively of those by the subtidal components (Table 1). Therefore, we need to consider only the subtidal transport in studying the characteristics of the estuary-subestuary exchanges of salt and DO in this case.

29.4 Temporal variability of mass exchanges

The subtidal current was variable with a dominant time scale of several days (Fig. 5c). Since the freshwater discharge into the Rappahannock River was fairly constant over the two months prior to and during the field measurements, the variability may be attributed to meteorological forcing, primarily the wind. The wind events not only exert variations in wind stress over the water surface, but also cause variations in salinity structure in the bay near the river mouth. The salinity structure in the bay may also vary in response to changes in discharge from the Susquehanna River, the major freshwater contributor of the bay. However, the time scale of variations resulting from this change should be much longer than several days. The magnitude, and its amplitude of variation, of the observed subtidal current are consistent with the values derived from the theoretical model of Hansen and Rattray (1965). They have derived an equation describing the vertical distribution of steady-state estuarine circulation. In the equation, the circulation consists of three modes: barotropic circulation forced by freshwater discharge, baroclinic circulation driven by density gradient and the wind-driven circulation. Based on the observed longitudinal salinity gradient, the model estimates that, under the condition of 5 m s^{-1} wind speed,

$$u_g / u_w = 0.32$$

where u_g and u_w are, respectively, the maximum density driven and wind-driven velocities in the lower layer of the water column. Application of the formula proposed by Hansen and Rattray (1966) gives

$$u_g = -3.3 \text{ cm s}^{-1}$$
$$u_w = -10.2 \text{ cm s}^{-1}$$

These values are consistent with the subtidal current presented in Fig. 5c.

Further evidence of wind-induced variability may be discerned by considering Figs. 5a, 5b and 5c. Increased landward (or westward) current near the river bottom (Fig. 5c) always occurred when a strong wind was blowing from the southwest quadrant. The wind drove surface water out of the river, and thus lowered the surface elevation (Fig. 5b) and caused a setup in the bay that drove the bottom water into the river.

There were characteristic differences in the properties of water masses transported into the river over different periods of time. On five occasions during the month of measurements, the incoming water masses were characterized by low DO and high salinity, typically 3 to 4 mg l^{-1} lower DO and 2 to 3 psu higher salinity than those of other periods. The import of low DO, high salinity water masses was always accompanied by increased landward current and southwest wind. Comparing Figs. 5d and 5e with Figs. 3 and 4, it is evident that the low DO, high salinity water masses were derived from the lower portion of the water column in the bay. In the absence of strong southwest wind, the shallow sill off the river mouth blocks the bottom water in the bay from entering the river, and therefore, the imported water is derived from the surface layer in the bay, with higher DO and lower salinity.

29.5 Mechanisms of wind induced mass exchange

As indicated in Fig. 5, the intrusion of the low DO, high salinity water into the Rappahannock River has a time scale of 2 to 3 days. This agrees with the time scale of the effect of local wind forcing in Chesapeake Bay. Wang and Elliott (1978) observed that the subtidal sea level fluctuations in Chesapeake Bay had spectral peaks at periods of 20, 5 and 2.5 days. Their analyses associated the 2.5 day fluctuations with the local wind forcing, while the lower frequency fluctuations were related to non-local wind forcing through the interaction with coastal sea. With local wind as forcing function, there are two possible mechanisms that contribute to the transport of the subpycnocline water of the bay into the Rappahannock River. For simplicity, we may look at the wind-driven circulation around the river mouth in two-dimensional space, one in the horizontal plane and the other in the vertical plane.

29.5.1 Horizontal plane

Fig. 6 depicts the forcing of bottom current around the river mouth. The left panel is for the wind from the southwest. Surface set-up in two transects across the bay is considered (Fig. 6b). The surface current in both transects is toward the east because of direct wind stress and Coriolis effect. The boundary on the east causes the water surface to rise and a return westward flow along the bottom. There is a difference in the degree of surface set-down on the western boundary between sections AA and BB. Owing to the surface out-flow from the river, the set-down at section AA will be less than that at section BB. This results in a barotropic flow away from the river mouth along the western boundary of the bay (Fig. 6c). Therefore, the bottom water transported into the river is mainly derived from the main stem of the bay where hypoxic conditions exist in the summer. On the other hand, if wind is blowing from the northeast quadrant the surface set-up at the river mouth will be less than that at section BB. This results in an along shore barotropic flow toward the river mouth. If the density-driven gravitational circulation at the bottom of the river mouth is stronger than the wind-driven current, the bay bottom water will still be transported into the river. However, this bottom water is derived from near shore regions to the north and south of the river mouth. Because of the shallow depths in these regions, the imported water mass has high DO and low salinity.

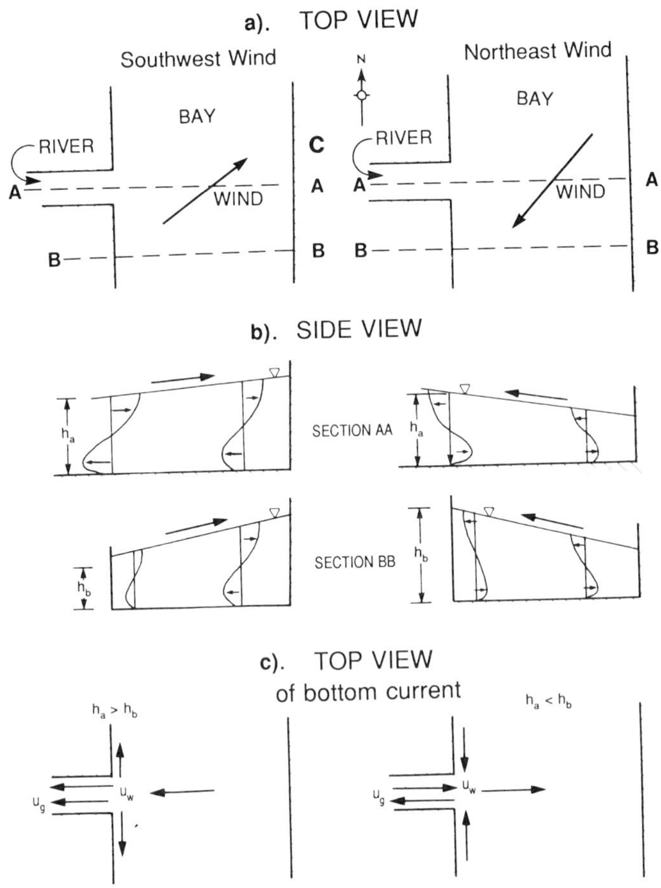

Figure 6: Directions of wind-driven currents around an estuary-subestuary junction.

29.5.2 Vertical plane

It was observed that the bay water is highly stratified, both in salinity and DO in summer months. Fig. 7 depicts the water surface and pycnocline in a transect across the bay (Section AA of Fig. 6). In the absence of wind, the pycnocline is nearly horizontal and is located at about the same height as the sill near the river entrance. Most of the water transported into the river by gravitational circulation is derived from the bay at a height above the pycnocline. In the time of strong southwest wind (Fig. 7b), the surface set-up and westward return flow near the bottom force the pycnocline to tilt upward on the western side of the bay. If the tilt is large enough to raise the pycnocline above the sill, it not only forces the subpycnocline water into the river, but also strengthens the bottom flow into the river. Therefore, strong inflow of low DO, high salinity water was observed (Fig. 5). Similar argument has been proposed to explain the observed intrusion of hypoxic water onto the shallow flanks of the bay (Tyler, 1984; Malone et al., 1986). On the other hand, if strong wind is blowing from the northeast, the pycnocline will be depressed below the sill on the western side of the bay (Fig. 7c). The water transported into the river will be that above the pycnocline, thus with high DO, low salinity.

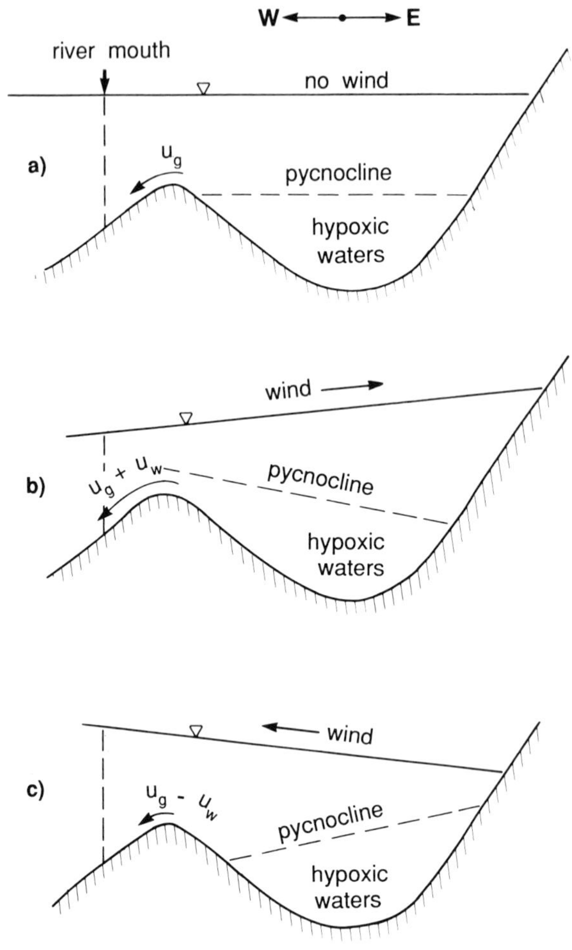

Figure 7: The tilting of pycnocline in a cross-bay section in response to wind.

29.6 Discussion

In this study, we have focused on the import of waters into the subestuary through the lower layer of the water column. Analyses of observed data indicate that transports of water, DO and salinity by subtidal components are an order of magnitude larger than those by tidal components of water movements. Since we have measurements at only one point in the water column no quantitative estimate of the amount of mass transport may be made. However, the temporal variability of mass transport may be assessed assuming similarity in spatial distribution over time.

Two distinct characteristic water masses were observed being transported into the subestuary, the low DO, high salinity subpycnocline water and the high DO, low salinity surface water. Through simultaneous comparison of wind data with surface elevation, subtidal current, salinity and DO, we can conclude that wind is the dominant factor influencing the properties of imported waters. The hypoxic water was transported into the subestuary as episodic events when strong wind was blowing from the southwest. During these periods, the pycnocline in the western side of the bay rose above the sill at the river mouth and subpycnocline water in the bay 'rushed' into the river along the river bottom. The landward flow of gravitational circulation was further enhanced by the wind-driven current and increased longitudinal salinity gradient.

Other studies also demonstrated the temporal variabilities of bottom-water intrusions over the entrance sills of estuaries or subestuaries. Cannon et al. (1990) reported that the onset of the intrusions into Puget Sound were associated with the combined effects of neap tides and the horizontal density gradient across the sill. The regular spring-neap cycle was the result of mixing effects over the sill; however, the episodic occurrence of bottom-water intrusion in winter was attributed to wind events on the Pacific Coast. Other causes have been proposed for the variability of the density gradient across the sills of fjords (e.g., Gade and Edward, 1980). These include seasonal variation in density and upwelling outside of the entrance sill. For a fjord estuary, bottom-water intrusion is considered to play a dominant role in the replacement of water below sill depth, and the flushing of contaminants. On the other hand, the intrusions of bottom water from Chesapeake Bay into its subestuaries could mean the intrusion of hypoxic or anoxic water in the summer season, adversely affecting the water quality in subestuaries. Sanford (1988) also reported that intrusion of lower-layer water from the bay into the Choptank River occurred as an episodic, wind-driven internal surge rather than as a slow, steady flow. Since the Choptank River is on the eastern side of Chesapeake Bay, the intrusion events occurred when strong wind was blowing from the northeast.

The episodic nature of subpycnocline-water intrusions often results in the reversal of the longitudinal salinity gradient in the lowest reach of the subestuary. This would pose some difficulties in specifying boundary conditions for numerical modelling of subestuaries. A priori knowledge of conditions in the main estuary is required, or the model domain would need to be extended to include portions of the main estuary.

Acknowledgements

This study was funded by the Commonwealth of Virginia as part of Chesapeake Bay Initiatives. The authors wish to express their appreciation to Mr. K. Curling for providing the data of the main stem of the bay, collected under the US EPA Chesapeake Bay Program.

29.7 References

Cannon, G. A., J. R. Holbrook, and D. J. Pashinski, 1990: Variations in the onset of bottom-water intrusions over the entrance sill of a fjord. Estuaries, 13(1), 31-42.

Elliott, A. J., 1978: Observation of the meteorologically induced circulation in the Potomac estuary. Estuarine Coastal Mar. Sci., 6, 285-299.

Flemer, D. A., G. B. Mackierman, W. Nehlsen, and V. K. Tippie, 1983: Chesapeake Bay: a profile of environmental change. U.S. Environmental Protection Agency, Washington, D.C., 199 pp.

Gade, H. G., and A. Edwards, 1980: Deep-water renewal in fjords. In H. Freeland, D. Farmer, and C. Levings (eds.), Fjord Oceanography, Plenum Press, New York, 453-489.

Godin, G., 1972: The analysis of tides. University of Toronto Press, Canada, 264 pp.

Hansen, D. V., and M. Rattray, Jr., 1965: Gravitational circulation in estuaries. J. Mar. Res., 23, 104-122.

Hansen, D. V., and M. Rattray, Jr., 1966: New dimensions in estuary classification. Limnol. Oceanogr., 11, 319-326.

Kuo, A. Y., and B. J. Neilson, 1987: Hypoxia and salinity in Virginia estuaries. Estuaries, 10(4), 277-283.

Malone, T. C., W. M. Kemp, H. W. Ducklow, W. R. Boynton, J. H. Tuttle, and R. B. Jonas, 1986: Lateral variation in the production and fate of phytoplankton in a partially stratified estuary. Mar. Ecol. Prog. Ser., 32, 149-160.

Newcombe, C. L., and W. A. Horne, 1938: Oxygen poor waters of the Chesapeake Bay. Science, 88, 80-81.

Sanford, L. P., 1988: Bathymetric control of tributary-mainstem interaction: the case of the Choptank River. EOS, Trans. Am. Geophys. Union, 69, 1256 (abstract).

Schroeder, W. W. (convener), 1985: The occurrence of hypoxic and anoxic conditions in estuaries and coastal environments. Abstract for the Eighth Biennial Estuarine Research Conference, Estuaries, 8(2B), 43A-47A.

Tyler, M. A., 1984: Dye tracing of a subsurface chlorophyll maximum of a red-tide dinoflagellate to surface frontal regions. Mar. Biology, 78, 285-300.

United Nations Educational, Scientific and Cultural Organization, 1983: Algorithms for computation of fundamental properties of seawater. UNESCO Technical Papers in Marine Sci., 44, Paris, France, 53 pp.

Wang, D. P., A. J. Elliott, 1978: Non-tidal variability in the Chesapeake Bay and Potomac River: evidence for non-local forcing. J. Phys. Oceanogr., 8, 225-232.

30

Interdisciplinary study on the tidal front in the Bungo Channel, Japan

T Yanagi, O Matsuda, S Tanabe and S Uye

Abstract

An interdisciplinary study was carried out around the tidal front in the Bungo Channel, Japan. Temperature and salinity differences across the tidal front are 3°C and 0.3, respectively. Nutrients are richer in the well-mixed region than in the surface water of stratified region where they have been consumed by phytoplankton. Man-made organic chemicals such as PCBs are concentrated in the surface water just on the front and they are accumulated in the plankton which are gathered there. Neuston such as marine skaters and *Vellela* lata are accumulated on the front. The tidal front plays as a barrier for some oceanic and neritic pontellid copepods.

30.1 Introduction

The remarkable tidal front is developed from May to August in the northern part of the Bungo Channel, the Seto Inland Sea, Japan (Yanagi and Koike, 1987). The detailed physical structure of this tidal front is already revealed and its temporal and spatial variations are well explained by the variable efficiency model (Yanagi and Tamaru, 1990).

However, the chemical and biological characteristics of this tidal front have not been clarified yet. Hence an interdisciplinary study was carried out in collaboration with physical, chemical and biological oceanographers.

30.2 Observations

The intensive and comprehensive field observations of the tidal front at the northern part of the Bungo Channel were carried out on board T & R/V Toyoshio-Maru of Hiroshima University from 16 to 19 June 1986, from 22 to 24 May 1987 and on 21 May 1988. The bottom topography and the map of the observation site are shown in Fig.1.

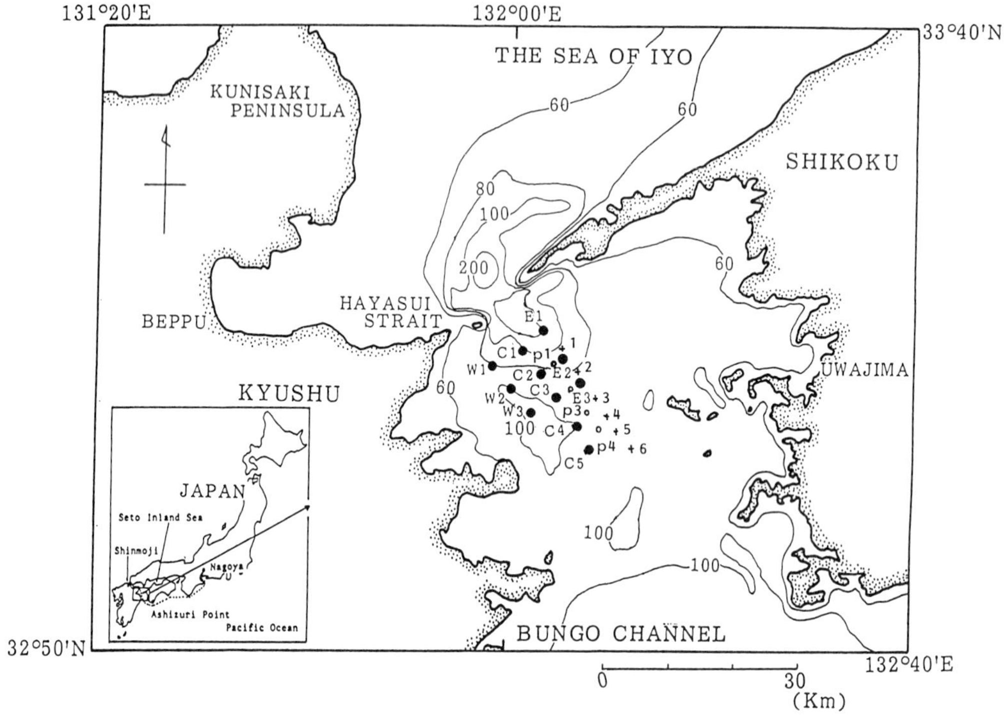

Figure 1: Observation stations in the Bungo Channel. Numbered contours show the depth in metres.

Vertical profiles of temperature and salinity were measured at 11 stations along three lines (W1-W3, C1-C5 and E1-E3) from 16 to 19 June 1986, at 4 stations along one line (P1-P4) from 22 to 24 May 1987 and at 6 stations along another line (1-6) on 21 May 1988 by using a CTD (ALEC Model AS-100). Water and plankton samples were collected at these stations with use of a pair of Van Dorn water bottles and plankton net, respectively. Water temperature, salinity and chlorophyll *a* were measured continuously at a depth of ca. 3 m along the above observation lines (water was pumped up from 3 m depth) using a multiple environmental senser (Martec Mark VI) and a fluorometer (Turner Designs) in order to detect the exact position and sharpness of the front.

30.3 Physical aspects

Water temperature and salinity abruptly changed from 18C and 33.8 to 22C and 34.3, respectively, across the frontal region with a width of ca. 500 m on 18 June 1986. Density also changed from 24.5 to 24.0 in sigma t across the tidal front (Fig.2).

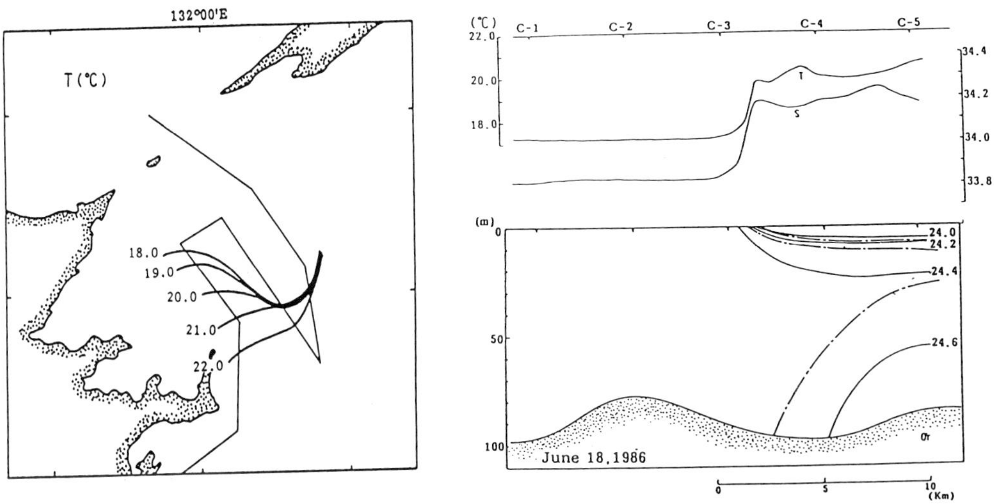

Figure 2: Horizontal distribution of water temperature ca. 3 m below the sea surface on 18 June 1986. Thin line shows the ship cruise track (left). Horizontal variations of water temperature, salinity and density 3 m below the sea surface and vertical distribution of density along the central observation line on 18 June 1986 (right).

The water in the northern part of this front was well mixed vertically due to the strong tidal current near Hayasui Strait while that in the southern part was stratified due to weak tidal current. The transparency (secchi disc depth) was 14 m in the northern part of the front and it was 16 m in the southern part.

The position of the front was noticed easily from the research vessel, as the front was accompanied by a narrow band of foam and various kinds of drifting material on the sea surface. Such accumulation of drifting materials suggests the existence of surface convergence at the tidal front. This basic physical structure was maintained consistently throughout our observation periods with only minor spatial changes according to tidal phase. The detailed physical characteristics of this tidal front are described by Yanagi and Tamaru (1990).

30.4 Chemical aspects

Representative vertical distributions of DIP (*D*issolved *I*norganic *P*hosphorus) and PP (*P*articulate *P*hosphorus) on 24 May 1987 are shown in Fig.3. The front existed exactly at Sta.P3, while Stas P1 and P2 were situated in the well-mixed region and Sta. P4 in the stratified region (Matsuda, 1990). The concentration of DIP is more important for the growth of phytoplankton than that of DIN because phosphate is a limiting factor in the Seto Inland Sea (Joh, 1986). DIP was about 0.3 µg-at/l in the well-mixed region and it abruptly decreased across the tidal front to about 0.15 µg-at/l in the surface layer of the stratified region. On the other hand, PP was low (about 0.1 µg-at/1) in both sides of the tidal front but high (about 1.15 µg-at/l) at the tidal front. This is associated with higher phytoplankton biomass as evidenced by higher concentrations of chlorophyll *a* and pheo-pigments just at the tidal front (Fig.4). The dissolved oxygen concentration was high in the surface layer at the tidal front (Fig.5). These facts indicates that the activity of phytoplankton is high at the tidal front.

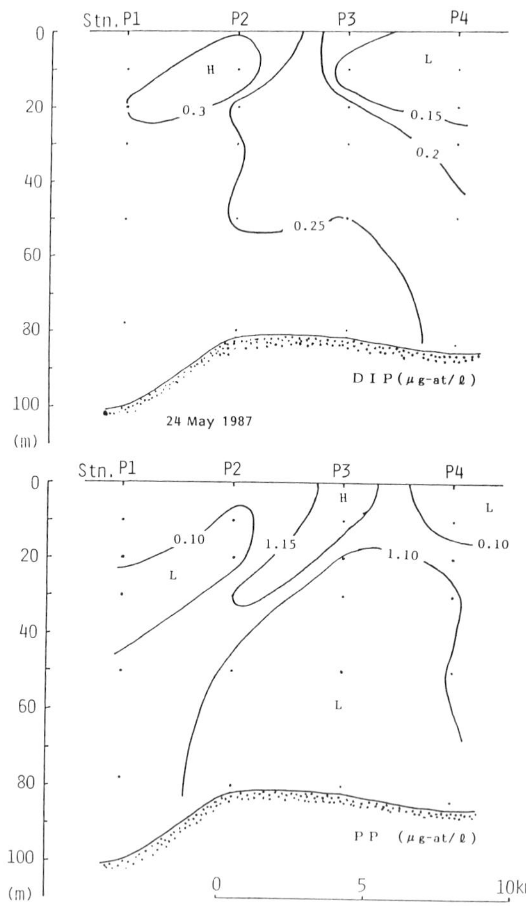

Figure 3: Vertical distributions of DIP and PP from Sta.P1 to Sta.P4 on 24 May 1987. Tidal front exists on Sta.P3.

Continuous observations of water temperature, salinity, chlorophyll *a* and beam transmittance of the surface water were carried out between Stations 1 and 6 on 21 May 1987 (Fig.6). On this occasion, chlorophyll *a* concentration was high and beam transmittance was low at the tidal front.

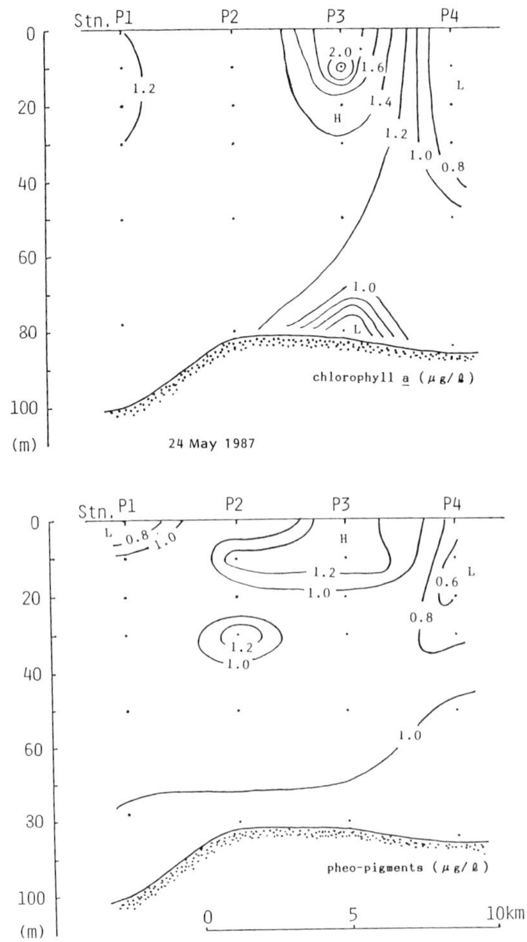

Figure 4: Vertical distribution of chlorophyll *a* and pheo-pigments from Sta.P1 to Sta.P4 on 24 May 1987. Tidal front exists on Sta.P3.

The reasons for the high concentration of chlorophyll *a* at the tidal front may be explained as follows. Phytoplankton can not grow successfully in both sides of the front. Because they can not stay in the shallow euphotic zone due to strong vertical mixing in the well mixed region, and nutrients are lacking in the deep euphotic zone of stratified region. On the other hand, phytoplankton can grow successfully near the front where the extent of vertical mixing and the supply of nutrients are moderate.

Table 1: Concentration of DDE and PCBs in plankton collected from the surface sea water in the northern part of the Bungo Channel (Tanabe et al., 1990).

Concentrations of p,p'-DDE and PCBs in plankton collected from surface seawater of Bungo Channel.

Sampling station	p,p'-DDE concentration (ng/g on dry wt. basis)	PCBs concentration (ng/g on dry wt. basis)
C-1	< 1.2	9.9
E-2	5.3	26.6
C-5	< 1.2	5.4

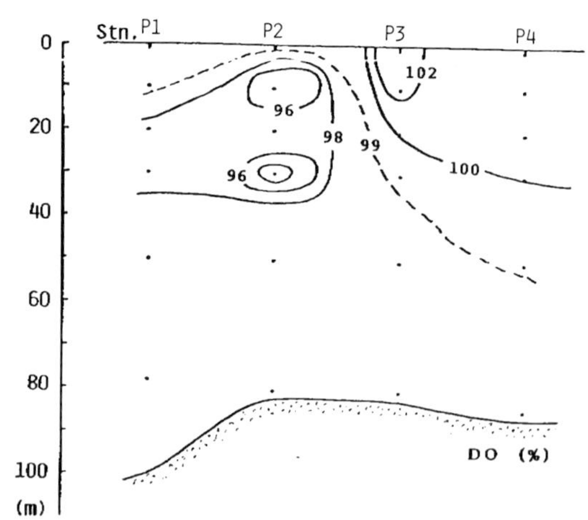

Figure 5: Vertical distribution of dissolved oxygen from Sta.P1 to Sta.P4 on 24 May 1987. Tidal front exists on Sta.P3.

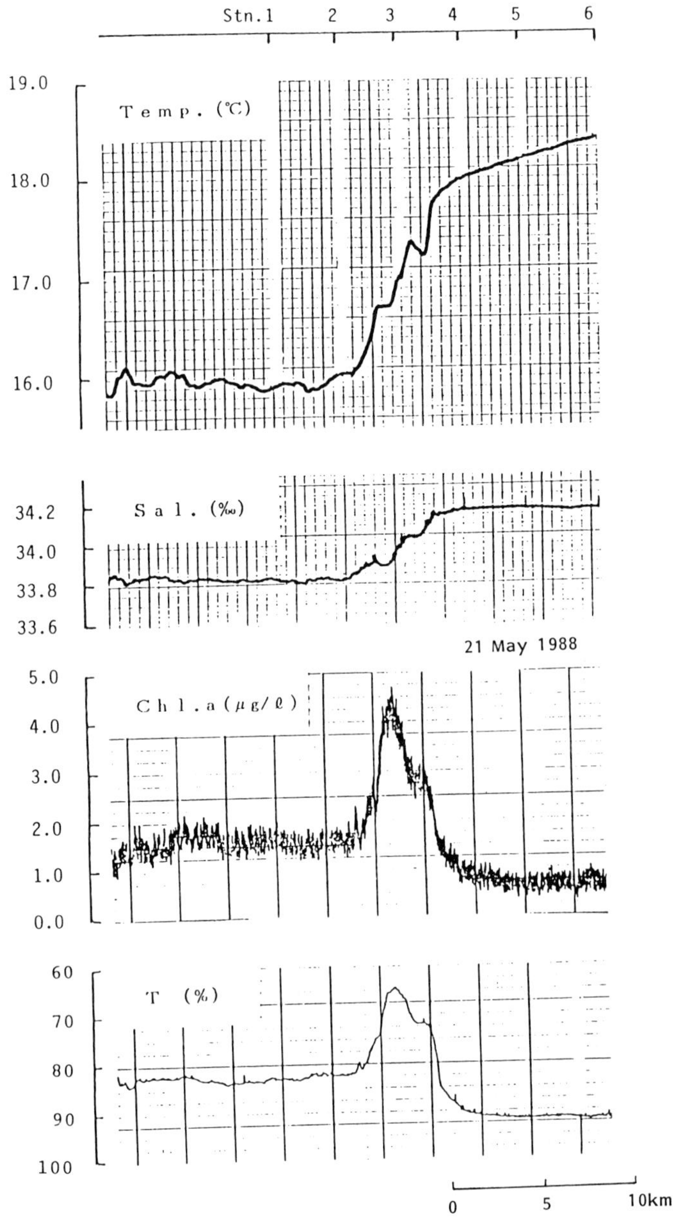

Figure 6: Horizontal variations of water temperature, salinity, chlorophyll *a* and turbidity (reciprocal of beam transmittance) ca. 3 m below the sea surface from Sta.1 to Sta.6 on 21 May 1988.

The concentrations of HCHs (Hexachlorocyclohexanes) were high in the surface layer of the tidal front as shown in Fig.7 (Tanabe et al., 1990). The difference of 0.2 ng/l in Fig. 6 is significant because variation in reproducibility is less than 10%. Our results indicate that the tidal front accumulates oil and particles in the surface layer due to the surface convergence. The surface convergence at the tidal front was not observed directly but the accumulation of foams and many kinds of drifting matter suggested its existence. The physicochemical properties of HCHs such as lipophilic and high particle affinity enables the enrichment of this contaminant at the tidal front. The concentrations of DDE and PCB in the plankton of the surface layer around the tidal front were examined and the results are shown in Table 1. The concentrations of DDE and PCB in the plankton were low in both sides of the front, Stas C-1 and C-5, but high at Sta.E-2 on the tidal front. These results also suggest that the man-made organics such as PCB and DDE which are enriched in the surface layer of the tidal front are bioaccumulated in the plankton.

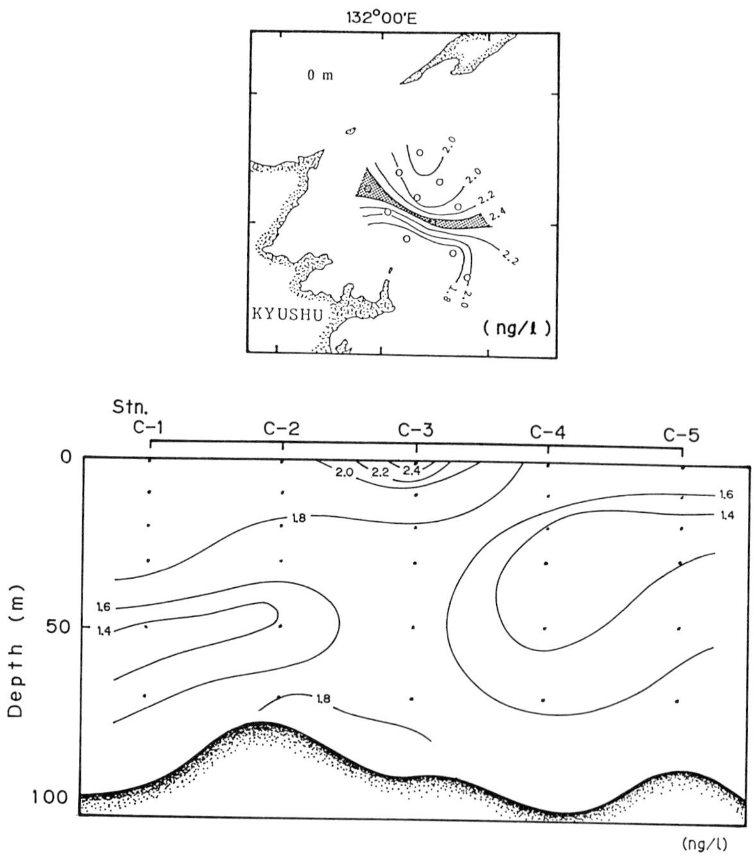

Figure 7. Horizontal distribution of HCH concentrations in the surface layer on 18 June 1986 (upper). Vertical distribution of HCH along the central observation line on 18 June 1986 (lower).

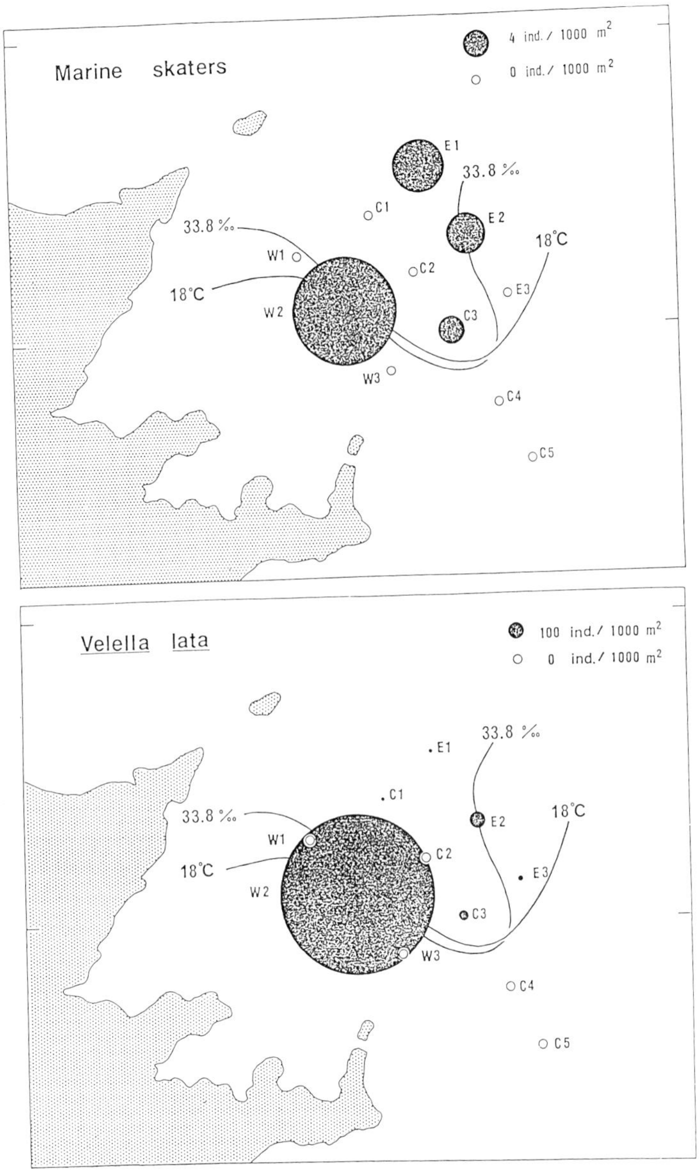

Figure 8: Horizontal distribution of marine skaters and *Velella lata* in the surface layer on 18 June 1986.

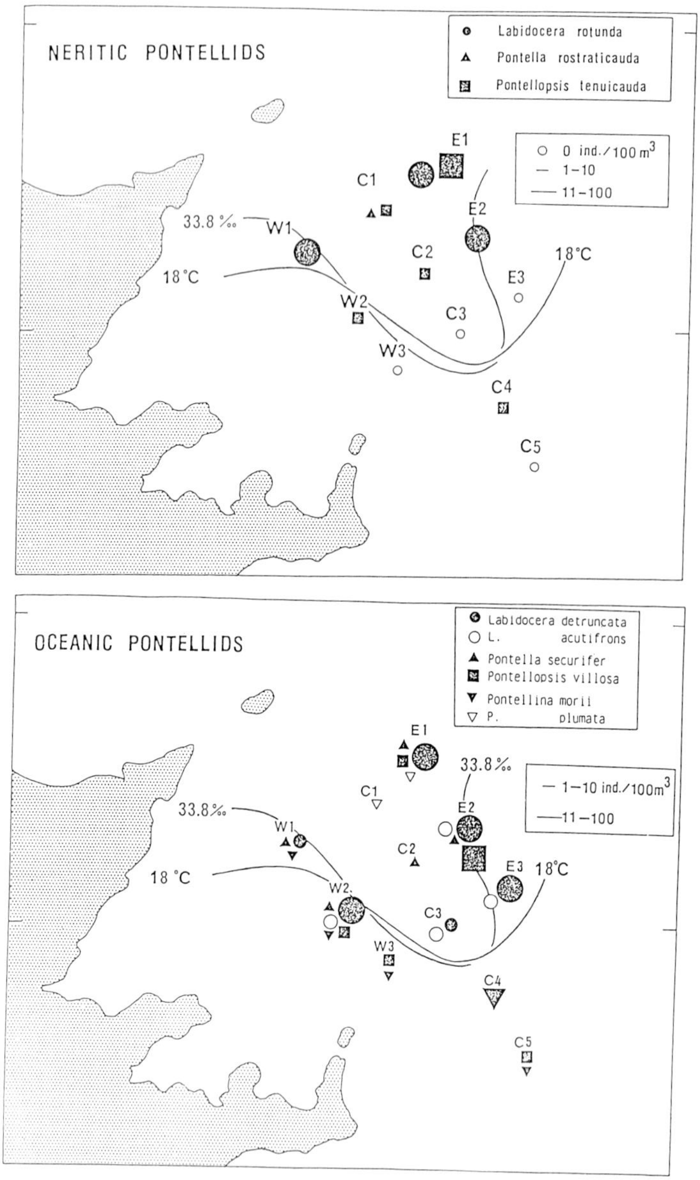

Figure 9: Horizontal distribution of neritic and oceanic *Pontellid* copepods in the surface layer on 18 June 1986.

30.5 Biological aspects

Epineuston (which lives on the sea surface) such as marine skaters whose species is not identified and pleuston (whose body exists across the sea surface) such as *Velella lata* are accumulated around the tidal front as shown in Fig.8 (Uye et al., 1990).

Both marine skaters and *Velella lata* are warm water species and they are considered to be advected from the Pacific Ocean and accumulated around the tidal front in the Bungo Channel due to the surface convergence. Although hyponeuston (which lives just below the sea surface) such as pontellid copepods were not accumulated around the tidal front, neritic pontellids were observed mainly in the northern part of the front and oceanic pontellids in the southern part of the front as shown in Fig.9. Such results mean that the tidal front plays a role of barrier of the distribution of hyponeuston.

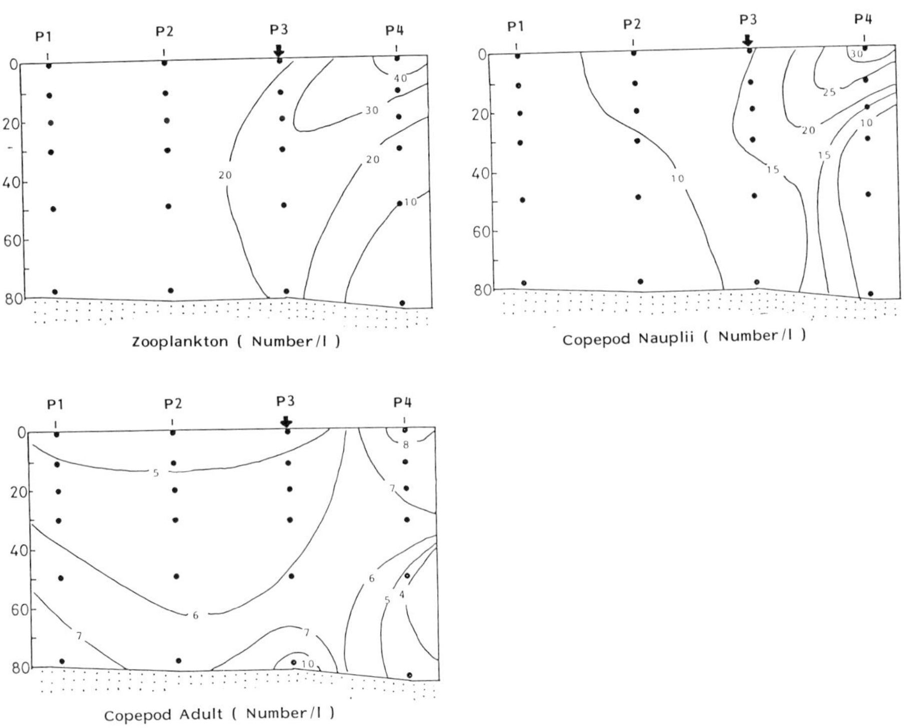

Figure 10: Vertical distribution of zooplankton, copepod nauplii and copepod adults from Sta.P1 to Sta.P4 on 24 May 1987. Tidal front exists on Sta.P3 which is shown by an arrow.

Figure 10 shows the vertical distributions of total zooplankton, copepod nauplii and adult copepods across the tidal front on 24 May 1987. Their population densities were high in the surface layer of the stratified region.

Egg production experiment of a copepod, *Palacalanus* sp. was carried out with use of sampled surface water around the frontal region and its result is shown in Fig.11. Egg production rate is positively related to chlorophyll *a* concentration. The same result is obtained at the frontal zone in the Northwest North Sea (Kiorboe et al., 1988). Egg production rate of a copepod, *Palacalanus* sp., is considered to be high on the tidal front where chlorophyll *a* concentration is high as shown in Fig.4. However, the concentration of copepods (nauplii and adults) was not necessarily high on the front. Further, Uye et al. (1990) demonstrates that the density of copepod nauplii is lower at the front than in both sides of the front. Such discrepancy suggests that copepod nauplii which might be produced at high rate at the front are intensively predated by carnivores such as *Velle lata* and jelly fish which are accumulated in the surface layer of the front.

Similar observational results of high concentration of chlorophyll *a* and low population density of zooplankton at the tidal front were obtained in the English Channel (Holligan et al., 1984) and in the Northwest North Sea (Kiorboe et al., 1988).

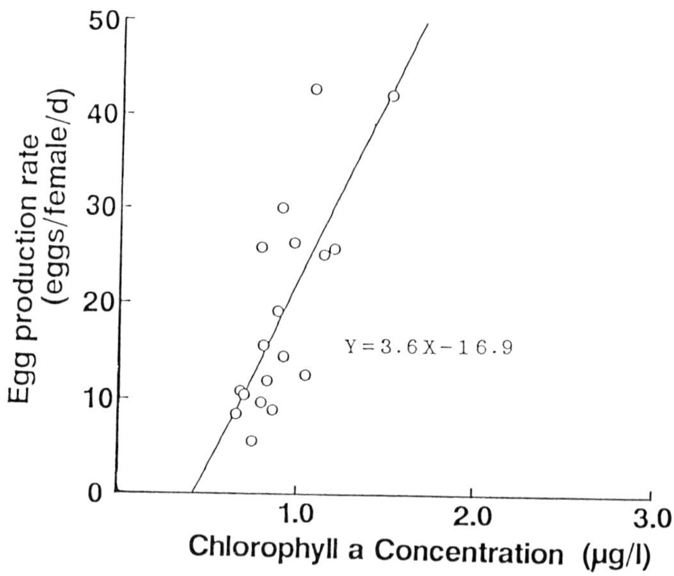

Figure 11. Relationship between egg production rate of the copepod *Paracalanus* sp. and chlorophyll *a* concentration.

30.6 Discussion

The tidal front is generated in a transition zone from the well-mixed region to the stratified one. Nutrients are rich in the well-mixed region and poor in the surface water of the stratified region. On the other hand, the biomass of phytoplankton is high at the edge of the stratified region because the vertical mixing is too strong in the well-mixed region for phytoplankton growth. The oleophilic chemicals are concentrated along the tidal front and bioaccumulated in the plankton which are also gathered along the tidal front due to the surface convergence.

Though the tidal front is a physical structure, it also affects the distribution of chemical compounds and planktonic organisms as schematically shown in Fig.12. It acts as a barrier for some zooplankton, i.e. oceanic species are found only in the southern part of the front and neritic species in the northern part. At the same time, the tidal front concentrates the particulate materials including living plankton along the frontal region in the surface layer where the chemical material are accumulated in the organisms of higher trophic level through the food chains.

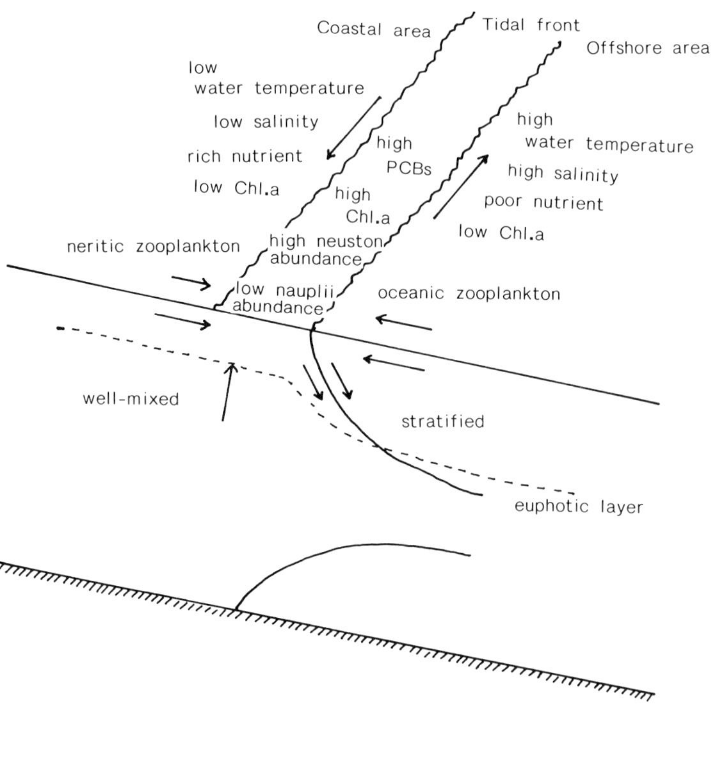

Figure 12. Schematic representation of the physical, chemical and biological structures of the tidal front.

The more intensive study on the higher trophic organisms such as jelly fish and fish around the tidal front will be necessary in the future. Though the physical structure of tidal front is consistent from late spring till late summer, its biological role seasonally changes. Therefore, the seasonal variation in the biological role of the tidal front needs to be revealed in order to clarify the ecological meaning of the tidal front in the coastal sea.

Acknowledgement

The authors express their sincere thanks to Prof.T.Onbe and Mr.S.Ohtsuka of Hiroshima University for their comments on zooplankton distribution and the crew of T & R/V Toyoshio-Maru for their kind cooperation during the field observations.

30.7 References

Holligan, P.M., R.P. Harris, R.C. Newell, D.S. Harbour, R.N. Head, E.A.S. Linley, M.I. Lucas, P.R.G. Tranter and C.M. Weekley (1984): Vertical distribution and partitioning of organic carbon in mixed, frontal and stratified waters of the English Channel. Mar.Ecol.Prog.Ser., 14, 111-127.

Joh, H. (1986): Studies on the mechanism of eutrophication and the effect of it on fisheries production in Osaka Bay. Bull. Osaka Prefectural Fisheries Experimental Station, 7, 1-174.

Kiorboe, T., P. Munk, K. Richardson, V. Christensen and H. Paulsen (1988): Plankton dynamics and larval herring growth, drift and survival in a frontal area. Mar.Ecol.Progr.Ser., 44, 205-219.

Matsuda, O. (1990): Chemical process around the tidal front. In "The Science of Siome" ed.by T.Yanagi, Koseisha-Koseikaku, Tokyo, 37-50.

Tanabe, S., S. Hanaoka, T. Yanagi, H. Takeoka and R. Tatsukawa (1990): Persistent organochlorines in the coastal front. Mar. Pollut. Bull., (in press)

Uye, S., T. Ohtsuka and T. Onbe (1990): Biological process around the tidal front. In "The Science of Siome" ed. by T. Yanagi, Koseisha-Koseikaku, Tokyo, 78-100.

Yanagi, T. and T. Koike (1987): Seasonal variation in thermohaline and tidal fronts, Seto Inland Sea, Japan. Continental Shelf Research, 7, 149-160.

Yanagi, T. and H. Tamaru (1990): Temporal and spatial variations of a tidal front. Continental Shelf Research, 10, 615-621.

31

Hydrodynamic modelling for a tidal power project

T L Shaw

Abstract

This paper reviews the programme of hydrodynamic studies carried out to date in connection with the development of the proposed Severn Barrage tidal power scheme. It also outlines the further work considered to be necessary before a decision to construct the barrage could be taken.

The tides and hence the processes which relate to these are fundamental to the performance of the project, both as an energy producer and in respect of the acceptability of its environmental effects. For example, salinity variations and the consequences of sediment transport determine much about the regime which the barrage would affect and create. The work done to date on these subjects is summarised.

The accuracy required of predictions and the availability of methods to achieve this are considered.

31.1 Introduction

The recent decisions of the UK Government to curtail, at least temporarily, the development of nuclear power stations and to expand support for renewable energy schemes have given added impetus to the possibility of exploiting wind, wave and tidal sources for commercial purposes. Each of these sources is capable of supplying more than 10% of present UK electricity demand, hence is regarded as potentially significant in national terms. However, only tidal power has a long track record and is capable of development in large units; the 'small is beautiful' concept is better suited to power production at a local rather than national scale, when it is organisationally cumbersome and economically dubious to promote.

The tides of the Severn Estuary are the highest in Europe and the cross-sectional profile of this estuary is generally well suited for power generation from this source in terms of the depth and width required to accommodate the number and size of turbines and other equipment required to harness the available energy. Either excessive or insufficient depth and/or width can introduce serious capital cost penalties to this type of project. Fig. 1 shows the location proposed for the Severn Barrage (D.En. et al, 1989); this is towards the seaward end of what is known as the Severn Estuary but which, as we shall see, has features which make much of it more akin to an open coastline than to a river estuary.

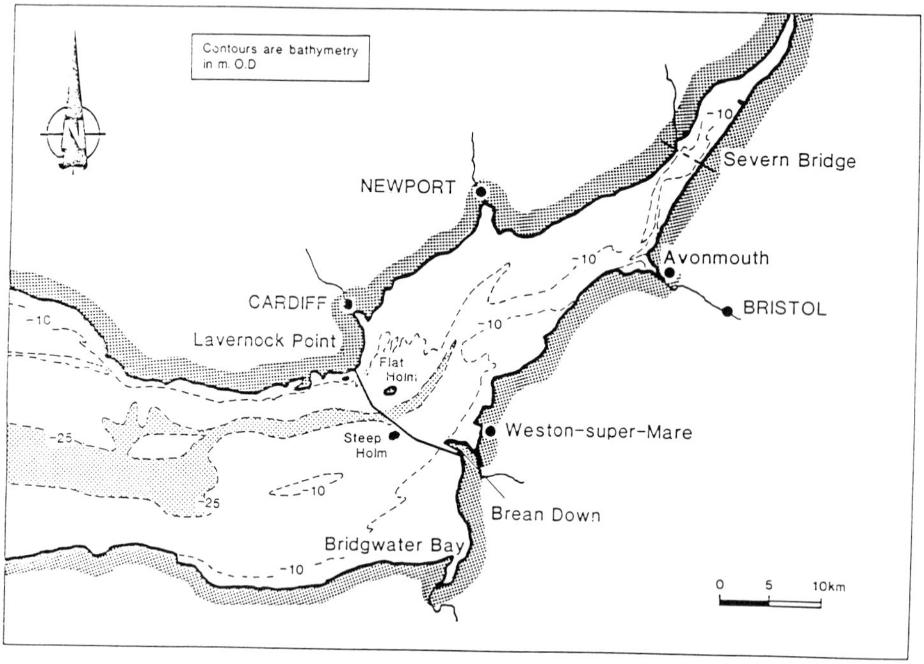

Figure 1: Barrage location

A barrage in this location would produce some 7-8% of the present electricity demand of England and Wales, equivalent to the output of 2-3 major thermal power stations. This would make it some ten times larger in size than the proposed Mersey Barrage, thirty times larger than the Rance Barrage in Brittany commissioned in 1966, and more than a hundred times larger than most possible UK tidal power schemes (Binnie and Ptnrs, 1989). Although tide range is a factor in these comparisons, the differences from site to site between these are small compared with the basin areas which each scheme would create, ie., with the live water volumes available to each set of turbines (volumes which largely determine the appropriate installed generating capacities). The basin formed by the Severn Barrage would have a high water surface area of over 500km^2.

This paper outlines the work done to determine the hydrodynamic conditions in which this barrage scheme would have to be constructed and in which it would operate. The large size of the estuary means that wave climate becomes an important consideration, in addition to tide range and associated currents.

31.2 Tide regime

Earlier studies of the proposed barrage (Severn Barrage Committee, 1981) made use of 2-D numerical models of the Severn Estuary and Bristol Channel, one of these models extending on to the Continental Shelf (IOS, 1980/1). These models confirmed the importance of good representation of coastal bathymetry and estuarine detail, also the seaward tide regime so that the full implications of the barrage and its method of operation were properly simulated.

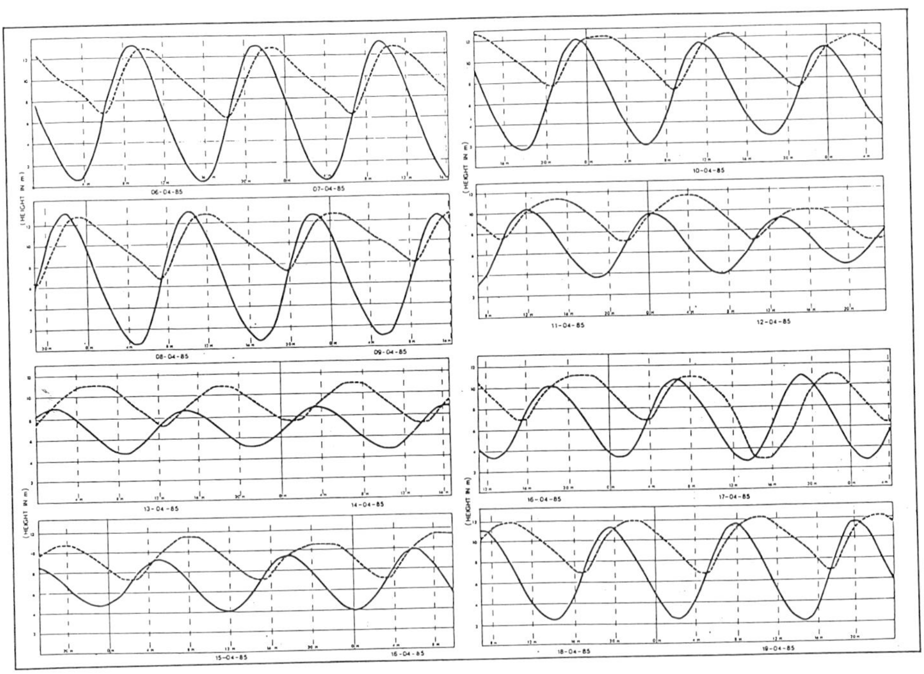

Figure 2: Representative tide and estuary water level curves - Rance Estuary

This information allowed a more targetted 2-D plan model to be developed for the recent study of this barrage (WRc, 1989a). The aim of this work related closely to the investigation of a range of issues not hitherto studied in detail, for example tidal flow patterns and their consequences during barrage construction, variations of salinity and the dispersion of effluents.

The study was less concerned with refining the details of the work done previously. The decision to proceed in this way was taken because of the need to improve awareness of a number of issues of particular environmental importance, not least because of the public concern now attached to this aspect of all new projects. It was also recognised that attempts to improve the accuracy of tide modelling could not sensibly progress until barrage operating procedures had been specified, in particular use of the turbine equipment in reverse as pumps around the time of high tide to raise the level of the basin above that which gravity refilling alone would achieve. Fig. 2 shows this operation as it is practised at the Rance Barrage in Brittany; the amount of pumping done clearly relates to tide range, being least on spring tides and more when the range is less than average.

Fig 3 shows the four levels of modelling definition used, these being selected to suit the requirements for information in each part of the estuary. The performance of the model was checked against simultaneous tide gauge data recorded at 24 stations. Fig 4 shows a selection of the comparisions achieved by the model when it had been developed to the form accepted for use throughout the investigation. The values of the bed roughness parameter varied over the whole estuary and with tide range by factors of 3-4, as would be expected (WRc, 1989a).

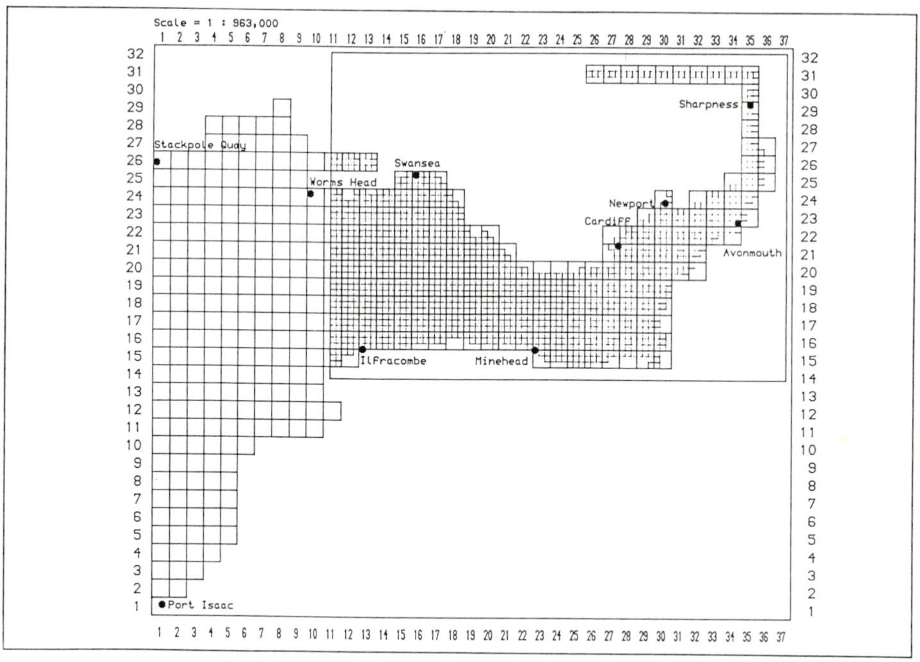

Figure 3a: 2-D tide model grids 4500m + 1500m

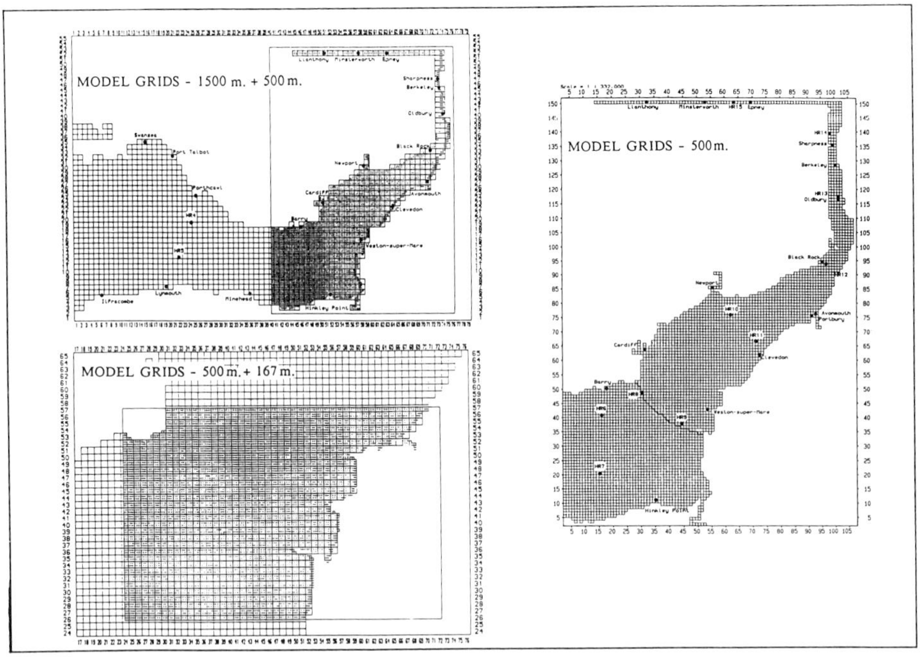

Figure 3b: 2-D tide model grids 1500m + 500m + 167m

It was recognised that the model assumed unchanged and unchanging bathymetry, not only over the period since the most recent surveys of each part of the estuary were done but also in respect of the consequences of the barrage for the distribution of sediment deposits. The model was therefore used to determine the extent to which local alterations in depths, whether by accretion or erosion, would significantly affect any aspect of the tide range. No discernible changes were noted within the considerable range of depth changes postulated to occur within the main Severn Estuary, though the effects of expected accretion in the river estuaries could be significant.

Another uncertain hydrodynamic factor necessarily specified in the model is the rate of dispersion of momentum. From a range of typically accepted values, a suitable number was chosen for use in all conditions, including those simulating the regime during barrage construction and operation. The sensitivity of the chosen value to the conditions simulated was acceptably small.

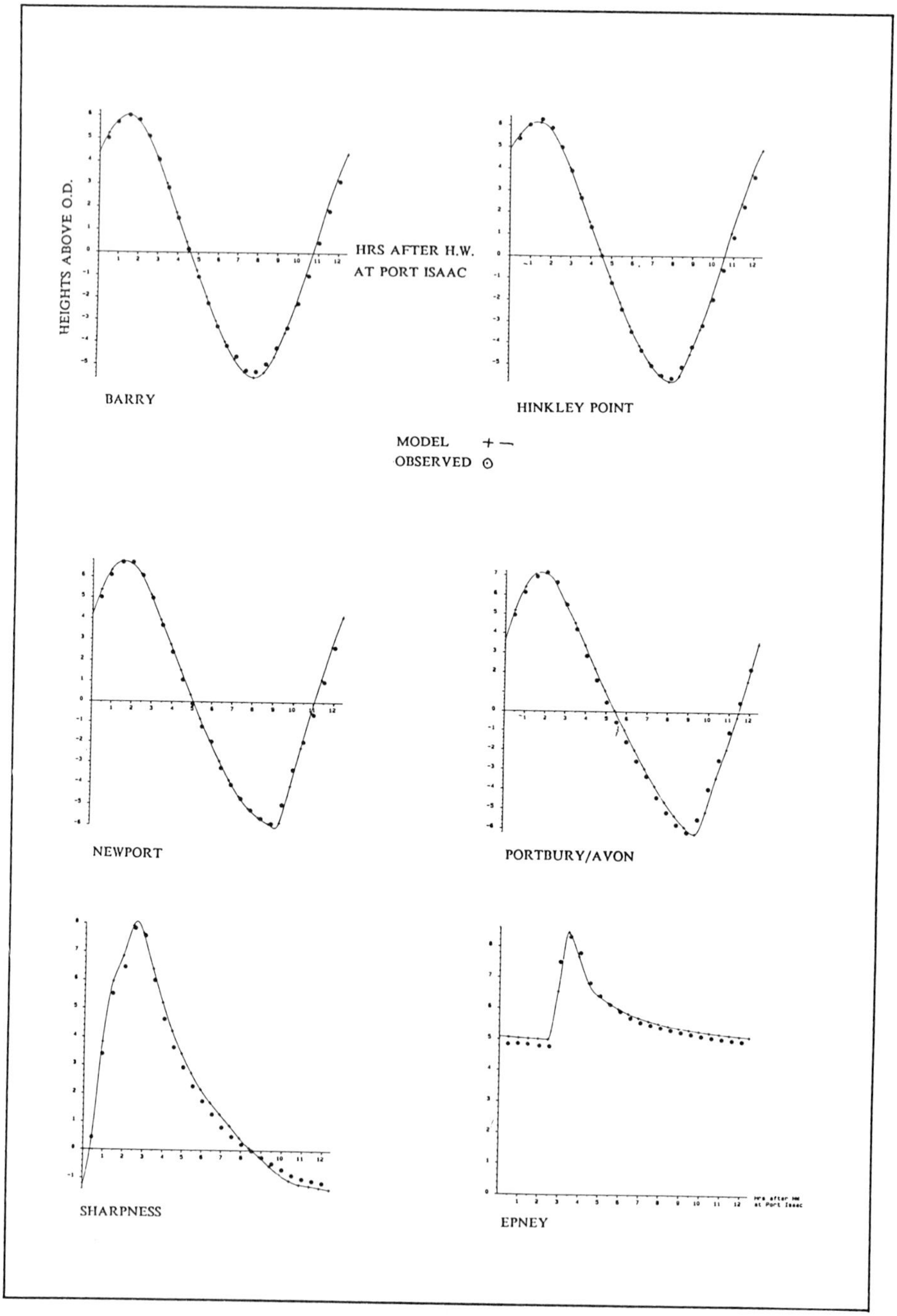

Figure 4: Spring tides - observed and model water level data

Tide modelling is usually employed to show features which occur during a single tide or as a result of a relatively short sequence of cycles. Many aspects of the effects of the barrage have been resolved in this way. However, the account of the evolution of the Severn Estuary (D.En. et al, 1989) which principally occurred in response to rising mean sea level since the last ice age, confirms the importance of also considering the longer timescale because of the change which the barrage would make to mean water level in the basin area and the effects which, by inference, this is likely to have. Fig 5 shows this change at representative locations about the Estuary.

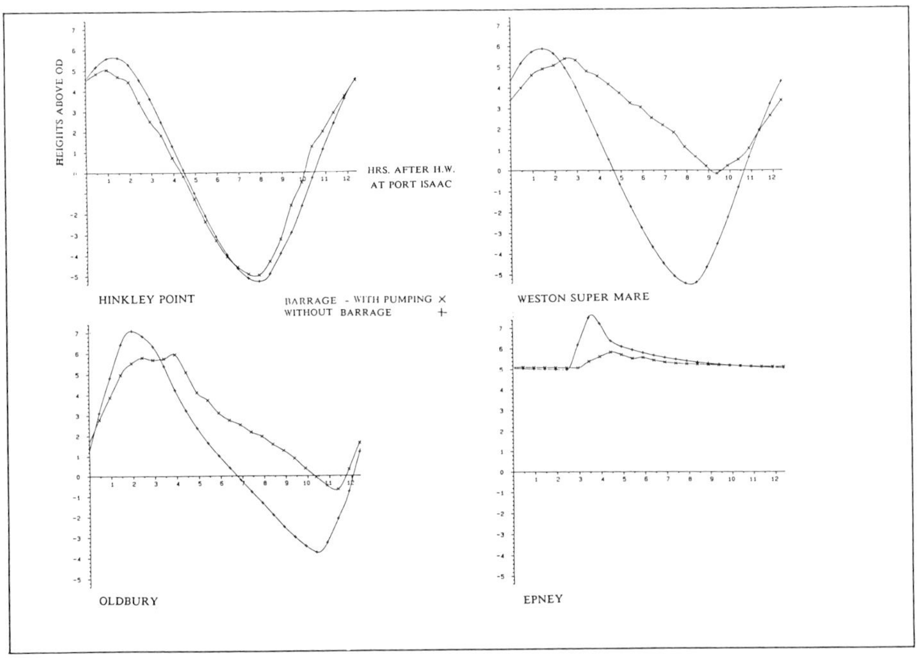

Figure 5: Spring tide curves - without and with barrage - with pumping

The fact that transformation from the full range (which modelling shows would be little changed during barrage construction) to a much reduced range would probably be made over a period of not more than a week or so means that many features of the estuary would quite suddenly be out of regime with the controlling tidal influence. A period of adjustment would follow, extending to decades and perhaps to centuries, and which would to some extent depend on how the barrage was operated (eg Fig 2). Possible methods of predicting these changes based on the use of regime concepts and field survey data have been considered (D. En., 1989). They differ significantly from conventional modelling methods and are still at the formative stage, mainly because they have not hitherto generally been needed for predictive purposes in connection with proposed engineering works. The widely anticipated but still to be proven world rise in mean sea level forecast to occur over the next 50-100 years is adding momentum to work in this area because of the effects it will have on coastal processes and sea defence provisions.

Earlier work on the proposed Severn Barrage (eg Severn Barrage Committee, 1981) gave more attention to tide conditions in the main Severn Estuary than in the river estuaries (Fig 1). However, many of the reasons for wishing to know more about the implications of the barrage now relate to its likely effects in the river estuaries, eg, changes in salinity, water quality, turbidity, sediment movement and flood risk. The present geometry of each river estuary is largely determined by the hydrodynamic factors of tide regime and river flow. Fine sediments are readily available and in some river estuaries there are geological controls on the otherwise unconstrained evolution of these interfaces between riverine conditions landward of the tidal limits and quasi-maritime conditions seaward in the main Severn Estuary. However, because the significance of these controls is only local, there is considerable similarity of form between these river estuaries, a consequence essentially of the common influence of high tide range.

For this initial study of the effects of the barrage on conditions in the river estuaries (D.En. et al, 1989), it was therefore appropriate to focus detailed work on one estuary and to seek to apply the lessons so learned to the others. In this way it proved possible to make general progress and to identify the more detailed effort required in due course. The geometry of the river estuaries allowed a 1-D model to be used, and this was applied to the simulation of turbidity and erosion/deposition (Wimpol, 1989). However, as previously explained, the consequences of the rise in mean water level in the basin area would encourage permanent accretion on a much longer timescale than could sensibly be simulated by a model of tide-by-tide processes. The techniques considered for this purpose and that applied in detail are described by the University of Liverpool (1989). A timescale of about 100 years to reach 95% of the accretion needed to give the ultimately stable bed form according to the changed tide regime and mean water level was deduced. The tide model was then run with this revised bathymetry to show how hydrodynamic conditions in the river estuaries would then compare with those which now exist. Those which would occur in the short term following barrage construction were also assessed.

The suitability of a 1-D model to simulate flows in the river estuaries, and of a 2-D model for the main Severn Estuary, respectively become increasingly questionable as the river estuaries broaden towards their mouths and as the main estuary deepens further seaward. In these outer regions, both geometric and density effects progressively influence the simpler regime assumed with justification to apply in each case further landward. Entending this simplicity to areas in which it is known not to be as strictly applicable gives rise to errors. These must be quantified in order to establish whether more complex models will be needed in future and, if so, what their specifications are likely to be, the field data needed to validate them and the costs of collecting those data and of developing and running the models, etc.

In the recent study of the Severn Barrage, an initial assessment was made of the specifications for and findings of 3-D models applied to the tide regime of the Severn Estuary and Bristol Channel (McLaren and Collins, 1989). Their results were compared with those of several other 2-D plan models. The results suggest that the overall structure of the currents in some shallow areas away from the main flows and in deeper areas could include vertical components. However, these appear to be generally weak compared with the main flows in plan, and it remains to be shown whether or not they justify further serious consideration in the context of the barrage. Present indications are that any effects will only be of secondary importance and are more likely to be significant in some specific situations rather than generally. If this is confirmed it may still mean that much of the Severn Estuary will at some stage have to be modelled in three dimensions.

31.3 Salinity regime

In the transition from freshwater to fully marine conditions, salinity concentrations of 5 and 18 parts per thousand (ppt) are commonly used to describe thresholds of tolerance for various species of plant and animal. A near-marine threshold of about 33 ppt is also identified for some species. However, the salinity at any location in an estuary varies throughout a tide cycle, the amplitude of change depending on tide range. Also, during periods of higher or lower than average incoming river flow, the mean salinity will be reduced or increased respectively, and in the nearshore zone will be affected by local runoff, eg., during prolonged heavy rainfall.

The above reference to the significance of specific values of mean salinity should therefore be read with reference to the extent to which they change on the shorter timescale of a tide cycle and on the longer timescale of varying river flows. However, there appears to be a lack of evidence on which to base this judgement. Indeed, in the case of higher tide range conditions such as those in the Severn Estuary, it must be questionable whether the biological regime is significantly determined by salinity or whether tidal exposure and submergence, strong currents, storm waves, high turbidity, mobile substrates, etc., either singly or collectively exert more influence. Clearly it would be wasteful to define the salinity (or any) regime in more detail than is necessary, which means that a specification for this has to be made. This remains to be done.

A significant step in this general direction was taken during the recent study of the Severn Barrage. It was argued that it would be instructive to know how the barrage would change the salinity regime throughout the estuary, both in terms of tide-averaged values according to tide range and river flow, and the amplitudes of variation per tide cycle. Earlier work done using a 1-D tide model (Severn Barrage Committee, 1981) suggested that average concentrations landward of the barrage would fall by at least 2-3 ppt, and by more in the river estuaries.

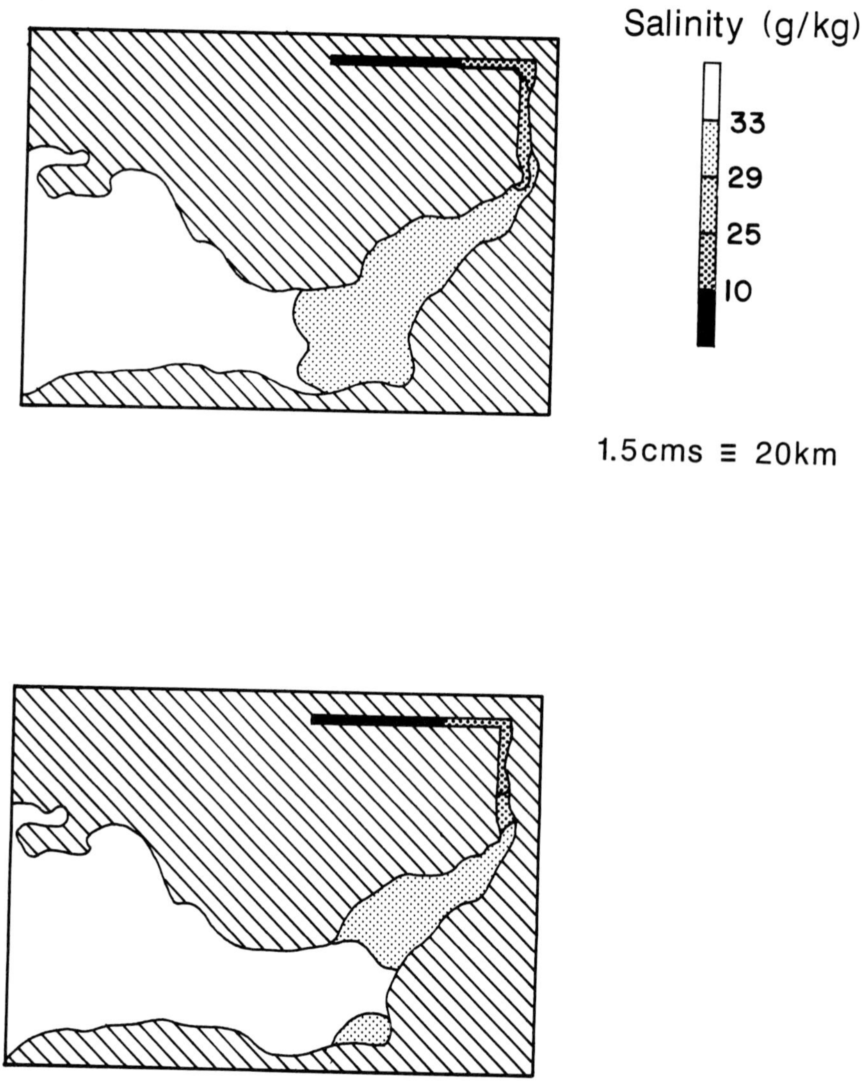

Figure 6: a) Severn - salinity contours for Spring tide, 10% flow, no barrage
b) Severn - salinity contours for Spring tide, 10% flow, barrage

The 2-D model developed for the recent study (WRc, 1989a) was used as the hydrodynamic basis for simulating salinity (WRc, 1989b). Typical examples of the output of this work are shown in Fig 6, in this case showing the concentration contours for the without and with-barrage conditions at low water of a spring tide. The full results suggest that the average concentration at any location in the main Severn Estuary on any tide would be increased slightly (up to about 1ppt), that the excursion per tide cycle would be reduced and that there would be a seaward movement of salinity (establishing fresher conditions) in the upper reaches of the river estuaries.

The reasons why the l-D and 2-D models differed in their forecasts of the direction of salinity change in the basin deserve to be resolved, though the absolute magnitude of the difference between the forecasts may not be too significant. The fundamental difference between the models is, of course, the way in which each represents lateral variations in water levels, currents, etc., across the estuary. Early models of tide propagation through the Bristol Channel and Severn Estuary (Fig 1) assumed lateral homogeneity and were able to demonstrate the principal details of the tide regime. However, lateral gradients of levels, currents, salinity, etc., are known to exist, hence there is a case for using 2-D representation for the present regime.

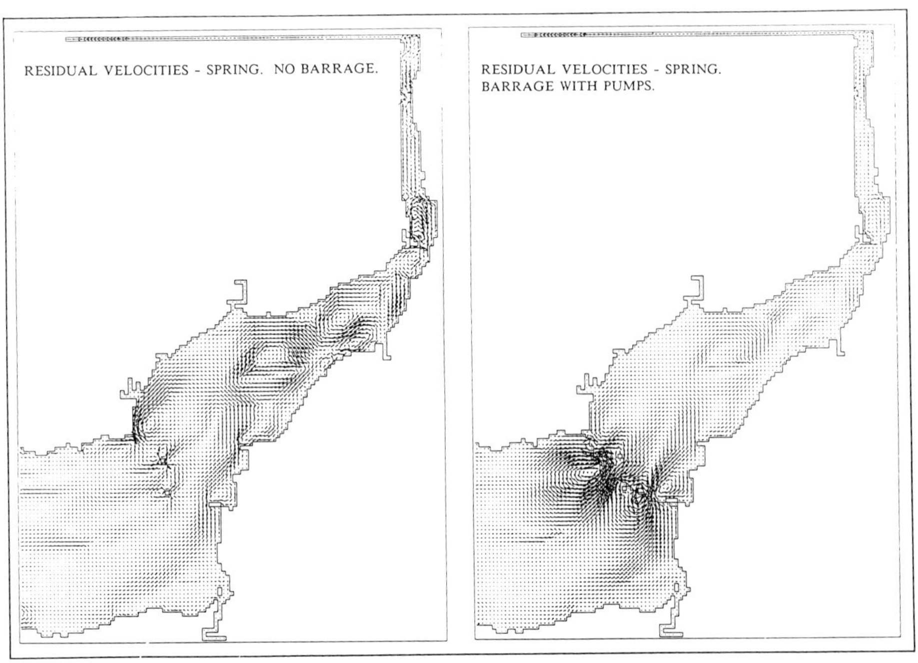

Figure 7: a) Residual velocities - Spring. No barrage.
b) Residual velocities - Spring. Barrage with pumps.

Furthermore, the 2-D model confirms that the layout of turbines and sluices along the barrage and the sequence in which these are used through a tide cycle would substantially increase the magnitude of these differential flows, particularly in the region of the barrage. (Fig 7 shows the residual currents for a spring tide without and with the barrage). The importance of the strong contra-rotating residuals set up by the barrage extends beyond salinity, for which it would ensure that a more rapid exchange of more saline water from seaward and fresher water for landward would occur. According to the results given by WRc (1989b), the net result of this motion is to alter the present salinity regime as indicated above.

Vertical gradients of salinity in the Severn Estuary have been shown to be generally small, less than lppt. Only in the deeper waters of the Bristol Channel do greater differences occur but even there they are believed to be of little consequence to the water quality and ecological regimes. Moreover, it is unlikely that they would be much altered by the barrage, though the possible occurrence of locally significant changes for some plant and animal communities, and the need to design the barrage to influence this, should not be ruled out.

31.4 Effluent dispersion

The effects of the differential currents set up by the turbines and sluices are made more clear by the way in which the patterns of dispersion of flows from individual outfalls would be affected. In order to quantify the change, two hypothetical outfall locations were selected (positions A and B on Fig l). The discharge of both degradable (eg, bacteria) and conservative (eg, metals) effluents were considered. Various bacteria lifetimes to 90% decay were considered in order to represent the somewhat uncertain effects of high turbidity in reducing typical decay rates. The 2-D tide model was also used for this study (WRc, 1989c).

Fig 8 compares the plumes of degradable effluent from the two outfalls without and with the barrage in place. The changes to the regime due to the barrage are particularly marked for outfall B, seaward of the barrage. In this case, the combined effect of the sluices permitting discharge landward through the barrage during the flood tide phase and the turbines further offshore releasing it seaward is to accelerate dispersion many-fold compared with that of the comparatively rectilinear flows of the tide alone.

The effect of the barrage on flows from outfall A, because of its greater distance from the barrage, would be less marked. Nevertheless, data from the model suggest that even close to this outfall, contaminant levels would be reduced. The effect of accelerated decay should the fall in turbidity levels be sufficient to increase light penetration significantly would be an added consequence of the barrage; as indicated above, this remains to be quantified.

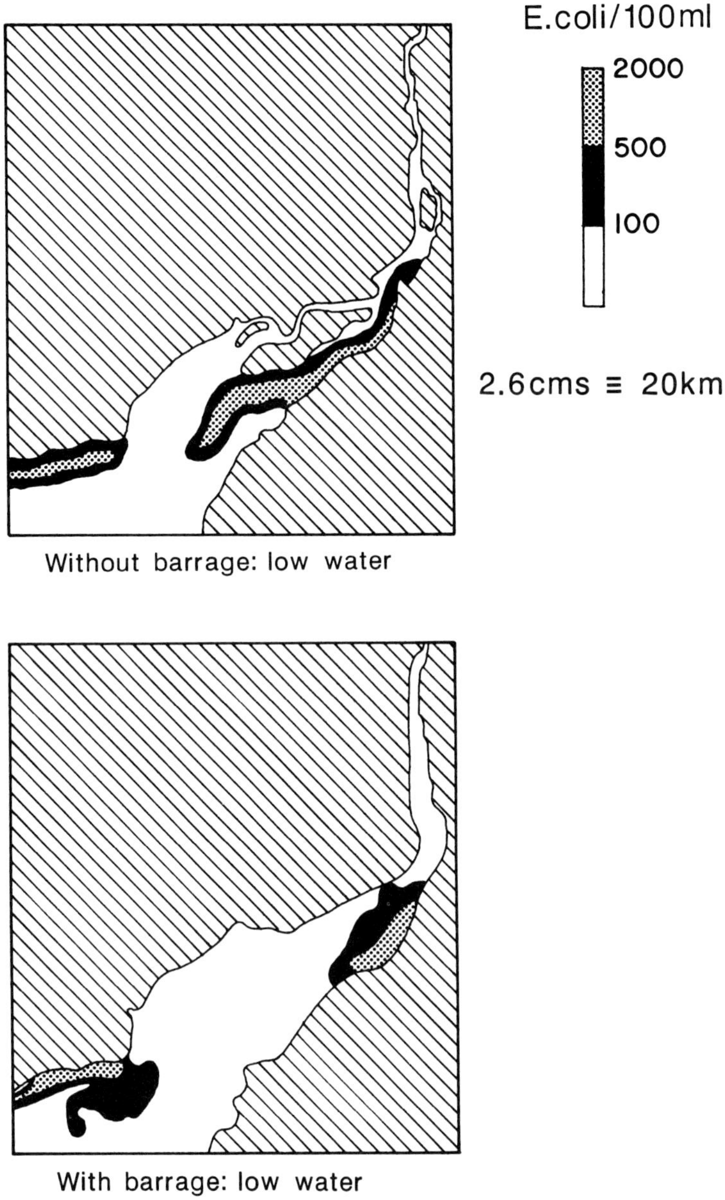

Figure 8: a) No barrage, Spring, T90 = 20 hrs, No wind, D. Coeff. 10.0 M*M/s
b) Barrage, Spring, T90 = 20 hrs, No wind, D. Coeff. 10.0 M*M/s

Based on the evidence given by WRc (1989c) for the Severn Barrage, the instinctive supposition that this type of project would lower the quality of the water in the basin is not correct. Hydrodynamic modelling of dispersion across the main estuary confirms that conditions may be expected to improve following barrage construction. However, in the river estuaries there may be specific discharge locations and phases, during lower range tides in particular, when this improvement may not occur. These conditions remain to be studied in detail; the combined effects of reduced currents, greater mean water depths, restricted lateral estuary widths and reduced peak turbidity need to be assessed before comments of either a general or site specific nature can be made. (The effect of long-term accretion would be to reduce depths and increase currents and hence turbidity; the net effect of these changes must also be considered).

31.5 Wave climate

The orientation and depth of the Bristol Channel permit waves having periods of up to 10 seconds to penetrate as far as the barrage (Fig 1). Sediment transport in this part of the estuary may therefore be expected to be influenced by wave action, though strong tidal currents are the major force at work. However, in the shallow coastal zone, this role is reversed; the risk of storm damage in some locations is high, as experiences often confirm.

The wave climate in the Bristol Channel does not penetrate far into the Severn Estuary. The coastlines seaward of Clevedon and from Cardiff to Newport are exposed to conditions which are in part of this origin but are also due to more locally raised seas. Further up-estuary and in the river estuaries it has been concluded that apart from the influences which reduced currents may have on refraction and its effects, the wave climate would be little affected by the barrage because these wave regimes are generated by local winds whose fetch would not be affected significantly, if at all (HRL, 1984).

The most uncertain influence of wave climate is its effects in shallow water and on coastlines as far as ecosystem productivity is concerned. Reference was made earlier to the information now available on the historic evolution of the Severn Estuary. Waves are believed to have played an important part in this, in conjunction with currents. Over the past few hundred years, the process has been altered by the construction of sea walls, and further changes may be expected to occur due to the presence of the barrage and how it would raise mean water level in the basin and reduce the currents (more so on the ebb phase than the flood).

The resultant effect on the stability of coastal sediments remains to be resolved and has a number of important implications. The wave climate further offshore appears to be less environmentally significant.

31.6 Conclusions

Sufficient basic work has been done to illustrate the principal hydrodynamic effects of the proposed Severn Barrage. By taking an objective approach to requirements, the recently completed study of the barrage has identified many significant effects which were not previously appreciated. The principal task now is to improve the level of scientific awareness in a number of hydrodynamically-related areas, including:

1 The significance for the effects of the barrage of three-dimensionality in the flow structure of the Bristol Channel, this being determined by topography and salinity.

2 The extent to which the flow structure set up by the barrage may need to be and may be modified by the design of the barrage, including the distribution and alignment of the turbines and sluices, the sequence of use of this equipment, the extent and regularity of its use including that for pumping, and the conditions which would occur during barrage construction according to the construction methods and programme adopted.

3 The significance, if any, of the duration of the period over which the tide regime is changed from that existing at the end of the barrage construction process, but when all sluices and turbines are still permanently open, to that required for power generation, including the time of year and any other factors relevant to the estuary's ecosystems.

4 How the rate of decay of degradable contaminants is determined by the extent to which sunlight penetrates into the water column (ie by turbidity), together with the influence of salinity.

5 How the stability of coastal sediments and hence the composition of this zone are influenced by the strength of currents, the availability of material, and the frequency and severity of wave action (information which must in practice be extended by data on the plant and animal colonisation of specific areas, these influencing stability).

6 How the barrage would, in the long term (decades), alter the distribution of sediments and how this would affect the overall regime.

This paper has not dealt specifically with sediments though their importance to the total regime of the estuary must be stressed. They have been mentioned in the context of the effects of turbidity and of the composition of the coastal zone. In their own way they are often regarded as the key to determining the productivity and hence the ecological character of an estuary, and it is clear from what is now known about the barrage that the project would alter this character of the estuary in a number of significant ways.

However, it will not be possible to determine these changes with confidence until some of the hydrodynamic implications of the scheme are more firmly specified. Whereas this task is not regarded as excessively demanding in scientific terms, it is in part urgent for resolution in order that attention may be given to certain implications.

The fact that the proposed barrage would create a tide regime which in many respects would be typical of that which occurs in many UK (and other) estuaries makes it important to look elsewhere for guidance as part of the process of forecasting the changes which the barrage would produce. Based on the findings of the recent study of the project (D.En. et al, 1989), these changes will be less extensive than has previously been suggested, and there appears to be considerable opportunity to influence them. A further concerted programme of more general study as outlined in this Paper is needed before estuary-specific conclusions about the environmental implications of the Severn (or any tidal power) barrage may be made. However, the results of the recent study have added considerable confidence to the viability and acceptability of the project in a number of hydrodynamic areas.

31.7 References

Binnie and Partners, 1989, "The UK Potential for Tidal Energy from Tidal Estuaries", Rpt. to ETSU, No. ETSU/TID 4048 Pt. 1.

Department of Energy, CEGB and STPG, 1989, "The Severn Barrage Project", Energy Paper No. 57, HMSO.

Hydraulics Research Ltd., 1984, "Waves Upstream of the Severn Barrage", Rpt. to the Severn Tidal Power Group, No. SD 2.4.

Institute of Oceanographical Sciences, 1980/81, "Tidal Barrage Calculations for the Bristol Channel", Rpts. to the Severn Barrage Committee, Nos. STP 15-17.

McLaren, P. and Collins, M.B., 1989, "Sediment Transport Pathways in the Severn Estuary and Bristol Channel", Rpt. to the Severn Tidal Power Group, No. SBP 18.

Severn Barrage Committee, 1981, "Tidal Power from the Severn Estuary", Energy Paper No. 46, 2 Vols, HMSO.

University of Liverpool, 1989, "Assessment of River Estuary Geometry", Rpt. to the Severn Tidal Power Group, N. SBP 10.

Water Research Centre, 1989a, "2-D Plan Hydrodynamic Modelling of the Bristol Channel and Severn Estuary Without and With Barrage", Rpt. to the Severn Tidal Power Group, No. SBP3.

Water Research Centre, 1989b, "Salinity Modelling of the Severn Estaury", Rpt. to the Severn Tidal Power Group, No. SBP 9.

Water Research Centre, 1989e, "The Effects of a Barrage on the Dispersion of Pollutants in the Bristol Channel and Severn Estuary", Rpt. to the Severn Tidal Power Group, No. SBP 34.

Wimpol Ltd., 1989, "1-D Mathematical Modelling Study of the Tidal Reaches of the River Usk", Rpt. to the Severn Tidal Power Group, No. SBP 5/5A.

Coastal and Estuarine Sciences

1. Coastal Upwelling, Francis A. Richards (Ed.) 529 pages. 1981.
2. Oceanography of the Southwestern U.S. Continental Shelf, L. P. Atkinson, D. W. Menzel, and K. A. Bush (Eds.) 156 pages. 1985.
3. Baroclinic Processes on Continental Shelves, Christopher N. K. Mooers (Ed.) 130 pages. 1986.
4. Three-Dimensional Coastal Ocean Models, Norman S. Heaps (Ed.) 208 pages. 1987.

Coastal and Estuarine Series

(formerly Lecture Notes on Coastal and Estuarine Studies)

Vol. 1: J. Sündermann, K.-P. Holz (Eds.), Mathematical Modelling of Estuarine Physics. Proceedings, 1978. 265 pages. 1980.

Vol. 2: D.P. Finn, Managing the Ocean Resources of the United States: The Role of the Federal Marine Sanctuaries Program. 193 pages. 1982.

Vol. 3: M. Tomczak Jr., W. Cuff (Eds.), Synthesis and Modelling of Intermittent Estuaries. 302 pages. 1983.

Vol. 4: H.R. Gordon, A.Y. Morel, Remote Assessment of Ocean Color for Interpretation of Satellite Visible Imagery. 114 pages. 1983.

Vol. 5: D.C.L. Lam, C.R. Murthy, R.B. Simpson, Effluent Transport and Diffusion Models for the Coastal Zone. 168 pages. 1984.

Vol. 6: M.J. Kennish, R.A. Lutz (Eds.), Ecology of Barnegat Bay, New Jersey. 396 pages. 1984.

Vol. 7: W.R. Edeson, J.-F. Pulvenis, The Legal Regime of Fisheries in the Caribbean Region. 204 pages. 1983.

Vol. 8: O. Holm-Hansen, L. Bolis, R. Gilles (Eds.), Marine Phytoplankton and Productivity. 175 pages. 1984.

Vol. 9: A. Pequeux, R. Gilles, L. Bolis (Eds.), Osmoregulation in Estuarine and Marine Animals. 221 pages. 1984.

Vol. 10: J.L. McHugh, Fishery Management. 207 pages. 1984.

Vol. 11: J.D. Davis, D. Merriman (Eds.), Observations on the Ecology and Biology of Western Cape Cod Bay, Massachusetts. 289 pages. 1984.

Vol. 12: P.P.G. Dyke, A.O. Moscardini, E.H. Robson (Eds.), Offshore and Coastal Modelling. 399 pages. 1985.

Vol. 13: J. Rumohr, E. Walger, B. Zeitzschel (Eds.), Seawater-Sediment Interactions in Coastal Waters. An Interdisciplinary Approach. 338 pages. 1987.

Vol. 14: A.J. Mehta (Ed.), Estuarine Cohesive Sediment Dynamics. 473 pages. 1986.

Vol. 15: R.W. Eppley (Ed.), Plankton Dynamics of the Southern California Bight. 373 pages. 1986.

Vol. 16: J. van de Kreeke (Ed.), Physics of Shallow Estuaries and Bays. 280 pages. 1986.

Vol. 17: M.J. Bowman, C.M. Yentsch, W.T. Peterson (Eds.), Tidal Mixing and Plankton Dynamics. 502 pages. 1986.

Vol. 18: F. Bo Pedersen, Environmental Hydraulics: Stratified Flows. 278 pages. 1986.

Vol. 19: K.N. Fedorov, The Physical Nature and Structure of Oceanic Fronts. 333 pages.1986.

Vol. 20: A. Rieser, J. Spiller, D. VanderZwaag (Eds.), Environmental Decisionmaking in a Transboundary Region. 209 pages. 1986.

Vol. 21: Th. Stocker, K. Hutter, Topographic Waves in Channels and Lakes on the f-Plane. 176 pages. 1987.

Vol. 22: B.-O. Jansson (Ed.), Coastal Offshore Ecosystem Interactions. 367 pages. 1988.

Vol. 23: K. Heck, Jr. (Ed.), Ecological Studies in the Middle Reach of Chesapeake Bay. 287 pages. 1987.

Vol. 24: D.G. Shaw, M.J. Hameedi (Eds.), Environmental Studies in Port Valdez, Alaska. 423 pages. 1988.

Vol. 25: C.M. Yentsch, F.C. Mague, P.K. Horan (Eds.), Immunochemical Approaches to Coastal, Estuarine and Oceanographic Questions. 399 pages. 1988.

Vol. 26: E.H. Schumann (Ed.), Coastal Ocean Studies off Natal, South Africa. 271 pages. 1988.

Vol. 27: E. Gold (Ed.), A Law of the Sea for the Caribbean: An Examination of Marine Law and Policy Issues in the Lesser Antilles. 507 pages. 1988.

Vol. 28: W.S. Wooster (Ed.), Fishery Science and Management. 339 pages. 1988.

Vol. 29: D.G. Aubrey, L. Weishar (Eds.), Hydrodynamics and Sediment Dynamics of Tidal Inlets. 456 pages. 1988.

Vol. 30: P.B. Crean, T.S. Murty, J.A. Stronach, Mathematical Modelling of Tides and Estuarine Circulation. 471 pages. 1988.

Vol. 31: G. Lopez, G. Taghon, J. Levinton (Eds.), Ecology of Marine Deposit Feeders. 322 pages. 1989.

Vol. 32: F. Wulff, J.G. Field, K.H. Mann (Eds.), Network Analysis in Marine Ecology. 284 pages. 1989.

Vol. 33: M.L. Khandekar, Operational Analysis and Prediction of Ocean Wind Waves. 214 pages. 1989.

Vol. 34: S.J. Neshyba, Ch.N.K. Mooers, R.L. Smith, R.T. Barber (Eds.), Poleward Flows Along Eastern Ocean Boundaries. 374 pages. 1989.

Vol. 35: E.M. Cosper, V.M. Bricelj, E.J. Carpenter (Eds.), Novel Phytoplankton Blooms. 799 pages. 1989.

Vol. 36: W. Michaelis (Ed.), Estuarine Water Quality Management. 478 pages. 1990.

Vol. 37: C.M. Lalli (Ed.), Enclosed Experimental Marine Ecosystems: A Review and Recommendations. X, 218 pages. 1990.

Vol. 38: R.T. Cheng (Ed.), Residual Currents and Longterm Transport. XI, 544 pages. 1990.

Vol. 39: M.I. El-Sabh, N. Silverberg (Eds.), Oceanography of a Large-Scale Estuarine System. X, 434 pages. 1990.

Vol. 40: D. Prandle (Ed.), Dynamics and Exchanges in Estuaries and the Coastal Zone. 648 pages. 1992.